Basic Notions of Condensed Matter Physics

Advanced Book Classics
David Pines, Series Editor

Anderson, P. W., *Basic Notions of Condensed Matter Physics*
Bethe, H. and Jackiw, R., *Intermediate Quantum Mechanics, Third Edition*
Nozières, P., *Theory of Interacting Fermi Systems*
Pines, D., *The Many-Body Problem*
Quigg, C., *Gauge Theories of the Strong, Weak, and Electromagnetic Interactions*

Basic Notions of Condensed Matter Physics

Philip W. Anderson
Princeton University
Princeton, New Jersey

Advanced Book Program

CRC Press
Taylor & Francis Group
Boca Raton London New York

CRC Press is an imprint of the
Taylor & Francis Group, an **informa** business

First published 1984 by Westview Press

Published 2018 by CRC Press
Taylor & Francis Group
6000 Broken Sound Parkway NW, Suite 300
Boca Raton, FL 33487-2742

ISBN 13: 978-0-201-32830-1 (pbk)

Visit the Taylor & Francis Web site at
http://www.taylorandfrancis.com

and the CRC Press Web site at
http://www.crcpress.com

Cover design by Suzanne Heiser

Editor's Foreword

Addison-Wesley's *Frontiers in Physics* series has, since 1961, made it possible for leading physicists to communicate in coherent fashion their views of recent developments in the most exciting and active fields of physics—without having to devote the time and energy required to prepare a formal review or monograph. Indeed, throughout its nearly forty-year existence, the series has emphasized informality in both style and content, as well as pedagogical clarity. Over time, it was expected that these informal accounts would be replaced by more formal counterparts—textbooks or monographs—as the cutting-edge topics they treated gradually became integrated into the body of physics knowledge and reader interest dwindled. However, this has not proven to be the case for a number of the volumes in the series: Many works have remained in print on an on-demand basis, while others have such intrinsic value that the physics community has urged us to extend their life span.

The *Advanced Book Classics* series has been designed to meet this demand. It will keep in print those volumes in *Frontiers in Physics* or its sister series, *Lecture Notes and Supplements in Physics*, that continue to provide a unique account of a topic of lasting interest. And through a sizable printing, these classics will be made available at a comparatively modest cost to the reader.

The informal monograph *Basic Notions of Condensed Matter Physics* is an obvious candidate for this series, containing as it does an authoritative and timeless account of the key concepts of condensed matter physics written by Nobel laureate Philip W. Anderson, who initiated so many of them. The topics he considers continue to be the focus of much of present-day research, so that his guided tour provides an invaluable introduction to contemporary condensed matter physics for everyone interested in the field, from the beginning graduate student to the experienced researcher. It gives me pleasure to welcome Professor Anderson to the *Advanced Book Classics* series.

David Pines
Urbana, Illinois
October 1997

CONTENTS

Quantum Solids

Renormalization Group

PREFACE AND ACKNOWLEDGMENTS

In the ten years since the first rough draft of these notes was begun (October 1973, in Port Isaac, Cornwall), far more people than the work deserves or than I can possibly acknowledge have helped with one aspect or another. First of all, the entire first draft was typed by Joyce Anderson, during a sabbatical term aided financially by both Bell Laboratories and Cambridge University. Later versions were typed in part by Clare Wilson at Cambridge, Grace Anderson at Princeton, and the most nearly complete of all by Mary Porreca at Bell Laboratories. Successive classes of graduate students sat patiently through the lectures based on these notes and often read or commented wisely upon them, both at Cambridge and Princeton. Final figures were mostly prepared by John Gomany and Morgan Phillips at Princeton. The final version owes much to the typing and editing of Marilena Stone and Karie Friedman, and especially to Clare Yu, who has gone through the whole and in many cases clarified my expression and even my ideas. The patience of Lore Heinlein at Addison Wesley was responsible for keeping the project alive these ten years.

June 1983
Murray Hill, New Jersey

Philip W. Anderson

CHAPTER 1

INTRODUCTION

A. MOTIVATION

Since 1973, in Cambridge and then in Princeton, I have given a series of graduate lecture courses based on the material from which this book is a selection. This material was written out in a preliminary but complete form starting in October 1973, and has been enlarged and revised many times since then. Parts of it have served as the basis for a set of published lecture notes; in particular, a chapter on disordered systems, which was not organically connected with the rest of the book was the basis for my Les Houches lecture notes of 1978 (in Ill-Condensed Matter, edited by R. Balian, R. Maynard, and G. Toulouse).

The concept of the book was that since the explosive growth of condensed matter physics as a field since 1945, it is no longer possible to write a conventional, comprehensive text in the field at anything but the most elementary level. It is even less possible to give a lecture course, which necessarily selects from the material in a textbook. There have been two or three solutions to this problem. One is specialization, often unconscious or at least unstated; the author covers the physics of metals, or of semiconductors, or bands and bonds, or some other subject fairly completely, and neglects the rest of the field.

A second solution is that chosen by Seitz as a follow-up to his marvelous 1940 text: to collect rather than select, and to publish series of reviews by different authors on their own specialties. At the advanced monograph level this is all that can be done, but it is no help to the student entering the field.

Finally, some books concentrate on the theoretical techniques which are common to all of many-body theory. Many of those who have given courses on many-body theory must have experienced some of the frustrations with teaching such a course which have led me to write this book. However much one may try to give a balanced course, emphasizing equally the physical and mathematical aspects of the subject, one finds the purely mathematical aspects of the field-theoretical perturbation techniques introduced in the fifties dominating the course. This subject seems to take on a life of its own. The material is lengthy; it has great appeal to the student because of its elegance and formal nature--that is, it does not force the student (or the lecturer) to think too deeply or to encounter any uncertainty; and, perhaps most important, it is the most extensively discussed part of the best texts on the subject, most particularly the justly revered "AGD" (Abrikosov, Gorkov, and Dzyaloshinskii, 1963). As a partial result, at least in my course, discussions of the rest of the physics tended to be confined to a few hasty illustrative examples at the very end. This left the students, I felt, under the mistaken impression that many-body physics was simply a matter of calling in two subroutines, one entitled "Green's functions" and the other "diagrams," and then feeding the appropriate data into the main program. There are, of course, texts which take a broader view, for instance, the early ones by Pines (1961), and Pines and Nozières (1966) but even these emphasize one aspect only and are now somewhat out-of-date. The most generally used modern text, Fetter and Walecka (1971), not only takes a less sophisticated view than Pines on many matters, but is also very formal and rather divided in its treatment between the different methodologies of solid-state and nuclear theory. I give below a list of several texts; my own favorites, in spite of various weaknesses, are AGD (1963), Pines (1961), Nozières (1964), and Thouless (1961).

My feeling is that behind the mathematics of AGD lies a much less formal, logically more complete, structure, associated most generally with the name of Landau; and few books seem to give much feeling for this. I therefore felt that it might be useful to write a book about all the rest of what I thought should have gone into my lectures, and in particular to emphasize the general principles of many-body physics, for which the marvelous formalism of perturbation theory provides the supportive mathematical apparatus. I would not advise the student to attempt to use this book to help with computations, but I would hope that it will help in many cases to understand what is being computed and why.

Another thing I have wanted to do for many years was to expand on some of the ideas that were presented very sketchily in the second half of my earlier book (Anderson, 1963). In fact, I have often found myself unconsciously following my earlier discussions. This indicates either that I am too old to have any new ideas or that the really basic principles of many-body physics are indeed fairly stationary in time; I hope the latter. Therefore to some extent this work can be considered an expansion and restatement of the second half of my earlier book. In the intervening years, apparently, no comparable discussion of these ideas has appeared, but instead most books have found themselves entrapped in diagrams and Green's functions.

I should like, in addition, to disavow any attempt at completeness; what this book contains is what might have gone into my courses instead of perturbation theory, and is thus idiosyncratic. Many important results of many-body theory--for example, some parts of the theory of liquids--are nowhere mentioned, and others--for instance, fluctuation-dissipation and other aspects of transport theory--are skimped, for no compelling reason except the need for finiteness.

Finally, as regards level: this book probably should not serve unaided as a basic text; it is meant primarily to supplement and deepen the student's or worker's understanding of what he first learned elsewhere. On the other hand, I will not assume a great degree of formal preparation. A knowledgeable experimentalist who has never studied diagrams or Green's functions should understand most of the material; I shall introduce diagrams only as a convenient descriptive device. Those who have a nodding acquaintance with second quantization and practical group theory can probably manage without further study. Since these two subjects are treated well in many texts, I don't try to teach them here. The discussion I give of second quantization in <u>Concepts in Solids</u> (Anderson, 1963) is the simplest; and Kittel (1971) covers group theory.

A great many of the ideas here presented are new points of view on old ideas, or in a few cases are wholly new. For this reason this book will sometimes be unsuitable for one of the most important uses of a true text, namely, to lead the student to the literature and teach him to understand it within its own context. In that respect, this resembles much more a monograph than a text. But writing a proper text of many-body physics is probably a task already beyond possibility; certainly it is beyond me.

B. GENERAL DESCRIPTION OF THE BOOK

In 1947, when I started my thesis research on pressure broadening of microwave spectra, I had not the faintest appreciation that what I did might come to be called many-body physics or that the methods I was about to use (which my adviser, Professor Van Vleck, insisted on calling "Schwingerian"), i.e., direct evaluation of the appropriate correlation functions, were soon to become a part of the routine apparatus of a new field. Those of us who were doing this kind of thing thought of it either as statistical mechanics--which has a fairly long history of dealing with the quantum mechanics of large systems--or as part of the new fields of chemical physics (like my work) or solid-state physics. Not only this fragmented nature of the subject, but a very real lack of confidence that any given problem was likely to be soluble, were characteristic of that time. We had, to be sure, the Debye theory of phonons, Wigner's and his students' work on the free-electron gas and the alkalies, and many other successes; but we were baffled, for instance, by antiferromagnetism, superconductivity, and superfluidity. At that time it was a challenging question whether the behavior of real macroscopic systems of atomic particles (of the order 10^{23} atoms in size) could be understood, starting from a knowledge of the laws that govern their motion on the microscopic atomic level. I think it is fair to say that in the last 25 or 30 years a quiet revolution has taken place in the methodology and in the confidence with which we approach this subject, of which a very definite but hard-to-define area has come to be known as many-body (as opposed to nuclear, solid-state, or low-temperature) physics. A revolution has also, we should hope, come in our predictive success, as well.

This success is based upon a much more rigorous habit of thought combined with a much more complete arsenal of ideas and methods than were available then. One set of methods which has come almost to dominate the field is the use of various forms of what is known as "many-body perturbation theory" involving Feynman diagrams and suited to different types of problems.

Other problems are solved by the use of the idea of elementary excitations, which stems from Landau, as well as the concepts of quasiparticles and of the Fermi-liquid theory (in its formal aspect also due to Landau). It is important to add also the idea of collective excitations in complex systems such as Fermi or Bose gases: phonons in helium, spin waves, plasmons, and zero sound in metals. Another line of thought that has been important stemmed from the fluctuation-dissipation ideas of Onsager and led through Callen's ideas and Kubo's and my work on dissipative effects in resonance spectra to the two-particle Green's-function methods now in use. I should mention also the concepts of broken symmetry, which began with Landau's ideas on phase transitions and with antiferromagnetism and which came into their own once the Bardeen-Cooper-Schrieffer theory of superconductivity appeared (Bardeen et al., 1957).

This is not a complete list of principles and concepts. I include a chapter on the renormalization group, for instance, a new idea that shows promise of being very basic, and my original plan was also to include a chapter on the new conceptual structure forming in the field of amorphous materials. But the list will serve to show that there is a large body of rather general ideas in many-body physics which it is appropriate to try to tie together into a more coherent conceptual structure. This book will attempt to present one possible way of putting together this structure.

The central purpose of this book is to set out what I believe to be the logical core of the discipline of many-body physics as it is practiced. This lies in the interaction of the twin concepts of broken symmetry and of adiabatic (or perturbative) continuity.

In Chapter 2 we shall examine broken symmetry, in Chapter 3 adiabatic continuity. It is hard to give this idea an adequate name; equally good would be renormalization, if that didn't have a different meaning in field theory, or, perhaps best of all, model-building.

Two additional chapters provide illustrative material—Chapter 4, on quantum solids, and Chapter 5, on the renormalization group. Chapter 4 I include because I think no adequate treatment of this problem yet exists and because it is at least a fertile source of speculative ideas. The chapter on the renormalization group (Chapter 5) is here because I believe that the renormalization group has the potential of providing an even more complete conceptual unification of the science of complex systems than has yet been realized, and that the common view of it as a technique useful only in phase transition theory and field theory is a great pity.

Remaining illustrative material which might be included--on magnetism, superconductivity, superfluidity, and metal-insulator transitions--is omitted at this time. Quantum magnetism is perhaps a bit too specialized a topic to include here; my original lecture notes now read like a bit of archaeology. The two (or three) superfluidities are large and complex subjects, and I felt they were best represented by reprinting some reviews and research papers which are as illustrative of the general ideas as my original lecture notes would have been. In particular, the new science of helium 3, and the whole flowering of the idea of topological classification of order parameters and their defects, are an illustration and outgrowth of the original conceptual structure of broken symmetry which would deserve addition to my basic material if I could not represent them well with reviews and research articles here.

Metal-insulator transitions and the whole subject of the magnetic state and strongly interacting fermions constitute a set of material which I found I could never finish without its continually being superseded by some major development. I suspect this area should simply be postponed for a totally separate book. It is an important, surprisingly recalcitrant, and exciting subject which has not yet gotten itself into an order that can be explained in any short space; I leave it out with regret.

Finally, I would like to describe the relevance of the choice of reprints. First, in addition to the articles mentioned above on superconductivity and my own articles on superfluidity and broken symmetry, which illustrate, to an extent, both Chapters 2 and 3, I include my own article with Stein (1980), which gives perhaps the most complete published discussion of the basic principle of broken symmetry.

To illustrate the continuity principle, in addition to Chapter 5, I have chosen not to reprint anything on Fermi-liquid theory, especially because of the excellent Nozières book on this subject (Nozières, 1964), as well as a good reprint collection by Pines (1961).

My own articles on superfluidity (Anderson, 1965 and 1966) are relevant to another continuation concept, the Bose liquid. The concepts here derive from early work I refer to there, by Penrose, Onsager, Gross, and Pitaevskii. But perhaps the greatest triumph of the concept of continuation--of an effective simple system replacing a complex one--has been in the general field of dynamical impurities in Fermi systems: the Kondo and Anderson models. I reprint here the important article by Nozières (1974) making this crystal clear for the Kondo problem. As an illustration both of this and of sophisticated uses of the renormalization group, I reprint Haldane's article (1978) on the generalized Anderson model.

Chapter 4 is itself an illustration of the general principles, but it has led to some ideas of present-day research interest, having to do with the properties of superimposed "quantum solids," i.e., density waves in conventional solids. A couple of basic reprints on this active area are included.

Finally, I felt it useful to give several reprints illustrative of the unconventional uses of the renormalization group which I advocate in Chapter 5. The Haldane work (1978) and mine with Yuval (Anderson 1961, 1970a and b; Anderson and Yuval, 1969, 1970, 1971, 1973; Anderson, Yuval, and Hamann, 1970a and b) on which it is based are one example; closely related is the work of Kosterlitz and Thouless (1973) on the two-dimensional xy model, which is here reprinted. Another subject where the method has been stretched past its conventional limits is localization--metal-insulator theory. In this area I give two seminal reprints, Abrahams et al. (1979), and McMillan (1981). Another extension is by Hertz, to quantum phase transitions at absolute zero (Hertz, 1976).

A note on format and units: in almost all cases I take $\hbar$ and k_B (Boltzmann's constant) = 1, i.e., I measure frequency, energy, and temperature in the same units. I will often also use atomic units where convenient.

In the text some sections are either parenthetical or excessively mathematical. These are indented.

CHAPTER 2

BASIC PRINCIPLES I: BROKEN SYMMETRY

A. WHAT IS BROKEN SYMMETRY?

For almost all practical purposes, space is homogeneous and isotropic. Even in the interior of a neutron star, the distortion of space due to gravitation can usually be treated as a relatively small perturbation, locally negligible, as certainly it is on earth. The laws that govern the behavior of the particles with which we usually deal--electrons and nuclei, or possibly individual nucleons and mesons in nuclear many-body physics--have a very high degree of symmetry: they are certainly translationally and rotationally symmetric, corresponding to the homogeneity and isotropy of space; and to the degree of accuracy which is relevant in all many-body problems, they also obey space inversion and time-reversal symmetry.

It has been speculated (Kirzhnitz and Linde, 1972; Weinberg, 1974) that in the elemental fireball, which preceded (or was) the cosmological "big bang," an even higher degree of symmetry obtained. It may be, it is supposed, that the vacuum and all the particles with their puzzling approximate symmetries of SU3, isospin, parity, etc., are the result of some cosmological phase transition which originated from a state with some very high overall symmetry which is not manifest to us. This kind of extension by field theorists and cosmologists of the broken-symmetry concept of many-body physics will be mentioned briefly in later chapters, but is a side issue here; the normal symmetries are quite adequate.

Matter at sufficiently high temperatures is in the gaseous state, and at even higher temperatures the molecules dissociate and the atoms ionize: the sun, for instance, is mostly a locally homogeneous and isotropic gaseous mixture of its elementary components. Such a state is clearly compatible with the basic symmetries of the controlling equations, and needs no explanation. But cold matter is a different case altogether, as Dr. Johnson forcefully demonstrated in his famous response to Bishop Berkeley (Boswell, 1791, entry for 6 August 1763). When you kick a stone, no doubt remains in your mind that it possesses a property that we call rigidity, which we recognize as the very epitome of broken translational invariance: not only does it occupy a fixed position in space, but translational symmetry is broken everywhere within it, in that the individual atoms all occupy fixed positions.

Mention of astrophysical situations already brings us to a separation between two fundamental types of symmetry breaking. Starting from a homogeneous, isotropic gaseous system--as in the elemental fireball--the nonlinearities of the gravitational equations and of the equation of state lead to what Thom has tried to describe mathematically in "catastrophe theory" (Thom, 1975): a more or less sudden breakup into a new regime, characterized by the onset of an instability whose precise realization is controlled by infinitesimal fluctuations already present in the medium. Such a gravitational instability broke up the original homogeneous gas first into the ancestors of galactic clusters, then again into galaxies, and finally into individual stars. A very similar symmetry breaking (Glansdorff and Prigogine, 1971) occurs in a homogeneous flowing fluid near the point of turbulent instability (again astrophysically a very important process). While the general umbrella of catastrophe theory is a common element between phase transitions and nonlinear instabilities, it is important to maintain the sharp distinction between them. One distinction is that, usually, in the case of the nonlinear driven instabilities the local equations of state and the local properties remain mostly symmetric and homogeneous; virtually no change takes place in the gas or liquid. This is the basic assumption, repeatedly stated, of the works of Prigogine _et al._ (e.g., Glansdorff and Prigogine, 1971; Haken, 1974) on such "symmetry-breaking instabilities." The phase transition to a solid, on the other hand, involves a complete change in the microscopic properties of the matter involved: microscopically the solid can no longer be described in the same terms, requiring not just an altered equation of state but a wholly new local, and even overall, description. (Of course, the condensation process often has all the complexities of the typical nonlinear instability as well.) The phase transition case usually exhibits great regularity and a form of stability which we will later emphasize as a generalization of rigidity; this is not known ever to be true of the nonequilibrium case (Anderson, 1981a).

It has been proposed that the laser is a representative of a third category of broken symmetry, more closely related to the condensation process than to the nonlinear instability which it at first resembles. In the laser there is a local, microscopic broken symmetry, which in this case is driven by an external pumping mechanism rather than being a consequence of thermodynamic equilibrium. It has recently been shown that the laser may be treated formally in a way very similar to phase transition theory (Haken, 1974). But the spontaneous regularity and rigidity properties are probably missing (Anderson, 1981a). Whether any "driven" cases of true broken symmetry exist (as, for example, in biological systems) is not yet clear (Kittel, 1970; Eigen, 1971; etc.).

I should caution the reader that broken symmetry is not the only way in which cold matter may change its behavior qualitatively. Although the cases in which the other alternative--a continuous phase transition to a qualitatively different behavior without change of symmetry--occurs are as yet relatively rare, they are quite fascinating. The Kondo effect is the best known: a continuous transition from magnetic to nonmagnetic behavior of a single impurity (Nozières, 1974). A number of systems of restricted dimensionality, such as one-dimensional metals or antiferromagnets, also show this kind of behavior (Lieb and Mattis, 1966); the physical realization of such systems is only approximate, of course, but it does seem that a number of real systems show quasi-one-dimensional behavior. We shall examine this fashionable new field in Chapter 5. There are also close relationships here to modern theories of elementary particles, both theories of "confinement" and "fully unified" theories.

Our experience shows us, then, that as matter cools down it usually no longer retains the full symmmetry of the basic laws of quantum mechanics which it undoubtedly obeys; our task here is to understand that the questions we must ask are "Why?", "In what sense?," and "What are the consequences?"

B. WHY BROKEN SYMMETRY? CRYSTALS AND THE BOSE LIQUID

Any discussion of why low-temperature phases exhibit broken symmetries must proceed with sufficient caution to allow for the fact that not all many-body systems do so, or at least not all artificial models of such systems. The prime example is the noninteracting Fermi gas; one might once have said here "the Fermi gas with repulsive forces," but it has been shown of the electron gas that it will eventually undergo an anisotropic BCS transition (Kohn and Luttinger, 1965), and very possibly the same is true for short-range repulsive forces (Layzer and Fay, 1971; Anderson and Brinkman, 1975). So the overwhelming majority of real physical systems with interactions between the particles--even purely repulsive ones--tend to exhibit the phenomenon of a lowest-energy state's not having the full symmetry of space or of the Hamiltonian's describing their interactions. The latter occurs in the case of various subsystems or model systems for which we create model Hamiltonians, such as the Heisenberg Hamiltonian of magnetic systems.

To make our problem clear, let us give some examples in detail. The purely noninteracting Fermi gas has, of course, the Hamiltonian (in second-quantized notation)

$$H = \sum_{k,\sigma} \varepsilon_k n_{k\sigma} \tag{2.1}$$

where the single-particle energies for free particles are

$$\varepsilon_k = k^2/2m.$$

σ labels the spin of the particle.

The lowest-energy state, occupied at absolute zero, is that in which all states below the Fermi energy $\varepsilon(k_F) = \mu$ are occupied:

$$\psi_0 = \prod_{\varepsilon_k < \mu} c^+_{k\sigma} \psi_{\text{vac}}, \tag{2.2}$$

where the total density N determines k_F: $k_F^3/3\pi^2 = N$ for two spin species.

At higher temperatures we have a randomly phased mixture in which states with energy ε_k are occupied according to the Fermi distribution

$$\langle n_{k\sigma} \rangle = \left(e^{(\varepsilon_k - \mu)/T} + 1\right)^{-1}. \tag{2.3}$$

These states clearly retain all of the invariance properties of the ordinary Boltzmann gas, which this system is when $T >> \varepsilon_F$, even though the ground state (2.2) has zero entropy.

If, on the other hand, we introduce interactions between our particles, even purely repulsive ones, the ground state changes. In the Coulomb case with a 1/r repulsive potential, the Coulomb forces dominate at very low density ($V_{coul} \sim N^{1/3}$, as compared to the kinetic energy $\sim k_F^2 \sim N^{2/3}$), and it has long been understood that the stable ground state as $r_s \to \infty$, where

$$\frac{4\pi}{3} r_s^3 = N^{-1} = \frac{3\pi^2}{k_F^3} \tag{2.4}$$

is the "Wigner solid," a regular (probably bcc) lattice array of electrons (Wigner, 1934; Coldwell-Horsfall and Maradudin, 1960). Equally, if one compresses a Fermi liquid with repulsive hard cores (roughly like ^3He), it will at high enough density form a regular solid, as ^3He indeed does (Palmer and Anderson, 1974).

The essential phenomenon in either case is that the lowest state of a potential energy of interaction between particles--for example, a pair interaction

$$V_{tot} = \sum_{ij} V(|r_i - r_j|)$$

--must occur for either a unique relative configuration of all the particles

$$C = \{r_1, r_2 \ldots r_N\}$$

(and all translations and rotations)

or, in artificial cases, perhaps for a highly restricted subset [as in, for example, order-disorder problems described by the Ising model on a triangular or magnetite lattice, which have a "zero-point entropy" (Wannier, 1950; Anderson, 1956). So far as I know, there exists no proof that among the lowest energy configurations C at least one is a regular lattice, but I for one would be very surprised if this weren't so.

One may work up a reasonable argument for it as follows: Let us take a relatively small box containing n atoms and consider its optimum configuration. There will be one minimum-energy configuration, all small displacements from which are described by a harmonic potential:

$$V - V_{min} = \frac{1}{2} \sum_{ij} V''_{ij} \delta r_i \delta r_j \tag{2.5}$$

(incidentally, we may allow small changes in the shape of our box to minimize the energy further). The effect of the smallness of the box may be minimized by using periodic boundary conditions.

For large displacements, however, there may be additional relative minima: for instance, but not typically, a single atom may have two possible potential wells within which it might sit, one lower than the other (see Fig. 2-1a). Such a second minimum

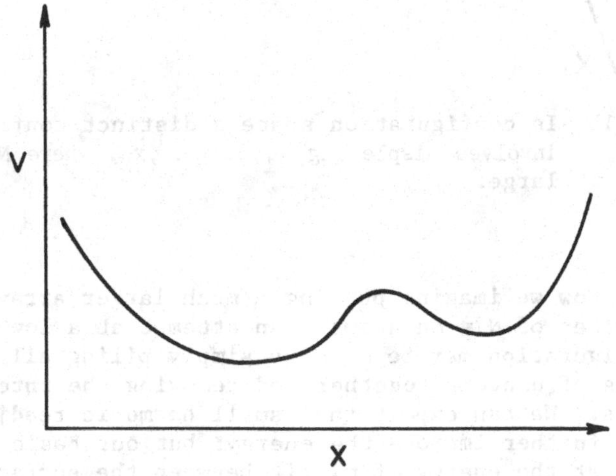

Figure 2-1a Local double minimum.

will have an energy only of order unity above the true minimum; but in general the other configurations will typically have energies ≈ n higher (as in the regular case, lattice energy differences for different lattices are proportional to the size of the crystal; even for an irregular array moving every atom in an essential way will change the energy by ≈ n).

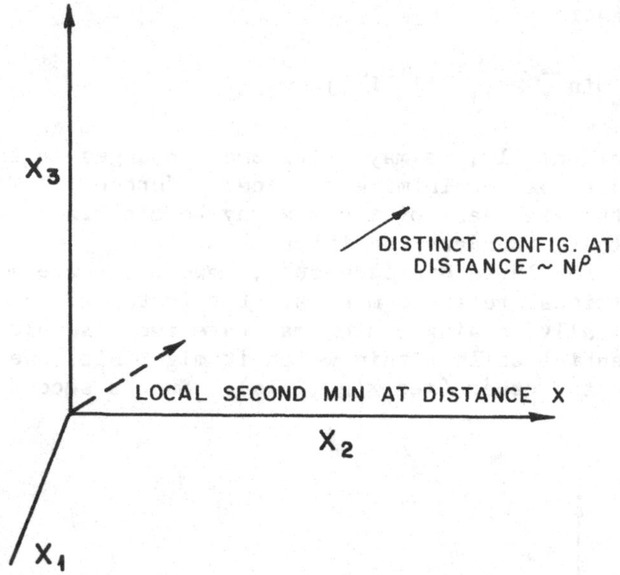

Figure 2-1b In configuration space a distinct configuration involves displacing $x_1, \ldots, x_N$, where N is very large.

Now we imagine putting a much larger array together of N = mn atoms. An attempt at a low-energy configuration may be made by simply piling all the boxes of n atoms together and removing the interior walls. We can expect that small harmonic readjustments will further improve the energy; but our basic argument is that the energy of misfit between the surfaces of the small pieces is a surface energy, of order $n^{2/3}$, while the energy necessary to change the interior configuration in an essential way will be ≈ n, so that there will be a cell size, n, beyond which it will not pay to modify the internal configurations; this then gives us a regular array.

The above is, of course, highly nonrigorous; clearly, the weak point is the definition of "essential" configuration changes. The result, however, seems hardly to be questionable. It is, then clear, that in any situation where the potential energy dominates kinetic energy and entropy (as in the two cases mentioned), a system of particles obeying a simple potential will take up a regular lattice structure.

That this regular, rigid lattice is a rather peculiar object from a quantum point of view is concealed from us by its familiarity. It seems a triviality that it is not permissible to treat the dynamics of the different parts of a rigid body independently, that there are long-range positional correlations throughout. But this means that there is a fundamental dichotomy between the mechanics of the rigid body as a whole and that of its individual parts. We have our choice of putting the body as a whole into an eigenstate of its full Hamiltonian, and in particular giving it a fixed total momentum $\bar{P}$. But then, of course, we would have no information about the position: the body would be in a wave packet of position with a spread $\Delta\bar{X} \propto \frac{1}{\Delta\bar{P}} = \infty$. But, of course, the relative position of the pieces of the body would be perfectly accurately known. Thus it is conventional, and the only reasonable choice, to specify the position of the body as a whole, and only then do we describe the whole and its parts consistently.

The weakly interacting Bose fluid breaks symmetry in a quite different way. We shall have to analyze this phenomenon in a much more detailed way later, but the point is that here even the kinetic energy alone does not allow for a symmetric ground state.

Imagine a box containing N free bosons. Again the energy levels for individual bosons are $k^2/2m$, and $k = 0$ is clearly the lowest one, separated in energy from the next by $\approx V^{2/3}$. Bosons can all occupy the lowest state (i.e., $\langle A_0^+ A_0 \rangle = n_0 = N$), and one is left at first glance with a highly symmetric situation.

In elementary statistical mechanics courses, it is shown that even at temperatures slightly above zero most of the atoms remain in the ground state ϕ_0. The mathematical reason is very simple; it is the convergence in three dimensions of the integral giving the occupation of all the other states:

$$\sum_{k\neq 0} n_k = \sum_k \frac{1}{e^{\frac{E_k-\mu}{T}} - 1} \leq \int \frac{k^2 dk}{e^{\frac{k^2}{2mT}} - 1} ,$$

so long as the chemical potential μ is below or at the energy level E_0. If μ is at E_0 the occupation of $\phi_0 \to \infty$; μ above E_0 is not meaningful because the denominator would be negative for some of the energy levels of the system. The integral is a function $f(T/T_B)$, where T_B is the Bose condensation temperature,

$$T_B \propto \frac{n^{2/3}}{m} ,$$

and $f(0) \to 0$, as $(T/T_B)^{3/2}$. Since the occupation of the other states decreases as the temperature decreases, eventually the ground state must become macroscopically occupied.

That this state is not truly symmetric in actual fact can be seen when we begin to try to subdivide our sample of bosons. In a gas or a normal liquid, and in fact in the Fermi gas as well, it is valid to a very good approximation to consider the state of a large system as simply the superposition of the states of different macroscopic-sized regions: this is what we mean by homogeneity of a system, and it is an implicit, but very real, assumption in our thinking about macroscopic objects. (Certainly, for instance, it is an implicit, if never explicitly stated, assumption of ordinary thermodynamics: it is part of the very definition of intensive versus extensive properties that the state must be specified by the local intensive variables.) But if in this case we divide our box into boxes of size N/m, and describe each box as a Bose-condensed gas of $n = N/m$ atoms, that does not in fact describe the total state at all. To show this, we need only introduce a_i and a_i^+ as the Bose amplitudes of the uniform state in each box, and, of course, assume $a_i^+ a_i = n_i$, $\langle n_i \rangle = N/m$.

But,

$$A_0 = 1/\sqrt{m} \sum_i a_i , \tag{2.6}$$

so that

$$\begin{aligned} N = \langle A_0^+ A_0 \rangle &= \langle \frac{1}{m} \sum_i n_i \rangle + \frac{1}{m} \sum_{i \neq j} \langle a_i^+ a_j \rangle \\ &= \frac{N}{m} + \frac{1}{m} \sum_{i \neq j} \langle a_i^+ a_j \rangle . \end{aligned} \tag{2.7}$$

Clearly, the second term must be much larger than the first, implying strong phase correlation of the different a_i. It is quite clear that the Bose-condensed state requires not only that each of the individual cells into which we subdivide the gas be itself condensed but that they have very strong phase relationships. It is for this reason that the definition of Bose condensation which is now more widely accepted assigns the Bose amplitude a finite mean value, not the occupation of the ground state. Defining the local field amplitude $\psi(r)$ and the ground-state amplitudes

$$A_0^+ = \frac{1}{\sqrt{V}} \int d^3r\psi^+(r) \tag{2.8}$$

$$a_i^+ = \sqrt{\frac{\overline{m}}{V}} \int_{(i^{th}\,\text{cell})} d^3r\psi^+(r) \,,$$

it is far more consistent--also with the behavior of superfluids under more general conditions--to define the ground state by

$$\begin{aligned} \langle\psi^+(r)\rangle &= \text{finite} \\ &= \sqrt{\rho}\, e^{-i\phi} \quad \text{for free particles;} \\ \langle\psi(r)\rangle &= \sqrt{\rho}\, e^{-i\phi} \,. \end{aligned} \tag{2.9}$$

Here I have inserted an arbitrary phase to demonstrate that in this kind of state it is gauge symmetry which is broken in the same sense that a lattice breaks translational symmetry. The phase angle is arbitrary, but it must be coherent throughout the system (Anderson, 1964, 1966; Josephson, 1974).

This definition has the advantage that it describes the states of the Bose system wholly in terms of local variables--a kind of generalization of the notion of an intensive variable whereby we feel we can describe any macroscopic phase in terms of T, P, ρ, etc., locally. Another advantage is that it automatically ensures--as is physically correct--that no part of the system can be ascribed a fixed number of Bose particles, because there is a number-phase uncertainty relationship which gives maximum uncertainty to N if ϕ is fixed and vice versa.

<u>Gauge symmetry</u>: whenever the Hamiltonian is such as to conserve the total number of particles of a par-

ticular sort--or, more generally, where there is a conserved "charge"-like quantity, such as lepton or baryon number, or electric charge itself--we shall find that the Hamiltonian will exhibit a gauge invariance property. In fact, this is simplicity itself: the appropriate field always enters into any interaction in the form $\psi^+(r)\psi(r) = \rho(r)$, so that multiplying ψ everywhere by $e^{i\eta}$ and ψ^+ by $e^{-i\eta}$ leaves H unchanged. Local gauge symmetry in which η may be a function of r is something else again: this occurs only in the case of electric charge, because in order to maintain gauge invariance it is essential to introduce a gauge field which couples to the conserved charge--in this case, the electromagnetic field. Attempts to introduce gauge fields for other conserved quantities are promising but inconclusive as yet. One of the most important motivations for the modern color-quark-gluon gauge theories of elementary particles is to restore this concept of local gauge symmetry connected to conserved charges.

To these two canonical types of broken symmetry--lattice formation and broken gauge symmetry--one may add broken time-reversal invariance, i.e., magnetic ordering, either ferro- or antiferromagnetic. In general, these phenomena also break an at least approximate spin rotation symmetry. Ferroelectricity is not in principle different; it is merely a lower crystal symmetry that involves formation of a polarity, i.e., broken inversion symmetry. Liquid crystals in the nematic phase exhibit a fourth type: broken local rotational symmetry. So does liquid ^{3}He, but this is intimately related to another form of phase transition, broken gauge symmetry, as in the Bose case. Superconductivity is also broken gauge symmetry.

We now have essentially answered the "Why?" of broken symmetry: under surprisingly general circumstances the lowest-energy state of a system does not have the total symmetry group of its Hamiltonian, and so in the absence of thermal fluctuations the system assumes an unsymmetrical state. The main remaining question is "What are the consequences?" These are fourfold:

(i) discreteness of phase transitions, and the resulting failure of continuation: disjointness of physical phases;

(ii) development of collective excitations and/or fluctuations;

(iii) generalized rigidity;

(iv) defect structures: dissipation and topological considerations.

C. DISCRETENESS AND DISJOINTNESS: FIRST-ORDER VERSUS SECOND-ORDER TRANSITIONS, FERROELECTRIC, $N \to \infty$ LIMIT

It was Landau (Landau and Lifshitz, 1958) who, long ago, first pointed out the vital importance of symmetry in phase transitions. This, the First Theorem of solid-state physics, can be stated very simply: it is impossible to change symmetry gradually. A given symmetry element is either there or it is not; there is no way for it to grow imperceptibly. This means, for instance, that there can be no critical point for the melting curve as there is for the boiling point: it will never be possible to go continuously through some high-pressure phase from liquid to solid. Landau suggested also the consequence that "second-order" phase transitions (of which he then held a slightly over-simplified view) could only occur between phases of different symmetry.

The liquid-gas transition is in fact typical of a symmetry-nonbreaking transition (Fig. 2-2). Another is the

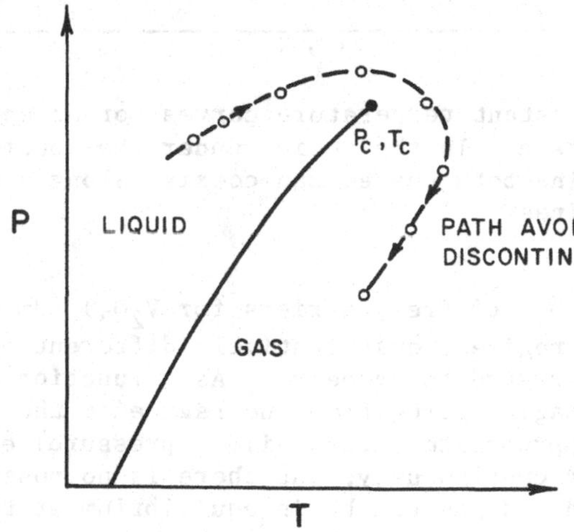

Figure 2-2a Phase diagram for a simple substance.

Mott transition in V_2O_3, which can be thought of as a free-electron liquid-to-gas transition (I shall have more to say about this case later) (Mott, 1974; McWhan et al., 1971). In either case, one of the parameters (density in the

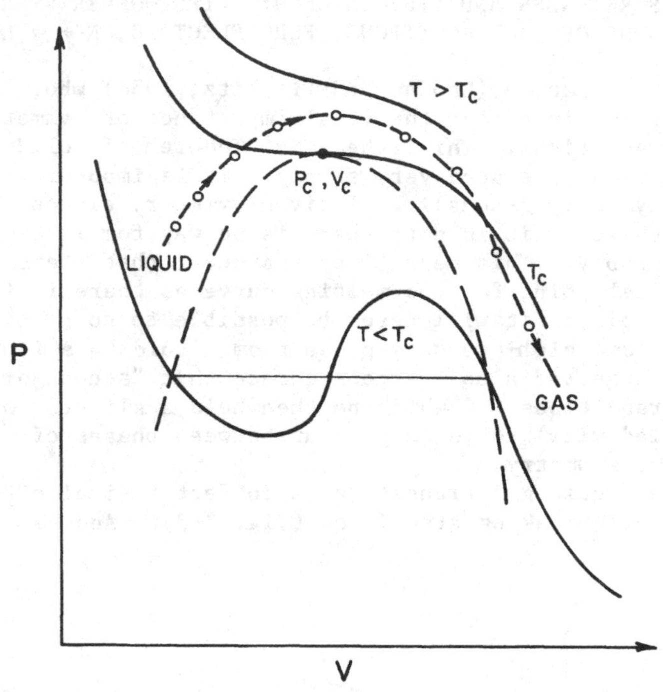

Figure 2-2b Constant temperature curves for an equation of state. In the region under the lower dashed line both phases can coexist along horizontal lines.

liquid-gas; number of free carriers for V_2O_3) admits of two very different regimes, quantitatively different but not different with regard to symmetry. As a function of temperature we can imagine going from one regime to the other (by applying the appropriate stress, i.e., pressure) either discontinuously or continuously; but there is no possibility that the liquid and gas can be in equilibrium at the same density except at an exceptional point on the boiling curve at which the discontinuous change disappears and the transition disappears entirely. This is because it is only the density parameter and other variables of state which distinguish the two.

In the other case of true broken symmetry, the unsymmetrical state is normally characterized by an "order parameter" η. By Landau's definition this is simply any parameter that is zero in the symmetric state and has a nonzero average uniquely specifying the state when the symmetry is broken; and, of course, normally there is no doubt what the correct definition of the order parameter is: such a variable is $\langle\psi\rangle$ in the Bose gas, $\langle\vec{M}\rangle$ in a magnet, etc. The concept has, however, a precise meaning: it is an additional variable necessary to specify the microscopic state in the lower symmetry state. In a nonmagnetic material, for example, in the absence of a magnetic field $\vec{H}$, $\vec{M}$ is zero by time-reversal symmetry, and the state is uniquely specified by the usual P, T intensive variables. In a normal, nonsuperfluid system, gauge symmetry ensures $\langle\psi\rangle \equiv 0$, as another example. Thus by broken symmetry a new variable is created. We shall see later how to look at the order parameter in an even more fundamental way, and, in particular, that it always has the important property of having a generalized phase as well as the amplitude which Landau emphasized in his statistical arguments. At the very least, the order parameter can take on two signs (as in the Ising model, for instance), but in general it has to specify as well one or several angles. The free energy F in general does not, by symmetry, depend on the phase angle variables, since different directions are connected by the original symmetry. This is the most fundamental broken-symmetry property, as we shall see.

Using the fact that once symmetry is broken, a new variable is necessary to specify the state completely, it is quite generally possible to predict that $F = -T\ln\langle e^{-\beta H}\rangle$ in the new system is a different mathematical function (even when the new variable is eliminated—say, by the minimum condition $\frac{\partial F}{\partial M} = H \equiv 0$) from that in the old. Let us, for instance, call the new variable ψ. Then we have in general from the (in principle) direct calculation of $\ln Z$ with ψ specified

$$F = F(V,T,\psi). \tag{2.9a}$$

We calculate F(V,T), which is what we observe by conventional means, by appending to this the equilibrium condition

$$\frac{\partial F}{\partial \psi} = f = 0 \ , \tag{2.9b}$$

where f is the generalized force variable (like the magnetic field H) corresponding to ψ. Above T_c, this is satisfied by symmetry; below, it is nontrivial, and when (2.9b) is substituted back into (2.9a), the result is a new function $F'(V,T)$, which is not analytic at T_c. We shall demonstrate this by an example shortly. This is Landau's essential insight.

A different sort of discontinuity is associated with first-order phase transitions. Suppose F is a single continuous, analytic function of T and V such that F plotted as a function of V at $T > T_c$ is an asymmetric double well with the minimum at V_1 lower than that at V_2. As we lower the temperature, the asymmetry is reduced until at $T = T_c$ the double well is symmetric with $F(T_c,V_1) = F(T_c,V_2)$ with $P = \left.\frac{\partial F}{\partial V_1}\right|_{T_c} = \left.\frac{\partial F}{\partial V_2}\right|_{T_c}$. Below T_c, $F(T,V_1) > F(T,V_2)$. Thus the system jumps discontinuously from V_1 to V_2. This, then, is a discontinuity in derivatives (in this case with respect to T), of the appropriate modified free energy

$$A = A(P,T) = F(V,T) - PV$$

as a function of the intensive variables P and T.

> It is possible to subsume the usual liquid-gas or two-phase solution critical point into the same structure, as chemical statistical mechanics does, by the "lattice gas" trick of introducing concentrations of "atoms" and "vacancies" into the two phases as a new intensive order parameter. In a way this is legitimate: segregation of some constituent is a new possibility which has some of the nature of a broken symmetry; but from our point of view perhaps this lattice gas ↔ broken symmetry analogy should be treated as a mathematically valid trick, not necessarily a fundamental insight.
>
> Until a short time ago I would have held, with Landau, that the absence of symmetry change precludes a higher-order transition--by definition, a point at which F and its first derivatives (e.g., entropy), which can be chosen to be intensive variables determining the state, are continuous. Only if a new variable, the order parameter, growing from zero, is introduced, can a higher-order transition occur. This has, however, become a very knotty question. Some--but not all--transitions to rigid, glasslike states, may

entail a hidden, microscopic order parameter which is not a macroscopic variable in any usual sense, and describes the rigidity of the system (Edwards and Anderson, 1975; Anderson, 1976b). This is the fundamental difficulty of the order-parameter concept: at no point can one be totally certain that one can really exclude a priori the appearance of some new hidden order, just as in antiferromagnetism the sublattice magnetization $\vec{M}$, or in superfluidity $\langle\phi\rangle$, were hidden variables for many years. It is probably best not to make any categorical or dogmatic statements, especially in view of the further complications caused by fluctuation effects at phase transitions, to which we shall return in Chapter 5. Another example is the insight by Kosterlitz and Thouless (1973) that an order parameter may be purely topological in nature, having arbitrarily large local displacements but retaining an overall topological integrity. In fact, one can almost characterize 1965-1975 as the decade in which, one after another, the value of Landau's great generalizations was first demonstrated, and then their precise validity destroyed; this is the essence of my last chapter.

Even in the Landau theory, symmetry changes can and often do take place discontinuously, i.e., as first-order transitions, as in the canonical liquid-solid case. One of the possible reasons is closely related to point (ii) which we mentioned at the end of the last section: at this transition certain response functions diverge, and as a result feedback affects the transition.

Barium titanate is a well-known case of this, which may be described mathematically in very simple fashion (Devonshire, 1949). At the ferroelectric transition a dielectric polarization P appears, destroying the inversion symmetry. A free energy which describes this adequately may be written

$$F = A(T-T_c)P^2 + BP^4 + CP^6 \quad , \tag{2.10}$$

where A, B, and C are positive constants, and (neglecting critical fluctuation effects) we at first expect P to vary as shown in Fig. 2-3. For $T > T_c$ the minimum free energy is F=0, which corresponds to no polarization, i.e., the order parameter P=0. But for T

$< T_c$, the first term is negative and F=0 is no longer

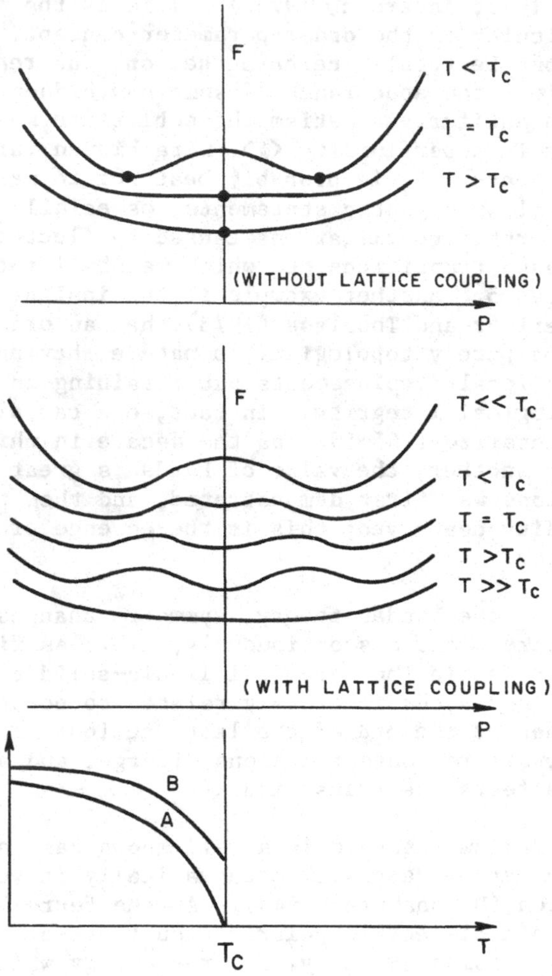

Figure 2-3a-c Ferroelectric phase transition.

the minimum free energy. To find the new minimum, we minimize F with respect to P^2 as follows:

$$\frac{\partial F}{\partial P^2} = 0 = A(T-T_c) + 2BP^2 + 3CP^4 + \dots \quad ,$$

so that to first order in P^2,

$$P^2 \simeq \frac{A}{2B}(T_c - T). \tag{2.11}$$

By substitution we have

$$F(T) = 0 \qquad T > T_c$$

$$F(T) = -\frac{A^2}{4B}(T-T_c)^2 + \dots \qquad T < T_c. \tag{2.11a}$$

This is a good example of the phenomenon discussed above, that F(T) is not the same analytic function above and below T_c, even in this "mean field" theory. There is, however, a complication: when the ions displace to form P, clearly they may (and do) distort the crystal as a whole, via a nonlinear elastic effect, causing a strain x, which in fact is the observed distortion towards tetragonal from cubic (see Fig. 2-3d). This gives us an $x\text{-}P^2$ coupling, which we write

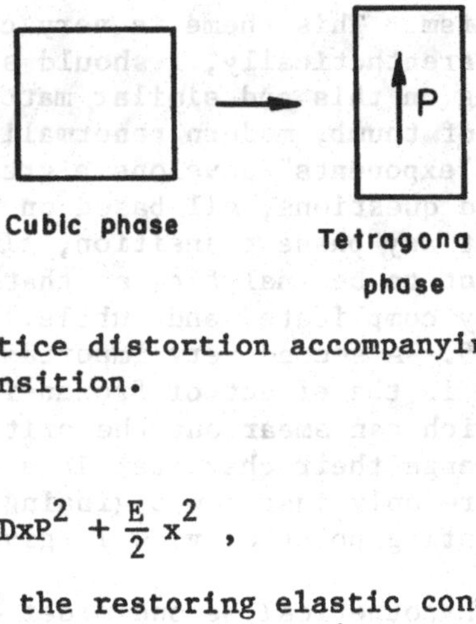

Figure 2-3d Lattice distortion accompanying ferroelectric transition.

$$F_{strain} = DxP^2 + \frac{E}{2}x^2 , \tag{2.12}$$

introducing the restoring elastic constant E corresponding to this strain (see Fig. 2-3b). Minimizing the free energy with respect to x, as well, we obtain

$$\frac{\partial F}{\partial x} = 0 = Ex + DP^2 ,$$

so that

$$F_{strain} = -\frac{D^2P^4}{4E}, \tag{2.13}$$

which, it turns out, outweighs the expected positive P^4 term in (2.10) and leads to a slightly first-order discontinuous appearance of P (see Figs. 2-3b,c).

In some cases symmetry forbids a second-order transition for reasons discussed by Landau (Landau and Lifshitz, 1958). The relevant question is whether the symmetry allows a so-called "third-order invariant," i.e., whether terms like $|P|^3$ can appear. In the case of ferroelectricity such a term doesn't satisfy reflection symmetry, but in many cases it is allowed. If no symmetry reason requires the P^3 term to vanish, a first-order transition always occurs, for an obvious reason: that $P = 0$ and $\frac{\partial F}{\partial P} = 0$ does not define a minimum. This occurs everywhere except possibly at a single point on a transition line. In either case the moral is the same: the second-order symmetry-breaking phase transition would occur except that it is preceded by a jump to the unsymmetric state, whose free energy is lowered by some additional mechanism. This theme is very common in solid-state physics. Parenthetically, I should say that while Landau's reasoning on this and similar matters is a good qualitative rule of thumb, modern renormalization group theory involving "exponents" develops a great number of subtle caveats and questions, all based on the fact that, actually, right at any phase transition, fluctuations cause the free energy not to be analytic, so that its behavior can be extraordinarily complicated and subtle. Such worries we defer to Chapter 5. A second very important caveat to the present reasoning is the effect of frozen-in impurities and other defects, which can smear out the critical phenomena remarkably and change their character in a number of ways. These phenomena are only just now beginning to be understood; one interesting point of view is given in Halperin and Varma (1976).

Perhaps the hypothetical second-order ferroelectric phase transition of $BaTiO_3$ will serve as well as any to illustrate the nature of a symmetry-breaking transition. The naive theory given here suggests that the symmetry-breaking order parameter should grow from zero according to $P \propto \sqrt{(T_c - T)}$ and that, therefore, $F \propto P^4$ exhibits a discontinuity in its second derivative with respect to temperature at T_c. In fact, the singularity in F is often found to be rather worse because of critical fluctuation effects. Ferromagnetic, antiferromagnetic, superconducting, and various other kinds of broken-symmetry transitions all behave in more or less the same way. The less symmetric state tends to be the lower one in temperature, simply because the more symmetric one is usually a distribution of thermal fluctuations among all the available values of the order parameter; in this case, it is a fact that microscopically $|P|$ is not zero in the higher-temperature state, but fluctuates in sign and direction to give a zero average. But this order of phases in temperature is not a general rule; ^{3}He, for instance, violates it because the solid has a greater nuclear paramagnetic entropy than the liquid, and at low temperatures the melting curve in the P-T plane has a negative slope. So in this temperature regime, the solid is the high-temperature phase and the liquid is the low-temperature phase.

In simplest terms, it is quite clear that the First Theorem requires that F have a boundary of singularities between the two regimes of symmetry, and that therefore analytic continuation between the two is not possible. The second-order phase transition is, as we have shown, a singular point of F; even where we do not observe it because the first-order transition intervenes, in any attempt to go continuously from the symmetric to the unsymmetric state via unstable configurations we would certainly encounter this strange point of nonanalyticity. When we examine renormalization group theory, we shall find this point expressed in a very clear way. It is, of course, quite clear that this kind of discontinuity is a result of the large size of the many-body system we consider. The free energy of a finite system is manifestly an analytic (even entire) function of the relevant parameters:

$$F/T = -\ln \mathrm{Tr} \exp\left[-\frac{H}{T} - \frac{\mu N}{T}\right]. \qquad (2.14)$$

Thus this analytic singularity must arise somehow in the limit of large systems: the $N \to \infty$ limit. How this does or does not take place may perhaps best be illustrated in terms of a simple model I have discussed elsewhere (Anderson, 1972b): the NH_3 molecule as a microscopic example of a system without inversion symmetry (see Fig. 2-4).

> Two of the five valence shell electrons in N prefer to lie in a rather deep, predominantly 2s

state. As a result, the three covalent bonds to the hydrogen atoms are predominantly p bonds, and in order to retain their orthogonality they form a triangular pyramid, leaving the N atom out of the plane of the three H's. The resulting effective potential for

Figure 2-4 Ammonia molecule.

motion of the N relative to the H triangle is a two-well potential. The lowest vibrational states are relatively little affected by the possibility of tunneling through the barrier, which in the lowest state leads to an energy splitting of 1 cm^{-1} in 10^3. Thus for most thermal purposes at ordinary temperatures, we may treat NH_3 as an asymmetrical pyramidal molecule. However, if we were to raise the temperature sufficiently high, the effect of the barrier would be very much reduced, since higher vibrational states first split and then reform in a well-known manner to the scheme appropriate for a single potential well.

There are two points to be noted here. First is that even in the ground vibrational state the actual lowest state of the molecule is symmetric with respect to the H_3 plane, since the two orientations are connected by a tunneling matrix element. In ND_3 this tunneling is greatly reduced, and in heavier pyramidal molecules practically not present at all. In principle, as a result, the polarizability of NH_3 would remain finite at absolute zero, while that of a heavier pyramid would obey a Curie law and diverge as T^{-1}, because it would effectively have a finite electric dipole in zero applied field. But any thermal agitation will cause the inversion transition to take place, so that there is no true "one-sidedness" at any temperature. When one gets to normal-size biological molecules, one finds that very often what are called "optical isomers" are quite stable: ordinary sugar (dextrose) is an example. But there are no sharp phase transitions in molecules.

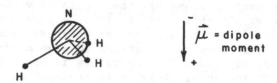

It is only as we go to macroscopic systems in which large numbers of particles cooperate that we find a sudden, rather than gradual, thermal transition. If each dipole moment acts to hold the others in the same direction via a "molecular field," the reversal of any one will not reverse the total but simply represent a local fluctuation. The probability that all fluctuate at once is

$$\propto \exp - \sqrt{N} \sim e^{-10^{12}},$$

and utterly negligible. When, on the other hand, a sufficient number are reversed, the system can quite suddenly transform with enormously high probability into a wholly different region of phase space, where each sign of the dipole moment occurs with equal probability. Thus there is a rather subtle dividing line [discussed for superconducting systems, for instance, by me (Anderson, 1964) and by Ambegaokar and Langer (1967)] between the microscopic unsymmetrical system and the macroscopic broken symmetry. One characteristic of this distinction is that one ceases to be correct in doing one's thermal average over all possible states of the system: essentially always, at least one aspect of a phase transition is an ergodic-nonergodic transition. Once $\vec{P}$ develops, one may no longer average Z over all configurations with both macroscopic $\vec{P}$ and macroscopic $-\vec{P}$; the latter are to be treated as unavailable. One uses (2.9a) with (2.9b), where (2.9a) is calculated with a fixed value of the order parameter ψ. The borderline comes in the shadowy area where quantum or thermal fluctuations are only just capable of carrying one over the barrier between two equivalent configurations. This borderline has been beautifully investigated experimentally by T. Fulton (1968) in the superconducting case.

The above is all part of elementary solid-state physics; nonetheless, its implications seem to be ignored rather often, and by very eminent scientists. The error into which one must not fall is to attempt to calculate the properties of one state of matter by methods suitable only to another. Between the two there must invariably exist a discontinuity if the system is large enough, and there is therefore no reason to expect convergence of the right method in the wrong domain. For instance, one of the founding fathers of solid-state physics, Eugene Wigner, used a continuous interpolation between the free-electron gas and the Wigner lattice to estimate the correlation energy for ordinary metals (Wigner, 1934). Quantitatively this may or may not be valid--some hypothetical liquid phase at low densities might have nearly the energy of the Wigner solid. But in principle these are two distinct analytic functions and no valid continuation procedures connect them; unfortunately, Wigner's idea, which was entirely reasonable at that early date, has become embalmed in the textbooks and is to my knowledge never questioned. I feel that nearly as false an extrapolation--from gas to liquid, which is not quite so bad--underlies much of modern nuclear matter theory.

But the most severe abuse by far of this principle has been the growth of the "quantum chemistry" approach to condensed matter physics: "If I can just compute the electronic state of some large but finite cluster of electrons and atoms, at absolute zero, with giant machines, that is all that I need to do." Such a calculation, however, will in general not demonstrate the existence or not of a broken symmetry, which in principle is ascertainable only in the $N \to \infty$ limit. Clearly, computation must come after understanding, not before. This question is closely related to my second general principle, that of continuity, and will be treated more fully in Chapter 3.

D. COLLECTIVE EXCITATIONS AND FLUCTUATIONS: SPIN WAVES, CRITICAL FLUCTUATIONS, PHONONS. DIMENSIONALITY OF ORDER PARAMETERS.

The Hamiltonian describing the internal interactions in our macroscopic system still obeys the initial symmetry group, whatever it may have been; the <u>state</u> of the matter involved does not, in some sense, at low enough temperatures; do any consequences flow from that? The answer is that some of the most important and interesting properties of matter follow from precisely this fact. It has two kinds of consequences: it controls much of the dynamics, affecting the spectrum of elementary excitations and fluctuations; and it causes new static properties of the sort I shall call "generalized rigidity," as well as permitting certain special solitonlike objects called "order-parameter defects." Chronologically, the former type of effect was the first to be understood in a general way, and we begin with the problem of low-frequency dynamics.

There are two possible cases. The new, unsymmetric state can be either an eigenstate of H or not.

The original Hamiltonian, in general, has a group of symmetries G. No dynamic effects at low temperatures follow unless the group is a continuous one--for instance, the

Ising model has no spin-wave-like excitations and no interesting low-temperature dynamics. G, if continuous and compact, may be described by a set of generators L_i, i.e., the group elements A belonging to G

$$AHA^{-1} = H$$

may be written as

$$A = e^{i\theta L_i} ,$$

where L_i is related to infinitesimal transformations (e.g., infinitesimal rotations $\delta\theta$). For an infinitesimal transformation ε

$$A = 1 + i\varepsilon L_i ,$$

and continuous groups may in general be built up by products of infinitesimal operators (in addition to, possibly, some group of discrete operators such as parity and time reversal). The spin rotation group, for instance, has the generators S_x, S_y, S_z, and a rotation about any angle can be expressed as exp $i(\vec{\theta}\cdot\vec{S})$, where $\vec{\theta}$ is a vector along the rotation axis.

The only unsymmetrical operators which commute with H are the various group generators L_i. These are therefore conserved quantities; and if the order parameter can be taken as one of them, it is conserved. This, of course, is the case for ferromagnets, where the order parameter may be taken as the total spin in some fixed direction Z. That may well be the only example of a simple, conserved order parameter, so it may be just as well not to try to create too general a theory for this case. The group may or may not have other generators which do not commute among themselves, i.e., it may be either Abelian, in which case all representations are of dimension 1 and the ground state is nondegenerate, or non-Abelian, like the rotation group, which is the isotropic ferromagnetic example: S_x, S_y, S_z are noncommuting. It is only in the latter case that group theory alone tells us the whole story: the order parameter may be taken as, say, S_z, and S_x and S_y commute with H but not with S_z. Thus if S_z is nonzero, there are other degenerate ground states ψ'_g formed by applying these generators to the original one ψ_g :

$$\psi'_g = (S_x \pm iS_y)^n \psi_g = (S^\pm)^n \psi_g \qquad n = 1, \ldots, 2S_{tot}$$

and we recognize these as the $(2S_{tot} + 1)$ degenerate ferromagnetic ground states of

$$H = -\sum_{ij} J_{ij} S_i \cdot S_j ,$$

obtained by lowering from the state $M_s = + S_{tot}$. Linear combinations of these are the ground states with spin oriented along different directions.

The Abelian case also occurs, of course: consider the axially symmetric ferromagnet

$$H = -\sum_{ij} \{J^z_{ij} S_{iz} S_{jz} + J^\pm_{ij}(S^+_i S^-_j + S^-_i S^+_j)\} ,$$

which has the rotation operator S_z^{tot} around the z axis as an Abelian group generator. If $J^z > J^\pm$, the ground state is unique: the state with $S = NS_0$ aligned along the z axis. If $J^\pm$ is bigger, we have the "x-y model," which is a typical example of the nonconserved order parameter that we will talk about shortly.

When--in the unique isotropic case--we have a group generator which generates the degenerate set of eigenstates, we can deduce the existence of a branch of excitations related to it: the spin waves. That is, we may consider the operation of rotating the spins in a local region.

This turns out to give us a set of excitations which are eigenstates, the frequencies of which approach zero as their momentum goes to zero, i.e., the spin waves first described by Bloch (1930). One can express the spin-wave excitations in two related ways. The operator $S^- = S_x - iS_y$ is a lowering operator for the spin S_z:

$$S^- \psi(S_z = S) = \psi(S_z = S-1) .$$

If S and S_z characterize the spin state of the sample, the "lowered" state is still an eigenstate, which may be shown to be a linear combination of states in which $\vec{S}$ is oriented along axes near z but rotated from it slightly:

$$S^- \psi \sim \int d\theta \; e^{i\theta} \; \psi(\theta).$$

We can also consider "lowering" the spin locally, over a region r, containing a fairly large number of individual atomic spins, by taking $S^-(r) = \sum_r S_i^-$. This lowered state is no longer degenerate with the ground state, because the spins i in r are coupled to other spins outside r. Lowering the total spin is done by summing over all regions,

$$S^-_{tot} = \sum_{\{r\}} S^-(r).$$

But we may lower different regions out of phase, e.g., by using

$$S^-(k) = \sum_{\{r\}} e^{ik\cdot r}\, S^-(r).$$

This is Bloch's way of producing spin waves; but it is just a linear combination of the classical Herring-Kittel way of producing a spin wave (Herring and Kittel, 1951), which is to allow a small spiral precession of the quantization axis of the total spin. Bloch showed that the resulting state is to an excellent approximation an eigenstate; we shall give a semiclassical derivation of this same result below. The essential idea is that since the total magnetization can rotate freely, $[H, S_x + iS_y] = 0$, the energy for twisting the magnetization about the z axis with a given wave number k must be proportional to k^2, i.e., the effective Hamiltonian must always be $\propto (\nabla S)^2$. Therefore a slow twist of S must be a very low energy excitation.

One actually perfectly rigorous way to approach spin waves is to note that, because their frequencies are very low, a quantum of excitation is very small and thus represents only small displacements of the system from its ground-state configuration with M parallel everywhere. Thus the restoring forces must be perfectly linear, and the potential harmonic and in this case classical equations of motion are perfectly correct. It is then correct to write

$$[H, M] = \frac{d\vec{M}}{dt} = \gamma\left(\vec{M} \times \vec{H}_{eff}\right) \qquad (2.15)$$

(we write $\vec{M} = \gamma\vec{S}$, γ being the so-called gyromagnetic ratio). Then we note that exchange leads to a potential

$$V_{exch} = \frac{A}{2} (\nabla \vec{M})^2. \tag{2.16}$$

To see this, note that

$$J_{ij} \vec{S}_i \cdot \vec{S}_j = J_{ij} S^2 \cos(\theta_i - \theta_j) \simeq J_{ij} S^2 [1 - \frac{(\theta_i - \theta_j)^2}{2}]$$

$$= \text{const} - J_{ij} S^2 (\vec{R}_{ij} \cdot \nabla\theta)^2$$

$$= \text{const} - J_{ij} S^2 (\vec{R}_{ij} \cdot \frac{\nabla M}{M})^2$$

where $\vec{R}_{ij} = \vec{r}_i - \vec{r}_j$. Therefore there is an effective field on the local magnetization

$$\vec{H}_{eff} = \frac{\delta V_{exch}}{\delta M} = A \nabla^2 \vec{M}. \tag{2.17}$$

We get

$$\frac{d\vec{M}}{dt} = \gamma A (\vec{M} \times \nabla^2 \vec{M}) , \tag{2.18}$$

which is the Landau-Lifshitz equation leading to spin waves with frequency

$$\omega = \gamma A k^2 \tag{2.19}$$

for waves with $\nabla^2 M = -k^2 M$, that is, $M_x + iM_y \propto e^{ik\cdot r - i\omega t}$.

It is only because the ground state violates time-reversal symmetry that the $\omega \propto k^2$ dispersion relation occurs: otherwise, as in antiferromagnets, ω cannot have a sense (i.e., right- or left-handed circular polarization) and the dispersion relation should depend only on ω^2 like that of phonons, which is

$$\omega^2 = s^2 k^2 \rightarrow \omega = s|k|.$$

Actual physical ferromagnets do not, in fact, obey perfect rotational symmetry, because they are normally crystalline; but very often, because of the weakness of spin-orbit coupling, the spin system of the ferromagnet comes rather close to doing so. It is characteristic that while spin-orbit coupling gives one a finite frequency for the longest wavelengths, the bulk of the spin-wave spectrum is unaffected. In fact, the effect of axial anisotropy can be determined by simply including in the Landau-Lifshitz equation an "anisotropy field" H_A, giving

$$\omega = \gamma\left(H_A + Ak^2\right). \tag{2.20}$$

This question is related to that of the "dimensionality of the order parameter," which plays a great role in critical fluctuation theory. When the violated symmetry is only that of a point group--a discrete symmetry--we speak of a "one-dimensional order parameter." An example is the Ising model for ferromagnetism, which is isomorphic symmetrywise, with the ferromagnet with axial anisotropy having the favored direction along the axis. In the Ising model, there is no question of collective modes, because there are no continuous degrees of freedom to support them, whereas in the axial ferromagnet, the modes are there but do not appear at zero frequency; they have a frequency proportional to the anisotropy. We recognize this difference as not being trivial, in spite of the symmetry situation, but rather fundamental; even though H does not obey full rotation group symmetry, the underlying dynamical objects do: they are spins, a consequence of the rotation group properties of the fundamental particles, only coupled together and to a lattice in such a way as not to exhibit their full symmetry. Nonetheless, the group generators play a role in their excitation spectrum, and it is easy to deduce from that the underlying objects. This is the same kind of role the famous "SU3" symmetry plays in fundamental particle physics; although not too well satisfied, its operators are important in classifying states.

Near a second-order T_c, no matter what the dimensionality and symmetry, there will be the finite temperature equivalent of zero-frequency modes, namely, critical fluctuations with divergent amplitude as $T \to T_c$. Critical fluctuations are, again, a wholly classical phenomenon, because they always occur at finite temperature and involve frequencies $\ll T$; thus it is fair to neglect dynamics. The classic "Ornstein-Zernike" (1914) treatment invokes the idea of a "Ginzburg-Landau-type" free-energy functional (Ginzburg and Landau, 1950). One imagines the coarse-grained average of the order parameter to be specified as a (necessarily very slowly varying) function of position: $\vec{M} = \vec{M}(r)$, and then asks for the free energy as a functional of this function, i.e., with $\vec{M}(r)$ constrained. It is clear that if an analytic expression exists near T_c for such an F, it must be approximated as

$$F = \alpha(T - T_c)M^2 + A(\nabla M)^2 + BM^4 + \ldots , \qquad (2.21a)$$

in close analogy to the observation about the energy at T = 0 and also to the Landau free energy of a ferroelectric discussed earlier. The $(T - T_c)$ term is the simplest way to ensure that M will grow from zero above T_c to a finite value below it--see the discussion of the ferroelectric. Then the probability of a given fluctuation M(r) will, of course, be

$$P \propto e^{-\frac{F(M)}{T}} , \qquad (2.21b)$$

which leads to fluctuation amplitudes

$$\langle M^2(k)\rangle = \frac{\langle M^2 e^{-F/T}\rangle}{\langle e^{-F/T}\rangle} \propto \frac{1}{\alpha(T-T_c) + Ak^2} , \qquad (2.22)$$

which in turn diverges as $k \to 0$ and $T \to T_c$ (we have Fourier-transformed M to momentum space).

This divergence in the fluctuation amplitudes in turn implies that the relevant frequencies approach zero in some manner. To see how, consider the following. It is clear that if we apply an external field H to interact with the order parameter, we introduce a term -M.H into the free energy. The susceptibility χ is given by

$$\chi = \frac{\langle M\rangle}{H} = \frac{1}{H}\frac{\langle M e^{\frac{M.H}{T} - \frac{F(M)}{T}}\rangle}{\langle e^{\frac{M\cdot H}{T} - \frac{F(M)}{T}}\rangle} = \frac{\langle M^2\rangle_{H=0}}{T} . \qquad (2.23)$$

Thus at least classically χ', the real part of the susceptibility near $\omega = 0$, and $\langle M^2\rangle$ are equivalent. But by the Kramers-Kronig relationship

$$\chi(\omega=0) = \frac{1}{\pi}\int_{-\infty}^{\infty} \frac{d\omega'}{\omega'} \chi''(\omega') \qquad (2.24)$$

where χ'' is the imaginary part of the susceptibility. χ'' has some spectral weight near $\omega = 0$ (there can be no absorptive part of a polarizability <u>precisely</u> at $\omega = 0$), thus implying excitations which become concentrated at low frequencies. (As far as I can see, this is not a rigorous argument, since χ'' could blow up in principle; but it does generally appear to be the case.) All of this argument is a rather trivial example of a use of the "fluctuation-dissipation theorem" of Callen and Welton (1951).

For many years the real difficulty with understanding critical-point behavior was the question of how the argument leading to (2.22) could possibly fail. Clearly the Landau function (2.21) must be the only possible result of the Gedanken-process of constraining $\langle M(r) \rangle$ to be some slowly varying function and then averaging over shorter-range fluctuations, if our ideas about such things as the very existence of a thermodynamic, $N \to \infty$ limit are to be correct. At some scale, in other words, it must be possible to treat the matter as though it were a medium with certain linear-response functions defined as functions of T and k, and in which, therefore, by Eqs. (2.21a and 2.21b) it has purely Gaussian fluctuations of the simple Ornstein-Zernike form. It is the great achievement of the scaling idea of Widom (1965) and Kadanoff (1964, 1966) and the Wilson-Fisher renormalization group (Wilson and Fisher, 1972) that they showed how these assumptions actually do fail right at the critical point, but nowhere else. The fluctuations do become of quasimacroscopic size and dimension at the critical point, as a consequence of the divergences predicted by (2.22).

I repeat that these fluctuations may or may not correspond to real collective modes. In some cases they do, and these are called "soft modes." In ferroelectrics, for instance (Anderson, 1960; Cochran, 1959), one of the optical modes drops to very low infrared frequencies. In other cases the fluctuations have a broad spectrum centered at zero ("diffusive case"), and in yet others there are both, for reasons not as yet entirely clear. In general, the frequency spectrum of the fluctuations may be (and often is) complicated.

In the case of a continuous symmetry group, the low-frequency collective excitations, which are necessarily there at absolute zero, are also present, of course, throughout the temperature range up to T_c and merge, more or less, with the critical fluctuation spectrum near T_c. These phenomena have been studied in much more detail by Halperin and Hohenberg. (For a review, see Hohenberg and Halperin, 1977.) At any temperature above zero, especially in a metal, the spin waves need no longer be sharp excitations, but will broaden to a greater or lesser degree into a spectrum.

In recent years, with the improvements in techniques for studying such hydrodynamic fluctuations--to a great extent because of laser scattering techniques--and with the interest in helium and its various kinds of sound--first, second, etc.--the understanding of these various low-frequency modes has grown apace. There are two dichotomies which must be understood: between "collisionless" and "hydrodynamic" modes, and the one already discussed between "diffusive fluctuations" and "soft modes."

Phonons in solids and the spin waves we have just been examining are examples of "collisionless" modes--essentially quantum excitations required by the symmetries, or the conservation laws in some cases, of the system. But at frequencies low compared to microscopic relaxation times, these may merge into "hydrodynamic" modes, which have the character of ordinary sound waves, compressional or other modes of vibration in which the matter remains in local thermodynamic equilibrium. Such hydrodynamic modes usually do <u>not</u> actually have collisionless analogues, as, for example, "second sound" in He, which is a hydrodynamic mode in the gas of collisionless excitations. Another example is ordinary sound in gases which have no collisionless excitations above $\omega \sim \tau^{-1}$. The hydrodynamic modes are themselves usually associated with a conservation law, in much the same way as phonons. The whole subject of modes near T_c is extraordinarily complex and is only now getting sorted out (see, again, Hohenberg and Halperin, 1977).

The nature of the low-temperature collisionless excitations is more interesting in the much more common case in which the order parameter is not a constant of the motion, and in which the low-temperature state does not belong to a representation of the symmetry group of the Hamiltonian. This case was first discussed in relation to antiferromagnetism, but in fact a far more familiar example offers itself: the crystalline state.

A crystal, as we have already remarked, manifestly does not have the symmetry of free space. Normally, in fact, we find ourselves forced to regard the crystal as essentially stationary and rigid. Once we locate a particular lattice plane in space, zero-point and even thermal fluctuations are so small as to allow us (if we know the mean lattice spacing well enough) to locate any atom to of order 10^{-9} cm, independent of its distance from the origin. If we are willing to wait long enough, the average atomic positions are rigidly connected to a far higher degree of accuracy than even that; long-range and long-term thermal fluctuations, though very small, represent the true limitations on our view of crystals as essentially perfectly rigid classical structures.

But the eigenstates of the pure translation group are not position eigenstates but momentum eigenstates, devoid of locality information. The same kind of homogeneity difficulty we encountered in the Bose gas occurs.

There is essentially perfectly precise correlation information for the relative positions of different macroscopic parts of the crystal, so that the only consistent description which treats the whole and its parts equivalently is to assign the entire crystal a precise location in space, or at least to locate it by a wave packet much smaller than the zero-point atomic motion.

In the absence of external forces, however, we know that different positions in space must be perfectly equivalent. The energy must depend not on absolute displacements u of the crystal but only on relative ones, i.e., on the gradient ∇u:

$$V = \frac{1}{2} \nabla u \cdot \underset{\approx}{c} \cdot \nabla u, \qquad (2.25)$$

where $\underset{\approx}{c}$ is the elastic stiffness tensor. This is indeed the standard expression for the elastic energy of a crystal, and is analogous to the $(\nabla M)^2$ energy we have already discussed. Only the symmetric part of ∇u matters in (2.25); the antisymmetric part $\nabla \times u$ simply denotes a local rotation, i.e., a reorientation of the fiducial lattice from which the displacement u_i is measured, and therefore the energy cannot depend on it.

It is important to understand the relation between u and the crystalline order: what is, in fact, the order parameter of a crystal? A crystal imagined at its inaccessible second-order transition point would undoubtedly first appear as a three-dimensional density wave,

$$\rho(\vec{r}) = \sum_{\vec{G}} e^{i\vec{G}\cdot\vec{r}} \rho_{\vec{G}} ,$$

with $\rho_{\vec{G}} = \rho^*_{-\vec{G}}$ ($\vec{G}$ denotes the reciprocal lattice vectors of the structure). It is interesting that a phenomenological Ginzburg-Landau free-energy expression would have to be of the form

$$F = \frac{1}{2}\sum_{\vec{G}} \alpha(|\vec{G}|)\left(T - T_c(\vec{G})\right)|\rho_{\vec{G}}|^2 + \text{higher-order terms.} \tag{2.26}$$

The orientation of the actual G's that appear is at first arbitrary; and the appearance of a set of three different ones coherently directed and phased in such a way as to form a three-dimensional crystal can occur only by virtue of "higher-order terms." There will in general be a third-order invariant $\rho_{\vec{G}_1} \cdot \rho_{\vec{G}_2} \cdot \rho_{\vec{G}_3}$ with $\vec{G}_1 + \vec{G}_2 + \vec{G}_3 = 0$, violating Landau's criterion for a second-order transition, except in one dimension.

Why is this third-order term and hence the relative phases of the density waves important? To answer this, take three density waves whose relative orientation is the same as that of the sides of an equilateral triangle (see Fig. 2-5). The lines in the figure represent the maxima of the

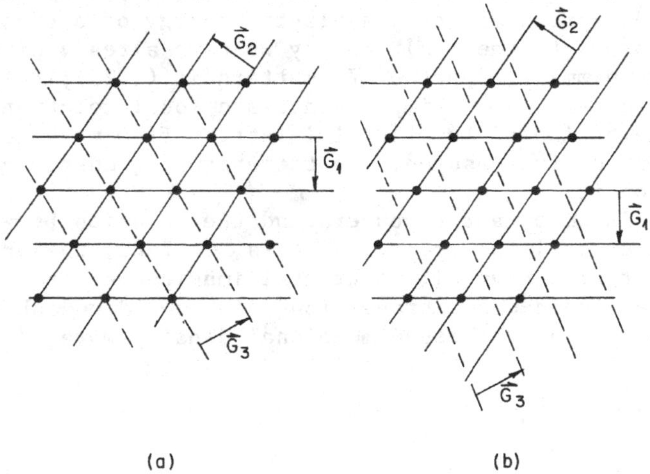

Figure 2-5 Three density waves form a two-dimensional lattice.

density waves. $\vec{G}_i$ gives the orientation of the ith wave as well as the spacing between maxima (or minima). Now fix the relative phase of two of the waves; i.e., superimpose two sets of lines. Where do we place the third set of lines? The phase of the third wave relative to the other two is not arbitrary because its maxima can fall on the lines of joint maxima of the other two (Fig. 2-5a) or between these lines of joint maxima (Fig. 2-5b). There is no reason why the energies of these two configurations should be the same. This proves that a two-dimensional lattice will always form, as, in fact, has been observed in the recent work on charge-density waves in TaS_2, etc. (Wilson et al., 1975). But in an isotropic case this construction is true in all planes and shows that a three-dimensional lattice will form. This, of course, is at least one reason why crystallization is normally first-order, but note that the question is far more complex than that: the topology of liquid crystals does indeed show a second-order transition to a one-dimensional density wave, but, vindicating Landau, this occurs only in the anisotropic nematic phase, where the third-order invariant isn't present! Alexander and McTague (1978) have recently pointed out that the bcc space lattice has the maximum possible number of these triangular third-order invariants, and have remarked that in fact many systems do crystallize into bcc lattices near the melting temperature T_m.

Be that as it may, we can consider the coupled set of $\rho_{\vec{G}}$'s which make up the crystal to be the order parameter, which, just as in the case of such order parameters as $\vec{P}$, $\vec{M}$, $\langle\psi(r)\rangle$, has an amplitude which is not related to invariance properties in any way, but which simply signifies the degree of order, and phases or directions, which are arbitrary and represent the symmetry-breaking parameters. In the crystalline case, we have two symmetry-breaking parameters: the direction of $\vec{G}$, which signifies the orientation of the crystal, and the phase of $\rho_{\vec{G}}$, which is related to the position variable $\vec{u}$. To see this, set

$$\rho_{\vec{G}} = |\rho_{\vec{G}}| e^{i\phi_{\vec{G}}}$$

$$\phi_{\vec{G}} = \vec{G} \cdot \vec{u},$$

and we have, by virtue of

$$\rho_{-\vec{G}} = \rho^*_{\vec{G}} ,$$

$$\rho(\vec{r}) = 2\sum_{\vec{G}} |\rho_{\vec{G}}| \cos[\vec{G} \cdot (\vec{r}+\vec{u})] \; . \tag{2.27}$$

Here $\vec{u}$ is a vector; hence we have three acoustic modes.

In this pure symmetry-breaking case, where the symmetry-violating part of the order parameter is not a constant of the motion, the equations of motion are more complex. The time dependence of the mean position X of the whole crystal doesn't vanish, but is just the corresponding velocity v:

$$i[H, X] = v = P_{tot}/M,$$

and there is a canonical commutation relation between X and its corresponding momentum P_{tot}, so that since we must make the uncertainty of X small, the uncertainty of P_{tot} is nearly infinite. However, the total mass $\underline{Nm}$ = M is so large that, of course, it leads to a very small v^2.

Again, it is valid to treat the long-wavelength collective coordinates (phonons) classically, so that it is an easy matter to work out their equations of motion. Define the Fourier transform of the displacement vector

$$u_q = \sum_n e^{iq\cdot r_n} u_n$$

Then the equations of motion are

$$\dot{u}_q = i[H, u_q] = \frac{\partial H}{\partial P_q} = P_q/M \tag{2.28}$$

$$\dot{P}_q = i[H, P_q] = -\frac{\partial V}{\partial u_q} = (q\cdot \underset{\approx}{c}\cdot q)u_q \tag{2.29}$$

so that we get a linear spectrum by substituting Eq. (2.28) into the derivative of (2.29):

$$\omega^2 = \frac{c}{M} q^2 \; . \tag{2.30}$$

This linear spectrum is characteristic of a number of systems: the Bose liquid and the antiferromagnet, particularly, as well as some excitations of the neutral Fermi superfluid such as ^3He. It is clear that the essentials of the derivation do not depend on the particular system. In fact, these ideas were generalized to a field-theoretic context by Goldstone, Salam, and Weinberg (Goldstone, 1961; Goldstone et al., 1962), whence this kind of mode has tended to be called a "Goldstone boson."

One special case in which the above argument does not hold is that in which there are long-range forces--specifically, Coulomb ones. In this case, although it is still correct to say that the energy does not depend on the uniform displacement $\vec{u}$, a longitudinal density wave leads to the accumulation of charge at the surface of the sample:

$$\rho = \frac{1}{4\pi} \nabla \cdot \vec{E} = ne\nabla \cdot \vec{u} \tag{2.31}$$

and the Coulombic restoring forces lead to a potential proportional to u_q^2:

$$V_{Coul} = \frac{1}{2} \frac{4\pi}{q^2} \rho_q^2 = 2\pi n^2 e^2 u_q^2. \tag{2.32}$$

Thus, in this case, the longitudinal collective mode has a finite frequency as $q \to 0$--in fact equal to the plasma frequency, as we see from

$$\dot{p}_q = -\frac{\partial V_{Coul}}{\partial u_q}$$

$$\dot{u}_q = \frac{p_q}{nm} .$$

Thus

$$\omega_p^2 = \frac{4\pi ne^2}{m} . \tag{2.33}$$

This plasma mode as a collisionless excitation is universal in charged condensed systems. For instance, it occurs in two important cases: the Wigner lattice, where only the two transverse modes behave as phonons, with $\omega \propto q$, and the superconductor, where there are no $\omega \to 0$ modes. Again this possibility has been picked up by the field theorists. Broken symmetry without Goldstone bosons is of interest to them because of the paucity of zero-mass particles (only neutrinos and photons, as far as is known) in the real world, so that if the vacuum is to be a broken-symmetry state the underlying theory should have long-range forces, as it does in the Weinberg models. I have elucidated this possibility in some detail (Anderson, 1963b), but a more complete discussion has been given by Higgs (1964); and the field-theoretic equivalent of a plasmon is now called a "Higgs boson."

In all cases, the zero-point motions of these collective modes play an important role in the energy and wave function of the ground state itself. The Debye-Waller theory is the earliest example of a theory relating the motions in the ground state to the collective excitations; the zero-point, as well as thermal vibrations, decreases the x-ray scattering amplitude from that which would result from stationary atoms. The quantum solids (He, H_2, the Wigner lattice) and also the Heisenberg antiferromagnet must basically be handled by calculating the zero-point motion, as we shall discuss in detail in Chapters 4 and 5.

The zero-point motion is central to the problem of restoring the broken symmetry. This was first realized in the antiferromagnetic case, where even the very existence of the antiferromagnetically ordered state postulated by Néel and Landau for the perfect Heisenberg antiferromagnet was in doubt, because of Hulthén's observation that the ground state must be a singlet (Hulthén, 1937). The $k = 0$ antiferromagnetic spin wave has essentially the equation of motion of a free, rather massive, rotor, the orientation of which is that of the sublattice magnetization. Thus the ground state belongs to $S = 0$, i.e., is isotropic; but there is a manifold of other states, degenerate in the $N \to \infty$ limit, which can be recombined to give a very stable wave packet with essentially the nature of the Néel-oriented sublattice state. This construction is discussed in some detail in the original paper, and then later on in some of my own work on superfluidity and superconductivity--for instance, Anderson (1966)--and is undoubtedly well understood by many of those who have thought reasonably deeply about broken symmetry. One finds the same kinds of arguments surfacing again in the modern "instanton" theory of gauge fields (Callen _et al._, 1978), in soliton theories of elementary particles (Dolan and Jackiw, 1974; Goldstone and Jackiw, 1975), and so on. But somehow this is one of those arguments that is, although very simple, and terribly important, not generally available, perhaps because everyone who has ever understood it thinks it too simple to write down.

The essential idea is to treat the system like the macroscopic, classical system it is, as far as the lowest-energy levels, infinitesimally close to the ground state, are concerned. For instance, in the original paper on antiferromagnetism, I showed that the Neel antiferromagnet behaves like a heavy rigid rotor as far as its lowest-energy levels are concerned. This follows from a very simple argument given by Leggett and Brinkman [for application to antiferromagnetic resonance see Halperin and Saslow (1977); see also Anderson and Brinkman (1975) reprinted in this volume]. This is that the susceptibility plays the role of a moment of inertia as follows: The angular velocity of uniform procession of the spins in the antiferromagnet is

$$\vec{\omega} = \frac{\partial E}{\partial \vec{S}},$$

where E is the energy of the system and $\vec{S}$ is the total spin angular momentum, which in turn determines $\vec{M}$, the sublattice magnetization, via the gyromagnetic ratio γ:

$$\vec{M} = \gamma \vec{S}.$$

But since

$$\partial E/\partial M = H = M/\chi,$$

where χ is the susceptibility,

$$E = \frac{1}{2}\frac{M^2}{\chi} = \frac{\gamma^2}{2\chi}\vec{S}^2.$$

So

$$E = \frac{1}{2} I_{eff}\omega^2$$

and

$$I_{eff} = \frac{\chi}{\gamma^2}$$

is the effective moment of inertia. This argument may be used to derive equations for antiferromangetic resonance as follows:

> In antiferromagnetic resonance, because of relatively weak anisotropy forces, the energy depends on the orientation of the sublattice structure, which we

characterize by a vector $\vec{d}$ (for difference of sublattice spins), lying perpendicular to the plane of the spins and giving an energy which may be schematized as $-\gamma\vec{d}\cdot\vec{H}_A$ ($\vec{H}_A$ the anisotropy field). $\vec{d}$ precesses with the angular velocity ω, i.e., $\vec{S}_{tot}$ is the generator for rotations of any vector including $\vec{d}$, and as a result one obtains the equations of motion for $\vec{d}$:

$$\dot{\vec{d}} = \vec{d} \times \left(\gamma\vec{H} - \frac{\vec{S}}{I_{eff}}\right) ,$$

while $\vec{S}$ changes according to the torque $\frac{d\vec{S}}{dt} = \vec{\tau}$

$$\vec{\tau} = -\frac{\delta E}{\delta\vec{d}} = -\left(\vec{d} \times \frac{\partial E}{\partial\vec{d}}\right) ,$$

where we used the chain rule and $\delta\vec{d} = \delta\vec{\theta} \times \vec{d}$. Then

$$\dot{\vec{S}} = \gamma\left(\vec{S} \times \vec{H} + \vec{d} \times \vec{H}_A\right) .$$

These give the famous antiferromagnetic resonance equation

$$\omega^2 = (\gamma H)^2 + \vec{H}_E \cdot \vec{H}_A ,$$

where $H_E = \frac{\vec{d}}{I_{eff}}$ is the "exchange field." Spin waves may also be conveniently calculated within this formal structure.

In the absence of an external orienting torque or magnetic field, however, we have simply a free rigid rotor with moment of inertia

$$I = \frac{\chi}{\gamma^2} \sim \frac{N\chi_S}{\gamma^2} ,$$

where χ_S is the susceptibility per spin (the susceptibility is nearly zero along the sublattice vectors, so the tensor

$\underset{\sim}{I}$ is that of a disc perpendicular to the sublattice). Thus the Hamiltonian is just

$$H = \frac{1}{2}\frac{(\vec{S}_{tot})^2}{I} = \frac{1}{2}\frac{S(S+1)}{I},$$

which is clearly diagonalized only by eigenstates of $\vec{S}_{tot}$, and the ground state is a singlet S = 0, totally isotropic. But the moment of inertia given above is of order N--very large--in fact equal to that of a copper sphere of about 10 microns radius. Thus the energy levels are enormously closely spaced--the lowest one being $\sim 10^{-20}$ degrees K above the ground state for this macroscopic sample. Thus extremely weak forces can pin down the orientation; and even in the absence of such forces, the correlation time for an oriented sample to lose its direction would be of the order of years.

Similar arguments, in many cases even simpler, hold in all of the broken-symmetry systems. The restoration of symmetry is a quantum fluctuation effect which is always small and usually utterly negligible. The only rational approach to a description of such a system is either to neglect or to treat as weak perturbations these symmetry-restoring effects, and to use a basic approach in which the state is and remains asymmetric. In the crystal, the construction of a total mass, momentum, angular momentum, and moment of inertia is obvious, and we are completely habituated to dealing with all of these as classical variables. In the superconductor, as shown in Anderson (1964), the momentum operator is the total charge, the "mass" is the capacitance to ground, and the classical description requires sufficiently soft charge fluctuations. In each case the construction is trivial and instructive.

I should note that, as I remarked in Anderson (1964), the singular spectrum encountered in the continuous broken-symmetry case always implies long-range interaction forces between entities which couple to the symmetry parameter--e.g., the long-range elastic forces between vacancies in crystals, the Suhl-Nakamura force between nuclear moments in ferromagnets, and the long-range forces between vortices in superfluids.

Perhaps the Suhl-Nakamura force is the simplest (Suhl, 1958; Nakamura, 1958). There is a hyperfine interaction $A_{HF}\vec{I}\cdot\vec{S}(r)$ between the nuclear spins I and the local ferromagnetic polarization S(r). A_{HF} is the coupling strength. The z component (if $\vec{M}$ is in the z direction) is ineffective, simply adding to the molecular field on S, but $A_{HF}I_x$ amounts to a local field on S in the x direction, which twists the ferromagnetic spin out of the z direction. To compute it we recognize that the effective transverse field on every spin must be zero except for the central one, so that the polarization around I goes according to $\nabla^2 M = 0$, or $M(r) \propto 1/r$. Dimensionally the coupling constant must be $(A_{HF})^2/A$, i.e., $(H_{HF})^2/H_{ex}$. A is the exchange coupling constant.

This very large and long-range field is the primary broadening mechanism of nuclear hyperfine resonance in ferromagnets. It does not actually have infinite range because of the anisotropy correction: the expression

$$\omega(k) \propto H_A + Aa^2k^2,$$

where a is the lattice spacing, gives the spin waves a "mass" in field-theory terms, which makes the interaction of Yukawa form

$$V \propto \left(e^{-\mu r}\right)/r \quad \text{where} \quad \mu a = \sqrt{H_A/H_E}$$

and a is the lattice spacing. This is an effect somewhat related to "generalized rigidity" but distinct from it. It is amusing that long-range forces in the background system destroy this long-range effect in the final system; this paradox is the foundation of the new Weinberg theories of elementary particles and fields.

One final caution about the general phenomenon of Goldstone boson excitations, as we shall call both of the two types of low-frequency excitations implied by broken symmetry. This is that there is not always a unique true eigenexcitation, in the case at least of metals and other Fermi liquids which have low-energy fermion excitations. Instead of the boson's making itself up (as it does in a metallic ferromagnet) as a bound state of a hole-electron pair, it may under some circumstances be either a broad resonance or only what is called a "diffusive" mode--a mode whose frequency is effectively pure imaginary, $\omega(q) \to if(q)$. (Its propagation then obeys a diffusion equation, like $\frac{\partial u}{\partial t} = \nabla^2 u$.) This appears to be the case for "orbit waves" in ^{3}He and possibly for antiferromagnetic spin waves in metals (Cross and Anderson, 1975; Lowde, 1974).

E. GENERALIZED RIGIDITY AND LONG-RANGE ORDER: MEASUREMENT THEORY

We are so accustomed to the rigidity of solid bodies—the idea, for instance, that when we move one end of a ruler, the other end moves the same distance—that it is hard to realize that such action at a distance is not built into the laws of nature except in the case of the long-range forces such as gravity or electrostatics. It is strictly a consequence of the fact that the energy is minimized when symmetry is broken in the same way throughout the sample: the phase and angle variables want to be uniform, so that the orientation and position of the lattice is the same everywhere. Of course, in general they are not quite the same, since the lattice can deform elastically: there is the elastic energy that is proportional to $(\nabla\vec{u})^2$ ($\vec{u}$ being the lattice displacement), which will allow the lattice to deform slightly under a force. But nonetheless the lattice transmits that force from one end to the other even in equilibrium and without having to flow constantly like a viscous liquid. To break down the rigidity completely, we must supply the condensation energy of a macroscopic piece of the sample, which is very large. We are so accustomed to this rigidity property that we don't accept its almost miraculous nature, that it is an "emergent property" not contained in the simple laws of physics, although it is a consequence of them. Perhaps our naive pre-freshman-physics apprehension that it <u>should</u> take work to exert a force (quite correctly for our bodies: our contractile cells are dissipative structures, not broken-symmetry ones) is more realistic after all than simple physics with its rigid rods and inextensible strings.

The generalization of this concept to all of the instances of broken symmetry is what I call generalized rigidity. It is responsible for most of the unique properties of the ordered (broken-symmetry) states: ferromagnetism, superconductivity, superfluidity, etc. For instance, permanent magnets are so because the magnetization cannot change a little at a time. That is, a whole macroscopic "domain" must reverse its magnetization at once in order to allow magnetization reversal; one spin at a time doesn't work. Equally, superconductivity is the phase rigidity of the electron pair fluid. The phase ϕ of the order parameter is held uniform by a phase rigidity energy $\frac{\Lambda(\nabla\phi)^2}{2}$, which, it turns out, responds to any accelerating force by causing a supercurrent $J_S \propto \Lambda\nabla\phi$, which in general cancels out the force. Here the rigidity idea is sufficiently subtle to have eluded both Landau [whose celebrated criterion for superfluidity as given in, for example, Khalatnikov (1976) (w.r.t. Landau, 1941), is simply incorrect] and Bardeen et al. (1957). Here what is transmitted is a current and not a force, but it seems more surprising to us (though of course perfectly permissible thermodynamically) because we are not as used to dissipationless transport of matter as we are of force.

I want to make one comment here about the rather deep relationships between broken symmetry, rigidity, and the theory of measurement. On one level it is obvious that there is a close relationship--namely, all of the canonical experiments to demonstrate the uncertainty principle, for instance, involve rigid apparatus and the interaction of the particles with bodies which are behaving essentially classically, i.e., broken-symmetry objects. Slits, magnets, etc., are all solid objects, and the very assumption of positional stability which goes into the standard discussion of interference experiments assumes that all the parts of the apparatus are solid and connected by rigid beams. Actually, even if an ingenious experimenter were to devise wholly quantum apparatus, in the end the observer himself and his observational instruments would necessarily still themselves be solid, broken-symmetry objects; otherwise, the pointers wouldn't point, the recorders wouldn't record, etc. At the most primitive level, the observer participates in the general broken time-reversal symmetry of the world (the "arrow of time"), and his measurements are necessarily recorded as part of a causal sequence involving irreversibility of the arrow of time. In addition, the very essence of measurement in the crucial cases is the decision between symmetry equivalent states, or at least between states of the same energy.

I would argue that this fact has, however, an even deeper significance, in that it places the observer and his apparatus in an essentially nonquantum world in which the phase information of wave functions is lost not just by inadvertence but essentially and irretrievably. The central problem of measurement theory is, of course, the contrast between the perfect determinism of Schrödinger's equation [$i\hbar\frac{\partial\psi}{\partial t} = H\psi$ determines ψ for all time if $\psi(t = 0)$ is known] and the actual uncertainties of measurements as typified by Heisenberg's uncertainty principle. The reason is the same reduction process we carried out in the Bose gas and the crystalline solid problem: we may place the system of macroscopic objects embodying the measurement and the measured entity as a whole in an eigenstate of its own Hamiltonian, but it is not valid to do so for any macroscopic subsystem, because the subsystems have essentially infinitely strong correlations with each other, which are best described by treating the whole system as though it were a classical object with fixed position. There is simply no possibility of preparing a subsystem, consisting of a macroscopic part of the apparatus plus a quantum particle, in a precise eigenstate, and of exhibiting the actual deterministic structure of quantum mechanics in this way, because the macroscopic bodies including the observer are undergoing or have undergone strong interactions among themselves which have left them in states which cannot be assumed to have independent phases. The correlation of phases is necessarily implied by the fact that the observer has "prepared" the apparatus (he has located it with respect to himself, and it has the same arrow of time as he does.)

I do not accept that this is a philosophical difficulty, or that it demonstrates anything like an impossibility in principle of treating macroscopic objects quantum-mechanically; merely that this discussion brings out the practical problems of determining all the relevant phases in a given system. It remains to be discovered whether or not someone more sophisticated and intrepid than the present students of measurement theory can work out a way of including the observer in the equations and achieving deterministic measurements. But this is perhaps not the most important test: as Landauer has remarked, uncertainty enters only once, and successive measurements contain no quantum noise: quantum uncertainty is not a noise source, and hence is physically irrelevant in any case.

F. MODES OF BREAKDOWN OF GENERALIZED RIGIDITY: VORTICES, WALLS, AND OTHER SINGULARITIES

To return from these philosophical convolutions of measurement theory to a thoroughly practical area of many-body physics, it is worth noting that, along with the other general phenomena associated with broken symmetry, there is a general rule that the breakdown of the generalized rigidity property, along with the resulting dissipation, is a consequence of the formation and motion of defect structures which are usually macroscopic in size. So far as I know, de Gennes was the first person to suggest that a general scheme relating defect structures to broken symmetry might be constructed, and he remarked only that there appeared to be few really general rules (de Gennes, 1973). Since then (and since this chapter was first written, in fact) there has grown up a much more systematic topological theory of these phenomena, which has been extensively summarized by Mermin (1979) in a review [another good review is by Poenaru (1979)]. This theory originated in papers by Toulouse and Kléman (1976) and Volovik and Mineev (1977), and the application to generalized rigidity and dissipation was made by Anderson and Toulouse (1977). The theory is, actually, not yet complete, since it applies unmodified only in cases with "local order parameters" [i.e., not, as shown by Stein (1979), to crystals or smectic liquid crystals, which are density waves where the order parameter has global significance] so that in fact the discussion given originally is not by any means superseded by the more mathematical modern work. I shall present the original arguments unmodified and then summarize the topological view and its difficulties in a succeeding section.

The twin models for such effects are the domain theory of ferromagnetism, due to Landau (1941), Bloch (1930) and others, and the dislocation theory of G. I. Taylor (1934), Burgers (1939, 1940), and others. These two cases demonstrate nicely the one general statement which will be justified by the topological theory: that a one-dimensional order parameter allows only two-dimensional ("wall") defects; and that a two-dimensional order parameter allows linear defects such as dislocations, but not necessarily always; while a three- or more-dimensional case may allow point defects.

Clearly in any broken-symmetry system we can imagine forces which would disorient the order parameter in one region relative to another: forces applied at the boundary, externally applied strains or fields, long-range forces of various kinds, or--a very important example indeed--simply history: different parts of the sample may have grown with different order parameters and eventually meet in the middle. In what ways can the system respond?

The most obvious and simplest is a boundary or "domain

wall": one simply has two or more regions, each locally homogeneous, separated by boundaries. In the case of the one-dimensional order parameter, there is only a discrete set of local equilibrium states (directions of $\vec{M}$ for a ferromagnet, for instance, or of $\vec{P}$ for a ferroelectric), and there must be a two-dimensional boundary (for a 3D sample) dividing the regions containing the different states. The shape of the boundary or other singularity is determined by competition between the forces of generalized rigidity--in the magnetic case essentially the exchange stiffness $A(\nabla M)^2$-- and other forces. In the boundary case, characteristically the wall is ≈ one lattice constant thick--essentially, there is no length parameter except the fundamental microscopic length. This is true of ferroelectric domain boundaries, twin boundaries in crystals, etc., and is dynamically important, because such thin boundaries cannot move continuously in space--they are located at a specific crystal plane and must overcome an activation energy to move to the next (see Fig. 2-6). In case there is an approximate higher symmetry,

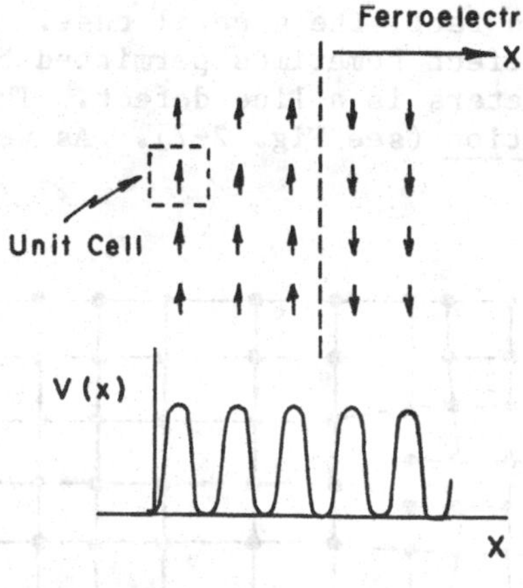

Figure 2-6 Pinning of ferroelectric domain boundary.

as in the Heisenberg Hamiltonian with anisotropy, one has the same length parameter μ as for the Suhl-Nakamura type of force, and the wall is in general thick and mobile, because its energy is nearly independent of position relative to the crystal planes. This kind of domain wall was discussed by Bloch (1930) and Landau (1941). This is the essential reason for the high permeability of pure, single-crystal ferromagnetic materials; it's amusing that the Bloch wall, perhaps the first such order singularity to be recognized, is the anomalous result of an approximate symmetry: there is an extra length parameter present, namely, the length at which the system can see that it is not isotropic. This is by no means a rare phenomenon--one encounters it again in ^{3}He, where the dipolar interaction coupling spin and orbit are weak, and in liquid crystals where different elastic constants can be sharply different in magnitude.

In general, the response to external forces tending to twist or reverse the magnetization must in the first instance come from the motion of these boundaries. Only if there is a sufficiently strong magnetic field applied to overcome the anisotropy field can the magnetization of a whole sample reverse in response to a reversal of H; but if a domain wall is present, it can move in the presence of an infinitesimal field, leading to very low coercive fields and high susceptibility. To make a permanent magnet, one can proceed in two ways: either to pin the existing boundaries, i.e., to make a very impure material, or to remove them, i.e., to make it of very small particles. Thus the response properties of magnets are wholly conditioned by the defect structures. This is, in fact, the general case.

A second type of defect sometimes permitted by multi-dimensional order parameters is a line defect. The original example was the dislocation (see Fig. 2-7). As we remarked

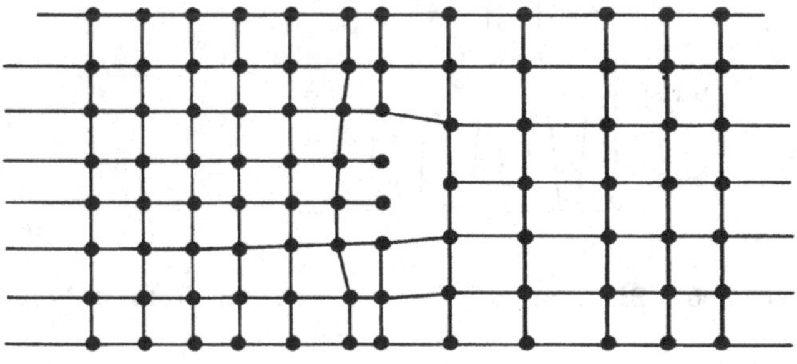

Figure 2-7 Edge dislocation (⊣) in two dimensions.

earlier, translation of a crystal can be expressed as a phase angle, and has in common with a phase angle that locally the states described by ϕ and by $\phi \pm 2n\pi$ are identical. Thus we can imagine following a closed path through a crystal along which the local phase varies gradually, corresponding to a small local strain, $\nabla u \propto \nabla\phi$, but then returns to the original phase changed by 2π. Such paths can clearly not be collapsed to zero length without encountering some region where ∇u becomes large; thus there must be associated a line of singularity where the lattice structure is essentially destroyed.

The two basic types of dislocations are edge dislocations, where an extra plane of atoms has been interpolated ending at a line singularity; and screw dislocations, where the lattice has the topology of a screw.

Like boundaries in magnetism, dislocations play a vital role in the strength and dynamic behavior of crystals. The easiest way in which a crystal can continuously slide against itself (continuously increase u, i.e., ϕ, in one region relative to another) is by the continuous flow of dislocations across a line between the two (Taylor, 1934). Again, there are two ways to make a strong crystal: the easy way is to pin the dislocations, one version of which is called work hardening; the hard way is to eliminate them altogether as in a "whisker" crystal (Herring and Galt, 1952).

Orientational misfit in a crystal cannot be healed by a line singularity, because rotation and translation don't commute: a rotation about one line is a large translation at another, so it is impossible to heal the crystal by simply returning to the same orientation. There is, however, the amusing fact that under some circumstances like dislocations may line up and form a plane, which, in fact, turns out to be an orientational boundary: this is the "dislocation theory of small-angle grain boundaries" [see Fig. 2-8; (Burgers, 1939; Read and Shockley, 1950).] But in general different orientations are separated by simple boundaries.

All of the phase-type order parameters lead to line singularities: quantized flux lines in superconductors, vortices in helium (4) II (the case of the ^{3}He superfluid is more complex), and, of course, dislocations as noted above. The relation between flux lines in hard superconductors and critical currents and fields is almost exactly analogous to that between dislocations and strength of materials. That is, when the flux lines move, a continuous variation of the relative phase is possible, and since one can show rather easily that $d\phi/dt = \mu$, the chemical potential (Anderson, 1966), this allows voltage to appear, and correspondingly leads to dissipation and production of heat. The strain or other disturbance about a line singularity cannot, in general, die off completely and rapidly as is true of a

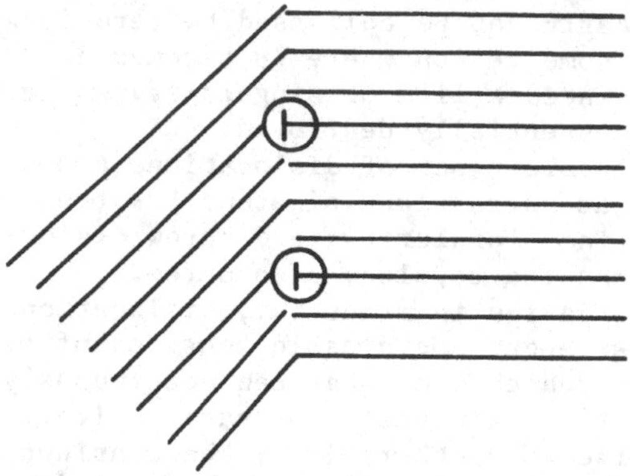

Figure 2-8 Grain boundary as a plane of edge dislocations.

boundary. The question of the range and nature of such a disturbance is again purely one of symmetry, very closely related to my earlier comments about long-range Goldstone boson-related forces. As already noted, for one-dimensional order parameters there need be no thickness to a boundary, just as there is no long-range force unless there is an approximate higher-dimensional symmetry. In the two-dimensional case with Goldstone bosons (solid, superfluid), the order parameter obeys $\nabla^2\psi = 0$, and about a line singularity ψ is a harmonic function, so that the phase ϕ (or u) varies as the angle θ about the line [giving v_s, or the strain $\nabla u_2 \sim 1/r$, and a logarithmically divergent defect energy $\int (\nabla u)^2 r dr$]. In the charged superfluid, lines are finite in extent because of the existence of the long-range electromagnetic screening. This, according to the particle theorists, is characteristic of cases of broken gauge symmetry involving a genuine dynamical gauge field. There is a new dimensional parameter, the Landau screening length or "penetration depth," of order 100 Å.

In at least one case other than dislocations, the problem of whether the singularities attract or repel arises: in the superconductor it is only in the so-called Type II case that isolated flux lines repel; in the Type I case they attract each other and coalesce into normal regions, so that one gets a second type of boundary phenomenon and a second type of domain structure, not at all precisely analagous to magnetic domains, but more related to phase equilibria.

At least one other variety of line singularity exists:

where a system has a "pure" orientation symmetry without a broken translation group, as in a nematic liquid crystal or in the anistropic superfluid phase of ^{3}He, a line singularity of orientation can occur, a line about which the "directrix" turns by π. This is called a disclination in nematics, and such a line has been named a disgyration in ^{3}He by de Gennes, who proposed it (de Gennes, 1973). But the properties of both entities are very dependent on the topological theory, and their discussion is better deferred.

In the case of these more complicated phases, yet another general type of order parameter singularity can occur: the point singularity. Points can occur wherever the broken symmetry is at least as complicated as rotations about a point--i.e., where it contains a subgroup isomorphic to the rotation group. In principle, for instance, points would appear in the isotropic magnetic phases as points where the magnetization came to a node; and in fact they do occur both in liquid crystals and in ^{3}He. Two kinds of points occur in nematics (see Fig. 2-9) and presumably also in ^{3}He-B, where

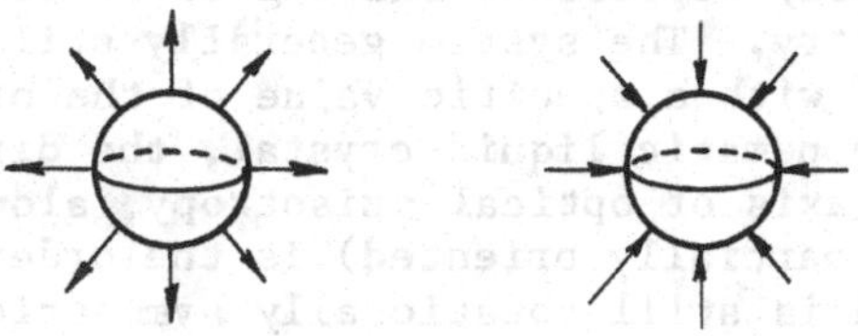

Figure 2-9 Point singularities for vector order parameter.

they have been studied by Brinkman (Anderson and Brinkman, 1975--reproduced here in reprints section). Points also occur interior to certain magnetic domain walls (Bloch points), specifically in forming a bubble domain (Thiele, 1970).

A point singularity is also allowed in a lattice, namely, a vacancy or interstitial. The displacement $\vec{u}$ is radial, and certain quantization conditions which I have not analyzed are obviously implied by the fact that there must be only a complete vacancy or interstitial. These point singularities seem to be the closest analogue of t'Hooft's theory of magnetic monopoles in Weinberg broken-symmetry theories. Again, especially in the lattice case, important mechanisms of relaxation of generalized rigidity occur due to vacancy migration.

G. TOPOLOGICAL THEORY OF DEFECT STRUCTURES: SUCCESSES AND LIMITATIONS

Founded in the earlier work of Friedel, Kléman, and others (Friedel, 1922; Kléman and Friedel, 1969) on liquid crystals, stimulated by the ideas of t'Hooft (1974) and Polyakov (1975) in quantum field theory, and by the need to understand the complicated situations that arise in the superfluid phase of ^{3}He, a new topological theory of defects and textures has grown up in the past few years. The first work was by Toulouse and Kléman (1976), but there are other important articles by Volovik and Mineev (1977), by Poenaru (1979) and by Mermin (1978, 1979). This theory, though still incomplete, has left us with a much clearer view.

The idea is basically extremely simple: it is to define and classify defect structures mathematically, in terms of mappings between real space and order-parameter space. As emphasized earlier, the order parameter (O.P.) almost invariably represents the degree of breaking of an original symmetry. The system generally still has some symmetry, even with a specific value of the order parameter ψ—thus in the nematic liquid crystal, the directrix $\hat{D}$ (which is the axis of optical anisotropy, along which the molecules are partially oriented) is the order parameter, but the liquid is still rotationally symmetric about $\hat{D}$, as well as translationally homogeneous. But the original symmetry group was larger, so that there is a group of operations on the order parameter ψ which leaves the energy invariant but not the state. This group is clearly always a "quotient" subgroup G/H of the original symmetry group G, divided by the remaining symmetry H. Equally, it defines the "space of the order parameter": the values over which the order parameter can range while remaining in one of the set of allowed equilibrium values. For simple gauge symmetry, or in He-II, for instance, the order parameter space is that of a single complex phase angle ϕ, leaving, of the full symmetry group, translation and rotation undisturbed.

For the nematic, $\hat{D}$ can clearly rotate over a whole spherical surface ("S_2"), except that the group still retains inversion (the sign of $\hat{D}$ is meaningless), so $\pm\,\hat{D}$ are equivalent, and we have for order parameter space the "projective" sphere "P_2," with opposite points being equivalent. ^{3}He-A is more complicated: the gauge phase and rotation are not independent, since the order parameter is two vectors, δ_1 and $i\delta_2$, perpendicular to each other and of identical magnitude, representing the two components of the gap function. Multiplication by a phase rotates δ_1 and δ_2 in the plane. The group is SO_3, the rotations of a rigid body in space, whose space is equivalent to the projective 3-sphere P_3 (rotations by angle θ about a vector $\hat{n}$ are plotted as $\theta(\hat{n})$ and clearly $\vec{\theta} = \pm\pi\hat{n}$ are equivalent). A hypothetical case often postulated to show the full possible complexity is the "biaxial nematic," which has full tensorial anisotropy but also reflection symmetry.

One aspect of order-parameter space is that it is not at all unusual to find approximate symmetries for nominally inequivalent different values of ψ. I have already mentioned the magnetic case, where there is an approximately perfect rotation symmetry in spin space, of which only axial rotation remains, so that the approximate order-parameter space is S_2; but at low temperatures especially, anisotropy occurs, and exact symmetry leaves only a few points on the sphere. Again, in ^3He to a good degree of approximation, orbital and spin rotations are independent, so that there is an approximate additional symmetry, namely, rotation of the spin degrees of freedom of the order parameter. These approximate symmetries affect the singularity structures. Their effect is often summarized by quoting a length of the Suhl-Nakamura type which we already described: for order-parameter variations at distances less than this length, the weaker energy may be neglected because it is outweighed by the gradient energies, while at fully macroscopic distances the fully broken symmetry holds.

A "defect" or "singularity" or (as we shall call them here) "defect structure" has as its raison d'être that the order parameter can take on an equilibrium value almost everywhere except possibly in a region which is topologically of smaller dimension than real space: a line, point, or plane, and the immediate vicinity thereof.

Outside of that region, the order parameter must be somewhere in its allowed space. It must take on one of the energetically equivalent values allowed by symmetry, and it must not vary rapidly, so that local regions are arbitrarily close to equilibrium, though if we move distances of order the size of the sample ($N^{1/3}$ atomic lengths $\ggg 1$), we are allowed to vary ψ. Gradient energies are then of order $\approx \nabla^2 \times \text{volume} \approx N^{-2/3} \times N \approx N^{1/3} \ll N$ which is very much smaller than the energy necessary to destroy order everywhere or even make a boundary: thus the system prefers to relax this way. We can therefore map the exterior region on the order-parameter space, in the sense that there is an allowed value of $\psi(r)$ everywhere outside the singularity: $\psi(r)$ is a unique mapping of r-space into ψ-space. But also, outside that region, the map must be continuous: as we go from one point to its neighbors in real space, we move only a small distance in O.P. space. It is only as we describe extended paths that we expect to leave our original value far behind; so the question is what kinds of distinct extended paths are allowed, given this restriction. This depends on the topology of this outside, or "surrounding" space. In particular, can we deform all paths continuously into a single point? If so, the system will do so, to reduce its gradient energy; small as it is, $N^{1/3}$ is still $\gg kT$.

This topology is clearly different for the different cases. For a boundary, the surround space is two disconnected regions. We may describe closed paths in either region, but we never go from one to the other. Thus the topology is just that of two points--we are at one point or the other. All paths in either region are trivial, in that they may be continuously shrunk to infinitesimal paths and thence to a point, and hence the order parameter might as well be uniform in each region. For a line, on the other hand, the topology is a circle: we may have paths surrounding the line as many times as we like, and continuous deformations can never eliminate that property, called the "winding number" invariant. Thus maps in the surrounding space are like maps on a circle, topologically. Similarly, a point defect has a surround space with the topology of S_2, the surface of a sphere.

There is a theory of mappings of this sort called homotopy group theory, which can tell us how many and what kind of topologically distinct maps of the surround space onto the order-parameter space (not the other way: else the order parameter at a point wouldn't be unique!) there are. It is not necessary to go into it in detail; many of the results of this sort of mapping are trivial once one has the idea. But some are not, otherwise. For an example of a nontrivial case, consider lines in nematics. Here the order-parameter space is P_2, and we want to find nontrivial paths between equivalent points in P_2. A simple rotation by 2π is, unexpectedly, not topologically stable, since it is just a circle on P_2 which can be deformed to a point (see Fig. 2-10). This is, as shown in Fig. 2-10, terribly obvious: for 2π rotations P_2 behaves just like S_2, which has no line defects, because one can obviously slip a rubber band off a ball. In fact, 2π lines in nematics are not stable and "escape in the third dimension" by standing up parallel to the z axis from the x-y plane, becoming an array of point defects, as seen by Cladis and Kléman (1972), point

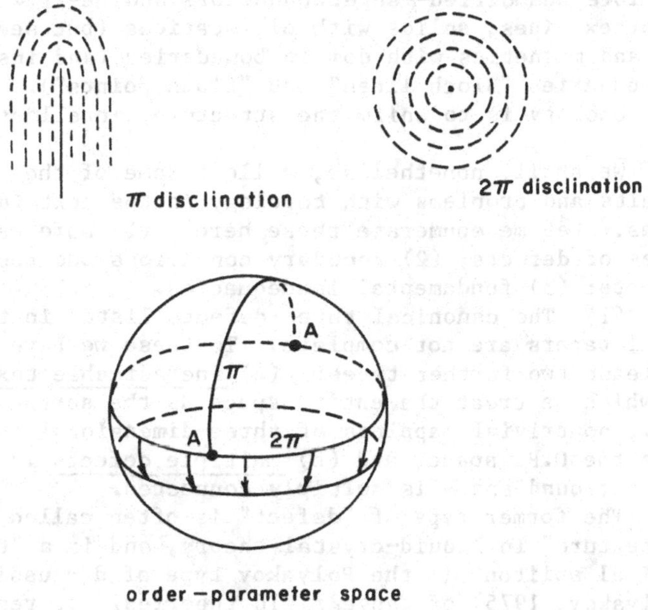

Figure 2-10 Π and 2π disclinations in a nematic liquid crystal.

defects which are stable: S_2 can map onto P_2 without coming unwrapped. On the other hand, a rotation by π (this object is an example of what is called a disclination) is stable, since it is a path from a point to its equivalent at the other pole, which can't be unwrapped.

One exciting result (Anderson and Toulouse, 1977) for ^{3}He-A is that the same phenomenon occurs for line defects in that case, except that 2π lines are stable, but 4π ones are not. The 2π line is a path from pole to pole (π to -π, if you like) which may be taken either way: a "disgyration" is not in principle distinct from a 2π vortex line because they can be transformed into each other by a continuous smooth rotation. But a 4π vortex line is not a singularity at all, and this leads to very strong modifications of superflow properties in this case (see also Mermin, 1977; Volovik and Mineev, 1977). The subject of flow in ^{3}He is still interesting, complicated, and controversial, but it appears clear that flow can drive very complicated motions of the anisotropic "texture," which are observed in some cases.

But in all the less esoteric cases, topology leaves ancient lore unmodified--superconductors and He-II with their vortex lines, solids with dislocations (but see below), and magnetics with domain boundaries (and inside these boundaries "Bloch lines" and "Bloch points"). The role of topology is to unify the structure, intellectually.

We shall, nonetheless, collect some of the further results and problems with topology in the next few pages. Let me enumerate these here: (1) more esoteric types of defects; (2) boundary conditions and boundary defects; (3) fundamental inadequacies.

(1) The canonical three defects listed in the usual papers are not complete. To these we have to add at least two further types: (a) ineradicable textures, in which we treat the entire space as the surround space, i.e., nontrivial mappings of three-dimensional space onto the O.P. space, and (b) multiple defects in which the surround space is multiply connected.

The former type of "defect" is often called simply a "texture" in liquid-crystal theory, and is a "topological soliton" in the Polyakov type of discussion (Polyakov, 1975) of gauge field theories. It requires that we specify boundary conditions; if these are merely that the order parameter takes on a single value at ∞, we have the soliton situation, and the "surround" space to be mapped is the three-dimensional sphere S_3. One such mapping is the nontrivial mapping of S_3 onto S_2 (Finkelstein, 1966), which would be an allowable texture for a ferromagnet except that it seems energetically unstable. A 4π vortex texture in ^{3}He-A in a torus is another example, i.e., a "4π vortex" ring. Both $S_3 \to S_2$ and the 4π vortex ring have energies that are linear in the overall scale, so that in equilibrium they will collapse to very small size, and presumably thereupon tunnel away to nothing when they reach the coherence length in size; but the 4π vortex texture rings at least play, very probably, an important role in dissipative effects [as foreshadowed by Volovik and Mineev (1977)]. More examples will come up in the next topic, since such textures can be connected to boundary defects. One way to visualize this kind of thing is to think of it as a point defect in a four-dimensional sample, which one has surrounded by a three-dimensional sphere which is our actual space. But, of course, our three-dimensional space does not contain the defect. This, however, is oversimplified, since real cases are usually complicated by boundary conditions.

The "surround space" need not necessarily be the simplest possible manifold of the given dimension. It can, for instance, be complicated by including more than one line (in the examples I know of). The individual lines must, of course, themselves be given by appropriate mappings, but they nonetheless may leave a more complex multiply connected surround space. The topology of such a space is appropriate for a noncommutative homotopy group, such as the quaternion group, which is possibly appropriate for lines in cholesterics (Kukula, 1977; Stein, 1979); and indeed cholesterics exhibit "focal conic" defects composed of two nonintersecting crossed lines. Presumably cholesterics could exhibit rather complicated "link" defects composed of several noncommuting lines.

(2) The effect of boundaries can be very interesting and crucial in the more complicated cases. In the first place, boundary conditions--or other overall constraints, as well--can require the presence of certain defects. We are very familiar with the long-range dipolar effects summarized by demagnetizing fields, which require most magnetic samples to be full of domain boundaries. Equally, in a rotating vessel He-II contains the well-known Feynman paraxial array of quantized vortices parallel to the rotation axis (Feynman, 1955). In liquid crystals and ^{3}He-A, boundaries constrain the order-parameter directions near the surface and in general enforce the presence of one or several defects interior to the sample. A particular example is the spherical geometry for ^{3}He-A, where $\hat{\ell}\perp(\vec{\delta}_1, \vec{\delta}_2)$ must be perpendicular to the surface and where the Euler number of the surface controls what defects are required topologically. In general, the correct prescription is as suggested by Anderson and Palmer (1977; see also Stein et al., 1978), i.e., to treat the surface as a two-dimensional manifold allowing line (boundary) and point (vortex) singularities, with a more limited order-parameter space as enforced by the boundary conditions. In ^{3}He-A and cholesterics, we find the possibility of a point defect on the boundary

attached to a mere texture in the bulk [a "boojum" (Mermin, 1977), or a "fountain" (Anderson and Palmer, 1977)].

(3) The failures of homotopy group theory have to do with the not infrequent situations where the order parameter is not really free to take on any value it likes locally. In a solid, for instance, the order-parameter space apparently should include not only displacements--a phase parameter u_1, u_2, u_3 having the topology of a three-dimensional toroidal surface in four dimensions--but also all rotations in space. Defects (also called "disclinations") have been actually proposed (and made it to Scientific American) involving rotating a cubic lattice, for instance, through 90 degrees in going around a line. But in fact such defects are meaningless and are never observed, because they have energies $\propto N$, not $N^{1/p}$, $p > 1$, as in the usual case. A grain boundary would be energetically far cheaper; the point being that a true rotation of one volume relative to another causes finite elastic strains almost everywhere in the crystal [except see the work of Halperin and Nelson (1978) discussed below].

Means of handling this situation have been discussed without firm conclusion. My own preference for the crystal is to define the order parameter as a three-dimensional phase field everywhere in space, and to introduce rotational symmetry by observing that it causes the energy not to depend on $\nabla \times \vec{u}$ (the antisymmetric part of the strain tensor). Thus rotation may be introduced as a strain which causes no elastic energy locally. The true order parameter, the object $\vec{u}$, with a d-dimensional toroid as order-parameter space, allows only line singularities; grain boundaries appear as sheets of dislocations as in the Burgers-Shockley-Read theory (Burgers, 1939; Read and Shockley, 1950). This does not fit into the general mathematical structure of making use of all symmetries of the order parameter, but it seems to fit all the facts.

A second failure, very similar but more severe, is the case of smectic liquid crystals. Again, we have a confusion of rotational and density wave-variables, but it is more serious: some of the topologically allowed lines do occur, but not all. Here a reasonable suggestion due to Mineev (Volovik and Mineev, 1977) is the introduction of a nonlocal "integral constraint" on the allowed configurations of the order parameter (see Fig. 2-11). We may define the order parameter in, say,

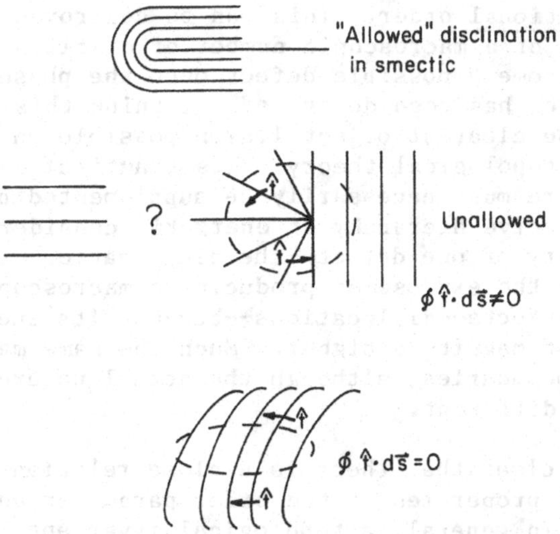

Figure 2-11 Mineev's integral constraint in smectics.

smectic A as the direction $\hat{t}$ perpendicular to the layers locally, and as a phase u at any given point which varies at a fixed rate along $\hat{t}$. The integral constraint on $\hat{t}$ is $\oint ds \cdot \hat{t} = 0$, which is simply the statement that a macroscopic number of layers may not be missing when we go around a circuit. Alternatively, we may use the phase field alone, itself, but then we have a local differential constraint $|\nabla u| \simeq$ const, and we predict only dislocations, not disclinations, some of which are allowed in a smectic. Apparently, even topologists are at a loss with this kind of system, which is called a "measured foliation," but which has little mathematical theory. Yet another defect that doesn't fit is the vacancy or interstitial. This may not be properly categorizable as a defect structure at all.

Defects also play an important role in some phase-transition phenomena. The best-known cases are the Kosterlitz-Thouless theory of two-dimensional xy problems and the Halperin-Nelson theory (Halperin and Nelson, 1978) of 2d melting. This latter is a fascinating case: it envisages a lower melting transition, mediated by the stability of a macroscopic number of dislocation points as in the Kosterlitz-Thouless theory, but dislocations (again because of the peculiar properties of the density-wave phases) do not destroy orientational order. This can be destroyed only by the freeing of a macroscopic number of disclinations which have become a possible defect once the phase-order parameter has been destroyed. I think this is, in a way, the clearest object lesson possible on the weakness of the topological theory: its beautiful logical structure must necessarily be supplemented by a quantitative hierarchy of energetic considerations. Stability of one defect--the disclination--is bought only at the expense of producing a macroscopic number of other defects--dislocations--because its energy is an order of magnitude higher. Much the same may be said of grain boundaries, although the actual nature of these is rather different.

It is clear that there is a close relationship between topological properties of the order parameter and generalized rigidity. In general, a topological invariant leads to a form of rigidity. The canonical case is the path invariant: the number of unit cells along a path in a crystal can't change without crossing that path with a dislocation, the supercurrent in a ring can't change witout passing a vortex across the ring, etc. But there are corresponding rigidities for point invariants (the relative sign of the magnetization $\vec{M}$ measures how many boundaries separate two points in a ferromagnet--whether an odd or even number, for instance) or surface invariants (no familiar cases). Thus the relaxation of these rigidities will always involve defect motion.

As a summary of this initial chapter on broken symmetry, it should help in finding our way through the bewildering variety of systems which I have discussed (and these are only a fraction of those which exist) if I present what I can of the systematics of broken symmetry in tabular form. In Table I, I have listed, for each type of transition, the identifying phenomenon, the high- and low-symmetry phases, the order parameter and its dimensionality, the common type of transition, the major type of fluctuation, the "Goldstone bosons" if any, the rigidity phenomenon, and finally, the types of singularities necessary to break down the rigidity. This table predates Sec. 2G and has not taken into account recent results in the homotopy theory of defects.

In later chapters, we shall examine in much greater detail many of the entries in this table; but first, I'd like to discuss the other main principle, continuity.

TABLE I: MASTER TABLE OF BROKEN-SYMMETRY PHENOMENA

Phenomenon	High Phase	Low Phase	Order Parameter (constant of motion)		Order-Parameter Dimensionality $T \to 0$	T_c	Common Transition Type (Can Always Be First Order	"Goldstone Bosons" (or "Higgs" Bosons)	Fluctuations	Collective Hydrodynamic Modes	Generalized Rigidity Phenomenon	Long-Range Forces Due to General Rigidity	Singularities
Ferroelectricity (Pyroelectricity)	Non-Polar crystal	Polar crystal	$\vec{P}$	(no)	1	1 or 3	2nd or 1st nearly 2nd	no (optical phonons)	Soft Modes	No	Ferroelectric hysteresis	No	Domain walls (thin)
Ferromagnetism	Paramagnet	Ferromagnet	$\vec{M}$	(yes)	1, often ≈ 3	1 or 3	2nd (time reversal)	Spin waves, one branch $\omega \propto k^2 (+\text{const})$	Spin waves	No ?	Permanent magnetism: hysteresis	Suhl-Nakamura	Domain walls (very mobile, ≈ 3 dim.)
Antiferromagnetism	Paramagnet	Antiferromagnet	$\vec{M}_{\text{sublattice}}$	(no)	1, often ≈ 3	1 or 3	2nd (or first)	Spin waves, 2 branches $\omega \propto \sqrt{k^2 + \text{const}}$ ("Fermions" in metal case?)	Spin waves (diffusion?)	No?	subtle effects in A.F. resonance	Suhl-Nakamura	Domain walls
Superconductivity	Normal Metal	Superconductor	$\langle \psi_\sigma^* \psi_{-\sigma}^* \rangle = Fe^{i\varphi}$	(no)	2	2	2nd (Gauge; no 3rd order terms)	no (plasmons)	diffusive fluctuations of gap	Mostly not	super conductivity	No: Penetration depth	Flux lines (or normal domains in Type I)
He II	Normal liquid	Superfluid	$\langle \psi \rangle$	(no)	2	2	2nd (Gauge; no 3rd order terms)	phonons (1 branch)	diffusive fluctuations of $\langle \psi \rangle$	2nd sound, especially	superfluidity	Yes, vortex lines unscreened	vortex lines

Nematic liquid crystal	Normal liquid	oriented liquid	$\overleftrightarrow{\lvert d \rvert}$ (directrix)	(no)	3 {complex topology}	3	2nd	not stable at T = O (yes in principle)	Yes	usually overdamped	various peculiar properties: orientation elasticity	Yes	disclinations, points
Cholesteric, Smectic liquid crystal	Nematic	Density wave	$\rho(Q)$	(no)	>3 {complex topology}	>3	2nd or 1st	″	Yes	″	″	Yes	disclinations, points and dislocations
Crystal	liquid	solid	ρ_G, all G on recip. lattice	(no)	3(2 orient, 1 phase) at least	3	1st	yes: 3 kinds 2 transverse 1 longitudinal	phonons	2nd sound in solid, etc.	rigidity	Yes: elasticity effects	dislocations, grain boundaries, points (vacancies, interstitials)
He 3	normal liquid	anisotropic superfluid	$d_{ij} = \langle \psi\psi \rangle_{M_L, M_S}$	(no)	at least 3 {complex topology}	18	2nd	yes, several kinds	yes, complex	probably	superfluidity and orientation elasticities	Yes	Vortex lines, disgyrations, point defects
CDW's	normal electron gas (2-dim.)	Incommensurate Density wave → commensurate	ρ_G G on triangular superlattice		2	2	1st	Yes, phasons		Yes	Yes, $NbSe_3$ sliding CDW's	?	discommensurations, dislocations

CHAPTER 3

BASIC PRINCIPLES II: ADIABATIC CONTINUITY AND RENORMALIZATION

A. FERMI SYSTEMS: THE FERMI-LIQUID THEORY

I think most people entering research find that by far the most difficult question is where to start, especially when confronted with something that is actually new. This, it seems to me, is the kind of question a book like this should be designed to answer. Many books are simply compendia of methods that have already been used or of techniques for calculating a little better something that is already understood. I am writing this immediately after the experience of having been confronted by the new phases of ^3He; faced with such a genuinely novel problem it is far more important to have some idea of what the relevant questions are than it is to do any one calculation with great accuracy or rigor. This is one of the reasons why I suggest that the two most important principles of condensed matter physics for our purposes are, first, broken symmetry, which tells us that what the order parameter is and what symmetry it breaks are the most vital questions; and, second, the continuity principle, which tells us to search for the right simple problem when confronted with a complicated one. To my way of thinking, detailed perturbation methods, and even Green's function and fluctuation-dissipation ideas, are somewhat less important, because they emphasize computation rather than understanding.

Let us first discuss Landau's Fermi-liquid theory--in principle, not in detail--as an example, perhaps the cleanest and best, of how continuity works. Landau's own "derivation" of this theory (1957a, 1957b, 1958) is a bit sketchy, even for him, but is at least clearer about what is going on than the normal derivations in terms of vertex functions, etc. It serves as something of a model for this type of theory.

The idea is to imagine oneself starting from a completely noninteracting Fermi gas and gradually turning on the interactions between the particles. Landau's argument is that there will be a one-to-one correspondence between states before and after this adiabatic turning-on process, and that states of similar symmetry do not cross, so that he can label the new states with the quantum numbers of the old and they will be in roughly the same order. The noninteracting gas can be described as having a distribution function n(k) (assumed to be very close to the T=0 Fermi distribution of the ground state), so Landau labels the new states by n(k) and asks how the energy can depend functionally on n(k). The fictional particles that occupy the states k are called "quasiparticles."

Before proceeding, let us try to clean up Landau's derivation a little. After working through the following, it should be clear from the two or three sentences Landau gives to this subject that he must have had more or less the same idea in mind.

In the limit of a very large system, energy levels get very close together, and it is not only impossible, but incorrect, to define the state with a given number of quasiparticles as the exact eigenstate with the same ordinal number counting up from the ground state.

This can be seen clearly in nuclear physics, for instance. It is not a bad approximation to describe nuclei as relatively weakly interacting Fermi-liquid droplets, and the ground state can often be assigned precise sets of single-particle quantum numbers (the "shell model"). A nucleon added in a nuclear reaction goes into an unoccupied single quasiparticle state at first (the so-called "giant single-particle resonances"), but this state is not an eigenstate. The density of actual eigenstates at the relevant high degrees of excitation is far higher than that of single-particle states, and the giant resonance spreads over many tens or hundreds of exact states and decays into one or another of them (Feshbach <u>et al.</u>, 1954). From the <u>detailed</u> point of view, the giant resonance appears to be just a phenomenon in the reaction cross section (see Fig. 3-1). There is no way of choosing a correct <u>exact</u> state to correspond to the quasiparticle state. This example points

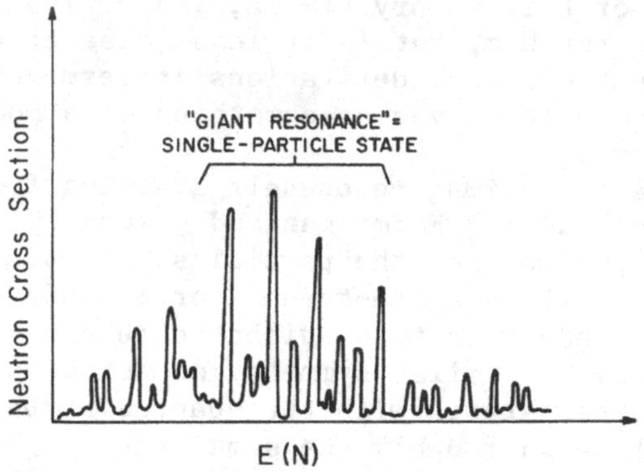

Figure 3-1 Neutron cross section illustrating giant resonance.

up the importance of time--equivalent to energy--resolution. On a short-time basis (i.e., low energy resolution), the single-particle picture is correct, while on a long-time detailed basis, only a statistical picture is possible.

Taking this hint [and following, actually, some ideas of Nozières (1964) rather closely], we shall suppose that the interactions are turned on, not infinitely slowly, but at a slow rate R chosen with two conditions in mind:

(1) If we are considering states of excitation of the system which correspond roughly to temperature T, i.e., excitations (holes below the Fermi surface, or particles above) having energies of order ~ T, R << T will ensure that our energy resolution is finer than the energies of interest. (2) On the other hand, if τ is the time that quasiparticle states take to decay into more complicated, exact eigenstates, we must have R >> $1/\tau$, or the quasiparticle states will lose their meaning and identity as distinct from the exact eigenstates.

It is easy to show that for low enough T--i.e., low enough degree of excitation--T >> $1/\tau$, so that conditions (1) and (2) are compatible with a correctly chosen R. This is the essential condition for the success of the Fermi-liquid theory. To demonstrate it, one uses perturbation theory, but only of the simplest kind, namely, the Golden Rule. The simplest decay process that can happen to a particle of momentum k is for it to excite a particle, either one out of the Fermi sea or one of the few excited particles, into an empty state, itself being scattered into a state k' = k + q by the potential matrix element

$$V(q) = \int d^3r\, e^{iq\cdot r} V(r). \tag{3.1}$$

We give the diagrammatic description of this process in Fig. 3-2.

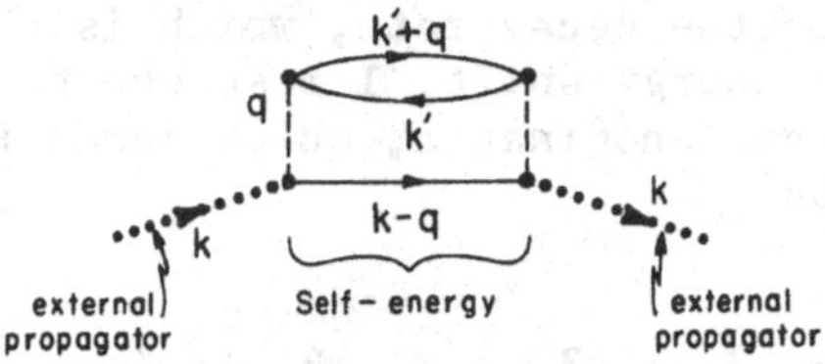

Figure 3-2 "Self-energy part": simplest case.

Diagrams: As K. G. Wilson has remarked (1974a), it simply doesn't matter which specific diagrammatic scheme one uses, so long as one works consistently within one's own scheme. Diagrams will be used here simply as a shorthand for the physical processes I am describing. (A somewhat more complete introduction to diagrams will be given in this chapter.) Thus the diagram in Fig. 3-2 is shorthand for the following statement: I am to calculate the effect of the scattering process, described above, on the energy of particle k. I need to include two matrix elements (vertices) of the scattering potential V (which I stretch out into dotted lines for simplicity), since it is a second-order process. I draw forward-pointing lines for particles and backwards ones for holes (the propagation of the absence of a particle being equivalent to a particle moving backwards in time; or, to put it another way, the state, having emptied at t, must be refilled at some later t'. I have "borrowed" an electron. It's just like selling short on the stock market). In space-time theory I'd include an energy denominator for every line, and integrate amplitudes over all intermediate energies and momenta, but that prescription would give the same answers as simple Brillouin-Wigner perturbation theory with a single energy denominator:

$$D = E_{k-q} + E_{k'+q} - E_{k'} - E_k . \tag{3.2}$$

To get the resulting energy shift, I integrate with the real part of energy denominator D^{-1},

$$\Delta E = \int d^3q \int d^3k' \frac{|V(q)|^2}{D} . \tag{3.3}$$

However, for the decay rate, which is the imaginary part of the energy shift, I must use the imaginary part of the energy denominator, which leads to the prescription

$$\begin{aligned} \frac{1}{\tau} &= \lim_{\varepsilon \to 0} \int d^3q \int d^3k' |V(q)|^2 \, \mathrm{Im}\left(\frac{1}{D-i\varepsilon}\right) \\ &= \pi \int d^3q \int d^3k' \ \delta(D) |V(q)|^2 , \end{aligned} \tag{3.4}$$

which is the Golden Rule. In calculating energies I must leave off the propagator of the original particle; only if I want the modification of the propagator of k do I leave in the incoming lines.

A diagram like 3-2 is called a self-energy diagram, and shows contributions to the energy shift (and inverse lifetime). More complicated processes such as those shown in Fig. 3-3 can also be imagined, and all of them can be represented by diagrams, e.g., scatterings against ground-state particles.

Let us now analyze the Golden Rule formula (3.4) for the decay rate of a quasiparticle at low degrees of excitation in the Fermi sea. First let us consider simply a single excited quasiparticle of excitation energy

$$E_k = \varepsilon_k - E_F = v_F(|k| - k_F) \tag{3.5}$$

taken to be much less than E_F:

$$E_k/E_F \simeq \frac{|k| - k_F}{k_F} \ll 1 . \tag{3.6}$$

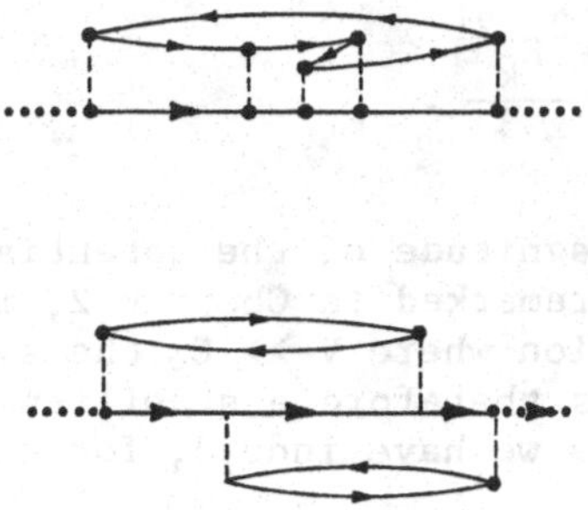

Figure 3-3 Higher-order self-energy diagrams.

We are making a linear approximation to the band energies near E_F.

For the δ function to give a finite result, all three of the excitation energies $E_{k'+q}$, E_{k-q}, and $-E_{k'}$, must be less than E_k, since in this case they are all positive. This means that in doing the integral over q and k' in Eq. (3.4) we are restricted in each case to an amount ~ $(E_k - E_F)$ of the available energy space (see Fig. 3-4), which in

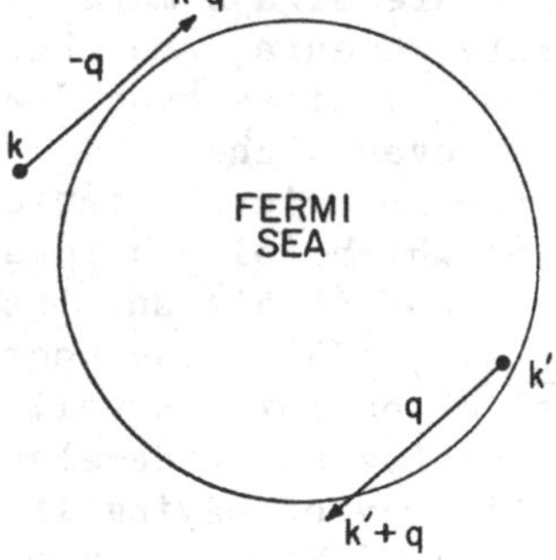

Figure 3-4 Momentum and energy requirements from the exclusion principle.

turn restricts us in momentum space. This gives us a factor

of E_k^2. The three internal propagators give roughly $\frac{1}{E_F^3}$. Thus we come to the conclusion that

$$\frac{1}{\tau} = \text{const} \times \frac{V^2 E_k^{\ 2}}{E_F^{\ 3}} \ . \tag{3.7}$$

Now the order of magnitude of the potential V is $\lesssim E_F$; this is because, as we remarked in Chapter 2, in any low-temperature situation where $V \gg E_F$ the system is not fluid but solid. This is therefore a stability condition on the Fermi liquid. Thus we have indeed, for this simple case, shown that

$$\frac{1}{E_k \tau} \lesssim \frac{E_k}{E_F} \ll 1 \ ,$$

and that therefore under the condition (3.6) it is possible to choose

$$E_k \gg R \gg \frac{1}{\tau} \ . \tag{3.8}$$

It is easy to see that so long as all excitations are to energies of order T only, the phase-space restrictions hold, and that lifetimes will still be of order T^{-2}. Even more negligible will be the effects of higher-order processes such as those represented by the diagrams of Fig. 3-3, because in these there are always more than two integrations over intermediate-state momenta, and thus the restriction to excited-state energies $\lesssim$ T bites even deeper. Such terms will be of order T^3 or even higher.

This, then, is the main justification for Landau's ideas, a justification which follows ideas put forward as early as 1951 by Weisskopf (1951) and later by Brueckner and others (Brueckner et al., 1954; Brueckner, 1959), in justifying the shell model of nuclei, as well as by Abrahams and Kittel (1953) in discussing the free-electron model of metals. A diagrammatic way of saying it is that, for sufficiently low-lying states of Fermi systems, it is always correct to neglect imaginary parts of self-energies. Note that there is absolutely no corresponding argument to allow neglect of the energy shift ΔE of Eq. (3.3), so that free-particle behavior does not imply an unshifted Fermi energy or Fermi velocity.

But Landau's theory is not only this simple neglect of $1/\tau$; it includes the idea that there is a unique one-to-one correspondence between the states of the free Fermi gas and--within a certain energy resolution, much smaller than the energies of interest when those are low--those of the strongly-interacting gas or "Fermi liquid." Thus one may label the two sets of states the same, and describe any low-lying state of the Fermi liquid as being made up of states with so-and-so many excited quasiparticles of each k. For instance, in principle one can construct quasiparticle operators q_k^+ and q_k which have the effect of creating and destroying quasiparticles of momentum k.

This is quite a different mode of thinking from conventional Green's-function perturbation theory. There one concentrates on the states that contain one excited "bare" particle over the ground state, states like $c_k^+ \Psi_g$, and observes that some of these states behave as though they were part of a "quasiparticle pole." One thinks, then, of expanding bare one-particle states in terms of exact eigenstates. Of the bare states created by C_k^+, only a fraction Z are in the appropriate quasiparticle pole, where Z is called the "wave-function renormalization constant." In Landau's idea Z plays no role; q_k^+ creates precisely one quasiparticle, which is made up of states with 3, 5, 7, etc. bare-particle excitations, as well as of the $c_k^+ \Psi_g$ state.

This derivation, which shows us the absence of certain renormalizations, allows us to see immediately several relationships that can be determined from the perturbation theory only with considerable difficulty. Any quantity that is conserved in every one of the individual scatterings which take place as the interactions are turned on will have exactly the same value for the actual state as it would have had in the free-particle state. It is obvious that the individual scatterings conserve charge, particle number, and momentum, because the interaction Hamiltonian does so, and there is no broken symmetry in the noninteracting state. This leads to three basic theorems: (1) quasiparticles transport as much current and momentum as real particles (Landau); (2) the volume of the Fermi sea is unchanged (Luttinger, 1960); (3) there is an identity due to Landau for the effective mass of the quasiparticle.

The first two are obvious from the above, but to explain this last theorem we have to go a bit more deeply into the details of Landau's theory. As we have just discussed, the quasiparticle states are eigenstates to an approximation increasingly good as we go to lower temperatures, and so the energy must be a function simply of their occupation numbers, to the same approximation:

$$E = E_g + \int d^3k \sum_\sigma \delta n(k,\sigma)\varepsilon_\sigma(k)$$

$$+ \frac{1}{2} \int d^3k \int d^3k' \sum_{\sigma\sigma'} f_{\sigma\sigma'}(\hat{k},\hat{k}')\delta n(k,\sigma)\delta n(k',\sigma') \, . \tag{3.9}$$

Both ε and f are unknown coefficients, not at all equal to their free-particle equivalents. In the spherically symmetric case the coefficient $\varepsilon_\sigma(k)$ is

$$\varepsilon_\sigma(k) = E_F + \frac{k_F}{m^*}\left(|k| - k_F\right) \, , \tag{3.10}$$

where m^* is defined by this equation; in ^{3}He, for instance, this is from four to six times the free mass. This is the "mass" which determines the linear specific heat $C = \gamma T$, coming from the constant density of states at the Fermi surface characteristic of Fermi systems.

Landau's identity follows if one changes to a frame of reference moving at a small velocity δv. The moving liquid must change energy by $\frac{Nm(\delta v)^2}{2}$, where m is the real mass and N the number of particles. This appears to contradict Eq. (3.9) because the linear term gives a coefficient of $(\delta v)^2$ involving $\frac{m^2}{m^*}$. One factor of δv comes from expressing the new n(k) in terms of the old one via $n_{new}(k) = n_{old}(k+m\delta v)$. The other factor comes from Eq. (3.10). The mean energy change in a given direction is

$$\overline{\delta\varepsilon} = \frac{1}{2}\frac{k_F}{m^*}\delta(|k|) = \frac{1}{2}\frac{m}{m^*}\delta v \cdot k$$

instead of the obvious $\frac{1}{2}\delta v \cdot k$ obtained by Galilean invariance. (The factor of 1/2 comes from averaging.) The difference must be made up from quadratic terms, which we can calculate using

$$\delta n(k) = m\delta v \cos(\hat{k}, \delta v)$$

for the change in the number of quasiparticles in any direction $\hat{k}$, being clear that, so long as they are very near the Fermi surface, their precise distribution in $|k|$ doesn't matter much. In fact, in Eq. (3.9) we have written the quasiparticle interaction term as a function only of the directions $\hat{k}$ and $\hat{k}'$ around the Fermi surface; corrections to the "bare," unrenormalized $f_{\sigma\sigma'}(\vec{k},\vec{k}')$ are, like the real part of the "self-energy," large but slowly varying because the exclusion principle and the Fermi surface play no crucial role so far as real energy corrections are concerned.

When we insert δn into the quadratic term of Eq. (3.9) all Fermi-liquid corrections drop out if and only if

$$\frac{1}{m} = \frac{1}{m^*} + k_F \mathrm{Tr}_{\sigma'} \int f_{\sigma\sigma'}(\theta) \cos\theta \, d\Omega, \qquad (3.11)$$

where θ is the angle between $\hat{k}$ and $\hat{k}'$; this is an identity for the $\ell = 1$ spherical harmonic in the expression of f.

It is in the realization that the energy should be expanded to second order in n, and no farther, that the practical value of the Landau theory lies. In almost any property, transport or equilibrium, there appears a correction due to the appropriate "Landau coefficient" [some average of f such as appears in Eq. (3.11)], which may even be quite large--a factor of 20, for instance, for the susceptibility of ^{3}He. Such corrections enter like molecular fields: they give the self-interaction of the quasiparticles with the field due to the surrounding atmosphere of other quasiparticles.

The exchange effect on spin susceptibility is a nice simple example of this. When we apply an external field H, up-spin and down-spin free particles are shifted in energy relatively, by μH. In equilibrium the Fermi levels must be the same, and therefore

$$\left(n_{\uparrow} - n_{\downarrow}\right)_{\text{non-int}} = \mu H \cdot \frac{dn}{dE} .$$

But in the interacting gas there is an exchange interaction J, essentially because parallel spin particles, under the antisymmetry requirement of the exclusion principle, see the "exchange hole" and are (in ^{3}He) less repelled by their neutral hard cores than anti-parallel spins. In other words, the spatial wave functions of parallel spin particles are orthogonal, and this reduces the hard-core repulsion. Thus there is an exchange field $\delta H = \delta n \cdot J$ in addition to the external one, and we may find χ self-consistently from

$$n_\uparrow - n_\downarrow = \mu \frac{dn}{dE}(H + \delta H)$$

$$(n_\uparrow - n_\downarrow) = \chi_0 (H + J\delta n)$$

$$= \frac{\chi_0}{1 - J\chi_0} H \quad .$$

The factor $1/(1 - J\chi_0)$ is often called the "Stoner enhancement factor" from a historical analogy to the old free-electron theory of ferromagnetism. The term of Eq. (3.9) which corresponds to J is $\int (f_{\uparrow\uparrow} - f_{\uparrow\downarrow}) d\Omega$, which is given various notations in the Fermi-liquid literature. The general idea is that the quadratic term in every case modifies the effective single-particle energy in response to changes in the distribution of other quasiparticles, leading to "mean-field" type corrections in the various properties.

B. GENERALIZATIONS OF FERMI-LIQUID THEORY

Landau's original "pure" version of the theory was aimed at the specific problem of ^{3}He and was concerned with deriving transport coefficients as well as equilibrium properties. To this end, Landau set himself a second restriction: that all quantities should be very slowly varying in space. Under this restriction, it is possible to define a local distribution function of quasiparticles $n(\vec{k}, \vec{r}, t)$ in space and time, and a local energy $E(\vec{k}, \vec{r}, t)$, which is the same functional of the local distribution as in the equilibrium case. In this way, changes in the distribution function with time can be treated, so long as they are slow compared to our turning-on time. The "residual interactions"--i.e., the terms which mix the different states labeled by quasiparticle number n--are, of course, the source of the scatterings which enter into the transport theory. That is, one writes a Boltzmann equation $\frac{dn}{dt} = \frac{\partial n}{\partial t} + \left(F \cdot \nabla_p + V \cdot \nabla_r\right)n =$ "collision term," and "collision term" is the result of quasiparticle interactions, which are inserted empirically in the Landau version and not related to the fundamental Landau parameters contained in f.

But the use of Landau's theory in transport processes is beyond the scope of the present discussion. What I wanted to do was to point out that there are several situations in which the standard Landau theory, including the restriction, doesn't apply, and yet the idea of quasiparticles and their reality as the "true" low-energy excitations of the system is just as valid. Let us take one of these, the case of nuclei, as an example: this one has been worked out in some detail by Migdal (1967). In this fairly typical case, the difficulty is that the noninteracting system will not go smoothly into the interacting one when the interactions are turned on adiabatically: the noninteracting nucleons won't even form a nucleus because they don't attract each other. What we clearly must do is to start with some kind of arbitrary mean single-particle potential well which gives us a fair approximation to the right density and energy-level structure, and gradually turn on, not the whole potential, but the difference between the real potential and this starting Hartree-type potential. It is hard to believe that there would not exist some potential which would give near enough to the right ordering of quasiparticle levels to allow the adiabatic continuation. We then have a set of Landau-type parameters giving the most important interaction effects; and again, as Migdal shows, the quadratic terms give a much better account of nuclear energy-level schemes than the linear terms alone, as well as allowing him to fit the various forms of nuclear collective excitation. Exactly the same modification is adequate for treating real metals (although here there are some fairly subtle corrections necessitated by the long-range Coulomb interactions of charged particles, involving careful treatment of $q \to 0$ and $\omega \to 0$ limits). The Fermi surface of a real metal is not the same as that of noninteracting electrons in the same potential, so that one must in some way allow for adjustments in the starting Hamiltonian as V is turned on. In this case the correct procedures have been checked by diagrammatic perturbation theory [as, I should mention, has the original Fermi-liquid theory by Abrikosov et al. (1963) and Luttinger and Nozières (1962)].

The same scheme, in effect, was reinvented for atomic

and molecular physics and called the "hyper-Hartree-Fock" theory by Slater (Slater et al., 1970). Here again the idea of single-particle configurations preexisted--the original shell theory is, after all, atomic--but the use of modified quadratic terms in the occupation numbers gives a much better fit with atomic data. In each of these cases the central point is that a low-excitation configuration specified by quasiparticle-state occupation numbers is a very good approximate eigenstate, and therefore, to get the real energy-level spectrum, it is necessary to mix only states which are degenerate in energy or nearly so. Because of the complexity of the spectrum of a many-electron atom or many-nucleon nucleus, there are often several of these, but they usually represent a manageable problem.

A very similar use of the quasiparticle idea is in justifying the straightforward use of BCS theory in superconductivity, treating the strongly interacting electrons as though they were nearly free. This case was analyzed clearly and rigorously by Leggett (1965b) in connection with the problem of BCS pairing in superfluid ^3He. Leggett's analysis makes use of the concept of the Fermi-liquid parameters as effective fields, and points out that as long as the BCS pairing is a weak coupling effect in itself it does not change the distribution of quasiparticles around the Fermi surface; it merely recombines them into new single-particle eigenstates. Fermi-liquid and pairing effects interact only via the effect of the latter on response functions such as $\chi_0 = \delta(n_{\text{quasi-particles}})/\delta F$ where F is some generalized force. Typically, the response function (e.g., the Pauli magnetic susceptibility) is severely modified by BCS pairing, and if the observed normal χ is strongly renormalized by f (e.g., Stoner enhancement), the modified one will be also, but, of course, to a different degree. Now, in place of χ_0 we will have χ_{BCS}.

This case, Khalatnikov's Fermi-liquid transport theory (Khalatnikov, 1965), and the two above all have in common that the quasiparticle scattering terms specifically neglected in the original theory are reintroduced by hand, to calculate effects which are smaller than the basic renormalization. That is, the quadratic Landau Hamiltonian is treated as an unperturbed H_0 and the corrections are brought in by, essentially, degenerate perturbation theory.

Another way to think of this kind of application is in terms of our original "turning-on" idea. The interactions which lead to decay of the quasiparticle states, and thus lead to the rate $1/\tau$, are retained and taken into account exactly. By definition their effects are much smaller than those of the full interactions that were turned on at the rate $R \gg 1/\tau$, and only mix nearly degenerate quasiparticle states with each other.

One is led, then, to the concept of a generalized Fermi-liquid theory (FLT). The generalized Fermi liquid is a system of interacting Fermi particles for which one can find a single-particle Hamiltonian with noninteracting particles which allows the usual adiabatic turning-on process. If such a Hamiltonian exists, and its energy levels are ε_n, with wave function ϕ_n, the quasiparticle lifetime can be estimated by the Golden Rule,

$$\frac{1}{\tau_n} = 2\pi \sum_{nn''n'''} |(nn'''|V|n'n'')|^2$$

$$\times\, \delta\left(E_n + E_{n'''} - E_{n'} - E_{n''}\right).$$

Again, if the density of energy levels is ρ, and if we take into account that the levels in the unperturbed system are filled up to a Fermi energy E_F, the lifetime τ_n is given by

$$1/\tau_n \sim \rho \langle V^2 \rangle \left(E_n - E_F\right)^2 .$$

This may be neglected in the limit $E_n - E_F \to 0$ of low energies: specifically, the ground and the lowest excited states will have the same topological structure as the noninteracting system.

Thus even extraordinarily impure metals, with mean free paths for scattering of order the interatomic distance, continue to act like noninteracting Fermi systems; it is only necessary to think of them, not in terms of the plane-wave, or Bloch, representation, but of the "n-representation" (Anderson, 1959b) of exact, but incredibly complicated, quasiparticle states ϕ_n, with an energy density that is, of course, renormalized according to the appropriate Landau parameter but is often not much different from that of the corresponding pure metal. The reason for this is that so long as such scattering is of reasonable strength, it mixes states only of order $1/\ell$ from the Fermi surface in momentum space (ℓ being the mean free path), which simply averages out properties over the Fermi surface. But these considerations show us that there is a limit to the validity of such a construction: if the eigenstates of the starting Hamiltonian are no longer states extending throughout the whole large system, but have a genuinely different character, the averaging of interaction matrix elements as in Eq. (3.9) is no longer correct.

This brings us to a difficult problem only now being solved. We must be prepared for the existence of systems where, if quasiparticle states exist, they are not at all like some randomly-phased average of plane-wave states, but may be localized by strong random potentials, or confined to the sites of a lattice as in a solid; and none of the above arguments are necessarily valid in that case. [But see the work of Fleishman (1978).] For the time being we note that, aside from these extreme cases, most of our experience with real metals--e.g., the "dirty superconductor theory" (Anderson, 1959b)--tells us that the concept of a generalized Fermi-liquid theory is very often valid.

I have elsewhere suggested (Anderson, 1973) that an appropriate background for such a generalized FLT may be renormalization group theory in the form typified by work on the Kondo problem, via what I like to call the "poor man's method" (Morel and Anderson, 1962; Anderson, 1970b; Wilson, 1974b; Nozières, 1974). The idea is to recognize that one is interested, usually, only in low-energy, low-temperature phenomena which rarely involve excitations far from the Fermi surface. This type of renormalization group theory simply eliminates progressively, by perturbation theory, all virtual transitions into states at high energies, leaving one with an effective Hamiltonian acting only between particles near the Fermi surface. Such an effective Hamiltonian will, of course, contain not only the all-important quasiparticle interactions of conventional FLT, but also presumably the residual scattering and pairing interactions as well.

Note added in proof: Very recent work--see for example Abrahams et al. (1981)--shows that scattering can have an unexpectedly profound effect on coupling to particle interactions in some instances, especially for lower dimensionality. This subject is too recent and too unsettled to include here, but most of the above remains essentially valid.

C. PRINCIPLE OF CONTINUATION

The Fermi-liquid theory, especially in its generalized

forms, serves as an almost perfect model for the idea I have in mind when I refer to a basic principle of continuation. FLT shows that in a very deep and complete sense we can refer back whole classes of problems involving interacting fermions to far simpler and fewer problems involving only noninteracting particles. While there can be great or small quantitative differences between a real metal, for instance, and a gas of noninteracting electrons, the essentials of their qualitative behavior--specific heat, T dependence of susceptibility, T dependence of various transport coefficients, etc.--are the same. This "referring back" has a very straightforward and complete meaning in the case of the fermion systems, namely, (1) the ground state can be labeled meaningfully in terms of the occupation numbers of a set of one-electron states; and (2) the excitation spectrum has the essential character, for low excitation energies, of a noninteracting Fermi gas. That is, the excitations have a finite density, are like single particles, and normally occur near a surface in momentum space. This brings us to the concept of elementary excitations, which I discussed at length in my previous book; here we simply state that for this wide class of systems the elementary excitation spectra have the same character as those of a Fermi gas.

One implication of such a continuity principle is that we can expect to be able to start from the noninteracting system with an unperturbed Hamiltonian and calculate the properties by perturbation theory. In the case of Fermi-liquid theory, this program has been formally carried out: one can produce diagrammatic expressions (Abrikosov et al., 1963) for all of the renormalizations, of the mass and of the various vertices, contained in the theory. In fact, another way to express the idea of continuation would be to say that a real system is in the continuum attached to a given simple problem if its properties can be computed, at least in principle, via perturbation theory starting from the simple problem. This is, however, probably too restrictive; we shall see that straight perturbation theory almost never converges, and we must allow certain resummations of infinite sets of terms. What is needed is "perturbation theory plus good sense." [As in almost the simplest of all problems, the high-density electron gas, the plasmon resummation (Bohm and Pines, 1951, 1952; Gell-Mann and Brueckner, 1956) is essential, but still leaves us with a Fermi liquid.]

Both this continuity principle and the idea of adiabatic variation of interaction parameters make it clear that one thing one can not do is continue past a symmetry boundary: the very obvious failure of analyticity of the free energy at such a boundary, which we mentioned in Chapter 2, shows that this cannot be done. It is also clear that two states of different symmetry cannot be reached by simple continuation from the same state.

D. EXAMPLES OF CONTINUATION

1. Bose and normal liquids

Before going on to a qualitative discussion of perturbation theory and renormalization, let me give a few additional examples of what I mean by continuation.

For Bose fluids (which so far, in practical fact, consist of ^{4}He liquid only), the free Bose-condensed gas represents a fair starting point. The free Bose gas has a ground state in which all the particles occupy the single lowest-energy state, with, as we discussed, a fixed mean amplitude $\langle\nabla_0\rangle$. The spectrum of excited states is that of a free-particle gas. Beliaev's (1958) theory of the weakly interacting Bose gas allows us to follow what happens as interactions are turned on. The excited states remain a set of Bose excitations as before, but the spectrum changes qualitatively from $p^2/2m$ to a linear, phonon-like form $E = sp$, where s is the "phonon velocity." [This behavior was already foreshadowed in the early theory of Bogoliubov (1947).] In this new spectrum the low-energy excitations decay even more slowly than fermions in a Fermi liquid because of the linear dispersion relation. Only when the spectrum curves upwards can there be any decay at all; otherwise, the density of momentum- and energy-conserving states vanishes. In ^{4}He itself this upwards curvature is so slight as still to be a subject of controversy (Dynes and Narayanamurti, 1974), and eventually one gets a downward curvature and goes into the well-known roton spectrum, which is therefore also nondecaying as $T \to 0$. Thus again, it is quite safe to assume a one-to-one correspondence between the spectra of the free-particle system and the interacting one, except for one proviso: as far as all physical parameters are concerned, the point at which the interaction V first becomes finite is a point of nonanalyticity. The properties of the system are not analytic in V as $V \to 0$. This is in fact a common case; as we pointed out, the electron gas has a similar nonanalyticity as $e \to 0$, which has no serious consequences. The reason, of course, is the sudden change in form of the spectrum (in the electron case to a finite-energy plasmon, in this case to a linear spectrum), although in either case no individual elementary excitation ever changes its energy discontinuously.

These longest-wavelength excitations are, as we saw in Chapter 2, closely related to the macroscopic response functions and, of course, to the correlation functions at very long distances. The minute even the smallest interactions are present in a Bose gas, it acquires a finite compressibility, and its density fluctuations are limited in the way we are used to in a true liquid: the density becomes macroscopically uniform, whereas, in a noninteracting gas, even boundary conditions affect the density because the lowest state ϕ_0 depends on them and is not necessarily uniform. In the charged Bose or Fermi gas, the compressibility effectively goes to zero as the plasma oscillation acquires a finite frequency, and density fluctuations become very small as $q \to 0$.

Momentum is conserved for the quasiparticle states; but the quasiparticles do not retain their single-particle character, as they do in the Fermi system, because of the kind of virtual scattering process shown in Fig. 3-5, in

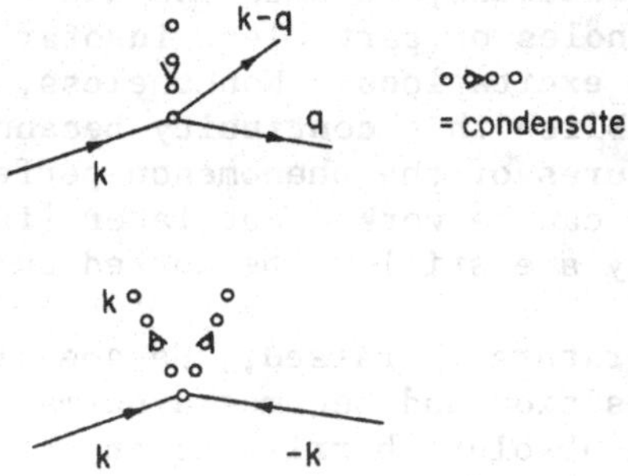

Figure 3-5 Bose diagrams: scattering against condensate.

which a quasiparticle can scatter against a condensate particle and convert itself into two quasiparticles. Thus quasiparticles in the Bose case are mixed with phonons and vice versa. Here, what we are calling phonons are simply density oscillations with no net particle character. We can, if we like, enshrine this as a general principle: The broken symmetry of the starting state infects the excitation spectrum in general, and especially the single-particle-like excitations. Here we find quasiparticles with indeterminate particle number, because it is particle number (gauge) symmetry that is violated in the starting state; in the BCS theory it will be the same, that is, there will be hole-particle mixing--as is demonstrated so strikingly by the Tomasch effect (Tomasch, 1965; McMillan and Anderson, 1966), which depends on double refraction and interference of hole- and particle-like states. In the case of solids, both electron and phonon excitations become imprecise eigenstates of momentum, the missing momentum being taken up by the lattice as a whole, i.e., the condensate, which breaks translational symmetry.

The only theorem that can be established by this approach to Bose liquids (and it may be more easily proved otherwise) is that the condensate at T = 0 is again unrenormalized, just as the volume of the Fermi sea doesn't renormalize in the Fermi liquid. It is valid to think of a condensate as consisting entirely of all of the quasi-particles, even though in real ^{4}He the actual delta function in the momentum distribution corresponding to the unperturbed k = 0 state is probably less than 10% (McMillan, 1965). Thus the entire fluid participates in superflow as $T \to 0$. But when there are excitations present, the number of quasiparticles in the condensate is not reduced by exactly the number of excitations, because the excitations can carry along condensate holes or particles, insofar as they are partly collective excitations. Nonetheless, it seems to me important to establish this continuity because it tells you the physical features of the phenomenon perfectly well; numerical details can be worked out later [in the case of ^{4}He, in fact, they are still to be worked out! (see Woo, 1976)].

As the temperature is raised, ^{4}He goes through a second-order transition and becomes a normal liquid. This is undoubtedly an absolute barrier to any kind of continuation from the Bose-condensed state. The normal Bose, high-temperature Fermi, and classical liquids seem to have no radically different features, precisely as one might expect from the corresponding noninteracting gases. Once T > T (degeneracy) ~ T_F, T_C^{B-E} statistics no longer play any appreciable role, even in the noninteracting gases. I would presume that turning on the interactions adiabatically (in the thermodynamic sense) would bring any of these three types of gases continuously to the corresponding liquid. This is the minimal sense in which one can understand the principle of continuity: that no singularity intervenes if one follows some adiabatic path from the simple state to a state including more or less strong interactions. Again, even with interactions present, one may manipulate pressure and temperature in such a way as to carry the dilute--and therefore nearly free--gases adiabatically over to the liquid states.

It is this basic similarity of the liquid states which lies behind de Boer's idea of the quantum law of corresponding states (de Boer, 1948a,b; Palmer and Anderson, 1974). For simple classical liquids with typical potentials of the Lennard-Jones type, a very accurate classical law of corresponding states exists, modeled more or less on the simple two-parameter van der Waals equation,

$$(P + a/V^2)(V-b) = T \qquad (3.12)$$

where P is the pressure, V is the molar volume, T is the temperature, a and b are the parameters, and the universal gas constant R is set equal to 1. That is, there exists a single expression for the free energy F [slightly different from the oversimplified version leading to Eq. (3.12)], which depends only on a length σ and on an energy parameter ε in the forces, fitting liquid and gas as well for all substances with van der Waals-type forces. [For the van der Waals equation (3.12), $b \propto \sigma^3$ and $a \propto \varepsilon\sigma^3$.] De Boer's idea was that this formalism could be extended to quantum liquids by introducing a third parameter measuring the "degree of quantum behavior,"

$$\Lambda^* = \left(\frac{h^2}{m\sigma^2\varepsilon}\right)^{1/2} , \qquad (3.13)$$

and writing a generalized three-parameter free energy

$$F = \varepsilon F^*\left(P^*, T^*, \Lambda^*\right), \qquad (3.14)$$

where

$$P^* = \frac{P\sigma^3}{\varepsilon}, \quad T^* = T/\varepsilon, \quad V^* = \frac{V}{\sigma^3} , \qquad (3.15)$$

which approaches the classical free energy in

the $\Lambda^* \to 0$ limit:

$$F(P, T, 0) = F^{\text{classical}}_{\text{corr. states}}$$

Correspondingly, one has a quantum equation of state,

$$V^* = V^*\left(P^*, T^*, \Lambda^*\right) . \tag{3.16}$$

In the case of the "normal" helium liquids, this works very well for both ^4He and ^3He, indicating that statistics play very little role in the normal state. The dynamical effect of quantum mechanics on the correlation functions--essentially spreading them out because of the uncertainty principle--is instead the major effect. We shall see later that precisely the same or even greater independence of statistics is to be expected in solid phases, where quantum mechanics primarily determines the zero-point energy of atomic motions.

But in fact, even in the genuine quantum fluids, the total free energy associated with statistics--the condensation energy of ^4He II, the free-energy modification due to Fermi-liquid effects in ^3He--is still not very large: a 10% effect or so, raising the melting pressure of ^4He, and lowering that of ^3He. These two liquids, diametrically opposite in their transport and thermal properties, unlike as night is to day in their phenomenology, fit so closely to the quantum law of corresponding states, which totally ignores their differences, that it is barely possible, with considerable effort, to make out the two opposite deviations.

The fact that liquid ^3He and liquid ^4He II are so similar in free energy to the corresponding classical liquids, while differing very much in other properties, brings out a principle of wide applicability in strongly interacting systems which obey the continuity principle. This is that the real particle correlation functions, in their crude outlines, are controlled mainly by the interaction forces. It is very hard to make a major change in, for instance, the near-neighbor correlation functions, because to do so will ruin the energy. Chemical species which nominally differ widely in valence or formal charge, for instance, will usually have much the same charge distribution around the atoms. This is why high-energy and high-momentum probes of atomic structure, such as x-ray and electron scattering, have to be very carefully refined and delicately interpreted before they can tell us much about the very striking macroscopic properties, such as superfluidity or metallic conductivity, of bodies.

Thus the system will very often obey the continuity principle only for its low-energy states of excitation and in the longest-range and lowest-energy part of its correlations, while the short-range, strongly interacting parts of the correlations go more or less their own way. Transport properties are different for liquid ^{3}He and ^{4}He because these depend on long-wavelength properties. However, the momentum distribution and correlation functions, which depend on short-range, strong interactions, are similar in both helium liquids to each other and to those of a classical liquid, except for the small, singular details which signify the type of state: a Fermi-surface discontinuity (see Fig. 3-6) for the Fermi liquid, a q = 0 delta

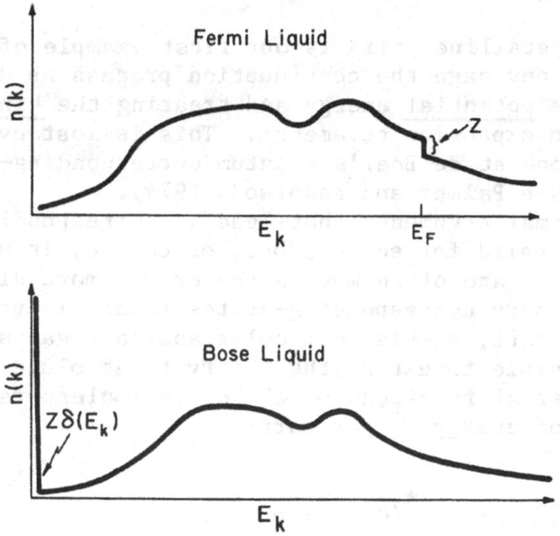

Figure 3-6 Fermi liquid - Bose liquid.

function for the superfluid. One finds this same kind of behavior in many other systems, particularly in magnetic metals, where the Kondo effect (Kondo, 1964) is a striking example.

In the Kondo effect, the underlying energetics determine that the correlation functions are those of a magnetic atom, so that the high-energy, high-temperature behavior is that of a magnetic atom. But the ground state can be thought of as a continuation from the ordinary impurity in a noninteracting gas, and so must be nonmagnetic (Anderson et al., 1970; Wilson, 1974b; Nozières, 1974). Often the division between these two energy regimes is at extremely low energies, of order T_K, which for Mn in Cu is ~ 10^{-3} °K. What to me is most impressive, perhaps, is not the constancy of this underlying structure, but the force of the continuation principle, which makes the system into a Fermi or Bose liquid or a into nonmagnetic impurity almost in spite of itself.

Note added in proof: Even more spectacular is the phenomenon of mixed-valence metals, where electronic quasiparticles of mass equal to several thousand electron masses are seen (Anderson, 1981b and other articles in the same book).

2. Solids, quantum and otherwise

The crystalline solid is our first example of a system in which we envisage the continuation process as diagonalizing the potential energy and treating the kinetic energy as an expansion parameter. This is most evident if we simply look at de Boer's quantum corresponding-states expansion (see Palmer and Anderson, 1974).

The formal arguments that lead to corresponding states are equally valid for solids; but, of course, in solids binding forces are often much stronger and more directional, so that ordinary corresponding-states theory is valid only for reasonably inert, simple, molecular and rare-gas solids. It is then possible to extend the theory to absolute zero and obtain universal functions of the dimensionless reduced volume V^* and energy U^* per particle,

$$V^*(P^*, \Lambda^*): \quad V^*(0, \Lambda^*) = V_0^* , \tag{3.17a}$$

$$U(P^*, \Lambda^*): \quad U^*(0, \Lambda^*) = U_0^* , \tag{3.17b}$$

which are plotted in Fig. 3-7. The liquid-solid energy difference in the heliums is quite small and hardly shows in

Fig. 3-7. Note that the ground-state U and V alone determine empirically ε and σ, from these curves, since we know the mass m.

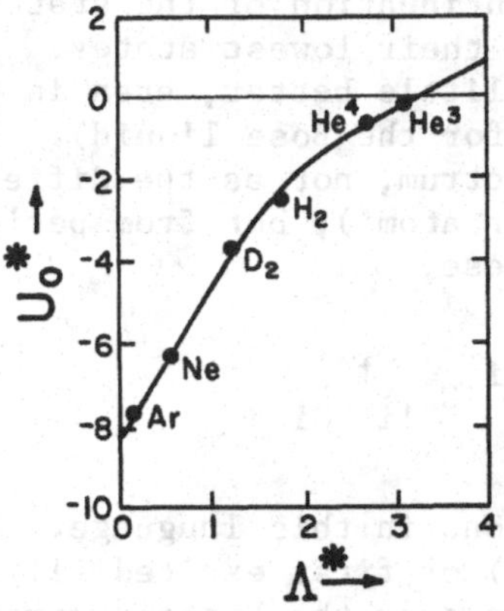

Figure 3-7 Energy of quantum liquids and solids as functions of Λ*.

One may derive from Debye theory the first few terms of an expansion for the ground-state energy of a solid:

$$U_0/\varepsilon = U_0^* = -8 + U_1\Lambda^* - U_2(\Lambda^*)^2 \dots , \tag{3.18}$$

where, you will remember, Λ^* is proportional to $(1/m\varepsilon)^{1/2}$ (note again that U_0 is not analytic as $1/m \to 0$, which is the essential parameter) (see Fig. 3-7). The first term is the energy of a body-centered cubic solid with atoms at fixed positions, the second the zero-point energy of a perfectly harmonic Debye-type solid, and the remainder rather difficult corrections. Though usually small, these are very large in the quantum-solid regime of the helium solids, which are stable only at pressures of a few tens of atmospheres.

As in the Bose liquid, it is useful to start from the very weak coupling--as opposed to zero coupling--regime. But we shall treat the almost classical solid not as a rigid lattice, whose vibrations we then quantize, but in the way envisaged in quantum-solid theory (Varma and Werthamer, 1976), that is, to very lowest order, we take it as a

Hartree self-consistent field problem in which each atom takes up a harmonic oscillator wave function in the sum of the self-consistent potentials of its neighbors. These localized wave functions are the quasiparticle states for this type of problem, and the ground state of any true solid is the analytic continuation of the state in which all oscillators are in their lowest states. It is, however, essential to do a little better, even in the very weak coupling case (as for the Bose liquid). One must make up the excitation spectrum, not as the different first excited, n = 1 levels (3 per atom!), but from periodic linear combinations of these,

$$b_k^\dagger = \sum_i e^{ik\cdot r_i} C_{1i}{}^\dagger C_{0i} , \tag{3.19}$$

which are the phonons in this language. The $C^\dagger$'s create atoms in lowest (0) or first excited (1) states. The lowest of these phonons do obey the broken-symmetry theorem and have $\omega \propto k$ as $k \to 0$. It is, incidentally, also essential to take into account properly their zero-point motion. These phonons, with their linear, down-curving spectra, are very-long-lifetime excitations ($1/\tau$ proportional to a high power of T) and thus by the usual construction make up a proper set of elementary excitations above the continued ground state.

The quantum solids and the Wigner solid, then, while in fact very far from the classical harmonic limit, are valid continuations of it and in principle have all their quasiparticles in localized orbitals. A problem we shall discuss at greater length later comes up when [as in the Wigner solid, solid ^{3}He, and many magnetic electronic systems (antiferromagnets)] the "atoms" are fermions and thus have at least the additional spin degrees of freedom.

This case has been rather carefully discussed by me (Anderson, 1959a) and by Herring (1960), but perhaps not wholly in the present context. [A less sophisticated discussion was later given by Kohn (1964).] The correct approach is to go a little farther than the unsymmetrized Hartree treatment which I have discussed so far and take into account the possibility of interchanges of the "atoms" (or electrons) on the different sites. There are n! ways of putting n atoms on n sites, and we must either symmetrize-in the Bose case-or antisymmetrize in the Fermi one. In the former, we are left with only one ground state, but in the latter we are left with an N/m-fold degeneracy for m up-spins out of N. This degeneracy of the ground state is lifted by exchange matrix elements. In simple cases these "exchange integrals" always favor antiferromagnetism. They contribute a term of a rather different form to $U_0^*(\Lambda)$ which is statistics-dependent, but is essentially quite small, because in the solid state interchanges of atoms take place only with difficulty, being rather severely obstructed by the presence of the other neighbors. We see then that there are three basic types of excitations, widely separated in energy: (1) spin reversals, of energy $\sim J$; (2) phonons, of energy $\sim \omega_D >> J$; (3) true free quasiparticles, of energy $>> \omega_D$--i.e., mobile interstitials and vacancies (J is the spin exchange coupling and ω_D is the Debye frequency). The scales of these three energies are easily visualized in terms of the depths of the Hartree wells, of order $U_0 \sim \varepsilon$, and the harmonic oscillator levels in the wells, with a frequency of order $\omega_D \sim \varepsilon\Lambda^*$. The exchange process involves tunneling out of the well, which is of order

$$J \sim \omega_D e^{-\text{const}/\Lambda^*}$$

so that the ratios are

$$J: \omega_D: U_0 \sim \Lambda^* e^{-(\text{const}/\Lambda^*)}: \Lambda^*: 1.$$

The Λ^* values > 3 of the He solids represent a very great departure from the "classical" harmonic solid--probably as distant in numerical terms as the He liquids are from free-particle systems. For instance, the pure Debye expression for the energy would give over an order of magnitude larger positive energy than the observed (and calculated, within ~ 20%) negative value. Nonetheless, the picture of stable localized atoms and, for ^{3}He, a thoroughly distinct degenerate subset of spin states seems--despite some quantitative problems in the case of ^{3}He--to be entirely valid. In all other solids, point defects (interstitials and vacancies) are not only expensive energetically but essentially immobile because of large lattice distortions; on the whole it appears that this situation extends even to the He solids. In summary, the quantum solid is a triumph of the continuity principle.

This conventional picture of a true solid is by no means the only possible one, nor, in fact, the only one to occur in nature. We could take a second point of view, similar to that discussed in Chapter 2, and look upon a solid as a three-dimensional expression of a giant density wave. As was pointed out there, a giant density wave in an isotropic medium automatically becomes at least two-dimensional and probably three. So let us ask to what extent a two- or three-dimensional density wave is equivalent to our usual concept of solidification.

First let us consider a Fermi system. The question of the occurrence of giant charge or spin-density waves in the electron gas is a recurring theme in the literature (Fröhlich, 1954; Peierls, 1955; Overhauser, 1960). One visualizes such a wave as leading to a self-consistent Hartree-Fock potential at a certain wave vector G:

$$V_{HF} = V_G \cos Gr.$$

This in turn modifies the quasiparticle wave functions by introducing a zone boundary at G/2, where approximately equal-energy states near $k = G/2$ will be mixed by this self-consistent potential to give a pair of states with energies

$$E_k = \varepsilon_{G/2} \pm \sqrt{(\varepsilon_k - \varepsilon_{G/2})^2 + V_G^2} , \qquad (3.20)$$

where $\varepsilon_k = \frac{k^2}{2m}$.

Finally, if the splitting term in Eq. (3.20) is sufficiently large and if G/2 is not too far from the Fermi momentum k_F, partial or total energy gaps may appear in the quasiparticle spectrum at certain points on the Fermi surface spanned by the vector G, and these gaps generate the original self-consistent potential by means of the resulting periodic density variation. Gaps at the Fermi surface are favored if they lower the total energy of the system.

This may be most simply understood, as suggested by Overhauser (1971), by starting from the much-discussed one-dimensional case. Leaving aside purely one-dimensional special problems, and confining ourselves to so-called "mean-field theory," this is a trivial exercise. In this one-dimensional case, the density wave is called a "Peierls distortion" (or sometimes "Peierls-Fröhlich"), and whether the interaction is attractive or repulsive such a distortion will appear: in the repulsive case, as a "spin-density wave" (SDW) (Overhauser, 1960, 1971); in the attractive case, as a "charge-density wave" (CDW) (Overhauser, 1968; Chan and Heine, 1973). The reason may be expressed in several ways, but perhaps most simply in terms of the response functions of the one-dimensional electron gas. If we apply a small potential V(q) of a given wave number q to a noninteracting gas,

$$V(r) = V(q)e^{iq\cdot r} + V^*(q)e^{-iq\cdot r} \quad ,$$

we will observe a corresponding linear response in the average density

$$\langle\rho(r)\rangle = \langle\rho(q)\rangle e^{iq\cdot r} + \langle\rho^*(q)\rangle e^{-iq\cdot r}$$

with

$$\langle\rho(q)\rangle = \chi_0(q)V(q)$$

where the susceptibility χ_0 is given by

$$\chi_0(q) = -\sum_k \frac{f_{k+q} - f_k}{\varepsilon_{k+q} - \varepsilon_k} \quad . \tag{3.21}$$

and f_k and f_{k+q} are Fermi distributions.

In the one-dimensional case, this is divergent as $T \to 0$, $q \to 2k_F$. To see this, notice that

$$\varepsilon_k \simeq -v_F\left(k_F + k\right)$$

$$\varepsilon_{k+q} \simeq v_F\left(k + q - k_F\right)$$

where $k < 0$ is on one side of the Brillouin zone and $(k+q) > ($ is on the other side. $k < -k_F$ implies $\varepsilon_k > 0$, and $(k+q) < k_F$ implies $\varepsilon_{k+q} < 0$. Thus

$$\varepsilon_{k+q} - \varepsilon_k \simeq 2v_F k + v_F q$$

$$= 2v_F\left(k + k_F\right) + v_F\left(q - 2k_F\right) \quad .$$

As $k \to -k_F$, and for $T = 0$, $q = 2k_F$,

$f_k - f_{k+q} = \text{sgn}(k + k_F)$ (and at $T \neq 0$ sgn

is replaced by the Fermi distribution).

Thus χ_0 diverges logarithmically, as a function of q:

$$\chi_0 \sim \rho(0)\ln\left|\frac{k_F}{q-2k_F}\right| \quad . \tag{3.22}$$

[One can also verify that at $q = 2k_F$ the divergence as a function of temperature goes as $\ln(E_F/T)$.]

The interactions of the particles may be written

$$V_{int} = \frac{1}{2}\sum_q \rho(q)v(q)\rho^*(q) \quad . \tag{3.23}$$

Thus, in simplest terms, a wave in the mean density at wave vector q,

$$\langle\rho\rangle = \rho_0(q) \quad ,$$

will lead to an effective mean-field potential on the other particles of

$$V_{eff}(r) = v(q)\left(\langle\rho_0(q)\rangle e^{-iq\cdot r} + \langle\rho_0^*(q)\rangle e^{iq\cdot r}\right) \quad . \tag{3.24}$$

The response to a given applied potential wave $V_{app}(q)$ must take this "effective field" or "mean field" into account, so that we have

$$\begin{aligned}\langle\rho(q)\rangle &= \chi_0(q)\left(V_{app}(q) + V_{eff}(q)\right) \\ &= \chi_0 V_{app} + \chi_0 v(q)\langle\rho(q)\rangle \quad .\end{aligned} \tag{3.25}$$

In (3.25) positive v is attractive, contrary to the usual

sign convention. If the effective potential is attractive, the density increases. We finally obtain

$$\langle\rho(q)\rangle = \frac{\chi_0 V_{app}}{1 - \chi_0 v(q)} = \chi^{eff} V_{app}$$

the well-known mean-field form. If χ_0 diverges as a function of T, and v is attractive, we see that there is some critical temperature T_c at which the denominator goes to zero, and χ^{eff} diverges. Below T_c a finite $\langle\rho\rangle$ appears with no V_{app}. Thus a charge-density wave will form. It is an easy matter to take spin into account, as we have not, and show that a spin-density wave will always form in the same way due to the exchange interaction if v is repulsive. Using the logarithmic dependence of χ, we see that T_c has the familiar BCS form--in this case $T_0 \sim e^{-1/(\rho|v|)}$ --and, in fact, the one-dimensional mean-field theory of the density wave can be seen to be in one-to-one correspondence with the BCS theory (Overhauser, 1971).

I remarked that there are certain specifically one-dimensional problems. These are (a) **Fluctuations.** Because fluctuations in one dimension diverge [for an obvious reason: the probability of an order defect's occurring somewhere is $\propto \exp[-E_{defect}/T] \times$ (length of chain) $\to \infty$ as length $\to \infty$], T_c is never sharp for a purely one-dimensional system. All natural one-dimensional systems are really weakly interacting parallel chains, and eventually the interchain interactions take over. However, the region above this "crossover" can be computed exactly for many one-dimensional models (Scalapino et al., 1972, and others), and it is usually clear in any given case which effective susceptibility tends to diverge as T $\to$ 0, thus signalling a mean-field transition. It is important to keep clearly in mind that one-dimensional fluctuations always reduce order, never increase it: fluctuations imply that the real T_c is less than the mean field T_c. (b) **Superconductivity.** Because there is a one-to-one correspondence between the logarithmic divergences in this theory and those in BCS theory, it usually turns out (Menyhard and Solyom, 1973; Luther and Emery, 1974) that an appropriate kind of superconducting fluctuation becomes non-negligible at the same time as the density wave, although the effect is almost never controlling. SDW's are accompanied by triplet superconductivity, CDW's by singlet (Luther and Peschel, 1973). As far as is so far known, this form of superconductivity never plays any real physical role, but just messes up the mathematics. (c) **Stoichiometry and commensurability.** In one dimension, the Fermi surface can be elsewhere than the zone boundary, or halfway up, only if there is nonstoichiometry and hence large random potentials, or if the bands of two different chains overlap, leading surely to strong interchain interaction phenomena. The commonest case is perfect alternation at a wavelength equal to twice the lattice spacing. Such a wave easily locks in to the underlying lattice, leading either to alternating pair bonds or to antiferromagnetism, and has fewer interesting properties. In many cases density-wave phenomena are strongly affected by commensurability effects (Lee *et al.*, 1973, 1974) (see below).

With these provisos, there is no doubt that true density-wave phenomena have been seen in many substances, not least KCP, a platinocyanide linear-chain molecular crystal (Comès, 1974), and the notorious organic TTF-TCNQ, which still remains controversial (Comès *et al.*, 1975). Recently, a rather considerable number of two-dimensional layer structures, many of them of the general class of 4d and 5d group dichalcogenides such as TaS_2, have also been shown to have two-dimensional CDW behavior, that is, two CDW's form so as to give a net triangular lattice (Wilson *et al.*, 1975). The complexity of these phenomena and the rapidity of growth of the field put them beyond the scope of this book, but there is no doubt that, as in the cases of KCP and TTF-TCNQ, these CDW's in many cases are incommensurate with the underlying lattices, although they exhibit with changing temperatures a bewildering sequence of commensuration transformations. The surfaces of many semiconductor crystals also exhibit long-period (7×7, 2×8, etc.), if commensurate, electron density waves (Lander and Morrison, 1962, 1963).

The most interesting general question of principle associated with these density-wave phenomena is that of the relative motions--of the density waves themselves, the electron gas as a whole, and the underlying lattice. By no means all the relevant problems have been solved, but some of them can at least be discussed.

Two important distinctions come up in such a discussion. The first is commensurability, the second has to do with the Fermi surface. A commensurate density wave has a wavelength equal to a lattice translation and is, quite generally, pinned to the underlying lattice. It does not have an independent transport behavior of its own; only its defects do so. Physically, this is obvious: it costs an energy of order some finite number × N to break it free everywhere, so that it can move only under a very large stress. Of course, there is a finite frequency vibration of the CDW relative to the underlying lattice, which may lead to an absorption peak in the infrared in some cases, but no possibility of "sliding conductivity" as suggested by Fröhlich (1954) and initially by Bardeen and co-workers (Allender *et al.*, 1974). Commensurability is generally enforced by the possibility of umklapp processes, in which the CDW at q couples to that at q + G or, in higher order, to that at 2q + G, etc. If, for instance, q + G = -q, this can lower the total energy appreciably. These commensurability effects have been nicely worked out in the case of the Chromium SDW (Motizuki *et al.*, 1968), and by Axe and co-workers (Moncton *et al.*, 1975) and McMillan (1976) in the plane example.

Given, then, an incommensurate CDW of wave vector q, how does it move, if at all, and what does it carry with it? The answer is easy if the energy gaps at k's satisfying

$$E_k = E_{k+q}, \quad \text{i.e.,} \quad k \cdot q = q^2/2$$

are such as to completely envelop and wipe out the Fermi surface. This is, of course, necessarily true in the mean-field theory of the one-dimensional electron gas, but in higher dimensions requires a strong density wave because of the dependence of E_k on the transverse wave vectors. Under those circumstances, the Brillouin zone of the density-wave superlattice is completely filled, and there is a true energy gap to excite any free particles relative to it. (This is a gap for single-particle excitations, not for the collective motion of the CDW. If the entire CDW slides, this gap is translated in momentum space.)

A filled Brillouin zone of Bloch functions is equivalent to a complete set of filled localized Wannier functions $W(r-R_i)$. The Bloch functions $\phi_k(r)$ are written as

$$\phi_k(r) = \frac{1}{\sqrt{N}} \sum_i e^{ik \cdot R_i} W(r-R_i) \ ,$$

and then it is easy to show that

$$\prod_k C_k^\dagger = \prod_i W_i^\dagger \ , \qquad (3.26)$$

where $W_i^\dagger$ is the creation operator for the Wannier function centered about R_i. Note that we may make the Wannier functions of up-spin coincide or alternate with those of down-spin; this is the only difference in principle between the CDW (coincidence, maximum charge oscillation) and SDW (alternation of spin, little charge oscillation) cases. The density superlattice then forms a true solid, just such as we have defined already, since all sites are filled and there is a gap to empty any of them. This true solid, if it moves (slides) through the background lattice, must carry all of the electrons involved in the original Fermi surface along with it, since each site is surely occupied and the sites move. Luttinger's theorem (1960) that interactions cannot change the volume enclosed by the Fermi surface can be reinterpreted in these cases. Effectively, the Fermi surface can be transferred from the underlying lattice to the density wave, so that a more precise expression of Luttinger's theorem would be that the mobile-fermion density cannot be changed, although any amount of this density can be taken up by a density wave. As Overhauser pointed out (1971), oscillations of the CDW or SDW exist with all the characteristics of phonons in an ordinary lattice. He called them "phasons," and they are the Goldstone bosons of this system, corresponding to the additional continuous broken symmetry in the incommensurate case (in the commensurate one, the broken symmetry is only discrete and no Goldstone boson is required). The fascinating possibility of "sliding conductivity" exists, but all best estimates suggest that, at least in one dimension, the two lattices are easily pinned to each other by random scattering potentials (Fukuyama et al., 1974; an argument of Larkin and Ovchinnikov, 1971, suggests this is valid in two and three dimensions as well), and this is borne out in the case of KCP, where the Goldstone mode shows up in the far-infrared spectrum as a broad bump at finite frequency.

It is much less clear what happens in cases like that of TaS_2, where there seems clearly to be a residual Fermi surface. It is my understanding that in this case, too, the superlattice, if it moves, carries a fraction of the charge density with it. That is, in this case, we can correctly speak of a two-(or three-) fluid model:

$$J = \rho_{DW} v_{DW} + \rho_n v_n (+ \rho_{latt} v_{latt})$$

$$\text{where } \rho_{DW} + \rho_n (+ \rho_{latt}) = \rho \ . \qquad (3.27)$$

DW refers to the density wave, n refers to normal electrons, and latt refers to the lattice ions. In other words, it is clear that if we displace the CDW Fermi surface in momentum space we are moving a certain amount of charge given by the volume enclosed by that Fermi surface, at a net velocity given by

$$v_n = \langle \delta k \frac{\partial^2 E}{\partial k^2} \rangle_{ave} \ .$$

The charge flows through the density wave like water through a pipe. We are not moving the underlying density wave (contrary to Fröhlich's ideas), any more than we move the metal as a whole, or its core electrons, when we set the electron gas in a metal in motion. Moving the density wave means changing its phase relative to the ionic lattice.

That we can do by the following indirect trick: we displace all the original electrons in momentum space before forming the density wave. Then we allow it to form; and then we allow the electrons on the remaining Fermi surface to relax back to zero v_n, i.e., we set them in motion at $-v_{DW}$ relative to the density wave. This will tell us ρ_{DW}, which is by definition, and using Eq. (3.27), precisely $\rho_{total} - \rho_n$.

The motion of this density ρ_{DW} will still be subject to Overhauser phasons and still can, in principle, contribute to sliding conductivity--it may in fact do so in TaS_2. But we have to recognize that what we have is very much a kind of "fluid solid": a solid some of the density of which can flow as a true fluid, presumably breaking up the density wave at one end and reforming it at the other. Such a solid is quite distinct from our conventional view of a solid, and to say that it is early to speculate on its properties is an understatement.

Leggett (1970) has considered the question of whether a Bose solid could also have this unlocalized character. It is, of course, perfectly possible to generate a three-dimensional density wave in a Bose fluid. One has merely to introduce an external periodic potential, and the resulting lowest-energy state $\phi_0(r)$ will vary periodically in space. Such a system will still have the characteristic Bose fluid behavior. But one would need a very special set of forces indeed to make such a density wave self-consistent, as we shall see in Chapter 4, and I am very much in doubt whether a Bose fluid will ever exhibit a density wave. This is because self-consistency would require attractive forces, but the Bose system with any net attractive forces would collapse.

It is worth mentioning that one of the liquid-crystal phases, "Smectic B," seems at this writing a likely candidate for a classical fluid solid. It has three-dimensional ordering at least in the sense of 2 + 1: There is a one-dimensional density wave of two dimensionally ordered sheets. Nonetheless it seems fluid. What is not known for sure is whether the two-dimensional sheets stack regularly in register or not. If they do, as seems almost necessary energetically--just as a commensurate density wave must surely be pinned--the only possible explanation for smectic B is that it is a density wave without an energy gap for creation of interstitials or vacancies, and with some non-integer number of molecules per unit cell. Then there is no activation energy for particle diffusion. This, then, is a classical equivalent of a free Fermi surface.

Note added in proof: This situation is now much clarified--see Halperin and Nelson (1978).

3. Superconductivity (and anistropic superfluidity)

I have left this case for last, not because it is a weaker example than the others, but because of its relative complexity. In fact, there is a beautiful completely renormalized theory of the BCS system due to Nambu (1960) and Eliashberg (1960), which also is one of the most accurate quantitative theories of many-body physics.

Here again, it is essential to take as one's model theory one in which the interactions already have modified the state qualitatively in a simple but radical way. You introduce a starting Hamiltonian in which the single-particle potential--to be thought of eventually as the equivalent of a Hartree or Hartree-Fock potential--already contains a nonclassical mixing term between electrons and holes:

$$H_0 = \sum_{k\sigma} \varepsilon_k n_{k\sigma} + \sum_k (\Delta_k^* c_{-k-\sigma} c_{k\sigma} + \text{H.c.}), \tag{3.28}$$

where Δ_k^* is the gap parameter. Because Eq. (3.28) is only quadratic in field operators, it can be diagonalized in terms of a new set of elementary excitations α_k^+, which have the standard BCS energy given by

$$E_k^2 = \varepsilon_k^2 + \Delta_k^2 . \tag{3.29}$$

(This is easy to see if we express H_0 as a 2 × 2 matrix with respect to the Nambu basis $\psi_{k\sigma}^+ = (c_{k\sigma}^+ \; c_{-k-\sigma})$. Then $H_0 = \sum_{k\sigma} \psi_{k\sigma}^+ \begin{pmatrix} \varepsilon_k & \Delta_k \\ \Delta_k^* & -\varepsilon_k \end{pmatrix} \psi_{k\sigma}$.) These α_k^+'s are hole-particle admixtures because of the "anomalous" second term of Eq. (3.28):

$$\alpha_k^+ = u_k c_{k\sigma}^+ + v_k c_{-k-\sigma} \tag{3.30}$$

with

$$\frac{2uv}{u^2 - v^2} = \Delta/\varepsilon . \tag{3.31}$$

Thus, as in the Bose-Einstein case, and as in the solid, the starting approximation already contains the broken symmetry, and the excitation spectrum at this stage already reflects the loss of symmetry. If we now turn on, as an interaction Hamiltonian, the difference between the true Hamiltonian and this one, we have the standard result that perturbation theory seems to encounter no further singularities.

With the starting H (3.28), the ground state is defined by $\alpha_k^+ \alpha_k \equiv 0$, which implies, when we transform back to the $c_{k\sigma}$'s, the existence of a mean anomalous amplitude

$$b_k^+ = \langle c_{k\sigma}^+ c_{-k-\sigma}^+ \rangle. \tag{3.32}$$

This amplitude is, to lowest order, maintained self-consistently if the potential matrix elements have a net attractive character,

$$V = -\frac{1}{2} \sum_{\substack{kk'q \\ \sigma\sigma''}} V_q \, c^+_{k\sigma} \, c^+_{-k'+q,\sigma'} \, c_{-k'\sigma'} \, c_{k+q,\sigma} \quad . \tag{3.33}$$

If the average amplitude (3.32) exists, this leads in lowest order to a term in V of the form in Eq. (3.28) if we assign

$$\Delta_k = \sum_{k'} V_{k-k'} \, b_{k'} . \tag{3.34}$$

It has been shown (Nambu, 1960; Eliashberg, 1960) that a consistent, fully renormalized theory can be based on the model theory of Eqs. (3.28) to (3.34). In particular, because the energy-level spectrum (3.29) has a gap, again the ground state can be thought of as a state where all quasiparticle states are occupied, precisely, and the low excited quasiparticle states have very precisely defined energies. It is also clear that there is no way to continue from the normal state to this one, because, of course, the normal state has no broken symmetry.

It is interesting to ask whether this state can also be thought of as a continuation of the Bose-Einstein condensed state of pairs; i.e., whether one could first bind the quasiparticle pairs together and then Bose-condense them, at least in principle. While there exists a transformation, originally due to Dyson (1957), that carries the BCS into the BE state, this contains a mathematical singularity that has never been analyzed properly. Many qualitative results can certainly be understood in this way--for instance, the fact that in He, at $T = 0$, $\rho_s = \rho$ indicates that all particles participate in the superflow.

In the case of the anisotropic superfluid phases of ^3He, this concept is especially useful. (See Anderson and Brinkman, 1975). In these phases the energy gap is not isotropic at the Fermi surface and is spin dependent. It is very convenient, therefore, and leads to results which usually agree qualitatively with detailed calculations, to think of binding $(^3He)_2$ molecules first (in, for example, $L = 1$ states with the resulting 9-fold degeneracy) and then allowing a BE condensation in one component of the multiplet. This leads to a rough understanding of the role of various anisotropies, of the superfluidity properties, etc. But at first sight it appears that this understanding is destroyed by the fact that the total orbital angular momentum, while it exists if the condensation is in a state for which $M_L = 0$, is not given by

$$\langle L_z \rangle = \frac{N}{2} M_L .$$

In fact, it is far smaller. It is, however, not clear if or in what sense orbital angular momentum would be conserved in the continuation process from the BE fluid, or if orbital angular momentum could be exchanged with the condensate (Cross, 1977).

Another problem pointed out by Leggett (private communication) in the case of the anisotropic ABM (Anderson-Brinkman-Morel) phase is the existence of gapless points, or sometimes lines, on the Fermi surface, which have no analog at all, even qualitatively, in the Bose molecule system. At any finite temperature, therefore, there is always a gas of normal fermions, which can carry on the order of $N^{2/3}$ more angular momentum than the condensate.

In all of these BCS-type theories, the BCS model based on the simple anomalous Hartree-Fock term (3.28) is a quantitatively satisfactory description of most low-energy or long-wavelength processes, even where its parameters, such as $V_{kk'}$, have little or nothing to do with the real interactions.

E. RENORMALIZED PERTURBATION THEORY: A QUALITATIVE DISCUSSION; MULTIPLE-SCATTERING THEORY AS AN EXAMPLE

The reason I have tried to soft-pedal renormalization via perturbation theory as the basis for the above discussion is that the idea of continuation is a more generally useful one. There are only two cases in which the full apparatus of perturbation theory is available, namely Bose and Fermi liquids (including BCS ones). But the idea of continuation applies to quantum solids as well as to ferro- and antiferromagnetism and may even have wider domains of applicability. Eventually, we may come to look upon the idea of continuation as based on something much more like the Wilson-Fisher renormalization group scheme (Wilson, 1971a, b; Wilson and Kogut, 1974) than perturbation theory, which is not as general as that scheme. Especially central is the idea of stable "fixed points"; for this interpretation see Chapter 5. In renormalization group theory, again, there is a classic, diagrammatic approach and a more useful, modern "scaling" interpretation, in which the operation of scaling is seen as carrying one "Hamiltonian" into another possibly simpler one, and again the use of diagrammatic perturbation methods is confining rather than liberating (Anderson, 1976a). Nonetheless, for the student who wants to read the literature, a rough familiarity with diagrams will be useful, and in this section we shall try to provide a dictionary of diagrams for him.

The diagram method in its full generality turns out to require, first, of course, that the starting approximation be expressible in terms of an unperturbed Hamiltonian H_0, and, second, that this unperturbed Hamiltonian be a linear operator, as well as that it define a simple, soluble problem. When H_0 is linear, the problem of inverting it to define unperturbed propagators for the "free" particles, which is a basic step of diagram theory, is a simple one. Unfortunately, not all the starting Hamiltonians which are perfectly satisfactory as "model" systems are linear; the spin Hamiltonians are a good example of this, or the large-U limit of the Hubbard Hamiltonian. The localized Hartree-type theory with which we start the theory of quantum solids is not even evidently a Hamiltonian theory.

For this reason the idea of renormalization must in many cases remain an ideal, or in a sense itself a "model" rather than a realistic approach. Nonetheless this model is, in the cases where it applies, so beautifully and completely worked out that an understanding of its central ideas in greater depth than the sketchy remarks of Section 3.A can be very valuable.

The formal structure of many-body perturbation theory is based very closely on the much simpler case of ordinary one-body perturbation theory in the presence of a scattering potential V. Imagine free particles, described by a Hamiltonian H_0, moving in the presence of the perturbing potential V,

$$H = H_0 + V . \qquad (3.35)$$

Many of the questions we want to ask may be answered by solving the Schrödinger equation

$$(E - H_0 - V)\Psi = 0,$$

but many more require a knowledge of the "Green's function" which is the operator solution of

$$[E - H_0 - V(r)]G(r,r') = \delta(r,r') . \qquad (3.36)$$

For instance, energy eigenvalues are the poles of G, while the residues at the poles tell us the eigenfunctions, as we can see from the equations

$$\begin{aligned} G &= (E - H_0 - V)^{-1} \\ &= \sum_n \frac{(r|n)(n|r')}{E - E_n} . \end{aligned} \qquad (3.37)$$

Many quantities can be calculated from G or generalizations thereof. The simplest is the density of states,

$$\begin{aligned} \rho(E) &= \sum_n \delta(E - E_n) \\ &= \sum_n \int d^3r(r|n)(n|r) \times \lim_{s\to 0} \frac{1}{\pi} \operatorname{Im} \frac{-1}{E + is - E_n} \\ &= \frac{-1}{\pi} \operatorname{Tr} \operatorname{Im} G_E . \end{aligned} \qquad (3.38)$$

Also, one may calculate a local density of states

$$\rho(r,E) = \frac{1}{\pi} \operatorname{Im} G_E(r,r).$$

Transport properties often involve "two-particle" Green's functions, which for this simple case are just products of two one-particle functions. For instance, the probability of transitions due to an operator X at a frequency ω (which is the spectral weight of the response function due to a force E acting on X) may be written

$$I(\omega) = \sum_{nn'} |(n|X|n')|^2 \delta(E_n - E_{n'} - \omega) , \qquad (3.39)$$

and we can use the famous identity

$$\begin{aligned} &\frac{1}{E_{n'}+\omega-E_n+is} \left(\frac{1}{E + i\frac{s}{2} - E_n} - \frac{1}{E - i\frac{s}{2} - E_{n'}-\omega}\right) \\ &= \frac{1}{E + is - E_n} \, \frac{1}{E - is - E_{n'}-\omega} \end{aligned}$$

to obtain

$$\delta\left(E_{n'}+\omega-E_n\right) = \lim_{s\to 0} \operatorname{Re} \int \frac{dE}{2\pi^2} \frac{1}{E+is-E_n} \frac{1}{E-is-E_{n'}-\omega}$$

and rewrite $I(\omega)$ (3.39) as

$$I(\omega) = \frac{1}{2\pi^2} \operatorname{Tr} \operatorname{Re} \int dE\left[G(E+is)\ X\ G(E-\omega-is)\ X\right] . \quad (3.40)$$

We may also insert, in this product form, occupation probability factors $f(E)\ [1-f(E+\omega)]$ to weight the transition probability for $E_n \to E_{n'}$ by the probability of occupation of the state $|n)$ $f(E_n)$ and the probability $(1-f(E_{n'}))$ that this state is empty. The resulting expressions are then very similar to the many-body generalization we shall discuss later.

In each case, the great value of the formal introduction of G is that it is a summation over all the eigenfunctions of H, no matter how complicated these may be, which contains relevant information about these functions. One hopes that by using G in a large system one may discard all the irrelevant details on a fine space and energy scale, and keep only the averages which experiment can measure. This is often the case; but in many other cases it is also a danger, in that one is often tempted to average before making some approximation rather than after, which can lead to qualitatively wrong results.

The perturbation series in ordinary single-particle quantum mechanics follows when we carry out the expansion

$$\begin{aligned} G &= \frac{1}{E-H_0-V} = \frac{1}{E-H_0} + \frac{1}{E-H_0} V \frac{1}{E-H_0} + \ldots \\ &= G_0 + G_0 V G_0 + G_0 V G_0 V G_0 + \ldots . \qquad (3.41) \\ &= G_0 + G_0 V G . \end{aligned}$$

The last equation is a way of writing "Dyson's equation" which is an integral equation for G:

$$G(r,r') = G_0(r,r') + \int d^3r'' G_0(r,r'')V(r'')G(r'',r')$$

or in momentum space

$$G(k,k') = G_0(k)\delta(k,k') + \int d^3k'' G_0(k)V(k-k'')G(k'',k'). \tag{3.42}$$

For free particles,

$$G_0(k) = \frac{1}{E - E_k} . \tag{3.43}$$

The Fourier transform of G from real to momentum space, as above, has an obvious meaning. Is the same also true of the time Fourier transform?:

$$G(t-t', r, r') = \frac{1}{2\pi} \int e^{-i\omega(t-t')} \frac{1}{\omega - H} d\omega . \tag{3.44}$$

We see immediately that this is very much the case. Since the exponential is an entire function, the zeros of the denominator are the only singularities, and if we choose to close the contour around the eigenvalues of H we get

$$G = ie^{-iH(t-t')},$$

i.e.,

$$G(r, r', t-t') = i \sum_n (r|n)(n|r')e^{-iE_n(t-t')}, \tag{3.45}$$

which is basically the amplitude for the particle to propagate from point r, t to point r', t', under the influence of the time-dependent Schrödinger equation. This is why G is often called a "propagator."

From this interpretation of G as a probability amplitude for propagation, we can easily visualize the Feynman interpretation of perturbation theory in terms of a sequence of "diagrams." That is, if we imagine that we can solve for G_0, this tells us exactly how a particle "propagates" under the influence of the unperturbed Hamiltonian H_0. Thus G_0 is called the "unperturbed propagator." We then visualize the process of propagation via G from r to r' in terms of propagation via G_0 from r to r_1, scattering at r_1 by $V(r_1)$, propagation from r_1 to r_2 via G_0, scattering at r_2, etc. in all possible ways ending up at r'. We must sum over all possible numbers of scatterings and over all possible intermediate scatterings at r_1, r_2 etc. This is the propagator interpretation of the second line in Eq. (3.41), as shown diagrammatically in Fig. 3-8. Alternatively, we can propagate via G_0 from r to r_1, scatter at r_1, and then propagate

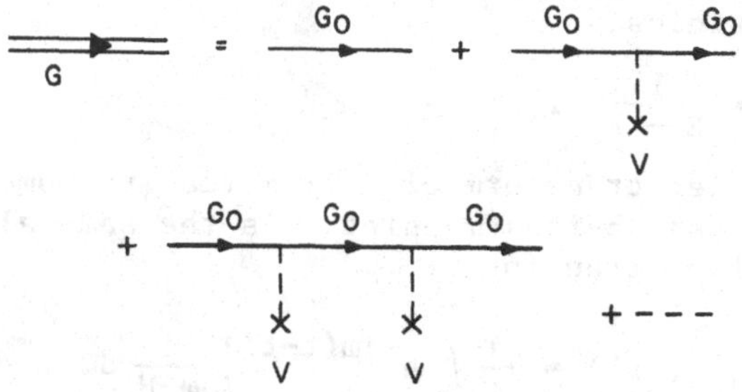

Figure 3-8 Perturbation series.

via G (all possible further scatterings) from r_1 to r, except that we must also add the unscattered amplitude. This is the meaning of Dyson's equation, as in Fig. 3-9.

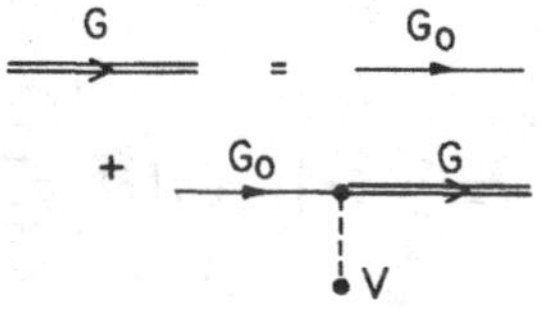

Figure 3-9 Dyson's equation: simplest case.

Of course, the choice of H_0 and V is somewhat arbitrary--for instance, it might be that one would want to treat the potential V(r) as H_0, in which case G_0 is sometimes called the "locator": it tells how the amplitude varies in time at the point r; and hopping from point to point is treated as the perturbation. This kind of series is important in disordered systems. It is convenient, especially with H_0, to carry out the integration in Eq. (3.44) precisely along the real axis and to give the eigenvalues ε_k small positive imaginary parts. This then causes a slow decay of $G_0(t)$, which is appropriate to the idea that particles slowly decay out of the system, i.e., it is the equivalent of outgoing boundary conditions, which imply that a particle propagating away is likely eventually to be scattered and not to return coherently.

An important case is that in which the particles are moving in a medium composed of many randomly arranged scatterers, so that we can neglect phase correlations among them:

$$V_{kk'} = \sum_\alpha V_\alpha(k-k')e^{iR_\alpha(k-k')} \tag{3.46}$$

where R_α is the random position of the αth scatterer.

If the scatterers are weak, we might at first think that we could confine ourselves to the first few terms of the perturbation-theory equation (3.41), as follows:

$$G(k,k) = G_0(k) + G_0(k) \sum_\alpha V_{\alpha kk} G_0(k) + \tag{3.47}$$

$$\sum_{k'\alpha\beta} G_0(k)V_\alpha(k-k')G_0(k')V_\beta(k'-k)G_0(k)$$

$$\times\, e^{i(k-k')(R_\alpha - R_\beta)} + \ldots.$$

$$\simeq \frac{1}{E-\varepsilon_k} + \frac{NV_{kk}}{(E-\varepsilon_k)^2} + \frac{N^2V_{kk}^2}{(E-\varepsilon_k)^3}$$

$$+ N \sum_{k'\neq k} \frac{1}{(E-\varepsilon_k)^2} |V_{kk'}|^2 \frac{1}{E-\varepsilon_{k'}} + \ldots. \tag{3.48}$$

In the second expression we have dropped the terms with random $\exp[-ik\cdot R_\alpha]$ phase factors, because these will be $\sim 1/\sqrt{N}$smaller, where N is the size of the sample, without a compensating factor of $V_{kk'} \propto$ (sample volume)$^{-1}$. Correspondingly, the off-diagonal elements of G always contain such phase factors and will be very small. They represent diffuse scattering, which is undoubtedly there--a corresponding amplitude will turn out to be absent from the diagonal part of G, which represents the main beam--but doesn't act in phase coherence.

We can see from Eq. (3.48) that our hope that small V would allow us to stop at lowest order is illusory. The second term in V_{kk} is of order $\frac{NV}{E - \varepsilon_k}$ relative to the first, which as $E \to \varepsilon_k$ is as large as we like, and all higher terms of similar form will be correspondingly bigger: the series as a whole is divergent.

The remedy for this is the simplest of a general class of techniques based on resummation of a divergent subclass of a diagram series. There is, so far as I know, no rigorous mathematical justification for it, but common sense certainly suggests that we had better "sum the most dangerous diagrams" before we get around to doing any others. In some few cases--e.g., Gell-Mann and Brueckner's (1957) treatment of Coulomb correlation energy--the series as a whole has been analyzed in powers of a physical parameter (in that case e^2), and it has been shown that summation of the "most divergent diagrams" actually works; but by no means does it always work.

In the present case the technique is known as a "self-energy summation." We observe that the offending terms always appear in geometric series, which we can derive in a diagrammatic way as follows. We take any diagram of the series for G_k and look for the first recurrence of G_k^o. To the right of that first recurrence can be any possible diagram starting and ending with G_k^o, including G_k^o itself (there may be no further recurrence). Thus to the right is G_k itself. To the left is, first, the original factor of G_k^o; then any diagram <u>within which</u> G_k^o does not occur. We call the sum of all such diagrams, defined as all those which cannot be split in two by cutting a G_k^o line, the <u>self-energy</u> and the individual pieces "self-energy parts." The first few diagrams of the self-energy $\Sigma(k,\omega)$ are shown in Fig. 3-10 where each scattering cross represents an impurity atom and carries a factor of N/V. Mathematically these diagrams correspond to

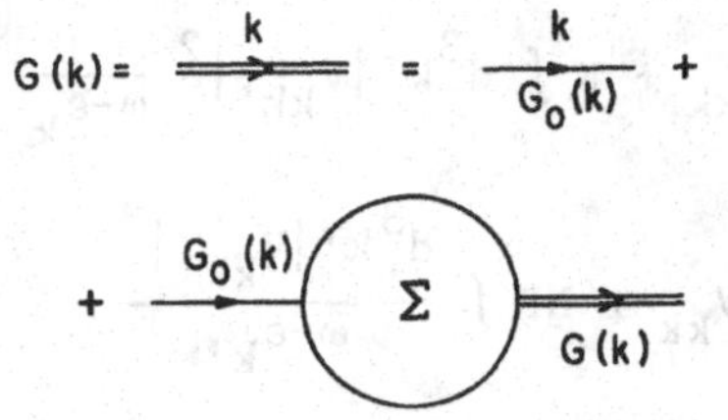

Figure 3-10 Dyson's equation: general case.

$$\sum(k,\omega) = NV_{kk} + N \sum_{k'\neq k} |V_{kk'}|^2/(\omega-\varepsilon_k) +$$

$$+ N \sum_{k'',k'\neq k} \frac{V_{kk'}V_{k'k''}V_{k''k}}{(\omega-\varepsilon_{k'})(\omega-\varepsilon_{k''})} + \ldots , \qquad (3.49)$$

and then we have a refined version of Dyson's equation (see Fig. 3-10)

$$G_k = G_k^o + G_k^{\,o} \sum(k,\omega) G_k \ ,$$

i.e.,

$$G_k = \frac{1}{\omega-\varepsilon_k-\Sigma(k,\omega)} \ . \qquad (3.50)$$

It is now clear why the series diverged. G has its pole, not at the point

$$\omega = \varepsilon_k \ ,$$

but at the point

$$\omega = \varepsilon_k + \sum(k,\omega) = E_k \ , \qquad (3.51)$$

and we cannot move a pole from one point to another with a nonsingular function.

We observe that $\Sigma(k,\omega)$ is not, in general, real. The second-order term is the one we already called attention to in Section 3.A, and gives the lowest-order term in the decay rate of the state k due to incoherent scattering:

$$\begin{aligned}\Sigma(k,\omega) &= NV_{kk} + N\int d^3k'|V_{kk'}|^2 \frac{1}{\omega-\varepsilon_k} \\ &= NV_{kk} + NP\int \frac{d^3k'|V_{kk'}|^2}{\omega-\varepsilon_{k'}} \\ &\quad + i\pi N\int d^3k'|V_{kk'}|^2\,\delta(\omega-\varepsilon_{k'})\ ,\end{aligned} \tag{3.52}$$

i.e.,

$$\begin{aligned}\Sigma &= \Delta E_k + \frac{i}{\tau} \\ \Delta E_k &= NV_{kk} + NP\int \frac{V^2}{\omega-\varepsilon} + \ldots \\ 1/\tau &= \pi\langle|V|^2\rangle\,\rho(\omega) + \ldots\end{aligned} \tag{3.53}$$

(P denotes the principal part.) This imaginary part of E has the same sign as the infinitesimal imaginary parts assigned to the ε_k. In general, then, there is no longer a pole at a real E_k, but a singularity somewhat off the real axis: a clustering of exact eigenstates and not a single state.

The basic scheme of multiple-scattering theory (Fig. 3-11) consists of the above with one improvement: it is observed that we can continue to disentangle completely the effects of different scatterers and arrive at an exact result as long as the density of scatterers is still small. If we continue the above procedure to fourth order, for instance, we shall find in Σ a term

$$\sum_{\alpha,\beta}\ \sum_{k',k''} V^{\alpha}_{kk'}\ G^{o}_{k'}\ V^{\beta}_{k'k''}\ G^{o}_{k''}\ V^{\beta}_{k''k'}\ G^{o}_{k'}\ V^{\alpha}_{k'k}\ , \tag{3.54}$$

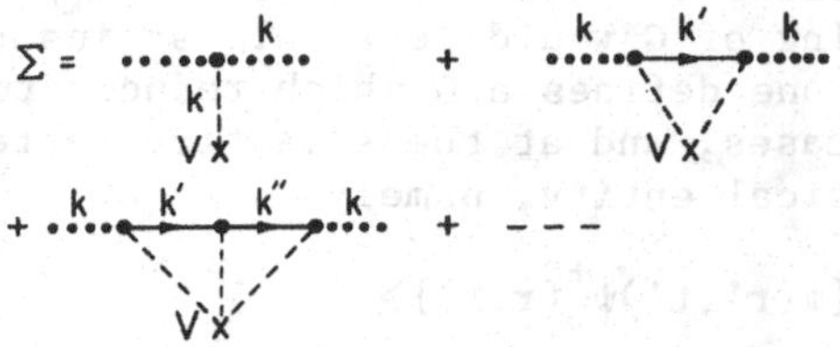

Figure 3-11 Self-energy: multiple scattering case.

in which scatterers α and β interfere with each other. But if the density of scatterers β is sufficiently small, this term is negligible compared to terms in which nothing but α appears, because the phase restriction on k' gives no sum on that variable and thus ensures that Eq. (3.54) is proportional to $(N/V)^2$. The corresponding term in which only α appears is

$$\sum_{k'k''} \sum_{k'''\alpha} \left(V^{\alpha}_{kk'}\, G^{o}_{k'}\, V^{\alpha}_{k'k''}\, G^{o}_{k''}\, V^{\alpha}_{k''k'''} G^{o}_{k'''} V^{\alpha}_{k'''k}\right),$$

which has no restrictions on k''' and is proportional to N/V only. Thus we may write

$$\begin{aligned} \Sigma\,(k,\omega) &\simeq \sum_{\alpha} \Sigma_{\alpha}\,(k,\omega) \\ &\simeq \sum_{\alpha} T_{\alpha}\,, \end{aligned} \qquad (3.55)$$

where Σ_{α} is the full series for the scatterer α alone, which in turn is just what is called the "scattering matrix" T_{α} for the single scatterer--the exact solution of the one-scatterer problem defined by the Hamiltonian

$$H_{\alpha} = T + V_{\alpha}\,.$$

A further considerable improvement to this approximation, which itself dates back to a number of wartime studies of propagation in scattering media (Lax, 1951), had to wait until much more recently. It is the "coherent potential approximation" of Soven (1966), Elliott and Taylor (1967), and others. But that work is rather far afield from our topic.

In generalizing perturbation theory to the many-body system, it is no longer possible to retain the definition of $G = (E-H)^{-1}$ because in this case H is so complicated that the mere labeling of G would be an almost insuperable problem. Instead, one defines a G which reduces to G_0 in the free-particle cases, and at the same time represents a manageable physical entity, namely,

$$G = -i\langle T\{\psi(r',t')\psi^{+}(r,t)\}\rangle \tag{3.56}$$

where T is the time-ordering operator, in the case of fermions given by

$$\begin{aligned} T\{A(t)B(t')\} &= AB \qquad t > t' \\ &= -BA \qquad t' > t \end{aligned}$$

and the $\psi(t)$'s are the Heisenberg representation of the relevant quantum fields. The average is a thermal one; in the important case of $T = 0$, it is the ground-state average. It is clear that there is a close analogy here to the "propagator" idea, in that Eq. (3.56) represents the amplitude for a particle to travel from r, t, where it is created, to r', t', where it disappears. It is also the quantity that reduces to G_0 of Eq. (3.43) when H_0 represents wholly free particles, and to the one-particle G we have been discussing when there are no particle interactions. We can see this by noting that, when H is quadratic in ψ's,

$$H = \int \psi^{+}(r')H(r,r')\psi(r)d^3r d^3r' \quad ,$$

the commutator of H with ψ^{+} is just $H\psi^{+}$, so that G obeys exactly the same equation of motion as the one-particle G:

$$i\frac{\partial G}{\partial t} = HG + \delta(r-r')\delta(t-t') \ . \tag{3.57}$$

But G is, in general, not a true Green's function in the 19th-century mathematician's sense, because the full time dependence is projected onto states in which only one particle is added or removed, and computed only in the equilibrium state. We shall see, and to a great extent understand already, that the latter is an absolutely necessary restriction; but the validity of quasiparticle methods leaves one wondering, with regard to the first restriction, if this is the only viable method.

It is in computing this Green's function that the standard diagrammatic methods come into play. The formalism looks almost precisely the same, but now diagrams must be drawn for more complicated processes. I have already shown in Fig. 3-3 diagrams for electron-hole pair creation, i.e., the result of scattering a particle out of the Fermi sea. In electronic systems, we must allow for the lattice vibrations, including processes of phonon emission, absorption, and scattering. In charged systems, we may include photons as well. We are restricted fundamentally, however, in that all entities must be fermions or bosons. I show a group of typical diagrams in Fig. 3-12. I am not even going to write

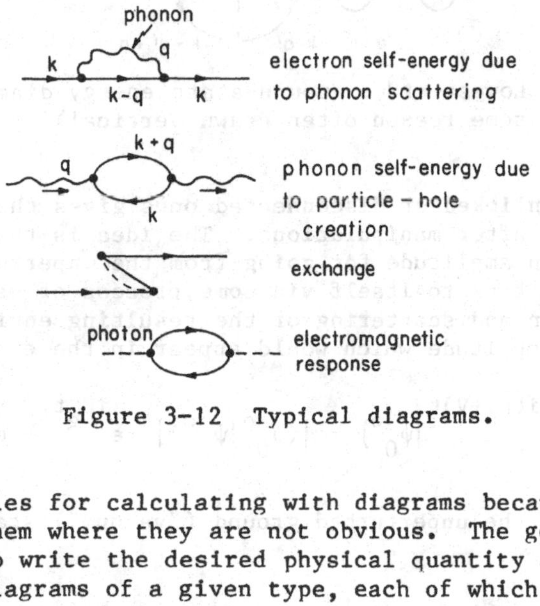

Figure 3-12 Typical diagrams.

out the rules for calculating with diagrams because we shall not need them where they are not obvious. The general principle is to write the desired physical quantity as a sum over all diagrams of a given type, each of which is to be evaluated by inserting unperturbed propagators for all lines and appropriate interactions at all vertices, and summing over all processes possible with the appropriate conserved momenta and energies at each vertex. Each diagram is multiplied by a sign factor (-1) for all closed fermion loops, and a combinatoric factor depending on the number of interactions in a simple way.

Let me now set down a brief glossary of terms we are likely to encounter, here or elsewhere.

Ground-State Energy Diagram, Vacuum-to-Vacuum Diagram. This is a diagram closed on itself, but otherwise calculated by the same rules as Green's function diagrams (see Fig. 3-13).

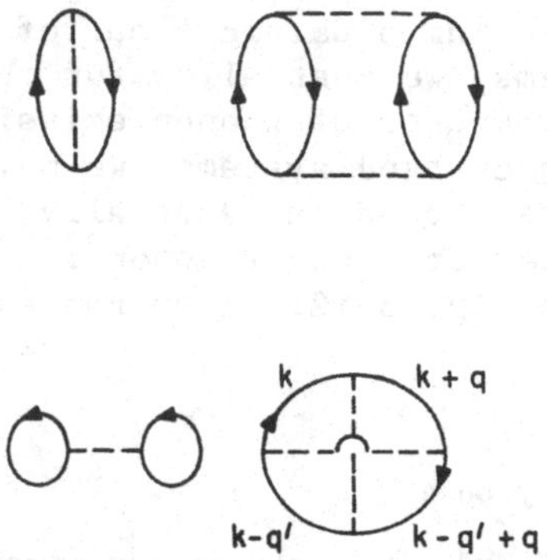

Figure 3-13 Lowest-order ground-state energy diagrams (for some reason often drawn vertical).

The sum of unlinked or disconnected ones gives the ground-state energy after manipulations. The idea is that such a diagram is an amplitude for going from the unperturbed ground state back to itself via some process of particle-hole creation and scattering of the resulting entities. It is thus an amplitude which would appear in the expansion of

$$(\psi_0{}^o|e^{i(H_0+V)t}|\psi_0{}^o) = |(\psi_0{}^o|\psi_0{}^o)|^2\, e^{iE_g t} + \text{other states.}$$

[Here $\psi_0{}^o$ is the unperturbed ground ("vacuum") state, ψ^o the exact one.]

Linked-Cluster Theorem (LCT). Some ways of deriving diagrams give one no clear directive not to add in all kinds of irrelevant vacuum-to-vacuum bits. The LCT is the observation that if one calculates E_g and not $\exp[iE_g t]$, these irrelevant bits disappear, i.e., they are absent from the log of the perturbation expansion. "Linked clusters" are connected diagrams containing only a single group of lines all attached to each other and to the external lines, and the LCT takes advantage of the fact that the rules (for large systems only) make the contribution of two connected pieces the product of the individual contributions; therefore, in the logarithm, they simply add up. This is clearly essential if E_g is to be an extensive quantity, proportional to the amount of substance present.

Self-Energy Summation. This is precisely as we have stated it for the noninteracting case: the procedure of finding the self-energy by adding up diagrams which cannot be disconnected by cutting across a single G_0 line. The self-energy plus the unperturbed energy gives the correct poles.) Usually this is done with a view to putting self-energy corrections into all of the lines of the diagrams, leaving one with only so-called "skeleton diagrams" (see Fig. 3-14).

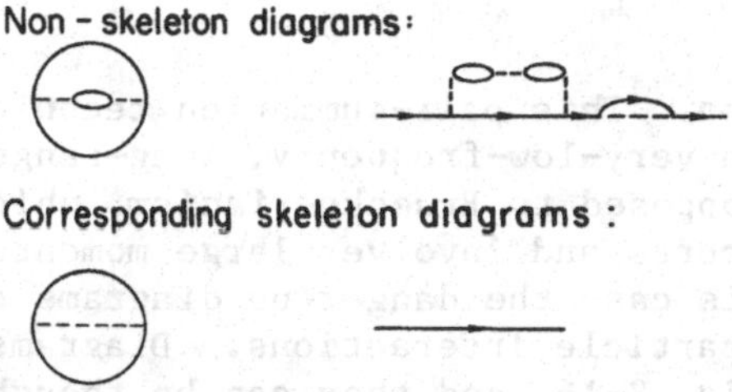

Figure 3-14 Non-skeleton diagrams: skeleton diagrams corresponding.

Ladder Summation. This is the first of a sequence of summations of pairs of particles propagating together. In this case, we sum all two-particle repeated scatterings by noting that the two-particle Schrödinger equation is often soluble and represents the result of all possible such two-particle scatterings. This is very like the multiple-scattering idea mentioned above: two-particle strong scattering can be managed by replacing V with the exact T matrix for the two (see Fig. 3-15). Again, one way to describe this summation is by a kind of Dyson equation which in this case is called the Bethe-Salpeter (Salpeter and Bethe, 1951) equation. This scheme was first used by Brueckner (1959; see also Thouless, 1961) to get rid of hard cores in nuclear matter calculations, and I like to refer to it as Brucknerianism. The relevant divergent diagrams come from repeated hard-core scattering. It is even more useful in ^{3}He. It also lies behind much of the lore of pseudopotentials, which tells us how it is used: one replaces V by T as the scattering potential and leaves out ladders [note the precise correspondence to multiple-scattering theory's Eq. (3.55)].

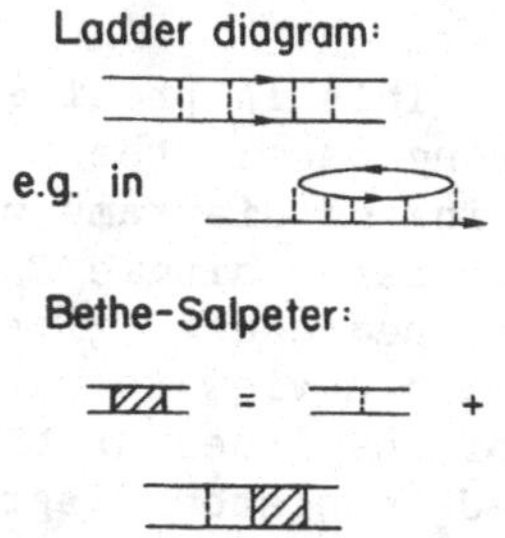

Figure 3-15 Ladder summation.

Bubble Summation. This pair-summation technique is ubiquitous when very-low-frequency, long-range effects are important--as opposed to Brucknerianism, which is often used for hard cores and involves large momentum and energy. In this case the dangerous diagrams come from repeated hole-particle interactions. Diagrams of this type are shown in Fig. 3-16, and they can be thought of as self-energy diagrams of the interaction lines (the dotted lines representing V in particle interactions), or phonon or

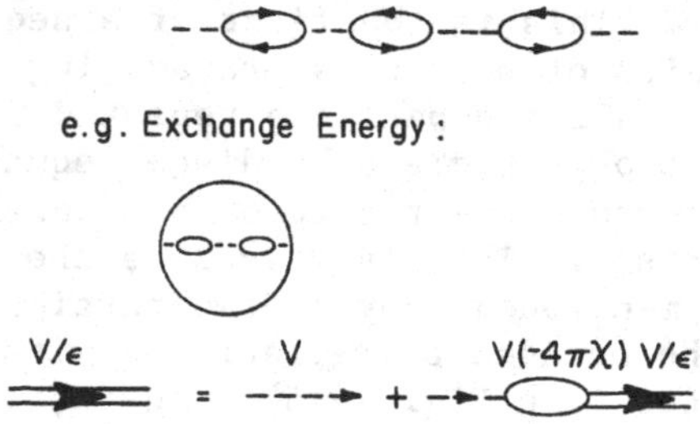

Figure 3-16 Bubble diagrams

photon lines (the Coulomb V can actually be thought of as a longitudinal photon if we like). These are then dielectric-constant-type corrections to the propagation of the interactions. Such corrections obviously modify the places where the poles of Green's functions describing the interaction occur, and they can be dangerously divergent for just the same reasons as self-energy diagrams. We know that, particularly in a metal, the dielectric constant can be effectively infinite at long wavelength and low frequency, so that is the region where bubble summation is important. "Dielectric screening" of the interactions is the old-fashioned word for summing bubbles. (See Fig. 3-16.)

Both bubble and ladder sums, but particularly the former, often lead to either a modification of an elementary excitation already present--e.g., renormalization of the phonon frequency--or a wholly new type of pole in the effective Green's function for pair propagation. Spin waves in metals, excitons, plasmons, and zero sound are all examples of bound hole-particle collective excitations which come out of bubble sums.

Vertex Corrections. These are mentioned most often to disavow that much can be done with them, but also to point out that vertices can be renormalized too, and often are in quantum electrodynamics. That is, a triple-interaction vertex (say, phonon, electron, hole, as in Fig. 3-17) can in principle be thought of as having added to it all the possible diagrams in which none of the incoming lines can be detached from either of the other two by cutting one line (an example is shown in this figure), just as we can think in terms of renormalized propagators containing all the

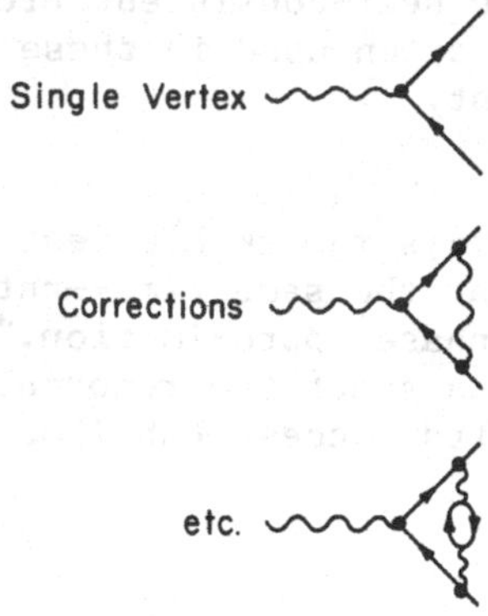

Figure 3-17 Vertex corrections.

self-energy corrections already made. Migdal's (1958) theorem states that, for phonons, vertex corrections can be neglected, but usually they are not negligible.

Self-Consistency: Hartree-Fock Theories. It appears to be a great step forward always to correct the internal propagators in, for instance, a self-energy diagram, for the self-energy as it will be later calculated. If we include only diagrams which describe the mean-field effects, and do this (see Fig. 3-18), that is the Hartree-Fock theory. This

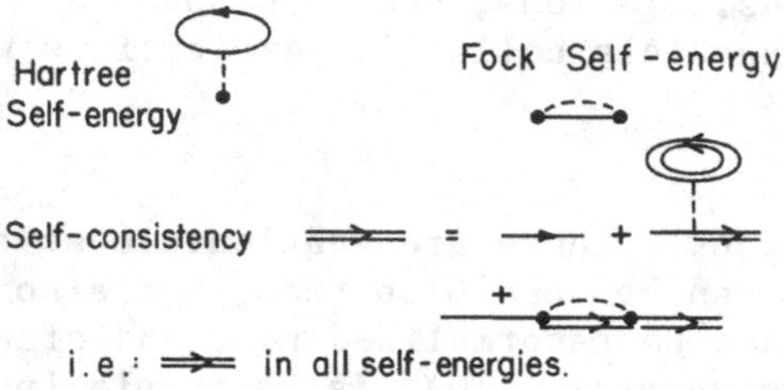

Figure 3-18 Hartree-Fock self-consistency.

"self-consistent" way of proceeding is often a great step forward, but in other cases self-consistency of this type can lead to double counting--counting some diagrams two or more times, always a danger in diagram games. But where we are careful not to double count, higher-order self-consistency is often useful. The self-consistent breaking of symmetry is a concept we have often used in these pages; without this idea, we would be lost.

Parquet Summation. This remarkable feat involves doing bubbles and ladders at the same time--rather like the old "generalized random-phase approximation." It works in special cases, but I suspect the renormalization group of always being behind its success (Abrikosov, 1965).

With this lexicon of techniques, a great many of the systems I have already referred to can be discussed in a diagrammatic, and therefore--to those used to these techniques--respectable manner. It is often essential, as in the BCS case, also to redefine the starting Hamiltonian. Especially notable are the treatments of the Fermi liquid, of normal metals and of superconductors with phonon effects, of correlation energy in electron and electron-hole systems, and of spin fluctuations in nearly magnetic systems. Promising inroads are now being made also on the problem of ferromagnetism in metals. But I should like to emphasize that a much wider class of problems can still be qualitatively and even quantitatively understood by the use of the continuity principle using a model system.

F. DISCRETENESS OF LARGE SYSTEMS: THE QUANTUM CHEMISTS' FALLACIES

Almost the very first calculation made in quantum chemistry, that of the binding energy of H_2, was made in two almost equally satisfactory but distinct ways. Heitler and London took the point of view that H_2 was to be made up of two approximately unperturbed H atoms, and that one should calculate the energy of a pair of such atoms put close together. One of the largest effects was the term which coupled the state

$$\psi_1(\uparrow)\,\psi_2(\downarrow) \quad \text{with} \quad \psi_1(\downarrow)\,\psi_2(\uparrow)\,,$$

the arrows representing the electronic spin directions. This term they called the exchange integral, and showed that it made the singlet state $(\alpha\beta - \beta\alpha)$ attractive, the triplet not. This came to be called Heitler-London valence bonding.

Hund and Mulliken, on the other hand, suggested that H_2 came about by adding another electron to H_2^+ in the symmetric wave function made up more or less of the two atomic 1s orbitals. The second electron had to be opposite in spin; it saw an attractive potential and caused extra binding. Thus was born the Hund-Mulliken "molecular orbital" method.

For H_2 there is in fact no difficulty in principle with taking either point of view. Either can be taken as a model system in our sense, and in a beautiful discussion Slater (1951) shows how we modify either to approach the other and that the truth lies in between.

To a lesser extent, the same approach may be taken for many reasonably small molecular systems. The point is that, in any sufficiently small system, there is <u>some</u> overlap between almost <u>any</u> trial wave function for the ground state and the real state, if the trial state is not chosen with deliberate stupidity, as for example choosing the wrong symmetry. Thus almost any starting point can be moved in the general direction of reality by judicious improvement. It is also characteristic of quantum-chemical methods to rely heavily on the variational theorem in improving one's approximation. The intuition which relies on these ideas is indeed soundly based for molecular problems.

The same independence--what you might call overlap of the radii of convergence--is almost unknown in large systems such as those we deal with here. In almost every case, a continuation method of any kind--perturbation, variational, etc.--will completely fail to give the right answer unless it starts from the right starting point. It is the mistaken assumption that any starting point gives the right answer which I call the "Quantum Chemists' Fallacy No. 1."

The reason is most easily seen if you think about the everyday facts of phase transitions. We are accustomed, in statistical mechanics, to the fact that when a system undergoes a phase transition it does not go to a unique eigenstate (except at absolute zero), but to a whole region of phase space, from which not even thermal fluctuations ever carry it back completely to the original region. When each of 10^{23} coordinates changes, even if the change is small, fluctuations of individual coordinates are wholly incapable of reversing the process. This is true even more strongly for quantum states. The field theorists have an expression for it which gives pretty much the right idea: a change of this sort puts the system, in the limit $N \to \infty$, into a wholly different, orthogonal Hilbert space, from which there is no easy continuous method of return.

This is the reason why self-consistent methods are so valuable in many-body physics: they do actually allow the coordinates of all particles to be changed simultaneously. But self-consistency is no guide, as we learned for superconductivity, if we are not allowing self-consistent fields which take us into the correct Hilbert space in which the ground state is. Also, very often a minimum principle can carry us to a metastable ground state: for example, the normal metal, when superconductivity is possible, is a perfectly good saddle point.

There is a large school of quantum-organic chemists who customarily approach the theory of aromatic compounds from the valence-bond point of view. The aromatic compounds are planar systems such as benzene, in which there is approximately a half-filled band of p electrons, and in which a considerable contribution of π bonding to the energy is expected. The now conventional Huckel theory, on the other hand, takes the point of view that would be automatic to the physicist, that the best starting point is molecular orbitals, and one fills these up to the Fermi surface with both spins just as one would in graphite, which is an infinite planar aromatic compound. Such an approach is, in general, very successful; but, oddly enough, it is only barely more successful than the valence-bond theory if one allows for resonance among different possible valence-bond pairings. (See Fig. 3-19 for a typical set of valence-bond diagrams

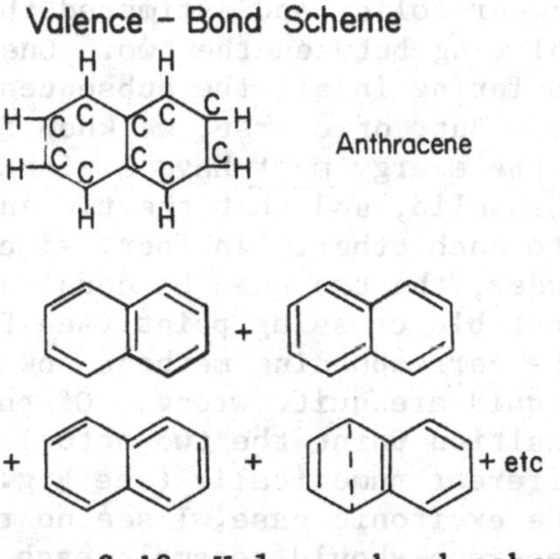

Figure 3-19 Valence-bond scheme.

for such a compound, anthracene.) For small molecules, in fact, the variational theorem suggests that the two should indeed both work. Emboldened by this success, not only have the practitioners carried this on to graphite itself--which is certainly a case where only band theory gives correct answers--but Pauling (1949) has attempted to create a theory of metallic bonding based on the same techniques. Again, this theory is not quite hopeless, so far as binding energies are concerned. At first one was tempted to take the tolerant view and say, "fine, perhaps they are both right"; but I think we must assume that the valence-bond type of wave function is really literally not in the same Hilbert space and consider the success of these methods as a triumph of parametric fitting.

I should, however, make one cautionary statement, related to what I said about the condensed phases of helium. It is, indeed, perfectly possible for a wave function that gets the short-range correlations right to give good energetics even if it has quite the wrong symmetry and transport properties. It is an empirical fact that the energy differences between different phases--different crystal structures, say, or even solid and liquid--are quite a bit smaller than the total binding energy. This is what makes quantitative predictions of phase transitions hard. Thus you may get pretty good energetics out of a qualitatively wrong state.

This is a slip of which even Wigner (but, of course, 50 years ago!) is guilty. In calculating the correlation energy of free electrons, he made an approximation at high density which was similar to present-day methods for Fermi gas systems (Wigner, 1933). At low density he computed the energy of the Wigner solid, and estimated the full energy curve by interpolating between the two. One finds this method of interpolating in all the subsequent literature and all of the texts. But, of course, we know from our general principles that the energy must have a singular point between the fluid and the solid, and that the two energies need have no resemblance to each other. In fact, since the transition will be first order, the two need be equal as a function of r_s only at an unstable crossing point (see Fig. 3-20). Therefore this and the corresponding methods now used for the electron-hole liquid are quite wrong. Of course, it is true that at the transition point the two actual energies may not be extremely different numerically (see Fig. 3-20); but, especially in the excitonic case, I see no reason why even the formal expressions should resemble each other.

"Quantum Chemists' Fallacy No. 2" is far less common and accepted, but I have come across it often enough that I think it worth warning about. It is based on the common experience of quantum chemists that energies are easy to calculate and wave functions hard, and a sense that only the variational principle is solid. (Not that field-theoretical types aren't just as stubborn in not accepting any reasoning not based on diagrams!) The fallacy is that no result is better than the degree of accuracy of the ground-state energy that it achieves. Thus, for instance, I have heard it claimed that the BCS theory is speculative because no one can calculate the ground-state energy of a metal as accurately as the BCS binding energy, by several orders of magnitude. The fallacy, of course, lies in everything I have said in the past two chapters: the importance and accuracy of the process of continuing from a model, and the vital importance of symmetry. These two principles together allow us to make exact statements of a wholly different, nonquantitative sort. Another way of putting it is that the word "semi-

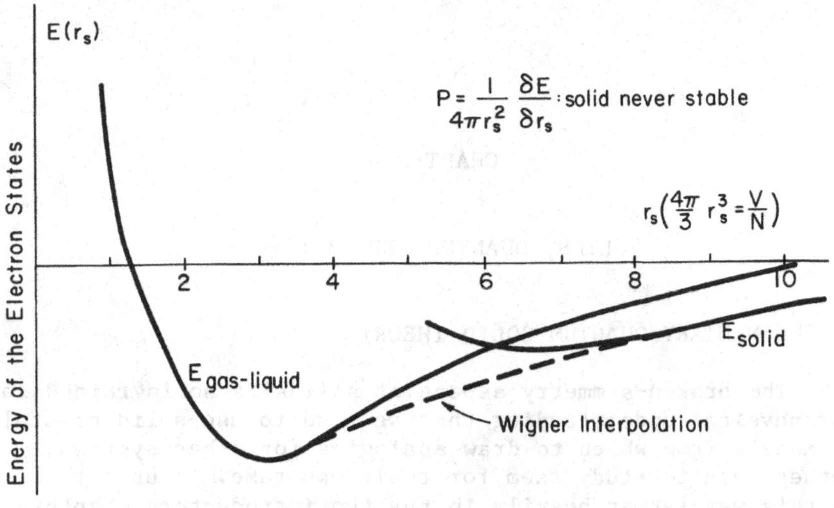

Figure 3-20 Energy of free-electron states.

empirical," which is a naughty word in quantum chemistry, perhaps because so many mistakes have been made in its name, can, with the help of these principles, often describe completely rigorous procedures, as rigorous as the Debye theory of solids or the phonon theory of superconductivity.

The following two chapters and, especially, the collection of reprints, illustrate these principles in more detail.

CHAPTER 4

SOLIDS, QUANTUM AND OTHERWISE

A. ELEMENTARY QUANTUM SOLID THEORY

The broken-symmetry aspect of solids is so ingrained in our physical understanding that we tend to use solid crystals as models from which to draw analogies for other systems, rather than to study them for their own sake. I used them in this way rather heavily in the two introductory chapters. What I should like to do now is to make a sketch of a quantum solid theory that is not in fact the one most successful in practice, in order to show how one might start from quantum mechanics and derive the properties of a solid, formally and with some intellectual rigor (which is not, to my mind, the same as mathematical rigor!). I shall then give a very brief description of the actual theory used for the solid heliums. My discussion is very much based on the important review article of Varma and Werthamer (1976). These discussions will include, but not emphasize, the broken-symmetry aspects of the theory.

Then I should like to cover at some length the special problem of exchange, symmetry, and the real nature of the solid state as viewed from a quantum point of view. Some of this discussion is anticipated in earlier chapters, but since it is unfamiliar it calls for expansion as well as repetition. Finally, I should like to say a few words about friction because it is unique to solids. I shall not discuss the defect theory of solids because it is relatively well known and I have already used it as a model.

Let us, then, start with a Hamiltonian for N particles,

$$H = \sum_{i=1} \frac{P_i^2}{2M} + \frac{1}{2} \sum_{ij} V(r_{ij}) \ . \qquad (4.1)$$

The question of interchange is left for later, so for the time being we consider the particles distinguishable.

If M is sufficiently big (again, a continuation!), the particles will take up some lattice structure so as to minimize V_{ij}. The vibrations of this classical lattice are the phonons, and, in general, near the minimum we can expand V about its minimum V_0 as follows:

$$V - V_0 = \sum_{i,j} \left(r_i - R_i\right) V_{ij} \left(r_j - R_j\right) \ , \qquad (4.2)$$

where R_i and R_j are the equilibrium positions of the particles i and j, respectively. V_{ij} is, of course, a tensor, but we use one-dimensional notation for simplicity; no confusion results. We set

$$u_i = r_i - R_i .$$

But if both particles i and j move the same amount, $u_i = u_j$, then $\Delta V = 0$ as far as their interaction $V(r_{ij})$ is concerned, so it is correct to combine the V_{ii}, V_{jj}, and V_{ij} terms in Eq. (4.2) into

$$\Delta V = -\frac{1}{2} \sum_{ij} \left(u_i - u_j\right) V_{ij} \left(u_i - u_j\right) = -\frac{1}{2} \sum_{ij} u_{ij} V_{ij} U_{ij} \qquad (4.3)$$

(V_{ij} is thus negative, since ΔV is positive definite). We see that equivalence of Eqs. (4.3) and (4.2) requires the identity

$$V_{ii} + \sum_{\substack{j \\ j \neq i}} V_{ij} = 0 \ . \qquad (4.4)$$

Now since translational symmetry of the lattice requires that $V_{ij} = f(i-j)$, Eq. (4.2) can easily be diagonalized by using phonons (lattice waves),

$$u_i = \frac{1}{\sqrt{N}} \sum_k e^{ik\cdot R_i} u_k , \tag{4.5}$$

and then H in turn is, in the harmonic approximation, diagonal:

$$H = \sum_k P_k^2/2M + \frac{1}{2} \sum_k V_k u_k^2 \tag{4.6}$$

where

$$V_k = \sum_j V_{ij} \exp[ik\cdot(R_i - R_j)]$$

$$= V_{ii} + \sum_{j\neq i} V_{ij} \exp[ik\cdot R_{ij}] = \sum_{j\neq i} |V_{ij}|(1-\cos k\cdot R_{ij}) , \tag{4.7}$$

using Eq. (4.4). The solution of Eq. (4.7) is the standard Debye-Born-van Karman model. The frequencies are given by

$$\omega_k^2 = \frac{V_k}{M} .$$

The corresponding wave function is easily written, since each of the oscillators u_k is in its ground state:

$$\psi = C \exp[-\sum_k M\omega_k u_k^2/2]$$
$$= C \exp[-\frac{M}{2} \sum_{ij} u_{ij}\Gamma_{ij}u_{ij}] , \tag{4.8}$$

where Γ_{ij} is the Fourier transform back into real space of the frequency spectrum. The ground-state energy is

$$E_o = V_0 + \sum_k \frac{\omega_k}{2} \ . \tag{4.9}$$

Debye was the first to carry this approach a bit beyond the perfectly harmonic assumption (4.2) by considering small anharmonicities, but in fact simple techniques of that sort are not adequate for dealing with the true quantum solids. I shall discuss the improvements necessary for quantum solid theory later, briefly, but first I want to take a step backwards in order to make it possible at least in principle to deal with indistinguishable particles and other quantum phenomena.

The Debye wave function (4.8) already takes into account rather subtle correlations between the particles, correlations which we have to discard when we go back to a self-consistent Hartree description. The assumption we make in such a description is that the true wave function still involves only particles localized at the different sites R_i, but that the particle at site i, instead of responding dynamically to the motion of a particle j as in Eq. (4.8), simply sees the average of its potential over that motion. This implies that the wave function is a product of single-particle functions

$$\Psi = \prod_i f(r - R_i) \ . \tag{4.10}$$

You will note that, on the other hand, Eq. (4.8) has the "Jastrow" form

$$\Psi = \prod_{ij} f(r_i - r_j) \ , \tag{4.11}$$

which contains far more correlations, although in this case the wave function f is restricted to a Gaussian.

When the product function (4.10) is optimized, one gets the Hartree self-consistent field equation,

$$-\frac{\nabla^2 f_i}{2M} + \Big(\sum_j \int V(r_{ij}) |f(r_j)|^2 \, d^3 r_j\Big) f_i = E f_i \ . \tag{4.12}$$

If the harmonic approximation (4.2) is made again, V can depend on r_i only quadratically, and every wave function is of the harmonic oscillator form. In fact, the u_j dependence in Eq. (4.2) cancels out, since

$$\int u_j d^3r_j |f(r_j)|^2 = 0 ,$$

and one has

$$-\frac{\nabla^2 f}{2M} + V_{ii}\,(r_i - R_i)^2\, f = Ef . \tag{4.13}$$

Thus this approach leaves us with an Einstein model of the solid, with every atom vibrating independently in its harmonic well. In this case the energy is V_0 plus the sum of all the oscillators,

$$E_0 = V_0 + \frac{N\omega_0}{2} , \tag{4.14}$$

and

$$\omega_0^2 = \frac{V_{ii}}{M} = \frac{\sum_{j\neq i} |V_{ij}|}{M} . \tag{4.15}$$

There is a fascinating resemblance between this situation and the antiferromagnetic one. The correspondence is close between the Néel and the Debye states. The ratio of the phonon frequencies is

$$\omega_k/\omega_0 = \left[\frac{\sum_{j\neq i} |V_{ij}|(1-\cos k\cdot R_{ij})}{\sum_{j\neq i} |V_{ij}|}\right]^{1/2} .$$

Suppose $V_{ij} = \begin{cases} V & \text{if } R_{ij} \text{ is between nearest neighbors,} \\ 0 & \text{otherwise.} \end{cases}$

Then

$$\frac{\omega_k}{\omega_0} = \sqrt{1 - \gamma_k} ,$$

where

$$\gamma_k = \frac{1}{z}\sum_{\delta} e^{ik\delta} ,$$

z is the number of nearest neighbors, and the vectors δ connect the atom i to its nearest neighbors. The ratio of the spin-wave frequencies is

$$\sqrt{1 - \gamma_k^2},$$

and there is the difference, therefore, that the spin-wave corrections all improve the energy; but, of course, we have the same divergences of zero-point motion, the same Goldstone theorem, etc., in the correct theory with collective motions properly taken into account.

On the other hand, the Hartree theory is far more convenient for thinking of other quantum phenomena. Thus to show how it can be corrected to give the Debye energy without losing its essential nature is not merely an exercise. To do this we have to correct the excitation spectrum by reanalyzing it in terms of particle pair excitations.

Each of the harmonic oscillator equations (4.13) has an excitation spectrum of the harmonic oscillator type, nondegenerate in the case of one dimension, triply degenerate for the lowest state in three. The energies are

$$E_n = (n + 1/2)\omega_0,$$

and the wave functions are given by

$$f_0(u) = Ce^{\left(-u^2/2u_0^2\right)} \qquad \left(u_0^2 = \frac{1}{M\omega_0}\right)$$

$$f_1(u) = \frac{u}{u_0} f_0$$

$$f_2(u) = \frac{1}{2}\left(\frac{u^2}{u_0^2} - \frac{1}{2}\right) f_0, \text{ etc.} \tag{4.16}$$

Let us now introduce second-quantized operators for the particles, rather than retaining the r_i labels; at this point this is just a simple counting device, since we shall for the time being never allow particles to leave their own potential wells. We thus introduce operators

$$c_{0i}^+,\ c_{1i}^+,\ c_{2i}^+,$$

which create particles in the zeroth, first, and second states, respectively, at atom i. Since as yet we are keeping M very large, the wave functions f_{0i} and f_{0j} do not overlap appreciably. It is therefore valid to do a Wannier orthogonalization on them. (By the time they overlap, we had better be taking symmetry into account anyway.)

The ground state, then, of the Einstein system is

$$\psi_0^E = \prod_i C_{0i}^+ \psi_{vac} , \tag{4.17}$$

and we can describe excited states in which the atoms all remain in place by pairs of field operators. For instance,

$$\psi(n_i = 1, \text{ all others } 0) = C_{1i}^+ C_{0i} \psi_0^E \ .$$

But this does not diagonalize even the perfectly harmonic Hamiltonian (4.1) with (4.2). Let us analyze the V_{ij} terms which we have neglected in the Hartree approximation. These are capable either of exciting or of deexciting the states (4.16), because they multiply these states by u:

$$u\, f_0(u) = u_0 f_1(u)$$

$$u\, f_1(u) = u_0\left[2f_2(u) + \frac{1}{2} f_0(u)\right].$$

Thus multiplying by u is equivalent to using the operator

$$u \to u_0\left\{C_1^+ C_0 + \frac{1}{2} C_0^+ C_1 + 2\, C_2^+ C_1 + \ldots\right\} . \tag{4.18}$$

In fact, we know that we can introduce a boson operator b_i^+ for each site which excites and deexcites its harmonic oscillator spectrum. Then using

$$b_i^+ + b_i = \frac{u}{u_0} \tag{4.19}$$

we can write the off-diagonal terms of (4.2) as

$$V' = \sum_{i \neq j} u_0^2\, V_{ij}\left(b_i^+ + b_i\right)\left(b_j^+ + b_j\right) . \tag{4.20}$$

In this form the problem is clearly the same as the Debye theory: we have a quadratic form to diagonalize in Bose operators for the individual harmonic oscillators, and clearly, because of the complete correspondence of classical and quantum theory for harmonic oscillators, the Hamiltonian will end up being diagonalized as

$$H = \sum_k (n_k + 1/2)\omega_k + V_0 , \quad (4.21)$$

where ω_k is the same as in the Debye theory. So are the zero-point motion and energy, etc. The formalism for diagonalizing this problem is the same as that given in Concepts in Solids, pp. 144ff, and is very similar to the one we have used for antiferromagnetic spin waves.

Thus it is actually quite trivial to get back and forth from Hartree-Einstein to Debye models. In fact, Hartree-Einstein with collective excitations is formally equivalent to Debye. Why should we bother? Obviously, because the corrected Hartree theory can be continued far outside the regime where the Debye theory fails because of large anharmonicities and symmetry effects. These are, in fact, the two important kinds of corrections, which are different in principle.

B. ANHARMONIC SOLIDS

The first type of correction, anharmonicity, is very important in trying to achieve a quantitative understanding of solid helium. In this case, the zero-point motion carries us very far from the minimum V_0 and far outside the region of harmonicity. Indeed, the actual lattice spacing is such that the atom sites would not even be local minima of the potential if the atoms were stationary (see Fig. 4-1). Nonetheless, it is still possible to set up a Hartree self-consistent potential well with a ground state and a ladder of excited states. In general, in the realistic case, the ladder will stop at a finite energy and be replaced by a continuum, and in fact, in the helium case, there are only one or two bound excited states at most. In spite of this, we may again set up a collective excitation theory that corrects the Einstein-Hartree theory back to a proper phonon one and restores the long-wavelength phonons.

Let me sketch how this goes. The potential energy, insofar as it affects particles in the ground states, can be expanded in terms of our set of particle operators which create the various excited states (now,

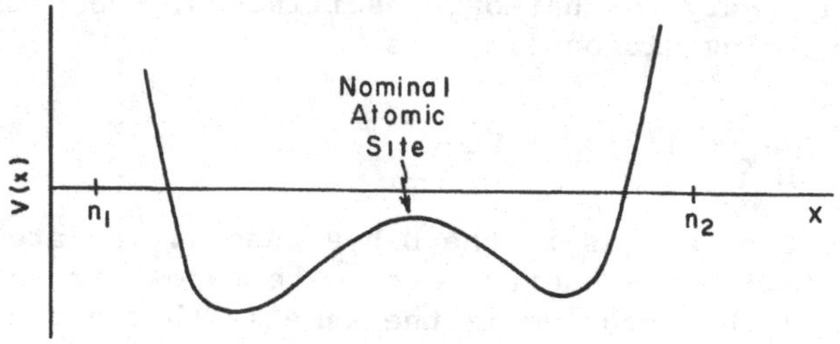

Figure 4-1 Potential for an atom in solid helium with fixed neighbors at n_1 and n_2.

of course, it does not simply step us up the ladder by one, but can convert us to any excited state):

$$V = \sum_{ij} V_{ij}$$

$$V_{ij} = \sum_{n,m\neq 0} \left(c_{0i}^{+}c_{ni} + c_{ni}^{+}c_{0i}\right)(0n|V^{ij}|0m) \times \left(c_{0j}^{+}c_{mj} + c_{mj}^{+}c_{0j}\right) + \sum_{n} \left(00|V^{ij}|nn\right) n_{nj} n_{0i} \quad , \tag{4.22}$$

where the matrix element is

$$\left(0n|V_{ij}|0m\right) = \int d^3r_i \int d^3r_j \phi_{ni}(r_i)\phi_{0i}(r_i)V(r_{ij}) \times \phi_{mj}(r_j)\phi_{0j}(r_j) \ . \tag{4.23}$$

The last term in Eq. (4.22) is the contribution of V_{ij} to the Hartree field. There are no "mixed" terms like $(00|V_{ij}|0m)$ because, since the states ϕ satisfy the mean-field Hartree equations, the mean potential is diagonal. It cannot excite just one atom. The fact that the set $\phi_{0,i}$ is overcomplete or nonorthogonal appears dangerous but is not. So long as all the states are orthogonalized to the set of ground states, no difficulties arise.

We are going to want to express a phonon as a collective excitation involving some linear combination of excited states of each site. The long-wavelength phonons are thus, in some real sense, simply displacements of the lattice with a slow variation from site to site, something like the conventional phonon

$$q_k = \frac{1}{\sqrt{N}} \sum_i e^{ik\cdot R_i} u_i \ .$$

Let us, then, write down the appropriate operator, which has the effect of multiplying the wave function at the site i by u_i:

$$u_i \phi_{0i} = \sum_n x_{n0} \phi_{ni} \ ,$$

i.e.,

$$u_i = \sum_n x_{n0} \left(C_{ni}^+ C_{n0} + C_{0i}^+ C_{ni} \right) . \qquad (4.24)$$

where

$$x_{n0} = \langle n|x|0\rangle .$$

In the harmonic case $x_{n0} = \delta_{n,1}$, but our system is anharmonic. (We include also all the matrix elements which can bring us back from a general state to the ground state, in order to keep our operators Hermitian.) There is also the displacement operator or momentum

$$\phi_0(u + \delta u) = \delta u \nabla \phi_0 + \phi_0$$

$$= \phi_0 + i\delta u \cdot \sum_n p_{n0} \phi_{ni} \ ,$$

and we know that, using the energies E_n of the excited states in the Hartree potential well,

$$p_{n0} = 2m\, i(E_n - E_0)x_{n0} ,$$

so that we can define an operator

$$p_i = 2m\, i \sum_n (E_n - E_0)x_{n0}(C^+_{ni}C_{0i} + C_{0i}C^+_{ni}) . \qquad (4.25)$$

Now we would argue that the total Hamiltonian must obey the standard identity

$$\dot{u}_i = i[H, u_i] = \frac{p_i}{m} \qquad (4.26)$$

because only the kinetic energy fails to commute with the operator u_i. Equally,

$$-i\dot{p}_i = [H, p_i] = +i\nabla_i V_i = i \sum_j \nabla_j V_{ij} . \qquad (4.27)$$

We see this, again, by referring back to the fact that p_i really <u>is</u> the operator that takes the gradient of functions based at site i. That is, the following sequence of functions is equivalent insofar as taking matrix elements in expansions like (4.22) is concerned:

$$f_j(r_j)[V(r_{ij}), i\nabla_i]f_i(r_i)$$

$$= f_j(r_j)[-i\nabla_i V(r_{ij})]f_i(r_i)$$

$$= f_j[i\nabla_j V_{ij}]f_i .$$

Let us now define our phonon variables,

$$q_k = \frac{1}{\sqrt{N}} \sum_i e^{ik\cdot R_i} u_i$$

$$p_k = \frac{1}{\sqrt{N}} \sum_i e^{ik\cdot R_i} p_i$$

$$\begin{aligned} p_k &= im[H, q_k] \\ &= m\dot{q}_k \end{aligned} \qquad (4.28)$$

[the last by (4.26)]. Using Eqs. (4.28) and (4.27), we now complete the chain by taking the time derivative of $\dot{p}_k$ to get the force,

$$-i\dot{p}_k = [H, p_k]$$

$$= \frac{i}{\sqrt{N}} \sum_{ij} e^{ik\cdot R_i} \nabla_j V_{ij}$$

$$= \frac{i}{2\sqrt{N}} \sum_{ij} e^{ik\cdot R_i} \left(\nabla_j V_{ij} - e^{i(k\cdot R_j - k\cdot R_j)} \nabla_i V_{ij}\right)$$

$$= \frac{i}{2\sqrt{N}} \sum_{ij} e^{ik\cdot R_i} \left(1 - e^{-ik\cdot(R_i - R_j)}\right) \nabla_j V_{ij}. \qquad (4.29)$$

The phase factors give us long-wavelength phonons. Now $\nabla_i V_{ij}$, unlike V_{ij}, can have matrix elements which excite single particles. If V were purely harmonic, in fact, $\nabla_i V_{ij}$ would just be $u_i \nabla^2 V_{ij}$ where $\nabla^2 V_{ij}$ is a constant coefficient, and hence it would have an expansion in terms of single bosons like (4.24) and (4.25). [By symmetry, ∇V will have no mean value or $(00|00)$ matrix element.] Thus we may linearize (4.29) by keeping only the one-excitation matrix elements:

$$\nabla_i V_{ij} \simeq \sum_n \left(\beta_n^i + \beta_n^{i+}\right)\left(0n|\nabla_i V_{ij}|00\right) + \left(00|\nabla_i V_{ij}|0n\right)\left(\beta_n^j + \beta_n^{j+}\right). \qquad (4.30)$$

Here the β's are pseudoboson operators

$$\beta_n^{i+} = C_{ni}^+ C_{0i} .$$

We expect that (4.30) will be roughly proportional to the expansion of u_i by (4.24), which would be correct if V were harmonic. In any case, with greater or lesser accuracy, we may set up linear equations of motion for the β's and extract from these a set of effective collective excitations or phonons p_k or q_k whose zero-point motion must be taken into account. This is all in the spirit of the "random-phase approximation," which works well in electron-gas problems: find equations of motion which assume the unperturbed ground state, and correct for the resulting zero-point motion.

The theory that is most completely worked out for the heliums is not this second-quantized, corrected Hartree theory, but a rather similar exercise called the "method of correlated basis functions." Taking advantage of the fact that symmetry effects are still very small in He, this method stays in coordinate space throughout, with distinguishable particles, and writes a wave function as the product of an explicit Jastrow part with a general functional form

$$\Psi_1 = e^{-\sum_{ij} f(r_{ij})}$$

and a phonon function which, as we pointed out, is also a Jastrow function but of a special sort,

$$\Psi_2 = \exp\left[-\frac{1}{2}\sum_{ij} u_{ij}\Gamma_{ij}u_{ij}\right].$$

This phonon function also has the effect of tying the atoms to particular sites R_i. Using various tricks--which mostly involve making sure that one is not doing the same job twice with Γ_{ij} and f--one arrives at a very good account of the binding energy, lattice constant, etc. The purpose of the f_{ij} is to do the job we have described as "ladder summation" or "Brucknerianism," that is, to renormalize the hard cores down to manageable proportions. (The phonon corrections we have just been discussing are, in terms of hole-particle pairs, a "bubble summation"; this is why the two parts Ψ_1 and Ψ_2 must be treated differently.)

e^{-f} is essentially a solution of the problem of two atoms interacting in the absence of the rest of the particles:

$$-\frac{\nabla^2}{2\mu}e^{-f} + Ve^{-f} = Ee^{-f}. \qquad (4.31)$$

where μ is the reduced mass. Of course, this solution is different depending on the energy E, but not very much in the region where V >> energy differences, i.e., where V may properly be regarded as a hard-core potential. One often chooses E = 0 and cuts V off at the hard-core radius. The effect of including f is primarily to give a modified effective potential V^+, which is, approximately,

$$V^+ \approx V - \frac{1}{2\mu}\nabla^2 f,$$

and in fact this is the motivation for using Eq. (4.31), which approximately minimizes V^+.

The computational details of these quantum solid theories are hair-raising, but they do work, even in cases where anharmonicity is severe. As I have said, the essential points are: (1) they give a ground-state energy that is indeed a function of Λ^* of roughly the form given in Fig. 3-7; (2) they all must incorporate the low-frequency correlations implied by the existence of phonons, and if not give very poor answers; (3) they also, in dealing with the hard-core problem, must in one way or another rely on pseudopotential techniques in which the strong part of the two-particle scattering is handled by solution of a Bethe-Salpeter scattering equation.

C. SYMMETRY CONSIDERATIONS--WHEN ARE SOLIDS SOLID?

The techniques we have just been describing seem to rely again and again on the very basic intuition we tend to have that, statistics notwithstanding, the atom in a solid near site R_i can be treated as distinguishable from that at site R_j. Particularly, the central dogma of phonon theory is that phonons are made up of excitations in which the atoms do not change places. An excitation $C_{ni}^+ C_{0i}$ is followed by $C_{0i}^+ C_{ni}$ or, if one insists on retaining strong anharmonic terms, by a chain that ends in C_{0i}^+, not in C_{0j}^+. This intuition must have a physical basis; it must in some real sense be one of the defining properties of solids as we know them.

As I remarked in Chapter 2, the existence of a one-dimensional or even three-dimensional density wave is the obvious criterion for "solidity," but doesn't seem necessarily to define what we mean by a solid, as we shall see in more detail in Sec. 4.E. For all practical purposes, glass is a solid, but it is not a density wave. (It contains a rigid framework, which in some sense plays the role of the density wave, but it is not clear why or how.) I would prefer to define a solid as a system in which elementary excitations involving the free motion of one atom relative to the background, or removal of an atom from a site and replacement elsewhere, involve a finite--usually large--energy gap. That is, my definition of a solid would be much like that of an insulator as defined by Mott. To expand upon this point of view the Hartree theory is the only approach, but of course in order to handle it properly we must put the Fock back into Hartree-Fock: we must retain exchange terms. In order to display the real complications immediately, let us take the Bose case first. We want to derive self-consistently a Bose ground state

$$\Psi = \Big(\prod_{\text{sites } i} c_i^+ \Big) \Psi_{vac} \tag{4.32}$$

where the c_i^+ refer to some set of localized self-consistent orbitals at the sites of a lattice:

$$c_i^+ = \int d^3r \phi_i(r) \Psi^+(r) \ . \tag{4.33}$$

where the wave function $\phi_i(r)$ can be taken to be real.

I derived in Concepts in Solids (Sec. 2.A2) the general Hartree-Fock scheme, which we now try to apply to the Bose case.

Let us first assume the self-consistent orbitals ϕ_i to be orthonormal. Then the equation of motion of the single-particle operators c_i is

$$\dot{c}_i = i[H, c_i] = - i\sum_j T_{ij} c_j - i\sum_{jk\ell} c_j^+ c_k c_\ell (V_{ijk\ell} + V_{jik\ell})$$

$$= - E_i c_i \tag{4.34}$$

and that for c_i^+ the Hermitian conjugate. The matrix elements here are

$$T_{ij} = -\int \phi_i^*\left(\frac{\nabla^2}{2m}\right)\phi_j d^3r$$

$$V_{ijk\ell} = \int d^3r_1 \int d^3r_2\ \phi_i^+(r_1)\phi_j^+(r_2)\phi_k(r_2)\phi_\ell(r_1) \times V(r_1-r_2). \tag{4.35}$$

When we apply Eq. (4.34) to the assumed ground-state function (4.32), the only way in which we can get back to a single-particle excitation is to have $j = k$ or $j = \ell$, so that we put one particle back where it came from. In the Bose case, there is a special argument about the $j = k = \ell$ element because, in contrast to the Fermi case, this element does not automatically drop out by the anticommutation rules (i.e., in the Bose case the exchange self-energy does not vanish). This vanishing reflects the fact that in Fermi statistics you cannot put two particles in state j, so that state's self-energy is unique. In the Bose case, however, we cannot <u>empty</u> state j twice, but we can fill it twice. Thus the equations for c_i and c_i^+ are different:

$$-E_i c_i = [H, c_i]_{H-F} = -\sum_j T_{ij}c_j - \sum_{j\neq k} \langle n_j\rangle c_k\left(V_{ij;jk} + V_{ji;jk}\right), \tag{4.36}$$

whereas the equation for adding a particle is

$$E_i c_i^+ = [H, c_i^+]_{H-F} = \sum_j T_{ji}c_j^+ + \sum_{\substack{j,k \\ \text{(all)}}} \langle n_j\rangle c_k^+\left(V_{jk;ij} + V_{jk;ji}\right) \tag{4.37}$$

and has no $j \neq k$ exclusion. This is very unfortunate for our original assumption of orthonormality. Equation (4.37) is the same for every i and thus has orthonormal solutions; but, because it is the same for every i and thus has the periodic symmetry of the lattice, it has only running-wave solutions, as is appropriate for an added particle in a regular lattice. Equation (4.36) could have localized solutions--because of the $j \neq k$ omission, the large self-energy $V_{ii;ii}$ never enters and there is a potential well at site i. But these solutions are, correctly, in general not orthogonal.

To see that this is, in fact, the correct outcome let us do a Slaterian description of a very simple case, a two-site solid. That is, suppose electrons had been bosons. How would H_2 bind?

We assume two "atoms" A and B with tightly bound and therefore rather inflexible orbitals a and b initially orthogonalized to each other. The "electrons" are assumed to have a rather short-range interaction so that

$$V_{aa,aa} = V_{bb,bb} = U \gg \text{all other } V\text{'s.}$$

(I am setting up, that is, a kind of Bose "Hubbard model." See Fig. 4-2.) However, ϕ_a and ϕ_b do overlap somewhat so

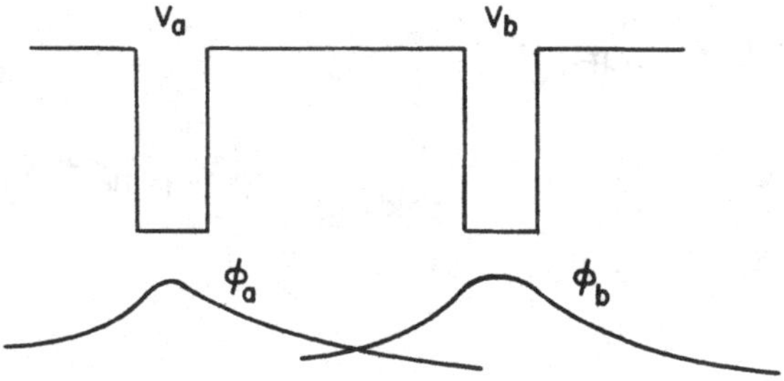

Figure 4-2 Bose H_2 model.

that there is a kinetic energy matrix element T between them. Thus the Hamiltonian may be written

$$H = E_0\left(n_a + n_b\right) - T\left(c_a^+c_b + c_b^+c_a\right) + U\left[n_a(n_a-1) + n_b(n_b-1)\right]. \qquad (4.38)$$

E_0 is arbitrary, so let us set it equal to zero. We now try to minimize the energy with respect to a modified pair of nonorthogonal single-site operators,

$$c_{a'} = uc_a + vc_b;\ c_{b'} = uc_b + vc_a\ , \tag{4.39}$$

where $u^2 + v^2 = 1$. The symmetry of Eq. (4.39) is a reasonable assumption. Note first that the state $c^+_{a'}c^+_{b'}\Psi_0$ is not even normalized:

$$|\Psi|^2 = |c^+_{a'}c^+_{b'}\Psi_0|^2 = \left\|\begin{matrix} u^2+v^2 & 2uv \\ 2uv & u^2+v^2 \end{matrix}\right\| = 1 + 4u^2v^2\ ,$$

where 2uv is the commutator of $c_{a'}$ with $c^+_{b'}$, i.e., the overlap of $\phi_{a'}$ and $\phi_{b'}$, and $\|\ \ \|$ denotes the "permanent," which is like the determinant but with all signs positive. Note that for bosons $\langle\psi_0|ccc^+c^+|\psi_0\rangle = 2$.

Now we apply Eq. (4.38) to our state,

$$\Psi = c^+_{a'}c^+_{b'}\psi_0 = \left[c^+_a c^+_b + uv\left(c^{+2}_a + c^{+2}_b\right)\right]\Psi_0\ ,$$

which gives

$$-T\left(c^+_a c_b + c^+_b c_a\right)\Psi = -T\left(c^{+2}_a + c^{+2}_b + 4uv\, c^+_a c^+_b\right)\Psi_0$$

$$U\left(n^2_a - n_a + n^2_b - n_b\right)\Psi = U\cdot 2uv\left(c^{+2}_a + c^{+2}_b\right)\Psi_0$$

so that

$$(\Psi, H\Psi)/(\Psi,\Psi) = \frac{-8Tuv + 8Uu^2v^2}{1 + 4u^2v^2}$$

$$= \left(-4Tx + 2Ux^2\right)/\left(1 + x^2\right) \quad (x = 2uv) \tag{4.40}$$

with the solution for a minimum

$$x = \frac{T}{U}\left(1 - \frac{T^2}{U^2} + \ldots\right) . \tag{4.41}$$

Thus, approximately, since $u^2 + v^2 = 1$,

$$c_{a'} = c_a\left(1 - \frac{T^2}{8U^2}\right) + c_b(T/2U)$$

and the modified energy from Eqs. (4.40) and (4.41) is

$$E_i = E_0 - \frac{2T^2}{U} \tag{4.42}$$

so that the Bose solid is stabilized to this extent relative to the naive assumption of orthogonality. Equation (4.41) shows that there is indeed a tendency for the two wave functions to become nonorthogonal. This is very like the result of an "unrestricted Hartree-Fock" calculation for fermions, but does not suffer from the same symmetry difficulties; this solution is correct even for the small system, if U is big enough.

Thus we have to work out a more sophisticated way of dealing with the self-consistent Hartree-Fock problem than simply using orthogonal local functions as for a filled band of fermions. I have never seen an attempt at this for Bose systems (the surprising arithmetical complication of even the trivial two-particle model above may indicate why) but, of course, one may exist without my knowledge. I think a proper approach may be the following.

Let us suppose that a ground-state wave function of the form (4.32) does indeed represent a good starting approximation, where the c_i's are not taken to be orthogonal (it may even be that, as for a Bose liquid, they are all nearly the same). Let us start from the "Koopman's theorem" aspect of Hartree-Fock theory, in which the equations of motion of the occupied states are considered as estimates of the single-particle elementary excitation energies. We rewrite the equation of motion of the holes and particles, keeping the general form

$$-E_i c_i = [H, c_i] = \int \frac{\nabla^2}{2m} \phi_i^+(r)\Psi(r)d^3r - \int d^3r' \int d\, rV(r-r')\phi_i^+(r)\Psi^+(r')\Psi(r')\Psi(r) \tag{4.43}$$

and

$$E_i c_i^+ = -\int \Psi^+_{(r)} \frac{\nabla^2 \phi_i(r)}{2m} d^3r +$$

$$\int d^3r \int d^3r' V(r-r') \Psi^+(r) \Psi^+(r') \Psi(r') \phi_i(r) \ , \tag{4.44}$$

where we used Eq. (4.33).

We now consider these equations applied to the assumed product wave function (4.32) on the right. Let us first consider Eq. (4.44), which we expect to be relatively well behaved (as was 4.37). Our philosophy will be that we wish to "factorize" the combination of three field operators in the interaction term, in all possible ways, into a ground-state average $\langle \Psi^+ \Psi \rangle$ times a single-particle operator. Diagonalizing the resulting equation will ensure that the Hamiltonian has no matrix elements causing single hole-particle excitations. This is both the physical essence of the Hartree mean-field idea and the essential condition for minimizing the energy of a state like (4.32).

On the right in Eq. (4.44) we get two terms,

$$\int d^3r \phi_i(r) \Psi^+(r) \bar{V}(r) + \int d^3r \int d^3r' \phi_i(r) \Psi^+(r') \bar{A}(r-r'), \tag{4.45}$$

where

$$\begin{aligned} \bar{V}(r) &= \int V(r-r') d^3r' \ \langle \Psi^+(r') \Psi(r') \rangle \\ &= \int V(r-r') d^3r' \rho(r') \end{aligned} \tag{4.46}$$

and

$$\bar{A}(r-r') = V(r-r') \ \langle \Psi^+(r) \Psi(r') \rangle \ . \tag{4.47}$$

The mean density $\rho(r)$ entering into Eq. (4.46) is not the same simple expression $\langle n_j \rangle$ entering into (4.37), but includes overlap charges. It is given by the rather complicated expression

$$\rho(r) = \langle\Psi^+(r)\Psi(r)\rangle = \frac{1}{\|S\|}\sum_{ij} \phi_i^+(r)\phi_j(r)\|S\|_{ij} \quad , \tag{4.48}$$

where as before $\|S\|$ is the permanent of the overlaps, and $\|S\|_{ij}$ is its ij'th minor. The same type of expression holds for the density operator $\langle\Psi^+(r)\Psi(r')\rangle$ that appears in Eq. (4.47): one simply substitutes $\phi_i^+(r)\phi_i(r')$. For our purposes it is adequate to expand to lowest order in overlaps, which gives

$$\langle\Psi^+(r)\Psi(r')\rangle = c[\sum_i \phi_i^+(r)\phi_i(r') + \sum_{ij} S_{ij}\phi_i(r)\phi_j(r')], \tag{4.49}$$

where $c = \frac{1}{\|S\|}$.

When we insert (4.45) into the equation (4.44), the effective potential becomes perfectly periodic, and there are therefore no localized solutions for added particles. If we are to find a self-consistent solid, we must hope that Eq. (4.43) has a different character, and it does. In this case the operator $\Psi(r)$ acts prior to the measurement of the density, and so the relevant terms like (4.45) are

$$\int d^3r\phi_i(r)\Psi(r)\overline{V_i'}(r) + \int d^3r\int d^3r'\phi_i(r)\Psi(r')\overline{A_i'}(r-r').$$

(Since ϕ_i can be regarded as a real function, we drop the dagger.) These terms differ in that V_i' is to be calculated, not in the actual ground state, but in one in which a particle has definitely been destroyed in state i, and A' is to be calculated in a state in which the operator $\phi_i(r)\Psi(r')$ has operated prior to the calculation of $\langle\Psi^+(r')\Psi(r)\rangle$.

It is easiest to visualize the modified density. Let us again confine ourselves to second order in overlaps; then, since precisely one particle has been removed, the change in ρ is

$$\Delta\rho_i(r) = -\frac{\left(|\phi_i|^2 + \sum_j S_{ij}\phi_i(r)\phi_j(r)\right)}{1 + \sum_j |S_{ij}|^2} \tag{4.50}$$

and the change in effective potential is

$$\Delta V_i = \overline{V_i'} - \overline{V} = \int d^3r' V(r-r')\Delta\rho_i(r'). \tag{4.51}$$

Similar equations hold for the exchange term A, which—incidentally—in the case of bosons always works in the same direction as the ordinary potential, and will therefore considerably strengthen the change in effective potential (as we should expect: in the fermion case the corresponding term wipes out the potential hole entirely, while here it has the opposite sign).

We have, then, an absolutely clear criterion for whether or not a Bose solid will form. This is that the extra potential well of Eq. (4.51) should, with the help of ΔA_i, be sufficiently attractive to bind at least one state ϕ_i. [Note that everything we say, even about overall stability, requires that the potential V(r-r') be repulsive on the average and at short distances. This does not preclude the possibility that the long-range attraction can lead to a finite binding energy, as in He.] We have in fact two equations. The first is for c^+ from (4.44),

$$E\phi(r) = -\frac{\nabla^2\phi(r)}{2m} + \overline{V}\phi(r) + \int d^3r'\overline{A}(r-r')\phi(r'); \tag{4.52}$$

since $\overline{V}$ and $\overline{A}$ are uniform or periodic, this equation has no bound solutions $\phi(r)$ and has, therefore, a continuous spectrum E with a lowest energy E_0. The second is the equation for the hole excitation c_i,

$$E_i\phi_i(r) = -\frac{\nabla^2\phi_i}{2m} + V\phi_i + \int d\, r'A(r-r')\phi$$
$$+ \overline{\Delta V_i(r)}\,\phi_i(r) + \int d^3r'\,\overline{\Delta A_i(r-r')}\,\phi_i(r'), \tag{4.53}$$

which has added a localized potential well $\overline{\Delta V_i}$ and $\overline{\Delta A_i}$ to (4.52) and therefore may or may not have a bound solution $E_i < E_0$. (See Fig. 4-3 for a graphic portrayal of this situation.) If it does, this implies that there is an energy gap,

$$E_g = E_0 - E_i \tag{4.54}$$

for excitation of a particle away from one of the sites of the lattice, and therefore one may, with some hope of self-consistency, occupy each of the orbitals ϕ_i in the ground state of the system. If this is not the case, it seems extremely difficult to see any motivation for a lattice to form in the Bose case.

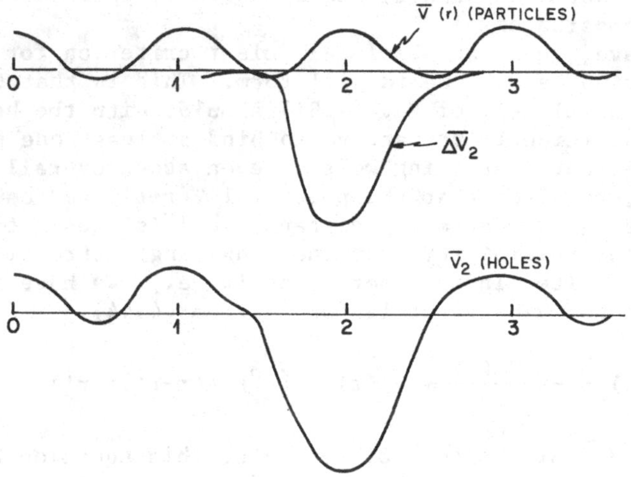

Figure 4-3 Effective potentials for particles and holes in Bose solid.

I have discussed this procedure purely within the confines of Hartree-Fock theory. But once we have the formal structure, it is easy to see that we could base a perturbation-theoretic hierarchy of diagrams upon it. We may sum ladders in order to renormalize the worst effects of the hard cores out of V, and the bubble sums which bring in the phonon collective excitations may be done equally well starting from our more rigorously defined local potential wells. In addition, in some higher approximation, the hole c_i must be capable of motion, but unlike the Fermi case this does not appear in Hartree-Fock approximation since the potential hole does not move with the atom. Apparently the hole becomes mobile only when three-boson excitations are taken into account. Nonetheless we visualize that the difference between hole and particle potentials persists and an energy gap will remain. In the absence of the Pauli principle, the conventional "Bloch" explanation of an energy gap is quite untenable for bosons. It is clear that <u>one</u> of the essential stabilities underlying the phenomenon of the solid is this energy gap.

In the absence of extremely unusual and artificial types of forces, if the running-wave solutions are the only stable ones, it seems certain that the lowest-energy one will be the $k = 0$ uniform wave function with no periodicity. (Such unusual forces have been postulated for pions in superdense nuclear matter; that, however, is not only a complicated system, but highly controversial. We are not, in any case, discussing Bose-like collective modes such as spin waves and, possibly, pions or isospin waves in nuclear matter.)

This, then, is the motivation for suggesting in a previous chapter that a Bose solid is not physically likely to be superfluid. In any case my definition of a true solid certainly excludes it. You will recognize that if free particles are to be produced only by thermal excitation across a gap, Bose condensation cannot take place. We shall find that the characteristic properties we ascribe to a solid also seem to be postulated on this energy gap in the spectrum.

The situation with regard to Fermi solids is somewhat different mathematically and brings in a new problem: exchange. Spinless particles obeying Fermi statistics would be much easier to deal with, so let us discuss this case first. The fermion Hartree-Fock equations are the same for particles and holes, and so the appearance of a gap comes in quite a different way, as an apparent consequence of the Pauli principle plus Bloch's theorem.

The Hartree-Fock equations in this case have the well-known structure given before. We postulate a product ground state as in Eq. (4.32),

$$\prod_{\text{sites } i}^{N} c_i^+ \Psi_{\text{vac}} ,$$

and the equation (4.33) for the creation operator. The

creation operator c_i^+ satisfies the equation

$$E_i c_i^+ = \sum_j T_{ji} c_j^+ + \sum_{j,k} \langle n_j \rangle c_k^+ \times \left(V_{jk;ij} - V_{jk;ji}\right) , \tag{4.55}$$

the matrix elements being just as in (4.35). Fortuitously--we now realize--we are not troubled by the special status of j = k; the exchange and direct matrix elements cancel, and the equations for holes and particles coincide. Thus Eq. (4.55) has only orthonormal solutions, which are bands of running-wave eigenstates. But Eq. (4.55) can lead to a gap just as could (4.43) and (4.44). This is because of the exclusion principle: if there are N sites in the periodic lattice, states will group into bands of just N levels, separated, in many cases, by energy gaps from the next band. If these gaps are large enough, the N different electronic states that must be filled may just fill up a band, leaving a gap to the first free-particle states.

The apparent difference from the Bose case is removed by the use of Wannier's theorem, which observes that any separated band of running-wave states ϕ_k is equivalent to N exponentially localized states $W(r-R_n)$ via the Bloch transformation

$$W_n = \frac{1}{\sqrt{N}} \sum_k \phi_k e^{-ik\cdot R_n} .$$

The Hartree-Fock equation (4.55) can then be transformed into a well-known equation for the localized function W:

$$HW(r-R_i) - \sum_{j\neq i} (W_i|H|W_j) W(r-R_j) = E_0 W_i , \tag{4.56}$$

where the artificially inserted term on the left-hand side has the effect of a repulsive pseudopotential confining the particle to site i. Other pseudopotential techniques developed originally by Adams and Gilbert (Adams, 1961; Gilbert, 1964), which I have modified and applied to some chemical problems (Anderson, 1968), can give a more localized function than W, satisfying an equation without the uniqueness difficulties of (4.56). For instance, it is satisfactory (Anderson, 1968) to replace H in the pseudopotential term simply by the potential energy V. We will, however, keep (4.56) for our formal discussion, recognizing that if necessary a better formalism exists.

In any case, we insert on the right-hand side of (4.56) the Hartree-Fock Hamiltonian, which itself depends on the functions W, since in this case

$$\langle \Psi^+(r')\Psi(r)\rangle = \sum_i W^+(r' - R_i)W(r - R_i). \qquad (4.57)$$

Then we have again an equation which may or may not be capable of exhibiting a bound-state solution separated from all possible free solutions:

$$H\phi(r) - \sum_{j\neq i} (\phi|H|W_j)W(r - R_j) = E\phi(r). \qquad (4.58)$$

All unoccupied bands are automatically solutions because HW_j is a linear combination of occupied states only. If the bands are not separated, Eq. (4.58) has no exponentially localized solution. On the other hand, (4.58) alone does not guarantee a solid, even within Hartree-Fock theory. E is not now the only relevant energy, and if the highest state in the band is above the lowest free state, we may lose our solid.

As I have pointed out at some length elsewhere, in spite of the fact that in this case the occupied and unoccupied states satisfy the same equation, there is still a large exchange self-energy contribution to the band gap (Inkson, 1971, 1973; Anderson, 1974). Of course, in the case of the quantum solid, this is the major component, since the gap must develop self-consistently. In the case of the occupied states, the repulsive self-energy

$$\int d^3r \int d^3r' V(r - r')|W(r - R_i)|^2|W(r' - R_i)|^2$$

is completely canceled out by the exchange term. For the unoccupied states, on the other hand, the cancellation is far less complete. The direct interaction with W_i for a wave function $\phi(r)$ is

$$\int d^3r \int d^3r' |W(r - R_i)|^2 V(r - r')|\phi(r')|^2 , \qquad (4.59)$$

while the canceling exchange term is

$$-\int d^3r \int d^3r' W(r-R_i)\phi^+(r)V(r-r')W^+(r'-R_i)\phi(r'). \quad (4.60)$$

Equation (4.60) is smaller than (4.59) because $\phi(r)$ is necessarily orthogonal to W, belonging as it does to a different band,

$$\int \phi(r)W^+(r-R_i)d^3r \equiv 0. \quad (4.61)$$

Thus the "exchange charge" distribution whose self-energy is (4.60) has zero average, as contrasted to the ordinary charge $|W|^2$ or $|\phi|^2$.

D. SPINS AND EXCHANGE EFFECTS

Now we have to introduce the complication of spin. As we saw from the Bose system, the formation and stability of a solid is not dependent at all upon the orthogonality of the localized functions. In fact, the Bose solid is probably intrinsically more stable than the spinless Fermi one because of its added stabilization energy [Eq. (4.42)] and because the holes do not move as easily. (But the solidification pressure of the Fermi solid will be lower, as in the case of ^3He and ^4He, because the Fermi liquid is less stable by the Fermi energy, a bigger effect. That is, both fermion phases are less stable, but the liquid more so.) Opposite-spin fermions can still be expected to occupy some kind of self-consistent local orbitals, which need no longer be perfectly orthogonal to each other and which thereby gain some degree of stabilization. This argument tells one the sign and, in a rough way, the magnitude of the resulting "exchange integrals."

Herring (Herring, 1966a) has given beautifully rigorous arguments for the nature of the interactions in such a case--aimed at systems of "electrons well separated on atoms," primarily, but suited to the present situation--which amount to a sophisticated use of our basic continuation principle. Let me sketch it out in a less sophisticated form.

In the extreme case of large mass M or of large repulsive interactions, there will be a very large gap to any excited state which involves the real--as opposed to virtual--excitation of a particle out of its potential well. (As we pointed out, even in the very classical limit the low-energy excitations within the wells, which continue into even-lower-frequency phonons, must be kept.)

But, in this limit, we must also keep account of the still-lower-energy excitations that involve the reversal of the particles' spins while leaving them in their respective wells. By the well-known argument of Van Vleck and Dirac, which leads to the "Heisenberg" Hamiltonian, this degeneracy is coupled to the near-degeneracy with respect to interchanges of the particles in their wells, so that in order to attain total antisymmetry with respect to interchanges of particles one must couple a space-symmetric function with a spin-antisymmetric one and vice versa. Thus, finally, an effective coupling Hamiltonian for the spins is obtained as

$$H_{eff} = -\sum_{ij} J_{ij}P_{ij} + \sum_{ijk} J_{ijk}P_{ijk} + \ldots \quad . \tag{4.62}$$

P_{ij} is the operator that permutes the spin labels on different sites treated as inequivalent, while the J's are the renormalized matrix elements of the Hamiltonian between permutations of Hartree-like states with the atoms in their respective potential wells. For instance J_{12} is the matrix element of (H_{tot}) between a state with atom 1 in well 1, atom 2 in well 2 (with an artificial barrier preventing 1 from occupying 2 and vice versa) and the corresponding permutation. As is well known,

$$P_{ij} = \frac{1}{2}\left(1 + \vec{\sigma}_i \cdot \vec{\sigma}_j\right) = \frac{1}{2} + 2\vec{S}_i \cdot \vec{S}_j \tag{4.63}$$

etc. Herring also shows that the J's can be represented by a kind of mutual tunneling expression, a current matrix element integrated over a surface in configuration space separating the two permuted configurations:

$$J_{12} = -\frac{1}{m}\int_{\left(\Sigma(P_{12})\right)} d\vec{S} \cdot \phi^{+}\nabla P_{12}\phi \ , \tag{4.64}$$

where $\Sigma(P)$ is a surface separating configuration (1 in 1, 2 in 2) from (2 in 1, 1 in 2), and ϕ is the pseudo-wavefunction of the artificially localized Hamiltonian. Similar expressions hold for the higher terms in Eq. (4.62), except that the relevant permutations are obviously more complicated.

Equation (4.64) shows us that the exchange series in (4.62) is, at the very least, a series in the overlaps between wave functions in neighboring potential wells. J_{12} clearly is of order $(S_{12})^2$ in this sense, in that the coordinates of two particles are involved, J_{123} of order $(S_{12})^3$ etc. In real solids another very important effect is the steric hindrance caused by neighboring atoms when atoms 1 and 2 try to tunnel past each others' hard cores. This is the reason why exchange integrals in ^3He decrease so rapidly as the density increases (see Fig. 4-4). One could easily,

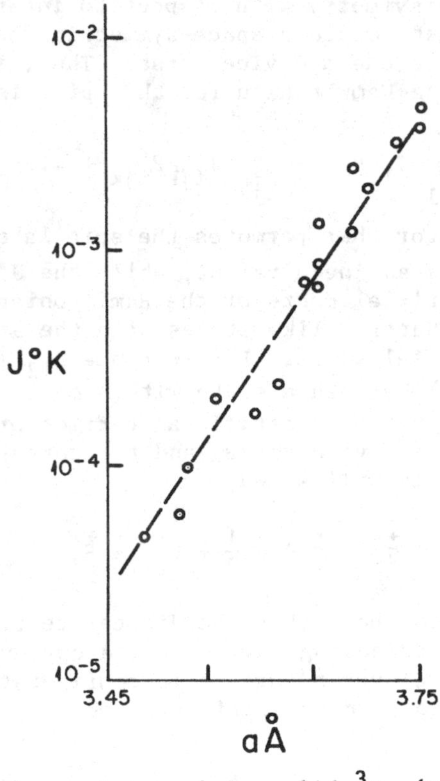

Figure 4-4 Exchange integral in solid ^3He (Panczyk and Adams, 1969).

in the case of He, imagine that at some density the steric factor could make ring interchange easier than pair exchange, as suggested by Thouless (1965), and in fact four-particle exchange seems to be one of the dominant modes experimentally. (Recent developments in the magnetic structure of solid ^3He underline this point but are too complex for inclusion here.)

A second obvious consequence of Eq. (4.64) is that normally J_{12} will be antiferromagnetic, simply because J_{12} is the solution of the two-particle problem, in which Wigner's theorem requires antiferromagnetism. That is, the most symmetric state seems sure to be lower in energy. In solids this is not subject to the exceptions expected in transition ion-electronic systems. Indeed, the gain in energy of the antiferromagnetic ground state is similar in origin to that of the Bose solid: antiparallel spin neighbors need not maintain an artificial orthogonality and can slop over into each other's Hilbert space to a slight extent. This is the oversimplified but physically correct view I took in my original study of "superexchange" in 1959 (Anderson, 1959a).

What happens as we continue this picture into the region of large overlaps and large anharmonicities? Phonon frequencies scale with the frequency of the excited states in the individual wells, while exchange scales with the wave function overlap; it seems intuitive that the former at first is considerably larger, and in fact a simple analysis suggests (using the notation of Chapter 3)

$$\frac{\omega_D}{\varepsilon} \sim \Lambda^* ,$$

whereas $J \sim e^{-\text{const}/\Lambda^*}$, a very much smaller effect at small Λ^*. In fact, even in ^3He the exchange integrals are $< 10^{-4}$ ω_D. Apparently exchange will remain manageably small even near the instability limit. It is still not entirely clear why, in this case, four-particle exchange is dominant, but plausible theoretical hypotheses can be proposed.

E. DENSITY WAVES, RIGIDITY, FRICTION

As I pointed out earlier, in the fermion case, it is not at all obvious that one cannot have "giant density waves" as opposed to "true" solids with an energy gap for particle excitation. This question depends on whether or not the energy gap cuts off the entire Fermi surface or leaves some behind. These two possibilities are familiar in the case of electrons in ordinary solids: we have every gradation from large-gap insulators to good metals with virtually no missing Fermi surface. The forces may, for instance, be such as to give a crystal structure that does not split certain degeneracies and therefore necessarily leaves some Fermi surface, which seems to be the typical case for many semi-metals. I do not see any reason to suspect such a case for simple hard-core van der Waals-type interactions such as the heliums exhibit, but no discussion of the Wigner solid I know of demonstrates to me the impossibility of such a phase (or a "spin-density wave") at intermediate densities; nor is it obviously ruled out for nuclear matter in some range. In fact, I believe that in these cases, as for excitonic systems, the transitions will usually go directly from true solid to liquid and "leap over" such phases; but they serve as an introduction to the question of rigidity and friction. I would argue that such phases do not show the properties we usually understand to be solid rigidity, but are essentially fluid.

To introduce this question, consider states in which the particles making up the crystal are moving and for simplicity start with the temperature at absolute zero. In the case of the "true solid" with an energy gap, there are two possible types of moving states, with a given total momentum P. The first is the state of the crystal moving as a whole, with velocity V, which has an energy

$$E = \frac{NMV^2}{2} = \frac{P^2}{2MN} \,.$$

Our definition of a true solid ensures that any state with the same momentum P obtained by exciting individual particles into motion must have an energy which is <u>linearly</u> proportional to P, with a coefficient at least equal to the energy gap E_g divided by the maximum momentum in the lowest excited band:

$$[E(P)]_{indiv} \geq E_g \times \left(\frac{P}{mv_{max}}\right) \gg \frac{P^2}{2NM} = [E(P)]_{coll} \,. \tag{4.65}$$

In particular, once the lattice is moving at some small velocity V, any frictional force attempting to slow it down by creating excitations has to provide an energy linearly proportional to the amount of slowing.

These trivial facts tell us two things. (1) If some clamping force is exerted on the lattice, the system as a whole cannot be accelerated by applying a body force or by exerting a pressure gradient. The few excited particles thermally present may move--in fact, it is well known that crystals do flow very gradually by thermal diffusion of vacancies and interstitials--but the crystal as a whole cannot. On the other hand, once the crystal is sliding, it cannot be brought to rest by stopping the individual particles, because that costs more energy than is available from the inertia.

If, on the other hand, there is a free Fermi surface, there are two kinds of states of motion. We can have uniform motion of the whole crystal, which again has energy $E = P^2/2MN$. But we can also allow the crystal lattice to remain at rest and merely move the Fermi surface, which is equivalent to taking particles from one side of it and replacing them on the other. Thus we gain $2p_F$ per particle, while costing an energy only proportional to the number of particles we excite.

If we minimize the resulting energy cost, we shall find ourselves simply displacing the Fermi surface as a whole by a momentum P/N' (N' the volume of the relevant piece of Fermi surface), and the resultant cost in energy is given by

$$E_{\text{displ}}(\text{Fermi}) = \frac{P^2}{2N'M^+}, \tag{4.66}$$

where M^+ is an effective mass per particle. I could find no sum rule requiring $N'M^+$ to be greater or less than NM.

This means that, even with the crystal lattice clamped (say by putting it in a rough container), a pressure on one end of the sample can cause it to accelerate and flow. Presumably the lattice re-forms downstream--such a solid might well have none of the usual kinetic obstacles to crystal growth. At the same time, a lattice of this sort sliding on a surface will be decelerated easily by scattering of quasiparticles around the Fermi surface. As far as I can see, such a material would have the characteristics of a liquid crystal, a "quantum liquid crystal" if you like.

The very old field, recently reactivated, of liquid crystals is a fascinating source of examples illustrating unexpected points in the theory of broken symmetry. The emphasis of this book is mostly on quantum many-body physics, and the liquid crystals are purely classical; that and my own lack of specialized knowledge are my excuses for bringing in so little on liquid crystals. One case, however, is especially relevant here: the smectic B phase, which from x-ray data appears to have a regular spacing of layers, each layer being two-dimensionally ordered. So far (1978) there is no conclusive evidence whether this is a very plastic solid or a true liquid crystal in some sense, i.e., whether even in the classical domain three-dimensional order and rigidity are not distinct phenomena. (As of 1982, both cases seem to have been proven to exist! Moncton et al., 1982.)

I describe these anomalous cases mainly to bring out my reasons for emphasizing that our usual intuitions about a solid depend far more on a one-to-one atom-to-site correlation than on the usual definition of crystalline regularity. I have asked the converse question--i.e., when are amorphous substances solid?--elsewhere (Anderson, 1979).

We can see from this argument also the reasons for the unique nature of solid friction. When the surface of one solid slides over that of another, two roughnesses encountering each other must either be completely overwhelmed by the forces driving the two, or stop the motion entirely; there is no way for a roughness simply to provide a continuous source of linear dissipation. Any roughness which causes only a purely elastic deformation leads to no dissipation at all, since the lattice exerts exactly as much force on the roughness, when rebounding, as the original deformation required. Thus we get the stick-slip nature of solid friction; either two surfaces stick entirely, or losses occur only by the continual establishment of sticking points and their complete breakdown, leading to the phenomenon of wear.

A beautiful illustration of this general principle has been worked out by Pippard and co-workers (1969) in the case of the "peak effect" of type-II superconductors. This is one of the few examples of one lattice sliding within another, the type-II flux-line lattice sliding within the crystalline lattice of the metal. It is observed that just before the second-order transition point H_{c2} where the flux lattice disappears it suddenly begins to move much more slowly through the underlying lattice and its impurities. This slow motion means that the voltage developed is much smaller, and that, as a result, one gets a large, sharp conductivity peak, hence the name "peak effect."

It seems clear that in the low-temperature resistive state the flux-line lattice moves fairly freely with little friction, presumably because most of the "pinning centers" are ineffective and cause only elastic deformations of the flux-line lattice. But as the temperature is raised, either the flux-line lattice becomes so plastic that it can accommodate locally to the pinning centers, or it becomes so soft that its motion is limited by its own sound frequency. In either case we lose the "suprafluidity" or "giant fluidity" of the stick-slip regime of the solid in favor of a more normal fluid behavior. [More recently, Lee (1979, 1981; see reprints) and others have begun to study the pinning phenomena for charge-density waves in great detail.]

It is amusing that only in these solid-friction situations do considerations allied to the "Landau criterion" for superfluidity have any genuine relevance. Here the question of velocities of available excitations plays a vital role, although generally not so simple a role as Landau described. (Landau's idea was that flow velocities would be limited by the velocity of available excitations that a perturbing center, such as a wall roughness or an impurity, could produce.) We see in the reprints that Landau's criterion is not relevant at all to superfluidity. Allender, Bray, and Bardeen (Bardeen, 1973; Allender _et al._, 1974) applying an old theory of Fröhlich (1954), have suggested that a number of organic "one-dimensional metals" are systems in which one observes giant "sliding" conductivity, caused by the more or less rigid sliding motion of an incommensurate electronic lattice within a normal molecular one. The one-dimensional case is very anomalous because of the characteristic large fluctuation effects, and it is as yet by no means clear whether the model describes the system or whether the theoretical suggestions are correct. More recent theories (e.g., Lee, Rice, and Anderson, 1973, 1974) suggest, in fact, just the opposite--that friction can be enhanced by density-wave formation. Unfortunately, many types of systems that might exhibit this mutual sliding behavior cannot easily be studied because the internal lattice does not transport any easily measured quantity. This is true, for instance, of such examples as Cr and the rare-earth spin spiral states. As one of the deepest as yet unsolved problems of many-body physics, this kind of phenomenon deserves much more experimental and theoretical attention. (As of 1983, in fact, it is receiving a great deal.)

CHAPTER 5

USES OF THE IDEA OF THE RENORMALIZATION GROUP IN MANY-BODY PHYSICS

A. INTRODUCTION

The renormalization group (RNG) idea was originally invented as a method of solving certain problems of "infrared divergence" in quantum electrodynamics (Bogoliubov and Shirkov, 1955). Because low-frequency photons are so small, a large number of them may be emitted in a single event such as acceleration of a particle near a nucleus ("bremsstrahlung"), even though no great amount of energy is involved. Calculation of such processes by normal diagram techniques failed, because of divergences in the photon number in the limit $\omega \to 0$ (the "infrared limit").

The principle behind the solution is all we need to describe here. Simple quantum electrodynamics is a purely dimensionless theory. The coupling constant $e^2/\hbar c$ is dimensionless, and for low-frequency processes no pairs are going to be created, so no masses enter. Thus we might well ask "low-frequency relative to what?" However, there is a hidden energy scale, in that the theory is actually not convergent, and so some kind of renormalization trick must be employed. One introduces a cutoff of all perturbation integrals at some very high cutoff energy Λ to handle ultraviolet divergences, and then also carries out a <u>renormalization:</u> recognizing that the "bare" quantities in the original Hamiltonian before all the high-energy processes are taken into account are not observable, one defines the "physical" quantities such as mass, charge, and wave function to be those at a specified energy μ or "subtraction point." Either μ or Λ set an energy or length scale for the problem (since we do not tamper with the velocity of light c, or with h, energy $\sim \omega \sim \ell^{-1}$.

Since electrodynamics is "renormalizable," the physical results at some energy ω depend only on the physical charge e and on the ratio of ω to the physical scale of energy μ, and are thus independent of Λ; as Λ changes we change the bare charge in such a way as to ensure this. This invariance of physical results to a group of transformations of Λ and e is the central idea of the "renormalization group." Since Λ also really sets the scale of energy, one may follow this group transformation with a scale transformation, moving Λ, μ, and ω all together. Combining these two transformations, we can see that processes at two different values of ω/Λ describe the same physics: we can shift ω about at will, but only if we also use a different coupling constant and a new scale of energy. A difficult process can be calculated at $\omega/\Lambda \approx 1$, where perturbation theory works, and only then transformed down to low frequency. Hence a combination of purely trivial scaling and the change in coupling constant gives the correct frequency dependence.

For various reasons, this idea languished in field theory for a number of years, although a great many formal results that are now very useful in many-body applications were gradually developed. Among these, the idea of the "fixed point," a point where the dimensionless coupling constant is independent of Λ and thus does not change under further scaling, deserves special mention, as does the idea of the "Lie differential equation," which controls the scaling relation among Λ, μ, and the coupling constant. The idea of the RNG is now very much more alive, both because of support from high-energy experiments of various sorts and for theoretical reasons; but these developments are not the business of this book. (It is interesting as an aside, however, that there is a remarkably close relationship between the concepts of the field-theoretic renormalization group and some of the new many-body applications we shall talk about in Sec. 5.C.)

The original field-theoretic RNG is basically too restrictive in practice, if not in principle, for our use. The difficulty is twofold: one is faced with both the relativistic prohibition on decoupling space and time, which severely restricts the allowable transformations, and the prejudice that field theories must be simple, i.e., that one must scale only between formally equivalent field theories with, say, only different values of $e^2/\hbar c$. The essential generalization that offered a way out of this problem was the contribution of K. G. Wilson--the idea of scaling the whole Hamiltonian containing a multiplicity of coupling constants, i.e., of proving the essential equivalence of problems of quite different structure (Wilson, 1971a,b), permitting, for instance, either a more complicated Hamiltonian to arise from a simple one, or vice versa (which, we shall see, is very common in systems with "universality" properties).

Such a theory is especially useful in the many-body problem, where we are usually most interested in the low-frequency, long-wavelength behavior of a system, i.e., its macroscopic response functions. As we implied in Chapter 3, what is needed is an effective Hamiltonian, or an effective theory of some kind, that will tell us something about a system's macroscopic responses. We are happy enough to scale out the irrelevant high-frequency, short-range correlations. Thus in general we look for "infrared-stable" fixed points, behavior characteristic of long times and large distances, that ceases to change as we look at it at such large scales.

The primary application of the RNG in many-body physics has been to the problem of phase transitions and critical points. Here a second vital contribution was made by Wilson--the idea that the scaling process leads in the "infrared limit" (actually the long-wavelength, large space-scale limit in the case of equilibrium statistical mechanics) to a special kind of fixed point when the system is precisely <u>at</u> a critical point. The properties of this fixed point determine the famous critical exponents which characterize the fluctuations at the critical point. We shall describe this "core" RNG Wilson-Fisher theory (Wilson and Kogut, 1974) in the next section. This theory has been extended to non-equilibrium processes by Hohenberg and Halperin (Hohenberg and Halperin, 1977), but, in keeping with our unjustifiable omission of most transport theory here, we shall not discuss that extension.

The Wilson theory was not chronologically the first application of scaling theory to many-body physics. Of course, that theory was foreshadowed by the scaling ideas of Widom (1965) and Kadanoff (1966), but I mean to refer to a totally different type of problem in which the scaling idea provides the key to a full solution, the first of these being the Kondo problem, the second that of the one-dimensional metal. Anderson, Yuval, and Hamann showed in 1969 (Anderson <u>et al</u>., 1970; the paper was delayed by referees) that the ferromagnetic Kondo problem leads to what would now be called an infrared-stable fixed point, and that the antiferromagnetic one can be solved, even though it corresponds to an infrared-unstable fixed point, by an indirect method based on scaling. These methods are also applicable to an anomalous one-dimensional classical thermodynamic problem with long-range interactions (Anderson and Yuval, 1970), to one-dimensional many-body theories (Fowler and Zawadowski, 1971; Zawadowski, 1973), and to certain two-dimensional long-range problems as well (Kosterlitz and Thouless, 1973). These developments will be described in Sec. C, not only because of their own intrinsic interest as physics, but because they seem to typify a qualitatively different behavior that may be more general than critical phenomena, and thus they suggest an opening up of many further possibilities inherent in the RNG.

Neither the Kondo problem nor that of the critical point would justify including an RNG chapter in this book. After all, materials are almost never near their critical points, and, if one takes the Wilson point of view, scaling to a trivial region always occurs even in that rare case. Moreover, Kondo and even one-dimensional metal problems are relatively esoteric questions. But it seems to me that both of these examples point to a more fundamental reason for understanding the RNG: that it affords at least one way of putting mathematical teeth into the basic concept of continuation we talked about in Chapter 3, and perhaps it is even the most fundamental way of doing so. For example, what the RNG idea has to say about the Fermi-liquid regime could be more complete and more basic than Fermi-liquid theory itself: it can include the Leggett (1965a,b) generalization to superconductivity and superfluidity, for instance. From this point of view we are more interested in the various types of "trivial" fixed points to which systems scale than in the "nontrivial" critical points, since the former classify the various physical regimes or phases which are possible. In Sec. 5.E, we shall state this point of view without proof, and also point out some historical first steps in this direction.

B. RNG THEORY OF CRITICAL POINTS

As we already emphasized in Chapter 2, phases of different symmetry cannot be continued into each other, and must be separated by some kind of nonanalytic boundary. Even where difference of symmetry is not obvious, one can encounter a difference in physical character that suffices to destroy continuity. It is not possible for phase separation to occur continuously, for instance, from gas to gas-liquid mixture; a macroscopically inhomogeneous system is really of different symmetry. In random dynamic systems we can see yet another type of critical point, which is a change in the physical nature of the system rather than in symmetry, i.e., localization or percolation (see, for example, Thouless, 1979). As we remarked, the overwhelming majority of such transitions are first order and not very interesting in themselves; but either by symmetry or by adjustment of parameters we can often so arrange things as to pass directly through the continuous but nonanalytic phase boundary or critical point.

A simple theory of such a phase boundary can easily be produced--the so-called mean-field theory (MFT), of which the Weiss molecular field theory of ferromagnetism was the first example, or perhaps van der Waals' equation of state for the liquid-gas transition. To pick a specific physical example, let us use the simple ferroelectric model, a macroscopic version of which I already used as one of the first examples in Chapter 2. (See also Anderson, 1960.) This has a scalar polarization p_i at each site i which, in a microscopic model, experiences the following interactions:

(1) A harmonic restoring force

$$E_H = \frac{K}{2} \sum_i p_i^2 ; \tag{5.1}$$

(2) A molecular-field interaction that favors ferroelectric polarization

$$\begin{aligned} E_{LF} &= -\frac{1}{2} \sum_{ij} L_{ij} p_i p_j \\ &= -\frac{1}{2} \sum_k L(k) p_k p_{-k} \end{aligned} \tag{5.2}$$

p_k being the Fourier transform of p_i

$$\frac{1}{\sqrt{N}} \sum_i \exp(ik\cdot R_i) p_i = p_k \tag{5.3}$$

(we shall assume that $L(k = 0)$ to be the largest local field coefficient); and

(3) An anharmonic restoring force that restricts the total fluctuations

$$E_{AH} = \frac{B}{2} \sum_i p_i^4 = \frac{B}{2N} \sum_{k_1+k_2+k_3+k_4 = 0} p_{k_1} p_{k_2} p_{k_3} p_{k_4} . \quad (5.4)$$

At absolute zero it is easy to see that this model will polarize if L(0) > K. The energy is minimized by

$$p_i(T = 0) = \left(\frac{L(0)-K}{2B}\right)^{1/2} . \quad (5.5)$$

At sufficiently high temperatures, on the other hand, it is pretty clear that the positive anharmonic term (5.4) will dominate both L and K and the system will be paraelectric $\langle p_i \rangle = 0$, $\langle Bp_i^4 \rangle \approx T$. Thus a symmetry change takes place and, as we pointed out in Chapter 2, a singularity must intervene between T = 0 and $T \to \infty$.

We are interested mainly in the k = 0 component p_0, related to the total polarization P_0 by $P_0 = \sqrt{N}\, p_0$. The inverse susceptibility is given by

$$\chi_0^{-1} = 2 \frac{\delta F}{\delta(P_0^2)}, \quad (5.6)$$

where

$$-\beta F(P_0) = \ell n \langle \exp -\beta\left(E_H + E_{LF} + E_{AH}\right) \rangle_{P_0} , \quad (5.7)$$

the average being taken with P_0 fixed. The simplest approximation is quasiharmonic, in which we simply average the P_0 dependence of the terms in E:

$$\begin{aligned} \frac{1}{2}\chi_0^{-1} = \frac{\delta F}{\delta(P_0^2)} &\approx \frac{1}{2}(K - L_0) + \frac{B}{2} \left\langle \frac{\delta(\Sigma p_i^4)}{\delta(P_0^2)} \right\rangle \\ &= \frac{1}{2}\left(K - L_0 + \frac{B}{N} \sum_{k \neq 0} \langle p_k p_{-k} \rangle + \frac{2B}{N} \langle p_0^2 \rangle\right) \\ &\approx \frac{1}{2}\left(K - L_0 + 3B \langle p_i^2 \rangle\right). \end{aligned} \quad (5.8)$$

Now $\langle p_i^2 \rangle$ is a regular function which decreases as temperature increases, and at some critical T_c

$$K - L_0 + 3B \langle p_i^2(T_c) \rangle = 0. \tag{5.9}$$

Above this T_c,

$$\chi_0^{-1} \simeq A\left(T - T_c\right) \tag{5.10}$$

where A is a constant, and below it P_0^2 takes on a finite value, at first linearly proportional to $T_c - T_0$. We have already discussed the corresponding typical mean-field behavior in Chapter 2. It is essentially contained in the resulting free energy as a function of P_0,

$$F_{MF} = \frac{A(T)P_0^2}{2} + \frac{B}{2N} P_0^4 \tag{5.11}$$

where

$$A(T) = K - L_0 + B \langle p_i^2(T) \rangle \tag{5.12}$$
$$\sim \left(T - T_c\right).$$

To pursue this regime a little further, let us look at the temperature-dependent fluctuation $\langle p_i^2(T) \rangle$ in detail. In principle, we could define a free energy keeping any of the p_k's fixed, which would define a susceptibility for each Fourier component p_k:

$$2 \frac{\delta F}{\delta p_k^2} = \left(\chi(k)\right)^{-1}. \tag{5.13}$$

Gibbs fluctuation theory tells us that

$$\frac{\langle p_k^2 \rangle}{2\chi(k)} = \frac{T}{2}, \tag{5.14}$$

i.e., the probability of a fluctuation is proportional to $e^{-\beta F}$. (We have set Boltzmann's constant equal to unity.) Within the same theory, we may calculate $\chi(k)$,

$$\chi^{-1}(k) = K - L(k) + 3B\langle p_i^2\rangle$$
$$= \chi_0^{-1} - \big(L(k) - L(0)\big) \tag{5.15}$$
$$\approx \chi_0^{-1} + \alpha k^2$$

when k is sufficiently small that L(k) may be expanded in a power series. Inserting this into Eq. (5.14) leads to the well-known Ornstein-Zernike form of the fluctuations,

$$\langle p_k^2\rangle = \frac{T}{A\big(T - T_c\big) + \alpha k^2} . \tag{5.16}$$

It is interesting to transform Eq. (5.16) into real space, where it gives a well-known form for the correlation of the order parameter p at long range,

$$\langle p(r)p(0)\rangle \propto \frac{1}{r}\exp\{-[\frac{A(T - T_c)}{\alpha}]^{1/2} r\}. \tag{5.17}$$

The coefficient in the exponent is often called the inverse range of order ξ^{-1}. ξ diverges as $T \to T_c$. If we are discussing a three-dimensional system, it is apparently perfectly legitimate to make Eqs. (5.8) and (5.16) self-consistent with each other, and hence to calculate the coefficient A of Eq. (5.10),

$$\langle p^2(T)\rangle = \text{const} \times \int \frac{k^2 dk\, T}{A\big(T - T_c\big) + \alpha k^2} . \tag{5.18}$$

This integral is convergent at all $T \geq T_c$, depending primarily upon the upper cutoff because of the k^2 convergence factor; and yet, in fact, the contribution of the long wavelengths leads to a sufficient singularity at criticality (which goes as $\tau^{1/2} = \big(T - T_c/T\big)^{1/2}$), to change the nature of the critical point. (The result is the same as that of the so-called "Gaussian model"--a third-order rather than second-order transition.)

What we have outlined here are the first stages in the process of deriving an effective "Ginsburg-Landau free energy" for this particular system. It is not at all clear why the concept is thus named. The Ginsburg-Landau paper (1950) does nothing at all with it which was not in Landau's general theory of phase transitions, and in fact Gibbs introduced the fluctuation expression equivalent to Eq. (5.14). The idea of eliminating short-wavelength modes is, incidentally, in my own "soft mode" paper given in the Soviet Union in 1958 (Anderson, 1960); it has nothing to do with Ginsburg-Landau and was formalized by Kadanoff in 1966 (Kadanoff, 1966). I shall refer to such an expression, where I use it, as a "Landau" free energy, since "Gibbs" is reserved to the macroscopic case. The Landau free energy, then, is a free-energy functional

$$F_L(p_k) = F(p(r)) \tag{5.19}$$

of an assumed slowly varying, continuous order-parameter function p, either in space or Fourier representation. F will in general have the same nature as the unperturbed energies (5.1) - (5.4),

$$\begin{aligned} F &= F_0 + F_{L-F} + F_{A-H} \\ &= \frac{A(T)}{2} p^2 + \frac{\alpha}{2} (\nabla p)^2 + \frac{B}{2} p^4 \quad , \end{aligned} \tag{5.20}$$

but it will contain a temperature dependence because it is assumed that all short-range, large-k thermal fluctuations have been integrated out. One may even develop a perturbation-theoretic diagram series for doing so, in our case simply a power-series expansion for $\delta F/\delta p_k^2$ in B, the anharmonicity coefficient, assuming the individual self-consistent $\langle p_k^2 \rangle$ as effective propagators. Diagrams such as those in Fig. 5-1 will contribute. The one we have already considered, Fig. 5-1a, gives

$$3B \sum_k \langle p_k^2 \rangle \quad ,$$

the next higher order, Fig. 5-1b, gives

$$\text{const} \times B^2 \sum_{k_1+k_2+k_3=0} \langle p_{k_1}^2 \rangle \langle p_{k_2}^2 \rangle \langle p_{k_3}^2 \rangle , \tag{5.21}$$

etc. When $T \neq T_c$ the Ornstein-Zernike form for $\langle p_k{}^2 \rangle$ causes no difficulty, and the individual terms of the series, at least, are convergent; in that case, the theory does, presumably correctly, lead to Gibbs fluctuation theory for the quasimacroscopic variable p(r). But as $T \to T_c$ it is

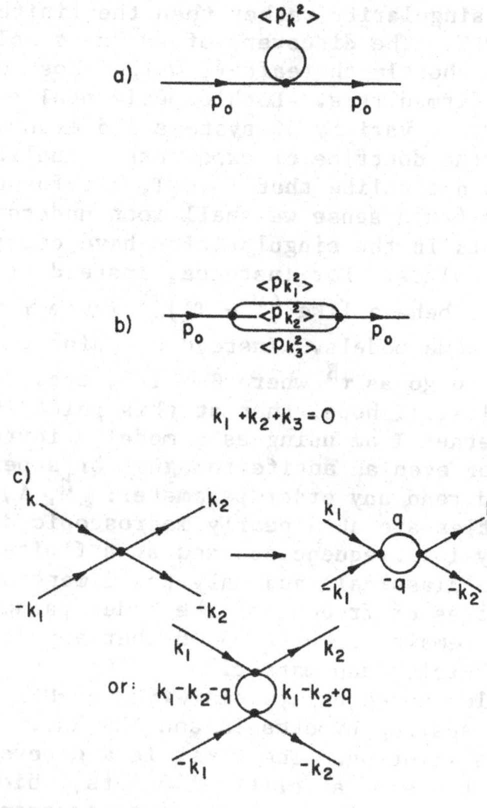

Figure 5-1 Diagrams for ferroelectric model.

easily seen that the terms diverge more and more severely as their order increases: with each additional vertex, there are two extra $\langle p^2 \rangle$ which can diverge as $k \to 0$, and only one extra integral $\int k^{d-1} dk$. (See Fig. 5-1c.) Thus our difficulties with the lowest term are compounded in higher order, and we must conclude that as $T \to T_c$ Gibbs fluctuation theory (or, as Wilson calls it, the Landau theory) fails, at least for all dimensionalities $d \leq 4$. For $d > 4$ the above theory converges and mean-field theory is valid, a result known to many of us after 1960 or so, but formalized by Patashinskii and Pokrovskii (1964).

The recognition of this theoretical difficulty gradually emerged from a combination of experimental and theoretical information. Perhaps earliest to question mean-field theory were Kramers and Wannier (1941), who showed that the two-dimensional Ising model had a logarithmic specific-heat singularity rather than the finite jump predicted in MFT. The discovery of an exact solution by Onsager (1944) shortly thereafter, with a host of non-MFT behaviors, confirmed this. Both experimental and numerical studies of a great variety of systems led eventually to the acceptance of the doctrine of exponents: qualitatively the singularity is not unlike that in MFT, but for each general type of system (in a sense we shall soon understand) the actual exponents in the singularities have certain unexpected but universal values. For instance, instead of changing like τ^{-1}, χ tends to behave like $(T - T_c)^{-\gamma}$ where γ tends to be about 4/3 for some models. Instead of going to zero as $\tau^{1/2}$, p tends to go as τ^{β} where $\beta \sim 1/3$, etc.

(It is clear, I hope, that at this point it is quite irrelevant whether I am using as a model a ferroelectric, ferromagnet, or even an antiferromagnet or superconductor. For p we could read any order parameter: $\vec{M}$, Δ, etc. The real difficulties are at a nearly macroscopic distance scale, at very low frequencies, and at a finite T. Thus the statistics are classical, and only the dimensionality and number of degrees of freedom of the order parameter are important; it remains to be seen whether any detailed aspects of the anharmonicity can matter.)

The problem posed by systems with non-MFT behavior was solved by the scaling hypothesis and the RNG. There are two aspects to the solution. The first is a general theory of exponents and behavior at critical points, which is extremely illuminating in general terms as to why exponents occur, and as to the nature of the phase boundary. We shall emphasize this aspect here. The second is the actual calculation of critical exponents, by expansions or numerically. This is remarkable for the mathematical ingenuity involved, and was initially vital from a pedagogical point of view, to make it clear that the RNG and scaling really <u>work</u>, but is otherwise of real interest only to specialists, and we shall omit it entirely.

Again, let us have recourse to the "ferroelectric"

model. The basic idea of scaling is simply to eliminate progressively the short-wavelength, microscopic degrees of freedom, not all at once as we tried to do--and failed--in the earlier part of this section, but continuously, and to try to develop a different effective energy, or Landau free energy, appropriate for use at each different scale.

We imagine our polarizations p_i on each site i, with their Fourier-transform variables as in Eq. (5.3):

$$p_k = \frac{1}{N^{1/2}} \sum_i^N e^{ik\cdot R_i} p_i \,.$$

Let d be the dimensionality of our system. We can imagine two possible ways of making a scale transformation. In the first, we simply take groups of neighboring sites in blocks of side b (say, b = 2) and define

$$p'_j = \frac{1}{b^{d/2}} \sum_i^{b^d} p_i \,, \tag{5.22}$$

the sum running over the b^d (say, 2^d) sites of the group. We have $b^d - 1$ additional internal variables, say ($\alpha = 1 .. b^d-1$),

$$p_\alpha = p_i - \frac{p'_j}{b^{d/2}} \,,$$

over which it is now our intention to average (see Fig. 5-2).

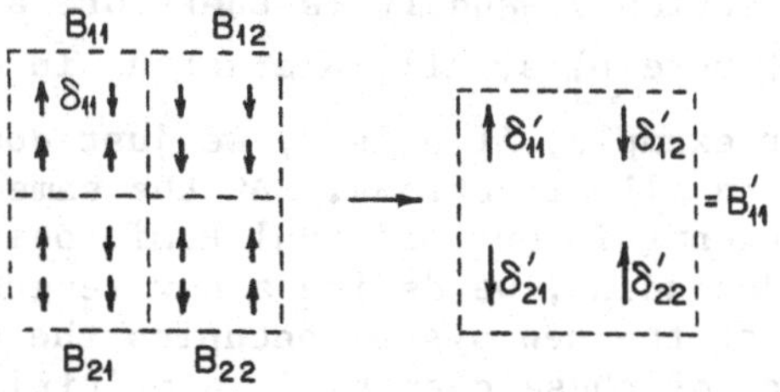

Figure 5-2 Block-spin process.

The free energy is as given in Eq. (5.7),

$$-\beta F = \ell n\langle \exp - \beta\left(E_H(p_i) + E_{LF}(p_i) + E_{AH}(p_i)\right)\rangle ,$$

and in RNG practice it is usual to define a dimensionless "Hamiltonian"

$$H = \beta E(p_i). \tag{5.23}$$

The first step of scaling is to do the average in Eq. (5.7) in two stages:

$$\begin{aligned} -\beta F &= \ell n\langle\langle \exp - \beta E(p_i)\rangle_{p_\alpha}\rangle_{p_j}, \\ &= \ell n\langle \exp - E'(p_{j'})\rangle_{p_j'} , \end{aligned} \tag{5.24}$$

defining $E'(p_j')$ as the average over the internal variables. The calculation of this average is, of course, extraordinarily difficult, but can in special cases be approximately done. $E'(p_j')$ is, however, not a suitable candidate for the scaled "Hamiltonian" H'. It does not refer to the same number of variables as H, because we have eliminated $N\left(1 - \frac{1}{b^d}\right)$ of them. Let us expand the size of the system before carrying out the next stage. We recognize that H' will certainly contain only local interactions like $p_j' p_{j+\tau}'$ where τ is a reasonably small number, since the individual moments summed to make p_j' interacted only locally. Thus the system is still in the thermodynamic limit (size large compared to range of interactions), and it is therefore allowable to add on $\left(N(1 - \frac{1}{b^d})\right)$ more p_j''s, all interacting in exactly the same way. For example, if b is 2, we just double the size of the system in all directions. At the same time, if any of the coefficients in our original Hamiltonian were explicitly r dependent, we define a new length variable $r' = r/b$ so that the new system occupies the same space. The only effect of these changes is a multiplicative factor in the definition of the actual free energy of the original system: from Eq. (5.24),

$$-\beta F = \ln\langle \exp - H'(p_j')\rangle_{\frac{Np_j' \text{'s}}{b^d}}$$

$$= \frac{1}{b^d} \cdot \left(-\beta F(H')_{Np_j' \text{'s}}\right) \; . \tag{5.25}$$

We have thus defined a transformation that is our fundamental group operator (actually a semigroup):

$$H \to H' \tag{5.26}$$

$$F(H) = \frac{1}{b^d} F(H') \; . \tag{5.27}$$

In this way we have developed a way of performing a unidirectional mapping in "Hamiltonian space," which we can in principle repeat indefinitely until we get to the macroscopic scale. Since at every stage the more complicated, short-range correlations are being eliminated, we would expect this mapping to go, at least after the first few stages, from more complicated to simpler H's, from special to universal. This turns out, in fact, to be always the case, but the way in which simplification occurs is qualitatively different at critical points from elsewhere.

Let me just indicate, so that it will not disturb the reader too much when I scale with differential rather than difference equations, the other method of scaling: in k-space rather than real space. Quite clearly, it is possible to start with $H(p_k)$ defined as a function of $N = (2k_m)^d$ p_k's, and eliminate a shell of p_k's defined by

$$\begin{aligned} k_m - dk_m < k_x &< k_m \\ k_y & \\ k_z & \end{aligned} \tag{5.28}$$

which removes

$$\frac{3dk_m}{k_m} \cdot N \text{ degrees of freedom.}$$

We can then rescale the system up to N degrees of freedom again and start over with the new Hamiltonian $H'(p_k)$.

Quite clearly, what we hope will happen is that after we have carried out either of these scaling procedures enough times, $H \to H'$, we will reach an invariant effective Hamiltonian, a fixed point of the sort we described earlier. In fact, this is sometimes too much to hope for; but what is always found is that as we carry out the transformation $H \to H' \to H'' \to$ etc., successive stages lead to similiar H's, Hamiltonians which can be scaled into each other by scaling the magnitudes of H itself and of the polarization variables, so that after appropriate rescaling (called in the field-theory analog "assigning anomalous dimensions"), we do aim at an invariant Hamiltonian H*. This is a general feature of such mapping transformations within a multidimensional space such as our "Hamiltonian space." The initial Hamiltonian H of Eq. (5.23) contains three parameters, $K - L_0$, α, and B essentially (called r, Δ, and u in the standard texts), if one wishes to ignore details of the local fields. In the course of scaling, a great many more parameters can and will appear--n^{th} order anharmonicities, with $n \geq 3$; nonlocal fourth-order anharmonicity, etc.--so that the actual Hamiltonian space is essentially an infinite-dimensional one. Nonetheless it is a general characteristic of mappings in such a space that they converge towards certain fixed points, points at which

$$\Delta H/\Delta(\ln b) = \frac{H' - H}{\Delta k_m/k_m} = 0 \ . \tag{5.29}$$

Let us first look at an ordinary point, the kind of point where simple Gibbs-Landau fluctuation theory is valid. Assuming we are above T_c, there will be some microscopic scale at which there is no longer any appreciable modification due to averaging over the internal variables, and at which the Landau free energy has become constant. In order to show that this can be possible, let us carry out the RNG in this trial case explicitly. We have to assume that the anharmonic terms are now unimportant, and show at least that that assumption can be self-consistent. Thus we start by assuming that, beyond this scale, H expressed in terms of p_k is constant and equal to

$$H(p_k) = \sum_k K_{eff}(k)p_k p_{-k}. \tag{5.30}$$

We may transform this back to real space at two different scales simply by defining

$$p_i = \frac{1}{\sqrt{N}}\sum_k^N p_k\, e^{ik\cdot R_i} \tag{5.31}$$

and

$$p_j' = \frac{1}{\sqrt{\frac{N}{b^d}}}\sum_k^{N/b^d} p_k e^{ik\cdot R_j} \tag{5.32}$$

where the number of independent R_j is, of course, only $1/b^d$ of the number of R_i. Note that indeed

$$\sum_{i=1}^{b^d} p_i = \frac{1}{\sqrt{N}}\sum_k p_k \sum_{i=1}^{b^d} e^{ik\cdot R_i} \tag{5.33}$$

$$\simeq \frac{b^d}{\sqrt{N}}\sum_k^{N/b^d} p_k e^{ik\cdot R_j} = \sqrt{b^d}\; p_j' \;.$$

Inverting Eqs. (5.31) and (5.32), we have

$$p_k = \frac{1}{\sqrt{N}}\sum_i^N e^{-ik\cdot R_i} p_i; \qquad p_k = \frac{b^{d/2}}{\sqrt{N}}\sum_{j=1}^{N/b^d} p_j'\, e^{-ik\cdot R_j} \tag{5.34}$$

so that

$$H(p_i) = \sum_{i,j} p_i p_j \frac{1}{N} \sum_k^N \exp[-ik\cdot(R_i - R_j)] K_{eff}(k)$$

$$= \sum_{i,j} p_i p_j V(R_i - R_j)$$

$$H(p_j') = \sum_{i,j}^{N/b^d} p_i' p_j' \frac{b^d}{N} \sum_k^{N/b^d} \exp[-ik\cdot(R_i - R_j)] K_{eff}(k)$$

$$= H(p_i) \ , \tag{5.35}$$

and we do indeed have a true fixed point. It is easy to see that, as b gets large, Eq. (5.33) implies that any quartic anharmonicity would rapidly disappear as b^{-d}. In this case the correct scaling of the polarization variable p is given without "anomalous dimensions" by Eq. (5.22). All of this is the obvious consequence of the fact that, if Eq. (5.10) holds, far enough from T_c the contributions from increasingly small k to fluctuation integrals such as (5.18) and (5.21) are increasingly negligible. In this kind of regime the RNG leads to a trivial result which might seem much more easily found otherwise. The point, however, is that the RNG theory can give another answer; and thus it can tell us when this trivial case breaks down.

Let us now postulate that there may exist some more general fixed point of the mapping

$$H \rightarrow H'.$$

It is most convenient either to use the dk_m momentum-cutoff scheme, or to subdivide the b scaling, so that we can consider an infinitesimal scaling transformation

$$b = 1 + \varepsilon$$

$$\varepsilon = d(\ln a) = -d(\ln k_m) \ , \tag{5.36}$$

where a is some length scale, i.e., a transformation in which the log of the scaling length changes by a small quantity ε. A fixed point is defined by

$$\frac{H' - H}{\varepsilon} = \frac{dH}{d\,\ell n\ a} = 0\ . \tag{5.37}$$

But of course, in general, our given physical Hamiltonian H will not be so convenient as to correspond precisely to the fixed-point Hamiltonian H*. It will differ in a number of ways. Therefore we must write

$$H = H^* + \delta H \tag{5.38}$$

and we must inquire how δH evolves as we carry out our mapping transformation. To elucidate the behavior it is adequate to assume that eventually the mapping approaches H* sufficiently well that a linear approximation in the small quantity ε may be used:

$$\delta H' - \delta H = \varepsilon \cdot \underset{\sim}{Y}\delta H \tag{5.39}$$

where $\underset{\sim}{Y}$ is a linear operator in δH space. An important role is played by the eigenvalues y_i and eigenoperators O_i of the matrix $\underset{\sim}{Y}$,

$$\underset{\sim}{Y}O_i = y_i O_i\ , \tag{5.40}$$

because when we expand H in terms of these eigenoperators,

$$\delta H = \Sigma \mu_i O_i\ , \tag{5.41}$$

we can follow the scaling process in terms of the expansion coefficients μ_i:

$$\frac{d\mu_i}{d\,\ell n\ a} = y_i \mu_i \tag{5.42}$$

or

$$\mu_i = a^{y_i}\ . \tag{5.43}$$

We immediately see that a crucial distinction between different possible components O_i of δH may be drawn, depending on the sign of y_i. All operators for which $y_i < 0$ are called <u>irrelevant</u> and disappear in the scaling process:

$$y_i < 0 \qquad O_i \text{ is irrelevant,}$$
$$\mu_i \propto a^{-|y_i|} \to 0 \quad \text{as} \quad a \to \infty \; . \tag{5.44}$$

The basic fact of universality--that all ferroelectrics behave the same, that exponents are the same for widely differing systems--shows that the great majority of all O_i are irrelevant. This seems to be not particularly obvious, but necessary for our conventional view of thermodynamically stable systems: that almost everywhere, except very near a critical point, thermodynamic fluctuations occur only at a microscopic scale. Thus the stable case of Gibbs fluctuation theory [Eqs. (5.30)-(5.35)] is the normal one. It is only at the boundary between two different stable, Gibbsian fixed points that the other case occurs, and it is not surprising that only one, or at most a very few, parameters of H exhibit instability.

This simplicity of thermodynamic fluctuation theory contrasts with the complexity being revealed by recent work in the case of dynamical, driven systems; the thermodynamic systems have only "fixed points" and not more complicated "attractors" such as "limit cycles."

Some of these O_i, of course, can be physically relevant operators, for which Eq. (5.43) gives a scaling law; of this more later.

Much more important are those few operators O_i which are or can be relevant:

$$y_i > 0 \quad O_i \text{ relevant.} \tag{5.45}$$

Clearly, if too many relevant operators exist, we shall never encounter this particular fixed point, since perturbations about it will grow unstably. But there are a number of relevant operators which can and must exist, on physical grounds, at any critical point. These have two functions, to set the scaling laws and to determine whether the system is or is not critical.

The first of these sets the scale for p_i. In our naive scaling, we have assumed Eq. (5.22), which is equivalent to a scaling law,

$$\frac{d \ln p}{d \ln a} = \frac{d}{2} \; .$$

This is not necessarily correct; it may well be that the transformation $H \to H'$ is a similarity transformation rather than an identity, i.e.,

$$H'(p_i) = H(\lambda p_i) \quad ,$$

where λ is some scale other than $b^{d/2}$. The appropriate operator O_p to express such a scale modification is

$$O_p = \frac{d}{d\lambda}\left(H(\lambda p_i)\right)\Big|_{\lambda=1}$$

$$= \sum_i p_i \frac{\partial H}{\partial p_i} \quad . \qquad (5.46)$$

This operator has a simple physical meaning, expressed by the vector

$$\vec{\nabla}_{p_i} H = \vec{h} \qquad (5.47)$$

where $\vec{h}$ is an appropriate field of force acting on the total polarization. That is, if we add a force term to H,

$$H \to H - \vec{h} \cdot \vec{p}_{tot} \quad , \qquad (5.48)$$

the field is determined by

$$\left\langle \frac{\partial \ln Z}{\partial \vec{p}_{tot}} \right\rangle = 0$$

$$= \vec{h} - \left\langle \frac{\partial H}{\partial \vec{p}_{tot}} \right\rangle$$

$$= \vec{h} - \left\langle \sum_i \frac{\partial H}{\partial p_i} \right\rangle \quad . \qquad (5.49)$$

Thus the operator (5.46) is essentially $\vec{h} \cdot \vec{p}_{tot}$. In order to keep track of the total scale of H, it is convenient to add the "counterterm" $\vec{h} \cdot \vec{p}_{tot}$ as in (5.48) and to adjust the scaling of the moment $p_i \to p_j'$ according to the exponent for this term. In other words, if

$$\frac{d \ln(\vec{h}\cdot\vec{p}_{tot})}{d \ln a} = y_h ,$$

the scaling of polarization p must be taken to be, instead of Eq. (5.22),

$$p_j' = b^{-y_h} \sum_i p_i . \tag{5.50}$$

This will now be our scaling assumption for p. Then, in the presence of the field term, the identity (5.49) will be maintained as we scale. The exponent y_h is, in terms of the traditional exponents of the critical-point problem, $(d/2 + 1 - \eta/2) > 0$. (η is usually fairly small.) It may easily be shown that this is the characteristic exponent of the correlation function in space at criticality:

$$\langle p_i \cdot p_j \rangle = \frac{\text{const}}{R_{ij}^{2y_h}} = \frac{\text{const}}{R_{ij}^{d-2+\eta}}$$

$$\langle p^2(k) \rangle = \frac{\text{const}}{k^{2-\eta}} . \tag{5.51}$$

A second type of relevant variable indicates those physical parameters which must be adjusted in order to make the system reach the unstable fixed point characteristic of critical behavior. We know that, typically, the temperature must be such a variable. To see this, we note that it is essentially the definition of a critical point that a singular behavior will take place with scaling, and that this will occur at a unique T. If there are no symmetry-conserving relevant operators, we are stuck with the regular, Gibbsian fixed-point type of behavior of Eqs. (5.30)-(5.35) [of course, a symmetry-breaking field term as in (5.48) is always relevant, since it changes the character of the effective Hamiltonian].

Let us call this symmetry-preserving relevant operator

O_E. Our initial Hamiltonian, again, may be expanded about the fixed point H*,

$$H = H^* + \mu_E O_E + \text{(irrelevant operators)}. \tag{5.52}$$

It is clear that if $\mu_E \neq 0$, H does not scale to H* because the second term changes as a^{y_E}, where y_E is the exponent solving Eq. (5.40) for O_E, and by assumption $y_E > 0$. Thus at the critical temperature $\mu_E = 0$ and by suitable normalization of O_E we may take

$$\mu_E \approx \frac{T - T_c}{T_c} = \tau. \tag{5.53}$$

Thus, if a is our measure of length scale, we have (adding the field term for completeness)

$$H(a,p) = H^*(p) + \tau a^{y_E} O_E + hp_{tot} a^{y_h} + \left(\sim a^{-y_i}\right) . \tag{5.54}$$

Figure 5-3a helps to visualize how the scaling goes in "Hamiltonian space" as we repeat the scaling process H → H'

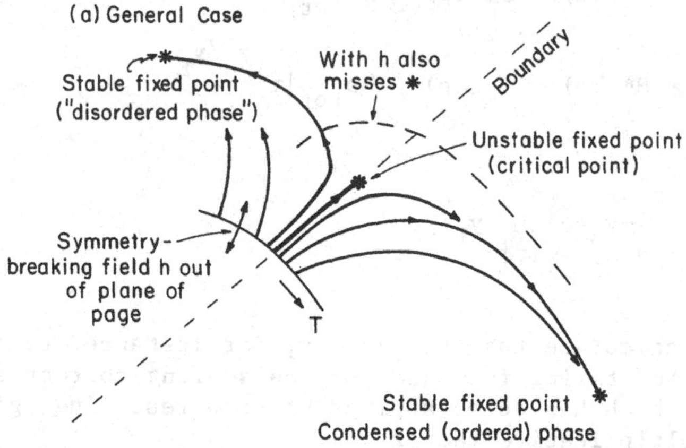

Figure 5-3a Mapping in Hamiltonian space.

→ H″ etc. When h = 0, only one point on the $\mu_E = \tau$ line scales into the critical fixed point H*; all others miss, by very little if $\tau \ll 1$, and eventually arrive at trivial fixed points like Eqs. (5.30)-(5.35). On the other hand, T_c ($\tau = 0$) represents a dividing line. It is the end point of a boundary of the mapping transformation, because in general the trivial fixed point above T_c (which we have discussed) will differ qualitatively from the equally trivial one below T_c, which represents the polarized state and in general has a quite different symmetry from that above. As we shall see, the effective Hamiltonian H^*_{trivial} [which is really just $\frac{1}{2}\Sigma\, K(k)p_k^2$ as in Eq. (5.30), above T_c] is the same in <u>form</u> for all $\tau > 0$, but its <u>coefficients</u> depend nonanalytically in a specific and characteristic way on τ as $\tau \to 0$. This is the characteristic behavior that I believe is generalizable to all broken-symmetry systems.

We see from the figure that, at least for h = 0, systems with different τ's eventually all follow the same trajectory, but at different values of the scale a. As a characteristic point at which the trajectories have adequately converged let us take $|\tau|_{\text{scale}} = 1$, i.e.,

$$|\tau|_{\text{scaled}} = |\tau| a^{y_E} = 1. \tag{5.55}$$

At this point the effective Hamiltonians of all the systems will just be

$$\begin{aligned} H &= H^*(\lambda p) + O_E(\lambda p) + hp_{\text{tot}}a^{y_h} \\ &= H^*(\lambda p) + O_E(\lambda p) + hp_{\text{tot}}|\tau|^{y_h/y_E} \end{aligned} \tag{5.56}$$

with

$$\lambda = a^{-y_h} = |\tau|^{-y_h/y_E}. \tag{5.57}$$

We must calculate the free energy, for instance, using Eq. (5.56), but taking into account the scaling correction (5.27), which has to be applied at each rescaling, giving a total multiplicative factor

$$F = F_{\text{scaled}}\left(\tau,\ h_{\text{scaled}}\right) \times \tau^{-d}$$

$$= \text{const} + |\tau|^{2-\alpha}\ F\left(h|\tau|^{-\Delta}\right) \qquad (5.58)$$

where

$$2 - \alpha = \frac{d}{y_E}$$

$$\Delta = \frac{y_h}{y_E}. \qquad (5.59)$$

Equation (5.58) is the basic homogeneity law of critical-point theory conjectured by Widom, and (5.59) gives explicit values for the two independent exponents of the Widom-Kadanoff laws, as derived from scaling theory. But, of course, Eq. (5.56) is more general, since it allows us to compute all fluctuation phenomena as well as static thermodynamics. As I mentioned earlier, extensions to dynamic phenomena are also possible (Hohenberg and Halperin, 1977).

This, then, is the key concept of the RNG theory of critical points. Figure 5-3b summarizes what, as one can

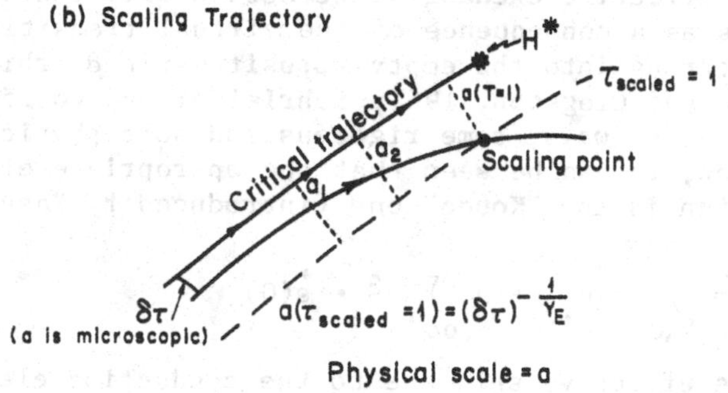

Figure 5-3b Mapping in Hamiltonian space.

see immediately, may be a very useful general way of viewing phase changes. Detailed values of exponents and the fascinating possibilities opened up by extending the number of dimensions in H-space--tricritical points, crossover phenomena, and the like--are really subjects for the specialist. The idea of scaling trajectories, however, while perhaps not of the "rigor and generality of equilibrium statistical mechanics" as once claimed by Fisher, is a core concept that may, in the end, be more widely, if not so rigorously, applicable.

C. KONDOESQUE SCALING THEORY

As I mentioned in the Introduction, scaling as a method in many-body physics first succeeded with an entirely different kind of system from the critical-point one. The original case was the so-called "Kondo problem," a Hamiltonian meant to describe a single additional magnetic impurity with a free spin, embedded in a metal and interacting with the metal electrons via an exchange scattering potential. This rather specialized and unphysical problem seems to contain the essential difficulties of handling the theory of a fairly wide class of magnetic impurities in metals. Characteristic "Kondo behavior" in a metal includes rather large scattering effects at moderate temperatures, anomalous temperature dependence of resistance, magnetoresistance, and thermopower, and often at low temperatures a recession to apparently nonmagnetic but very strong scattering behavior. (Some typical data are shown in Fig. 5-4.)

The physics of this phenomenon are clear enough. A magnetic impurity will normally experience an antiferromagnetic effective exchange interaction with the free electrons as a consequence of the virtual transition of the free electrons into the empty opposite-spin d orbital (Anderson and Clogston, 1961; Schrieffer and Wolff, 1966). In a number of ways, some rigorous and some physically convincing, it can be seen that the appropriate effective Hamiltonian is the "Kondo" one (introduced by Kasuya, 1956),

$$H = \sum_{k,\sigma} \varepsilon_k n_{k\sigma} + J \sum_{\sigma\sigma'} \vec{S} \cdot \vec{s}(0) \tag{5.60}$$

where the effective spin due to the conduction electrons at the impurity site ($\vec{r}=0$) is given by

$$\begin{aligned} \vec{s}(0) &= \frac{1}{N} \sum_{k'k\sigma\sigma'} c^+_{k\sigma} c_{k'\sigma'} (\vec{\sigma})_{\sigma\sigma'} \\ &= \sum_{\sigma\sigma'} \psi^+_\sigma(0) \psi_{\sigma'}(0) \vec{\sigma}_{\sigma\sigma'} \ . \end{aligned} \tag{5.61}$$

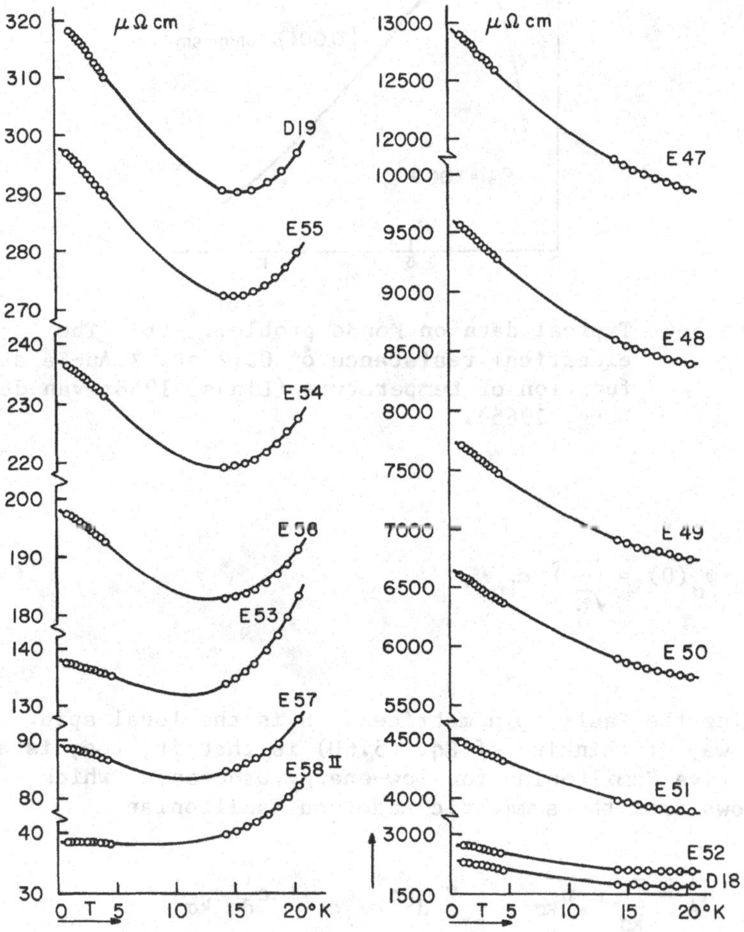

Figure 5-4 Typical data on Kondo problem. (a) ρ vs. T curves for a series of copper-iron wires. The concentration increases from 0.0005 to 0.123 at. % (Knook, 1962; Knook et al., 1964; van den Berg, 1965).

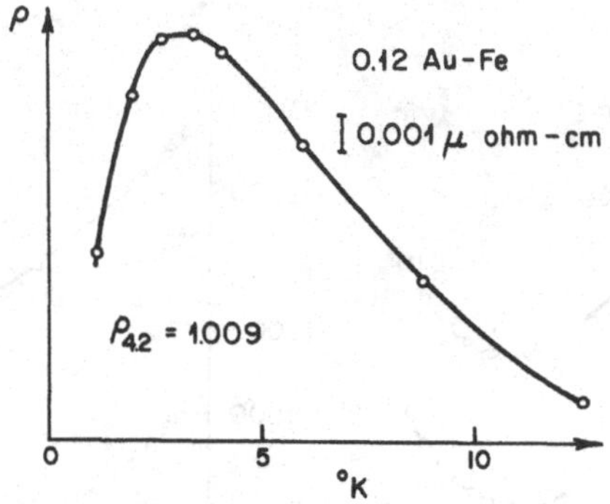

Figure 5-4 Typical data on Kondo problem. (b) The electrical resistance of 0.12 at. % Au-Fe as a function of temperature (Linde, 1958; van den Berg, 1965).

Here

$$\psi_\sigma(0) = \frac{1}{\sqrt{N}} \sum_k c_{k\sigma} , \tag{5.62}$$

$\vec{\sigma}$ being the Pauli spin matrices. S is the local spin. One good way of thinking of Eq. (5.60) is that it, too, is an effective Hamiltonian for low-energy processes, which follows from the symmetric Anderson Hamiltonian

$$H_A = \sum_{k\sigma} \varepsilon_k n_{k\sigma} + \sum_{k\sigma} V_{kd}\left(c^+_{k\sigma} c_{d\sigma} + c^+_{d\sigma} c_{k\sigma}\right)$$

$$+ U n_{d\uparrow} n_{d\downarrow} - \frac{U}{2}\left(n_{d\uparrow} + n_{d\downarrow}\right) \tag{5.63}$$

when we eliminate processes with enough energy (~ U/2) to empty or to fill doubly the local d state. Thus it, too, may be thought of as the outcome of a suitable scaling process. V_{kd} is the transition amplitude between the local d state and the conduction band. More detail is given in the Haldane (1978) article reprinted in this volume. The sum over k in Eq. (5.60) is taken over a band of energies ε_k about the Fermi surface (we set $E_F = 0$), with a cutoff E_c in energy space which, consistently with the source of (5.60) in the Anderson Hamiltonian, is of order U/2, where U is the intra-atomic exchange energy. We may, if we like, think of this cutoff as a form factor for the wave function $\psi(0)$ [Eq. (5.62)] of electrons interacting with the local spin, but that is not its real physical source.

The antiferromagnetic interaction J, taking advantage of the constant density of states ε_k down to $\varepsilon_k = 0$, can form a bound state between the local spin and one made up from the free-electron states, as one may demonstrate variationally if necessary. The binding energy of such a state is of order

$$E_B = E_c e^{-1/J\rho} \tag{5.64}$$

where ρ is the density of states. Thus the ground state is essentially nonmagnetic, and there is no free spin left. On the other hand, when $T > E_B$ there <u>is</u> a free spin, and its magnetic properties include a Curie-Weiss susceptibility and spin-flip scattering. Since the Kondo Hamiltonian is essentially a zero-dimensional one, involving only a single additional entity with a single degree of freedom, it is pretty clear that no phase transition can separate these two regions. The problem of how to connect them is the canonical "Kondo problem," and a few years ago I would have felt, probably with very general agreement, that far too much time and effort had been expended on this apparently physically simple problem of interpolation. It is not now at all clear that such an attitude is correct. "Kondoesque" renormalization has already been established as relevant to a collection of other types of problems, perhaps artificial but entirely physical, of which one or two (especially the Kosterlitz-Thouless two-dimensional dislocation theory and the theory of one-dimensional metals) are promising to lead still further. Finally, it is not unlikely that some otherwise mysterious physical phenomena such as, perhaps, the continuous metal-insulator transitions of certain rare-earth compounds, may be further examples.

We shall depart here almost totally from the historical approach to the Kondo problem. I have described that history elsewhere (Anderson, 1971) as a "Wilderness" (the 1864 Civil

War battle connotation being perhaps closer than the biblical one). The RNG scaling approach is a much simpler, as well as more correct, point of view, and yet it appeared only as the last stage in the solution, rather than the first. (As of 1982, yet another stage, that of "exact" solutions, has been added; "simplification" is not the word for these, and the reader is not directed to this work.) Even in the post-RNG history, after the solution had been obtained by scaling (Anderson and Yuval, 1969; Anderson, Yuval, and Hamann, 1970) other physicists did not accept the final verdict, and it was only after explicit computer calculation of the crucial crossover region by Wilson (1974b, 1975) that the correct picture gained total acceptance.

Like the field-theory original of the RNG, the Kondo Hamiltonian can be made dimensionless and contains no intrinsic scale of energy. This is provided only by the cutoff, which does not appear explicitly in the Hamiltonian (see Anderson and Yuval, 1971). There are three parameters with the dimensions of energy: J, the inverse density of states in energy $\rho^{-1} = \left(\frac{dn}{dE}\right)^{-1}$, and the cutoff E_c. (A fourth at finite temperature is, of course, T.) If we scale the normalization volume Ω, we obtain

$$\Omega \to \lambda\Omega$$

$$\frac{dn}{dE} \to \lambda \frac{dn}{dE} \; ; \quad J \to J/\lambda \tag{5.65}$$

[the last because ψ(0) goes as $\frac{1}{\sqrt{N}}$], so that the dimensionless ratio Jρ of J to the energy-level spacing remains constant, whereas neither J nor ρ alone has any intrinsic meaning independent of sample size and E_c. It is convenient, therefore, to use a normalization volume Ω for which

$$\frac{dn}{dE} = E_c^{-1} = \tau . \tag{5.66}$$

This equation also defines a cutoff time τ, which can be thought of as the time for an electron to pass the local spin. All shorter times are meaningless. Ω is approximately the atomic volume. We have, therefore, two parameters, one dimensionless, $J\rho = J\tau$, and τ itself, the latter setting an atomic time scale; if we wish to calculate some property at a temperature β^{-1}, this must then be a function $f(J\rho, \beta/\tau)$. Of course, such properties as resistivity depend on additional dimensional parameters such as v_F, but usually can be reduced to dimensionless form (such as the generalized scattering phase shift δ in the case of resistance, or the extra entropy in the case of specific heat).

It is not unexpected that many physical parameters depend on the cutoff τ and would diverge if τ were allowed to be zero; this is the conventional "ultraviolet catastrophe" of many field theories. It poses no problem for us because all condensed matter problems contain natural cutoffs at atomic sizes. For instance, the total energy modification due to the spin is, in second order,

$$\Delta E \sim J^2\rho \sim \frac{(J\tau)^2}{\tau} \to \infty \quad \text{as} \quad \tau \to 0. \tag{5.67}$$

But in lowest order the magnetic scattering is just proportional to the square of the scattering phase shift, which is cutoff independent,

$$\sigma \simeq \frac{1}{k_F^2}\sin^2\delta \sim \frac{1}{k_F^2}(J\tau)^2 ,$$

and transport effects, thermal effects, etc., appear to be perfectly well behaved. However, as Kondo showed (1964), in the next higher order of the Born approximation for the spin-flip scattering amplitude one runs into terms like

$$J^2\left(S_zS_+ - S_+S_z\right)\sum_{k''}\frac{f\left(\varepsilon_{k''}\right) - \frac{1}{2}}{\omega - \varepsilon_{k''}} ,$$

(f = Fermi function $\left[\exp(\varepsilon_k/T) + 1\right]^{-1}$) which diverge logarithmically at the upper cutoff E_c and as $T \to 0$, i.e., as $\ln(E_c/T)$. Kondo showed that this leads to a scattering amplitude which goes as

$$\left(f_{kk'}\right)_{Sf} \sim J\rho + (J\rho)^2 \ln\frac{E_c}{T} + \ldots \tag{5.68}$$

and thus increases rapidly at low temperatures, explaining

the well-known "resistance minimum" and other phenomena as shown in Fig. 5-4. The reason for this is obvious with antiferromagnetic exchange--the $S_z s_z$ scattering amplitude is attractive for antiparallel spins, pulling the opposite-spin electrons closer to the impurity. This increases the spin-flip scattering amplitude in the next order; but since the scattering is actually rotationally invariant, it also enhances the original effect leading to an essentially cooperative effect. The successive dots in Eq. (5.68) represent successively higher powers of $(J\rho)\ell n(E_c/T)$, clearly indicating that (5.68) probably fails at an energy of order (5.64).

This kind of divergence should immediately have suggested the use of the RNG, but it was actually not until about 1970 (Anderson, 1970b; Fowler and Zawadowski, 1971; see also Shiba, 1970), that the very simplest interpretation in terms of scaling was given. It seems almost obvious that, for purposes of calculating low-energy phenomena such as transport properties, our main concern should be to eliminate the irrelevant higher-energy scattering processes that are causing the trouble. The "poor man's method" of doing this is simply to eliminate by perturbation theory all scattering processes to energies between the cutoff E_c and some lower cutoff $E_c - dE_c$. The first example of this procedure in many-body theory was the effective potential U of superconductivity theory (Morel and Anderson, 1962). The process of elimination is described in detail in Armytage's thesis (1973) and in the Haldane (1978, 1979) references. We give a brief description here with no details; the reader who is not interested in the formal nature of this method (which is slightly different from the conventional RNG) may skip to the end of the indented section. The equations that result are self-explanatory.

> Physical properties of a system may all be calculated from the "resolvent operator" or full Green's function (as opposed to the one-particle Green's function of conventional theory),
>
> $$G(\omega) = (\omega - H)^{-1} \tag{5.69}$$
>
> [H is the full Hamiltonian (5.60)]. We wish to calculate this in terms of the unperturbed $G_0 = (\omega - H_0)^{-1}$ arising only from the kinetic energy term
>
> $$H_0 = \sum_{k\sigma} \varepsilon_k n_{k\sigma} \; .$$

The equation for G in terms of G_0 defines the scattering matrix T,

$$G(\omega) = G_0(\omega) + G_0\ T\ G_0, \tag{5.70}$$

which obeys Dyson's equation

$$T = V + VG_0\ T, \tag{5.71}$$

V being the scattering term. [Equations (5.70) and (5.71) must be supplemented by information concerning whether the poles of G_0 lie in the upper or lower half of the ω-plane or on the real axis, but that is here irrelevant.]

To scale, let us now suppose that we want to calculate the scattering matrix T only within a subspace of one-electron states of H_0 with a lower cutoff

$$|\varepsilon_k| < E_c - dE_c\ , \tag{5.72}$$

for which we introduce a projection operator Q:

$$\begin{aligned} Q\Psi\big(E_c > |\varepsilon_k| > E_c - dE_c\big) &= 0 \\ &= Q\Psi_{out} \end{aligned} \tag{5.73}$$

$$Q\Psi_{in} = \Psi_{in}\ .$$

We want to find a new interaction $\tilde{V}$ for which Eq. (5.71) holds <u>within the "in" subspace</u> only:

$$QTQ = Q\tilde{V}Q + Q\tilde{V}QG_0QTQ\ , \tag{5.74}$$

whereas the actual projected equation (5.71) is

$$QTQ = QVQ + QVG_0TQ. \tag{5.75}$$

We write

$$
\begin{aligned}
V &= QVQ + (1-Q)V(1-Q) + \Delta V \\
&= QVQ + V_{out} + \Delta V
\end{aligned}
\tag{5.76}
$$

where $\Delta V = QV(1-Q) + (1-Q)VQ$ connects "in" and "out" subspaces. Inserting this in Eq. (5.71) we get

$$
\begin{aligned}
T = {} & QVQ + V_{out} + \Delta V \\
& + (QVQ + V_{out} + \Delta V)G_0 T ,
\end{aligned}
\tag{5.77}
$$

whereas in Eq. (5.75) it gives

$$
QTQ = QVQ + QVQG_0QTQ + Q\Delta VG_0TQ \tag{5.78}
$$

[using $QV_{out} = 0$ and $(1-Q)G_0Q = 0$]. We iterate the last term of Eq. (5.78), using (5.77),

$$
\begin{aligned}
Q\Delta VG_0TQ = {} & Q\Delta VG_0\Delta VQ + Q\Delta VG_0V_{out}G_0TQ \\
& + Q\Delta VG_0\Delta VG_0QTQ
\end{aligned}
\tag{5.79}
$$

and again the middle term of this equation,

$$
\begin{aligned}
Q\Delta VG_0V_{out}G_0TQ = {} & Q\Delta VG_0V_{out}G_0(\Delta VQ) \\
& + Q\Delta VG_0V_{out}G_0(V_{out} + \Delta V)G_0TQ,
\end{aligned}
\tag{5.80}
$$

and again the last ΔV term ends in QTQ, and the middle one must continue to be iterated. Combining Eqs. (5.78), (5.79), and (5.80), we see that we can reproduce (5.74) if we use the following very simple perturbation series for $\tilde{V}$:

$$
\begin{aligned}
Q\tilde{V}Q = {} & QVQ + Q\Delta VG_0\Delta VQ \\
& + Q\Delta VG_0V_{out}G_0\Delta VQ + Q\Delta VG_0V_{out}G_0V_{out}G_0\Delta VQ \\
& + \cdots
\end{aligned}
\tag{5.81}
$$

[In Eqs. (5.79) and (5.80), the first term gives a contribution to the first term of (5.74), the last to the second.] We thus can only use ΔV twice, once in he first scattering from "in" to "out" and once in the last from "out" to "in". The expansion (5.81) is, in this case, an expansion in powers of $J\tau$, and converges well only when $J\tau$ is small; V_{out} is, unfortunately, not necessarily small in any sense, since the electron scattered "out" by ΔV need not be that scattered by V_{out}.

The first two terms of (5.81) may be written

$$\tilde{V}_{kk'} = V_{kk'} + \sum_{E_c > |\varepsilon_{k''}| > E_c - dE_c} (\Delta V)_{kk''} \, G_0(E_c) \, (\Delta V)_{k''k'} \, , \tag{5.82}$$

and the second, like all subsequent terms, is proportional to an integral over dE_c and thus may be written

$$\tilde{V} = V + \frac{dV}{dE_c} dE_c \, ,$$

where

$$\frac{dV}{dE_c} = \Delta V \, G_0(E_c) \, \Delta V + \Delta V G_0(E_c) V_{out} G_0(E_c) \Delta V + \cdots \tag{5.83}$$

Equation (5.83) is the "Lie differential equation," which is central to the conventional field-theory RNG, but we see here a slightly different interpretation. In particular, we suggest that the integration of Eq. (5.83),

$$\tilde{V}(E_c') = V + \int_{E_c}^{E_c'} dE_c \frac{dV}{dE_c} \, , \tag{5.84}$$

gives us a suitable Hamiltonian to be employed at any stage in the scaling process where we may wish to stop. For instance, we clearly do not wish to scale beyond E_c = a few × T, because electrons important in the various transport processes have energies of order T.

One additional formal aspect of the scaling procedure is important. In general, the new V will have a form somewhat similar to the old, and it is the assumption of RNG theory that most possible terms in V are irrelevant variables that disappear if scaling is carried out over a wide enough range. One new term, however, cannot be avoided. This is the possible existence of a constant in V. Such a constant is simply a correction to the ground-state energy

$$\begin{aligned}\tilde{V} &= V + \frac{dE_g}{dE_c} dE_c + \frac{dV}{dE_c} dE_c \\ &= V' + \frac{dE_g}{dE_c} dE_c ,\end{aligned} \tag{5.85}$$

where the second term is just a constant. The effect of this term is to shift the effective origin of the frequency variable ω in G,

$$G = \frac{1}{\omega - H_0 - V} \rightarrow G = \frac{1}{\omega - \Delta E_g - H_0 - V'(E_c)} , \tag{5.86}$$

so that if we want to calculate the total energy we must add into any calculation using $V'(E_c')$ the shift

$$\Delta E_g = \int_{E_c}^{E_c'} dE_c \frac{dE_g}{dE_c} . \tag{5.87}$$

This shift, as we remarked earlier [see Eq. (5.67)], diverges as $E_c \rightarrow \infty$, but such an ultraviolet divergence is rendered totally harmless by Eqs. (5.86)-(5.87). The only important feature of Eq. (5.86) to recognize is that ΔE_g is itself a function of ω. The true E_g is that at the ground-state pole, i.e., $E_g(E_g) = E_g(\omega' = 0)$ when we shift origin to the new ground-state energy, $\omega' = \omega - E_g(E_c)$. But when calculating physical properties we shall be looking for poles of G and their residues, which will be modified by the standard renormalization effect,

$$G \approx \left(\omega \left(1 - \frac{\partial \Delta E_g}{\partial \omega}\right) - H_0 - V\right)^{-1} ,$$

so that the residue for states near the ground state, which is the new renormalization of the wave function, is modified by

$$Z = \left(1 - \frac{\partial \Delta E_g}{\partial \omega}\right)^{-1} . \tag{5.88}$$

It is necessary to change to unit renormalization, which just has the effect of multiplying V by Z:

$$V'' = ZV'. \tag{5.89}$$

V'' is the new correct effective potential for calculating with the new cutoff E_c. We see that in general ΔZ will also be of order J^2, so this gives a J^3 correction to V, which is not so large for small J as the leading terms but does lead to a marginally relevant term in the Kondo temperature, as shown in Fowler and Zawadowski (1971).

The appropriate scaling equations which relate the appropriate effective Hamiltonian, for a given value of the cutoff E_c, to that for $E_c - dE_c$ are

$$\frac{dE_g}{dE_c} = -\frac{1}{2}(J\tau)^2 S(S+1)\ln 2 - (J\tau)^2 S(S+1)\left(\frac{\omega}{E_c}\right) + J^3(\ ) + \dots \tag{5.90}$$

$$\frac{d(J\tau)}{dE_c} = -\frac{(J\tau)^2}{E_c} + \frac{(J\tau)^3}{2E_c}\left(1 - S(S+1)\right) + J^4(\ ) + \dots , \tag{5.91}$$

the second equation <u>not</u> including the renormalization factor Z of Eq. (5.89). When we include that factor, i.e.,

$$\frac{dV}{V} = dZ = \frac{\partial}{\partial \omega}\left(\frac{dE_g}{dE_c}\right) dE_c ,$$

we find that Eq. (5.91) becomes

$$\frac{d(J\tau)}{dE_c} = -\frac{(J\tau)^2}{E_c} + \frac{(J\tau)^3}{2E_c} + J^4(\ ) + \ldots \quad . \tag{5.92}$$

The important equation is (5.92): how does the coupling vary with E_c? So long as $J\tau$ is small, the first term suffices, and we have

$$\frac{d(J\tau)}{(J\tau)^2} = -\frac{dE_c}{E_c}; \quad \frac{1}{J\tau} - \frac{1}{(J\tau)_0} = \ln\left(E_c/E_c^o\right) \tag{5.93}$$

or

$$(J\tau)(E_c) = \frac{(J\tau)_0}{1 - (J\tau)_0 \ln\left(\frac{E_c^o}{E_c}\right)} \quad . \tag{5.94}$$

Equation (5.94) is the essence of the Kondo effect. As E_c is carried down to low energy, $J\tau$ increases if $(J\tau)_0$ is positive (antiferromagnetic), but decreases if $(J\tau)_0$ is ferromagnetic. That is, the system has an infrared-stable fixed point if ferromagnetic, and that case is therefore trivially soluble: as $T \to 0$, the logarithmic divergences add up to give weaker and weaker scattering and a steadily decreasing resistivity. Thus that case is essentially solved by Eq. (5.94). At the same cutoff energy and thus temperature $T_k = E_c^o \exp[-1/(J\tau)_0]$ at which the antiferromagnetic difficulties occur, the ferromagnetic case begins to have appreciably weaker coupling.

On the other hand, for the antiferromagnetic case, the only apparent fixed point is "ultraviolet stable." Unfortunately, we know that scaling towards higher energies makes no sense in a solid-state problem. There is every likelihood that there will be a large number of relevant operators, which will grow (in fact, as we already remarked, the Kondo problem itself is a low-energy simplification of a more complicated problem, the Anderson Hamiltonian, which is one of the many more complicated problems to which such an "antiscaling" procedure could lead).

The next term in the series (5.92) is of no help,

although it is interesting that it actually has a large physical effect, which was unknown until scaling was introduced. It is more nearly correct to invert the series in (5.92) and write it as a series in ascending powers of $(J\tau)$, since it is based on perturbation theory:

$$d(J\tau) \times \left(-\frac{1}{(J\tau)^2} - \frac{1}{2(J\tau)} + \text{const} + \ldots\right) = \frac{dE_c}{E_c}, \text{ i.e.,}$$

$$\frac{1}{(J\tau)} + \frac{1}{2}\ln(J\tau) + \text{const} \times J\tau + \ldots = \ln E_c + \text{const.}$$

$$(5.95)$$

Both of the first terms on the left-hand side diverge as $(J\tau) \to 0$, so that both contribute large factors to the Kondo temperature T_k, and if we arbitrarily define T_k by

$$\left(J\tau(T_k)\right) = 1, \qquad (5.96)$$

we get

$$T_k = \frac{E_c^o}{\sqrt{J(\tau)_0}} \exp\left[-\frac{1}{(J\tau)_0}\right] . \qquad (5.97)$$

Thus in the limit $(J\tau)_0 \to 0$, which is the true Kondo limit, a large correction factor appears in T_k. On the other hand, the fact that the first two terms of (5.92) appear to define an infrared fixed point at $(J\tau) = 2$ is quite irrelevant, because (5.95) is the correct series. Figure 5-5 from Armytage's thesis shows the scaling using the naive Eq. (5.94), the first two terms of (5.92) as suggested by Zawadowski, and the "correct" scaling which we shall demonstrate later. Fowler (1972) was the first to point out the correct solution--that $J\tau \to \infty$ as $\tau \to \infty$, $E_c \to 0$; we shall show that in fact J = const., so the line of slope unity is correct in Fig. 5-5.

Of course, if initially $\beta \gg \tau_i$ as defined by Eq. (5.95), we must stop our scaling near T, and (5.95) gives us the appropriate value of $(J\tau)$ to use in perturbation theory with a low cutoff (and therefore no difficult large logarithmic terms). Then the resulting $(J\tau)$ will be small and

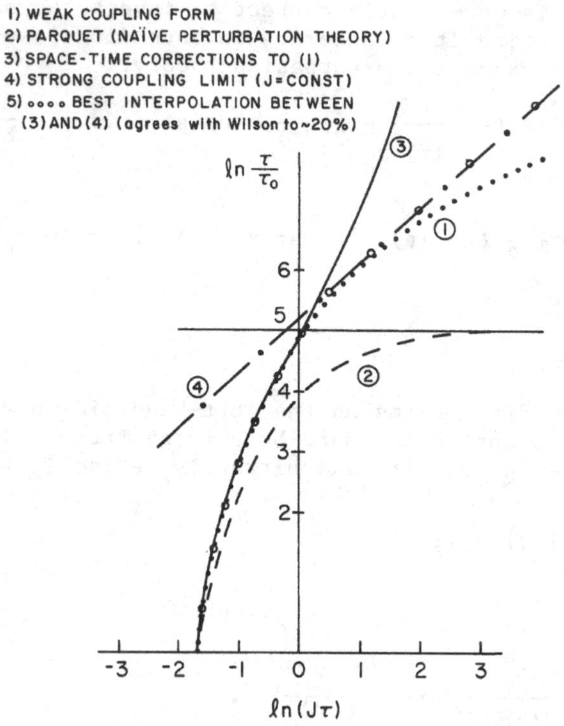

Figure 5-5 Results of various scaling procedures for Kondo problems.

finite, and perturbation theory will work well. But the simple scaling method fails near T_k because the series (5.95) no longer is useful. Several methods, all of which are of some interest and have possible applicability elsewhere, have succeeded in carrying the problem beyond this point.

The last, and in some ways most conclusive, of these is a superb tour de force of calculation carried out by K. G. Wilson (1974b, 1975), in which he succeeded in an actual numerical computation of the scaling to the final low-energy behavior, starting from a cutoff E_c well above the

"Kondotemperature" T_k. This calculation* is as yet the only source of accurate numerical results on the few percent level, for some of the "experimental" data, specific heat and susceptibility in particular. It is also very interesting from a general RNG point of view.

The trick Wilson uses to make the problem computable is a special form of discretization in which the hopping integrals decrease rapidly with distance from the spin:

$$H = E_c\Big\{ \sum_{\substack{n=0\\ \sigma}}^{\infty} \Lambda^{-n/2}\left(a^+_{n\sigma}a_{n+1\sigma} + a^+_{n+1\sigma}a_{n\sigma}\right) + (J\tau)\sum_{\sigma\sigma'} a^+_{0\sigma}\,(\vec{\sigma})_{\sigma\sigma'}\,a_{0\sigma'}\cdot\vec{S}\Big\} \tag{5.98}$$

(Λ is a number ~ 2). The Kondo problem is one dimensional in any case, since the spin only interacts with $\ell = 0$ spherical waves. The one-electron terms correspond to a one-dimensional chain with nearest-neighbor hopping. The peculiar variation of the hopping integral $\Lambda^{-n/2}$ with position in Eq. (5.98) is in fact merely a special, and rather strange, choice of the form factor for the cutoff in energy, which Wilson argues is probably justifiable. It has the effect of enormously simplifying the computational problem.

Wilson's scaling process for (5.98) is done in what appears at first to be a backwards way. He uses a discrete box size, i.e., he solves, instead of (5.98),

$$H_N = E_c\Big\{ \sum_{n=0}^{N} \Lambda^{-n/2}\left(a^+_{n\sigma}a_{n+1\sigma} + a^+_{n+1\sigma}a_{n\sigma}\right) + (J\tau)a_0^+\overleftrightarrow{\sigma}a_0\cdot\vec{S}\Big\} \tag{5.99}$$

---

* An exact solution now has been found which reproduces the Wilson results. This solution was found by Andrei (Andrei, 1980; Andrei and Lowenstein, 1981; Rajan, Lowenstein and Andrei, 1982) and independently by Wiegmann (1980). The solution leads to no physical enlightenment, as far as I can see, but is a mathematical tour de force.

and calculates numerically the lowest eigenvalues of Hamiltonians with successively larger values of N. These low eigenvalues, as N increases, are essentially sampling the energy levels of electrons of successively lower energies of order $\Lambda^{-N/2}$. It is the transformation $H_N \rightarrow H_{N+1} \rightarrow H_{N+2}$ which he considers to be the RNG operation (with appropriate rescaling); this is quite different in principle from other versions of the RNG, where it is the energy cut-off rather than the scale of examination which is moved down, but it seems to work equally well. The actual numerical values of the low energy levels are then fitted to an appropriate pseudo-Hamiltonian, which is the $H^* + \delta H$ of this system.

He notes that this transformation has two obvious fixed points. $J\tau = 0$ is one. Then the spin is decoupled and one has the soluble problem of the "hopping Hamiltonian,"

$$H_0 = E_c \Sigma \Lambda^{-n/2} a^+_{n+1} a_n + cc \ , \tag{5.100}$$

whose lowest eigenvalues just scale as $\Lambda^{-N/2}$, once N is large enough. Equally, $(J\tau) = \infty$ is an obvious fixed point. In this case the $(J\tau)$ term must be diagonalized first. The lowest eigenvalue is obtained when the state ψ_0 is singly occupied and coupled to a singlet with the local spin. The other eigenvalues involving zero or double occupation of ψ_0 are so much higher that the occupation of the state may not change; thus the state a_0 is decoupled from the hopping Hamiltonian, and there is just one less hopping integral: the logarithmic one-dimensional chain starts with a_1 and not a_0. But scaling doesn't change that state of affairs either.

By actual computations Wilson shows that the scaling process moves <u>away</u> from the $J\tau = 0$ fixed point very much according to the perturbation law (5.92), until a crossover region near T_k is reached; from this point on, the system steadily approaches the $J\tau = \infty$ limit. This is not surprising, if we really believe in the "bound state" concept. In order to bind one electron at a fixed energy T_k, we must have a fixed effective exchange interaction $J \simeq T_k$, and therefore as E_c decreases $J/E_c \simeq J\tau$ must go to infinity.

In order to calculate physical properties we need to understand also the lowest-order <u>irrelevant</u> operators. It is clear enough why. As we scale according to either the cutoff scaling concept or the Wilson "box expansion" idea, we are looking effectively at electrons in a bigger and bigger box. The effect of the spin or of any impurity center becomes smaller and smaller proportionally to the box size. A finite phase shift δ leads to an energy shift $\propto \delta/R$ falling off inversely with box size R. This would be as true of the utterly simple problem of a nonmagnetic impurity as of a Kondo spin. But the physical effects themselves are also inversely proportional to box size. They are connected with the impurity alone. Thus they can depend on irrelevant operators that behave just like ordinary scattering.

As Nozières (1974) has emphasized in the article that we reprint, the Wilson numerical results and effective H can be interpreted in just this way. They give one an effectively <u>ordinary</u> impurity which is equivalent to the dominant irrelevant operators as $T \to 0$ (the only relevant operator, $J\tau \to \infty$, drops totally out of the problem as we explained above). In fact, these operators are shown by Wilson to be two: a modified hopping integral

$$O_1 = T^1\left(f_0^+ f_1 + f_1^+ f_0\right) \tag{5.101a}$$

and a local, repulsive, Hubbard-like interaction

$$O_2 = U\left(n_0 - 1\right)^2 \tag{5.101b}$$

where $n_0 = a^+_{0\uparrow} a_{0\uparrow} + a^+_{0\downarrow} a_{0\downarrow}$. It should be recognized that the orbital labeled zero here is <u>not</u> the one in direct contact with the spin, because that has been eliminated by the infinite exchange coupling, but the nearest-neighbor one.

D. SPACE-TIME SCALING IN THE KONDO PROBLEM AND PHASE TRANSITIONS AS DEFECT CONDENSATIONS

The Kondo scaling problem was actually first solved by a quite different method, by Yuval, Hamann, and myself (Yuval and Anderson, 1970; Anderson and Yuval, 1969; Anderson, Yuval, and Hamann, 1970). This method has led to interesting generalizations of some importance in the understanding of many phase transitions.

Yuval and Anderson were able to transform the Kondo problem into a sum over dynamical histories of the local spin. Such a dynamical history is shown in Fig. 5-6: the spin can be only up or down, and any history consists of a recurring succession of flips at times t_i. The Hamiltonian

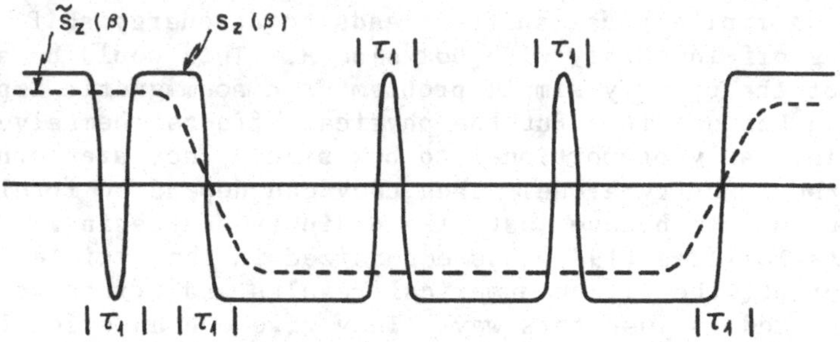

Figure 5-6 Dynamical history of impurity spin. Dashed line is the result of averaging out "quick" flips as prescribed by space-time scaling.

terms $J\left(S^+s^- + S^-s^+\right)$ produce such flips and lead to an interaction E, quite long-range in time (in fact logarithmic), between each pair of flips $E(t_i - t_j)$, while after some rather subtle calculations on the interference of waves of flipped electron spins one may show that the JS_zs_z term modifies this interaction. Since, in this representation, the two parts of the interaction are not explicitly symmetric, it is useful to allow the two to be different and study a more general, asymmetric interaction term

$$
\begin{aligned}
H_{int} &= J_zS_zs_z + J_\pm\left(S_xs_x + S_ys_y\right) \\
&= J_zS_zs_z + J_\pm\left(S_+s_- + S_-s_+\right) . \qquad (5.102)
\end{aligned}
$$

Naturally, any correct scaling law must keep $J_\pm = J_z$ if that holds initially--i.e., Kondo problems scale into each other, by symmetry--but otherwise we have a new coupling constant to play with.

With the two interactions, the sum over histories becomes

$$\operatorname{Tr} \exp\left(-\beta H_{Kondo}\right) = \sum_N \left(J_{\pm}\tau\right)^{2N} \int_0^{\beta} dt_{2N} \int_0^{t_{2N}-\tau} dt_{2N-1} \cdots$$

$$\cdots \int_0^{t_2-\tau} dt_1 \exp\left[\sum_{i>j} (-1)^{i-j} f(J_z\tau) \ln\left|\frac{t_i - t_j}{\tau}\right|\right] \tag{5.103}$$

where $f(J_z\tau)$ is linear for small $J_z\tau$ and all integrals cut off at small t/τ. The expression on the right is the sum over states for a one-dimensional problem in statistical mechanics involving interacting defects. These "defects" are the points at which the spin reverses in time. They have the character of "boundaries" between regions with constant spin. Clearly we can visualize the defects as spin reversals in time (as in the original case) or as domain boundaries in an equivalent Ising model; or we could visualize them as constituting the basic entities themselves, as a kind of "charge" that can be + or - in a model of charged points on a chain. Each of these problems leads to the same expression (5.103). The "defects" or charges interact by a long-range (in fact, logarithmic) potential, proportional to f(J) where f starts linearly with J, while each has a "fugacity" proportional to $(J_{\pm}\tau)$ for the Kondo model.

The transcription into different, equivalent models suggests that we need not treat the two parts of the Kondo interaction symmetrically, and, for instance, in the equivalent Ising model the two coupling constants are quite independent, one representing a short-range ferromagnetic spin coupling, the other a long-range one (proportional to $1/r_{ij}^2$, in fact). The interesting discovery of Anderson, Yuval, and Hamann (1970) was that this unsymmetrical problem had a phase transition, whose nature was first guessed by Thouless (1969). The defects have a logarithmic repulsion between similar sign defects, a logarithmic attraction between opposites. One would suppose that the logarithmic repulsion would require that defects appear only in bound pairs, whose long-range interactions cancel out. But at high enough temperatures--small enough J_z, in effect--the entropy gained per defect, which is also logarithmic in total defect density, outweighs the energy loss, and a finite density of defects ensues.

The calculation of the scaling diagram for this phase transition is perhaps the simplest of any known case (it may also have been the first such diagram ever drawn) and seems to have very general relevance, as we shall see. Thus it may be amusing to sketch it here.

Perhaps the simplest way to get at the actual phase transition is via the "poor man's scaling" expression (5.91), which we repeat here:

$$\frac{d(J\tau)}{dE_c} = -\frac{(J\tau)^2}{E_c} + \frac{(J\tau)^3}{2E_c} + \dots \quad . \tag{5.91}$$

In the first place, the calculation (5.83) <u>et seq.</u> can be easily adapted to an anisotropic J, and Eq. (5.91) becomes two equations:

$$\frac{d(J_z\tau)}{dE_c} = -\frac{(J_\pm\tau)^2}{E_c} + \dots$$

$$\frac{d(J_\pm\tau)}{dE_c} = -\frac{(J_z\tau)(J_\pm\tau)}{E_c} + \dots \quad . \tag{5.104}$$

For the phase transition problem, as distinct from the Kondo problem proper, the higher-order term isn't very important. (In any case, the symmetry of the Kondo problem assures us that the symmetric case will remain so under scaling, so that the line $|J_z\tau| = |J_\pm\tau|$ will be a trajectory of the mapping.) To express the consequences of Eq. (5.104) most straightforwardly, let us divide the two equations. We obtain

$$\frac{d(J_\pm\tau)}{d(J_z\tau)} = \frac{J_z\tau}{J_\pm\tau} \quad ,$$

which is solved by a set of hyperbolas,

$$(J_z\tau)^2 - (J_\pm\tau)^2 = \text{const.} \tag{5.105}$$

This is, of course, exact only where both $J\tau$'s are

small, but that is enough to tell us what the unique fixed point of (5.104) is: $J_{\pm}\tau = J_z\tau = 0$, and the geometry near that fixed point, which we see in Fig. 5-7 (from Anderson and Yuval, 1971). By looking back at Eq. (5.104), we can ascribe a mapping direction to the "scaling trajectories" (5.105), shown by the arrows on Fig. 5-7. "Const" = 0 is the isotropic case $J_z = J_{\pm}$, with the ferromagnetic case J_z negative (the sign of $J_{\pm}$ can be reversed by a simple 180° rotation of $\vec{S}$ and is irrelevant) scaling directly into the fixed point of zero interaction, while J_z positive scales towards the strong coupling case $J_z\tau = J_{\pm}\tau \gg 1$. All cases with $J_z < -J_{\pm}$ scale to the $J_{\pm} = 0$ line, with no flips and hence ferromagnetic, while $J_z > -J_{\pm}$ cases all scale towards strong coupling.

These results are the basis of what may be the earliest and simplest actual solution of a statistical model by the RNG equations. Since the same methods give the well-known Kosterlitz-Thouless solution of a less trivial problem, we sketch the ideas here. The space-time integral (5.103) is, as we noted, the same as a partition function for a one-dimensional statistical problem, equivalent either to 2-d Coulomb charges on a 1-d ring or to a 1-d Ising model with long-range

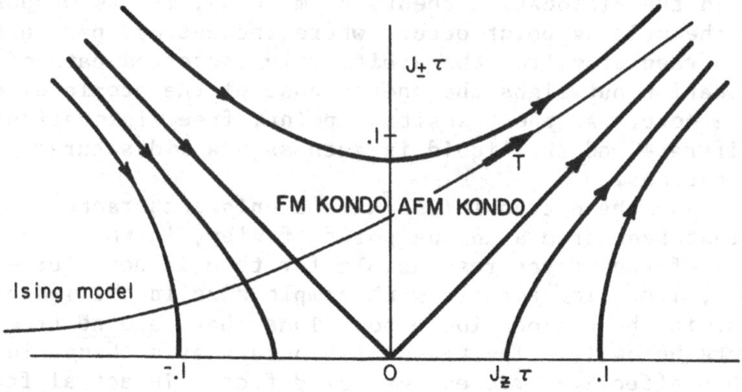

Figure 5-7 Scaling trajectories for Kondo and 1-d Ising problem.

$1/r^2$ interactions, a classical problem earlier studied by Dyson. In this problem the long-range interaction strength is proportional to $(2 - 2J_z\tau)$, while the short-range interaction (in the Ising model) or the fugacity y (in the charge case) is related to $(J_\pm\tau)$ in our formalism. Taking the Ising model as an example, the trajectory of the long- and short-range interactions as T is increased runs across the scaling trajectories from lower left to upper right, hence, at a particular T_c, crossing the $J_\pm = J_z$ line which scales into the singular point. This then is the temperature at which the scaling trajectory changes qualitatively and is the critical point of the phase transition in the Ising model.

This phase transition plays only a minor role in the Kondo problem proper. All antiferromagnetic Kondo problems are above it, all ferromagnetic ones below. But it was the first completely worked out case of a type of phase transition which had been suggested many times in the past, but never really proven to exist, a phase transition due to the sudden proliferation of defect structures.

Two such older suggestions were the dislocation theories of melting of Shockley (1952), Nabarro (1967), and Cotterill (1975) and the idea, sometimes ascribed to Bloch, that the λ point of helium is caused by the evaporation of an infinite number of quantized vortices. In fact, my own present best guess is that neither conjecture is correct.

In the dislocation theory of melting, it was proposed that the melting point occurs where the entropy per unit length resulting from the arbitrarily contorted path of a dislocation outweighs the energy cost of the atomic disorder at the core. At the transition point, free dislocations proliferate and the liquid is seen as a solid saturated with dislocations.

These phase transitions have a unique character, or are at least seen from a unique point of view, in that the free energy of the defect responsible for them is not microscopic, i.e., logarithmic with sample size in the 1-d case, linear in the dislocation case. Thus there are no true defects below T_c. The transition occurs as a change in sign of this effective free energy per defect. In actual fact, one has bound pairs of defects, or bound finite configurations such as dislocation loops in the liquid case, which increase in size and number as one approaches T_c. One may, in fact, if one likes, see the standard Ising-model ferromagnetic transition as an unbinding of domain walls, or more conventional melting theories as unbinding of grain boundaries, so this view may really be a universal point of view on phase transitions.

The transition can easily be made first order via the interaction of the density with the core volume of the dislocations. But Nelson and Toner (1981) have observed that single dislocations alone do not destroy orientational order, only positional order, and hence cannot lead to a true liquid; melting must be a more thoroughgoing state change than dislocations can provide.

The vortex theory of the λ point is less easy to refute, but also rather difficult to express in any very clear fashion.

Thouless and Kosterlitz (Kosterlitz and Thouless, 1973) realized that the mathematics of the Yuval-Anderson one-dimensional Ising model was easily generalized to the two-dimensional xy model, and this important insight has led to a busy industry in the study of two-dimensional systems, both physical and mathematical, as well as of a number of one-dimensional field theories. We reproduce the Kosterlitz paper in the reprints section. Recent work has made it virtually certain that at least the phase transition of the liquid He film goes in this way, and probably also the melting of some two-dimensional crystals.

E. FURTHER EXTENSIONS OF THE RENORMALIZATION GROUP: POSSIBLE NEW DIRECTIONS

In the preceding sections we have sketched perhaps the most successful of the extensions of the renormalization group idea. We also include in the reprints section a group of articles central to these ideas. Even viewed as a statistical-mechanical method for the understanding of critical points, the original Wilson-Fisher ideas have furnished us with many valuable insights to which I have hardly done justice. For instance, I have left out the idea of "crossovers." Where there are two interactions highly dissimilar in magnitude, it can occur that the scaling procedure causes the weak one to grow and become a relevant variable partway through the critical region, causing a "crossover" from one universality class to another. This is in fact usual in three-dimensional systems with weak interactions in one or two dimensions, since scaling will always weaken the lower-dimensional strong interactions until the weak ones can take over. But the concept of "crossover" is much more general than that simple case. Wilson, for instance, sees the Kondo phenomenon as a "crossover" from weak to strong J behavior. The crossover phenomenon invariably involves the presence of an unstable fixed point; in fact, even conventional critical behavior can be thought of as a "crossover" from critical to conventional behavior, occasioned by the close proximity of the unstable fixed point at the critical point.

I have, as already mentioned, also ignored critical transport phenomena, which can be straightforwardly--but with great complication in some cases--subsumed under the general scaling theory. The variables being scaled (like the polarizations p_i in Sec. 5.B) experience a fictitious fluctuating force or "Langevin force," and the scaling is carried out including these terms. As in the conventional theory of noise in linear systems, the fluctuating force is assumed to have a white-noise spectrum and is normalized by assuming Gibbs fluctuation theory. Dissipation arises via mode-mode coupling, the fluctuations of higher modes to which the macroscopic variables are coupled serving as a damping term via terms in the equations of motion like $\dot{p}_0 \propto p_0 \sum_k p_k^2$.

This theory has had many successes and some failures. Overall, I find it not particularly satisfying. The Wilson-Fisher theory of statistical equilibrium proceeds from a precise microscopic theory with no extraneous assumptions--in fact, one might employ much the same techniques to derive statistical mechanics itself from first principles, so that even statistical mechanics is not really a basic precondition. But the transport theory is far more artificial-feeling. One would be happier if it proceeded directly from fluctuation-dissipation theory, like the Kubo formula. For instance, I see no way at all that it could be extended to purely quantum transport processes such as metallic conduction or spin diffusion. It is a deeply classical construct. For these reasons I do not seriously regret omitting it here.

It seems particularly regrettable that no attempt has been made to give a general derivation of macroscopic transport equations such as those of hydrodynamics, of electrical conduction in solids, or of plasma physics from a RNG point of view. One would like to see the coefficients of such transport equations appear as the only relevant variables near some fixed point of a microscopic-to-macroscopic mapping process.

The conceptual structure of the renormalization group

certainly has an inviting similarity to the general concept we have called "continuation" in Chapter 3, so inviting that one is led immediately to the idea that each of the simplified model systems referred to there may represent a fixed point of a renormalization group process: the Fermi liquid, the Bose liquid, the quantum solid, etc. A start at discussing this kind of problem has been made by J. Hertz (1976) in the paper reprinted herewith. As I suggested a number of years ago (Anderson, 1973, Goteberg Nobel Symposium) and as Hertz foresees, one may think of various regions of such systems (spin fluctuations, valence fluctuations, etc.) as more or less close approaches to unstable fixed points of the Hamiltonian.

A unique use of RNG techniques has been possible in the problem of quantum transport under strong scattering, the so-called "Anderson localization" problem. While Wegner (1976, 1979, 1980) and others have been able to use mathematical strategems to relate the localization problem to certain field theory problems, with considerable success, I feel it is more significant to carry out the renormalization directly on the random Hamiltonian that leads to this quantum transport process, as was done by Abrahams *et al.* (1979) and more rigorously in a special case by Anderson (1981c). Again, we have a direct mapping in Hamiltonian space, where the significant factor is that the Hamiltonian is a member of a random distribution of similar Hamiltonians, and it is the distribution P that is renormalized and that has a fixed-point behavior, involving an unstable fixed point and a transition between two regimes in some cases. Several reprints relating to this interesting and possibly fundamental new way of looking at macroscopic stochastic systems are presented herewith.

It has recently become fashionable to borrow certain concepts based on the topological theory of mapping and its relationship to nonlinear differential equations--the so-called "catastrophe theory"--for application to all kinds of biological and even economic and sociological systems. Surely the simplest equations of population dynamics have bifurcations and instabilities, just as do the constitutive equations of matter about which we have been talking. Such ideas are fruitful in population biology, where simple dynamical equations may often be good approximations, but perhaps not so much so for more complicated systems such as human social ones.

For such systems, it seems to me that by the time a set of determinate, dynamical equations are written down, oversimplification to the point of nonsense has usually already taken place. It may be that a kind of "real-space renormalization group" approach could be much more fruitful: an attempt to look at larger and larger groupings of animals, people, or whatever, as averages or distributions over their internal degrees of freedom, and then to allow the larger groupings to interact, etc. The stochastic element surely must be left in as a factor in a rational treatment of real systems of this sort.

In either case, it is surely a hopeful sign that for the first time we, as theoretical physicists, have begun to develop methods for dealing with complex systems, methods that are sufficiently powerful to allow us to speculate about how far beyond the simple statistical phase transition it may be possible to go. One may hope that the exercise of going farther is carried out with a proper appreciation of how little has really been achieved even in the simple physical systems from which we start.

REFERENCES

Abrahams, E., P. W. Anderson, P. A. Lee, and T. V. Ramakrishnan, 1981, Phys. Rev. B 24, 6783.
Abrahams, E., P. W. Anderson, D. C. Licciardello, and T. V. Ramakrishnan, 1979, Phys. Rev. Lett. 42, 673.
Reproduced in Reprints section.
Abrahams, E., and C. Kittel, 1953, Phys. Rev. 90, 238.
Abrikosov, A. A., 1965, Physics 2, 5.
Abrikosov, A. A., L. P. Gor'kov, and E. Dzyaloshinskii, 1963, Methods of Quantum Field Theory in Statistical Physics (Prentice-Hall, Englewood Cliffs).
Adam, G., and J. H. Gibbs, 1965, J. Chem. Phys. 43, 139.
Adams, W. H., 1961, J. Chem. Phys. 34, 89.
Adams, W. H., 1962, J. Chem. Phys. 37, 2009.
Adkins, K., and N. Rivier, 1974, J. Phys. (Paris) 35, Colloq. C4, 237.
Alexander, S., and J. P. McTague, 1978, Phys. Rev. Lett. 41, 702.
Allender, D., J. W. Bray, and J. Bardeen, 1973, Phys. Rev. B 7, 1020.
Allender, D., J. W. Bray, and J. Bardeen, 1974, Phys. Rev. B 9, 119.
Ambegaokar, V., and J. M. Langer, 1967, Phys. Rev. 164, 498.
Anderson, P. W., 1952, Phys. Rev. 86, 694.
Anderson, P. W., 1956, Phys. Rev. 102, 1008.
Anderson, P. W., 1958a, Phys. Rev. 109, 1492.
Anderson, P. W., 1958b, Phys. Rev. 110, 827.
Anderson, P. W., 1958c, Phys. Rev. 110, 985.
Anderson, P. W., 1959a, Phys. Rev. 115, 2.

Anderson, P. W., 1959b, J. Phys. Chem. Solids 11, 26.

Anderson, P. W., 1960, Izdatel'stvo. Akad. Nauk., Moscow, Vsesouznaia Konferentsia po fizike dielektrikov. 2d (Conference Proceedings "Physics of Dielectrics" 1958), 290.

Anderson, P. W., 1961, Phys. Rev. 124, 41.

Anderson, P. W., 1963a, Concepts in Solids (Benjamin, New York).

Anderson, P. W., 1963b, Phys. Rev. 130, 439.

Anderson, P. W., 1964, in Lectures on the Many-Body Problem, edited by E. R. Caianiello (Academic Press, New York), Vol. 2, p. 113. See also Proceedings of the 1963 Midwest Conference on Theoretical Physics (University of Notre Dame, Notre Dame, Indiana), p. 124.

Anderson, P. W., 1965, in Some Recent Definitions in the Basic Sciences, edited by A. Gelbart, Annual Science Conference Proceedings (Belfer Graduate School of Sciences, Yeshiva Univ., New York, 1969), p. 21.

Anderson, P. W., 1966, Rev. Mod. Phys. 38, 298. **Reproduced in Reprints section.**

Anderson, P. W., 1968, Phys. Rev. Lett. 21, 13.

Anderson, P. W., 1970a, Comments Solid State Phys. 2, 193.

Anderson, P. W., 1970b, J. Phys. C. 3, 2436. **Reproduced in Reprints section.**

Anderson, P. W., 1971, Comments Solid State Phys. 3, 153.

Anderson, P. W., 1972a, Nature Phys. Sci. 235, 163.

Anderson, P. W., 1972b, Science 177, 393.

Anderson, P. W., 1973, in Proceedings of the Nobel Symposium, edited by S. and B. Lundqvist (The Nobel Foundation, Stockholm), Vol. 24, p. 68.

Anderson, P. W., 1974, in Elementary Excitations in Solids, Molecules, and Atoms (Part A), edited by J. T. Devreese, A. B. Kunz, and T. C. Collins (Plenum Press, New York), p. 1.

Anderson, P. W., 1976a, in Gauge Theories and Modern Field Theory, Proceedings of a Conference held at Northeastern University ... 1975, edited by R. Arnowitt and P. Nath (MIT, Cambridge, Mass.), 311.

Anderson, P. W., 1976b, in Amorphous Magnetism II, edited by R. A. Levy and R. Hasegawa (Plenum, New York, 1977), p. 1.

Anderson, P. W., 1979, in Ill-Condensed Matter/La Matière mal condensée, Les Houches École d'été de physique théorique, Session XXI, 1978, edited by R. Balian, R. Maynard, and G. Toulouse (North-Holland, Amsterdam), p. 159.

Anderson, P. W., 1981a, in Proceedings of the XVII Solvay Conference on Physics, Brussels, 1978, edited by G. Nicolis, G. Dewel, and J. W. Turner (Wiley, New York), p. 289.

Anderson, P. W., 1981b, in Proceedings of the International Conference on Valence Fluctuations in Solids, Santa Barbara, California, edited by L. M. Falicov, W. Hanke, and M. V. Maple (North-Holland, Amsterdam), p. 451.

Anderson, P. W., 1981c, Phys. Rev. B 23, 4828.

Anderson, P. W., and W. F. Brinkman, 1973, Phys. Rev. Lett. 30, 1108.

Anderson, P. W., and W. F. Brinkman, 1975, in The Helium Liquids: Proceedings of the 15th Scottish Universities Summer School in Physics, 1974, edited by J.G.M. Armitage and I. E. Farquhar (Academic, New York), 315. Also in The Physics of Liquid and Solid Helium, edited by K. H. Bennemann and J. B. Ketterson (Wiley, New York, 1978), Part II, p. 177. **Reproduced in Reprints section.**

Anderson, P. W., and Clogston, A. M., 1961, Bull. Am. Phys. Soc. 6, 124. (See also Anderson, P. W., unpublished talk given at 1959 Oxford Conference on Magnetism.)

Anderson, P. W., and P. Morel, 1961, Phys. Rev. 123, 1911.

Anderson, P. W., and R. Palmer, 1977, in Quantum Fluids and Solids, edited by S. B. Trickey, E. Adams, and J. Duffy (Plenum, New York), 23.

Anderson, P. W., and D. L. Stein, to be published in Self-Organizing Systems: The Emergence of Order, edited by F. E. Yates (Plenum, New York). **Reproduced in Reprints section.**

Anderson, P. W., D. J. Thouless, E. Abrahams, and D. S. Fisher, 1980, Phys. Rev. B 22, 3519.

Anderson, P. W., and G. Toulouse, 1977, Phys. Rev. Lett. 38, 508. **Reproduced in Reprints section.**

Anderson, P. W., and G. Yuval, 1969, Phys. Rev. Lett. 23, 89.

Anderson, P. W., and G. Yuval, 1970, Phys. Rev. B. 1, 1522.

Anderson, P. W., and G. Yuval, 1971, J. Phys. C. 4, 607.

Anderson, P. W., and G. Yuval, 1973, in Magnetism, edited by H. Suhl (Academic, New York), Vol. V, p. 217.

Anderson, P. W., G. Yuval, and D. R. Hamann, 1970, Phys. Rev. B 1, 4464. **Reproduced in Reprints section.**

Andrei, N., 1980, Phys. Rev. Lett. 45, 379.

Andrei, N., and J. Lowenstein, 1981, Phys. Rev. Lett. 46, 356.

Armytage, J., 1973 Ph.D. Thesis (Cambridge University).

Arnowitt, R., and P. Nath, 1975, Eds., Gauge Theories and Modern Field Theory, Proceedings of a Conference held at Northeastern University...1975 (MIT, Cambridge, Mass., 1976).
Ashcroft, N. W., and D. Mermin, 1976, Solid State Physics (Holt, Rinehart and Winston, New York).
Balian, R., and N. R. Werthamer, 1963, Phys. Rev. 131, 1553.
Bardeen, J., 1973, Solid State Commun. 13, 357.
Bardeen, J., L. N. Cooper, and J. R. Schrieffer, 1957, Phys. Rev. 108, 1175.
Becker, R., 1932, Phys. Z. 33, 905.
Beliaev, S. T., 1958, Sov. Phys. JETP 37, 289 [Zh. Eksp. Teor. Fiz. 34, 417 (1958)].
Blandin, A., and J. Friedel, 1959, J. Phys. Radium 20, 160.
Bloch, F., 1930, Z. Phys. 61, 206.
Bloch, F., 1932, Z. Phys. 74, 295.
Bogoliubov, N. N., 1947, Izv. Akad. Nauk SSSR, Ser. Fiz. 11, 77.
Bogoliubov, N. N., and D. V. Shirkov, 1955, Dokl. Adad. Nauk SSSR 103, 203, 391.
Bohm, D., and D. Pines, 1951, Phys. Rev. 82, 625.
Bohm, D., and D. Pines, 1952, Phys. Rev. 92, 609.
Bohm, D., and T. Staver, 1951, Phys. Rev. 84, 836.
Boswell, J., 1791, The Life of Samuel Johnson Dilly, London).
Brinkman, W. F., and T. M. Rice, 1970, Phys. Rev. B 2, 4302.
Broadbent, S. R., and J. M. Hammersby, 1957, Proc. Cambridge Philos. Soc. 53, 629.
Brueckner, K. A., 1959, in The Many-Body Problem/Le Problème à N Corps, Cours donné a l'Ecole d'été de physique théorique, Université de Grenoble, Les Houches--Session 1958, edited by C. DeWitt (Wiley, New York; Methuen, London; Dunod, Paris), p. 47.
Brueckner, K. A., C. A. Levinson, and H. M. Mahmoud, 1954, Phys. Rev. 95, 217.
Burgers, J. M., 1939, K. Ned. Akad. Wet. 42, 293.
Burgers, J. M., 1940, Proc. Phys. Soc. London 52, 23.
Callan, C. G., R. Dashen, and D. J. Gross, 1978, Phys. Rev. D 17, 2717.
Callen, H. B., and T. A. Welton, 1951, Phys. Rev. 83, 34.
Canella, V., and J. A. Mydosh, 1972, Phys. Rev. B 6, 4220.
Chan, S.-K., and V. Heine, 1973, J. Phys. F 3, 795.
Chui, S. T., and P. A. Lee, 1975, Phys. Rev. Lett. 35, 315.
Cladis, P., and M. Kleman, 1972, J. Phys. 33, 591.
Cochran, W., 1959, Phys. Rev. Lett. 3, 412.
Coldwell-Horsfall, R. A., and A. A. Maradudin, 1960, J. Math. Phys. 1, 395.

Comès, R.,1974, Solid State Commun. 14, 98.
Comès, R.,S. M. Shapiro, G. Shirane, A. F. Garito, and A. J. Heeger, 1975, Phys. Rev. Lett. 35, 1518.
Cotterill, R.M.J., 1975, Phil. Mag. 32, 1283.
Cross, M. C., 1977, J. Low Temp. Phys. 26, 165.
Cross, M. C., and P. W. Anderson, 1975, in Proceedings of the 14th International Conference on Low Temperature Physics, Otaniemi, Finland, edited by M. Krusius and M. Vuorio (North-Holland, Amsterdam and American Elsevier, New York), Vol. I, p. 29.
de Boer, J., 1948a, Physica (The Hague) 14, 139.
de Boer, J., and R. J. Lunbeck, 1948b, Physica (The Hague) 14, 520.
de Gennes, P. G., 1973, in Proceedings of the Nobel Symposium, edited by S. and B. Lundqvist (Nobel Foundation, Stockholm), Vol. 24, p. 112.
Devonshire, A. F., 1949, Philos. Mag. 40, 1040.
Dolan, L., and R. Jackiw, 1974, Phys. Rev. D 9, 3320.
Doniach, S., and S. Engelsberg, 1966, Phys. Rev. Lett. 17, 750.
Dynes, R., and V. Narayanamurti, 1974, Phys. Rev. Lett. 33, 1195.
Dyson, F. J., 1957, private communication.
Edwards, S. F., and P. W. Anderson, 1975, J. Phys. F 5, 965.
Eigen, M., 1971, Naturwissenschaften 58, 465.
Eliashberg, G. M., 1960, Sov. Phys. JETP 11, 696.
Elliott, R. J., and D. W. Taylor, 1967, Proc. R. Soc. London Ser. A 296, 161.
Feshbach, H., C. E. Porter, and V. F. Weisskopf, 1954, Phys. Rev. 86, 448.
Fetter, A. L., and J. D. Walecka, 1971, Quantum Theory of Many-Particle Systems (McGraw-Hill, San Francisco).
Feynman, R. P., 1954a, Phys. Rev. 91, 1291.
Feynman, R. P., 1954b, Phys. Rev. 91, 1301.
Feynman, R. P., 1955 in Progress in Low Temperature Physics, edited by C. J. Gorter (North-Holland, Amsterdam), Vol. 1, Chap. 2.
Finkelstein, O., 1966, J. Math. Phys. 7, 1218.
Fisher, M. G., 1974, Rev. Mod. Phys. 46, 587.
Fleishman, L., 1978, Ph.D. Thesis (Princeton University).
Fowler, M., 1972, Phys. Rev. B 6, 3422.
Fowler, M., and A. Zawadowski, 1971, Solid State Commun. 9, 471.
Friedel, M. G., 1922, Ann. Phys. (Paris) 18, 273.
Fröhlich, H., 1954, Proc. R. Soc. London A 223, 296.
Fukuyama, H., T. M. Rice, and C. M. Varma, 1974, Phys. Rev. Lett. 33, 305.

Fulton, T., 1968, Phys. Lett. A 27, 521.
Gell-Mann, M., and K. A. Brueckner, 1957, Phys. Rev. 106, 364.
Gilbert, T. L., 1964, in Molecular Orbitals, a Tribute to R. S. Mulliken, edited by P. O. Löwdin and B. Pullman (Academic, New York), p. 405.
Ginzburg, V. L., and L. D. Landau, 1950, Zh. Eksp. Teor. Fiz. 20, 1064.
Glansdorff, P., and I. Prigogine, 1971, Structure, Stability and Fluctuations (Wiley-Interscience, New York).
Goldstone, J., 1961, Nuovo Cimento 19, 1.
Goldstone, J., and R. Jackiw, 1975, Phys. Rev. D 11, 1486.
Goldstone, J., A. Salam, and S. Weinberg, 1962, Phys. Rev. 127, 965.
Gor'kov, L. P., 1958, Sov. Phys. JETP 34, 505 [Zh. Eksp. Teor. Fiz. 34, 735 (1958)].
Gutzwiller, M. C., 1965, Phys. Rev. 137, A1726.
Haken, H., 1974, Synergetics (Springer, New York).
Haldane, F.D.M., 1978, Phys. Rev. Lett. 40, 416. **Reproduced in Reprints section.**
Haldane, F.D.M., 1979, Ph.D. Thesis (Cambridge University).
Halperin, B. I., and D. Nelson, 1978, Phys. Rev. Lett. 41, 121; errata, 41, 519.
Halperin, B. I., and W. M. Saslow, 1977, Phys. Rev. B 16, 2154.
Halperin, B. I., and Toner, 1982,
Halperin, B. I., and C. M. Varma, 1976, Phys. Rev. B 14, 4030.
Herring, C., 1952a, Phys. Rev. 85, 1003.
Herring, C., 1952b, Phys. Rev. 87, 60.
Herring, C., 1966a in Magnetism, edited by H. Suhl and G. Rado (Academic, New York), Vol. IIB, p. 1.
Herring, C., 1966b, Magnetism, Vol. IV: Exchange Interactions among Itinerant Electrons (Academic, New York).
Herring, C., and J. K. Galt, 1952, Phys. Rev. 85, 1060.
Herring, C., and C. Kittel, 1951, Phys. Rev. 81, 869.
Hertz, J. A., 1975, in AIP Conference Proceedings No. 24 (American Institute of Physics, New York), p. 298.
Hertz, J. A., 1976, Phys. Rev. B 14, 1165. **Reproduced in Reprints section.**
Higgs, P. W., 1964, Phys. Lett. 12, 132.
Higgs, P. W., 1964b, Phys. Rev. Lett. 13, 508.
Hohenberg, P. C., and B. I. Halperin, 1977, Rev. Mod. Phys. 49, 435.
t'Hooft, G., 1974, Nucl. Phys. B 79, 276.
Hubbard, J., 1959, Phys. Rev. Lett. 3, 77.

Hubbard, J., 1963, Proc. R. Soc. London, Ser. A 276, 238.
Hubbard, J., 1964a, Proc. R. Soc. London, Ser. A 277, 237.
Hubbard, J., 1964b, Proc. R. Soc. London, Ser. A 281, 401.
Hulthéin, L., 1937, Ph.D. Thesis (University of Leiden).
Inkson, J. C., 1971, Ph.D. Thesis (Cambridge University).
Inkson, J. C., 1973, J. Phys. C 6, 1350.
Josephson, B. D., 1974, Nobel Lecture, Rev. Mod. Phys. 46, 250.
Kadanoff, L. P., 1964, in Lectures on the Many-Body Problem, edited by E. R. Caianiello (Academic, New York), Vol. 2, p. 77.
Kadanoff, L. P., 1966, Physics 2, 263.
Kadanoff, L. P., and G. Baym, 1962, Quantum Statistical Mechanics--Green's Function Methods (Benjamin, New York).
Kanamori, J., 1963, Prog. Theor. Phys. 30, 275.
Kasuya, T., 1956, Prog. Theor. Phys. 16, 45.
Kauzmann, W., 1948, Chem. Rev. 48, 219.
Khalatnikov, I. M., 1965, Introduction to the Theory of Superfluidity, 2nd Ed. (Benjamin, New York).
Khalatnikov, I. M., 1976, in The Physics of Liquid and Solid Helium, edited by K. H. Bennemann and J. B. Ketterson (Wiley, New York), Vol. 1, p. 1.
Kirzhnitz, D. A., and A. D. Linde, 1972, Phys. Lett. B 42, 471.
Kittel, C., 1970, Am. J. Phys. 40, 60.
Kittel, C., 1971, Introduction to Solid State Physics, revised edition, first published in 1953 (Wiley, New York).
Kléman, M., and J. Friedel, 1969, J. Phys. (Paris) 30, Suppl. 43.
Knook, B., 1962, Ph.D. Thesis (University of Leiden).
Kohn, W., 1964, Phys. Rev. 133, A171.
Kohn, W., 1968, in Many-Body Physics, Lecture notes and summaries of courses presented at the Ecole d'été de physique théorique of the Université de Grenoble, 1967 session, edited by C. DeWitt and R. Balian (Gordon and Breach, New York), p. 353.
Kohn, W., and J. M. Luttinger, 1965, Phys. Rev. Lett. 15, 524.
Kondo, J., 1964, Prog. Theor. Phys. 32, 37.
Kosterlitz, J. M., 1974, J. Phys. C. 7, 1046.
Kosterlitz, J. M., and D. J. Thouless, 1973, J. Phys. C 6, 1181. **Reproduced in Reprints section.**
Kramers, H. A., and G. H. Wannier, 1941, Phys. Rev. 60, 252, 263.
Kukula, J., 1977, Senior Thesis (Princeton University).

Landau, L. D., 1941, J. Phys. USSR 5, 71.
Landau, L. D., 1947, J. Phys. USSR 11, 91.
Landau, L. D., 1957a, Sov. Phys. JETP 3, 920 [Zh. Eksp. Teor. Fiz. 30, 1058 (1956)].
Landau, L. D., 1957b, Sov. Phys. JETP 5, 101 [Zh. Eksp. Teor. Fiz. 32, 59 (1957)].
Landau, L. D., 1958, Sov. Phys. JETP 8, 70 [Zh. Eksp. Teor. Fiz. 35, 97 (1958)].
Landau, L. D., and E. M. Lifschitz, 1958, Statistical Physics, Course of Theoretical Physics, Vol. 5 (Pergamon, London).
Lander, J. J., and J. Morrison, 1962, J. Chem. Phys. 37, 729.
Lander, J. J., and J. Morrison, 1963, J. Appl. Phys. 34, 1403.
Langer, J. M., and V. Ambegaokar, 1967, Phys. Rev. 164, 498.
Larkin, A. I., and Y. N. Ovchinnikov, 1972, Sov. Phys. JETP 34, 651 [Zh. Eksp. Teor. Fiz. 61, 1221 (1971)].
Lax, M., 1951, Rev. Mod. Phys. 23, 287.
Layzer, A. J., and D. Fay, 1971, Int. J. Magn. 1, 135.
Lee, P. A., 1981, Nature 291, 11. **Reproduced in Reprints section.**
Lee, P. A., and T. M. Rice, 1979, Phys. Rev. B 19, 3970. **Reproduced in Reprints section.**
Lee, P. A., T. M. Rice, and P. W. Anderson, 1973, Phys. Rev. Lett. 31, 462.
Lee, P. A., T. M. Rice, and P. W. Anderson, 1974, Solid State Commun. 14, 703. **Reproduced in Reprints section.**
Leggett, A. J., 1965a, Phys. Rev. Lett. 14, 536.
Leggett, A. J., 1965b, Phys. Rev. 140, A1869.
Leggett, A. J., 1970, Phys. Rev. Lett. 25, 1543.
Licciardello, D. C., and D. J. Thouless, 1975, Phys. Rev. Lett. 35, 1475.
Lieb, E., and D. C. Mattis, 1966, Mathematical Physics in One Dimension (Academic Press, New York).
Linde, J. O., 1958, in Proceedings of the Fifth International Conference on Low Temperature Physics (University of Wisconsin, Madison), p. 402.
Lowde, R. D., 1974, private communication.
Luther, A., and V. J. Emery, 1974, Phys. Rev. Lett. 33, 589.
Luther, A., and L. Peschel, 1973, Phys. Rev. B 9, 2911.
Luttinger, J. M., 1960, Phys. Rev. 119, 1153.
Luttinger, J. M., and P. Nozières, 1962, Phys. Rev. 127, 1423.
Mattis, D. C., 1967, Phys. Rev. Lett. 19, 1478.
McMillan, W. L., 1965, Phys. Rev. 138, A442.
McMillan, W. L., 1976, Phys. Rev. B 14, 1496.

McMillan, W. L., 1981, Phys. Rev. B 24, 2739. **Reproduced in Reprints section.**
McMillan, W. L., and P. W. Anderson, 1966, Phys. Rev. Lett. 16, 85.
McWhan, D. B., J. P. Remeika, T. M. Rice, W. F. Brinkman, J. P. Maita, and A. Menth, 1971, Phys. Rev. Lett. 27, 941.
Menyhard, N., and J. Solyom, J. Low Temp. Phys. 12, 520, 547.
Mermin, N. D., 1977, in Quantum Fluids and Solids edited by S. B. Trickey, E. D. Adams, and J. W. Dufty (Plenum, New York), p. 3.
Mermin, N. D., 1978, J. Math. Phys. 19, 1457.
Mermin, N. D., 1979, Rev. Mod. Phys. 51, 591.
Migdal, A. B., 1958, Sov. Phys. JETP 7, 996 [Zh. Eksp. Teor. Fiz. 34, 1438].
Migdal, A. B., 1967, Theory of Finite Fermi Systems and Applications to Atomic Nuclei (Pergamon, London).
Moncton, D. E., J. D. Axe, and F. J. di Salvo, 1975, Phys. Rev. Lett. 34, 734.
Moncton, D. E., R. R. Pindak, S. C. Davey, and G. S. Brown, 1982, Phys. Rev. Lett. 49, 1865.
Mook, H. A., R. Scherm, and M. K. Wilkinson, 1972, Phys. Rev. A 6, 2268.
Morel, P., and P. W. Anderson, 1962, Phys. Rev. 125, 1263.
Motizuki, K., A. Shibatani, and T. Nagamiya, 1968, J. Appl. Phys. 39, 1098.
Mott, N. F., 1974, Metal-Insulator Transitions (Taylor & Francis, London).
Nabarro, F.R.N., 1967, Theory of Dislocations, (Oxford University, Oxford).
Nakamura, T., 1958, Prog. Theor. Phys. 20, 542.
Nambu, Y., 1960, Phys. Rev. 117, 648.
Nambu, Y., and G. Jona-Lasinio, 1961, Phys. Rev. 122, 345.
Nelson, D. R., and B. I. Halperin, 1979, Phys. Rev. B 19, 2457.
Nelson, D. R., and J. Toner, 1981, Phys. Rev. B 24, 363.
Nozières, P., 1952, Le Problème à N Corps (Dunod, Paris; also in translation, Benjamin, New York).
Nozières, P., 1964, Theory of Interacting Fermi Systems (Benjamin, New York).
Nozières, P., 1974, J. Low Temp. Phys. 17, 31. **Reproduced in Reprints section.**
Nozières, P., and C. De Dominicis, 1969, Phys. Rev. 178, 1097.
Onsager, L., 1944, Phys. Rev. 65, 117.
Ornstein, L. S., and K. Zernike, 1914, Proceedings K. Ned. Akad. Wet. 17, 793.

Orowan, E., 1934, Z. Phys. 89, 634.
Overhauser, A. W., 1960, Phys. Rev. Lett. 4, 462.
Overhauser, A. W., 1962, Phys. Rev. 128, 1437.
Overhauser, A. W., 1963, J. Applied Phys. 34, 1019.
Overhauser, A. W., 1968, Phys. Rev. 167, 691.
Overhauser, A. W., 1971, Phys. Rev. B 3, 3173.
Palmer, R. G., and P. W. Anderson, 1974, Phys. Rev. D 9, 3281.
Panczyk, M. F., and E. D. Adams, 1969, Phys. Rev. 187, 321.
Patashinskii, A. Z., and V. L. Pokrovskii, 1964, Sov. Phys. JETP 19, 677 [Zh. Eksp. Teor. Fiz. 46, 994 (1964)].
Pauling, L., 1949, Proc. R. Soc. London, Ser. A 196, 343.
Peierls, R. E., 1955, Quantum Theory of Solids (Oxford University, Oxford), p. 108.
Pines, D., 1961, The Many-Body Problem (Benjamin, New York).
Pines, D., and P. Nozières, 1965, Theory of Quantum Liquids (Benjamin, New York).
Pippard, A. B., 1969, Phil. Mag. 19, 217.
Poenaru, V., 1979, in Ill-Condensed Matter/La Matièremal condensée, Les Houches Ecole d'été de physique théorique, Session XXI, 1978, edited by R. Balian, R. Maynard, and G. Toulouse (North-Holland, Amsterdam), p. 263.
Polyakov, A., 1975, Sov. Phys. JETP 41, 988.
Polanyi, M., 1934, Z. Phys. 89, 660.
Rajan, V. T., J. H. Lowenstein, and N. Andrei, 1982, Phys. Rev. Lett. 49, 497.
Read, W. T., and W. Shockley, 1950, Phys. Rev. 78, 275.
Ruderman, M. A., and C. Kittel, 1954, Phys. Rev. 96, 99.
Salpeter, E. E., and H. A. Bethe, 1951, Phys. Rev. 84, 1232.
Scalapino, D. J., M. Sears, and R. A. Ferrell, 1972, Phys. Rev. B 6, 3409.
Schrieffer, J. R., 1964, Theory of Superconductivity (Benjamin, New York).
Schrieffer, J. R., and P. A. Wolff, 1966, Phys. Rev. 149, 491.
Sherrington, D., 1975, J. Phys. C 8, L208.
Sherrington, D., and B. W. Southern, 1975, J. Phys. F 5, L49.
Shiba, H., 1970, Prog. Theor. Phys. 43, 601.
Shockley, W., 1952, in 9th Solvay Conference, edited by R. Stoop (Solvay Institute, Brussels), p. 431.
Slater, J. C., 1950, Phys. Rev. 78, 748.
Slater, J. C., 1951, J. Chem. Phys. 19, 220.
Slater, J. C., J. B. Mann, T. M. Wilson, and J. H. Wood, 1970, Phys. Rev. 184, 672.
Solyom, J., and F. Zawadowski, 1974, J. Phys. F. 4, 80.

Soven, P., 1966, Phys. Rev. 151, 539.
Star, W. M., 1969, in Proceedings of the 11th International Conference on Low Temperature Physics, St. Andrews, Scotland, 1969, edited by J. F. Allen, D. M. Friedlayson, and D. M. McCall (University of St. Andrews, Scotland), pp. 1250, 1280.
Stein, D., 1979, Ph.D. Thesis (Princeton University).
Stein, D. L., R. D. Pisarski, and P. W. Anderson, 1978, Phys. Rev. Lett. 40, 1269.
Stratinovich, R. L., 1958, Sov. Phys. Dokl. 2, 416.
Suhl, H., 1958, Phys. Rev. 109, 606.
Taylor, G. I., 1934, Proc. Roy. Soc. A 145, 362.
Thiele, A. A., 1970, J. Appl. Phys. 41, 1139.
Thom, R., 1975, Structural Stability and Morphogenesis (Benjamin, New York).
Thouless, D. J., 1961, The Quantum Mechanics of Many-Body Systems (Academic, New York).
Thouless, D. J., 1965, Proc. Phys. Soc. London 86, 893.
Thouless, D. J., 1969, Phys. Rev. 187, 732.
Thouless, D. J., 1979, in Ill-Condensed Matter/La Matière mal condensée, Les Houches Ecole d'été de physique théorique, Session XXI, 1978, edited by R. Balian, R. Maynard, and G. Toulouse (North-Holland, Amsterdam), p. 1.
Tomasch, W. J., 1965, Phys. Rev. Lett. 15, 672.
Toulouse, G., 1969, C. R. Acad. Sci. 268, 1200.
Toulouse, G., and M. Kléman, 1976, J. Phys. (Paris) Lett. 37, 149. **Reproduced in Reprints section.**
van den Berg, G. J., 1965, in Low Temperature Physics LT9 (Part B), Proceedings of the Ninth International Conference on Low Temperature Physics, Columbus, Ohio, 1964, edited by J. G. Daunt, D. O. Edwards, F. J. Milford, and M. Yaqub (Plenum, New York), p. 955.
Varma, C. M., and N. R. Werthamer, 1976, in The Physics of Liquid and Solid Helium, edited by K. H. Bennemann and J. B. Ketterson (Wiley, New York), Vol. 1, p. 503.
Volovik, G. E., and V. P. Mineev, 1977, Zh. Eksp. Teor. Fiz. Pis'ma Red. 72, 2256.
Wannier, G. H., 1950, Phys. Rev. 79, 357.
Wegner, F. J., 1976, Z. Phys. B 25, 327.
Wegner, F. J., 1979a, Phys. Rev. B 19, 783.
Wegner, F. J., 1979b, Z. Phys. B 35, 207.
Wegner, F. J., 1980, Z. Phys. B 36, 209.
Weinberg, S., 1974, Phys. Rev. D 9, 3357.
Weisskopf, V. F., 1951, Science 113, 101.
Widom, B., 1965, J. Chem. Phys. 43, 3898.

Wiegmann, P. B., 1980, Zh. Eksp. Teor. Fiz. Pis'ma Red. 31, 392 [JETP Lett. 31, 364, (1981)].
Wigner, E. P., 1933, Phys. Rev. 43, 252.
Wigner, E. P., 1934, Phys. Rev. 46, 1002.
Wigner, E. P., 1938, Trans. Faraday Soc. 34, 678.
Williams, C., P. Peranski, and P. E. Cladis, 1972, Phys. Rev. Lett. 29, 90.
Wilson, J. A., and A. D. Yoffe, 1969, Adv. Phys. 18, 193.
Wilson, J. A., F. J. di Salvo, and S. Mahajan, 1975, Adv. Phys. 24, 117.
Wilson, K. G., 1971a, Phys. Rev. B 4, 3174.
Wilson, K. G., 1971b, Phys. Rev. B 4, 3184.
Wilson, K. G., 1972, in AIP Conference Proceedings No. 10 (American Institute of Physics, New York), p. 843.
Wilson, K. G., 1974a, in Proceedings of the Nobel Symposium, edited by S. & B. Lundqvist (Nobel Foundation, Stockholm), Vol. 24, p. 14.
Wilson, K. G., 1974b, in Proceedings of the Nobel Symposium, edited by S. & B. Lundqvist (Nobel Foundation, Stockholm), Vol. 24, p. 68.
Wilson, K. G., 1975, Rev. Mod. Phys. 47, 773.
Wilson, K. G., and M. E. Fisher, 1972, Phys. Rev. Lett. 28, 240.
Wilson, K. G., and J. Kogut, 1974, Phys. Rep. 12C, No. 2.
Woo, C. W., 1976, in Physics of Liquid and Solid Helium, edited by K. H. Bennemann and J. B. Ketterson (Wiley, New York), Vol. 1, p. 349.
Yoshino, S., and M. Okazaki, 1977, J. Phys. Soc. Japan 43, 415.
Yosida, K., 1957, Phys. Rev. 106, 896.
Zawadowski, F., 1973, in Proceedings of the Nobel Symposium, edited by B. and S. Lundqvist (Nobel Foundation, Stockholm), p. 76.
Ziman, J. M., 1962, Electrons and Phonons (Clarendon, Oxford).

REPRINTS

COHERENT MATTER FIELD PHENOMENA IN SUPERFLUIDS

P. W. Anderson

BELL TELEPHONE LABORATORIES, MURRAY HILL, NEW JERSEY

It is ironic that during the same years that the elementary-particle physicists, at the highest energy end of the spectrum, have been doing their best to discard as many as possible of the properties of the quantum field operator, at the very lowest energy end of the spectrum, in the two low temperature phenomena of superconductivity and superfluidity, we have been provided with a series of very direct demonstrations which seem to bring the quantum particle field almost into the ordinary, tangible, macroscopic realm. This will be the burden of what I will be trying to say today. In these phenomena the quantum particle field plays a role very similar to the roles, with which we are familiar, of the classical fields, the electromagnetic and gravitational fields, in our ordinary macroscopic experience. This has been shown by a series of experiments of various kinds on coherence in quantum fluids. I would like here to emphasize more the basic meaning of these experiments than to describe in detail their results and their experimental mechanisms.

It is interesting that nowhere in this discussion will I have to treat in very much detail the microscopic theory of either of these two superfluids, either the superfluid, helium, or the superconducting electron gas in metals. In fact, the foundations of the subject of quantum coherence were laid long before there were any acceptable microscopic theories of these phenomena. One could even argue that there is no acceptable microscopic theory of helium to this day. These foundations were laid by Landau, Penrose, Onsager, London and many others. Even the two types of macroscopic quantization, the quantization of vorticity and the quantization of magnetic flux, were proposed by 1949, long before there were any microscopic theories. The first approach to the properties of superfluids and superconductors which contained the possibility of understanding these phenomena formally and theoretically was, however, proposed in 1951 and 1956 by Penrose and Onsager [1]. This is still, in a sense, equivalent to the accepted way of discussing these phenomena, and since it is historically the first and possibly the best known, it seems to be a good starting point for our discussion today.

The proposal they made was that one could essentially define superfluidity of a Bose system such as liquid He as a state in which the density matrix of the system factorized in a certain special way. They said that the density matrix $\rho = \rho(r, r') = \langle\psi^*(r)\psi(r')\rangle$, which is defined as the average of the product of the field creation operator at point r and the destruction operator at point r', could be factorized into $f^*(r)f(r')$, a function of r times a function of r' plus small terms. The density matrix $\rho(r, r')$ of any ordinary system tends to vanish as the points r and r' attain macroscopic distances from each other. The "small terms" are normal, i.e., go to zero as $|r - r'|$ goes to infinity. The term which has no dependence on the relative positions of r and r' is characteristic of superfluidity. So the statement is that this is equivalent to superfluidity; I hope I will be able to make that point clear in the rest of the talk. I should say, of course, that at absolute zero the average is taken in some hypothetical ground state of the system; on the other hand, at finite temperatures one does not use ground states but a thermal average.

Beliaev [2] extended this concept of factorization to the time-dependent Green's functions. He observed that the density matrix, which is the average of the time independent field operators, is a special case of the general Green's function, which is the average of the Heisenberg field operators $\langle\psi^*(r, t)\psi(r', t')\rangle$. He assumed that that could be factorized into $f^*(r, t)f(r', t')$ + small terms. Almost immediately and in a practically succeeding issue of the *Journal of Experimental and Theoretical Physics*, Gor'kov [3] observed that the same kind of theory could be generalized to apply to the electrons in superconductivity. It was not possible to have this kind of theory for just the single field operators because of the exclusion principle, but if one substituted instead the two-particle Green's function, which has a pair of field operators ψ^*, and a pair of ψ's, one could do the same thing for the fermions. One could talk about the fermion gas in the metal in terms of pairs of electrons which then have sufficiently similar commutation relations to bosons so that one can carry out the same kind of thing.

So Gor'kov introduced a function $F^*(X_1, X_2)$ of two variables X_1 and X_2 (including both time and space variables) and factorized his two-particle Green's function, a function of four variables, into pairs of two variable functions:

$$G(X_1, X_2, X_3, X_4) = \langle \psi^*(X_1)\psi^*(X_2)\psi(X_3)\psi(X_4)\rangle$$
$$= F^*(X_1, X_2)F(X_3, X_4) + \text{small terms}$$

Again the important term is not correlated between variables of the two different F's, but now one has a strong correlation between X_1 and X_2. One thinks of F^* as a kind of a two-particle wave function, and the two bound electrons in such a wave function should remain close together. Gor'kov showed that this theory, if one took F to be homogeneous in space, was completely equivalent to the then just developed Bardeen–Cooper–Schrieffer (BCS) theory of superconductivity, which now, of course, has come to be recognized as the correct microscopic theory of superconductivity [4]. As far as the superfluid (helium) version is concerned all reasonably useful microscopic theories that have been proposed have led to this kind of behavior but, as I say, no really satisfactory microscopic theory yet exists.

In superconductivity there already existed a theory of Ginsburg and Landau [5], which seems to describe superconductivity from a phenomenological point of view very satisfactorily. This theory describes the properties of superconductors in terms of a function which they called the order parameter function $\Psi(r)$, but the basic equations of the theory were remarkably similar to the equations for ordinary one-particle quantum mechanics in terms of a single-particle wave function $\Psi(r)$. So this function came to be known as the wave function although its meaning was not at that time understood. Gor'kov very shortly showed that in the appropriate limiting cases he could derive the Landau–Ginsburg theory from his theory [6], in which he identified $\Psi(r)$ with the function $F(X, X)$ at identical values of the variables. So he gave the function Ψ a physical meaning. It was only later that the corresponding theory for superfluidity, which is not quite as useful, was proposed by Gross and Pitaevskii [7].

Coherence phenomena can be treated perfectly satisfactorily using this Penrose–Onsager "off-diagonal long-range order" scheme. Unfortunately, all of those who until recently have chosen to work with this theory have made an additional assumption which is not necessary, and which makes the amount of work one has to go through in order to derive results considerably greater. This assumption is, that one should work always with states (in making the averages with which one forms either the density matrices or the Green's functions), with fixed definite total particle number N, i.e., total number of helium atoms or total number of pairs of electrons. In the case where one does the thermal average over a grand canonical ensemble, it is sometimes convenient, of course, to use many values of the particle number, but even there it was assumed that there was no coherence between states of different particle number. Now, so long as one is talking about a single isolated superfluid system, a single hunk, say, of superconductor or a single bucket of superfluid with a cover on it so that nothing evaporates, then, of course, the number of helium atoms or the number of electron pairs is perfectly fixed and the Hamiltonian must commute with the number operator. It seems extremely reasonable that, since the number of particles is a perfectly good quantum number, we should insist, in taking our average, on keeping the total number of particles fixed.

Unfortunately, it begins to be a little more difficult if we start trying to do interference experiments. In doing the kind of experiment we are going to talk about, we will be thinking very characteristically of systems which consist of two separate parts between which we have some kind of connection, and the basic feature of that connection is that it can transfer particles coherently from one side of the system, or one piece of the system, to the other. We will, in other words, be talking about systems such as those in which we have two regions of space separated by "slits" as you do in an electromagnetic interference experiment. What is more, it will be very hard, in discussing these coherence experiments, as it would be very hard in discussing for example electromagnetic interference experiments, to keep the slits open at all times. You would often like to compare what happens when you close off a slit with what happens when the slits are open. You would like to be able to turn off or on switches, superconducting switches or superfluid switches, here and there in your system. And under those circumstances, there is no reason whatever to expect that you are going to get satisfactory results by working with states in which the numbers of particles on the two sides are absolutely fixed and in which there are not allowed to be coherent superpositions of states in which the particles are differently partitioned between the two sides.

Let me suppose I have a system composed of two pieces connected by such a switch, and that I start with each piece having ODLRO, i.e., being described by

$$G(x, x') = f^*(x)f(x'),$$

but each having just $N/2$ particles. It turns out that there is no way that I can make up from this description the correct description of the system with the switch open, i.e., with ODLRO embracing the full composite system, in which x or x' may range over both halves. Thus no separate piece of a superfluid system is adequately described by a single state with ODLRO. Of course, it is possible to do things properly, by starting from a supermacroscopic system which has definite fixed particle number, and which contains every piece that I am ever going to attach to my system, i.e., every superconductor that I may ever bring into contact with my system or every reservoir of liquid helium that I may be interested in. But that is just inconvenient.

It is much easier to do it in a different way. The different way is to satisfy our original requirement of factorization by taking as our basis states (confirming afterward that what we do is all right), states in which there is an average of the particle operator itself.

$$\langle \Psi | \psi^*(r,t) | \Psi \rangle = \langle \psi^*(r,t) \rangle = f^*(r,t)$$

Clearly, when we do this our original assumption of ODLRO is satisfied:

$$\langle \psi^* \psi \rangle = \langle \psi^* \rangle \langle \psi \rangle + \text{other terms},$$

but equally clearly, the reverse is not necessarily true.

In other words, this implies that I still have my original assumption but I have made an extra assumption in addition. Clearly also these are states which do not have a fixed number of particles because I have destroyed a particle between one Ψ and the other, so at least the state has components with N and with $N-1$ particles, i.e., with two different numbers of particles. These states that I am going to use are very similar to, except for certain special features, the coherent states of the electromagnetic field with which people have begun to realize it is best to work in quantum optics. On the other hand, the states that we are confronted with in superfluidity and superconductivity do not have the simple complete coherence that one can get in the electromagnetic analogy. The maximum value that the mean value of Ψ can have is the square root of the density, because $\langle \psi^* \psi \rangle = \rho$, but in the helium case, it was shown in the original paper of Penrose and Onsager (and the estimate has been refined recently by MacMillan) that one gets of the order of $0.1\sqrt{\rho}$ for the absolute value. In the superconductivity case, the situation is even less coherent, one gets of the order of 10^{-4} of the maximum possible value. Nonetheless, these numbers are clearly recognizable as numbers or order unity rather than numbers of order 10^{-23}, and therefore the coherence is macroscopic in magnitude rather than microscopic.

Another way of putting it is that we all recognize that when we deal with macroscopic systems such as tables, chairs, solar systems and so on, it is extremely inconvenient to work with eigenstates of the conserved quantities such as total momentum, total parity, or anything like that;

it is much more convenient to work with wave packets of the eigenstates which have the property of localizing reasonably closely the macroscopic variables of the system, and that is exactly what we are doing here. One can make wave-packet transformations back and forth and demonstrate that we are working with the same kind of system.

The only condition we have is that someone sometime has to sit down with the equations and decide that the zero-point motion that we are ignoring in forming our wave packet, the diffusion of the wave packet, is quantitatively negligible compared with other fluctuations in the system. You have to decide, when you are talking about the solar system, that the uncertainty in the position of the sun is utterly negligible. In this case, you have to decide when you are talking about a macroscopic superfluid that the uncertainty in definition of the wave function of the quantum field is essentially negligible and it will not diffuse. I have done that calculation [8] and it does indeed turn out that even in the worst possible cases the zero-point motion is negligible compared to thermal fluctuations.

Now after a long prologue we come to the basic idea of what I am going to do today, which is to deal with the mean value of the quantum particle field as though it were one of the macroscopic thermodynamic and dynamical variables of the system, just as in, say, elasticity of crystals we deal with the position of a particular part of the crystal as a macroscopic thermodynamical variable, or in ferromagnetism we deal with the magnetization as a macroscopic variable. Both are, in fact, variables which destroy symmetry principles but are well known by our daily experience to be usable as thermodynamic variables. As in these cases, it is perfectly possible to define the variable not only for the total system but also for individual pieces of the system, small cells macroscopic from the atomic point of view but microscopic from our point of view. In other words, one can deal with coarse-grained averages as well as with macroscopic averages.

It is quite important that f, the mean value of the wave field, is a complex order parameter. It contains an absolute value and a phase:

$$f = |f| \, e^{i\varphi}$$

The two real parameters $|f|$ and φ turn out to be two different kinds of thermodynamic variables. In the thermodynamics of order–disorder systems, for example, we have a kind of thermodynamic variable, the long-range order in that case, which is characterized by the fact that while it is extremely convenient in discussing the thermodynamics and in keeping our concepts straight, there is no way we can get in there with a force and act on that particular variable. It is hard to find a force that pulls differently on copper and gold atoms. Similarly in this case, the magnitude of the mean value of the quantum field is not a variable that we can get in and manipulate. So this is not of much value to us in discussing macroscopic dynamics. The phase φ of the mean field, on the other hand, has the opposite property. It is a real dynamical variable such as the ones I gave as examples previously, the magnetization or the position of a piece of crystal. It is a variable corresponding to which there are forces, and thus is simultaneously a dynamical and a thermodynamical variable. The fact that it is a dynamical variable follows from the fact that the number operator N_i which corresponds to any one of our partial systems is the conjugate dynamical variable to the phase operator φ_i for the corresponding system. That is, the commutator of these two operators is the standard commutation relationship of conjugate dynamical variables, just like the P and Q of particle dynamics: $[N, \varphi] = i$. This is easily verified in a number of ways. Perhaps the simplest possible way is to ask oneself how one would go about forming, from wave functions with fixed numbers of particles, wave packets with fixed phase. That is fairly easy: we take our wave functions with fixed number, we sum over them in order to form a wave packet, we multiply by some real coefficient a_N which has a peak for some value of N and which is non-vanishing over a sufficiently broad region in order not to get into serious trouble with the uncertainty principle, and then we multiply by the coefficient $e^{i\varphi N}$:

$$\Psi(\varphi) = \sum_N a_N e^{i\varphi N} \Psi(N).$$

It is clear that in forming the mean value

$$\langle\psi\rangle = (\Psi(\varphi), \psi\Psi(\varphi)),$$

the term on the left with phase $e^{-i\varphi(N-1)}$ connects only with that on the right with phase $e^{i\varphi N}$, so that

$$\langle\psi\rangle = |f|\, e^{i\varphi}.$$

Thus the transformation matrix between phase and number operators is $e^{i\varphi N}$, just as for p and q. This implies the commutator, and also that we may write the operator equivalences

$$N = i\, \partial/\partial\varphi \qquad \varphi = -i\, \partial/\partial N.$$

There are limitations on the usability of these commutation relations but those are relevant only for very small systems, and we are talking specifically about a large system, so that we can really say this is a precise relationship. In addition this gives us an illustration of how to form wave packets for our system.

Now the essential dynamical equations for the whole phenomenon of superfluidity follow from these two ideas: that we are going to take states in which there are mean values of the particle field, and that the phase and the number operators are conjugate dynamical variables. The equations those lead to are, of course, the equations of motion of these two dynamical variables. We can write down the equation of motion of the number operator for one of our pieces of a dynamical system, connected by a superfluid connection to another system. Let the number and phase operators of the two systems be N_1, N_2 and φ_1, φ_2.

The equation of motion of the number N_1 is given by the standard quantum mechanical equation of motion, which, using the fact that these are conjugate dynamical variables, can be written

$$i\hbar\dot{N}_1 = [\mathscr{H}, N_1] = -i\,\partial\mathscr{H}/\partial\varphi_1,$$

and correspondingly we have the other Hamilton's equation for this pair of conjugate variables

$$i\hbar\dot{\varphi}_1 = [\mathscr{H}, \varphi_1] = i\,\partial\mathscr{H}/\partial N_1.$$

These equations are essentially equivalent to two equations which are familiar in the theory of superconductivity. The first is more or less equivalent to the London–Ginsburg–Landau current equation ([5], [9]), the equation which determines the supercurrent in a superfluid system. The second is an equation which has had a shorter history, although a similar equation has occurred in both the theory of superfluidity and the theory of superconductivity. It is equivalent to London's acceleration equation [9], telling how the system responds if we apply forces. As I have written them down, these are equations for operators, but they hold, of course, for mean values; so these are equally good equations for the mean value of the number operator and for the mean value of the phase in our wave packets.

The current equation is historically the first, and the most straightforward of the interference experiments that I promised to talk about are based on it. Of course, for an isolated system, as I said, the Hamiltonian commutes with the number operator, i.e., the Hamiltonian must obey gauge invariance, and be independent of the total phase. That just says, of course, that the number of particles moving into or out of an isolated system is zero. The situation is quite different if we talk now about two separate pieces of a superfluid system. Now imagine that we have two separate pieces of a superfluid system connected by a superfluid bridge (we must always remember that when I say superfluid I mean either helium or superconductors, in which case I am talking about pairs of electrons). Our very definition of superfluidity has said that we have a mean value of the quantum field operator in this whole system. If the energy were independent of the relative phase of the quantum field operator on the two sides I would not have such a mean value because on the average the two phases would fluctuate completely independently. So by my very definition of superfluidity I have said that there must be a dependence of the total energy on the relative phase on the two sides; there must be an energy which must be a minimum where the phases are equal, and must increase on each side of that minimum. Incidentally, of course, this tells me why superfluidity and superconductivity are low-temperature phenomena, because as thermal fluctuations increase, eventually at high enough temperatures the system is going to overcome this energy (which can be traced in every case to the quantum zero-point energy of the individual particles).

If, then, there is a dependence of the energy on $(\varphi_1 - \varphi_2)$:

$$U = U(\varphi_1 - \varphi_2),$$

we can have supercurrents between the two subsystems:

$$\frac{dN_1}{dt} = -\frac{dN_2}{dt} = \frac{1}{\hbar}\frac{\partial U(\varphi_1 - \varphi_2)}{\partial(\varphi_1 - \varphi_2)}.$$

We can equally well apply this to two small, semimacroscopic neighboring cells within an individual bucket of superfluid and we have again the same equation but in that case we get that the current will be the variational derivative of the energy with respect to the gradient of the phase:

$$J_s = \frac{1}{\hbar}\frac{\delta U}{\delta(\nabla\varphi)}.$$

This is the course-grained equivalent of the previous equation. Again we expect that the energy as a function of the gradient of the phase must be a minimum when the gradient is zero, and that it will be a quadratic function for small enough values of the gradient:

$$U = n_s(T)(\hbar^2/2m)(\nabla\varphi)^2.$$

As the coefficient it is convenient to put in $\hbar^2/2m$ times a perfectly arbitrary function $n_s(T)$ of the temperature and the internal state of the system, and then the current is the variational derivative of this with respect to the gradient of the phase and so may be written

$$J_s = n_s(T)(\hbar/m)\,\nabla\varphi.$$

Well, it is clear now why I have written the coefficient the way I did. I wanted to get a quantity which looked as though it was a velocity, a superfluid velocity $v_s = \hbar/m\, \nabla\varphi$ multiplying a superfluid number of particles. At this point, this is just a definition of the superfluid velocity. At absolute zero, in a perfectly homogeneous system, we can prove, in fact, that this is the velocity of the superfluid and that $n_s(T)$ is equal to the total number of particles or of electron pairs in the system. You prove that very simply by taking the perfectly homogeneous system, and translating it at a certain velocity, which must then be the superfluid velocity. I am aware of no proofs that v_s defined in this way means anything else in any real system, either superconducting or superfluid. In any real bucket of helium, any experimentalist like myself who has worked with it will tell you that there are always counterflows of normal and superfluid particles, and in any real superconductor there are impurities. Even at absolute zero in the presence of impurities n_s is not identically equal to N. Thus this is an idealization and one's reasons for calling this v_s are reasons basically of convenience rather than necessity. It is sometimes but not always a good approximation of the actual particle velocity in the system. But it is the current equation which is the basic equation.

In the case of superconductivity, another somewhat more familiar route may be taken to this equation. In the case of superconductivity you can recognize immediately that I have told you a whopper when I said that the energy is a functional of the gradient of the phase of the mean value of the wave field operator because that, of course, is not a gauge invariant statement. In the case of superconductivity my particles are charged; there is an electromagnetic field present and I can change the phase of the wave function arbitrarily if I at the same time change the value of the vector potential of that electromagnetic field. So in order to maintain gauge invariance I have to write instead that U is a functional of $(\hbar\, \nabla\varphi - 2e\mathbf{A}/c)$ where $\mathbf{A}$ = vector potential. $2e$ occurs in the factor rather than e because φ is the phase of a two-particle wave function. Then, of course, I take the well-known identity that the current is equal to $1/c$ times the variational derivative of the energy with respect to the vector potential and that gives me the electromagnetic current in this case.

At this point I would like to make two things clear. The first is that this discussion of the supercurrent already makes it very plausible (and I believe probably it can be made more than plausible by someone more patient than I) that superfluidity as we understand it, the existence of superflows in the absence of driving forces, and this type of coherence within the wave function or the off-diagonal long-range order concept to which it is equivalent, are both necessarily and sufficiently related to each other. Thus we suggest it is impossible to have supercurrents without off-diagonal long-range order and vice versa. The one seems plausible because there is only one variable, A, or the gradient of the phase, to which the current operator is conjugate, and therefore it does not seem that we can get supercurrents unless we have a dependence of the energy on the appropriate variable. Conversely, if we have a dependence of the energy on this variable we must have supercurrents. Now this seems absolutely trivial except that there are papers in the literature during the past year which violate both directions of the implication. There are papers producing hypothetical phases with off-diagonal long-range order which do not have any supercurrents, and there are papers producing hypothetical superfluid phases which have no off-diagonal long-range order, so that it may be worth pointing out that they seem to be necessarily and sufficiently connected.

The second importance of this part of the discussion is that this is the point at which macroscopic quantization begins to enter the picture. The first point at which macroscopic quantization enters is that the $\langle\psi\rangle$ is to be taken as a macroscopic variable of the system, and therefore it is a physical quantity. Thus it is not by any means to be taken as multiple valued. Therefore, as you follow this quantity around some circuit within which you have, say, two bridges between two superconducting samples (i.e., some multiply connected circuit) the value of that quantity must come back to the same value when we return to the same point. The phase, however, need not return to the same value: it can return to the same value $+2\pi$ or 4π or any number of integers times 2π. So we immediately get that there is a kind of quantization of a circuit integral. As we go around any closed curve within our superfluid, the integral of the gradient of the phase must be equal to $2n\pi$. And that is the basic quantization statement we have to make. Now in the case of superfluidity, as I have said, we often take the superfluid velocity to be $h/m\,\nabla\varphi$. So we immediately get that the line integral of the superfluid velocity around a closed curve is equal to nh/m, where n is an integer. So we get quantization of the circulation of the superfluid velocity. The question as to whether actual physical velocities are quantized in this way is a question of how good this assumption about the superfluid velocity is, and it turns out to be a pretty good one. But as far as I know there is no reason to say that it is exact.

The quantization of the magnetic flux in a superconductor comes about in a similar way but not exactly the same. We again take the formula for J_s which, using the similar quadratic formula for the energy, is $J_s = n_s e v_s$ and $v_s = \hbar/m\,\nabla\varphi - 2e/mc\mathbf{A}$. Realizing that in most macroscopic superconducting circuits the superfluid velocity inside the material is practically zero because of electromagnetic shielding (i.e., the London penetration effect) we can usually set it equal to zero. Then we go through our circuit integration again and we find that the line integral

$\oint A\, dl$ is equal to $hc/2e$. This line integral in turn is easily shown to be the magnetic flux. Again this is an approximation because we have had to use a circuit in which there is no supercurrent and in many instances, which have actually been checked out experimentally, the sample can be made so small that there is supercurrent at every point. That inexactness in the macroscopic quantization is thus one which can actually be proved experimentally. So it is important to remember that the quantization in the sense of the line integral of $\nabla\varphi$ is, of course, exact but the quantization of the usual quantities that are discussed is rather inexact.

Now, let us return to talk about the actual interference experiments. In order to do this, we have to go back from these more sophisticated continuum results involving U as a functional of the gradient of the phase to the trivial question of what happens when we have two superfluid or superconducting samples connected by a superconducting bridge, which can be represented by an energy which depends on the relative phase on the two sides. By general principles you can see that the dependence of the energy on the relative phase must be periodic because there is no way the system can tell whether the phase difference is zero, 2π, etc. as long as there is only the single bridge. Therefore, in this case, we expect the function $U(\varphi_1 - \varphi_2)$ to be periodic. Now there are two ways in which this periodic function can be constructed. The way in which ordinary macroscopic samples behave is very disappointing. There is one energy parabola $U \sim (\varphi_1 - \varphi_2)^2$ at the origin, another centered at $\varphi_1 - \varphi_2 = 2\pi$, etc.:

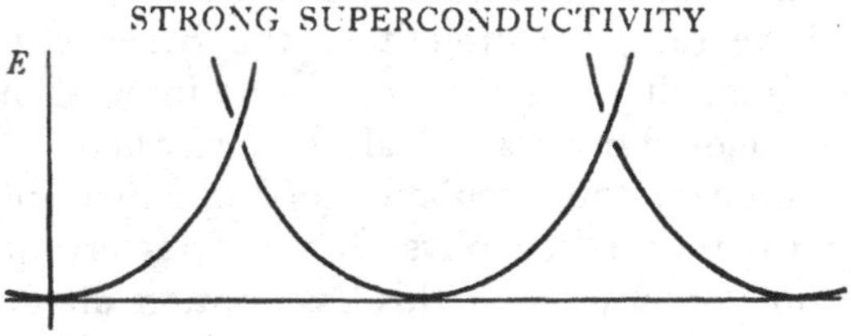

Figure 1

For reasons which are easy enough to understand but that I do not want to go into here, when you actually start working with this system, it tends almost always to pass through the intersections between the parabolas without paying any attention to the other parabola, so that it is not possible to get from one of these states to another easily. The tremendous contribution that Josephson [10] made to the subject is the fact that he discovered a way in which this periodic energy curve could be made to be simply sinusoidal. It did not then have any of these annoying metastable states up at high energy so that the system would follow the sinusoidal behavior as one changed the phase. The original Josephson junction was simply a tunneling barrier between two pieces of superconductor separated by an extremely thin film of insulator, but now it is realized that simply extremely thin wires or extremely thin bits of superconducting film behave in much the same way. Almost all the macroscopic interference experiments make use of this kind of junction between two pieces of superfluid, except for the original macroscopic flux quantization experiment by Doll–Näbauer [11] and Deaver–Fairbank [11].

What is the simplest way in which this could be used? We have two pieces of superconductor connected by an insulating film, and in order to arrive at this energy curve as a function of phase, I have had to assume that the phase difference is a certain fixed value everywhere in the junction. That need not be so, however, if I insert a magnetic field through the insulating barrier and parallel to it. Then the field is given by the curl of the vector potential so I can say that there is a vector potential perpendicular to the barrier.

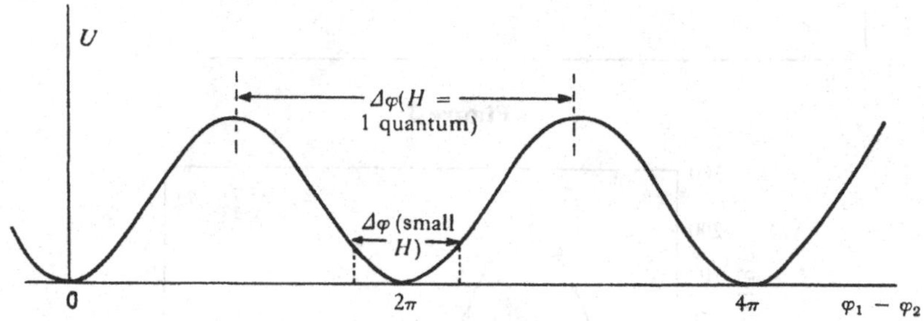

Figure 2

Thus the phase difference is a linear function of x. (Figure 3) Then if we draw the sinusoidal phase-energy curve we see that for small enough $H(\ll 4\lambda we/hc$, it turns out, where w is the width of the junction and λ the penetration depth) the phase difference is smeared only over a small portion of a cycle and the total energy is still sinusoidal; but when H is big enough to cover a full cycle, there is no longer any total dependence of energy on phase; it encompasses a whole 2π of the phase axis and then the energy clearly does not depend on where I am along the curve. (see figure 2) Therefore, since the total current is the derivative of the energy with respect to the phase, the total current will be exactly zero. In fact you can follow through the argument and convince yourself that the total current follows a perfect oneslit interference pattern. This experiment was the first check by Rowell and myself [12] that we were really seeing this kind of junction.

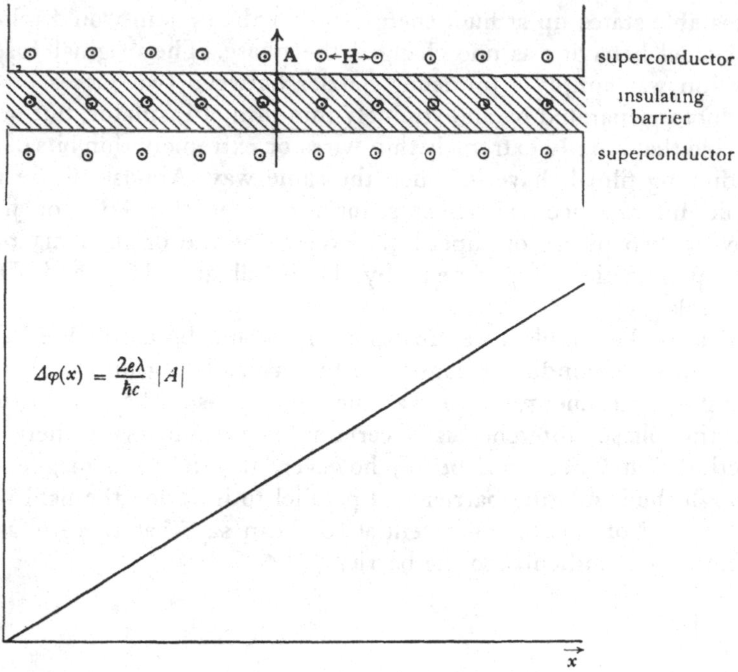

Figure 3

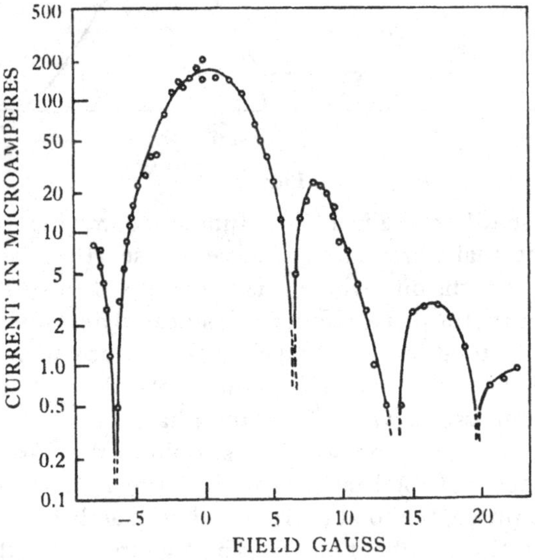

Fig. 4. The field dependence of the Josephson current in a Pb–I–Pb junction at 1.3°K.

Once you are convinced that you can do that then you can immediately see—we should have but Mercereau did—the next experiment to do. The next experiment to do is to make two of these things and separate them by a hole. In other words, I connect two superconducting reservoirs by two Josephson junctions and have a hole in the middle through which I pass a magnetic field; of course, the magnetic field also

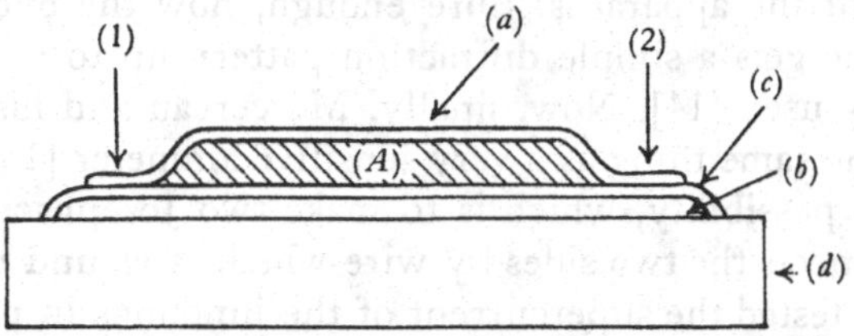

Fig. 5. Cross-section of a Josephson junction pair vacuum-deposited on a quartz substrate (*d*). A thin oxide layer (*c*) separates thin (~1000 Å) tin films (*a*) and (*b*). The junctions (1) and (2) are connected in parallel by superconducting thin film links forming an enclosed area (*A*) between junctions. Current flow is measured between films (*a*) and (*b*).

had to be in the two Josephson junctions. In that case, I get the original pattern for each of the junctions but the phase at one is correlated with the phase at the other depending on how much of the vector potential comes from the magnetic field in the hole. So as I increase the magnetic field, since the hole is much bigger than the junction I get a two-slit interference pattern with a one-slit interference pattern as its envelope. This was checked by Mercereau *et al.* [13].

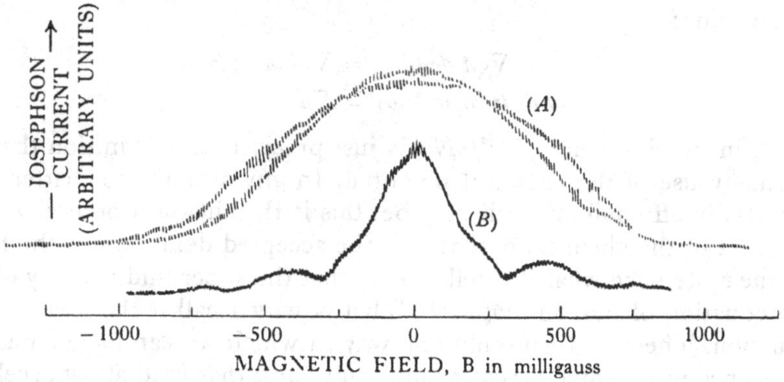

Fig. 6. Josephson current versus magnetic field for two junctions in parallel showing interference effects. Magnetic field applied normal to the area between junctions. Curve (*A*) shows interference maxima spaced at $\Delta B = 8.7 \times 10^{-3}$ G, curve (*B*) spacing $\Delta B = 4.8 \times 10^{-3}$ G. Maximum Josephson current indicated here is approximately 10^{-3} A.

The next experiment, also carried out by Mercereau *et al.*, was to check the idea of Bohm and Aharonov that the vector potential plays a special role in quantum mechanics relative to classical mechanics. They wanted to check that one could get an interference phenomenon without there being any actual magnetic field acting on the electrons so they made the same device but now enclosed the magnetic field in a solenoid, and insofar as possible there was no magnetic field from the solenoid in any other parts of the apparatus. Sure enough, now the one-slit pattern is gone, but one gets a simple diffraction pattern up to as large magnetic fields as they used [14]. Now, finally, Mercereau and his co-workers at Ford used the same thing as a very sensitive ammeter [15]. They carried out the final possibility, which is to make two Josephson junctions but now one connects the two sides by wire which is wound noninductively. So now they tested the supercurrent of the junctions in parallel in order to test the relative phase of the two junctions. This then is testing the Josephson current as a function of the superfluid velocity, i.e., as a function of $\boldsymbol{\nabla}\varphi$, where $\boldsymbol{\nabla}\varphi$ is caused by a superfluid velocity rather than by a vector potential. Sure enough one gets a nice interference pattern, in fact one can get several orders of magnitude higher sensitivity as a micro-ammeter than with any other device known to man.

It is now interesting to go on to the other of these two equations, the acceleration equation. So far I have not said anything about what really makes the currents flow. How do I accelerate currents in the superfluid? Of course, the answer must be this equation. What is the first thing this equation tells us? Let us again suppose that we are talking about a set of neighboring microscopic cells in the uniform sample. Let us take the gradient of this equation, and a macroscopic, thermodynamic mean value:

$$\nabla\langle\hbar\, d\varphi/dt\rangle = \nabla\langle\partial\mathscr{H}/\partial N\rangle$$
$$m\, d/dt\,(\hbar/m\;\nabla\varphi) = \nabla\mu$$

Well, in the first place $\langle\partial\mathscr{H}/\partial N\rangle$ is just precisely the definition that we normally use of the chemical potential. In any system, no matter how dreadfully off equilibrium it may be, this is the chemical potential. The gradient of the chemical potential is the accepted definition of the force on the system, so what this tells you is that the superfluid velocity obeys the equation of motion $m\dot{v}_s = F$. That is why I call it the acceleration equation. There is, then, only one way in which we can have an actual difference in potential across a superfluid, and that is to allow acceleration to take place. We can do that very directly: for instance, suppose I had two buckets of liquid helium with an opening between them, I could make the difference in head on the two sides finite but I would know then that unless something else happened, the liquid helium would accelerate and one would get just U-tube oscillations; one would never get any dissipation.

Now fortunately, or perhaps unfortunately, that is not the only way in which acceleration can take place.

The other possibility is the relatively new idea of phase slippage by vortex motion [16]. In order to have a phase difference, a change in the phase around a circuit, I have to have a hole in the middle of it, otherwise I am going to have a point which is a branch point of the macroscopic particle field. I do not like that, but I have to assume that there may be situations in which one has to introduce branch lines of the quantum particle field. In fact, the simplest such arrangement I can introduce in order for the phase not to be single valued within a single macroscopic sample of superfluid is a line at which you take the mean value of the particle field to be zero, and then around that branch line I can allow the phase to have a variation of 2π: around that branch line I have $\oint \nabla\varphi \, dl = 2\pi$. This object is called a vortex core and occurs both in superconductivity and superfluidity. In superconductivity there is often a quantum unit of flux associated with it [17]. In superfluidity it is the well-known quantized vortex [18].

Now the concept of phase slippage comes from the fact that I can get a time-dependent difference in phase from one point in a sample of superfluid to another by the following procedure. Let us take two points in the superfluid, point 1 and point 2, and a path between them. If I have a quantized vortex line on one side of the path, the phase difference from 1 to 2 may have one value. Now I move my vortex line to the other side of the path; and because I have a branch line of the phase, I now know that my phase integral,

$$\int_1^2 \nabla\varphi \, dl = \varphi_2 - \varphi_1$$

is increased by 2π. If now I am fortunate enough to have another vortex that I can pass through the path, I can increase the phase by 2π again and so on. In other words I can keep the phase varying continuously if I am willing to pass vortices indefinitely through the path between the two points; so that I can have a chemical potential difference between point 1 and point 2 given by

$$\mu_1 - \mu_2 = h\langle dn/dt \rangle,$$

where $\langle dn/dt \rangle$ is the average rate of passage of vortices through the path.

Now, this tells us two things. One: it tells us the conditions under which dissipation can occur in superfluids: it can occur only if there are vortex lines. It turns out that in liquid helium it is very easy to create vortex lines and so one almost always has dissipation of some sort. There are two kinds of superconductors. The first is the so-called soft superconductors, which are very resistant to the presence of vortex lines, and pass supercurrents without any dissipation whatever. The second is the hard superconductors, which contain vortex lines and always have an associated dissipation [19].

Second, and more interesting from the point of view of interference experiments, is the idea of using this expression to connect a potential difference with a frequency. You can see this is like Schrödinger's equation $i\hbar\, \partial\psi/\partial t = H\psi$, i.e., $(i\hbar/\psi)\, \partial\psi/\partial t$ equal to an energy. So I can associate a frequency with an energy difference simply by Einstein's frequency condition, which when it is applied in superconductivity is called Josephson's frequency condition. A typical simplified experiment, not far from the experiment that we actually carried out, is the following. I can take two buckets of helium with a partition between them, with a little hole in the partition, and then I can make a level difference so that the liquid is trying to flow through the hole. As the liquid flows through the hole it turns out that it probably blows smoke rings. In any case it creates vortices and these vortices move continuously out of the hole so if I make a path from the liquid surface in one bucket to the other, vortices pass across that path at a rate determined by the head difference. How can I tell whether they are passing at that rate? I can tell by introducing an ultrasonic transducer which modulates the flow through the hole at the appropriate rate, i.e., I can introduce an a.c. flow and synchronize the motion of those vortices. When the a.c. flow is synchronized with the height difference I would expect a singularity to occur in the flow and that indeed is exactly what we have observed [20]. This is interesting because it is the only one of these interference experiments so far which have been carried out in a precisely analogous way in both superconductivity and superfluidity [21].

What is the conclusion to be drawn from all this? In the first place, probably the presently most interesting conclusion is that our ideas about superconductivity and superfluidity seem to be right. That is particularly interesting in liquid helium, where the microscopic theory is not in as strong condition as the microscopic theory is in superconductivity. In the second place, something which probably is much more important for the more distant future is that there are here the germs of a series of devices which act on an entirely different principle from anything which has appeared on the macroscopic level previously. It is so different, in fact, and so exciting, that, to anyone who is familiar with what is going on, it is intensely frustrating that we have not been able to make more use of them than we have. One of the reasons why I think we have not is that somehow these things are so different from what we are used to in our technology that we do not even yet have the sense to realize what we need them for, but I am sure that somewhere we need them.

QUESTION. Would it be possible for several types of off-diagonal long-range order, several types of condensation to occur simultaneously?

ANSWER. Well, that is an interesting question. That was proposed by Gor'kov and Galitskii. The question is very simple. What they suggested is essentially that you might also have a second term in the factorization, $\langle\psi(r)\psi(r')\rangle = f(r)f(r') + g(r)g(r')$. It does not seem too important in the superfluid case but in the superconducting case $F(r, r')$, which is a function of two variables, might be a d state and the corresponding G might be an s state and you might have a superposition of d and s, say. There was quite a hassle about whether this was possible. It never was completely resolved except by recourse to actually directly soluble models. It is also interesting because it is a question of whether the whole use of the coherence concept the way I did it is correct or whether it might be possible to have a ring with simultaneously two types of long-range order: two values of quantized flux, say. I think the answer is no. There are only two ways in which this could occur. You can, first, have not this but $[f + g(r)][f + g(r')]$. That is what happens if you make a coherent superposition, but that is not interesting because that is just a new kind of coherence. It is a different coherence, not a superposition of the original ones but a different coherent state. The other possibility is, you might actually make a wave function which is a linear superposition of the two kinds of order and that is the one that is harder to shoot down. I know of no mathematical proof that such a thing can be shot down but I know the physics of it. The physics is that really these two wave functions are very far apart from each other in Hilbert space. One has a macroscopic number of particles quantized in one state, the other has a macroscopic number of particles quantized in another. Therefore you have to change the states of an infinite number of particles to get from one to the other. Therefore you can never have any coherence between the two. This is very like Schrödinger's old paradox about the live cat and the dead cat. You can make a wave function which is a linear superposition of a live cat and a dead cat, but it does not mean anything.

REFERENCES

1. O. Penrose, *Phil. Mag.* **42**, 1373 (1951); O. Penrose and L. Onsager, *Phys. Rev.* **104**, 576 (1956). It has been pointed out to me that the concept of ODLRO is actually mentioned in the paper of V. L. Ginsburg and L. D. Landau, in 1950, which as we will shortly discuss gives the phenomenological theory of superconductivity. These authors failed to observe the limitation that fermions could not exhibit ODLRO except as pairs. It is clear, however, that by 1958, when the papers of Beliaev and Gor'kov to which we later refer appeared, the concept was clearly understood by the Russian group. It is unfortunate that the paper of Yang, *Rev. Mod. Phys.* **34**, 694 (1962), which has had great influence in this country, did not acknowledge the Russians' priority.
2. S. Beliaev, *J. exp. theor. Phys.* **34**, 417 (1958).
3. L. P. Gor'kov, *J. exp. theor. Phys.* **34**, 735 (1958).
4. J. Bardeen, L. N. Cooper, and J. R. Schrieffer, *Phys. Rev.* **108**, 1175 (1957).
5. V. L. Ginsburg and L. D. Landau, *J. exp. theor. Phys.* **20**, 1064 (1950).

6. L. P. Gor'kov, *J. exp. theor. Phys.* **36**, 1918 (1959).
7. E. P. Gross, *Nuovo Cim.* **20**, 454 (1961); L. P. Pitaevskii, *Sov. Phys. JETP* **13**, 451 451 (1961).
8. P. W. Anderson, in *Lectures on the Many-Body Problem* (E. R. Caianello, Ed.) Vol. 2, p. 113. Academic Press, New York, 1964.
9. F. London, *Proc. Roy. Soc.* **152A**, 24 (1935).
10. B. D. Josephson, *Physics Lett.* **1**, 251 (1962).
11. R. Doll and M. Näbauer, *Phys. Rev. Lett.* **7**, 43 (1961); B. S. Deaver and W. Fairbank, *Phys. Rev. Lett.* **7**, 43 (1961).
12. P. W. Anderson and J. M. Rowell, *Phys. Rev. Lett.* **10**, 230 (1963); J. M. Rowell, *Phys. Rev. Lett.* **11**, 200 (1963).
13. R. C. Jaklevic, J. J. Lambe, A. H. Silver, and J. E. Mercereau, *Phys. Rev. Lett.* **12**, 159 (1964).
14. R. C. Jaklevic, J. J. Lambe, A. H. Silver, and J. E. Mercereau, *Phys. Rev. Lett.* **12**, 274 (1964).
15. R. C. Jaklevic, J. J. Lambe. A. H. Silver, and J. E. Mercereau, *Phys. Rev.* **140**, A1628 (1965).
16. P. W. Anderson, *Rev. Mod. Phys.* (1966) to be published.
17. The theoretical discovery of the superconducting flux line is certainly due to A. A. Abrikosov, *J. exp. theor. Phys.* **32**, 1442 (1957).
18. R. P. Feynman, *Prog. of Low Temp. Physics* (C. J. Gorter Ed.), Vol. 1, Ch. II (1955).
19. P. W. Anderson, *Phys. Rev. Lett.* **9**, 309 (1962); Y. B. Kim, C. F. Hempstead, and A. R. Strnad, *Phys. Rev. Lett.* **9**, 306 (1962); *Phys. Rev.* **131**, 2486 (1963).
20. P. L. Richards and P. W. Anderson, *Phys. Rev. Lett.* **14**, 540 (1965).
21. P. W. Anderson and A. H. Dayem, *Phys. Rev. Lett.* **13**, 195 (1964).

Reprinted from REVIEWS OF MODERN PHYSICS, Vol. 38, No. 2, 298–310, April 1966
Printed in U. S. A.

Considerations on the Flow of Superfluid Helium*

P. W. ANDERSON
Bell Telephone Laboratories, Murray Hill, New Jersey

First, we show that the most important equations of the dynamics of the two types of superfluids, He II and superconductors, follow quite directly from the simple assumption that the quantum field of the particles has a mean value which may be treated as a macroscopic variable. The background of this ansatz is also discussed. Second, we apply these equations to various physical situations in He II, notably the orifice geometry and the superfluid film, and show how they, and particularly the idea of phase slippage accompanying all dissipative processes, can be applied and what kinds of macroscopic interference phenomena may be expected. The effect of synchronization in the ac interference experiment is discussed.

I. INTRODUCTION

The material of the first part of this article covers really much the same areas of basic physics which are treated by Martin and by Nozieres in their articles at the same conference at which this was presented. Nonetheless the reader will find that the emphasis is sharply different. The striking macroscopic interference phenomena, the observation of which has been stimulated by Josephson's remarkable discovery,[1] call out for a description in terms of a definite wave function with a definable phase ϕ in every part of the system, while quantum hydrodynamics as pioneered by Landau[2] has tended to emphasize the superfluid velocity v_s and its equations of motion. The identification of $v_s = (\hbar/m)\nabla\phi$ has its limitations in relating these points of view, especially in Josephson junctions; while the opposite point of view, that ϕ exists, makes theorists uncomfortable because it breaks the gauge symmetry. Nonetheless I shall choose to take the latter path, assuming either that the reader has read such a discussion as Ref. 3 which gives the reason why this broken symmetry is possible, or that he will understand that the superfluid system is to be attached at some point to a large superfluid reservoir, with respect to which the phase is to be measured.

The idea of off-diagonal long-range order was introduced by Penrose and Onsager.[4] They suggested that superfluidity be described as a state in which the density matrix

$$\rho(r, r') = \langle \psi^*(r)\psi(r') \rangle$$

could be factorized:

$$\rho(r, r') = \psi^*(r)\psi(r') + \text{small terms}. \tag{1}$$

Beliaev[5] extended this for helium to a Green's function theory in which ψ was explicitly time-dependent, and first observed that the chemical potential determined the time dependence. Sortly thereafter Gor'kov[6] observed that superconductivity theory could be cast into the same form by substituting electron pair field operators $\psi\psi$ for the He atom bose field. In superconductivity there already existed a set of phenomenological equations proposed by Ginzburg and Landau[7] which dealt with an "order parameter" η, and it was soon recognized that this order parameter was the same as the factorized ψ of the Gor'kov theory.[8] It was only much later that Gross[9] and Pitaevskii[10] proposed similar sets of equations for liquid helium.

The notion that it was possible to regard the function which appears in these treatments as essentially the mean value of the quantum particle field has long been accepted in both helium[5] and superconductivity (there the first explicit discussion of the transformation between the "ODLRO" and mean field points of view was given by Anderson[11]) but only in the case of a homogeneous system; apparently the general case has not been discussed until recently even for superconductivity.[3] The basic idea is that it is as legitimate to treat the quantum field amplitude as a macroscopic dynamical variable as it is the position of a solid body; both represent a broken symmetry which, however, cannot be conveniently repaired until one gets to the stage of quantizing and studying the quantum fluctuations of the macroscopic behavior of the system.

Here we are going to discuss less the microscopic background of this ansatz than a number of its most important consequences for He II, many of which follow without further microscopic assumptions and are therefore of fundamental interest. Only some of what we will have to say is new, in the sense that many of the

* This paper was presented at the Brighton Symposium on Quantum Fluids at the University of Sussex, England, 18 August 1965. The full proceedings of this Conference will be published by the North-Holland Publishing Company in 1966, including the present article as well as those referred to above, in which a more conventional approach to superfluid dynamics is employed. Many of the other contributions have of course also been published in various journals.

[1] B. D. Josephson, Phys. Letters **1**, 251 (1962).

[2] L. D. Landau, J. Phys. USSR **5**, 71 (1941).

[3] P. W. Anderson, in *Lectures on the Many-Body Problem*, edited by E. R. Caianello (Academic Press Inc., New York, 1964), Vol. 2, p. 113.

[4] O. Penrose and L. Onsager, Phys. Rev. **104**, 576 (1956); see also O. Penrose, Phil. Mag. **42**, 1373 (1951).

[5] S. T. Beliaev, Zh. Eksperim. i Teor. Fiz. **34**, 417 (1958) [English transl.: Soviet Phys.—JETP **7**, 289 (1958)].

[6] L. P. Gor'kov, Zh. Eksperim. i Teor. Fiz. **34**, 735 (1958) [English transl.: Soviet Phys.—JETP **7**, 505 (1958)].

[7] V. L. Ginsburg and L. D. Landau, Zh. Eksperim. i Teor. Fiz. **20**, 1064 (1950).

[8] L. P. Gor'kov, Zh. Eksperim. i Teor. Fiz. **36**, 1918 (1959) [English transl.: Soviet Phys.—JETP **9**, 1364 (1958)].

[9] E. P. Gross, Nuovo Cimento **20**, 454 (1951).

[10] L. P. Pitaevskii, Zh. Eksperim. i Teor. Fiz. **40**, 646 (1961) [English transl.: Soviet Phys.—JETP **13**, 451 (1961)].

[11] P. W. Anderson, Phys. Rev. **112**, 1900 (1958).

appropriate equations have been written down (see the papers of Martin and Nozieres). The basic idea was previously stated in a Letter,[12] and some of the consequences have been explored in subsequent Letters by Donnelly[13] and by Zimmerman.[14] We will first discuss a particularly simple point of view on the derivation of the basic equations, following more closely than usually the corresponding ideas in superconductivity.[3,15] The emphasis will be on what can be shown to follow more or less rigorously from using this ansatz as a general semimicroscopic definition of superfluidity. Only the parameter values in the resulting equations require any other knowledge of the microscopic system. Then we will discuss the dynamical consequences for the orifice experimental geometry, as well as some more general situations such as film flow. In Appendix B we discuss briefly the interesting connection with classical ideal fluid hydrodynamics, where the basic "Josephson" theorem turns out to have a classical analogue which has not to our knowledge been clearly stated previously.

II. BASIC EQUATIONS

We take as our definition of a superfluid that it is a fluid in which the particle field operator ψ has a macroscopic mean value, in a sense which is defined shortly.

$$\langle\psi(r,t)\rangle=f(r,t)\exp[i\phi(r,t)]. \tag{2}$$

Here we allow slow (on the atomic scale) space and time variations; the essential point is that $\langle\psi\rangle$ has a mean value in the thermodynamic, quasi-equilibrium sense. One may think of the situation here as completely analogous to the definition of temperature in a non-equilibrium state; it is possible if there are equilibrating processes of shorter range in time or space than the coarse-grained scale of our averaging, which in turn is to occur over regions smaller than the macroscopic scale of physical measurements. Of course, in every system one can define a temperature and entropy if the average is allowed to extend over a long enough time, while only certain special systems will give a stable limiting value to $\langle\psi\rangle$. More explicitly, we visualize averaging ψ over some small region of space–time; if the region is small enough compared to the rates and ranges of microscopic fluctuations, we will obtain some finite value. As we increase the size ΔV of the region $\langle\psi\rangle$ will decrease to zero in a normal system very rapidly, in times of the order $h/$(mean kinetic energy) and ranges of order interparticle distances. In a superfluid system, on the other hand, we assume that at an intermediate, "coarse-grained" scale $\langle\psi\rangle$ approaches a limiting finite value, not changing until we reach a scale on which it varies because of the presence of macroscopic perturbations such as macroscopic fields and flows. (In Appendix A we discuss this definition a little more deeply.)

The whole problem of superfluid dynamics, then, reduces to the question of how to deal with $\langle\psi\rangle$ (often denoted the superfluid order parameter) as a thermodynamical variable. It is quite important that $\langle\psi\rangle$ is a *complex* order parameter; it has both an amplitude f and a phase ϕ. There is actually rather a sharp distinction between the two real thermodynamic variables f and ϕ, and it is the behavior with respect to ϕ which is responsible for specific superfluid properties. The distinction is that ϕ is coupled, like a polarization or a strain, to external forces, where f is merely an internal order parameter in the sense, for instance, of the original long-range order parameter of order–disorder systems, or of the antiferromagnetic sublattice magnetization. In principle a corresponding force might exist but in practice it does not, so f simply manifests itself as a convenient tool by which to describe the condensation process.

The point of ϕ, then, is that it is not only a thermodynamic but also a dynamical variable. The latter fact comes from its being the dynamically (not thermodynamically) canonically conjugate variable to N, the total number of particles in the system described by ϕ. (The limitations of this statement for systems of few particles[16] are irrelevant here.)

We illustrate this fact by forming wave packets in many-body wave-function space, just as in discussing the relationship of p and q it is useful to form wave packets. Let us write the wave function of one of our coarse-grained cells of volume ΔV as

$$\Psi(\Delta V)=\sum_N a_N\Psi_N(\Delta V) \tag{3}$$

where, since our cell is only a part of the superfluid, it is essential to realize that the state is a superposition of states Ψ_N with different numbers of particles N, with coefficients a_N large in some range of values $\Delta N\sim N^{\frac{1}{2}}$. An important simplifying assumption concealed in (3) is that the cell may be made big enough so that a_N and Ψ_N do not depend very much parametrically on the variables of the rest of the system: this is what is meant (see Appendix A) by a "satisfactory" local description.

We postulate that $\langle\psi\rangle$ has a limiting mean value, which must be of order $(N/\Delta V)^{\frac{1}{2}}$; we calculate this value from (3):

$$\frac{1}{\Delta V}\int_{\Delta V}d\tau\,\langle\psi(r)\rangle=\frac{1}{\Delta V}\sum_{N,N'}a_{N'}{}^*a_N\int(\Psi_{N'},\psi(r)\Psi_N)\,d\tau$$

$$=\sum_N a_{N-1}{}^*a_N\int\frac{(\Psi_{N-1},\psi(r)\Psi_N)\,d\tau}{\Delta V}. \tag{4}$$

[12] P. L. Richards and P. W. Anderson, Phys. Rev. Letters **14**, 540 (1965).
[13] R. J. Donnelley, Phys. Rev. Letters **14**, 939 (1965).
[14] W. Zimmermann, Phys. Rev. Letters **14**, 976 (1965).
[15] P. W. Anderson, N. R. Werthamer, and J. M. Luttinger, Phys. Rev. **138**, A1157 (1965).
[16] W. H. Louisell, Phys. Letters **7**, 60 (1963); L. Susskind and J. Glogower, Physics **1**, 49 (1964).

In the case of the Bose-condensed perfect gas, one may take

$$\Psi_N=(c_0^+)^N\Psi_{\text{vac}}, \tag{5}$$

where

$$c_0=\frac{1}{(\Delta V)^{\frac{1}{2}}}\int\psi(r)\,d\tau$$

$$c_0\Psi_N=N^{\frac{1}{2}}\Psi_{N-1}, \tag{6}$$

so that

$$\langle\psi\rangle=\sum_N a_{N-1}{}^*a_N(N/\Delta V)^{\frac{1}{2}}. \tag{7}$$

This is the maximum possible value of the matrix element. (Here we have taken advantage, as we will always, of our freedom to choose the phase of Ψ_N to make the matrix element real in order to keep the phase factors—which of course are *not* irrelevant—in the a_N's.) In real superfluid systems and at finite temperatures (the case of superconductivity is of course quite similar if for a single Bose ψ or c we read Fermion pair operators) the matrix element does not take on its maximum possible value—in He, for instance, as McMillan has shown,[17] the value is about 11% of the maximum when ΔV is reasonably large. Even if

$$M=(\Psi_{N-1},\psi\Psi_N) \tag{8}$$

is large, however, $\langle\psi\rangle$ will not be large unless the a_N preserve phase coherence. For example, we may form a wave packet using

$$a_N=(2\pi\Delta N)^{-\frac{1}{4}}\exp[-\tfrac{1}{4}(N-\bar{N})^2/(\Delta N)^2]\exp(i\phi N). \tag{9}$$

In this case if $\Delta N\gg1$,

$$\langle\psi\rangle\cong|M|\exp(i\phi) \tag{10}$$

whereas if the phase factors are arbitrary $\langle\psi\rangle$ will be very much smaller, even if every individual Ψ_N represents a pure Bose-condensed state of the volume element ΔV. We will very shortly discuss the reason why wave packets like (9) actually occur in superfluid systems; first, however, let us dispose of a few formal preliminaries.

Any linear combination of the set of fixed-number states Ψ_N may be written as a linear combination of our basic wave packets (3)–(9) [which we will call $\Psi(\phi)$]. In particular, the number eigenfunction Ψ_{N_0} may itself be written

$$\Psi_{N_0}\propto\int_0^{2\pi}d\phi\exp(-i\phi N_0)\Psi(\phi) \tag{11}$$

$$(=\sum_N a_N\,\delta(N,N_0)\Psi_N).$$

The operator $-i\partial/\partial\phi$ acting on $\Psi(\phi)$ has the same effect as multiplying Ψ_{N_0} by N_0:

$$\int_0^{2\pi}d\phi\exp(-i\phi N_0)\left(-i\frac{\partial}{\partial\phi}\right)\Psi(\phi)$$

$$=N_0\int_0^{2\pi}d\phi\exp(-i\phi N_0)\Psi(\phi)=N_0\Psi_{N_0}$$

so that we may take

$$N\leftrightarrow-i(\partial/\partial\phi) \tag{12a}$$

and correspondingly it may be shown that

$$i(\partial/\partial N)\leftrightarrow\phi \tag{12b}$$

in the limit that N may be considered a continuous variable. Thus, as we stated, N and ϕ are conjugate dynamical variables.

In general, also, N and ΔN are sufficiently large that the system's dynamics may be treated reasonably well classically and the uncertainty in ϕ is not excessive—of course (12) implies $\Delta N\Delta\phi\sim1$.

In any case the equations of motion of N and ϕ are

$$i\hbar\dot{N}=[\mathcal{H},N]=i(\partial\mathcal{H}/\partial\phi)$$

$$i\hbar\dot{\phi}=[\mathcal{H},\phi]=-i(\partial\mathcal{H}/\partial N).$$

Taking the mean values of these two equations and assuming that the wave packets are such that $\Delta N/N$ and $\Delta\phi$ are both small (quasi-classical case) is the source of the two equations which essentially characterize superfluidity: the equation for superfluid flow, corresponding to London's equation or the Ginsburg–Landau current equation in superconductivity, and the "Josephson" frequency equation, which is related to the acceleration equation for superfluid flow in both He and superconductivity. The second is somewhat simpler though less generally known: it is just

$$\hbar(d\phi/dt)=\partial E/\partial N=\mu. \tag{13}$$

Here we have used the standard definition of the chemical potential μ; if the fluid is in motion so that the particles have kinetic energy $\frac{1}{2}mv_s^2$ that is to be included in μ as well as the internal energy and any external forces. Obviously the partial derivative holds S fixed. In an isothermal, assumed incompressible bath of liquid helium, with free surface at height h in a gravitational field g,

$$\mu=m(p/\rho)+mgh+\tfrac{1}{2}mv_s^2. \tag{14}$$

In the nonisothermal case (14) should contain the thermomechanical term.

Equation (13) is of the utmost importance in understanding superfluidity. It says two things: first, that if the state of the superfluid is constant in time, because ϕ is a thermodynamic variable it will be constant and μ must be constant everywhere: there can be no potential differences in the truly steady state. Second, if there

[17] W. L. McMillan, Phys. Rev. **138**, A442 (1965).

is any potential difference $\mu_1-\mu_2$ between two elements in the superfluid $\phi_1-\phi_2$ must change in time. This can happen in two ways. The simplest is acceleration. If we take the gradient of (13), we obtain

$$(d/dt)(\hbar\nabla\phi)=\mathbf{F}, \tag{15}$$

where $\mathbf{F}$ is the total force on the particles, and making the identification (which we shall discuss shortly) of

$$\hbar\nabla\phi=p_s=mv_s, \tag{16}$$

this is the statement that the superfluid may undergo acceleration without frictional damping by whatever external forces act upon it. A potential difference may lead to continuous acceleration, then.

Slightly more subtle and much more physically important is the concept of phase slippage by quantized vortex motion. As we have emphasized, ϕ is the phase of the thermodynamic variable $\langle\psi\rangle$, which is of course necessarily single-valued. ϕ, however, being a phase, need not be single-valued in a multiply connected system such as a toroid; it need merely return to its original value $\pm 2n\pi$ on traversing a path around a nonsuperfluid obstacle:

$$\oint\nabla\phi\cdot d\mathbf{s}=2n\pi. \tag{17}$$

In terms of the superfluid momentum p_s this expresses the idea of the quantization of angular momentum in units of $\hbar$ or of vorticity $\omega=\frac{1}{2}\nabla\times v_s$ through a closed curve in units of $h/2m$. As we shall see, however, the quantity v_s is not necessarily a measurable particle velocity and so (17) is somewhat more fundamental than the usual concept of vorticity quantization.

A bucket of superfluid may be made multiply connected not only by the presence of actual solid obstacles but by the introduction of one-dimensional regions of nonsuperfluidity within the liquid itself: lines where $\langle\psi(r)\rangle=0$. These are "vortex cores" and may of course move according to the usual laws of hydrodynamics along with the surrounding fluid. Around such a line there can be a circulation of one or, less usually, an integral number of quanta, according to (17).

Equation (17) shows that the integral of $\nabla\phi$ along a path on one side of a vortex core differs from that on the other by 2π (see Fig. 1). Thus when a vortex core moves across the line between points 1 and 2, this may cause a time rate of change of $\phi_1-\phi_2$. Mathematically, we may write

$$\langle\mu_1-\mu_2\rangle_{Av}=T^{-1}\int_0^T dt\hbar\frac{d(\phi_1-\phi_2)}{dt}=\frac{\hbar}{T}\int_0^T dt\frac{d}{dt}\int_1^2\nabla\phi\cdot dl, \tag{C}$$

where $\langle\ \rangle_{Av}$ denotes time average, where we consider the limit T very long, and C is any path from 1 to 2. If the liquid is assumed to be in a reasonably steady state, so that at 0 and T the positions of the various vortices do not differ very much, the integral is just equal to 2π times the number of vortices which have crossed C in this time interval. Thus we have

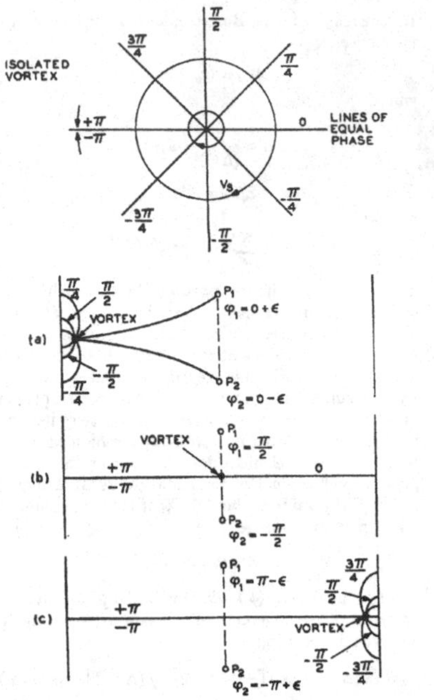

FIG. 1. Illustration of phase changes at two points P_1 and P_2 in a channel as a vortex moves between them. From a to b to c the vortex moves across from left to right; as it moves from one wall to the other the relative phase changes by 2π.

$$\langle\mu_1-\mu_2\rangle_{Av}=h\langle dn/dt\rangle_{Av}, \tag{18}$$

where $\langle dn/dt\rangle_{Av}$ is the average rate of motion of vortices across a path from 1 to 2. This is the "phase slippage" concept which is used to explain the various "ac Josephson"-like experiments.[18,19] Of course, this is not incompatible with the acceleration equation (15), so in a sense this phenomenon too is merely a manifestation of the fact that potential differences occur only in an accelerated superfluid; but it is a point of view which had not previous to Ref. 12 found its way into the He literature, nor until recently, with the discovery

[18] S. Shapiro, Phys. Rev. Letters 11, 80 (1963).
[19] P. W. Anderson and A. H. Dayem, Phys. Rev. Letters 13, 195 (1964).

of the Josephson effect and of flux creep and flow, into that of superconductivity.

Let us now take up the current equation, the theory of which is somewhat more complex. There are two aspects to this. In the first place there is the quasi-rigorous Eq. (16) the meaning of which becomes a bit vague with more careful consideration. If we simply suppose that the state of a volume element ΔV moving with velocity v_s is obtained from that of a stationary element by a pure Galilean transformation, multiplying ψ by $\exp(imv_s x/\hbar)$, this equation is trivially valid. That is the usual derivation of it, in one form or another. I know of no acceptable proof of (16), however, in the sense of showing v_s to be a real particle velocity, in physically important situations such as counterflow of normal and superconducting fluids, or where v_s is varying reasonably rapidly, as near a vortex core. Since (16) is the statement which leads to vorticity quantization, this means that that concept, often claimed to be exact, is apparently not so.

The statement has been made in the literature[13,14] that the results—specifically the "ac Josephson effect"—which follow from (13) or the phase slippage concept can equally well be "derived" from vorticity quantization plus perfect fluid hydrodynamics for the superfluid. Neither of these latter ideas, however, has at the moment a very quantitative experimental background, while theoretically as we have just seen the phase has a much more secure theoretical meaning than v_s; it seems to us that (13) is a much more rigorous and complete theoretical statement than the hydrodynamic equations, which are derived via (15) and (16) from it. Normal fluid counterflow and dissipation do not affect it. In a perfect classical fluid, for instance, the vortices cannot move across stream lines, so clearly He II is not such a fluid.

Perhaps a more rigorous general reason for using (16) at least as a *definition* for v_s (other than that it allows us to use ϕ as a velocity potential for the superfluid flow) is (15), which shows that $m\,dv_s/dt$ does then give the rate of exchange of momentum per superfluid particle with external forces, an excellent operational definition of v_s.

The superflow, however, is best determined not from the expression for v_s but from the other of the two Hamilton equations for the conjugate variables N and ϕ:

$$\hbar(dN/dt)=\partial E/\partial\phi. \tag{19}$$

Gauge invariance assures us that in fact the energy of an isolated bit of superfluid is independent of ϕ, so that in the absence of a coupling to its neighbors $dN/dt=0$. Consider, on the other hand, a pair of neighboring volume elements ΔV_1 and ΔV_2. Our definition of superfluid implies that $\langle\psi\rangle$ has a tendency to constancy, so that the mean phases ϕ_1 and ϕ_2 of the two neighboring elements will be coupled by some energy which is a minimum when $\phi_1-\phi_2=0$:

$$E=U(\phi_1-\phi_2).$$

Assuming the effects of other neighboring elements can be treated independently (suitable arguments can be found for this step) we see that the flow across the boundary between ΔV_1 and ΔV_2 is given by

$$J_{\text{tot}}=\frac{dN_1}{dt}=-\frac{dN_2}{dt}=\frac{1}{\hbar}\frac{\partial U(\phi_1-\phi_2)}{\partial(\phi_1-\phi_2)}. \tag{20}$$

This is the expression used in the theory of the Josephson current[1,3] across a barrier between two macroscopic pieces of superconductor; at this point we are carrying the same reasoning over to the continuous interior of a superfluid or superconductor. As in that case, we may pause now to point out that it is this coupling energy which enforces the phase coherence of each individual volume element of the superfluid. The kinetic energy matrix element which transfers particles across the boundary can cause transitions like

$$\{\Psi_1^N\to\Psi_1^{N-1};\,\Psi_2^{N'}\to\Psi_2^{N'+1}\}, \tag{21}$$

with a matrix element we may call M_{12}. (Here $\Psi_{1,2}$ is the many-body wave function of $\Delta V_{1,2}$.) If the wave packets (3) have coefficients a_N which, like (9), have coherent phase relationships, all transitions $N'\to N'+1$ can occur coherently with all transitions $N\to N-1$. Mathematically, the energy due to transitions like (21), inserting wave packets like (3), is

$$(\Psi_1\Psi_2,\,(\text{K.E.})\Psi_1\Psi_2)=\Big(\sum_N a_N{}^*a_{N-1}\Big)\Big(\sum_{N'} a_{N'}{}^*a_{N'+1}\Big)M_{12}. \tag{22}$$

As we see, this energy is orders of magnitude larger for coherent wave packets like (9), for which ϕ is determinate, than in the incoherent case.

In the interior of the superfluid, it is more convenient to go over to a continuum representation of the coupling energy which maintains phase coherence. We write U as a functional of the gradient of ϕ (as well as f, of course)

$$U=\int U[(f),\nabla\phi]\,d\tau$$

$$\simeq\int E(f,S)(\nabla\phi)^2\,d\tau \tag{23}$$

and then by considering a cell of wall area A and thickness d we find

$$\frac{J}{\text{unit area}}=\frac{dN_1}{dt}\Big/A=\frac{\delta U}{\hbar\delta\nabla\phi}=\hbar^{-1}E\nabla\phi. \tag{24}$$

If we use the pseudo-identity

$$v_s=\hbar\nabla\phi/m,$$

we may write

$$J_s = n_s(f, S)v_s$$
$$= [\hbar n_s(T)/m]\nabla\phi, \qquad (25a)$$

so that we can identify the parameter E of (24):

$$E = \delta^2 U/\delta(\nabla\phi)^2 = \hbar^2 n_s(T)/m. \qquad (25b)$$

Thus the current equation contains a completely arbitrary parameter n_s. In pure, homogeneous systems at absolute zero Galilean invariance can be used to show that $n_s = n$, the total number of particles; but in impure systems at $T=0$, or any system at $T \neq 0$, no such identity holds, though in general n_s is of the same order of magnitude as n.

The supercurrent (24) exists even if the phase ϕ is completely stationary in time, which as we have shown implies the absence of accelerating forces. In fact, in the presence of accelerating forces and time-dependent ϕ we must expect additional quasi-particle currents, in general; the system will exhibit a two-fluid hydrodynamics, the complexities of which need not concern us here. So far as I know their proper treatment requires more knowledge of the actual physical system than we are assuming here.

One very important point about (24) taken together with (15) is that it makes it at least highly probable that there is both a necessary and a sufficient connection between the existence of supercurrents and our definition of superfluidity (1) (which is essentially equivalent to what Yang[20] has named ODLRO). Namely, $\langle\psi\rangle$ and therefore ϕ will not exist if the energy is not such as to maintain spatial coherence of ϕ, so $\delta^2 U/\delta(\nabla\phi)^2$ must exist and be positive in a superfluid in this sense, meaning necessarily supercurrents by (24). (15) shows that they flow with zero forces and are therefore supercurrents. Hypothetical phases with $\langle\psi\rangle$ but no supercurrents (Cohen[21]) seem to ignore this half of the argument.

Conversely, the only dynamically conjugate variable to N is the phase as we have defined it, so that the existence of a dN/dt in a stationary state implies a $\delta U/\delta\phi$, which implies that ϕ is a meaningful variable. Various hypothetical superconducting phases (e.g., Frohlich[22]) do not satisfy *this* half of the argument.

This concludes our general discussion of the basic equations of superfluidity. We restate the conclusions: the phase equation (13) and the corresponding equation of phase slippage are exact in the "integrated" sense that they give the phase difference between two distant points in undisturbed regions of superfluid. The existence of the order parameter alone guarantees the existence of quantized vortices, and according to (24) these are indeed vortices in that they contain a superfluid circulation. However, the quantization of vorticity in any true sense is dependent on the imprecise assumption (15) that the phase is the velocity potential with fixed coefficient. This need not be true unless we treat (15) merely as a definition of v_s, for instance near a vortex core, just as the quantization of flux in superconductivity is not necessarily precise. Operationally, for example, the measurement of h/e by the ac Josephson effect, or of h/mg by the helium counterpart, is more precise in principle than by flux or vorticity quantization. In practice, of course, present-day methods are not capable yet of distinguishing these niceties, but it is of value to have a clear idea of the theoretical assumptions behind the various equations, since it is foreseeable that the most precise measurements of many important physical quantities will involve quantum coherence.

[20] C. N. Yang, Rev. Mod. Phys. **34**, 694 (1962).
[21] M. H. Cohen, Phys. Rev. Letters **12**, 664 (1964).
[22] H. Frohlich and C. Terreaux, Proc. Phys. Soc. (London) **86**, 233 (1965).

III. SOME DYNAMICAL CONSEQUENCES

The macroscopic quantum interference effects promised by the existence of $\langle\psi\psi\rangle$ in superconductivity have been relatively easy to observe for a number of reasons: the light mass of the electron, permitting weak superfluid connections to be made easily by use of the tunneling phenomenon, the coupling to the electromagnetic field which leads to flux quantization, and most particularly the fact that that coupling provides a second parameter, the penetration depth λ, which in screening out the current and magnetic field from the interior of the superconductor creates the Meissner effect, that is ensures that every superconductor exhibits a finite critical magnetic field H_{c_1} below which it contains no vortices and thus essentially exhibits constant ϕ. In He, $\lambda \to \infty$ so that the corresponding $\omega_{c_1} = 0$; no rotation of a sufficiently large He sample is too small for vortices to be energetically favorable. Indeed, experiments show that few samples are ever free of vorticity. Even worse, the coherence length—the length given by the ratio of $\delta U/\delta|\psi|^2$ to $\delta U/\delta|\nabla\psi|^2$, which determines how rapidly the order parameter can vary and thus how large a vortex core is—is of order a few Å, so that no channel through which He can flow is too small to contain a vortex. All this means that the useful idealization in superconductivity of the "ideal Josephson junction," a weak link between two reservoirs having constant phase, is probably not relevant to the helium case. Any barrier in which the flow occurred only by quantum-mechanical tunneling would have to be of subatomic dimensions, especially in thickness, to permit any measurable current to flow, and could not be supported mechanically. Any attempt to replace the ideal junction by a thin channel, on the other hand, runs into exactly the same difficulties that are encountered with long thin bridges in the case of superconductivity,[23,19] namely, that the device acts like a

[23] R. D. Parks and J. M. Mochel, Rev. Mod. Phys. **36**, 284 (1964).

sequence of weak links in series and one is never sure exactly where the phase rigidity is breaking down. The closest approximation we could imagine to a single definable "weak superfluid junction" was the orifice geometry, which is analogous to the "short" thin film bridge of Anderson and Dayem[19] in superconductivity.

Let us analyze this system in some detail, as we did the Josephson junction previously.[3] First, a note as to the driving term to be inserted to represent any externally applied conditions. In the case of normal systems it is natural to place the ends of a specimen in contact with reservoirs at different chemical potential levels μ_1, μ_2, giving an "applied" potential gradient $\nabla\mu$, and to describe any situation in terms of solutions of the resulting applied field problem; we calculate J as a function of the pressure gradient or field, even though actually we may be driving the system with a constant current generator. The microscopic theory is done by inserting μN terms into the Hamiltonian, as is well-understood in the calculation of resistance, for instance, or thermoelectric power.

It is precisely the nature of superfluids that they cannot assume a stationary state under a field or pressure gradient, but will, as explained by Anderson, Werthamer, and Luttinger,[15] have to be described by phases with a time dependence obeying (13). This condition follows when we insert the appropriate μN terms, as there described; but that does not lead to a way to discuss the equally interesting case of an imposed current. One obvious technique is to impose a fixed phase difference by postulating an infinitely tight coupling to reservoirs consisting of large superfluids of fixed phase, but that is often quite unphysical. We have used without discussion[3] the technique of inserting a term

$$\Delta E = J_{12}(\phi_1 - \phi_2)$$

in the energy to describe the effect of a fixed supercurrent J_{12} between regions 1 and 2. In superconductivity a rather rough physical derivation of this can be given in terms of the electromagnetic interaction between J and the magnetic flux of a vortex line. It is analogous to, but more general than, the technique of inserting a $\mathbf{p}\cdot\mathbf{v}_s$ term sometimes used to derive the hydrodynamic equations.

In the case of He we may return very simply to Eq. (19). We showed that the particle accumulation rate dN_1/dt in a volume element ΔV_1 is given by:

$$dN_1/dt = \hbar^{-1}(\partial U/\partial\phi_1).$$

If no net accumulation is to occur, and if the current leaving ΔV_1 to neighboring elements is given in terms of the coupling energy between them by

$$J_{12} = dN_1/dt)_{\text{into superfluid}} = \hbar^{-1}(\partial U_{\text{coupling}}/\partial\phi_1)$$

there must be a compensating term

$$dN_1/dt)_{\text{current generator}} = -\hbar J_{12}\phi_1.$$

Similarly, to make the corresponding current leave element (2) we must have $dN_2/dt)_{\text{c.g.}} = \hbar J_{12}\phi_2$ so we may represent a constant current generator by

$$\mathcal{H}_{\text{gen}} = \hbar J_{12}(\phi_2 - \phi_1). \tag{26}$$

Using this let us discuss the orifice problem in the presence of a constant driving current. Current acceleration can be important only on the time-scale of the U-tube oscillations, which is usually longer than the time necessary to create a vortex or otherwise change the phase but can be included if desired. Also, it will greatly simplify one's thinking without falsifying any important physical features to assume $T \to 0$; i.e., only incompressible, superfluid flows.

Under these conditions $\hbar/m\nabla\phi = \mathbf{v}_s$ and

$$\nabla\cdot\mathbf{v}_s = 0$$

so

$$\nabla^2\phi = 0.$$

Thus we can solve for the flow in the absence of vortices by a simple potential calculation. Let the radius of the orifice be a. The equipotentials and streamlines are along coordinate surfaces in a set of oblate spheriodal coordinates, defined in terms of cylindrical coordinates r, θ, z through the axis of the orifice by

$$\begin{gathered}(r^2/a^2\cosh^2 u) + (z^2/a^2\sinh^2 u) = 1;\\ (r^2/a^2\sin^2 v) - (z^2/a^2\cos^2 v) = 1;\\ \theta = \theta;\\ -\infty \le u \le \infty, \qquad 0 \le v \le \pi/2, \qquad 0 \le \theta < 2\pi;\end{gathered} \tag{27}$$

or

$$\begin{gathered} r = a\cosh u \sin v\\ z = a \sinh u \cos v.\end{gathered} \tag{28}$$

The potential (i.e., phase) is

$$\phi(r, z) = C\int \frac{du}{\cosh u} = 2C\tan^{-1} e^u, \tag{29}$$

so that the total phase difference between the two large reservoirs separated by the orifice is

$$\phi_1 - \phi_2 = \pi C = \phi(+\infty) - \phi(-\infty). \tag{30}$$

The velocity is in the "u" direction and is

$$v_s = \frac{\hbar}{ma}\,\frac{C}{\cosh u(\sinh^2 u + \cos^2 v)^{\frac{1}{2}}}. \tag{31}$$

Some special values of the velocity field are:

Along the axis:

$$r = 0, \qquad \sin v = 0, \qquad z = a\sinh u$$

$$v_s = \frac{\hbar C}{ma}(\cosh u)^{-2} = \frac{\hbar C}{m}\,\frac{a}{a^2 + z^2}. \tag{32a}$$

In the orifice:

$$z=0, \quad r=a\sin v, \quad u=0$$

$$v_s=(\hbar/m)[C/(a^2-r^2)^{1/2}]. \qquad (32b)$$

On the plane:

$$r=\pm a\cosh u, \quad \cos v=0$$

$$v_s=(\hbar C/mr)[a/(r^2-a^2)^{1/2}]. \qquad (32c)$$

The total particle flow is

$$J=\frac{\rho}{m}\int v_s\cdot dS=\frac{2\pi\hbar Ca\rho}{m^2}=\frac{2\hbar a\rho}{m^2}(\phi_1-\phi_2), \qquad (33)$$

[using (30)] and the total kinetic energy is

$$K=\tfrac{1}{2}\rho\int v_s^2\,d\tau$$

$$=\tfrac{1}{2}\rho\frac{\hbar^2}{m^2}\int(\nabla\phi)^2\,d\tau$$

$$=\tfrac{1}{2}\rho\frac{\hbar^2}{m^2}\left(\int dS_{+\infty}\cdot-\int dS_{-\infty}\cdot\right)\phi\nabla\phi$$

$$K=\tfrac{1}{2}J\hbar(\phi_1-\phi_2)=(\rho a/m)[\hbar^2(\phi_1-\phi_2)^2/m]. \qquad (34)$$

Thus this energy is a quadratic function of the phase difference $(\phi_1-\phi_2)$. However, it is essential to realize that this is only one branch of a multiple-valued function, because by gauge invariance K must be a periodic function of $\phi_1-\phi_2$ with period 2π. That is, if we were to pin down the phases in the two reservoirs by coupling to reservoirs of fixed relative phase $(\phi_1-\phi_2)_0$ we could satisfy the boundary conditions by any flow

$$J=(2\hbar a\rho/m^2)[(\phi_1-\phi_2)_0\pm 2n\pi]$$

and the corresponding kinetic energy

$$K=(\rho a/m)(\hbar^2/m)[(\phi_1-\phi_2)_0\pm 2n\pi]^2. \qquad (34)$$

Equation (34) must also be used, then, in the presence of a current generator (27); thus the total energy is

$$E=(\rho a\hbar^2/m^2)[(\phi_1-\phi_2)\pm 2n\pi]^2-\hbar J_{12}(\phi_1-\phi_2). \qquad (35)$$

This energy considered as a function of $\phi_1-\phi_2$ has an infinity of points of metastable equilibrium where

$$J=J_{12}$$

$$(\phi_1-\phi_2)=2n\pi+(m^2J_{12}/2\hbar a\rho).$$

$$E=-J_{12}[nh+(m^2J_{12}/4a\rho)]. \qquad (36)$$

The situation is diagrammed in Fig. 2. The first drawing assumes a small current generator, $J_{12}<ha\rho/m^2$, which is the current value when $\phi_1-\phi_2=\pi$, the value at which the parabolas cross. For this current, although none of the energy minima are truly stable each is at least the lowest energy state for a given fixed phase difference (this has not been proved but seems obvious).

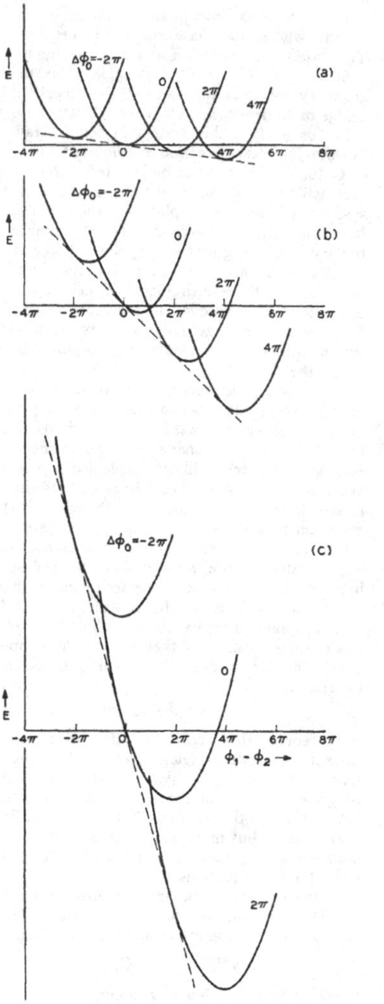

FIG. 2. Energy parabolas for potential flow in an orifice as a function of relative phase for different constant current generators J_{12}. $(\Delta\phi)_0$ is the relative phase "slippage" defined by $E(\Delta\phi_0)=J_{12}\Delta\phi_0$. (a) Low current; (b) $J_{12}\simeq v_c^0$; (c) J_{12} approaching observed critical current.

The system can still absorb energy from the current generator by "running downhill" in phase, but only if some fluctuation or external perturbation lifts it over the energy barriers between parabolas. In principle, if $J_{12} > h\alpha\rho/m^2$ the third drawing is correct, and there is no true energy minimum; this corresponds to a "zeroth-order" critical velocity (we take the mean value over the area of the orifice)

$$(v_s)^0 = 2\hbar/ma \tag{37}$$

which is $\frac{1}{4}$ cm/sec for $a = 10\ \mu$. Actually, the crossing points between parabolas do not represent possible transitions, because the two parabolas represent entirely different "sheets" of the energy connected only by passage of a vortex across the orifice, and we must consider the activation energy problem for creation of a vortex, as has been discussed by Vinen.[24] At the very least a length of vortex line of the order $2a$ must be created, which has energy of order

$$(E_{\text{vortex}})_{\min} = (\pi\rho\hbar^2/m^2)\ln(a/\xi)\times 2a,$$

where ξ is the coherence length, ~ 1 Å. This must be compared to the energy gained when a vortex line is halfway across the orifice. which is of order

$$\pi\hbar J_{12} = (\pi^2\rho a^2\hbar/m)v_s.$$

The result is another "critical" velocity

$$v_s^{(1)} = (2/\pi)(\hbar/ma)\ln(a/\xi)$$
$$\cong 7.5\hbar/ma. \tag{37'}$$

This is also far smaller than observed critical velocities, indicating as discussed by Vinen that the great difficulty in forming vortices in most situations is probably nucleating them at the walls.

Yet another "critical" velocity may be estimated if we suppose that the mechanism for phase slippage is the most plausible one in a simple orifice geometry, that of blowing vortex rings out on the downstream side of the orifice, of approximately the size of the orifice itself. The energy of a vortex ring of radius a is

$$E_{\text{ring}} = (2\pi^2\rho a\hbar^2/m^2)\ln(a/\xi)$$

which can be produced from an energetic point of view only if the energy available from the current source in each cycle, $2\pi\hbar J_{12}$, is equal to E_{ring}. From this we get

$$v_s^{(2)} = (\hbar/ma)\ln(a/\xi) \cong 11.5(\hbar/ma). \tag{37''}$$

[Note that the momentum conservation equation, as opposed to these energy considerations, simply gives us the frequency condition (13) as expected from (15)].

The essential physical point here is that all of these "critical" velocities are much less than real observed superflow velocities, indicating that in all cases the generation and motion of vortices is controlled by large random fluctuations, presumably either in the generation near the walls or in the motion of vortices already present. In general, the working point of an orifice is found to be not near the lowest intersection of two energy curves, where

$$v_s = 2\hbar/ma \quad \text{and} \quad (\phi_2 - \phi_1)_0 = \pi,$$

but at a phase-difference of the order of 10π, where the energy available is much greater than that necessary to form a vortex and thus we may expect rather irregular and unstable behavior. When vortices are created under such conditions they are accelerated rather strongly and give up considerable energy to the normal excitations.

It is because of this large value of the phase difference that the orifice geometry—and, correspondingly, to a lesser extent superconducting thin film bridges—are more difficult to demonstrate spacial interference effects with than the Josephson tunnel junction, for which the system moves adiabatically from one energy minimum to the next, 2π away, whenever that is energetically possible. Incidentally, it is clear that since v_c is not very dependent on channel length, the total phase difference for a long channel at v_c is even greater than for a short one, and as a result even more randomness in the creation of vortices and even less sensitivity to the precise value of phase difference is to be expected for long channels.

It is probably for these reasons that of all the macroscopic quantum interference effects, only the driven ac experiment has as yet succeeded in He. This experiment depends on the principle of synchronization, in which a strong external ac signal is introduced which can override the internal fluctuations.

First let us consider a free-running orifice connecting two reservoirs with a height difference Z. This height difference will decrease at a rate

$$\frac{dZ}{dt} = \frac{J\rho}{mA} = v_s\left(\frac{\pi a^2}{A}\right), \tag{38}$$

where A is the area of the surface in the smaller reservoir. J also determines the rate of generation of vortices; if J is greater than the observed J_c, vortices will be generated very rapidly, and conversely; but actually of course there is a functional dependence of the rate on the current:

$$dn/dt = f(v_s) \tag{39}$$

which will be rather steep, as shown in Fig. 3(a), but at least finite at very low v_s. Finally, we have the Josephson relation (13):

$$mgZ = -h(dn/dt) = -hf[(A/\pi a^2)(dZ/dt)]. \tag{40}$$

Equation (40) neglects the possibility of phase change by acceleration, i.e., it really contains a term in dJ/dt or d^2Z/dt^2, which will have no effect in the mean but does lead to the U-tube oscillations. If f were a step

[24] W. F. Vinen, *Proceedings of the International School of Physics "Fermi" (Varenna) 1961*, edited by G. Careri (Academic Press Inc., New York, 1963), p. 336.

function, dZ/dt would be fixed and the height difference would decay linearly, but presumably f is somewhat "soft," and as Z decreases the decay rate will do so also, but perhaps less slowly. Figure 3(b) shows a hypothetical decay curve which fits qualitatively with the vortex generation rate shown in 3(a).

Now suppose that as the height difference drops, we are causing an ac flow to be superimposed on the dc. During half the cycle we will be increasing the tendency to form vortices, during the other half decreasing it. When the height difference is such that $dn/dt=h\nu$, one vortex per cycle will be formed—presumably in the positive half-cycle, with quite high probability. When formed it uses up some considerable fraction of the available energy so another cannot be formed immediately; thus there will be a strong tendency for exactly one vortex per cycle to be formed, since the second half-cycle is not available. Because of (40) this will mean a tendency to fixed Z, i.e., a plateau in dn/dt [Fig. 3(a)] and thus in Z. Another way of putting it is that when the vortex formation is in a definite phase relationship with the ac signal, power can be transferred from the ac generator to the system as a whole, enough power to appreciably change the flow rate. If the alternating current is larger than the dc—as in our experiment[12]—clearly it is quite possible to stop the dc flow entirely, because we can control vortex formation wholly with the ac.

I have found a very simple mechanical analogy useful for understanding the ac Josephson effect. (See Fig. 4.) The relative phase of the two reservoirs I think of as the angular position coordinate ϕ of a set of locomotive wheels, and the velocity of the locomotive is the height difference Z; the ratio of position to phase is then the radius of the wheels corresponding to h/mg. The equation (40) for the rate of generation of vortices as a function of the flow (acceleration) may be inverted to give an effective nonlinear frictional force on the "locomotive"

$$\frac{A}{\pi a^2}\frac{dZ}{dt}=-f^{-1}\frac{mgZ}{h}=-f^{-1}\left(\frac{dn}{dt}\right). \qquad (41)$$

If f is a step function, this gives a constant frictional

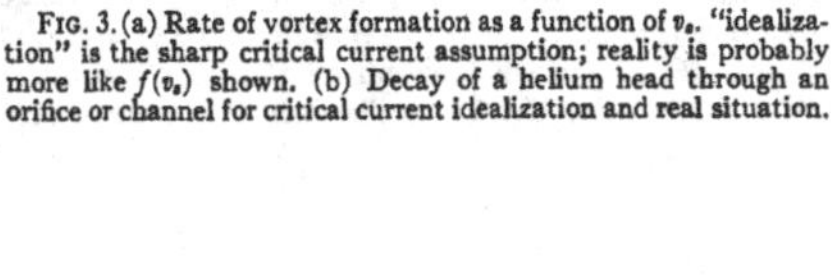

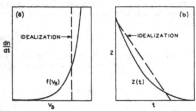

FIG. 3. (a) Rate of vortex formation as a function of v_s. "idealization" is the sharp critical current assumption; reality is probably more like $f(v_s)$ shown. (b) Decay of a helium head through an orifice or channel for critical current idealization and real situation.

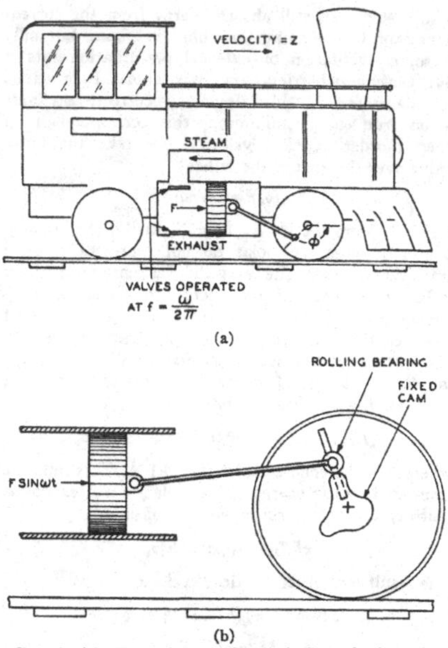

FIG. 4. (a) Illustration of "locomotive" mechanism of ac Josephson-type effect. (b) "Locomotive" system which could be driven at submultiple velocities.

force—i.e., the height (velocity) decreases linearly to zero. If f is as in Fig. 3, we get a "braking" action which leads to a decay between linear and exponential. It is an oversimplification to think of this as a constant force in time. Think of the locomotive as having square wheels and rusty bearings, so that the losses occur in some definite but not simple way during each cycle.

Now let us introduce the driven alternating current. This gives us an energy proportional to the phase coordinate, and alternating at frequency ω; it may be schematized by attaching a piston through a simple linkage to our locomotive wheels, and applying a force on the piston by (for instance) admitting steam intermittently at a rate $\omega/2\pi$:

$$F=F_0\cos(\omega t+\phi_0). \qquad (42)$$

(See Fig. 4.)

As we very well know, such an alternating force is capable of keeping the locomotive going at a steady velocity Z if it is large enough, and if

$$\omega=2\pi(dn/dt),$$

i.e., the force is applied once per revolution of the wheels. Also, as is less well-known but obvious, this mechanism can act as a brake or an accelerator at this

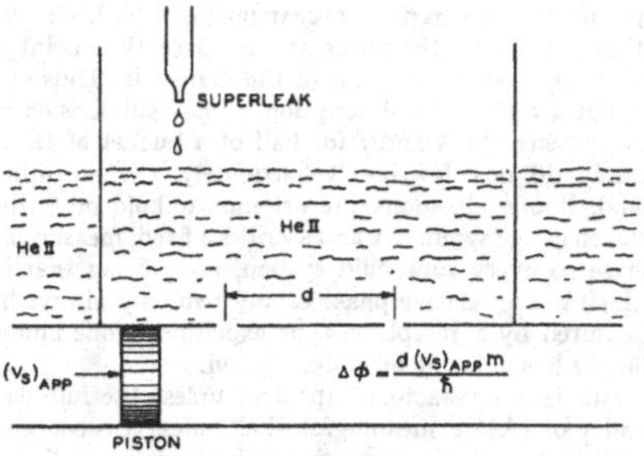

FIG. 5. Suggested space interference experiment with He II.

velocity. The system will attempt to synchronize itself, even where the power available is not adequate to hold it in synchronization.

Another possibility is to run the wheels at n times the frequency of the valves which admit the steam. Clearly if the velocity of the locomotive is not perfectly uniform, or the valves do not regulate the steam harmonically, the two can run in synchronism and drive the locomotive. This kind of harmonic ($V=n\hbar\omega/2e$) is often observed in the true ac Josephson effect.

In helium and in the thin film bridges, the phase slippage takes place by means of vortices. This means that the motion to which the driving ac is coupled is highly anharmonic (square wheels). Another valid schematization of this is to make the piston linkage highly anharmonic—let it roll on a queer-shaped cam, for instance [see Fig. 4(b)]. Then not only can the wheels be driven faster than the frequency of the force, but also they can be driven at subharmonics, since, for example, the vortex may be formed only every nth cycle of the driving current. Thus we expect and observe both harmonics and subharmonics, in the Anderson–Dayem and Richards–Anderson experiments.[12,19]

Both of these experiments, for final quantitative description, must wait for quantitative theoretical treatments of the generation and motion of vortices. But the basic principle of phase synchronization of the external signal with the relative phase of the order parameter in the reservoirs is independent of detailed mechanism.

As for spacial interference experiments, the most promising seems to be the orifice analog of the Mercereau "current" experiment.[25] Here one drives a supercurrent through two orifices in parallel, and at the same time causes a superflow past the orifices on one side (see Fig. 5).

The second superflow enforces a phase difference $\Delta\phi$ at the two orifices on one side so that both orifices cannot simultaneously be at metastable minima of their energy curves (see Fig. 2), unless that phase difference is $2n\pi$. Another way of putting it is that the circulation through the two orifices must be quantized, leading to an additional circulating current which may aid breakdown (vortex creation) at one or both orifices. This effect would be periodic in the phase. Unfortunately, it is very sensitive to fluctuations and instabilities.

A phenomenon in which the idea of phase slippage must play an important role is the superfluid creep of films. It has been suggested that vortices form at the critical velocity with axes parallel to the film and perpendicular to the flow.[13] That is almost certainly correct—it gives dimensionally the correct critical velocity, which again is of order $\sim 10\hbar/md$. However, I would speculate somewhat differently on certain details.

First, the motion of the vortices. Examination of typical estimates indicates that the frictional forces on He vortices allow them to move with a velocity component parallel to the Magnus force of about 1% of the flow velocity. Thus vortices of the type postulated above will be generated at the solid surface and move out of the film into the free surface only a few thousand Å downstream (or vice versa, but this seems less likely). This will be the predominant dissipative mechanism if it occurs. It is hard to believe that vortex flow into the bulk fluids at either end can play an important role in a direct fashion.

There is, however, a somewhat more subtle question to be considered. If we are to take seriously the usual vortex creation critical velocity expression, h/md, it is not obvious that the smaller dimension of the film is really the value of d which must be considered. Why do not vortices form perpendicular to the film and move across it from one edge to the other? While at velocities $\sim h/mW$, where W is the width of the film ~ 1 cm (velocities $\sim 10^{-3}$ cm/sec) one would expect the formation of such vortices to be difficult dynamically, at velocities of 10^2–10^3 times that, still small compared with critical film velocities, there appears to be no such process at work.

It is suggested here that the predominant mechanism in film flow is the *pinning* of such vortices by surface flaws and thin spots in the film. Thus the film is a "hard superfluid": its flow is maintained by a pinning effect rather than by an absence of suitable vortices.

Vortices of the parallel type can also become pinned at either end and retard the motion of other vortices by their mutual repulsion. This is probably a mechanism which increases the critical velocity for rough substrates. Finally, the interaction of the pinned perpendicular vortices and the moving parallel vortices can lead to quite complicated effects: such things as the vortices reattaching themselves after a crossing in such a way that the pinned end attaches itself to the parallel vortex can occur, and become a mechanism for pinning of parallel vortices which may increase the pinning, and thus the critical velocity, as a function of the vorticity flowing into the film from the bulk liquids. Another mechanism which may play a role is motion of the free end of a pinned perpendicular vortex under

[25] R. C. Jaklevic, J. Lambe, J. E. Mercereau, and A. H. Silver, Phys. Rev. **140**, A1628 (1965).

the Magnus force until it becomes a pinned parallel vortex. This could be a copious source of vorticity.

In conclusion, then, the fundamental point made here is that in helium II, as in superconductivity, the Josephson equation and the associated concept of phase slippage are the most fundamental and exact consequences of our present theoretical understanding of superfluidity. Where phase slippage in superconductivity can occur in the absence of identifiable flux quanta, in helium with present technology it will always involve vortex lines because their core size is only a few Å, and no tunneling medium is available. Thus the crucial problem in helium flow is to find the vortex lines and study how they move across the flow path into the walls or disappear into the bulk. The complicated dynamics of vorticity is beyond the scope of this paper; we have merely speculated, with little quantitative study, in order to present concrete examples of the central ideas.

ACKNOWLEDGMENTS

I have benefitted throughout from the close collaboration of P. L. Richards. Discussions with P. C. Hohenberg, J. M. Luttinger, and W. L. McMillan were of value. A suggestion of D. J. Scalapino was reworked into the form of the space-interference experiment using orifices suggested in Sec. III. Questions asked by F. Reif and P. A. Wolff stimulated the first part of the work.

APPENDIX A. ODLRO VS MACROSCOPIC PARTICLE FIELDS

As explained in the introduction, Penrose's initial definition of the order parameter[4] in terms of a large eigenvalue of the density matrix was later extended by Beliaev,[5] by Gor'kov[6] (in terms of Green's functions), and a generalization conveniently named "off-diagonal long-range order" (ODLRO) by Yang.[20] That is, one writes

$$\langle 0_N | \psi^*(x)\psi(x') | 0_N \rangle = \sum_n \lambda_n f_n^*(x) f_n(x') \quad \text{(A1)}$$

and "ODLRO" is present when $\lambda_1 \sim N$, giving a contribution to the sum comparable to the sum of all others. One may then define a "ground state" $\langle 0_{N-1}|$ of the system with a different number of particles so that f_1 becomes a matrix element

$$[\lambda_1 f_1(x)]^{\frac{1}{2}} = \langle 0_{N-1} | \psi(x) | 0_N \rangle. \quad \text{(A2)}$$

In this way the necessity for dealing with states which are coherent mixtures of states with different numbers N of particles in the system is avoided, apparently, and for this reason most of the above authors prefer this scheme.

We argue that this approach is physically unnecessary, though valid, and occasionally inconvenient. For one, this definition does not permit convenient subdivision of a system. The over-all phase of f_1 is quite arbitrary—as it correctly should be for an isolated system with no particle exchanges permitted. On the other hand, once the phase at any space–time point is fixed the phase of the rest of the system is. Thus one cannot use the same description for any subdivision of the system; the λ and f for half of a bucket of He II simply do not describe it adequately. On the other hand, if one abandons the attempt to hold on to the broken gauge symmetry and ascribes a fixed, measurable phase to every superfluid system, recognizing that in principle the relative phase of any two may always be measured by a Josephson-type experiment, one immediately has a usable local description.

This is a satisfactory expedient unless the full generality of (A1) is meaningful—i.e., unless it is conceivable that more than one eigenvalue λ_1 is "large" and more than one intermediate state $|0_{N-1}\rangle$ is involved. That is, we may ask whether the system may ever in any sense be a superposition of several distinct types of ODLRO. An attempt at such a theory was made by Gor'kov and Galitskii[26] for the d-state BCS theory, and proven invalid by various groups.[27] The question enters in many other cases—even, for example, in discussing flux quantization, one must be assured that one type of ODLRO only is present.

The most generally applicable argument here is that made by the author[28] in the Gor'kov–Galitskii case. It is that the two intermediate states $|0_{N-1,1}\rangle$ and $|0_{N-1,2}\rangle$ are demonstrably in entirely distinct Hilbert spaces in the limit $N\rightarrow\infty$, in the sense that $\sim N$ different particle states must be changed a finite amount to get from one to the other. Thus the $|0_N\rangle$ state must be simply a superposition of a $|0_{N,1}\rangle$ state communicating with $|0_{N-1,1}\rangle$ and a $|0_{N,2}\rangle$ state, and no interference effects whatever can connect the two types of states. In particular, every measurable quantity—energy, current, etc.—is the simple linear superposition of the two values. Then such a state is no more meaningful than Schrödinger's famous superposition of the quantum states of a dead cat and a live cat: a possible mathematical description of a physical absurdity.

APPENDIX B. A "NEW" COROLLARY IN CLASSICAL HYDRODYNAMICS?

Euler's equation of motion in a classical ideal fluid is

$$(\partial v/\partial t)+\nabla[(v^2/2)+\mu]=v\times\nabla\times v. \quad \text{(B1)}$$

μ is an appropriately defined chemical potential per unit mass. We now consider a general flow and draw a path C entirely inside the fluid—otherwise general—between two points P_1 and P_2 in the fluid. Points P_1 and P_2 are to be thought of eventually as being in reasonably quiet regions where the flow is steady over a long time T.

---

[26] L. P. Gor'kov and V. M. Galitskii, Zh. Eksperim. i Teor. Fiz. **40**, 1124 (1961) [English transl.: Soviet Phys.—JETP **13**, 792 (1961)].

[27] D. Hone, Phys. Rev. Letters **8**, 370 (1963); R. Balian, L. H. Nosanow, and N. R. Werthamer, *ibid.* **8**, 372 (1962).

[28] P. W. Anderson, Bull. Am. Phys. Soc. **7**, 465 (1962).

Let us now perform two integrations on (B1): first, along C from P_1 to P_2

$$\int_{P_1(C)}^{P_2} \frac{\partial v}{\partial t}\, dl + (\tfrac{1}{2}v^2+\mu)_{P_1}{}^{P_2} = \int_C (v \times \nabla \times v)\, dl. \quad \text{(B2)}$$

[It was brought out in the discussions of the conference that (B2) is even more general than I had thought, in that most types of viscosity terms which might be added to (B1) involve gradients, so that if viscosity is not acting at P_1 and P_2 they cancel out.]

Second, we integrate over a very long time interval T and divide by T, thus taking a time mean value as is done in the virial theorem:

$$\int_{P_1}^{P_2} dl\, T^{-1} \int_0^T \frac{\partial v}{\partial t}\, dt + \left\langle \left(\frac{v^2}{2}+\mu\right)_{P_2} \right\rangle_{Av} - \left\langle \left(\frac{v^2}{2}+\mu\right)_{P_1} \right\rangle_{Av}$$

$$= T^{-1} \int_0^T 2 \int dl \cdot (v \times \omega). \quad \text{(B3)}$$

We have defined ω, the vorticity, as $\frac{1}{2}\nabla \times v$ and written the time mean value at the points of steady flow in an obvious notation. The first term on the left-hand side is

$$T^{-1}\left[\left(\int_C v \cdot dl\right)_T - \left(\int_C v \cdot dl\right)_0\right].$$

We define a "quasi-steady" flow as one in which this difference increases less rapidly than T; almost any turbulent flow one wishes to consider, or periodic flow, etc. will satisfy this condition. Then as far as time mean values are concerned we arrive at the basic corollary of Euler's equation:

$$\langle (\tfrac{1}{2}v^2+\mu)_{P_2} \rangle_{Av} - \langle (\tfrac{1}{2}v^2+\mu)_{P_1} \rangle_{Av} = \left\langle 2\int_{P_1}^{P_2} dl \cdot (v \times \omega) \right\rangle_{Av}. \quad \text{(B4)}$$

It is easy to interpret the quantity on the right-hand side. Writing v as dr/dt, the particle velocity, this is

$$2\int_{P_1}^{P_2} (dl \times dr/dt) \cdot \omega,$$

which is the *rate at which vorticity is being carried across the curve* C by the particle motion. Thus

$$\langle \Delta[(v^2/2)+\mu] \rangle_{Av} = \langle 2\text{``}(d\omega/dt)\text{''} \rangle_{Av}. \quad \text{(B5)}$$

We see immediately that this equation is far more important in a superfluid, where vorticity is conserved and quantized, than it is in ordinary fluids, where in a laminar flow, for instance, the right-hand side has little or no special significance. In helium, in fact, by turning to the integrated form of (B1) involving the potential we get the detailed Josephson equation without the special assumptions necessary here.

A number of somewhat surprising consequences immediately appear. One example is that the Pitot tube,[29] for instance, must involve transport of vorticity and thus motion of vortex lines in liquid He II. Ordinary aerodynamic lift and drag also would do so if the surface condition were $v=0$, but of course it is not; the vorticity there can be thought of as all in the surface layer outside the superfluid and thus not quantized.

I have tried at length to find a clear statement of (B4–5) in the classical literature, including the voluminous works of Rayleigh and Lamb, but have so far failed to find anything but corollaries and lemmas related to it. I will be pleased to hear from any reader who can find the theorem stated in this form; one can only assume that it was understood by the "classics" but is of no value in classical hydrodynamics so was never stated.

[29] J. R. Pellam, Phys. Rev. **78**, 818 (1950).

BROKEN SYMMETRY, EMERGENT PROPERTIES, DISSIPATIVE STRUCTURES, LIFE: ARE THEY RELATED?

by

P. W. Anderson ♯ * and D. L. Stein
Princeton University **
Princeton, New Jersey 08544

The more theoretical physicists penetrate the ultimate secrets of the microscopic nature of the universe, the more the grand design seems to be one of ultimate simplicity and ultimate symmetry. Since all of the interesting parts of the universe - at least those of interest to us like our own bodies - are markedly complex and *unsymmetric*, the first, correct conclusion one draws from this statement is that the deep probing of the nature of matter (on which physicists expend great effort and greater sums of money) is becoming more and more irrelevant to *us*. But that isn't really an adequate retreat for any scientist who hopes to achieve the ultimate goal of science, which we take to be real understanding of the nature of the world around him from first principles. It is essential for him to explain the real world in terms of the ultimately simpler constituents of which it is made. In fact, he must thank his stars that the world becomes simpler as each underlying level is discovered - the opposite case would make his task difficult indeed.

The simplicity to which we refer is, of course, the recent success of the elementary particle theorists in reducing the equations of the fundamental constituents of matter to perfectly symmetrical ones, in which all constituents initially enter in exactly the same way, and in which all the interactions themselves are derived from a principle which *itself* is a manifestation of an especially perfect kind of symmetry. But those who are not acquainted with these developments need have no fear that what we say will depend on them in any way. we wish merely to make the point that there is a sharp and accurate analogy between the breaking-up of this

---

♯ This material is essentially that given in this author's Cherwell-Simon Memorial Lecture, Oxford, May 2, 1980. I am grateful to the Clarendon Laboratory for their hospitality on this occasion.

* Also at Bell Laboratories, Murray Hill, New Jersey 07974.

** The work at Princeton University was supported in part by the National Science Foundation Grant No. DMR 78-03015, and in part by the U.S. Office of Naval Research Grant No. N00014-77-C-0711.

ultimate symmetry to give the complex spectrum of interactions and particles we actually know, and the more visible complexities we will shortly discuss.

There has been gradually arising during the past twenty years or so a set of concepts related to the ways in which complexity in nature can arise from simplicity, some of which are quite rigorously and soundly based in the theoretical physics of large and complex systems, while others extend all the way to the speculative fringe between physics and philosophy.

The most basic question with which such a conceptual structure might hope to deal would be placing life itself within the context of physics in some meaningful way: to relate the emergence of life itself from inanimate matter to some general principle of physics. Can we understand the existence or even the origin of life in some purely physical context?

We approach this question in four steps. Clearly, we are trying to look at life as what the philosophers call an *emergent property:* a property of a complex system which is not contained in its parts. So we start from the very simple question of whether such properties exist at all.

The most rigorously based, physics-oriented description of the growth of complexity out of simplicity is called the theory of *broken symmetry.*

> Can properties emerge from a more complex system which are not present in the simpler substrate from which the complex system is formed?

The theory of broken symmetry gives an unequivocal "yes" answer to this question: In equilibrium systems containing large numbers of atoms, new properties - such as rigidity or superconductivity - and new stable entities or structures - such as quantized vortex lines - can emerge which are not just nonexistent but meaningless on the atomic level.

Unfortunately, the emergent properties we are most seriously interested in are not these simple ones of *equilibrium* systems: specifically, we need to know whether life, and then consciousness, can arise from inanimate matter, and the one unequivocal thing we know about life is that it always dissipates energy and creates entropy in order to maintain its structure. So we come to a second deep question:

> Are there emergent properties in dissipative systems driven far from equilibrium?

The answer is yes: dynamic instabilities such as turbulence and convection are common in nature and their source is well-understood mathematically. When they occur these phenomena exhibit striking broken symmetry effects which very much resemble the equilibrium structures which exist in condensed matter systems. These have been called "dissipative structures". Examples are convection cells or vortices in turbulent fluids. But these seem always very unstable and transitory: can they explain life, which is very stable and permanent (at least on atomic time scales)?

> Is there a theory of *dissipative structures* comparable to that of equilibrium structures, explaining the existence of new, stable properties and entities in such systems?

Contrary to statements in a number of books and articles in this field, we believe there is *no such theory* and it may even be that there are no such structures as they are implied to exist by Prigogine, Haken and their *collaborators.*[1,2] What does exist in this field is rather different from Prigogine's speculations and is the subject of intense experimental and theoretical investigation at this time.

Thus the answer to the fourth deep question

> Can we see our way clear to a physical theory of the origin of life which follows these general lines?

is already evident: No, because there exists no theory of dissipative structures. The best extant theoretical speculations about the origin of life, those of *Eigen*[3], are only tenuously related to the idea of dissipative structures, and instead are sui generis to the structure of living matter; still, it may be that they contain deep problems of the same sort which destroy the conventional ideas about dissipative structures. We are setting out to study this question in detail.

The above is the basic outline of what we have to say: now we would like to set it all out in more specific and detailed terms.

Initially, we must learn some things about the real physics: What and why is "broken symmetry"? - more properly called "spontaneously broken symmetry". The answers to both questions are so simple that we almost miss their depth and generality. First, what is it? Space has many symmetries - it is isotropic, homogeneous, and unaware of the sign of time, at the very least. Correspondingly, the equations which control the behavior of all particles and systems of particles moving in space have all of these symmetries. But Nature is not symmetric: "*Nature abhors symmetry.*" Most phases of matter are unsymmetric: the crystals of which all rocks are made, for instance, are neither homogeneous nor isotropic, as Dr. Johnson forcefully pointed out. Molecular liquids are often not *isotropic* but form liquid crystals (see Fig. 1); magnets such as iron or rust (which is antiferromagnetic) are not invariant under time-reversal. Superfluids break one of the hidden symmetries of matter, the so-called gauge symmetry, allowing the phase of quantum wave-functions to be arbitrary, related to the laws of conservation of charge and number of particles.

Second, why? Fluctuations, quantum or classical, favor symmetry: gases and liquids are homogeneous, magnets at high temperatures lose their magnetism. Potential energy, on the other hand, always prefers special arrangements: atoms like to be at specific distances from each other; spins like to be parallel or antiparallel, etc. Thus we define *spontaneously broken symmetry*.

Definition. Although the equations describing the state of a natural system are symmetric, the state itself is *not*, because the unsymmetric state can become unstable toward the formation of special relationships among the atoms, molecules, or electrons it consists of.

So far the idea is purely descriptive: it becomes meaningful when we find that it relates and explains many apparently different and unrelated phenomena. Initially, the concept was introduced by *Landau*[4] to solve a series of problems related to the nature and meaning of thermodynamic phase transitions, but it also relates and explains many other properties of broken symmetry phases. In the course of this, he introduced the single most important concept of the whole theory: the idea of the *order parameter*.

The order parameter is a *quantitative measure* of the loss of symmetry. The canonical one is $\vec{M}$ in a ferromagnet: the mean moment $<\mu_i>$ on a given atom. Others are:

1. director $\vec{D}$ of the nematic liquid crystal;

2. amplitude ρ_G of the density wave in a crystal;

3. mean pair field in a superconductor $<\psi\ (r)>$.

Landau: the loss of symmetry requires a new thermodynamic parameter whose value is *zero* in the symmetric phase - for instance, the magnetization of a ferromagnet (see Fig. 2), the director in a nematic liquid crystal, etc. The magnitude of the order parameter η measures the degree of broken symmetry.

This has many implications. For instance, the appearance of a wholly new thermodynamic variable is a necessary condition for a continuous (so-called second-order) phase transition. It can, and often does, appear discontinuously, but it need not - see M(T) in Fig. 2, and for contrast ρ_G (T) in Fig. 3 for a typical crystal. Also, since there is an extra parameter, the free energy and all the thermodynamic properties can never be the same mathematical functions in the two phases of different symmetry, so the phases are always separated by a sharp phase transition - unlike, for instance, water and steam.

The thermodynamic consequences which flow from the amplitude of η alone are ample excuse for the broken symmetry concept - but even more important consequences follow from another property of the order parameter which Landau never formalized but sometimes used: *it is a quantity which always has a phase: the free energy $F\ (\ T, |\eta|\ )$ is a function of its magnitude $|\eta|$*

but must not be so of its "phase" or direction, because of the existence of the original symmetry. For instance, the energy may not depend on the *direction* of the director since space is isotropic; nor may it depend on the orientation or position of a crystal, nor on the phase ϕ, of the superfluid wave function. Another way to say it is that the order parameter has a space within which it is free to move without changing the energy. (In quantum-mechanical terminology, the ground state is highly degenerate in the broken symmetric phase, which in a way is a remnant of the original symmetry of the Hamiltonian (which remains unchanged). This is connected to some of the dynamical consequences of broken symmetry, such as Goldstone modes and the Higgs phenomenon, as will be shortly discussed.)

A second property of η is obvious if we see it as a physical thermodynamic parameter: it may vary over macroscopic distances in the sample, and $\eta(r)$ may be defined locally (just as we can define a *local* temperature or pressure in a sample not too far from equilibrium, if these do not vary too rapidly.)

From this follow three major emergent properties of spontaneously broken symmetry:

1. Generalized rigidity
2. New dynamics
3. Order parameter singularities and their role in dissipative processes.

All of these are very interesting, since most of the important properties of solids depend upon them, but time and our subject mean that we can only afford to discuss the first, which is the simplest and most general.

Again we use the idea that η is a physical thermodynamic parameter to which we can by one means or another apply a force. This is clear in the case of $\vec{D}$ - which couples to boundary orientation - or of $\vec{M}$ or crystal orientation θ , ϕ; but it can be a little more w esoteric for "hidden" order parameters like sublattice magnetization in the antiferromagnet or ψ the superfluid order parameter. Nonetheless it is always possible to grasp hold of η at any point in the system. While F is not a function of the phase angles of η, it is naturally a function of the *gradient* of these phases, because otherwise arbitrarily large relative fluctuations of the phase would destroy the existence of the order parameter. Thus we must have

$$F = F(|\eta|, |\nabla\eta|^2, \ldots)$$

and

$$\frac{\partial^2 F}{\partial(\nabla\eta)^2} > 0$$

a positive stiffness for variations of η.

This is enough to ensure that if we exert a force on η (r) at *one* end of a sample, $\eta(r')$ will respond at the other. We can essentially use η as a crankshaft to transmit forces from one point to another, that is, to exert action at a distance (see Fig. 4). We emphasize that this rigidity is a true *emergent property:* none of the forces between actual particles are capable of action at a distance. It implies that the two ends cannot be decoupled completely without destroying the molecular order over a whole region between them.

Rigidity of solids, then, is a model for a wide class of other rigidity properties, including permanent magnetism, ferroelectricity, and superconductivity and superfluidity. These last two have, since the discovery of the Josephson effect, been understood to be the phase rigidity of the order parameter $\psi(r)$[5].

The other two major emergent properties are also consequences of this phase freedom in broken symmetry systems: the existence of long-wavelength collective motions of the order parameter, such as phonons and spin waves, which are the models for the Goldstone and Higgs phenomena of elementary particle physics; and the existence and classification of singularities and textures of the order parameter: the possible order parameter fields which are allowed when we permit lower-dimensional regions to be excluded from our order parameter field $\eta(r)$ - vortex lines and dislocations, domain boundaries, singular points, etc. Broken symmetry gives rise to the appearance of new length scales that did not exist in the symmetric phase.

Now let us return to the main theme of our discussion: that there does *not* exist a corresponding theory of the dissipative case. First, let us describe the kinds of experiments which seem at first to lead to very similar types of broken symmetry in the dissipative case and

have been so described. The canonical example is the Benard instability: the layer of fluid heated from below, which, once a critical heating rate is exceeded, exhibits very regular-appearing "rolls" of convection, arising spontaneously with a rather fixed size or wavelength (see Fig. 5).

Other examples abound, such as the Couette instability of a viscous fluid between rotating cylinders (Fig. 6), or even the familiar laser exhibiting a periodic wave of excitation density (Fig. 7).

Clearly all of these systems exhibit spontaneously broken symmetry in the simple sense, in that, for instance, in each case the sign is arbitrary, and also an initially homogeneous state changes suddenly into an inhomogeneous one. The initial transition is often continuous, just like the typical second-order transition, and it has often been suggested that there is some kind of deep analogy between these two types of systems. There is indeed one mathematical respect in which there is at least a similarity, in that both are examples of dynamic instabilities, for which there exists a general mathematical theory described by *Thom*[6] called "catastrophe theory" and much elaborated in recent years by a large number of mathematicians of whom perhaps *Ruelle*[7] should be specially cited. But the thermodynamic phase transitions invariably present only the simplest kind of catastrophe, the so-called "bifurcation", and the simplest type of state, the so-called "fixed point", while dynamical instabilities seem always to evolve, - even oversimplified mathematical models of them - towards more and more complex types leading eventually to completely chaotic behavior. The evolution of chaos in such systems has been beautifully described by *Gollub*[8] in a number of articles and by R. Abrahams and J. *Marsden*[9] in a well-known book. It is a pity that I cannot describe here in detail the beautiful work described in these sources in following the successive instabilities from classical to steady rolls to singly-periodic dynamic to multiply periodic and finally to total chaos.

Experimentally the situation is even more complex. *Ahlers*[10] particularly has shown that even the complicated behavior seen by *Gollub*[8] and predicted by the mathematicians may be an artifact of an over-constrained system heavily influenced by its boundary conditions: they find

finer-scale chaos or near-chaos even in the apparently quiescent region of the Benard system . We have tried to show that this is inevitable and that dissipative structures in a real, physical, open system unconstrained by artificial boundary conditions will inevitably be chaotic and *unstable*[11,12]. (For instance, the laser can be persuaded to oscillate in a single mode only with the utmost artificiality and difficulty. This depends on the proper placement of endplates or mirrors so that here broken symmetry is strongly dependent on externally applied boundary conditions. Lasers occurring naturally in nature (for example, from astrophysical sources) seem to show no mode selection.)

Prigogine and his school have made a series of attempts to build an analogy between these systems and the Landau free energy and its dependence on an order parameter, which leads to the important properties of equilibrium broken symmetry systems. The attempt is to generalize the principle of maximum entropy production which holds near equilibrium in steady state dissipative systems, and to find some kind of dissipation function whose extremum determines the state. As far as we can see, in the few cases in which this idea can be given concrete meaning, it is simply incorrect. In any case, it is clearly out of context in relation to the observed chaotic behavior of real dissipative systems.

Thus we conclude that there is no analogy visible between the stability, rigidity and other emergent properties of equilibrium broken symmetry systems, and the properties of dissipative systems driven far from equilibrium. The latter types of systems have never been observed to, nor can any mathematical reason be found why they should, exhibit the rigidity, stability, and permanence which characterizes the thermodynamically stable broken symmetry systems. (A case in point of a driven system that might have exhibited broken symmetry but failed to do so is described in Refs. 11 and 12.)

The reason this is unfortunate is that many authors have chosen to use such systems as the laser and the Benard instability as models for the nature and origin of life itself, as an emergent property of inanimate matter. It is indeed an obvious fact, noted since Schrödinger's 1940 *book*[13], that life succeeds in maintaining its stability and integrity, and the identity of its genetic material, at the cost of increasing the rate of entropy production of the world as a whole. It is at least in that sense a stable "dissipative structure" - i.e., an existence proof by example.

Turing[14] long ago observed that a fertile source of dynamic instabilities was the autocatalytic chemical reaction in which reaction products serve as catalysts as well. The base-pairing mechanism of DNA is an obvious and good example. *Eigen*[3] in particular has tried to develop a theory of autocatalytic instabilities in the primeval soup as a detailed explanation of the origin of life. It is a glorious picture to imagine the growth of an "order parameter in molecular information space," driven by a dynamic autocatalytic instability and self-stabilized in some mystic way by the magical power of Darwinian evolution. This may in fact be the way it happened - one can hardly assume it did not! - but there are reasons to be skeptical of the claim that we have yet found the full story. Why should dynamic *instability* be the general rule in all other dissipative systems, except this supremely important one? We are attempting a computer simulation of a model of the origin of information-carrying macromolecules which is already producing quite interesting results in terms of the spontaneous generation of complex molecules. (See Appendix)

Let us then conclude by reiterating our main point: we still believe, since in fact we understand the process in all details, in the reality of emergent properties: the ability of complex physical systems to exhibit properties unrelated to those of their constituents. But we do not believe that stable "dissipative structures" maintained by dynamic driving forces can be shown to exist in any inanimate system, and thus we do not see how speculations about such structures and their broken symmetry can yet be relevant to the still open question of the origin and nature of life.

Appendix

In the simple picture we are using, we begin with a "soup" of monomers of two different varieties, A and B, and an externally applied energy flux which drives the system toward formation of strings of monomers, according to a simple set of rules for lengthening and shortening chains. This process relies on temperature cycling: in the low temperature phase two strands (or a strand and monomers) attach weakly via A-B attraction (as in hydrogen bonding between a purine-pyrimidine pair). While held together in this fashion, stronger bonding may take place between two adjacent, previously unattached strands. In the high temperature phase the hydrogen bonds break, but the stronger bonds along the length of the strands do not, and the newly created (or lengthened) strands separate until the next cycle. There is a slightly higher probability for strong bonding between dissimilar monomers than for similar ones. There is also a certain chance that, in the high temperature phase, a strong bond may be broken (and a strand thereby shortened) because of, say, interaction with an energetic cosmic ray. In addition to these "birth" and "death" rates (more accurately, lengthening and shortening processes) there is also a small error probability; that is, in the low-temperature phase, an A monomer may mistakenly hydrogen bond to another A, rather than a B as it should. These birth, death, and error rates thus form a complete prescription for building up many lengthy strands starting from a sea of lone monomers and a single strand of two or three monomers.

We wish to see if, from this very simple picture, a polymer with nontrivial information content will be selected from the near infinite number of possibilities, selection (if it exists) being implicit in the strong nonlinearity of the problem. Clearly, if most chains are of the form ABABABAB .... (or AAAAA... or BBBBB...) nothing very interesting has happened. On the other hand, if many chains with irregular sequences, such as ABBABABAAABBA... are formed but no pattern appears to predominate, again little of interest has occurred.

In looking for patterns that may predominate, we have found it most useful to search for "triplets", by which we mean the following. Suppose we are given the strand that appears above:

$$A_2B \wr BAB_6ABA \wr A_1 \wr A_2B \wr B_2A$$

Whenever two like monomers appear adjacent, we draw squiggly lines separating them as pictured. We then count the respective lengths of the purely alternating sequences that make up the polymer (these are the numbers that appear in the picture above). A "triplet" is then the triplet of lengths of three adjacent alternating sequences; in the example above, we have a (2, 6, 1), (6, 1, 2), and (1, 2, 2). Note that the mirror image of the polymer would give the same result. We therefore wish to see if certain triplets make up the bulk of most long polymers. This seems to us to be more useful than trying to select an entire polymer itself as the prototype of what should be selected.

Our preliminary results indicate that for certain choices of bonding probabilities selection of a number of triplets occurs and can in fact be quite strong (as well as persistent over many cycles, which is a requirement if we are to say selection has occurred). It is also amusing, and somewhat unexpected, that a small error probability is necessary for selection to occur in the cases studied so far.

Many questions remain unanswered, the most prominent of which is, how does one assign a meaningful information content to a polymer? So far we've only discussed necessary, but not sufficient, conditions for symmetry breaking in "information space" to occur. One would guess that, in some sense, structure and function are intimately related (in that in the real world, DNA serves as a blueprint for manufacture of proteins, some of which act as enzymes in replication and other processes governing the DNA molecule itself). Is there any way in which this can be seen in the simple model presented here? This is one of the most fundamental problems in understanding the origin of life, a not so subtle variant of the "chicken and egg" *problem*[3]. We are not attempting to answer this problem at this stage, but rather the somewhat less ambitious problem of whether one can relate the issues of symmetry breaking discussed earlier to the problem of the origin of life (specifically, a primitive genetic code in this instance), and in what context this is possible and meaningful.

References

Some of the ideas which appear here can be found in P. W. Anderson, Science, 177, 393 (1972). A lengthier discussion of the relation between broken symmetry, generalized rigidity, Goldstone modes, etc., in the spirit of this article appears in P. W. Anderson, unpublished lecture notes.

1. P. Glansdorff and I. Prigogine, Thermodynamic Theory of Structure, Stability and Fluctuations (Wiley, New York, 1971); G. Nicolis and I. Prigogine, Self-Organization in Non-Equilibrium Systems (Wiley, New York, 1977).

2. H. Haken, Synergetics: An Introduction (Springer- Verlag, Berlin, 1977).

3. M. Eigen, Naturwissenschaften, 58, 465 (1971). For recent work along these lines, see M. Eigen and P. Schuster, The Hypercycle: A Principle of Natural Self-Organization (Springer-Verlag, Berlin, 1979)

4. L. D. Landau and E. M. Lifschitz, Statistical Physics (Pergamon Press, Ltd., Elmsford, N.Y. 1969).

5. For a discussion of some related points, see P. W. Anderson, Rev. Mod. Phys., 38, 298 (1966).

6. R. Thom, Structural Stability and Morphogenesis (W. A. Benjamin, Reading, Mass., 1975).

7. D. Ruelle and F. Takens, Comm. Math. Phys. 20, 167 (1971).

8. J. Gollub and H. L. Swinney, Phys. Rev. Lett., 35, 927 (1975).

9. R. Abraham and J. Marsden, Foundations of Mechanics (W. A. Benjamin, New York, 1967).

10. G. Ahlers and R. P. Behringer, Phys. Rev. Lett., 40, 712 (1978).

11. P. W. Anderson, "Can Broken Symmetry Occur in Driven Systems?", Remarks at Solvay Conference, Brussels, 1978 (to be published).

12. D. L. Stein, J. Chem. Phys., 72, 2869 (1980).

13. E. Schrödinger, What is Life? (University Press, Cambridge, 1945).

14. A. M. Turing, Philos. Trans. Roy. Soc.,London Ser. B, 237, 37 (1952).

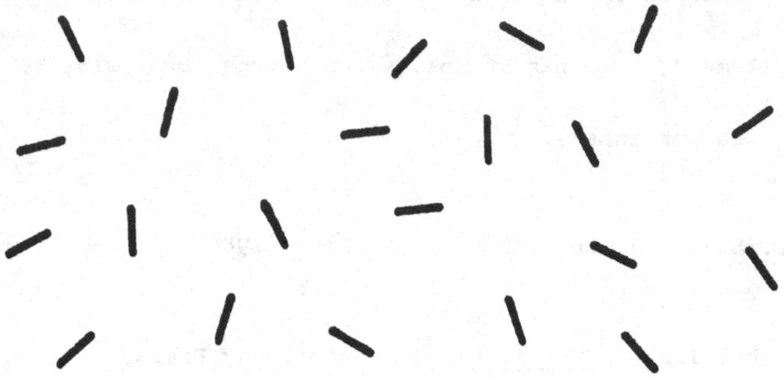

1a. Nematic liquid crystal in the disordered state. The line segments represent the rodlike molecules of the nematic. Averaging molecular orientations over macroscopic distances yields zero.

1b. For a suitable choice of thermodynamic parameters, the nematic enters the ordered state, with the appearance of a macroscopic order parameter (the director $\vec{D}$). The system is no longer isotropic, but has chosen a special direction: rotational symmetry has been broken.

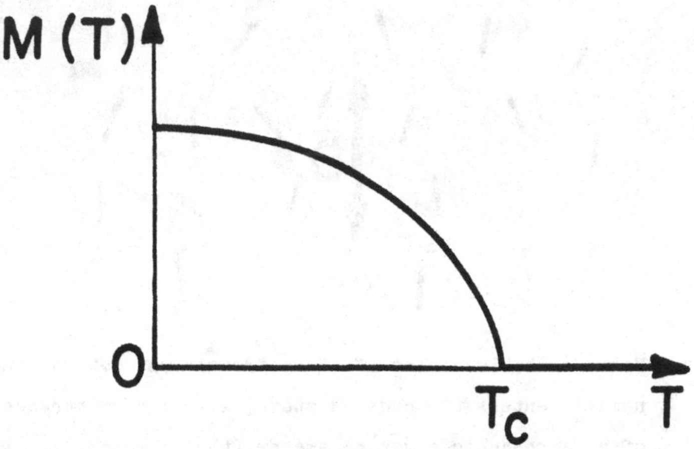

2. Variation of magnetization M with temperature T in a simple ferromagnet. This is a typical second-order phase transition, in which the order parameter grows continuously from zero as T is lowered below a critical temperature T_c.

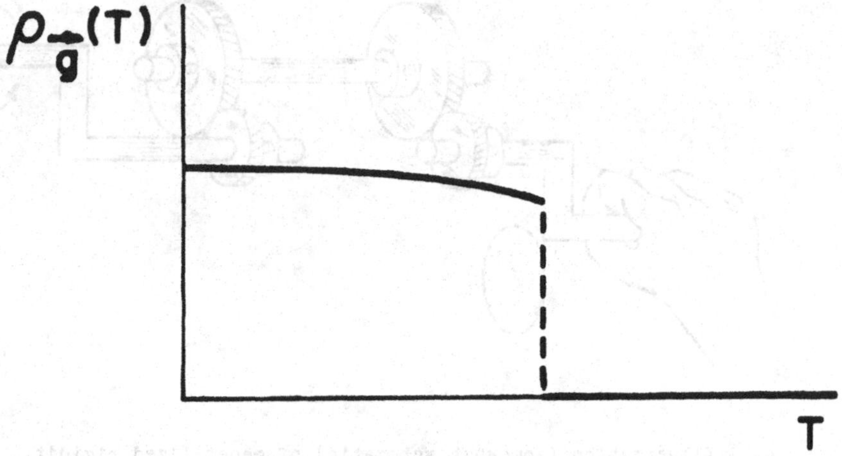

3. In a first-order transition, such as the liquid to solid crystal transition shown here, the order parameter will exhibit a discontinuous jump at the transition with an associated release (or absorption) of latent heat.

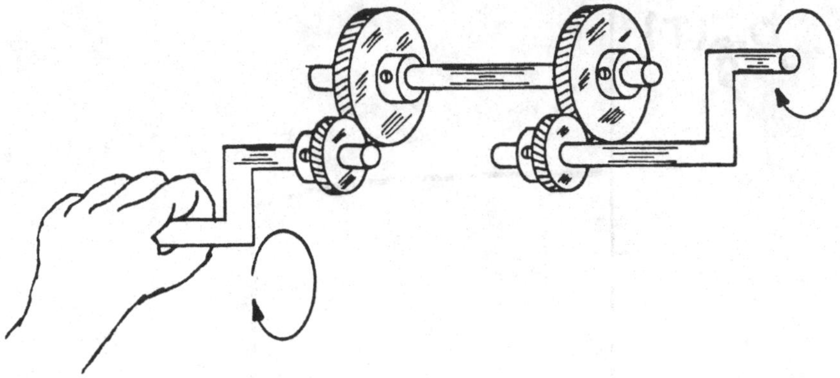

4. Illustration (somewhat schematic) of generalized rigidity. An external force (the crank) couples to the order parameter at one end of the system, represented as a gear. A change in the order parameter at any point in the ordered system is transmitted to all other parts of the system (first gear turns the second gear). The second gear turns the second crank: a force has been transmitted from one end of the system to the other via the order parameter.

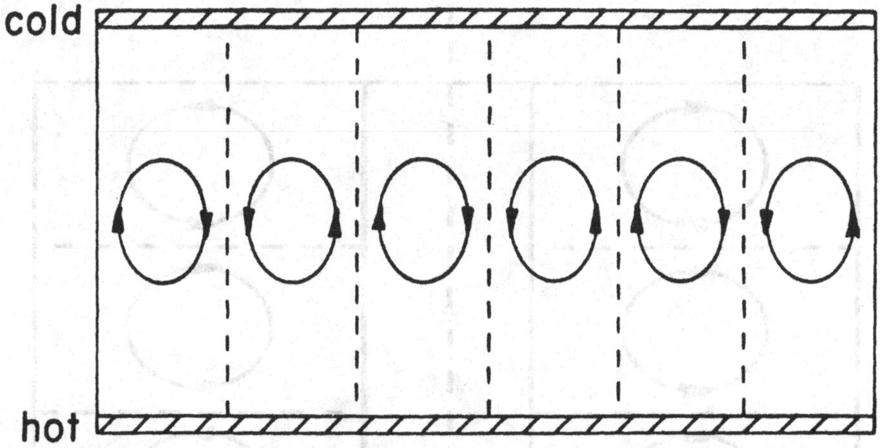

5. The Bénard instability in rectangular geometry. A layer of fluid between two horizontal rectangular plates is heated from below. When a sufficient thermal gradient is reached between top and bottom plates, convection arises in the form of rolls. In this cutaway edge-on view, the arrows represent the fluid velocity.

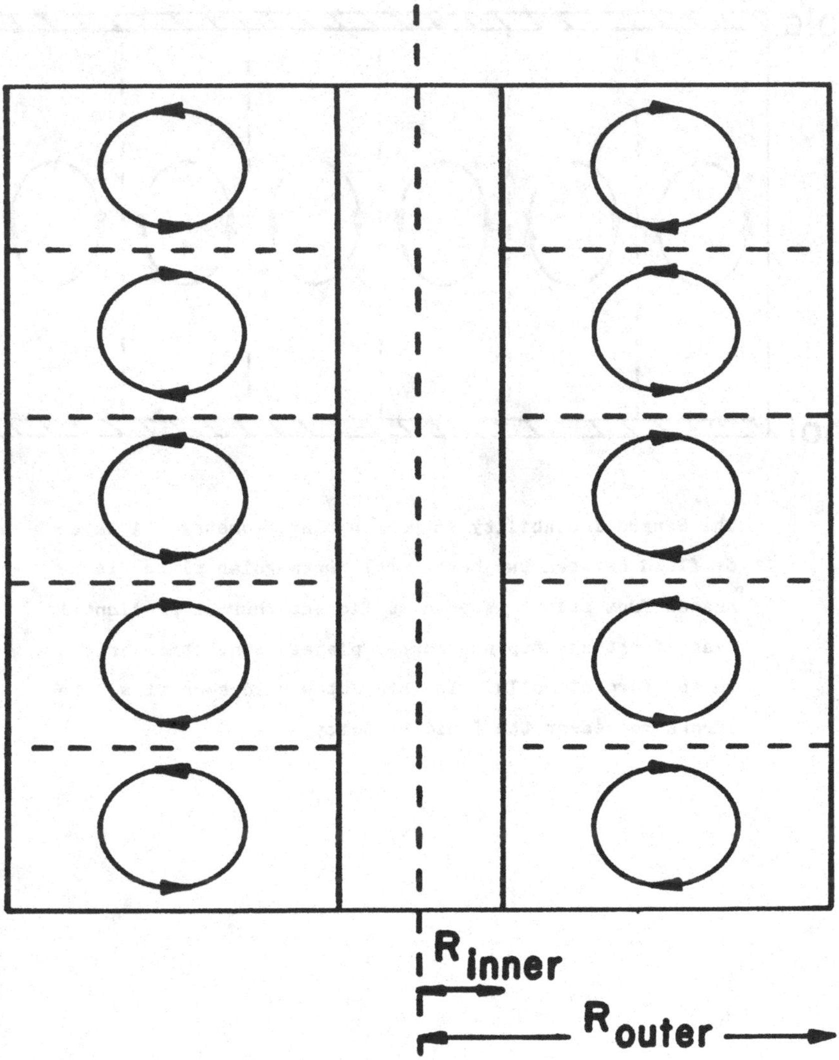

6, Couette flow: A fluid is placed between two cylinders with different rotational velocities about their axes. When the velocity gradient exceeds a critical value, rolls of vortices form. In this view the cylinder is cut along its length.

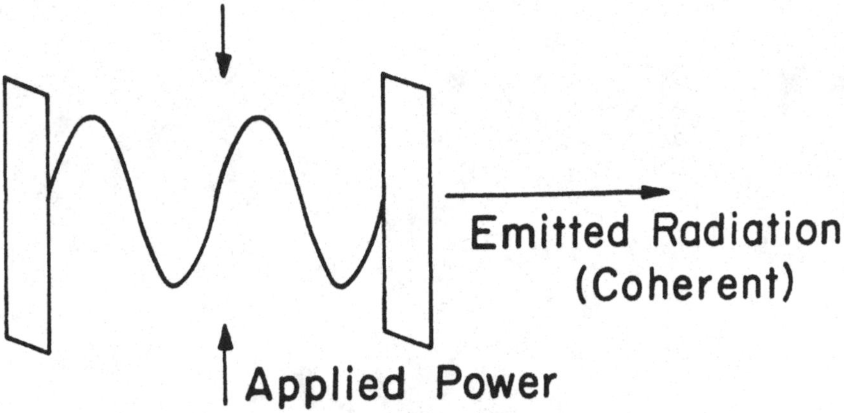

7. In a laser, a standing wave of excitation density is set up between two end plates, or mirrors, resulting in emission of a beam of coherent radiation.

THEORY OF ANISOTROPIC SUPERFLUIDITY IN He^3

P. W. ANDERSON

Bell Laboratories, Murray Hill, New Jersey 07974, USA

also

Cavendish Laboratory, Cambridge, England

and

W. F. BRINKMAN

Bell Laboratories, Holmdel, New Jersey 07733, USA

also

H. C. Ørsted Institute, Copenhagen, Denmark

I. HISTORICAL INTRODUCTION

1. PREFACE

From a microscopic point of view helium is the simplest of all condensed substances: the atoms may be treated as structureless particles (except for the convenient nuclear spin of $\frac{1}{2}$ carried by He^3) interacting via an interatomic potential which is quite accurately known. Nonetheless both solid and liquid phases of both isotopes exhibit properties which seem as complex as those of any inorganic system, and which in all cases are beyond the present capacity of many-body theory to calculate quantitatively, and in many cases to understand except in qualitative terms. Of all these varied phenomena, perhaps the most challenging intellectually and most complex phenomenologically are the recently discovered anisotropic superfluid phases of He^3. These phases are superfluid liquid crystals exhibiting ferromagnetism; their nuclear magnetic resonance behaviour has its closest analogue in antiferromagnetism; they exhibit the totally unique phenomenon of orbital supercurrents, as well as various new kinds of structural singularity, such as the disgyrations of de Gennes and the point defects proposed by Brinkman. It is the task of the theorist to show how Nature can construct, using the basic principle of broken symmetry, all this richness of phenomenology out of such a totally simple microscopic Hamiltonian.

Why should the physicist go to the very considerable trouble necessary to study in detail this minimicrocosm, all of which occurs well below 3 mK? Partly, of course, for the challenge: "it's there". This study stretches to their absolute limits his intellectual and experimental resources. More, perhaps, the motivation must be the expansion of the human consciousness. But one can by no means assume that at least the theoretical lessons learned will be of no value elsewhere in science. The understanding of superconductivity introduced new perspectives into the theory of the elementary particles, for instance, and we have no way of knowing whether this far more complex case of broken symmetry may not be of even greater value. Certainly, as well, advances in our capability of understanding Fermi systems can be important both to nuclear and to electronic physics.

2. EARLY (PRE-1972) HISTORY

In spite of uncertainties and difficulties which remain even today, the fact that a fair amount of still useful theoretical work was done on this system in the years 1959–1970, while the experimental discovery dates from 1972, correctly suggests a remarkable triumph of theoretical many-body physics. Theory, in the absence of experiment, often went off down blind alleys;† but also many of the most important phenomena were first suggested from the theoretical side.

Almost immediately after the BCS theory[1] several authors proposed anisotropic generalizations of the BCS phenomena. (Thouless[2] and Fisher[3] may well not be a complete list.) But it was not until the general realization that He^3 is a genuine Fermi liquid, which realization took place around 1958–1960, that these generalizations were taken seriously. Pitaevskii[4] was perhaps the first to propose an anisotropic BCS state for He^3, followed very shortly by Brueckner, Soda, Anderson and Morel[5] and by Emery and Sessler.[6]

Cooper, Mills and Sessler[7] had previously attempted unsuccessfully to find a conventional BCS transition in He^3. The reason for their failure was the nature of the He^3 interparticle potential (see Fig. 1). The repulsive hard core cannot be renormalized out completely enough, apparently, so that the $l = 0$ isotropic interaction potential at the Fermi surface is almost cer-

† We will try not to refer to papers which seem to have made no contribution to the advancement of the subject. We assure the reader that any impression of steady progress in the positive direction which he obtains is to some extent a result of selection.

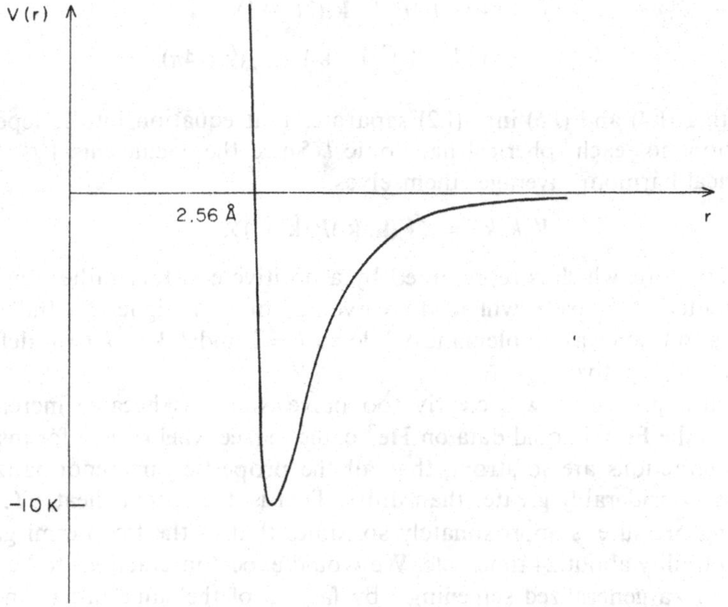

FIG. 1. A plot of the interparticle potential as a function of separation.

tainly repulsive and can lead to no pair condensation. The reason for trying $l \neq 0$ was in order to avoid this hard core. All of these calculations really amounted to rather naively introducing the bare interatomic potential

$$V(\mathbf{k}, \mathbf{k}') = \int \exp\left[i(\mathbf{k} - \mathbf{k}')\cdot\mathbf{r}\right] \mathrm{d}^3 r \, V(r) \tag{I.1}$$

into the straightforward BCS transition temperature equation

$$\Delta(\mathbf{k}) = \sum_{\mathbf{k}'} \frac{V(\mathbf{k}, \mathbf{k}')\Delta(\mathbf{k}')}{\varepsilon(\mathbf{k}')} \tanh \beta_c \varepsilon(\mathbf{k}'). \tag{I.2}$$

Then one looks for a solution

$$\Delta(\mathbf{k}) = \Delta_0(\varepsilon_{\mathbf{k}}) Y^l_m(\hat{\mathbf{k}}) \tag{I.3}$$

having the symmetry of a spherical harmonic of order l. Since (I.2) is linear in Δ and V is rotationally symmetric, V may be separated out into its spherical harmonic components

$$V(\mathbf{k}, \mathbf{k}') = \sum_l \bar{V}_l \tag{I.4}$$

where

$$\begin{aligned}\bar{V}_l &= V_l(k, k')P_l(\hat{\mathbf{k}} \cdot \hat{\mathbf{k}})(2l + 1) \\ &= V_l(k, k') \sum_m Y^l_m(\hat{\mathbf{k}}') \, Y^l_{-m}(\hat{\mathbf{k}}') \, (4\pi).\end{aligned} \tag{I.5}$$

Inserting (I.4) and (I.5) into (I.2) separates that equation into independent equations for each spherical harmonic l. Since the coefficients $V_l(k, k')$ are spherical harmonic averages themselves,

$$V_l(k, k') = \langle V(\mathbf{k}, \mathbf{k}')P_l(\hat{\mathbf{k}} \cdot \hat{\mathbf{k}}') \rangle,$$

the hard core, which is represented by a positive constant rather uniformly distributed in k-space, will tend to average out for higher l's. Indeed, for $l = 1$ a solution was problematic, while for $l = 2$ and 3, $V_l(k, k')$ was definitely attractive (negative).

Such a procedure was clearly too naive, since, as became increasingly clear as the Fermi-liquid data on He^3 came in (see Abel *et al.*[8] for instance) the interactions are so strong that all the properties are renormalized by factors considerably greater than unity. That is, the specific heat, γT, at the melting pressure is approximately six times that of the free Fermi gas, the susceptibility about 24 times, etc. We would expect interactions to be renormalized—a generalized screening—by factors of the same order, and with $T_c = \beta_c^{-1}$ depending exponentially on the coupling constant according to the standard formula $T_c \propto \exp[-N(0)V_l(k_F, k_F)]^{-1}$, the T_c values were merely wild guesses. Nonetheless this parameter is the most vital one to experimentalists and the point that the range of possible values included experimentally accessible ones was important to make.

The first papers seriously attempting a description of what the resulting state might be like were by Anderson and Morel.[9, 10] The first of these called attention to two aspects of these phases which have turned out to be of importance. These were, first, that a number of inequivalent solutions belonging to a single l existed, and that these different phases had to be tested for relative stability: different spherical harmonics led to inequivalent solutions. Second, they observed that a favourable type of solution in many cases might be complex, exhibiting internal orbital currents: as they named it, "orbital ferromagnetism", or aligned orbital angular momentum.

The reasoning behind these statements becomes clear if we look at the BCS gap equation below T_c—say, at absolute zero. As Anderson and Morel showed, so long as the coupling is reasonably weak (i.e. Δ considerably less than E_F) different l's continue to separate to a good approximation, so we may write with good accuracy (we omit spin considerations for the time being)

$$\Delta_l(\mathbf{k}) = -\sum_{\mathbf{k}'} \frac{V_l(k, k')P_l(\hat{\mathbf{k}} \cdot \hat{\mathbf{k}}')\Delta_l(\mathbf{k}')}{E(\mathbf{k}')}.$$

where

$$E^2(\mathbf{k}') = |\Delta(\mathbf{k}')|^2 + \varepsilon_{\mathbf{k}'}^2.$$

Because of the dependence of E on Δ, this is a nonlinear equation, and therefore different Y^l_m's will give different solutions. For instance, for $l = 1$ either $\Delta \propto \cos\theta$ or $\Delta \propto \sin\theta e^{i\varphi}$ will be solutions, but they have different energies. It is easy to see intuitively that this simple weak-coupling theory favours the most uniformly distributed $|\Delta|^2$ possible. Thus in fact of the two solutions above $\sin\theta e^{i\varphi} = \hat{k}_x + i\hat{k}_y$ is favoured: the complex one. While the various solutions are inequivalent, they must all have the same T_c, since the equation for T_c is a linear one.

Anderson and Morel borrowed from the BCS paper a device which is of great value in visualizing these states. This is the idea of the extra correlation or "correlation hump" caused by the condensation. This "correlation hump" is defined as

$$\Delta\langle nn(\mathbf{r})\rangle = \langle n(\mathbf{0})n(\mathbf{r})\rangle - \langle n\rangle^2.$$

$\Delta\langle nn\rangle$ may be calculated from a quality $b(\mathbf{r})$ which takes the formal place of a wave function of condensed pairs and is given by

$$b(\mathbf{r}) = \int d^3k \exp(i\mathbf{k}\cdot\mathbf{r})\frac{\Delta(\mathbf{k})}{E(\mathbf{k})}.$$

Then

$$\Delta\langle nn\rangle = |b(\mathbf{r})|^2.$$

Clearly, if $\Delta(\mathbf{k})$ has a certain spherical harmonic form $\Delta = \Delta(k)Y^l(\hat{\mathbf{k}})$, the "wave function" b will very nearly be proportional to the same harmonic.

One could imagine a gas of $(He^3)_2$ molecules, in an orbital state of given L, undergoing a Bose condensation. Such a Bose-condensed gas of molecules would also have a "correlation hump" since each atom would be correlated with its molecular partner. The total magnitude of that correlation hump would be almost precisely unity:

$$\int [\langle n(\mathbf{0})n(\mathbf{r})\rangle - n^2]_{B-E\,\text{gas}}\, d^3r \simeq 1 - na^3,$$

where n is the density of molecules and a the molecular diameter. In contrast, the "correlation hump" of a BCS condensed Fermi gas is very small:

$$\int d^3r[\langle n(\mathbf{0})n(\mathbf{r})\rangle - n^2] \simeq \frac{\Delta}{E_F} \ll 1.$$

This is the basic difference between the two cases: in this sense they belong to the same continuum but are opposite extreme cases. It is often useful in thinking about the system to keep the molecular picture in mind.

If a particular solution $\Delta(\mathbf{k})$ is a complex spherical harmonic such as $\hat{k}_x + i\hat{k}_y$, the resulting b will be complex and, in this case, the corresponding molecular state could have a net angular momentum $M_L^z = +1$. In precisely the same way the correlation hump will have a number-current correlation calculated, again, as though b were a wave function:

$$\langle n(\mathbf{0})\mathbf{j}(\mathbf{r})\rangle = \frac{\hbar}{m}(b^*\nabla b - b\nabla b^*).$$

In the given case, this will lead to a net orbital angular momentum of the sample, but, as Anderson and Morel showed, one may have $\langle n\mathbf{j}\rangle$ correlation without net angular momentum.

A word should be said about spin: just as for the molecule, a symmetric orbital function implies a spin singlet, an antisymmetric one a triplet. For even l, then, the spin behaviour will be just as in BCS, but for odd l there will be triplet pairing. Anderson and Morel considered only parallel spin pairing, ignoring the third $(\uparrow\downarrow + \downarrow\uparrow)$ component of the triplet. This error unfortunately has a qualitative effect on some results, as we shall see. All of the theory we have so far discussed assumes a single spin wave function, either up and down spins independently paired parallel, and l odd, or BCS type singlet pairing as one always has for l even. The magnetic behaviour in either case is simple: for odd l the spin susceptibility of independently paired up and down spin systems is unchanged, since the relative Fermi levels may shift at will. For even l a curve similar to Yosida's[11] as calculated for BCS is correct, giving a χ which vanishes as $T \to 0$ because shifting the Fermi surfaces relatively breaks up the $\uparrow\downarrow$ pairs.

In their final paper Anderson and Morel worked out a number of results with some care. Unfortunately, the greatest effort went into the case $l = 2$ which, at the time, was considered to be the most favourable.

They made the point that the higher l states would be superfluid even though the energy gap vanishes at several points on the Fermi surface. This was the first example of "gapless superfluidity" belying the Landau criterion for superfluidity and showing that the real criterion is the existence of the anomalous self-energy Δ. They gave the presently used formula for superfluid density in weak coupling and observed that in many cases this would be anisotropic. A final remark which is relevant to the modern theory is that anisotropic BCS states would be very sensitive to all forms of scattering—impurities, walls, etc., since all such would be "pair-breaking", in modern terms, like magnetic scattering in superconductors.

A few years later Balian and Werthamer[12] gave the first proper treatment of all spin components and worked out what is still the accepted formalism of the weak-coupling theory. Since they were concerned not with liquid He^3, primarily, but (despite developing the remark about scattering made by Anderson and Morel) to try to explain discrepancies in the spin susceptibility of normal superconductors in terms of an $s = 1$ state, they concentrated on the $l = 1$, $s = 1$ case, fortunately for the later development. (The singlet case is not modified, of course.) Since we will have occasion to use the BW (Balian–Werthamer) formalism at length in the body of this review, we shall not present it in detail here. Suffice it to say that upon including the third $(\uparrow\downarrow + \downarrow\uparrow)$ pairing component, one has available the possibility of using each of the three $l = 1$ spherical harmonics $m_l = \pm 1, 0$ (or k_x, k_y, k_z) with a different $s = 1$ spin state. One thereby can derive a pseudoisotropic, essentially real state with properties very like the standard $l = 0$ BCS state. In weak coupling this is clearly preferable to the complex AM (Anderson–Morel) state and apparently scotches the chance for orbital ferromagnetism in real systems. The BW formalism for the spin susceptibility shows that this state follows an intermediate course, extrapolating to $\chi_S/\chi_N = \frac{2}{3}$ at $T = 0$.

A final major early advance in discussing the properties of these hypothetical states was made by Leggett[13, 14] in 1965. Leggett observed that even if the transition temperature were so low as to validate the weak-coupling assumption so far made, at the very least Landau Fermi-liquid-parameter corrections would have to be made to the properties—such as χ_S/χ_N, ρ_S, etc.—of the anisotropic state. His results again will be used here repeatedly and thus will be treated in more detail later. Therefore we quote only the most striking ones: that the exchange correction to the spin susceptibility would bring the BW value down to $\frac{1}{3}$ or so at $T = 0$, and that the temperature dependence of the superfluid density is drastically changed because of large mass enhancement.

In the meantime, a new and fruitful line of discussion of the problem of the transition temperature was opening up. Emery[15] appears to have been the first to remark that magnetic renormalization effects might very appreciably favour odd-l, triplet states *vis-a-vis* even-l, singlet ones. The quantitative working out of this idea stemmed from the application by Berk and Schrieffer[16] of spin-fluctuation theory to the problem of superconducting transition temperatures.

The spin-fluctuation theory applies to Fermi systems of the type known as "nearly magnetic" in which the renormalization of the spin susceptibility by exchange effects is large (Pd, Pt, Ni among the metals; and, as we have observed, He^3). In such materials the spectrum of single-particle pairs has an enhanced low-frequency, long-wavelength region which can be thought of as a spectrum of a non-propagating type of collective excitation, the "spin fluctuation" or paramagnon. Berk and Schrieffer showed that the exchange of such fluctuations leads to a repulsive effective potential between the members of a singlet pair, and thus depresses the superconducting T_c. Although quantitatively this theory is never reliable, this and another qualitative result—that exchange of spin fluctuations enhances the linear specific heat γT and leads to thermal anomalies—are well borne out by experiment. One can think of the depression of T_c in another way, which is equally valid and gives one a good feeling for the physics: that the strong reduction which singlet BCS causes in the spin susceptibility makes the propagation of paramagnons more difficult, and reduces the paramagnon free energy. This interference of the two disfavours BCS superconductivity. Applied to He^3 by Brinkman and Engelsberg,(17) even quantitatively the specific heat theory was not a total failure.

Layzer and Fay(18) seem to have been the first to attempt a calculation of the effect of paramagnons on the T_c for anisotropic BCS states in He^3. One sees immediately that the effect must be very large, and indeed in this and later papers the major theoretical difficulty remains simply that. Not only is the even-l, $s = 0$, T_c very severely depressed, one finds, qualitatively because of the very different spin susceptibility of the $s = 1$ states, a strong enhancement of T_c for them. While quantitatively the effect on T_c cannot be calculated with any confidence at all even yet, Layzer and Fay's work strongly indicated that if anisotropic superfluidity were found, odd l was the correct choice. Incidentally, it is still not clear why it has so far not been discovered in paramagnon metals such as Pt and Pd as well.

3. POST-EXPERIMENT HISTORY—LATE 1972 TO PRESENT

Here we will give only a quick sketch of the developments as an introduction to the remainder of this chapter. Once correctly interpreted, the experiment of Osheroff *et al.*(19, 20) called out strongly to all those aware of the theoretical background for an interpretation as anisotropic BCS superfluidity. A low T_c/E_F, second-order phase transition, and approximately correct specific heat jump suggested BCS. The existence of more than one low phase, the A and the B, with what had to be equal or similar T_c's, is, as we have seen, a characteristic of these states. The magnetic susceptibility behaviour could be relatively easily fitted into the AM–BW structure, and militated against any alternative possibility such as some spin-density wave or liquid-crystal state. Finally, the shifts of the NMR frequency in the A phase required, rigorously, an anisotropic $\langle n(\mathbf{0})n(\mathbf{r})\rangle$ correlation.

The first to fit some of these results into the anisotropic BCS picture correctly was Leggett.(21, 22, 23) Leggett's theory of the resonance shifts has been remarkably successful in interpreting both the original data and subsequent measurements which have played the major role in identifying the A and B phases.

To briefly recapitulate the early experiments: these revealed a second-order phase transition along the melting curve at about 2·6 mK, into a phase which had unchanged nuclear susceptibility but a large NMR shift obeying

$$\omega_R^2 = (\gamma H)^2 + \Delta\omega_T^2(T).$$

At a lower T, about 2·0 mK, there was a first-order transition into a state with $\frac{1}{2}$ to $\frac{1}{3}$ of the nuclear susceptibility.

Leggett showed that any axial, odd-l phase such as those postulated by Anderson and Morel in 1960 would show such a shift. His theory will be discussed in detail in our section on resonance effects. His early work was marred by two slight difficulties: an incorrect computation of the shift in the BW phase, which is correctly zero; with this, and an understandable unwillingness to accept the stability of an AM type phase, he hesitated to identify the phases correctly.

On the basis of a less accurate theory of shifts Anderson[24, 25] reproduced Leggett's frequencies in the high-field regime and correctly identified the A phase as an $l = 1$, AM phase, the B phase as BW. Where they differ Anderson's theory is wrong, except possibly very near to T_c.

To continue with the resonance story, Leggett's full paper in 1973 correctly calculated shifts for all possible $l = 1$, $s = 1$ phases, predicted a "longitudinal resonance" which has now been observed at the correct frequency in both phases (Osheroff and Brinkman,[26] Osheroff[27]), and produced a complete theory of orientation effects by dipolar interactions. These theories have been among the most useful results in phenomenologically identifying the phases.

A second confirmation of the triplet BCS nature of the transitions came from the small, but characteristic, effect of magnetic fields on the second-order "A" transition (Ambegaokar and Mermin;[28] Varma and Werthamer[29]). In a magnetic field there is a linear displacement of the two Fermi levels and an accompanying change $\pm(\mathrm{d}^2n/\mathrm{d}\varepsilon^2)\mu H$ in the densities of states of both up and down spins. Thus T_c for a pure ↑↑ pairing, say, is slightly, and linearly, raised, leading to a small region of the so-called "A_1" phase. The theory does not give the correct, observed effect in detail, but the presence of this effect is a clear signal of a triplet BCS phase.

Early calculations of the attenuation of ultrasound (Patton,[30] Wölfle[31,32]), when compared with that measured by Richardson and Lee[33] confirmed, to a considerable extent, Anderson's assignment of the phases.

Although all the experimental data (magnetic susceptibility, resonance, ultrasound) pointed in the direction of the AM ↔ A, BW ↔ B assignment, it was clearly in contradiction with microscopic BCS theory and one was very reluctant to accept it whole-heartedly. The reason for this inversion of the stability of the two phases was found by Anderson and Brinkman[34] in a feedback effect of the susceptibility difference upon the spin-fluctuation interactions. The feedback calculation gives rise to a coupling between up and down spin pairs which favours parallel orbital angular momentum which is a specialization of the AM state; therefore this phase is now called the ABM (Anderson–Brinkman–Morel) state.

The many experimental consequences of this idea, as worked out in subsequent papers (Brinkman and Anderson;[35, 36] Brinkman *et al.*;[37] Kuroda[38, 39]) have been confirmed in considerable detail, especially by the series of phase diagrams and thermodynamic measurements carried out by Wheatley *et al.*[40, 41, 42, 43] and the NMR measurements in the A_1A_2 region by Osheroff and Anderson.[44]

The superfluidity properties have not been as accessible experimentally, but have attracted a great deal of theoretical attention. The formal calculation of superfluid density was carried out by Saslow.[45] demonstrating the anisotropy of ρ_s, and Legget's considerations on Fermi-liquid renormalization were applied to the anisotropic ρ_s. Unfortunately, the viscosity measurements of Alvesalo *et al.*,[46, 47] the anomalous heat conductivity measurements of Wheatley's group[41] and the fourth-sound measurements of Wheatley's group[48] and the Cornell groups,[49] while adding up to a very conclusive demonstration of superfluidity, do not settle many quantitative questions.

A remarkable new phenomenon has been predicted theoretically by de Gennes.[50, 51, 52] This is the phenomenon of "orbital supercurrents", resulting from cross-coupling of superflow and the angular properties. These effects are intrinsically dependent upon the complex wave function and their observations will present direct proof of its existence. The original speculations of de Gennes have been modified and completed by Wölfle[53] and Blount, Brinkman and Anderson.[54]

Another area which has been fertile for theoretical discussion but has not had direct experimental confirmation is that of collective modes of motion. Pre-1972 speculations were, on the whole, found to be infertile, and it is only from the detailed equations of motion of Anderson[24, 25] and Leggett,[22, 23, 55] the detailed calculations of Maki and Ebisawa,[56] Combescot[57, 58] and Brinkman and Smith.[59] and the current equations of de Gennes[50, 51] and Wölfle[53] that a coherent picture has arisen. Anderson's "orbit waves" are apparently to be expected at quite low frequencies in the ABM–A phase, while spin waves (Brinkman and Smith;[59] Maki and Tsuneto;[60] Combescot[57, 58]) will be present in both phases.

Finally, de Gennes[50, 51] has opened the discussion of "textures" (in the liquid crystal sense), singularities and surface orientation effects. Brinkman, Smith, Osheroff and Blount[61] have given a first computation of resonance effects in the BW–B phase in terms of textures and the weak orientation effects in the B phase suggested by Leggett.[21, 22, 23, 55]

At the present moment one may summarize the situation as follows: The theoretical structure based on Leggett's equations of motion and the spin-fluctuation stabilization of the ABM state has had a series of remarkable successes in correlating the static and NMR properties of these phases. The theoretical suggestions of de Gennes and others on the dynamical side are adequate to foreshadow a much greater richness of phenomena than demonstrated either by He^4II or by superconductors, and thus perhaps it is not surprising that on that side theory and experiment, while both very active, are yet to be reconciled in detail.

II. Properties of He^3 in the Normal State

The properties of He^3 as a normal Fermi liquid have recently been reviewed by Baym and Pethick.[62] Here we shall only briefly summarize the features which are relevant to our considerations of the superfluid phases. Using Landau's theory of a normal Fermi liquid the low temperature properties of He^3 can be parameterized in terms of the interaction function between quasiparticles defined as

$$f_{\alpha\beta}(\mathbf{k}, \mathbf{k}') \equiv \frac{\delta^2 E}{\delta n_{\mathbf{k}\alpha} \delta n_{\mathbf{k}\beta}} \qquad \text{(II.1)}$$

where E is the total energy of the Fermi liquid and $n_{\mathbf{k}\alpha}$ is the occupation number of the quasiparticle state with momentum $\mathbf{k}$ and spin α. Since in most experiments quasiparticles are always excited near the Fermi surface, $|\mathbf{k}| \simeq |\mathbf{k}'| \simeq k_F$ and the $f_{\alpha\beta}(\mathbf{k}, \mathbf{k}')$ depends only on the angle θ between $\mathbf{k}$ and $\mathbf{k}'$. Therefore f can be expanded in Legendre polynomials. Using the spin symmetry of the function one is left with two sets of parameters

$$F_l^s \equiv \frac{(2l+1)}{2} N(0) \int_{-1}^{1} \mathrm{d}\cos\theta\, P_l(\cos\theta)(f_{\uparrow\uparrow}(\theta) + f_{\uparrow\downarrow}(\theta)) \qquad \text{(II.2)}$$

$$F_l^a \equiv \frac{(2l+1)}{2} N(0) \int_{-1}^{1} \mathrm{d}\cos\theta\, P_l(\cos\theta)(f_{\uparrow\downarrow}(\theta) - f_{\uparrow\downarrow}(\theta)). \qquad \text{(II.3)}$$

In principle, one should be able to determine all of these parameters from experiment and thus give a complete characterization of the Fermi liquid. In practice, one can only determine the values of F_0^s, F_1^s, F_0^a and crude estimates of F_1^a and it turns out that even if we knew all the Fermi-liquid parameters we would not have the essential information necessary to determine T_c and other properties of the superfluid and normal states. However, considerable information on the nature of the state of normal He^3 is available in the values of the known parameters. The experimental values[63] of F_0^s are large and positive, varying from 10·8 at the saturated vapour pressure to

≈ 100 near the melting curve. The values of F_1^s, which determines the effective mass, i.e. $m^*/m = 1 + F_1^s/3$, of the quasiparticles, vary from 6·3 at the s.v.p. to ≈ 15 on the melting curve. Finally, F_0^a is determined from the so-called Stoner enhancement factor defined from the ratio of the measured susceptibility to the non-interacting Pauli susceptibility

$$\frac{\chi}{\chi_P} = \frac{m^*/m}{(1 + F_0^a)}. \tag{II.4}$$

It is found that F_0^a is negative and only slightly pressure dependent, varying from $\approx -0{\cdot}67$ at s.v.p. to $-0{\cdot}74$ near the melting curve. The large positive values of F_0^s imply that those particle–hole pair excitations which are of the density-fluctuation type are strongly coupled together into the higher energy zero-sound mode, while the negative value of F_0^a implies that the low energy particle–hole pair excitations of the spin-fluctuation type are strongly enhanced. Therefore, the dynamic spin susceptibility $\chi(q, \omega)$ will be dominated by the low energy excitations while the density response function will be dominated by the zero-sound collective mode. These facts led Doniach and Engelsberg[64] to show that deviations of the specific heat of He^3 from the simple linear temperature dependence given by the Landau theory could be explained in terms of the low-frequency long-wavelength spin fluctuations. In particular, they showed that the singular nature of the fluctuations at long wavelengths and low energies gives rise to an enhancement of the effective mass and to a $T^3 \ln T$ correction to the specific heat whose coefficient is sufficiently large to explain the observed anomalous behaviour.

Shortly thereafter, Rice[65,66] showed that the spin-fluctuation model also leads to corrections to the transport coefficients that differed by only one power of temperature from the low temperature limiting forms. A $T^3 \ln T$ correction was also found in the temperature dependence of the susceptibility.[67] These results were all discovered by studying a simple model with a contact interaction I which was assumed to be determined from the susceptibility enhancement $\chi/\chi_p = (1 - IN(0))^{-1}$, where $N(0)$ is the bare-particle density of states at the Fermi surface $mk_F/2\pi^2$. More recently Pethick,[68] Emery,[69] Amit, Kane, and Wagner[70] and others have shown that spin-fluctuation-model results can be placed into the context of Fermi-liquid theory and that such corrections are a general property of any Fermi liquid, not just almost ferromagnetic liquids for which the Stoner enhancement is large.

In the spin-fluctuation model one uses the parameter $\bar{I} \equiv IN(0)$ to describe the entire susceptibility enhancement including the effective mass corrections, i.e.

$$\frac{\chi}{\chi_P} \equiv \frac{1}{1 - \bar{I}} \equiv \frac{m^*/m}{(1 + F_0^a)}. \tag{II.5}$$

Thus in this model He^3 "looks like" an almost ferromagnetic system in that $\bar{I}$ becomes closer to one as the pressure is increased. However, from a Fermi-liquid point of view, as we already mentioned, F_0^a is almost independent of pressure and the variation of m^*/m gives rise to the pressure dependence of χ/χ_P. Therefore, one would not say that He^3 is almost ferromagnetic. In fact, this variation of F_0^s, F_1^s, and F_0^a as a function of pressure is qualitatively what one expects for a Fermi system approaching localization in the Mott[71] sense. Indeed, He^3 does localize at 33 atm when it becomes a solid. To see this we compare the behaviour of He^3 with the Gutzwiller[72] results on the Hubbard model with one particle per site. In this model I is the intrasite repulsive interaction between the particles. Let us assume that increasing pressure on He^3 is equivalent to increasing the repulsive interaction in the model. Brinkman and Rice[73, 74] showed that the Gutzwiller trial-wave-function approach to this problem gives rise to a critical value of the interaction I at which the system becomes localized, each particle occupying a single site. As $I(P)$ approaches the critical value the Fermi liquid has the following properties

$$\frac{m^*}{m} = \left\{1 - \left[\frac{I(P)}{I_c}\right]^2\right\}^{-1}, \qquad \text{(II.6)}$$

$$F_0^a = IN(0)\frac{1 + I/2I_c}{(1 + I/I_c)^2} \qquad \text{(II.7)}$$

and

$$F_0^s = IN(0)\left(2 + \frac{I}{I_c}\right)^{-1}\left(1 - \frac{I}{I_c}\right)^{-2}. \qquad \text{(II.8)}$$

Here I_c is the critical interaction strength at which localization sets in. Varying $I(P)$ to fit the m^*/m versus pressure data we find that F_0^a is almost constant, while F_0^s varies roughly as the square of the mass. These results qualitatively agree with the experimental facts, so that the properties of He^3 liquid are undoubtedly related to its incipient localization rather than to its being almost ferromagnetic. However, this does not mean that the spin fluctuations which characterize the behaviour of $\chi(q, \omega)$ are unimportant, but indicates rather that they are a consequence of the incipient localization. We will find that these fluctuations are very important in understanding the stability of the superfluid A phase.

Since the Landau parameters offer a means of parameterizing a normal Fermi liquid, one would like to be able to estimate their values from first-principle calculations. In practice, such work has not proved very successful. The most recent work is that of Babu and Brown,[75] who attempt to calculate the interaction function as a self-consistency problem in the virtual excitation of quasiparticle–hole pairs. Although the approach is interesting it does not appear to give good quantitative results. In the final analysis it is found necessary to introduce the spin-fluctuation concepts. This problem has also been attacked using variational wavefunction techniques by Feenberg[76] and his collaborators with slightly more quantitative success. Older work using the Brueckner K-matrix approach[77] has been shown by several authors to lead to poor results for the effective interaction. However, it has been shown by Østgaard[78] that by including the so-called rearrangement terms one obtains reasonable estimates of the Landau parameters from the Brueckner scheme. These calculations are simply pointed out because their sophistication makes one appreciate the difficulty of calculating the effective quasiparticle interaction.

III. Higher Angular Momentum Pairs and Estimates of T_c

The superfluid transition of a Fermi system can be determined by examining the stability of the normal state with respect to the formation of bound states of pairs of quasiparticles with total momentum zero. The transition temperature is determined by a divergence in the scattering amplitude of two quasiparticles at the Fermi surface.[6] The scattering amplitudes satisfy an integral obtained by summing the particle–particle ladder diagram shown in Fig. 2.

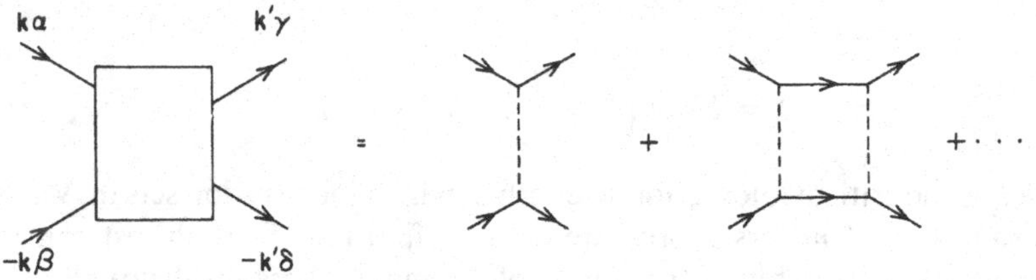

FIG. 2. The ladder diagrams that must be summed to calculate the stability of the normal state with respect to the creation of bound Cooper pairs.

Since each quasiparticle has spin $\frac{1}{2}$, the scattering states can be divided into states with total spin angular momentum 0 or 1. The two spin channels satisfy separate integral equations

$$A_\sigma(\mathbf{k}, \mathbf{k}') = v_\sigma(\mathbf{k}, \mathbf{k}') - \int \frac{d^3k''}{(2\pi)^3} v_\sigma(\mathbf{k}, \mathbf{k}'') \frac{(1 - 2f_{k''})}{2\varepsilon_{k''}} A_\sigma(\mathbf{k}'', \mathbf{k}'). \qquad \text{(III.1)}$$

We have assumed that the total momentum and energy of the pair is zero. The f_k is the occupation number of the quasiparticle state k of energy ε_k. Energy is measured relative to the Fermi momentum. The variable σ indicates total spin (=0.1). The $v_\sigma(\mathbf{k}, \mathbf{k}')$ is the matrix element of the interaction potential. For a system invariant under spin rotations

$$v_1(\mathbf{k}, \mathbf{k}') = \tfrac{1}{2}\langle \mathbf{k}\uparrow, -\mathbf{k}\uparrow | v | \mathbf{k}'\uparrow, -\mathbf{k}'\uparrow\rangle, \tag{III.2}$$

where

$$|\mathbf{k}\sigma, -\mathbf{k}\sigma'\rangle = a^\dagger_{\mathbf{k}\sigma} a^\dagger_{-\mathbf{k}\sigma}|\psi\rangle$$

and $|\psi\rangle$ is the state with no quasiparticles.

Since the scattering of an up spin with a down spin must be half the sum of the triplet and singlet matrix elements we have

$$v_0(\mathbf{k}, \mathbf{k}') = \tfrac{1}{2}[2\langle \mathbf{k}\uparrow, -\mathbf{k}\downarrow | v | \mathbf{k}'\uparrow, -\mathbf{k}'\downarrow\rangle - \langle \mathbf{k}\uparrow, -\mathbf{k}\uparrow | v | \mathbf{k}'\uparrow, -\mathbf{k}'\uparrow\rangle]. \tag{III.3}$$

It is easy to see that $v_1(\mathbf{k}, \mathbf{k}')$ is an odd function of $\mathbf{k}$ and $\mathbf{k}'$, while v_0 is an even function. This in turn implies that A_0 and A_1 are even and odd respectively. The early papers on He^3 concentrated on the solution of (III.1) when $v(\mathbf{k}, \mathbf{k}')$ is chosen to be the bare interaction potential between two He^3 atoms.

To do this we write

$$v(\mathbf{k}, \mathbf{k}') = v(k, k', \theta) \tag{III.4}$$

$$= \sum_{l=0}^{\infty} (2l+1) v_l(k, k') P_l(\cos\theta), \tag{III.5}$$

where θ is the angle between $\mathbf{k}$ and $\mathbf{k}'$ and where the P_l are the Legendre polynomials. By making the same expansion for $A_\sigma(\mathbf{k}, \mathbf{k}')$ we obtain an integral equation for states of given orbital angular momentum

$$A_l(k, k') = v_l(k, k') - \int_0^\infty \frac{k''^2 dk''}{(2\pi^2)} v_l(k, k'') \frac{(1 - 2f_{k''})}{2\varepsilon_{k''}} A_l(k'', k'). \tag{III.6}$$

We use a single index l, since its value implies the spin value; i.e. odd l implies spin triplet and even l spin singlet. This equation is usually solved in two steps by introducing an artificial cut-off ω_c and first solving (III.6) with the integration over $\mathbf{k}''$ restricted to be outside the region of momentum space defined by $|\varepsilon_{k''}| < \omega_c$. If the solution to the resulting equation is defined to be $K_l(k, k')$ then one can show that

$$A_l(k, k') = K_l(k, k') - \int_{|\varepsilon_k| < \omega_c} \frac{k''^2 dk''}{(2\pi^2)} K_l(k, k'') \frac{(1 - 2f_{k''})}{2\varepsilon_{k''}} A_l(k'', k'). \tag{III.7}$$

By doing this one treats the problem of the hard-core repulsion in solving for K_l and then studies (III.7) for the superfluid transition temperature. By a proper choice of ω_c it is possible to make K_l relatively independent of k and k'. In this case the solution to (III.7) is

$$A_l = \frac{K_l}{\left\{1 + K_l N(0) \int_0^{\omega_c} \frac{d\varepsilon}{\varepsilon} \tanh\left(\frac{\beta\varepsilon}{2}\right)\right\}}. \tag{III.8}$$

The transition temperature is thus given by the vanishing of the denominator,

$$1 = -K_l N(0) \int_0^{\omega_c} \frac{d\varepsilon}{\varepsilon} \tanh\left(\frac{\beta\varepsilon}{2}\right), \tag{III.9}$$

or defining

$$\lambda_l = -N(0)K_l \tag{III.10}$$

for negative K_l we have

$$T_c^l = 1{\cdot}14\omega_c e^{-1/\lambda_l}. \tag{III.11}$$

Although somewhat different procedures have been developed in order to treat separately the hard core, the principle behind them is the same as in the above treatment. Estimates by Brueckner *et al.*[5] and by Emery and Sessler[6] based on the 6–12 potential of de Boer and other suitable bare potentials give the result that the $l = 2$ *d*-wave pairs should become unstable at temperatures in the region of 0·01 K although the latter number was very uncertain. The *p*-wave pairs were found to have a small K_l whose sign was uncertain.

Pitaevskii[4] in his early work on $l \neq 0$ pairing examined carefully the consequences of the long-range nature of the van der Waals force and was able to show that v_l was necessarily attractive for sufficiently large l even if one includes many-body polarization effects. This problem was re-examined by Gor'kov and Pitaevskii[79] who attempted to extrapolate the large l results to $l \approx 2{\cdot}3$. Recently Bookbinder[80] showed that solutions for the function K_l, which is the Brueckner K-matrix, do not approach the large l limit until $l > 20$, so that in practice the exact result of Pitaevskii, although interesting, is not very useful.

Since liquid He^3 is not a low-density gas one cannot expect the estimates made from the two-particle K-matrix to be very accurate. This point was made clear in 1964 when Emery[15] investigated the use of the scattering amplitude calculated from the bare interaction as a means of determining the Landau parameters of the normal state. As mentioned in the previous section, he showed that this approximation leads to a poor description of these parameters and argued that it is quite important to include the effect of other He^3 atoms on the effective interaction. He thus noted that the values of the Landau parameters suggested that *p*-wave pairing would be most favourable. Since it has not yet proven possible to calculate this amplitude from first principles we proceed in a phenomenological fashion.

The scattering amplitude needed is that between two quasiparticles, so that v_l is the four-point vertex irreducible with respect to particle–particle scattering. We can gain some insight into its structure by considering the spin-fluctuation model, as was first done for singlet s-wave pairing by Berk and Schrieffer[16] and for triplet pairing by Layzer and Fay.[81]

In this model one assumes that the spin fluctuations are much like phonons in the case of strong-coupling superconductivity. The interaction between two quasiparticles is assumed to be due to the attraction of one quasiparticle by the spin-polarization cloud of the other. Clearly the force will be attractive only if the spins of the two quasiparticles are aligned. To see this first consider a spin-dependent interaction of the general form[82]

$$v = \frac{1}{2}\sum_{\substack{\mathbf{kk}' \\ \mathbf{q}}} [\Gamma_\rho(\mathbf{q})\delta_{\alpha\beta}\delta_{\gamma\delta} + \Gamma_s(\mathbf{q})\boldsymbol{\sigma}_{\alpha\beta}\cdot\boldsymbol{\sigma}_{\gamma\delta}]\, a^\dagger_{\mathbf{k}+\mathbf{q}\alpha} a^\dagger_{\mathbf{k}'-\mathbf{q}\gamma} a_{\mathbf{k}'\delta} a_{\mathbf{k}\beta}. \qquad \text{(III.12)}$$

Taking the matrix elements in (III.2) and (III.3) we find that

$$v_1(\mathbf{k}, \mathbf{k}') = \tfrac{1}{2}[\Gamma_\rho(\mathbf{k} - \mathbf{k}') + \Gamma_s(\mathbf{k} - \mathbf{k}') - (\mathbf{k} \to -\mathbf{k})] \qquad \text{(III.13)}$$

and

$$v_0(\mathbf{k}, \mathbf{k}') = \tfrac{1}{2}[\Gamma_\rho(\mathbf{k} - \mathbf{k}') - 3\Gamma_s(\mathbf{k} - \mathbf{k}') + (\mathbf{k} \to -\mathbf{k})]. \qquad \text{(III.14)}$$

If Γ_s is negative, it enhances triplet or odd-l pairing while suppressing singlet even-l pairing. In the spin-fluctuation model using the contact interaction I the Γ's are obtained from the summation of the diagrams shown in Fig. 3.

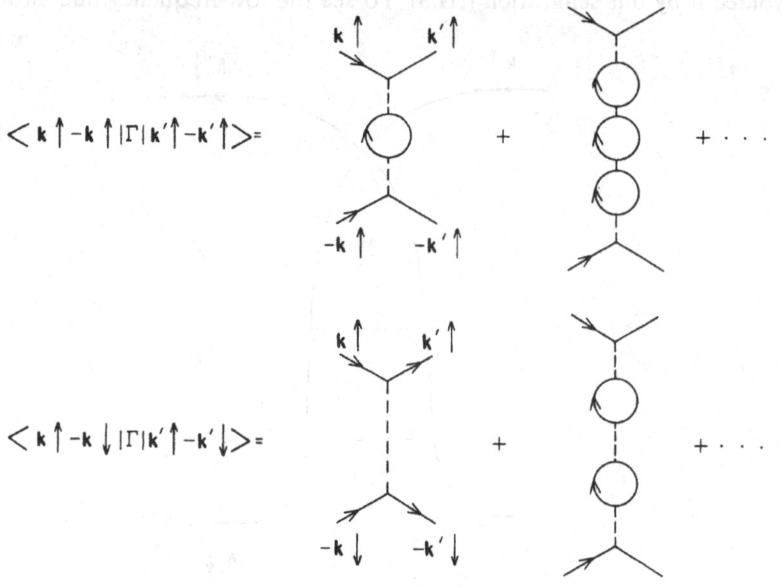

FIG. 3. The diagrams that must be summed to obtain the spin-fluctuation contributions to the effective interaction Γ_s. The bubble represents $\chi^0(q, \omega)$.

These diagrams can be summed to give

$$\langle \mathbf{k}\uparrow, -\mathbf{k}\uparrow | v | \mathbf{k}'\uparrow, -\mathbf{k}'\uparrow \rangle = -\frac{I^2\chi^0(\mathbf{k}-\mathbf{k}',0)}{\{1-[I\chi^0(\mathbf{k}-\mathbf{k}',0)]^2\}} + (\mathbf{k}\to -\mathbf{k}), \qquad \text{(III.15)}$$

$$\langle \mathbf{k}\uparrow, -\mathbf{k}\downarrow | v | \mathbf{k}'\uparrow, -\mathbf{k}'\downarrow \rangle = +\frac{I}{\{1-[I\chi^0(\mathbf{k}-\mathbf{k}',0)]^2\}}. \qquad \text{(III.16)}$$

The χ^0 is the noninteracting susceptibility

$$\chi^0(\mathbf{q},\omega) \equiv \int \frac{d^3k}{(2\pi)^3} \frac{(f_{\mathbf{k}+\mathbf{q}} - f_{\mathbf{k}})}{\omega - \varepsilon_{\mathbf{k}+\mathbf{q}} + \varepsilon_{\mathbf{k}}}. \qquad \text{(III.17)}$$

Equating (III.15, 16) to (III.13, 14) via (III.2, 3) we find

$$\Gamma_\rho(\mathbf{k}-\mathbf{k}') = \tfrac{1}{2}\frac{I}{[1+I\chi^0(\mathbf{k}-\mathbf{k}',0)]}, \qquad \text{(III.18)}$$

$$\Gamma_s(\mathbf{k}-\mathbf{k}') = -\tfrac{1}{2}\frac{I}{[1-I\chi^0(\mathbf{k}-\mathbf{k}',0)]}. \qquad \text{(III.19)}$$

The Γ_ρ describes the exchange of density fluctuations in the model. Since these fluctuations are high energy fluctuations we will ignore them for the present discussion. It should be remarked that the diagram in Fig. 4 must be considered if one desires to examine all the matrix elements. We have avoided it by our separation (III.3). To see the low-frequency fluctuation in

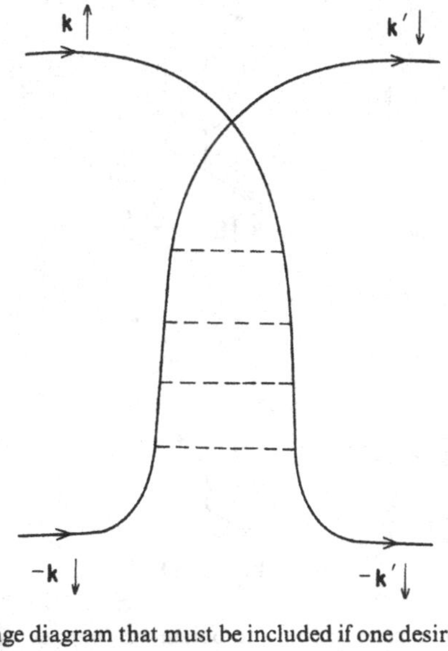

FIG. 4. The particle-exchange diagram that must be included if one desires to obtain the complete symmetry of the effective interaction.

Γ_s we calculate the noninteracting susceptibility χ^0 in the long-wavelength limit

$$\chi^0(q, \omega) = N(0)\left(1 - \frac{q^2}{12k_F^2} + \frac{i\pi}{2}\frac{\omega}{qv_F}\right). \tag{III.20}$$

Inserting this expression into (III.19) and taking the imaginary part, a peak is found in $\operatorname{Im}\Gamma_s$ at $\omega_q = (1 - \bar{I})(2/\pi)v_F q$ where $\bar{I} = IN(0)$. If $\bar{I}$ is close to one, ω_q is considerably less than $v_F q$. The situation is illustrated in Fig. 5. The

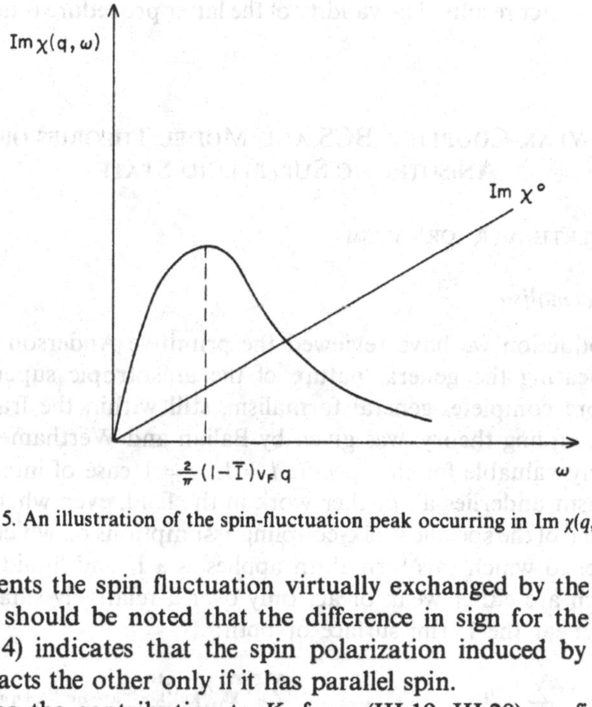

FIG. 5. An illustration of the spin-fluctuation peak occurring in Im $\chi(q, \omega)$.

peak represents the spin fluctuation virtually exchanged by the two quasi-particles. It should be noted that the difference in sign for the Γ_s term in (III.13, III.14) indicates that the spin polarization induced by one quasi-particle attracts the other only if it has parallel spin.

Calculating the contribution to K_l from (III.19, III.20) we find that the spin fluctuations give rise to a contribution

$$\lambda_l^{\mathrm{SF}} \simeq \frac{-3}{2}\ln\left(\frac{1 - 2/3\bar{I}}{1 - \bar{I}}\right) \tag{III.21}$$

for triplet pairing. For even-l pairing λ_l^{SF} is -3 times this result. (III.21) holds only for relatively small l since we have set $P_l(\cos\theta) = 1$ in doing the integrals. If we use $\bar{I} = 0{\cdot}75$ we obtain a number close to one. This is presumably cancelled to a large extent by the repulsive density-fluctuation term. What is essential here is that the calculation illustrates the importance of the spin fluctuations. Since one expects the result (III.21) to be most applicable for p-wave pairing one expects that the p-wave state will be most stable. Layzer and Fay[18] have attempted to calculate T_c^1 by using the sum of the spin-fluctuation interaction and the bare interaction for $v(\mathbf{k}, \mathbf{k}')$. In this way they obtained rather high values of $T_c^1 \approx 0{\cdot}1$ K. Recently, Layzer and Fay[83] have attempted to improve on their early work by examining the strong coupling equations used by Berk and Schrieffer for the $l = 0$ case. They find that the quasiparticle renormalization effects included in these equations tend to strongly lower T_c^1 and that its pressure dependence is the opposite of experiment. They then suggest a scaling of the renormalization factors which appears to give better results. The validity of the latter procedure is not presently clear.

IV. Weak-Coupling BCS and Model Theories of the Anisotropic Superfluid State

1. BALIAN–WERTHAMER FORMALISM

(a) *General formalism*

In the Introduction we have reviewed the primitive Anderson–Morel formalism indicating the general nature of the anisotropic superfluid state. A much more complete, general formalism, still within the framework of BCS weak-coupling theory, was given by Balian and Werthamer,[12] which is particularly valuable for the specific $l = 1$, $s = 1$ case of interest in He^3. This formalism underlies all further work in the field, even where one must abandon some of the specific weak-coupling assumptions on which it is based.

The model to which this formalism applies is a Fermi liquid with interactions which are either weak or act only over a relatively small region of phase space near the Fermi surface or both:

$$\mathscr{H} = \sum_{\mathbf{k}\sigma} \varepsilon_{\mathbf{k}} n_{\mathbf{k}\sigma} + \sum_{\substack{\mathbf{kk'q} \\ (|\varepsilon_k|,|\varepsilon_{k'}|<\omega_s)}} \sum_{\sigma\sigma'} V_{\mathbf{kk'}} a^\dagger_{\mathbf{k}\sigma} a^\dagger_{\mathbf{k'}\sigma'} a_{\mathbf{k'}-\mathbf{q}\sigma'} a_{\mathbf{k}+\mathbf{q}\sigma} \tag{IV.1}$$

where ω_s, the cutoff energy, is $\ll E_F$ and we measure the kinetic energy $\varepsilon_{\mathbf{k}}$ from the Fermi surface. We shall see that (IV.1) is far less restrictive than it appears. To a great extent, the BW formalism, like the BCS theory, can serve as a *model* even when it is not a valid microscopic *theory*. We will discuss the degree to which these restrictions may be removed, and the types of modifications which result, in later sections. The basic restriction, however, which makes BCS-type theories workable, cannot be abandoned: $T_c/E_F \ll 1$. This restriction ensures among other things that the part of the interaction which is responsible for the pairing condensation is in some real sense small (since $T_c/E_F \approx e^{-1/\lambda}$, where λ is the coupling constant). More importantly, the pairing interaction, and the pairing phenomenon, does not cause major changes in the basic correlations among the particles, because pairing only involves a few particles $\approx T_c/E_F$ near the Fermi surface. The bulk of the binding energy, the correlation function behaviour at most energies and wavelengths, and the major part of the Fermi-liquid renormalizations, and thus the normal quasiparticle self-energies, are not much affected.

With this restriction, and only with this, we can treat the pairing phenomenon as an independent additive contribution. The truncated hamiltonian formalism used by Bardeen, Cooper and Schrieffer and by Balian and Werthamer is then a perfectly satisfactory model, but we will use the more modern idea of introducing an anomalous self-energy which connects particle and hole excitations, and may be derived self-consistently from the existence of an anomalous amplitude or "pair amplitude"

$$\tfrac{1}{2}x^{\mathbf{k}}_{\sigma\sigma'} = \langle a_{\mathbf{k}\sigma} a_{-\mathbf{k}\sigma'} \rangle. \tag{IV.2}$$

Let us start the self-consistency process with this anomalous amplitude, which as you can see is already assumed to be constant in space, i.e. we for the present treat only pairs with equal and opposite momenta.

In the presence of an anomalous amplitude such as (IV.2) the fermions will experience an anomalous self energy

$$\Delta^{\mathbf{k}}_{\sigma\sigma'} = -\tfrac{1}{2} \sum_{\mathbf{k}'} V_{\mathbf{k}\mathbf{k}'} x^{\mathbf{k}}_{\sigma\sigma'} \tag{IV.3}$$

which will have the effect of mixing particles and holes in the equations of motion, by simple calculation from (IV.1):

$$\frac{\hbar}{i}\frac{\partial}{\partial t} a^{\dagger}_{\mathbf{k}\sigma} = [\mathscr{H}, a^{\dagger}_{\mathbf{k}\sigma}] = \varepsilon_{\mathbf{k}\sigma} a^{\dagger}_{\mathbf{k}\sigma} + \sum_{\sigma'} \Delta^{\mathbf{k}}_{\sigma\sigma'} a_{-\mathbf{k}\sigma'} + \ldots \tag{IV.4}$$

The + ... in (IV.4) are the many-body corrections, which may be subsumed into many-body corrections to the ordinary and anomalous self-energies.[84, 85] The anomalous term in (IV.4) requires us, in searching for elementary excitations, to make a canonical transformation involving both spins and both particles and holes:

$$a_{\mathbf{k}\sigma} = \sum_{\sigma'} u^{\mathbf{k}}_{\sigma\sigma'} \alpha_{\mathbf{k}\sigma'} + v^{\mathbf{k}}_{\sigma\sigma'} \alpha^{\dagger}_{-\mathbf{k}\sigma'} \tag{IV.5}$$

where the α's are the quasiparticle elementary excitations. Balian and Werthamer introduce a convenient 4-component matrix notation (following Eliashberg[86])

$$\hat{a}^{\mathbf{k}} \equiv \begin{pmatrix} a_{\mathbf{k}\uparrow} \\ a_{\mathbf{k}\downarrow} \\ a^{\dagger}_{-\mathbf{k}\uparrow} \\ a^{\dagger}_{-\mathbf{k}\downarrow} \end{pmatrix} \tag{IV.6}$$

and (IV.5) becomes

$$\hat{a}^{\mathbf{k}} = \begin{pmatrix} \hat{u}^{\mathbf{k}} & \hat{v}^{\mathbf{k}} \\ \hat{v}^{-\mathbf{k}*} & \hat{u}^{-\mathbf{k}*} \end{pmatrix} \hat{\alpha}^{\mathbf{k}} = \hat{U}^{\mathbf{k}} \hat{\alpha}^{\mathbf{k}}. \tag{IV.7}$$

$\hat{u}$ and $\hat{v}$ are 2×2 matrices, $\hat{U}$ is a 4×4 matrix. For (IV.7) to be canonical, $\hat{U}$ must be unitary. Clearly in this theory (as opposed to BCS) we cannot allow $\hat{U}^{\mathbf{k}}$ and $\hat{U}^{-\mathbf{k}}$ to be independent, and $\hat{U}$ satisfies the symmetry relationship

$$\hat{U}^{\mathbf{k}} = \begin{pmatrix} 0 & 1 \\ 1 & 0 \end{pmatrix} \hat{U}^{-\mathbf{k}*} \begin{pmatrix} 0 & 1 \\ 1 & 0 \end{pmatrix}. \tag{IV.8}$$

The equation of motion (IV.4) may be written in 4-component notation as well:

$$\frac{\hbar}{i} \frac{\partial (\hat{a}^{\mathbf{k}})^{\dagger}}{\partial t} = \hat{\mathscr{E}}^{\mathbf{k}} (\hat{a}^{\mathbf{k}})^{\dagger} \tag{IV.9}$$

where

$$\hat{\mathscr{E}}^{\mathbf{k}} = \begin{pmatrix} \varepsilon_{\mathbf{k}} & \hat{\Delta}_{\mathbf{k}} \\ -\hat{\Delta}^{*}_{-\mathbf{k}} & -\varepsilon_{\mathbf{k}} \end{pmatrix} = \begin{pmatrix} \varepsilon_{\mathbf{k}} & \hat{\Delta}_{\mathbf{k}} \\ (\hat{\Delta}_{\mathbf{k}})^{\dagger} & -\varepsilon_{\mathbf{k}} \end{pmatrix}. \tag{IV.10}$$

$\hat{\Delta}_{\mathbf{k}}$ is a 2×2 matrix in spin space given by the components (IV.3). The symmetry relation for $\hat{\Delta}$ is, by the anticommutation relations,

$$\hat{\Delta}_{\mathbf{k}} = -\tilde{\hat{\Delta}}_{-\mathbf{k}} \tag{IV.11}$$

as is true for $\hat{\mathscr{E}}$, and this has been used in the second equality of (IV.10). What we need to do is to diagonalize the equation of motion with the quasiparticle transformation $\hat{U}$ of (IV.7):

$$\hat{E}^{\mathbf{k}} = \hat{U}^{-1} \hat{\mathscr{E}}^{\mathbf{k}} \hat{U}, \tag{IV.12}$$

where $\hat{E}$ is a diagonal matrix:

$$\hat{E}^{\mathbf{k}} = \begin{pmatrix} E_{\mathbf{k}+} & & & 0 \\ & E_{\mathbf{k}-} & & \\ & & E_{-\mathbf{k}+} & \\ 0 & & & -E_{-\mathbf{k}-} \end{pmatrix}. \tag{IV.13}$$

When we have done that, the α's will be eigen-excitations whose thermal averages obey

$$\langle \alpha_{\mathbf{k}\sigma}\alpha^{\dagger}_{\mathbf{k}\sigma}\rangle = 1 - f(E_{\mathbf{k}\sigma}/kT), \tag{IV.14}$$

f being the Fermi function. This can be rewritten as a matrix equation

$$\langle \alpha^{\mathbf{k}}\alpha^{\mathbf{k}\dagger}\rangle = \begin{pmatrix} 1 - f(E_{\mathbf{k}+}) & & & 0 \\ & 1 - f(E_{\mathbf{k}-}) & & \\ & & f(E_{-\mathbf{k}+}) & \\ 0 & & & f(E_{-\mathbf{k}-}) \end{pmatrix}$$

$$= \tfrac{1}{2}(1 + \tanh \tfrac{1}{2}\beta \hat{E}^{\mathbf{k}}). \tag{IV.15}$$

From $\langle \alpha\alpha^{\dagger}\rangle$ we can, by transforming back to the a's, calculate the mean values of the real, bare-particle operators a. We write these in matrix form as well, defining matrices $\hat{w}$ and $\hat{x}$ by

$$\langle a^{\mathbf{k}}a^{\mathbf{k}\dagger}\rangle = \tfrac{1}{2}\begin{pmatrix} 1 + \hat{w}^{\mathbf{k}} & \hat{x}^{\mathbf{k}} \\ -\hat{x}^{-\mathbf{k}} & 1 - \hat{w}^{-\mathbf{k}} \end{pmatrix} \equiv \tfrac{1}{2}(1 + \hat{W}^{\mathbf{k}}). \tag{IV.16}$$

Here $\hat{x}$ is the matrix of anomalous amplitudes we started from, (IV.2), and the elements of the 2×2 matrix $\hat{w}$ are defined by

$$\langle a_{\mathbf{k}\sigma}a^{\dagger}_{\mathbf{k}\sigma'}\rangle = \tfrac{1}{2}(\delta_{\sigma\sigma'} + \hat{w}^{\mathbf{k}}_{\sigma\sigma'}). \tag{IV.17}$$

Initially we defined the transformation $\hat{U}$ to carry the a's into the α's and vice versa; thus

$$\langle \hat{a}^{\mathbf{k}}\hat{a}^{\mathbf{k}\dagger}\rangle = \hat{U}^{\mathbf{k}}\langle \alpha^{\mathbf{k}}\alpha^{\mathbf{k}\dagger}\rangle(\hat{U}^{\mathbf{k}})^{-1} \tag{IV.18}$$

or, using (IV.15),

$$\hat{W}^{\mathbf{k}} = \hat{U}^{\mathbf{k}}(\tanh \tfrac{1}{2}\beta\hat{E}^{\mathbf{k}})(\hat{U}^{\mathbf{k}})^{-1}$$
$$= \tanh \tfrac{1}{2}\beta\hat{\mathscr{E}}^{\mathbf{k}}. \tag{IV.19}$$

Since $\hat{x}^{\mathbf{k}}$ is determined by $\hat{W}^{\mathbf{k}}$, and in turn $\hat{\Delta}^{\mathbf{k}}$ is defined from $\hat{x}^{\mathbf{k}}$ by (IF.3), this gives us the essential self-consistency relationship. (IV.19) with (IV.3) is the gap equation in 4×4 matrix form.

Balian and Werthamer immediately specialize to a particular form of $\hat{\Delta}$ in order to simplify this equation. This is not necessary as yet, if we observe that the matrix

$$(\hat{\mathscr{E}}^{\mathbf{k}})^2 = (\hat{\varepsilon}_{\mathbf{k}} + \hat{\Delta}_{\mathbf{k}})^2$$

is block diagonal:

$$\hat{\mathscr{E}}^2 = \begin{pmatrix} (\varepsilon_{\mathbf{k}}^2 \hat{1} + \hat{\Delta}_{\mathbf{k}}^\dagger \hat{\Delta}_{\mathbf{k}}) & 0 \\ 0 & (\varepsilon_{\mathbf{k}}^2 \hat{1} + \hat{\Delta}_{\mathbf{k}} \hat{\Delta}_{\mathbf{k}}^\dagger) \end{pmatrix}. \tag{IV.20}$$

Thus the eigenvalues always appear in $\pm$ pairs, corresponding to quasiparticle and quasihole. Any function of $\hat{\mathscr{E}}^2$ will have the same diagonal form, as will

$$F(\hat{\mathscr{E}}^2) = \frac{1}{\hat{\mathscr{E}}} \tanh \tfrac{1}{2}\beta\hat{\mathscr{E}} \tag{IV.21}$$

which may be expanded as a power series in $\hat{\mathscr{E}}^2$. This allows us to break up (IV.19) into two 2×2 equations

$$\begin{aligned} \hat{w}^{\mathbf{k}} &= \varepsilon_{\mathbf{k}} F[(\hat{\mathscr{E}}^{\mathbf{k}})^2] \\ &= \varepsilon_{\mathbf{k}} \left[\frac{1}{(\varepsilon_{\mathbf{k}}^2 + \hat{\Delta}^\dagger\hat{\Delta})^{\frac{1}{2}}} \tanh \tfrac{1}{2}\beta(\varepsilon_{\mathbf{k}}^2 + \hat{\Delta}^\dagger\hat{\Delta})^{\frac{1}{2}} \right] \\ \hat{x}^{\mathbf{k}} &= \hat{\Delta}_{\mathbf{k}} F(\varepsilon_{\mathbf{k}}^2 + \hat{\Delta}_{\mathbf{k}}^\dagger \hat{\Delta}_{\mathbf{k}}), \end{aligned} \tag{IV.22}$$

where in (IV.22) F signifies the 2×2 block of F in which the $\hat{\Delta}^\dagger\hat{\Delta}$ order is taken. From now on except where noted all matrices will be 2×2 rather than 4×4, with spin indices only.

It is useful to note the theorem that

$$\tilde{\hat{x}}^{\mathbf{k}} = \tilde{\hat{\Delta}}_{\mathbf{k}} F(\varepsilon_{\mathbf{k}}^2 + \tilde{\hat{\Delta}}_{\mathbf{k}}^\dagger \tilde{\hat{\Delta}}_{\mathbf{k}}). \tag{IV.23}$$

That is, (IV.22) retains the transposition symmetry of $\hat{\Delta}$. Thus the full gap equation, which we may now write, does as well, since V is symmetric in its spin indices:

$$\begin{aligned} \Delta_{\sigma\sigma'}^{\mathbf{k}} &= (-\tfrac{1}{2}) \sum_{\mathbf{k}'} V_{\mathbf{k}\mathbf{k}'} [\hat{\Delta}_{\mathbf{k}'} F(\varepsilon_{\mathbf{k}'}^2 + \hat{\Delta}_{\mathbf{k}'}^\dagger \hat{\Delta}_{\mathbf{k}'})]_{\sigma\sigma'} \\ &= (-\tfrac{1}{2}) \sum_{\mathbf{k}'} V_{\mathbf{k}\mathbf{k}'} [\hat{\Delta}_{\mathbf{k}} \hat{\mathscr{E}}_{\mathbf{k}'}^{-1} \tanh \tfrac{1}{2}\beta\hat{\mathscr{E}}_{\mathbf{k}'}]_{\sigma\sigma'} \end{aligned} \tag{IV.24}$$

where

$$\hat{\mathscr{E}}_{\mathbf{k}}^2 = \varepsilon_{\mathbf{k}}^2 + \hat{\Delta}_{\mathbf{k}}^\dagger \hat{\Delta}_{\mathbf{k}}. \tag{IV.25}$$

The first of the equations (IV.22) simply determines $\hat{w}$ given $\hat{\Delta}$; (IV.24) is the basic equation of the system, or "gap equation".

(b) *Singlet–triplet separation and spin rotation symmetry*

Now we can separate out the singlet and triplet parts of (IV.24). The symmetry (IV.23) shows us that symmetry in interchange of σ and σ' is a good quantum number: a symmetric $\hat{\Delta}$ on the right-hand side leads self-consistently to a symmetric one on the left, and vice versa (though we see no proof that mixed spins *cannot* occur!). Thus we have the two cases

$$\hat{\Delta} = -\tilde{\hat{\Delta}} = \begin{pmatrix} 0 & d_0 \\ -d_0 & 0 \end{pmatrix}$$

(Singlet)

and

$$\hat{\Delta} = \tilde{\hat{\Delta}} = \begin{pmatrix} d_v + id_u & -id_w \\ -id_w & d_v - id_u \end{pmatrix}. \tag{IV.26}$$

(Triplet)

We have chosen the matrix elements in (IV.26) to conform to the BW convention

$$\hat{\Delta} = \sum_{i=0}^{3} \sigma_i \sigma_v d_i, \tag{IV.27}$$

where σ_u, σ_v, σ_w are Pauli spin matrices along a set of axes u, v, w in spin space, and σ_0 is $-i$ times the unit matrix. d_u, d_v and d_w are amplitudes of the pairs in the $m_s = 0$ quantum state along the corresponding axis, e.g. d_w has the symmetry of a wave-function w belonging to $s = 1$, where we have quantized our spins along $\hat{w}$.

The rotational properties of $\hat{\Delta}$ are peculiar, but the d's transform just like u, v and w under rotations of the spin axes. The unitary transformation on particle operators $a_{\mathbf{k}\sigma}$ which rotates the spin system through the Eulerian angles θ, φ, ψ is given by [87]

$$U(\theta, \varphi, \psi) = \begin{pmatrix} \cos[\theta/2] \exp[-\frac{1}{2}i(\varphi + \psi)] & -\sin[\theta/2] \exp[\frac{1}{2}(\psi - \varphi)] \\ \sin[\theta/2] \exp[\frac{1}{2}i(\varphi - \psi)] & \cos[\theta/2] \exp[\frac{1}{2}i(\psi + \varphi)] \end{pmatrix}.$$

The gap matrix $\hat{\Delta}$, by its definition, transforms not unitarily but according to

$$R(\hat{\Delta}_{\mathbf{k}}) = \hat{U}(R)\hat{\Delta}_{\mathbf{k}}\tilde{\hat{U}}(R), \tag{IV.28}$$

which in turn gives a conventional orthogonal rotation transformation to the d-vector:

$$R(d_u, d_v, d_w) = \hat{O}(R)\mathbf{d}.$$

The unusual rotation transformation leads to some counterintuitive results—such as, for instance, that it is the antisymmetric part of $\hat{\Delta}$ which is spin rotation-invariant. Because of the further symmetry of the commutation relationship (IV.11), (IV.26) implies that (as one would expect from Fermi statistics) the singlet corresponds to a $\hat{\Delta}(k)$ even in k, the triplet to an odd one.

The two sets of states see essentially different potentials

$$V^S_{\mathbf{kk}'} = \tfrac{1}{2}(V_{\mathbf{k}+\mathbf{k}'} + V_{\mathbf{k}-\mathbf{k}'})$$
$$V^T_{\mathbf{kk}'} = \tfrac{1}{2}(V_{\mathbf{k}+\mathbf{k}'} - V_{\mathbf{k}-\mathbf{k}'}). \tag{IV.29}$$

In actual fact, the real potential in He^3 is a result of complicated many-body screening and renormalization effects, and has a spin dependence which is not necessarily derivable from a real potential $V_{\mathbf{kk}'}$. It is then convenient to give the potential an explicit spin dependence: replace the potential in (IV.1) by

$$\tfrac{1}{2}\sum_{\mathbf{kk}'}\sum_{\sigma_i} a^\dagger_{\mathbf{k}\sigma_1} a^\dagger_{-\mathbf{k}\sigma_2} a_{-\mathbf{k}\sigma_3} a_{\mathbf{k}\sigma_4} V^{\sigma_1\sigma_2;\sigma_3\sigma_4}_{\mathbf{kk}'}. \tag{IV.30}$$

Here, if the system is isotropic in spin space we may write

$$V = V_1\delta(\sigma_1, \sigma_4)\delta(\sigma_2, \sigma_3) + V_2(\sigma)_{\sigma_1\sigma_4}\cdot(\sigma')_{\sigma_2\sigma_3}$$
$$= V_1 + V_2\sigma\cdot\sigma'. \tag{IV.31}$$

The triplet sees the potential $V_1 + V_2$, the singlet $V_1 - 3V_2$. We shall see that in the anisotropic BCS theory of He^3, (IV.31) is not quite adequate and we have to use the general expression (IV.30). The gap equation now reads

$$\Delta^{\sigma_1\sigma_2}_{\mathbf{k}} = \sum_{\sigma_3\sigma_4}\sum_{\mathbf{k}'} V^{\sigma_3\sigma_4;\sigma_1\sigma_2}_{\mathbf{kk}'} \chi^{\sigma_3\sigma_4}_{\mathbf{k}'} \tag{IV.32}$$

but the property of V that

$$V^{\alpha\beta;\gamma\delta} = V^{\gamma\delta;\alpha\beta}$$

still ensures the singlet–triplet separation.

(c) *Orbital angular momentum separation*

The gap equation may be further separated, to a very good approximation, according to orbital angular momentum l. At T_c the energy Δ is very small and

$$F(\hat{\mathscr{E}}^2_{\mathbf{k}}) = \frac{1}{\varepsilon_{\mathbf{k}}}\tanh\frac{\beta_c\varepsilon_{\mathbf{k}}}{2}, \tag{IV.33}$$

which is independent of Δ and isotropic in $\hat{\mathbf{k}}$. The gap equation (IV.24) then becomes a linear eigenvalue equation for the critical temperature β_c. Since the interaction V is isotropic, we may make the standard spherical harmonic separation of it as given in the Introduction

$$V_{\mathbf{kk}'} = \sum_l \overline{V}_l$$
$$\overline{V}_l = V_l(k, k')P_l(\hat{\mathbf{k}}\cdot\hat{\mathbf{k}}')(2l + 1)$$
$$= \overline{V}_l(k, k')(4\pi)\sum_m Y^m_l(\hat{\mathbf{k}})Y^{-m}_l(\hat{\mathbf{k}}'). \tag{IV.34}$$

If we assume a spherical harmonic solution of (IV.24)

$$\hat{\Delta}_{\mathbf{k}} = \hat{\Delta}_l(k) Y_l(\hat{\mathbf{k}}),$$

where Y_l is any linear combination of spherical harmonics of order l, the orthogonality of spherical harmonics of different l will eliminate all terms but Y_l,

$$\int \mathrm{d}\Omega_{\hat{k}} Y_l(\hat{\mathbf{k}}) P_{l'}(\hat{\mathbf{k}} \cdot \hat{\mathbf{k}}') = \delta_{ll'} Y_l(\hat{\mathbf{k}}') 4\pi/(2l+1), \tag{IV.35}$$

and in fact will reproduce Y_l on the left-hand side. Thus all spherical harmonics of order l are solutions at the appropriate T_c for l, and all therefore have the same T_c.

This is given by the standard BCS theory as in Section III:

$$\Delta_0 Y^l(\hat{\mathbf{k}}) = \frac{V_l}{2} \sum_{\mathbf{k}'} (2l+1) P_l(\hat{\mathbf{k}} \cdot \hat{\mathbf{k}}') \Delta_0 Y^l(\hat{\mathbf{k}}') \varepsilon_{\mathbf{k}}^{-1} \tanh \frac{\beta_c \varepsilon_{\mathbf{k}}}{2}$$

or

$$1 = N(0) V_l \int_0^{\omega_s} \frac{\mathrm{d}\varepsilon_{\mathbf{k}}}{\varepsilon_{\mathbf{k}}} \tanh \frac{\beta_c \varepsilon_{\mathbf{k}}}{2}$$

$$T_c \simeq 1{\cdot}14 \omega_s \exp\left(- \frac{1}{N(0) V_l} \right),$$

where $V_l \equiv V_l(k_F, k_F)$.

Below T_c $F(\mathscr{E}^2)$ is no longer independent of Δ or isotropic in k; but at most energies (if T_c is small) $\varepsilon_{\mathbf{k}} \gg \Delta$, and it is approximately so. Since the T_c's for different l's will in general be very different, the amount of coupling between, say, $l = 1$ and 3, or 0 and 2, is quite small—Anderson and Morel estimated a few percent at most. Thus, as they showed, it is an excellent approximation to neglect l–l' coupling, and to retain only V_l at all temperatures in the gap equation.

(d) *Special cases: Singlet $l = 2$ and free energy minimization*

Only two cases have been explored in any detail in the literature: $l = 1$ and $l = 2$. There has been some discussion of $l = 3$ but no definitive treatment. It is now rather certain that $l = 1$ describes He^3, so that detailed treatments of general l are pointless; but it may be interesting, for orientation, to summarize $l = 2$.

In the singlet case

$$\hat{\Delta}_{\mathbf{k}} \hat{\Delta}_{\mathbf{k}}^{\dagger} = \hat{\Delta}_{\mathbf{k}}^{\dagger} \hat{\Delta}_{\mathbf{k}} = \begin{pmatrix} |d_0|^2 & 0 \\ 0 & |d_0|^2 \end{pmatrix}$$

$$E_{\mathbf{k}}^2 = \varepsilon_{\mathbf{k}}^2 + |d_0|^2 \tag{IV.36}$$

and matrix notation is irrelevant. The gap equation becomes precisely the BCS equation, but with an $l = 2$ interaction

$$d_0(\hat{\mathbf{k}}) = V_2 \sum_{\mathbf{k}'} \frac{(2l+1)}{2} P_2(\hat{\mathbf{k}}\cdot\hat{\mathbf{k}}')d_0(\hat{\mathbf{k}}') \frac{\tanh \frac{1}{2}\beta E_{\mathbf{k}'}}{E_{\mathbf{k}'}}. \tag{IV.37}$$

Anderson and Morel showed that symmetry considerations controlled, to a great extent, the allowable solutions of (IV.37). A good way to form a solution is to take a subgroup S of the rotation group R, such as D_∞ or the cubic group T. If we can find an $l = 2$ spherical harmonic such that $|d_0|^2 = |Y_l|^2$ belongs to the identity representation of S, d_0 will automatically be a solution, since the kernel FP_2 will be invariant under S. This may be done in two obvious ways for $l = 2$: (1) $|Y_m^l|^2$ for any m belongs to the identity of D_∞, so Y_m^l, $m = 0, 1, 2$ is a solution. (2) We choose (with $\hat{\mathbf{k}} \equiv \hat{k}_x\hat{\mathbf{x}} + \hat{k}_y\hat{\mathbf{y}} + \hat{k}_z\hat{\mathbf{z}}$)

$$d_0^r = \hat{k}_z^2 - \tfrac{1}{2}(\hat{k}_x^2 + \hat{k}_y^2)$$

$$d_0^i = \frac{\sqrt{3}}{2}(\hat{k}_x^2 - \hat{k}_y^2); \tag{IV.38}$$

i.e. d_0^r and d_0^i are the two members of the 2-dimensional $l = 2$ representation of the cubic group.

To choose among these solutions (and any others: there appears, for instance, to be an unfavourable one based on the orthorhombic group) we must compute the total free energy of each solution and choose the lowest. Within the weak-coupling approximation, the following is the appropriate procedure, since we can assume that the total modifications of background contributions to the free energy, such as collective excitations, are negligible (primarily because the response functions are modified only for $\hbar\omega \lesssim kT_c$, $q \lesssim (1/\xi_0) = k_F(\Delta/E_F)$, which is a tiny portion of phase space).

We assume that for each direction $\hat{\mathbf{k}}$ the gap matrix $\hat{\Delta}_{\mathbf{k}}$ is independent of $|\mathbf{k}|$, and as a function of $\hat{\mathbf{k}}$ represents an independent variable. The free-energy calculation is best done following the method of Eilenberger[88] which we shall need in later work as well. The scheme is to try to find a best single-quasiparticle hamiltonian

$$\mathscr{h} = \sum_{\mathbf{k}\sigma} \varepsilon_{\mathbf{k}} n_{\mathbf{k}\alpha} - \tfrac{1}{4}\sum_{\mathbf{k}} \mathrm{Tr}(\hat{\Delta}_{\mathbf{k}}\hat{\mathscr{X}}_{\mathbf{k}}^{\dagger} + \mathrm{H.C.}) \tag{IV.39}$$

where $\hat{\Delta}_{\mathbf{k}}$ is a given trial value of the energy gap to be optimized as a function of $\hat{\mathbf{k}}$. In more general terms, $\hat{\Delta}$ can be thought of as a trial self-energy and the scheme fits within the more general Luttinger–Ward[89] technique.

We calculate the trial free energy g given $\mathscr{h}$ (IV.39)

$$g = -kT \ln \mathrm{Tr}\, e^{-\beta\mathscr{h}} \tag{IV.40}$$

(averages are to be taken with the trial hamiltonian $\mathscr{h}$) and correct g by subtracting the mean value of the artificial interaction and replacing it with the real one:

$$F = g + \tfrac{1}{4}\langle \mathrm{Tr} \sum_{\mathbf{k}} (\hat{\Delta}_{\mathbf{k}} \hat{x}^{\dagger}_{\mathbf{k}} + \mathrm{H.C.})\rangle - V_l \sum_{\mathbf{k}\mathbf{k}'} P_l(\hat{\mathbf{k}}\cdot\hat{\mathbf{k}}')\langle \hat{x}^{\dagger}_{\mathbf{k}}\hat{x}_{\mathbf{k}'}\rangle \frac{2l+1}{2}. \qquad \text{(IV.41)}$$

Let us add and subtract the quantity

$$\frac{1}{2(2l+1)} \frac{N(0)}{V_l} \int d\Omega_{\hat{\mathbf{k}}} \int d\Omega_{\hat{\mathbf{k}}'} P_l(\hat{\mathbf{k}}\cdot\hat{\mathbf{k}}') \,\mathrm{Tr}\,[\hat{\Delta}(\hat{\mathbf{k}})\hat{\Delta}^{\dagger}(\hat{\mathbf{k}})]$$

$$= \frac{N(0)}{(2l+1)V_l} \int d\Omega_{\hat{\mathbf{k}}}\, \mathrm{Tr}(\hat{\Delta}^{\dagger}(\hat{\mathbf{k}})\hat{\Delta}(\hat{\mathbf{k}}))$$

and we obtain

$$F = g + \frac{1}{(2l+1)V_l} \int d\Omega_{\hat{\mathbf{k}}}\, \mathrm{Tr}\,\hat{\Delta}^{\dagger}(\hat{\mathbf{k}})\hat{\Delta}(\hat{\mathbf{k}})$$

$$- \frac{\mathrm{Tr}}{2(2l+1)V_l} \int d\Omega_{\hat{\mathbf{k}}} \left(\hat{\Delta}(\hat{\mathbf{k}}) - \frac{(2l+1)}{2} V_l \sum_{\mathbf{k}'} P_l(\hat{\mathbf{k}}\cdot\hat{\mathbf{k}}')\hat{x}_{\mathbf{k}'}\right)$$

$$\times \left(\hat{\Delta}^{\dagger}(\hat{\mathbf{k}}) - \frac{(2l+1)}{2} V_l \sum_{\mathbf{k}'} P_l(\hat{\mathbf{k}}\cdot\hat{\mathbf{k}}')\hat{x}^{\dagger}_{\mathbf{k}'}\right). \qquad \text{(IV.42)}$$

F is now to be minimized by differentiating with respect to $\hat{\Delta}(\hat{\mathbf{k}})$ functionally. By the optimal properties of the free energy, this must give us the correct gap equation (IV.24), by virtue of which the last term above vanishes. Since this term is negative definite, it is equally valid to use as the free energy for minimization the far simpler expression obtained by removing this term, which must be greater than or equal to F:

$$F' = g + \frac{1}{(2l+1)V_l} \mathrm{Tr} \int d\Omega_{\hat{\mathbf{k}}} \hat{\Delta}^{\dagger}_{\hat{\mathbf{k}}} \hat{\Delta}_{\hat{\mathbf{k}}}. \qquad \text{(IV.43)}$$

This is the form which corresponds to the free energy used in the generalized Ginsburg–Landau theories (see also Werthamer,[90]) and which results from perturbation theory.

Now g may be calculated directly from

$$g = \langle \mathscr{h} \rangle - T\langle S \rangle,$$

where contributions from each **k** add in each term, and S_k is given by Balian and Werthamer:

$$\langle S \rangle = -\sum_{\mathbf{k}} \mathrm{Tr}\,\{\tfrac{1}{2}(1 \pm \hat{W}^{\mathbf{k}}) \ln \tfrac{1}{2}(1 \pm \hat{W}^{\mathbf{k}})\}. \qquad \text{(IV.44)}$$

The result for g takes on especially simple forms in two limits, although of course it may be integrated out in general. In the limit $T \to 0$, TS vanishes and one need only do $\langle \hbar \rangle$. We will follow the procedure of integrating radially over $\varepsilon_{\mathbf{k}}$ and leaving the angular integrations to be done later.

$$\langle \hbar_{\mathbf{k}} \rangle = 2N(0) \int_0^{\omega_s} d\varepsilon_{\mathbf{k}} \operatorname{Tr} \hat{w}^{\mathbf{k}} - \frac{N(0)}{4} \int_{-\omega_s}^{\omega_s} d\varepsilon_{\mathbf{k}} \operatorname{Tr} \langle \hat{\Delta}_{\mathbf{k}} \hat{\chi}_{\mathbf{k}}^{\dagger} + \text{H.C.} \rangle. \quad \text{(IV.45)}$$

The somewhat arbitrary cutoff ω_s must be introduced.[10] At $T = 0$, $\hat{\chi}_{\mathbf{k}} = (E_{\mathbf{k}})^{-1} \hat{\Delta}_{\hat{\mathbf{k}}}$ and the integral on the right is

$$\frac{\Delta_{\hat{\mathbf{k}}}^{\dagger} \hat{\Delta}_{\hat{\mathbf{k}}}}{(\varepsilon_{\mathbf{k}}^2 + \Delta_{\hat{\mathbf{k}}}^{\dagger} \Delta_{\hat{\mathbf{k}}})^{\frac{1}{2}}}.$$

Thus all quantities are functions of $\Delta_{\hat{\mathbf{k}}}^{\dagger} \Delta_{\hat{\mathbf{k}}}$ and may be simultaneously diagonalized; all traces become, then, simply sums over the two eigenvalues of this quantity. We give the result in the appropriate weak-coupling limit $\Delta \ll \omega_s$:

$$\langle \hbar_{\mathbf{k}} \rangle = \frac{N(0)}{4} \operatorname{Tr} \left(-\hat{\Delta}_{\hat{\mathbf{k}}} \hat{\Delta}_{\hat{\mathbf{k}}}^{\dagger} \ln \frac{2\omega_s}{\Delta\Delta^{\dagger}} - \hat{\Delta}\hat{\Delta}^{\dagger} \right) \quad \text{(IV.46)}$$

so that the total "free energy" (IV.43) becomes

$$F_0 = \frac{N(0)}{4} \int \frac{d\Omega_{\hat{\mathbf{k}}}}{4\pi} \operatorname{Tr} \left[-\hat{\Delta}\hat{\Delta}^{\dagger} \ln \frac{2\omega_s}{\Delta\Delta^{\dagger}} - \hat{\Delta}\hat{\Delta}^{\dagger} \left(1 - \frac{4}{(2l+1)V_l} \right) \right]. \quad \text{(IV.47)}$$

Inserting the gap equation, which we obtain by differentiating with respect to $\hat{\Delta}\hat{\Delta}^{\dagger}$, this becomes the expression first given by Anderson and Morel:

$$E_0 = -\frac{N(0)}{8\pi} \int d\Omega_{\hat{\mathbf{k}}} \operatorname{Tr} \hat{\Delta}_{\hat{\mathbf{k}}} \hat{\Delta}_{\hat{\mathbf{k}}}^{\dagger}. \quad \text{(IV.48)}$$

The ground-state energy is proportional to the total normalization of the gap.

In the "Ginzburg–Landau" limit $T \to T_c$ we can expand all quantities in power series in the gap parameter. The coefficients may be deduced from the corresponding coefficients in a power-series expansion of the gap equation (IV.24), if we can determine the normalization. This may be done by observing that the entropy, which is proportional to $\tanh(\beta\mathscr{E}/2) - \tanh(\beta\varepsilon/2)$, converges exponentially to zero at large $\varepsilon_{\mathbf{k}}$, so that the logarithmic terms must all come from (IV.45), and can be shown to be of the form

$$(F(T))_{\text{logarithmic}} \simeq -\frac{N(0)}{2} \int_{-\omega_s}^{\omega_s} \varepsilon^2 \, d\varepsilon \operatorname{Tr} \frac{\hat{\Delta}\hat{\Delta}^{\dagger} \tanh(\beta E/2)}{E^3}$$

$$= -\frac{N(0)}{2} \operatorname{Tr} \hat{\Delta}\hat{\Delta}^{\dagger} \left[\ln \frac{\omega_s}{T} + \text{const.} \right]; \quad \text{(IV.49)}$$

and since the remainder of the integrals are all convergent they can only depend on $(\Delta\Delta^\dagger)/T^2$, so that we have

$$F(T) = -\frac{N(0)}{2}\int\frac{d\Omega_{\hat{k}}}{4\pi}\,\mathrm{Tr}\left\{\hat{\Delta}_k\hat{\Delta}_k^\dagger \ln T_c/T + T^2 f\left(\frac{\hat{\Delta}\hat{\Delta}^\dagger}{T^2}\right)\right\} \qquad \text{(IV.50)}$$

where, since any quadratic term can be included in $\ln T_c$, f must start with fourth-order terms. Explicitly, from well-known expansions (see, e.g. Ambegaokar and Mermin,[28] Anderson,[25] Brinkman, Serene and Anderson,[37] Kuroda[38, 39]) we have

$$F = -N(0)\int\frac{d\Omega_{\hat{k}}}{4\pi}\tfrac{1}{2}\,\mathrm{Tr}\left\{\hat{\Delta}_k\hat{\Delta}_k^\dagger \ln T_c/T + \frac{(\hat{\Delta}\hat{\Delta}^\dagger)^2}{T^2}\frac{7\zeta(3)}{8\pi^2} - \frac{(\hat{\Delta}\hat{\Delta}^\dagger)^3}{T^4}(0{\cdot}007735) + \ldots\right\}. \qquad \text{(IV.51)}$$

Again, if we stop at fourth order and use the gap equation

$$\frac{\partial F}{\partial(\Delta\Delta^\dagger)} = 0,$$

the fourth-order term becomes $-\frac{1}{2}$ times the second-order one and we have

$$F(T \simeq T_c) \simeq -\frac{N(0)}{4}\ln\frac{T_c}{T}\int\frac{d\Omega_{\hat{k}}}{4\pi}\mathrm{Tr}\,\hat{\Delta}_{\hat{k}}\hat{\Delta}_{\hat{k}}^\dagger. \qquad \text{(IV.52)}$$

Again, the free energy is proportional to the normalization of Δ^2.

Specialized to the singlet case, the question is then for which state the gap eqn (IV.37) allows the largest $\langle d_0^2\rangle$. The general rule may be understood by realizing that $(\tanh \beta E_k/2)/E_k$ is a concave function of E^2 and hence of d_0^2. Integration over ε cannot remove this property: the curvature will remain upwards. Multiplying (IV.37) on the left by $d_0(k)$, and using the normalization of $P_2(\hat{\mathbf{k}}\cdot\hat{\mathbf{k}}')$, we recognize that it is in an equation of the form

$$\langle d_0^2(k)\rangle = \int d\Omega_{\hat{k}}\, d_0^2(k')\Psi(d_0^2(k')) = \langle d_0^2\Psi(d_0^2)\rangle$$

where

$$\frac{d^2\Psi(x)}{dx^2} > 0 \qquad \text{(IV.53)}$$

everywhere. This means that more uniform solutions are favoured, because the right-hand side will allow a bigger $\langle\Psi\rangle$. We see easily from the free-energy expressions (IV.47) and (IV.51) that the function Ψ near $T = 0$ is essentially $\propto -\ln|d_0^2|$, while near T_c it must clearly be $|d_0|^2$. The solution

(IV.38) is favoured near $T = 0$, while as Mermin and Stare[91] have pointed out the Y_1^2 and Y_2^2 solutions have the same $\langle d_0^4 \rangle$ and are *a priori* acceptable at T_c. It seems likely, however, that the more uniform "cubic" solution will be favoured at all $T < T_c$ and is the only stable solution in the weak-coupling limit.

(e) *Special cases: Triplet* $l = 1$

(i) *Nonunitary case.* The $l = 2$ case is only of academic interest. What occurs in the $l = 1$, $s = 1$ case which almost certainly represents He^3?

The eigenvalue spectrum is given by diagonalizing (IV.20) or (IV.25) with (IV.26)

$$\hat{\Delta}_{\mathbf{k}}^{\dagger}\hat{\Delta}_{\mathbf{k}} =$$
$$\begin{pmatrix} |d_u|^2 + |d_v|^2 + |d_w|^2 + i(d_u d_v^* - d_v d_u^*) & -i(d_v d_w^* - d_w d_v^*) + d_w d_u^* - d_u d_w^* \\ i(d_v d_w^* - d_w d_v^*) + d_u d_w^* - d_w d_u^* & |d_u|^2 + |d_v|^2 + |d_w|^2 - i(d_u d_v^* - d_v d_u^*) \end{pmatrix} \tag{IV.54}$$

i.e.

$$(E_{\mathbf{k}}^2 - \varepsilon_{\mathbf{k}}^2 - |\mathbf{d}|^2)^2 = |i\mathbf{d}^* \times \mathbf{d}|^2. \tag{IV.55}$$

The eigenvalues of $\hat{\Delta}\hat{\Delta}^{\dagger}$ are thus

$$|\mathbf{d}|^2 \pm |i\mathbf{d}^* \times \mathbf{d}|. \tag{IV.56}$$

The quantity $\mathbf{d}^* \times \mathbf{d}$ represents a kind of spin angular momentum in the sense that in the pictorial description in terms of $(He^3)_2$ molecules, it would give the molecular spin moment. We can see this, for instance, from (IV.26) by setting

$$d_v = i d_u = d$$
$$d_w = 0$$

which gives a Δ only for $\uparrow\uparrow$ pairing, $m_s = 1$, while in this case $i\mathbf{d} \times \mathbf{d}^* = +2|d|^2\hat{w}$. Then general spin moments follow from the fact that $\mathbf{d}$ transforms like a vector under rotations. In the presence of $(i\mathbf{d}^* \times \mathbf{d})$, time-reversal symmetry is broken and the Kramers degeneracy removed. $(i\mathbf{d}^* \times \mathbf{d})$ does not, however, imply the presence of a total spin angular momentum, both because it may average out over the Fermi surface, since d_u, d_v and d_w are functions of $\hat{\mathbf{k}}$, and because moments can and do also arise from displacements of the Fermi surface. The correlation-hump picture mentioned in the Introduction clarifies this point: the existence of a $(\mathbf{d}^* \times \mathbf{d})_w$ implies a spin-polarized correlation hump $\langle n\uparrow(\mathbf{0})n\uparrow(\mathbf{r})\rangle \neq \langle n\downarrow(\mathbf{0})n\downarrow(\mathbf{r})\rangle$ but that does not imply an average $\langle n\uparrow - n\downarrow\rangle$. In fact, in the most obvious physical phase where $\mathbf{d}^*\times\mathbf{d}$ exists, the A_1 phase which occurs in a magnetic field near T_c, there is a moment because of Fermi-surface displacement and also a tiny $\mathbf{d}^* \times \mathbf{d}$ moment resulting from hole-particle disymmetry (see (IV.98) following).

Calculating the free energy of solutions with $\mathbf{d}^* \times \mathbf{d}$ finite is complicated even near T_c or 0. These are called "non-unitary" because, as (IV.54) and (IV.55) show, in their absence $\hat{\Delta}^\dagger\hat{\Delta}$ is proportional to a unit matrix and $E_\mathbf{k}$ is given by the standard BCS gap expression

$$\begin{aligned} E_\mathbf{k}^2 &= \varepsilon_\mathbf{k}^2 + |d_u|^2 + |d_v|^2 + |d_w|^2 \\ &= \varepsilon_\mathbf{k}^2 + |\mathbf{d}|^2. \qquad (\text{"unitary"}) \end{aligned} \tag{IV.57}$$

Our general free-energy formalism clearly applies even in the non-unitary cases, and tells us that we must minimize the variation of $\hat{\Delta}\hat{\Delta}^\dagger$ *including* the variation caused by $\mathbf{d}^* \times \mathbf{d}$. That is, if we call the two eigenvalues of $\hat{\Delta}$ (IV.56). Δ_1 and Δ_2, we want to minimize

$$T \to 0 \qquad \frac{\langle -|\Delta_1|^2 \ln|\Delta_1|^2 - |\Delta_2|^2 \ln|\Delta_2|^2\rangle}{\langle |\Delta_1|^2 + |\Delta_2|^2\rangle} \tag{IV.58}$$

or

$$T \to T_c \qquad \frac{\langle |\Delta_1|^4 + |\Delta_2|^4\rangle}{\langle |\Delta_1|^2 + |\Delta_2|^2\rangle^2}.$$

A non-unitary solution can make Δ more uniform by making $|\mathbf{d}|^2$ more uniform: Re $\mathbf{d}$ can be made up of a different set of functions from Im $\mathbf{d}$. This is not, however, possible for $l = 1$, since Re $\mathbf{d}$ can saturate the set of only 3 independent $l = 1$ functions.

For $l = 3$ a non-unitary solution will improve $|\mathbf{d}|^2$ since there are 7 functions to choose from, but in actual fact [contrary to a previous remark (Anderson[24, 25])] this does not seem to compensate for the extra variation $|\mathbf{d}^* \times \mathbf{d}|^2$.

Thus non-unitary solutions occur only in the presence of magnetic fields or in boundaries. The one well-known case is the A phase near T_c in a large magnetic field, which will be discussed as an interesting NMR phenomenon.

(ii) *Unitary Solutions: BW versus AM.* For the discussion of $l = 1$ solutions it is very convenient to introduce a "bivector" notation in which we expand the spin-space vector $\mathbf{d}$ in orbit-space components:[91, 35, 36]

$$d_u(\hat{\mathbf{k}}) = \sum_\alpha d_{\alpha u}\hat{k}_\alpha \text{ etc.} \tag{IV.59}$$

$d_{\alpha i}$ transforms like a vector in orbit space as regards its first component, in spin space as regards its second and thus like a tensor under full rotation. Except for the weak dipolar interactions the energy is invariant under both types of transformation.

The general principle given for $l = 2$ allows us to construct at least three solutions. That of Balian and Werthamer[12] comes from using the full rotation group as our "subgroup", under which k_x, k_y and k_z determine a 3-dimensional representation, which we may use for d_u, d_v and d_w. The BW solution is then given by all spin and space rotations of

$$\text{BW: } d_{\alpha i} = d\begin{pmatrix} 1 & 0 & 0 \\ 0 & 1 & 0 \\ 0 & 0 & 1 \end{pmatrix}. \qquad \text{(IV.60)}$$

A second subgroup is the axial group, and we may construct solutions based on either Y_0^1 or $Y_{\pm 1}^1$ relative to a given axis z. The solution based on $Y_0^1 = \hat{k}_z$ is the most non-uniform possible function, and, although it could be favoured for extreme spin fluctuations, it does not occur and we will not further discuss it. Anderson and Morel pointed out that we can use $Y_{\pm 1}^1 = \hat{k}_x \pm i\hat{k}_y$ and pair up spins and down spins independently. In terms of the gap matrix $\hat{\Delta}$ this leads to

$$\hat{\Delta} = \Delta_0\begin{pmatrix} \hat{k}_x + i\hat{k}_y & 0 \\ 0 & \hat{k}_{x'} + i\hat{k}_{y'} \end{pmatrix}, \qquad \text{(IV.61)}$$

where we observe that in weak coupling the $\uparrow\uparrow$ and $\downarrow\downarrow$ pairs may be totally decoupled and so we can allow the $\uparrow\uparrow$ and $\downarrow\downarrow$ orbital directions to be different. In fact, there are two explicitly unitary and axially symmetric solutions. The first, chosen more or less by chance by Anderson and Morel, is the parallel one now known as the ABM solution, which in terms of **d** is

$$\text{ABM: } \mathbf{d} = \begin{pmatrix} d & 0 & 0 \\ id & 0 & 0 \\ 0 & 0 & 0 \end{pmatrix} \begin{matrix} x \\ y; \\ z \end{matrix} \qquad \text{(IV.62)}$$
$$\quad u \quad v \quad w$$

and the second is the "real" solution suggested by Balian and Werthamer, in which $y' \to -y$:

$$\text{A}'\text{: } \mathbf{d} = \begin{pmatrix} d & 0 & 0 \\ 0 & d & 0 \\ 0 & 0 & 0 \end{pmatrix}. \qquad \text{(IV.63)}$$

This solution is very likely to be the limit of BW near a vessel wall, where the component of k perpendicular to the wall will be severely depaired. Although ABM and A′ have the same energy in weak coupling, their response functions are very different, as we shall see, and as a consequence their higher-order free energies can be very different. (IV.63) never occurs in equilibrium.[91]

In weak-coupling theory there was never any doubt that the BW solution (IV.60) is the minimum one, as Balian and Werthamer proved. The gap parameter is given by $(\mathbf{d})^2 = d^2$, which is totally uniform and cannot be improved on under any criterion. In fact $\overline{d^2_{\mathrm{ABM}}} = 0{\cdot}88\ \overline{d^2_{\mathrm{BW}}}$ at $T = 0$, $\overline{d^2_{\mathrm{ABM}}} = \frac{5}{6}\overline{d^2_{\mathrm{BW}}}$ as $T \to T_c$. Most of the thermal properties of BW are identical with those of BCS. It was thus rather a puzzle that the "A" stable state first discovered could clearly not be BW, experimentally.

In attempting to discuss this problem, Mermin and Stare[91] and Brinkman and Anderson[35, 36] developed a general scheme for the Ginzburg–Landau regime near T_c, which has been of value in later work. In the absence of a theory of the phases they investigated by group theory what quantities might enter the free energy (IV.50) or (IV.51) to second and fourth order, taking into account rotation symmetry for both spins and orbits. Clearly to second order only the invariant

$$I_2 = \mathrm{Tr}\int \frac{d\Omega_{\hat{k}}}{4\pi}\hat{\Delta}_{\mathbf{k}}\hat{\Delta}^{\dagger}_{\mathbf{k}} = d^*_{\alpha i}d_{\alpha i} \qquad \text{(IV.64)}$$

(using the usual summation convention for repeated indices) can enter, but in fourth order there are five invariants:

$$A_1 = \Big|\sum_{\alpha, i}(d_{\alpha i})^2\Big|^2; \qquad A_2 = d_{\alpha i}d_{\alpha j}d^*_{\beta i}d^*_{\beta j};$$

$$A_3 = d^*_{\alpha i}d^*_{\beta i}d_{\alpha j}d_{\beta j}; \qquad A_4 = \Big(\sum_{\alpha i}|d_{\alpha i}|^2\Big)^2 = (I_2)^2;$$

$$A_5 = d^*_{\alpha i}d^*_{\beta j}d_{\alpha j}d_{\beta i}. \qquad \text{(IV.65)}$$

They proposed that the coefficients of these five invariants be considered as arbitrary parameters of the theory:

$$F = \alpha I_2 + \sum_i a_i A_i \qquad \text{(IV.66)}$$

and observed that the correct weak-coupling form of the coefficients is

$$\mathrm{BW}: a_2 = a_3 = a_4 = -a_5 = -2a_1 \qquad \text{(IV.67)}$$

where $a_1 = (N(0)/T_c^2)7\zeta(3)/120\pi^2$ (Ambegaokar and Mermin[28]). In a general strong-coupling theory α remains the same but the a_i's become parameters to be calculated, as we shall see in Section V.

2. STATIC AND LOW-FREQUENCY PROPERTIES OF ANISOTROPIC SUPERFLUID MODEL PHASES

In this subsection we will discuss various low-frequency and static phenomena most of which can be calculated reasonably directly from old-fashioned BCS theory and do not require the apparatus of temperature perturbation theory or Landau–Ginzburg–Gor'kov methods, and on the whole we will not indulge in these sophisticated techniques. We are essentially looking at the ways in which the simple properties of the anisotropic superfluid differ from those of a classic BCS superconductor. It is a special feature of He^3 that because it is a rather strongly coupled Fermi liquid, many of these results will be modified by Landau Fermi-liquid factors. This aspect of the problem we reserve for the next section, where we will see that the BCS calculation is a necessary preliminary to that one.

(a) *Correlations*

As we remarked in the Introduction, many of the properties follow, and more can be understood, from a consideration of the additional correlations which are introduced by the existence of the anomalous amplitudes $x^{\mathbf{k}}_{\sigma\sigma'}$. Except for negligible corrections to the Wigner exchange hole, these follow directly by simply factorizing the appropriate two-particle correlation into the relevant x's. For instance,

$$\begin{aligned}\Delta(\langle n_\sigma(\mathbf{r})n_{\sigma'}(\mathbf{r}')\rangle &= \langle\Psi^\dagger_\sigma(\mathbf{r})\Psi^\dagger_{\sigma'}(\mathbf{r}')\rangle\langle\Psi_\sigma(\mathbf{r}')\Psi_\sigma(\mathbf{r})\rangle \\ &= b^\dagger_{\sigma\sigma'}(\mathbf{r}-\mathbf{r}')b_{\sigma'\sigma}(\mathbf{r}'-\mathbf{r})\end{aligned} \qquad \text{(IV.68)}$$

where

$$\begin{aligned}b_{\sigma\sigma'}(\mathbf{r}-\mathbf{r}') &= \sum_{\mathbf{k}}\exp[i\mathbf{k}\cdot(\mathbf{r}-\mathbf{r}')]\langle c_{\mathbf{k}\sigma}c_{-\mathbf{k}\sigma'}\rangle \\ &= \tfrac{1}{2}\sum_{\mathbf{k}}\exp[i\mathbf{k}\cdot(\mathbf{r}-\mathbf{r}')]x^{\mathbf{k}}_{\sigma\sigma'}.\end{aligned} \qquad \text{(IV.69)}$$

Anderson and Morel, using methods suggested by the BCS paper[1], computed this quantity for absolute zero and for $k_F r \gg 1$ in terms of the gap matrix $\hat{\Delta}$, and found

$$\hat{b}(r) = \frac{N_0\cos k_F r}{ik_F r}K_0\left(\frac{r|\hat{\Delta}(\hat{\mathbf{r}})|}{\hbar v_F}\right)\hat{\Delta}(\hat{\mathbf{r}}) \qquad \text{(IV.70)}$$

(here it is unnecessary to go beyond the unitary case). By $\hat{\Delta}(\hat{\mathbf{r}})$ we mean $\hat{\Delta}(\hat{\mathbf{k}})$ with $\hat{\mathbf{k}}\|\hat{\mathbf{r}}$. The integration technique used by Anderson and Morel and by Bardeen, Cooper and Schrieffer is actually even easier in the opposite limit† $T \to T_c$ and gives us

$$\begin{aligned} \hat{b}(r) &= \hat{\Delta}(\hat{\mathbf{r}}) \frac{N_0 \cos k_F r}{i k_F r} \ln \coth\left(\frac{\pi k T_c}{2\hbar v_F}\right) r \\ &= \hat{\Delta}(\hat{\mathbf{r}}) \frac{2N_0 \cos k_F r}{i k_F r} \ln \coth\left(\frac{r}{3{\cdot}5\zeta_0}\right), \end{aligned} \tag{IV.71}$$

defining the average coherence length ξ_0 in terms of T_c. The correlation hump is an oscillating function with exponentially decaying amplitude at large r, with a range of the order of the coherence length, $\approx 10^3$ Å. Its total amplitude may be calculated by integrating over r, or, what is equivalent because of completeness, by summing over k:

$$\int \Delta\langle n_\sigma(0) n_{\sigma'}(\mathbf{r})\rangle \, d^3 r = n_c = \tfrac{1}{4}\sum_{\mathbf{k}} |x^{\mathbf{k}}_{\sigma\sigma'}|^2. \tag{IV.72}$$

As calculated by Anderson and Morel for temperatures near absolute zero, this is

$$\begin{aligned} n_c(T=0) &= \tfrac{1}{4}\sum_{\mathbf{k}} \frac{|\Delta^{\sigma\sigma'}_{\mathbf{k}}|^2}{E^2_{\mathbf{k}}} \\ &= \frac{\pi N(0)|\Delta^{\sigma\sigma'}_{\mathbf{k}}|^2}{4} \int \frac{d\Omega}{4\pi} (\mathrm{Tr}\, \Delta_{\mathbf{k}} \Delta^{\dagger}_{\mathbf{k}})^{-\frac{1}{2}} \\ &\approx \frac{\pi N(0)}{4} \Delta. \end{aligned} \tag{IV.73}$$

For temperatures near T_c the result is

$$\begin{aligned} n_c(T_c) &= |\Delta^{\sigma\sigma'}_{\mathbf{k}}|^2 \frac{N(0)}{4} \int \frac{d\varepsilon}{\varepsilon^2} \tanh^2 \frac{\beta\varepsilon}{2} \\ &= |\Delta^{\sigma\sigma'}_{\mathbf{k}}|^2 \frac{N(0)}{T_c} \frac{7}{\pi^2} \zeta(3). \end{aligned} \tag{IV.74}$$

Note that the result is of order Δ/E_F or about 10^{-3} in He^3.

Among effects which follow directly from the correlation hump itself, is first, of course, the condensation energy

$$\int V(\mathbf{r} - \mathbf{r}')|b(\mathbf{r} - \mathbf{r}')|^2 \, d^3 r \, d^3 r',$$

† The relevant integral is[92]

$$\int_0^\infty \cos \frac{r\varepsilon}{\hbar v_F} \frac{d\varepsilon}{\varepsilon} \tanh \frac{\beta\varepsilon}{2} = \ln \coth \frac{\pi r}{2\hbar v_F \beta}.$$

which we have computed in the last section. The anisotropic spin–position correlation leads to a non-zero nuclear spin–spin dipolar interaction, which is vital to the discussion of orientation effects. Another effect which may eventually be experimentally important is the effect of this correlation on dielectric constants and index of refraction.[93] This comes about through the modification of local field corrections. From the magnitude of the magnetic dipolar orientation energies (see later) we find that $\langle 1/r^3 \rangle \approx 10^{20}\,\mathrm{cm}^{-3}$, so that we can expect a change in the local field correction $\Delta L \approx 1/nr^3$, i.e. of order 10^{-3}. This will lead to an anisotropy of the dielectric constant ε of order $(\varepsilon - 1)^2 \Delta L \approx 10^{-6}$. This will give a barely observable birefringenece and an appreciable orienting effect in an electric field of order 10^4–10^5 volt cm^{-1}, in the ABM state, at which field the effect is comparable to the dipolar interaction.

In addition to the static density and spin correlations, Anderson and Morel pointed out the existence of density–current correlations $\langle n\mathbf{j} \rangle$, and Balian and Werthamer generalized this to spin–current correlations in the pseudoisotropic BW state.

Again, in the most direct model theory such correlations may be computed by factorizing the product of one-particle operators which defines it:

$$\langle n_\sigma(\mathbf{r})\mathbf{j}_{\sigma'}(\mathbf{r}')\rangle = \frac{i\hbar}{2m}\big[\langle \Psi_\sigma^\dagger(\mathbf{r})\nabla\Psi_{\sigma'}^\dagger(\mathbf{r}')\rangle\langle \Psi_{\sigma'}(\mathbf{r}')\Psi_\sigma(\mathbf{r})\rangle - \langle \Psi_\sigma^\dagger(\mathbf{r})\Psi_{\sigma'}^\dagger(\mathbf{r}')\rangle\langle \nabla\Psi_{\sigma'}(\mathbf{r}')\Psi_\sigma(\mathbf{r})\rangle\big]. \qquad \text{(IV.75)}$$

Obviously this too follows the general rule that b may be manipulated as though it were the wave function of the pairs:

$$\langle n_\sigma(\mathbf{0})\mathbf{j}_\sigma(\mathbf{r})\rangle = \frac{i\hbar}{2m}\langle \nabla b^\dagger(\mathbf{r})b(\mathbf{r}) - b^\dagger(\mathbf{r})\nabla b(\mathbf{r})\rangle. \qquad \text{(IV.76)}$$

Since in (IV.76) the r-dependent factors have constant phase, the current correlation is entirely azimuthal and determined by the phases of the components of $\hat{\Delta}$. In the ABM state,

$$\hat{\Delta}^\dagger(\hat{\mathbf{r}}) = \hat{\Delta}^\dagger\, e^{i\psi}$$

where ψ is the angle in the x–y plane: as a result

$$\langle n\mathbf{j} \rangle = \hat{\boldsymbol{\psi}}\frac{\hbar}{mr}\langle nn \rangle, \qquad \text{(IV.77)}$$

or the density–density correlation rotates with a net angular momentum $\hbar$ about the z axis. The total angular momentum of this motion is, then, $\int \mathbf{r} \times \langle n\mathbf{j} \rangle$ which is equal to $\hbar$ times the total number of correlated particles. This, from (IV.73) or (IV.74), gives a total orbital angular momentum in the correlation hump of order $\hbar\Delta/E_F \approx 10^{-3}\hbar$.

This total angular momentum must, as Anderson and Morel emphasized, imply a net orbital angular momentum of the sample as a whole. As in the conventional discussions of diamagnetism or ferromagnetism, in a uniform sample the correlation currents at any interior point will average to zero, but at a surface there will be a net surface current. Equally, if the state is slowly-varying but not uniform in space, we can expect a net current related to this orbital angular momentum density:

$$\mathbf{J}_L = \nabla \times \mathbf{L}. \tag{IV.78}$$

The existence of $\langle L \rangle$ is indirectly fairly important to a number of physical phenomena such as the orbital supercurrents, and directly to the possible existence of "orbit waves" with a k^2 dispersion relation. $\mathbf{J}_L$ should appear as part of the general supercurrent structure,[49, 50, 53] which we will describe later in this article. Relative to other supercurrent terms this first appears in order Δ^2/E_F^2. This is one order smaller than given by Anderson and Morel, the difference being the result of a cancellation due to hole–particle symmetry.[94] For this reason orbit waves may not have quite as low a frequency as has been proposed.[24, 25, 53]

A convenient formal expression for $\langle L \rangle$ has been suggested by de Gennes.[50, 51] We write the gap function in terms of the d bivector $d_{\alpha i}$ (IV.59)

$$d_i(\hat{\mathbf{k}}) = d_{\alpha i} \hat{k}_\alpha$$

and observed that the α component of the density–current correlation (IV.76) may be written, using (IV.70) or (IV.71), as

$$\langle n_\sigma \mathbf{j}_{\sigma'} \rangle^\alpha = \frac{\hbar}{2m} \frac{f(r)}{\rho} (\sigma_\nu \sigma_i)_{\sigma\sigma'} (\sigma_j \sigma_\nu)_{\sigma'\sigma} (d_{i\alpha} d^*_{j\beta} - d^*_{i\beta} d_{j\alpha}) x_{\beta},$$

where $f(r)$ is the numerical r-dependent factor of b in (IV.70) or (IV.71) and

$$\int r^2 \, \mathrm{d}r \int \mathrm{d}\Omega (m(\hat{\mathbf{r}} \times \langle n\mathbf{j} \rangle)_\alpha) = \langle L \rangle_\alpha$$

$$\int r^2 \, \mathrm{d}r \int \frac{\hbar f(r)}{\rho} (\sigma_\nu \sigma_i)_{\sigma\sigma'} (\sigma_j \sigma_\nu)_{\sigma'\sigma} \varepsilon_{\alpha\beta\gamma} (d_{i\alpha} d_{i\beta} - d_{i\beta} d_{i\alpha})$$

$$= \int \frac{\hbar}{6} \frac{f(r)}{\rho} r^2 \, \mathrm{d}r \, (\text{spin factors}) \times (\mathbf{d} \times \mathbf{d}^*)_\alpha; \tag{IV.79}$$

i.e. when integrated over r the total orbital angular momentum is proportional to $\mathbf{d} \times \mathbf{d}^*$, where in this case the cross-product refers to the orbital, not spin, components of the d bivector. As in the case of the spin, $\mathbf{d}^* \times \mathbf{d}$ is the nominal angular momentum of the pairs, and the total real angular momentum has a rather complex relation to this quantity which is not yet wholly understood. For the ABM state, both spins have the same orbital shape of d; but in the pseudoisotropic state, as Balian and Werthamer indeed pointed out, or in the A′ state, there is an orbital angular momentum for each spin direction but it cancels out on the average when summed over spins.

(b) *Susceptibility and orientation effects*

The nuclear spin susceptibility plays a great role in many phenomena in He^3 and will be derived here in some qualitative detail. First, however, we would like to clarify the confusing remark which one finds in the literature that this susceptibility is not "really" anisotropic. This is no more meaningful than it would be to say that the optical or elastic properties of crystalline quartz are "really" isotropic. The phases of He^3 are broken symmetry states[95] for which the only practical method of discussion is to assume that the spin and orbital axis systems, as well as the complex phase of the order parameter, are macroscopic variables of state, equivalent in nature to the orientation and position of a crystal (and in fact, in resonance phenomena the orbital axis system plays precisely the mathematical role of the crystal axes in antiferromagnetic resonance). Thus we calculate in every case under the assumption of a fixed numerical value and orientation of the $d_{\alpha l}$ bivector. Under some circumstances the d bivector will orient itself so as to optimize the nuclear susceptibility in a magnetic field, and χ will appear isotropic; but experimentally this is by no means always the case.

Here we calculate the susceptibilities purely within the weak-coupling theory, assuming non-interacting quasiparticles. As we have remarked, these must, and will in the next subsection, be corrected severely for Fermi-liquid interactions. There is clearly a question whether that is the only correction which need be applied. Again, it seems certain that the universal justification, that pairing simply modifies too small a region of phase space to be significant, is adequate to guarantee that.

This being the case, the calculation of ω, $k = 0$ susceptibilities becomes almost trivial (the far more complicated problem of general q and ω is carried through in the spin-fluctuation section). The total susceptibility will be the sum of those for each k, and therefore, of course, also for each direction $\hat{\mathbf{k}}$. In the sub-hamiltonian h_k for each $\mathbf{k}$ we insert the effect of the magnetic field on the one-particle energy $\varepsilon_{\mathbf{k}}$

$$\varepsilon_{\mathbf{k},\sigma} \to \varepsilon_{\mathbf{k}} + \mu\boldsymbol{\sigma}\cdot\mathbf{H} \tag{IV.80}$$

and recalculate the eigenvalues $E_{\mathbf{k}}$, and from them we can calculate the susceptibilities and add them up.

Even this need not be done in general. Since each of these susceptibilities is a tensor, we can pick the three principal axes of that tensor and then rotate to calculate a general direction.

For any unitary odd-l state, and for any Fermi-surface direction $\hat{\mathbf{k}}$, $d_i(\hat{\mathbf{k}})$ is a vector whose components all have the same phase, and thus defines a real vector $\mathbf{d}_i$ in spin space. Clearly the susceptibility axis must be $\mathbf{d}_i$. For this direction, the $\mathscr{E}_{\mathbf{k}}$ tensor becomes

$$\mathscr{E}'_{\mathbf{k}} = \begin{pmatrix} \varepsilon_{\mathbf{k}} + \mu H & 0 & 0 & -i|\mathbf{d}| \\ 0 & \varepsilon_{\mathbf{k}} - \mu H & -i|\mathbf{d}| & 0 \\ 0 & i|\mathbf{d}| & -\varepsilon_{\mathbf{k}} - \mu H & 0 \\ i|\mathbf{d}| & 0 & 0 & -\varepsilon_{\mathbf{k}} + \mu H \end{pmatrix}.$$

This separates into two sub-matrices in which the effect of the magnetic field is identical to that in BCS theory itself: essentially the pairing is between opposite spins quantized along the magnetic field, and the susceptibility in that case was calculated by Yosida[11]:

$$\chi_{\parallel} = \chi_N Y(\beta|\mathbf{d}|) \tag{IV.81}$$

where Y is the "Yosida function"

$$Y = \tfrac{1}{2}\beta \int_0^\infty d\varepsilon \operatorname{sech}^2 [\tfrac{1}{2}\beta(\varepsilon^2 + d^2)^{\frac{1}{2}}]. \tag{IV.82}$$

On the other hand, we note that when $\mathbf{H} \perp \mathbf{d}$ the Δ matrix is diagonal, and we have strictly the "equal-spin pairing" (ESP) case of Anderson and Morel in which

$$\chi_{\perp} = \chi_N, \tag{IV.83}$$

since the $\uparrow$ and $\downarrow$ spin Fermi surfaces may still move perfectly independently of each other in the magnetic field. (Incidentally, we may remark on the other two cases of interest. Even l: all pairing is opposite-spin and (IV.81) is correct in general. Odd l, non-unitary, is complicated in general (an expression is given by Balian and Werthamer[12]), but in fact the only important case is the high-field regime near T_c in the "A" phase, which is purely ESP, and (IV.83) is correct.)

Now χ of the system as a whole depends on the relationship of the $\mathbf{d}$ vectors for different $\hat{\mathbf{k}}$ directions, i.e. on $d_{\alpha i}$. Since the Yosida function (IV.82) depends in a complicated way on the magnitude of $\mathbf{d}$, this is a little complicated except in the two limits $T \to 0$ ($Y(0) = 0$) and $T \to T_c$, where

$$Y \simeq 1 - \tfrac{7}{8}\frac{\zeta(3)d^2}{\pi^2 T^2} \cdots$$

As $T \to 0$

$$\frac{\chi_{ij}(T=0)}{\chi_N} = \delta_{ij} - d_{\alpha i} d_{\alpha j} \tag{IV.84}$$

whereas as $T \to T_c$

$$\frac{\chi_{ij}}{\chi_N} = \delta_{ij} - \tfrac{7}{8} \frac{\zeta(3)}{\pi^2 T^2} d_{\alpha i} d_{\alpha j} \ldots \tag{IV.85}$$

There are three important cases. For BW χ is isotropic and equal to

$$\frac{\chi_{BW}}{\chi_N} = \tfrac{2}{3} + \tfrac{1}{3} Y\left(\frac{\Delta}{T}\right). \tag{IV.86}$$

For ABM all of the principal axes are equal and along the axis we have chosen as w, so that

$$\left(\frac{\chi_{ABM}}{\chi_N}\right)_u = \left(\frac{\chi_{ABM}}{\chi_N}\right)_v = 1$$

$$\left(\frac{\chi_{ABM}}{\chi_N}\right)_w = \langle Y(\beta d)\rangle \tag{IV.87}$$

while, finally, $\chi(\mathrm{A}')$ has one "ESP" axis

$$\left(\frac{\chi_A}{\chi_N}\right)_w = 1$$

$$\left(\frac{\chi_{A'}}{\chi_N}\right)_u = \left(\frac{\chi_{A'}}{\chi_N}\right)_v = \tfrac{1}{2} + \tfrac{1}{2}\langle Y(\beta d)\rangle.$$

As will be seen, A′ and ABM will be very strongly oriented in a magnetic field because of this anisotropy, the orientation energy being of the order $\chi_N H^2 = \mu^2 H^2/E_F$ per atom at low temperatures; this in He^3 is of order 1 erg cm^{-3} for 10^3 gauss. The A′ phase will be uniquely oriented as far as its spin axes are concerned, but w for the ABM or "A" phase can lie anywhere in a plane perpendicular to **H**.

For orientation of **L**, and for all orientation effects in the BW or "B" phase, we must go on to study the nuclear dipole–dipole interaction. In every possible phase, the spatial correlation for a given spin configuration is anisotropic, which means that (as opposed to the ordinary atomic forces, which depend on spin only through statistics and thus are completely invariant to spin orientation) the nuclear dipole–dipole interaction

$$V_{dd}(\mathbf{r}) = \frac{(\boldsymbol{\mu}_1 \cdot \mathbf{r})(\boldsymbol{\mu}_2 \cdot \mathbf{r})}{r^5} - \frac{3\boldsymbol{\mu}_1 \cdot \boldsymbol{\mu}_2}{r^3} \tag{IV.88}$$

depends on relative orientation.

As shown by Leggett[22, 23, 55] the correlations derived in Section IV.2a imply a nuclear dipole–dipole interaction which may be written

$$V_{dd} = \frac{2}{3}\frac{\mu^2}{4}\sum_{\mathbf{kk}'}[\mathbf{x_k}\cdot\mathbf{x_{k'}} - 3\mathbf{x_k}\cdot(\mathbf{k}-\mathbf{k}')\mathbf{x_{k'}}\cdot(\mathbf{k}-\mathbf{k}')]. \tag{IV.89}$$

Here we have defined an **x**-vector in analogy with the **d**-vector

$$\hat{x} = \begin{pmatrix} x_v - ix_u & ix_w \\ ix_w & x_v + ix_u \end{pmatrix}.$$

Note however that in (IV.89) for the first time vectors in k and in spin space are related to each other and we are no longer free to use a separate u, v, w coordinate system for spins. At this point we equate the axis systems (u, v, w) and (x, y, z).

In any unitary state $\mathbf{x_k}$ is proportional to $d_{\mathbf{k}}$:

$$\mathbf{d(k)} = -\tfrac{1}{2}\sum_{\mathbf{k}'} V_{\mathbf{kk}'}\mathbf{x_{k'}} \tag{IV.90}$$

We now express $d_{\mathbf{k}}$ in terms of the bivector $d_{\alpha i}$, and do the angular averages in (IV.89) (Engelsberg, Brinkman, and Anderson[96]):

$$V_{dd} \simeq \frac{2\pi\mu^2}{12}(d_{ii}d_{jj}^* + d_{\alpha i}^* d_{i\alpha})/(2\lambda\omega_c)^2. \tag{IV.91}$$

Here λ is the "coupling constant" $N(0)V_l[(2l + 1)/2]$. We have neglected the dependence of $F(E)$ on angles which then allows us to use the gap equation to evaluate the k-sum approximately. There is no approximation in the angle-dependence. The magnitude of the coefficient is of order 10^{-4} erg cm^{-3}, or a little larger if evaluated directly from resonance data (5×10^{-27} erg per atom).

The orientation effects caused by this dipolar interaction are straightforward for the ABM phase. This phase has two unique axes, which we have called w and z, and rotating, say, w away from z into the y direction we obtain

$$d_{i\alpha} = \begin{pmatrix} 0 & 0 & 0 \\ d\sin\theta & id\sin\theta & 0 \\ d\cos\theta & id\cos\theta & 0 \end{pmatrix}$$

and

$$d_{ii}^* d_{jj} + d_{i\alpha}^* d_{\alpha i} = 2|d|^2 \sin^2\theta$$

which is minimized with $\sin\theta = 0$: w and z must be parallel, and in a magnetic field both perpendicular to **H**.

In the A′ phase, again, there are two unique axes, but the parallel state written down by Balian and Werthamer is the least favourable. The most favourable appears to be a 90° rotation about the z axis to obtain

$$d_{i\alpha} = \begin{pmatrix} 0 & -d & 0 \\ d & 0 & 0 \\ 0 & 0 & 0 \end{pmatrix}$$

with $d^*_{ii}d_{jj} + d^*_{i\alpha}d_{\alpha i} = -d^2$. As we shall see, this resembles the BW case and will naturally arise from it when one component is depaired, as near a wall.

By definition, there is always a pair of sets of coordinate axes x, y, z and u, v, w in which the $d_{\alpha i}$ of the BW phase is

$$d_{\alpha i} = \begin{pmatrix} d & 0 & 0 \\ 0 & d & 0 \\ 0 & 0 & d \end{pmatrix}.$$

Two such sets of axes define an axis of rotation z and a rotation angle θ about z which carries one into the other. If we set our coordinate system along this axis, the two coordinates are related by the simple rotation through θ. Since in the original system d was a unit matrix, total rotations leave it so and only when we make the relative rotation u, v into x, y do we modify d. Thus in this system its matrix is

$$d_{i\alpha} = d\begin{pmatrix} \cos\theta & -\sin\theta & 0 \\ +\sin\theta & \cos\theta & 0 \\ 0 & 0 & 1 \end{pmatrix}$$

and

$$d^*_{ii}d_{jj} + d^*_{i\alpha}d_{\alpha i} = 4d^2(\cos\theta + 2\cos^2\theta),$$

which is minimized when $\theta = 104{\cdot}5° = \cos^{-1}(-\frac{1}{4})$. This condition (since other orientation effects in BW are weak, it is likely to be well satisfied in most regions of the liquid) leaves the BW "B" phase with the symmetry of an axial vector, the rotation $\hat{\mathbf{n}}\theta$. The vector $\hat{\mathbf{n}}$ defines the local state in the analysis of "textures".

The magnetic field orients the BW phase weakly through the following effect.[22, 23, 55, 96] Since it is the component of $d_{\alpha i}$, d_w, along the field $\mathbf{H}$ (taken in the w direction) which reduces the susceptibility, clearly there is a term in the free energy $-\frac{1}{2}\Delta\chi H^2$ with $\Delta\chi$ a function of d_w. For temperatures near T_c this term was written down by Ambegaokar and Mermin[28] and by the above authors:

$$\Delta F = c_1|\mathbf{M}\cdot\mathbf{d}_\alpha|^2 \tag{IV.92}$$

with c_1 an approximate constant, at least near T_c. Thus in the presence of a magnetic field, the free energy will be minimized by a d-matrix of the form (IV.91) but in which the z–z matrix element is increased and the x and y

portions decreased by a factor. Then the dipolar energy (IV.92) will in turn depend on the field. The appropriate expression is given by Engelsberg, Brinkman and Anderson:[96]

$$F_A \simeq \frac{0{\cdot}5(\hbar\omega_0)^2}{E_F}\left(\frac{\mu \mathbf{H}\cdot\hat{\mathbf{n}}}{kT_c}\right)^2, \tag{IV.93}$$

where the observed longitudinal frequency shift is $\omega_0[(T_c - T)/T_c]^{\frac{1}{2}}$, and is a measure of the dipolar anisotropy. The resulting orientation is with $\hat{\mathbf{n}}$ parallel to **H**.

Note that F_A is independent of T, approximately, hence, though small, finite near T_c. In general, as we shall see later, textures in the BW "B" phase are controlled by wall free energies, or in some cases by interactions with heat or fluid flow, etc., all of which are probably proportional to Δ^2 and hence to $T - T_c$. Thus as Webb *et al.*[97, 98] have observed, only extremely near to T_c will one have good orientation in weak magnetic fields.

(c) *Specific heats*

In subsection (1) we derived expressions for the condensation free energy F It is in principle not a separate question to calculate the specific heat

$$\Delta C = -T\frac{d^2\Delta F}{dT^2}. \tag{IV.94}$$

What we shall do here is briefly to mention the salient diagnostic features in which the anisotropic case differs from BCS and brings out the relevance of measuring the specific heat.

First we consider the discontinuity near T_c. Here we use the expression (IV.51) to obtain

$$\frac{dF}{dT} = \frac{dF}{\partial T} + \frac{\partial F}{\partial \Delta}\frac{d\Delta}{dT} = \frac{\partial F}{\partial T}$$

$$= \frac{N(0)}{T}\frac{\partial}{\partial T}\int \frac{d\Omega_k}{4\pi}\operatorname{Tr}\left[\hat{\Delta}\hat{\Delta}^\dagger - \frac{7\zeta(3)}{8\pi^2}\frac{(\hat{\Delta}\hat{\Delta}^\dagger)^2}{2T^2} + \dots\right]$$

and the second term is of order $(T - T_c)^2$ so may be neglected. Picking out the discontinuity, then, we have

$$\Delta(C)|_{T_c} = -N(0)\int \frac{d\Omega_k}{4\pi}\frac{d}{dT}\operatorname{Tr}\hat{\Delta}\hat{\Delta}^\dagger. \tag{IV.95}$$

We see already from (IV.95) an obvious fact: that the discontinuity in C is precisely proportional to the relative condensation free energy of the different states.

For any given solution in this limit, we can (as we did towards the end of subsection 1) parameterize the free-energy expression (IV.50) or (IV.51) and find an effective $|\Delta|^4$ term. For example, for the ABM or A_1 solutions[91]

$$\frac{\langle(\Delta\Delta^\dagger)^2\rangle_{\mathrm{ABM}}}{(\langle\Delta\Delta^\dagger\rangle)^2} = \frac{6}{5}\frac{\langle(\Delta\Delta^\dagger)^2\rangle_{\mathrm{BW}}}{(\langle\Delta\Delta^\dagger\rangle)^2}. \tag{IV.96}$$

Using $\partial F/\partial\langle\Delta\Delta^\dagger\rangle = 0$, this gives that $(\Delta C_{sp})_{\mathrm{ABM}}$ is only $\frac{5}{6}$ as great as for BW, which is in fact the same as for BCS, since the two are thermally totally equivalent.

When, in the early measurements, it became clear that C_{sp} had a greater jump experimentally than for BW–BCS, it became certain that the weak-coupling structure of the theory had to be abandoned. Any exotic phase was as certain to have a lower C_{sp} as to have a higher free energy. We will see that this is the case in Section V. Ordinary BCS theory, in fact, for most metals gives too low a C_{sp} (Bardeen and Schrieffer[99]). This is the result of the effects of strongly coupled phonons, and similar effects may be expected here, in addition to those we discuss in Section V.

A brief footnote to the specific heat discontinuity question: in a magnetic field the transition splits, both experimentally and theoretically, for reasons given by Ambegaokar and Mermin[28] and Varma and Werthamer.[29] The reason is the presence of two extra free-energy terms in the magnetic field; the one equivalent to the susceptibility decrease due to opposite-spin pairing already mentioned a few pages ago:

$$\begin{aligned} F_{\Delta\chi} &= C_1 H^2 |\Delta_{\uparrow\downarrow}|^2 \\ &= C_1 |\mathbf{H}\cdot\mathbf{d}_\alpha|^2 \end{aligned} \tag{IV.97}$$

and a second effect due to (presumably and primarily) the change in density of states at the Fermi level caused by the shift $\pm\mu H$ in E_F:

$$F_{\Delta E_F} = \eta H(|\Delta\uparrow|^2 - |\Delta\downarrow|^2) = i\eta\mathbf{H}\cdot(\mathbf{d}_\alpha \times \mathbf{d}_\alpha^*). \tag{IV.98}$$

(The change in density of states will of course modify T_c and hence $\ln T_c/T$.)

As observed, two transitions appeared, with unequal specific heat jumps adding up to the total in the absence of a field. Low-field theory gives two equal jumps, at higher fields [since (IV.97) is bigger than (IV.98) at most fields] there are three: A_1 ($\uparrow\uparrow$ pairing only), A (ESP), and finally B = BW. This phenomenon will be analysed in detail for the real ABM case in Section V.

At low temperatures Anderson and Morel pointed out (note however that their formal specific heat expression is wrong) that all phases expect $l = 1$ BW will have algebraic rather than exponential specific heats, because of the existence of nodes of the gap. For instance, in the ABM state the effective single-particle density of states as $T \to 0$ is

$$N(E) = N(0)\int \frac{d\Omega}{4\pi}\left[E^2 - \Delta_0^2(\hat{k}_x^2 + \hat{k}_y^2)\right] \simeq N(0)\frac{E^2}{\Delta_0^2}. \tag{IV.99}$$

This leads to a specific heat as $T \to 0$

$$(C_{sp})_{ABM}|_{T\to 0} \simeq 3{\cdot}5(\gamma T_c)(T/T_c)^3. \tag{IV.100}$$

The entropy also varies as T^3. The entropy of states with gap nodes, then, will of course tend to be higher, but not, of course, sufficiently so to make them stable in weak coupling.

(d) *Superfluid and normal fluid densities*

Anderson and Morel gave a correct expression for the superfluid density as normally defined in the weak-coupling, perturbation limit. In fact the supercurrent structure is considerably more complicated than that of an ordinary superfluid, as we shall see in Section VI, but for many practical purposes we expect that the straightforward approach followed by Anderson and Morel,[9, 10] Balian and Werthamer,[12] and later Saslow[45] will suffice.

The idea is to exploit the BCS perturbation-theory calculation of the electromagnetic response of a superconductor to a vector potential **A**. This is formally simple, so long as one neglects the Fermi-liquid corrections which we have so far omitted and will replace in the next subsection, but conceptually somewhat tricky, and in fact strictly speaking incorrect. In essence, as Saslow makes clear, this is a calculation of the normal fluid density ρ_n: we impose a velocity $\mathbf{v}_n$ on the quasiparticles, calculate $E(\mathbf{v}_n)$ and hence

$$\mathbf{P}_n = \frac{\partial E}{\partial \mathbf{v}_n} = \overleftrightarrow{\rho}_n \mathbf{v}_n. \tag{IV.101}$$

Then we deduce ρ_s by the standard Galilean invariance arguments which give $\rho_s = \rho - \rho_n$, i.e. the condensate has velocity

$$\mathbf{v}_s = \frac{\hbar}{m}\nabla\phi \tag{IV.102}$$

and density ρ_s, while ϕ is the phase of the condensate wavefunction and obeys the Landau acceleration–Josephson frequency equation

$$h\nabla\dot{\phi} = \nabla\mu. \tag{IV.103}$$

All of this is indeed true (although existing proofs are incomplete, Cross[100] has completed the logical argument) if the condensate has a unique velocity $\mathbf{v}_s$. For $l = 1$, unfortunately, there are nine components $d_{\alpha i}$ of the condensate and there is no guarantee that there is a single ϕ or a single v_s. Thus the condition for validity of the AM–Saslow theory is that $d_{\alpha i} \propto (d_{\alpha i})^0 e^{i\varphi}$, i.e. $(v_s)_{\alpha i} \neq f(\alpha, i)$. In any case the calculation of ρ_n is strictly correct, and is, in analogy to the corresponding BCS expression

$$\overleftrightarrow{\rho}_n = \sum_{\mathbf{k}} \mathbf{kk} \frac{\partial f(E_{\mathbf{k}})}{\partial k} \tag{IV.104}$$

where f is the Fermi factor $(e^{\beta E} + 1)^{-1}$. In the ABM state ρ_n is anisotropic, with deviations from ρ differing by a factor 2 near T_c:

$$\begin{aligned} \rho_n^{\parallel}/\rho &= \left(1 - \frac{7}{10\pi^2} \zeta(3) \frac{\langle \Delta^2 \rangle}{T_c^2}\right) \\ \frac{\rho_n^{\perp}}{\rho} &= \left(1 - \frac{7}{5\pi^2} \zeta(3) \frac{\langle \Delta^2 \rangle}{T_c^2}\right) \\ T &\to T_c \end{aligned} \tag{IV.105}$$

where $\parallel$ is aparallel and $\perp$ perpendicular to the orbital axis z. Thus ρ_s as defined within this theory is anisotropic by a factor 2:

$$\rho_s^{\parallel}/\rho = \tfrac{1}{2}\rho_s^{\perp}/\rho = \frac{7}{10\pi^2} \zeta(3) \frac{\Delta^2}{T_c^2} + \dots \tag{IV.106}$$

As $T \to 0$, however, $\rho_n \to 0$, although differently for $\parallel$ and $\perp$; so $\rho_s = \rho$, as required for Galilean invariance.

These quantities are then, one imagines, (after Fermi-liquid corrections if necessary) to be inserted into a two-fluid model on the HeII pattern, in which the kinetic free energy is

$$F_{\text{kin}} = \tfrac{1}{2}(\rho_s v_s^2 + \rho_n v_n^2) \tag{IV.107}$$

with v_s obeying (IV.102, IV.103), and v_n pinned to the walls in narrow channels. v_n carries the entropy of the fluid, since it represents the quasiparticle gas, etc. From this one may, as Saslow(45) and Wölfle(53) have suggested, derive an anisotropic fourth sound (ρ_s alone moving), second sound (a hydrodynamic counterflow oscillation), etc.

Except for fourth sound, it appears to the present authors that direct use of these two-fluid concepts will be at best unwise and at most misguided or incorrect. (See Wölfle(53) for a very sophisticated discussion.) Aside from the difficulty of the unique v_s, one also has different relative orders of magnitude of relaxation times, coherence lengths, diffusion lengths, etc. The many more degrees of freedom will allow many more collective modes, both hydrodynamic and collisionless, about which a massive literature is already growing. Since there is available a basically almost exact microscopic theory it seems usually possible, and far wiser, to trust for quantitative work to direct microscopic calculation rather than two-fluid heuristics until these can be sorted out.

Meanwhile, however, there are a few measurements which do have their natural discussion in the two-fluid picture: fourth sound[48, 101] and, particularly, early heat transport experiments[41] which demonstrated clearly a two-fluid counterflow. In these very interesting experiments non-linear instabilities were observed in the A phase highly suggestive of the expected interaction between anisotropic superflow and gap orientation.

3. FERMI–LIQUID CORRECTIONS TO STATIC AND LOW-FREQUENCY EFFECTS

Section 2 describes the very strong interaction effects between quasiparticles in normal He^3, which severely modify most of that liquid's properties from those of a free Fermi gas, which is the model we have used so far for the pairing phenomena. How can we treat the system in weak coupling when these strong interactions exist?

Leggett[13, 14] was the first to study the interaction of pairing and Fermi-liquid effects, and to develop a scheme for handling them together. The quantitative effects on many properties—in particular χ and ρ_n—are fairly large, but in a very specific sense the qualitative picture of handling the pairing independently is unchanged. We will follow Leggett's remarkably clear and complete discussion here. (It is striking that the susceptibility of the B phase is still compared with a figure given by Leggett[13] at that time)—see Fig. 6.

The philosophy of Landau is that states of the interacting Fermi system at temperatures $\ll T_F$ can be referred back and enumerated as adiabatic continuations of weakly-excited states of a non-interacting Fermi sea: i.e. labelled in terms of occupation numbers $n(\mathbf{k}, \sigma)$ of "quasiparticles" near the Fermi surface. The residual interactions of these quasiparticles may not be neglected but their total effects are weak: the large parts of the interactions are taken into account in renormalization effects.

Landau uses an effective hamiltonian containing the renormalized single-particle energy

$$H_k = \sum_{\mathbf{k},\sigma} \varepsilon(\mathbf{k}) n(\mathbf{k}, \sigma) \tag{IV.108}$$

plus a residual quasiparticle interaction between changes δn in quasiparticle occupation at the Fermi surface:

$$H_L = \tfrac{1}{2} \sum_{\mathbf{k}\mathbf{k}'\sigma\sigma'} f(\mathbf{k}, \mathbf{k}'; \sigma, \sigma') \delta n(\mathbf{k}, \sigma) \delta n(\mathbf{k}', \sigma'). \tag{IV.109}$$

This was expanded in Section II (II.2) in terms of direct and exchange-like

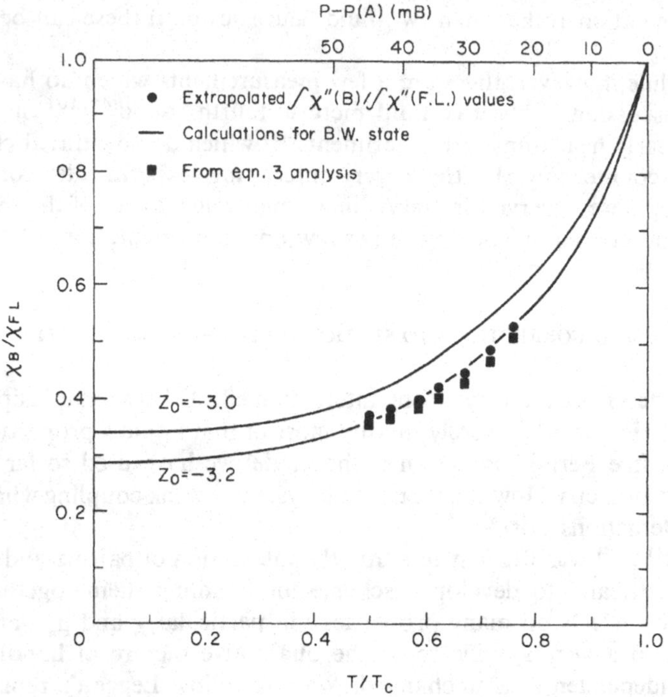

FIG. 6. A plot of the susceptibility versus reduced temperature for the B phase. Two values of $z_0 = 4F_0^a$ were used. The two sets of data are simply two slightly different ways of analyzing the saturation data discussed in Section VIII. See also Fig. 17.

parts F^s and F^a and in terms of spherical harmonics of $(\hat{\mathbf{k}}.\hat{\mathbf{k}}')$. This interaction represents the forward scattering amplitude of the quasiparticles and has the effect of modifying the effective single-particle energy $\varepsilon(\mathbf{k})$ proportionally to changes in the occupations of other states. As a result each quasiparticle moves in a "mean field" created by all the others.

A second type of residual interaction which cannot be neglected is the pairing interaction which causes the BCS transition (IV.1):

$$V_p = \sum_{\mathbf{kk}'\sigma\sigma'} V_{\mathbf{kk}',\sigma\sigma'} a^\dagger_{\mathbf{k}\sigma} a^\dagger_{-\mathbf{k}\sigma'} a_{-\mathbf{k}'\sigma'} a_{\mathbf{k}'\sigma'}. \tag{IV.110}$$

Although Leggett's basic work (like the later work by Betbeder-Matibet and Nozières[102] and Wölfle[53] along the same lines) proceeds from a formal diagrammatic approach, the principle may be seen from a much less formal treatment given in his Section IV. In this section he points out that in a real sense the Fermi-liquid interactions do not interfere with the pairing phenomenon because pairing does not change the net occupation for a given direction on the Fermi surface:

$$\delta n(\mathbf{k}, \sigma) = N(0) \int \mathrm{d}\varepsilon_{\mathbf{k}}\, \delta n(\mathbf{k}, \sigma) \simeq 0, \tag{IV.111}$$

if, as is always the case in weak coupling, approximate hole–particle symmetry is maintained. Thus in fact the Fermi-liquid interactions remain unchanged upon pairing and the pairing proceeds according to the sum of H_k and V_p. Thus, for instance, all the thermal phenomena are unaffected by H_L.

The pairing does, however, affect most of the characteristic Fermi-liquid transport, response, and collective-mode effects, because pairing changes precisely the long-wavelength response functions which are relevant to these. It is the essential result of the work of Leggett, Nozières, and Wölfle that one form or another of mean field theory can be used to calculate these effects. In rather general terms, we can suppose that, due to some applied field, the one-particle energies ε undergo a small change:

$$\varepsilon \to \varepsilon + \delta\varepsilon(\mathbf{k}, \sigma). \tag{IV.112}$$

We calculate the response of the non-interacting Fermi gas to this perturbation (as in subsection 2 of this section IV):

$$n(\mathbf{k}', \sigma) = \sum_{\mathbf{k}, \sigma} \chi^0(\mathbf{k}', \sigma'; \mathbf{k}, \sigma)\delta\varepsilon(\mathbf{k}, \sigma) \tag{IV.113}$$

and, finally, we used the Fermi-liquid relation (II.1)

$$\delta\varepsilon(\mathbf{k}, \sigma) = (\delta\varepsilon)_{\mathrm{ext}} + \sum_{\mathbf{k}', \sigma'} f(\mathbf{k}\mathbf{k}'; \sigma\sigma')\delta n(\mathbf{k}', \sigma') \tag{IV.114}$$

to obtain a renormalized response which can be schematically written

$$\chi = \frac{\chi^0}{1 + f\chi^0}. \tag{IV.115}$$

In general, the self-consistent solution of (IV.112–IV.114) will involve an integral equation and also must be carried out dynamically, but in most cases it is much simpler than that. If, with Leggett, we ignore higher spherical harmonics than $l = 1$ in the series of F's (II.2), we obtain for the spin susceptibility exactly the standard Fermi-liquid "Stoner" result equivalent to (II.3)

$$\chi = \frac{\chi^0}{1 + [\chi^0/N(0)]F_0^a} \tag{IV.116}$$

and for the normal fluid density

$$\rho_n = \frac{\rho_n^0}{1 + (\rho_n^0/Nm^*)\cdot\frac{1}{3}F_1}. \qquad \text{(IV.117)}$$

Here χ^0 is the susceptibility calculated in the last subsection including pairing only, and ρ_n^0 is the normal fluid density so calculated; F_0^a is the Stoner-type exchange Landau parameter (often called $Z_0/4$ or $-\bar{I}$) and F_1 the effective mass parameter, for which

$$\frac{m^*}{m} = 1 + F_1/3. \qquad \text{(IV.118)}$$

These expressions are very successful experimentally, as may be seen from Fig. 6 for the susceptibility (Osheroff;[27] one should note that prior susceptibility measurements[103] in the B phase were in a confused state) and Fig. 7 for the superfluid density $\rho_s = \rho - \rho_n$.

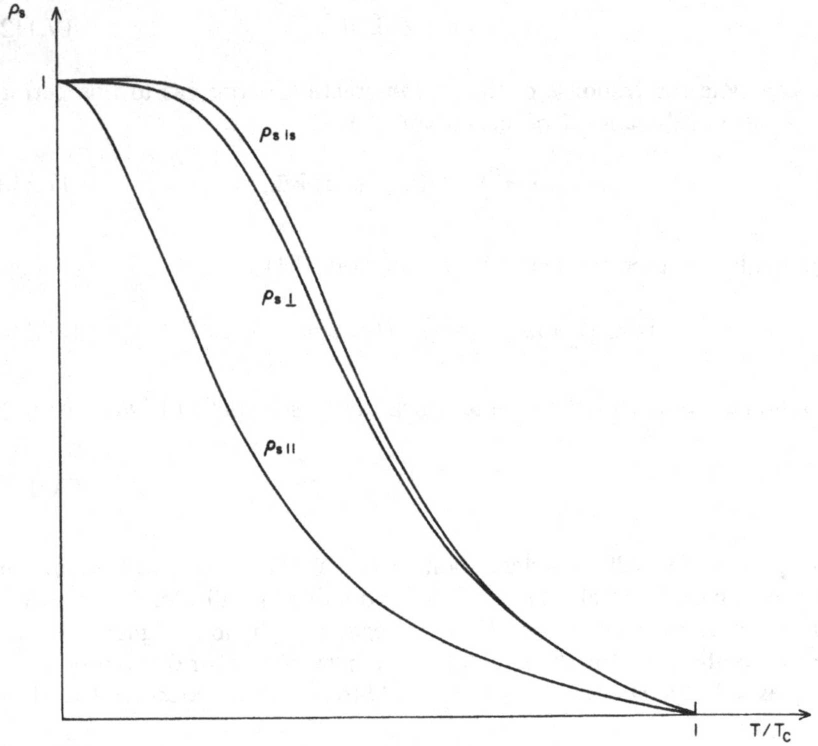

FIG. 7. A plot of ρ_s including the Fermi-liquid corrections. Note that the actual ρ_s is considerably smaller than the weak-coupling value and that it is quite anisotropic.[57]

The effect of these corrections may well be understood from the effective-field picture. When there is a large negative Z^0 (large enhancement) the reduction of the susceptibility by the pairing reduces the enhancement factor severely as well, and the susceptibility drops rapidly. The normal fluid density, on the other hand, is reduced rather than enhanced, so reaches its zero-temperature value more slowly at first.

Leggett himself already gave the type of expression which is correct if higher Z's and F's are not negligible.

$$\chi(T) = \int \frac{d\Omega_{\hat{k}}}{4\pi} \frac{\chi^0(T, \hat{k})}{1 + \int [\chi^0(T, \hat{k}')/N(0)] F^a(\hat{k}, \hat{k}')(d\Omega_{\hat{k}'}/4\pi)}. \qquad \text{(IV.119)}$$

As he pointed out, although (IV.119) reduces to (IV.117) (even as to slope) as $T \to T_c$, for the BW state there is no reason to be sure that $F^a(\hat{k}, \hat{k}')$ is not severely cut off in $|\hat{k} - \hat{k}'|$, which would have the effect of changing low-temperature susceptibilities rather severely—one would guess, incidentally, in quite the opposite direction to that suggested by Vuorio:[104] i.e. nearer for the BW state, not farther. The reason for this effect is that then the electrons on the different parts of the Fermi surface would act almost independently and of order $\frac{2}{3}$ of them would see no change in their effective fields at all.

For this reason we suggest that in fact the effective fields are, for whatever reason, of quite short range in space and of long range in k-space, and that higher Fermi-liquid corrections are empirically not very big. We await with interest more data on the temperature dependence of ρ_s.

4. SOUND IN ANISOTROPIC SUPERFLUID He^3

Although sound, as a collective mode, properly might be considered to belong in a later section, both chronologically and from a theoretical point of view it seems appropriate to discuss it here—chronologically, because zero-sound attenuation was a very early measurement which gives a striking peak at the A transitions which was used as a convenient signal in early Cornell measurements[105] and theoretically, because it is closely related to Fermi-liquid theory and because the relevant theory is not highly dependent on the anisotropic effects, or on the spin-fluctuation theory.

Normal He^3, of course, at high temperatures and low frequencies has only ordinary "hydrodynamic" sound with velocity given by the classical formula

$$c_0^2 = \frac{p_F^2}{3mm^*}(1 + F_0^s), \qquad \text{(IV.120)}$$

where the quantity

$$\frac{p_F^2}{3m^*}(1 + F_0^s) = \kappa \tag{IV.121}$$

is the modulus of compressibility per atom. F_0^s is quite large and $c_0 \gg v_F$. As we lower the temperature and raise the frequency we observe a peak in attenuation and a transition to "zero sound" which is a collisionless collective mode of the Fermi system in which the quasiparticles do not come into thermal equilibrium internally but respond only dynamically. This mode satisfies the integral equation for the zero-sound dispersion relation

$$(\omega - \mathbf{k}\cdot\mathbf{v}_0)\nu(\hat{\mathbf{k}}) = \mathbf{k}\cdot\mathbf{v}_0 \int F^s(\hat{\mathbf{k}},\hat{\mathbf{k}}')\nu(\hat{\mathbf{k}}')\frac{d\Omega_{\hat{k}'}}{4\pi}, \tag{IV.122}$$

where $\mathbf{v}_0$ is a vector of magnitude v_F along the propagation direction, and $\nu(\hat{\mathbf{k}})$ is the modification of the distribution function in the $\hat{\mathbf{k}}$ direction. Because $\omega/k = c$ is large compared to v_F, because of the high compressibility modulus, c for zero sound is only 4% larger at $T = 0$ than c_0.

Wölfle[32] has given a very clear discussion of sound propagation in the anisotropic case, restating in part some of the results of Leggett.[106] Earlier Wölfle[31] had also given a full treatment of the rather complicated problem of sound attenuation.

Briefly, the physical effect goes back to the simple observation of Anderson[107] and Bogoliubov[108] that at absolute zero the BCS superfluid should exhibit an ordinary sound mode with the normal velocity c_0 (IV.120) rather than zero sound. The physical reason is the fact that the Fermi continuum of hole–particle states $a^\dagger_{\mathbf{k}+\mathbf{q}}a_{\mathbf{k}}$ of energies up to $q \cdot v_F$ is removed by pairing, and thus the sound mode itself must saturate the sum rule: it must be a uniform compressional excitation because there are no quasiparticle modes at low frequency which can distort the Fermi surface, i.e. the mode is not repelled by a lower-frequency continuum (see Fig. 8).

Since the dispersion law changes as the mode passes through the region of the gap (well above the gap frequency, of course, zero sound will return) this implies a strong absorption, which in fact peaks very severely near T_c, and is entirely due to simple quasiparticle pair creation. In the normal state, momentum conservation prevents hole–particle pair excitation and only allows the scattering of excited particles, but the presence of a gap breaks this conservation law. As long as there is a region with $\Delta < \omega$ hole–particle creation can occur, although this is obviously a very anisotropic effect for the sound wave vector $\mathbf{q}$ with respect to the positions of the nodes of the gap in k-space. For the BW state, $\omega > \Delta$ only for ordinary ultrasonic frequencies and only very near T_c and one expects a very narrow peak (see Fig. 9), but for ABM there are always nodes and one can understand the observed long tail of absorption for $T < T_c$.[109]

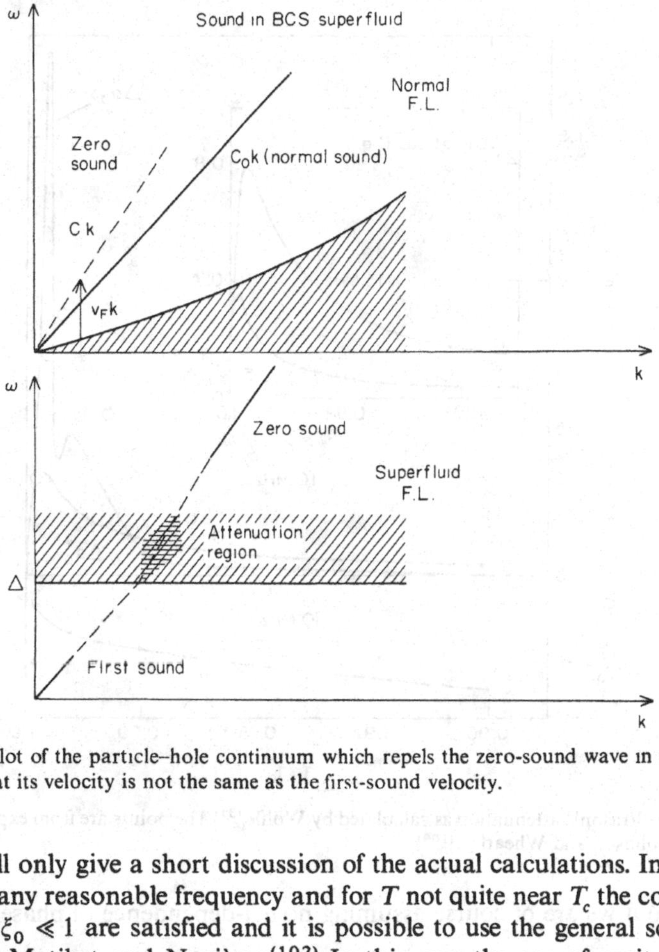

FIG. 8. A plot of the particle–hole continuum which repels the zero-sound wave in the normal state so that its velocity is not the same as the first-sound velocity.

We will only give a short discussion of the actual calculations. In the first place, at any reasonable frequency and for T not quite near T_c the conditions $\omega \ll \Delta$, $q\xi_0 \ll 1$ are satisfied and it is possible to use the general scheme of Betbeder-Matibet and Nozières.[102] In this case the gas of excited quasiparticles can be treated just precisely similarly to the Landau theory of normal Fermi liquids, which obey the kinetic equation (essentially (IV.122))

$$(\omega - \mathbf{q}\cdot\nabla_{\mathbf{k}}E_{\mathbf{k}})\nu_{\mathbf{k}} + \mathbf{q}\cdot\nabla_{\mathbf{k}}E_{\mathbf{k}}N(0)\delta E_{\mathbf{k}}(\mathbf{q},\omega) = I_{\mathbf{k}}(\nu)$$

where $I(\nu)$ is the collision integral and $\delta E_{\mathbf{k}}$ is the change in quasiparticle energy due to external forces and to the modified distribution ν itself:

$$\delta E_{\mathbf{k}} = \frac{\varepsilon}{E}\left(\frac{F_0^s}{N_F}\delta\rho - \delta\mu\right) + \frac{F_1^s}{3}\left(\frac{m}{m^*}\right)\mathbf{k}\cdot\frac{\overleftrightarrow{\rho}_n}{\rho}\cdot(\mathbf{v}_n - \mathbf{v}_s) + \mathbf{k}\cdot\mathbf{v}_s + \frac{\delta|\Delta^2|}{E_{\mathbf{k}}}. \quad \text{(IV.123)}$$

The tensor $\overleftrightarrow{\rho}_n$ is the normal fluid density we already derived in the last section. $\mathbf{v}_s$ obeys the acceleration equation we also derived in the last section (and

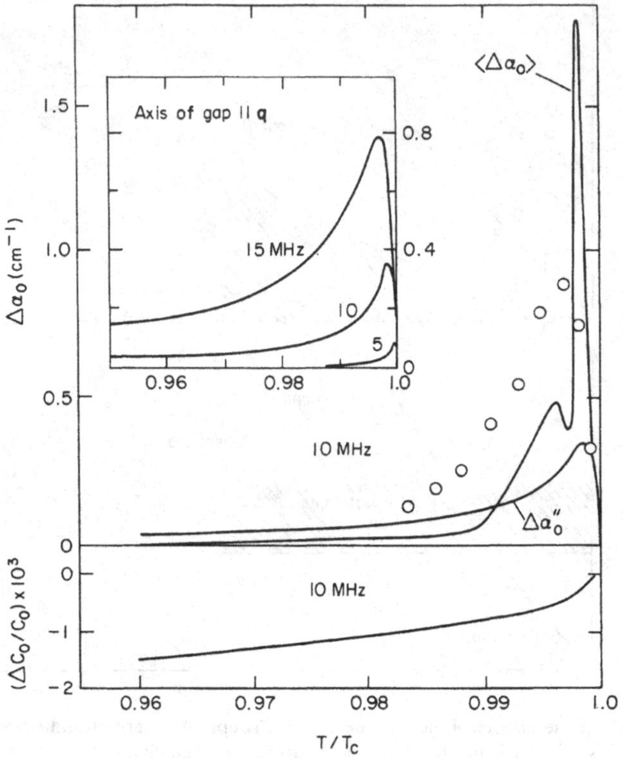

FIG. 9. The ultrasonic attenuation as calculated by Wölfle.[31] The points are from experiments of Paulson, Johnson and Wheatley.[109]

throughout we are of course assuming no α, i-dependence of phase)

$$m\frac{\mathrm{d}\mathbf{v}_s}{\mathrm{d}t} = \hbar\nabla\delta\mu$$

and the continuity equation

$$\frac{\partial\delta\rho}{\partial t} = \nabla\cdot(\rho_n\mathbf{v}_n + \rho_s\mathbf{v}_s). \qquad \text{(IV.124)}$$

$\mathbf{v}_n$ is defined in the obvious way as the mean quasiparticle velocity. Finally, the change in density is

$$\delta\rho = \frac{\Sigma(\varepsilon/E)\nu + N(0)\langle\lambda_k\rangle\delta\mu}{1 + F_0\langle\lambda\rangle}. \qquad \text{(IV.125)}$$

$\langle\lambda_k\rangle$ is a kind of effective average BCS superfluid density for the direction $\hat{\mathbf{k}}$

$$\lambda(\hat{\mathbf{k}}) = \int_{-\infty}^{\infty} d\varepsilon_{\mathbf{k}} n(E_{\mathbf{k}}) \frac{|\Delta_{\mathbf{k}}|^2}{E_{\mathbf{k}}^3} \tag{IV.126}$$

which varies from 0 at T_c to 1 at $T = 0$. For the "collisionless" sound velocity, Wölfle obtains

$$c^2 = c^2(T = 0) + (3 + F_1)\left[-\langle\lambda(\hat{\mathbf{k}})(\hat{\mathbf{k}}\cdot\hat{\mathbf{q}})^4\rangle + \frac{\langle\lambda(\hat{\mathbf{k}})(\hat{\mathbf{k}}\cdot\hat{\mathbf{q}})^2\rangle^2}{\langle\lambda\rangle}\right], \tag{IV.127}$$

which has the appropriate behaviour but makes the zero-to-first-sound transition different for different orientations.

The hydrodynamic regime $\omega\tau < 1$ is very limited in extent, because of the presence of the gap and the resulting low quasiparticle density, to either extremely low frequencies or $T \simeq T_c$. In this limited regime first sound is normal, and a second sound exists with a highly anisotropic velocity

$$u_2^2 = v_F^2\left(\frac{m^*}{m}\right)\frac{T}{2E_F}\left[\rho\mathbf{q}\cdot\overleftrightarrow{\sigma}\cdot\overleftrightarrow{\rho}_n^{-1}\cdot\overleftrightarrow{\sigma}\cdot\mathbf{q} - (\mathbf{q}\cdot\overleftrightarrow{\sigma}\cdot\mathbf{q})^2\right], \tag{IV.128}$$

where $\overleftrightarrow{\sigma}$ is a tensor entropy density of the normal fluid fraction, defined by the same sum (IV.104) as $\overleftrightarrow{\rho}_n$ multiplied by $\varepsilon_{\mathbf{k}}$. u_2 is very small, $<10^{-2}\, v_F$. For fourth sound, with the normal fluid pinned, Wölfle's results agree with those of Saslow[45] and de Gennes,[50, 51] which we have already discussed.

As for attenuation, Wölfle's earlier calculation is quite complex because it was necessary to go entirely outside the effective Landau model of Nozières and Betbeder-Matibet and use the general self-consistent dynamical theory which Leggett first gave in detail (it is amusing that in essence this is identically the "generalized RPA" scheme of Anderson[107]). This is no serious difficulty in principle, but one must take into account self-consistently modifications $\delta\hat{\mathscr{E}}_{\mathbf{k}}$ and $\delta\hat{W}_{\mathbf{k}}$ in the Balian–Werthamer matrices. Results for the attenuation peaks are shown in Fig. 9.

V. Spin-fluctuation Stabilization of the "A" Phase

As discussed in the previous section on weak-coupling theory, Balian and Werthamer[12] have shown that their state is the lowest free-energy state at all temperatures. Clearly, there must be a strong coupling effect which tends to stabilize the A phase, which is now fairly conclusively established to be the Anderson–Morel state with the orbital angular momentum of both up and down spin pairs aligned. Since the estimates made in Section III indicate that the spin fluctuations are important in determining T_c we examine a possible feedback mechanism involving these fluctuations which, as we shall see, stabilizes the ABM state.[34] The attractive interaction via spin fluctuations is similar to the attractive interaction via the virtual exchange of phonons in electronic superconductors. In the case of superconductors, because the phonons are mostly a property of the ions plus some high screening coming from the electrons, one does not expect large changes in the phonon spectrum unless one has a very strong coupling situation. On the other hand, in He^3 the spin fluctuations are part of the Fermi liquid itself and, as such, may change below the superfluid transition temperature. This is especially true at long wavelengths where the spin fluctuations are at very low frequencies. As was discussed in the previous section, the static susceptibilities χ^0 become anisotropic and undergo large changes in the superfluid state. Since the effective interaction for zero momentum transfer depends on these susceptibilities, we expect the interaction to change on going into the superfluid state.

To see how such an effect occurs we examine the effective interaction more carefully. In the superfluid state the dynamic susceptibility is not isotropic, so that below T_c each component of the triplet experiences a different effective interaction. This effect can be described by making $\Gamma_s(\mathbf{q})$ dependent on the components of the spin operators in (III.12), i.e. we write the exchange term as

$$\frac{1}{2}\sum_{\substack{\mathbf{k}\mathbf{k}'\\ \mathbf{q}}} \Gamma_s^i(\mathbf{q})\sigma^i_{\alpha\beta}\sigma^i_{\gamma\delta}a^\dagger_{\mathbf{k}+\mathbf{q}\alpha}a^\dagger_{\mathbf{k}'-\mathbf{q}\gamma}a_{\mathbf{k}'\delta}a_{\mathbf{k}\beta}. \tag{V.1}$$

We have not written down the most general form but rather one with which we can easily visualize the effect. In fact, we assume that $\Gamma_s^x = \Gamma_s^y \neq \Gamma_s^z$. Since the total spin is no longer conserved we must calculate the interaction for each component of the triplet. Using (III.2, III.3) we find

$$v_{m_l=\pm 1}(\mathbf{k}-\mathbf{k}') = -\Gamma_s^z(\mathbf{k}-\mathbf{k}') \tag{V.2}$$

$$v_{m_l=0}(\mathbf{k}-\mathbf{k}') = \Gamma_s^z(\mathbf{k}-\mathbf{k}') - 2\Gamma_s^x(\mathbf{k}-\mathbf{k}'). \tag{V.3}$$

Here m_l denotes the three components of the triplet. If the weak-coupling susceptibilities have the same form as Γ_s, i.e. $\chi^0_{xx} = \chi^0_{yy} \neq \chi^0_{zz}$, then summing the diagrams in Figs. 3 and 4 leads to the results that

$$\Gamma_s^i(\mathbf{q}) = \frac{1}{2}\frac{I}{(1 - I\chi^0_{ii}(\mathbf{q},0))}$$

so that

$$v_{m_l=\pm 1}(\mathbf{q}') = -\frac{1}{2}\frac{I}{(1 - I\chi^0_{zz}(\mathbf{q},0))} \tag{V.4}$$

and

$$v_{m_l=0}(\mathbf{q}') = \frac{1}{2}\left(\frac{I}{(1 - I\chi^0_{zz}(\mathbf{q},0))} - \frac{2I}{(1 - I\chi^0_{xx}(\mathbf{q},0))}\right). \tag{V.5}$$

The feedback effect is very easily seen in these equations. At $q = 0$ we can use the results for the static values of χ^0_{ij} from the previous section. If we consider the BW state, the three susceptibilities are equal and reduced. Consequently, the effective interactions are also equal and reduced. If fed back into the gap equation, the value of the gap will be reduced and the free energy raised. However, in the ABM state, only χ^0_{zz} is reduced and only $m_l = 0$ pairs have a non-zero amplitude, so that as χ^0_{zz} decreases the relevant interaction *increases*. Thus the magnitude of the energy gap is increased and the free energy of the ABM state is lowered. Therefore, if the spin-fluctuation term is large enough, one can expect the ABM state to be energetically more favourable than the BW state. Of course, one cannot simply look at the values of the susceptibility at $q = 0$. In the first paper(34) on this effect, the values of $\chi^0_{ij}(\mathbf{q}, 0)$ were guessed from earlier work on the superconductors by Anderson and Suhl,(110) and it was shown that this feedback mechanism gives rise to fourth-order terms in the free energy which are such that if the spin fluctuations are sufficiently strong the ABM state can be stabilized.

More recently, Brinkman, Serene, and Anderson(37) and Kuroda(38, 39) have extensively investigated the feedback effect by calculating properly the susceptibility tensor for finite q and ω and then calculating the free-energy changes. In order to do this they studied the change in the spin-fluctuation contribution to the free energy induced by the transition to the superfluid state. The spin-fluctuation contribution to the free energy can be written as

$$\Delta F = \frac{T}{2}\,\mathrm{Tr}\sum_m \int \frac{d^3q}{(2\pi)^3}\left[\ln\{1 - I\chi^0_{ij}(\mathbf{q}, i\omega_m)\} + I\chi^0_{ij}(\mathbf{q}, i\omega_m)\right], \qquad \text{(V.6)}$$

where ω_m are the Matsubara frequencies $2m\pi T$ and the logarithm is defined in terms of its expansion. The trace is taken over i and j. ΔF represents the sum of the diagrams in Fig. 10. The susceptibilities are defined as

$$\chi^0_{ij}(\mathbf{q}, i\omega_m) \equiv \int_0^\beta \exp(i\omega_m\tau)\langle T_\tau[M_i(\mathbf{q}, \tau)M_j(-\mathbf{q}, 0)]\rangle\, d\tau \qquad \text{(V.7)}$$

where

$$M_i(\mathbf{q}, \tau) = \frac{\sigma^i_{\alpha\beta}}{2}\sum_{\mathbf{p}} \exp[(H_0 - \mu N)\tau]a^\dagger_{\mathbf{p}+\mathbf{q}/2,\alpha}a_{\mathbf{p}-\mathbf{q}/2,\beta}\exp[-(H_0 - \mu N)\tau]. \qquad \text{(V.8)}$$

T_τ designates the usual time ordering and H_0 is the reduced hamiltonian about which we do perturbation theory. In order to calculate the change in ΔF we define $\delta\chi^0_{ij}$ to be the difference between the normal-state susceptibility functions and those of the superfluid, i.e.,

$$\delta\chi^0_{ij} = \chi^0_{ij}|_N - \chi^0_{ij}|_S \qquad \text{(V.9)}$$

where N and S stand here for normal and superfluid respectively. We define

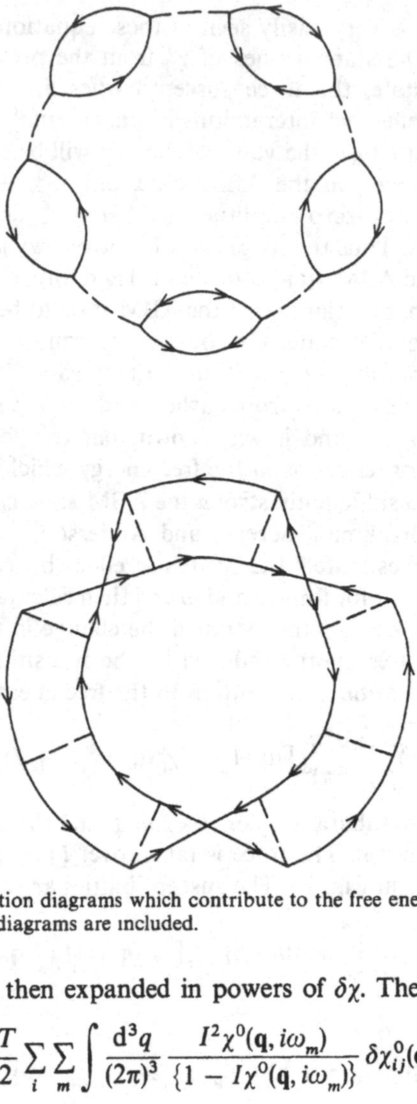

FIG. 10. The spin-fluctuation diagrams which contribute to the free energy. Note that both the normal and anomalous diagrams are included.

$\chi^0_{ij}|_N = \delta_{ij}\chi^0$. ΔF is then expanded in powers of $\delta\chi$. The first-order term is

$$\Delta F_1 = \frac{T}{2}\sum_i\sum_m \int \frac{d^3q}{(2\pi)^3} \frac{I^2\chi^0(\mathbf{q}, i\omega_m)}{\{1 - I\chi^0(\mathbf{q}, i\omega_m)\}} \delta\chi^0_{ij}(\mathbf{q}, i\omega_m). \qquad \text{(V.10)}$$

The physical interpretation of ΔF^1 is that it represents a contribution to the free-energy difference between the superfluid and normal states when the effective interaction is taken to be that obtained by virtual exchange of normal-state spin fluctuations. We assume that the effect of ΔF^1 can be incorporated into the attractive interaction of the unperturbed weak-coupling calculation. This assumption has recently been investigated by Serene and Rainer[111] who find that although there are some non-negligible corrections to the results presented here they do not change the overall result that the ABM state becomes stable.

The term second order in $\delta\chi^0$ is

$$\Delta F^2 = -\frac{T}{4}\sum_{ij}\sum_m \int \frac{d^3q}{(2\pi)^3}\left[\frac{I\delta\chi_{ij}^0(\mathbf{q}, i\omega_m)}{1 - I\chi^0(\mathbf{q}, i\omega_m)}\right]^2. \qquad \text{(V.11)}$$

One can convince oneself that the higher-order terms in the expansion of ΔF are negligible compared to ΔF^2 because the important region of integration with respect to q is for $2k_F\, q \gg \xi^{-1}$, where ξ is the temperature-dependent coherence length. In this region it can be shown that $\delta\chi \propto N(0)/(q\xi)$, so one is expanding in powers of $\bar{I}/(k_F\xi)(1 - \bar{I})$ which is much less than one.

Examining equation (V.11) we see that the change in the spin-fluctuation free energy is always negative since $\delta\chi(\mathbf{q}, i\omega_m)$ is real. The principle we therefore use to study the feedback effect is the maximization of the change in the spin-fluctuation free energy.

The simple discussion in terms of the effective interaction is not entirely correct as it implies that the BW state will increase in energy via the feedback. This can, however, be rectified if one studies the gap equation in a strong-coupling formulation of the problem.

The calculation of $\delta\chi_{ij}^0$ is fairly complicated[37] so we shall not repeat it here. Rather we simply discuss the results of the calculations of the free energies in the various regimes. To do this we first rewrite (V.11) so that we can separate out the dependence of ΔF^2 on various parameters. As already mentioned, the integral over q is dominated by the $q > \xi^{-1}$ region and in this region we can write

$$\delta\chi_{ij}^0 = \frac{N(0)}{qv_F}\,\delta\bar{\chi}_{ij}^0(\hat{\mathbf{q}}, i\omega_m), \qquad \text{(V.12)}$$

i.e. we separate out the explicit dependence on the magnitude of $\mathbf{q}$. Then we can write

$$\Delta F^2 = -\frac{T}{4}\frac{1}{2\pi^2}\frac{1}{v_F^2}\int_0^{2k_F} dq\left(\frac{\bar{I}}{(1 - \bar{I} + \bar{I}q^2/12k_F^2)}\right)^2 \times \sum_{ij}\sum_m \int \frac{d\Omega_{\hat{\mathbf{q}}}}{4\pi}[\delta\bar{\chi}_{ij}(\hat{\mathbf{q}}, i\omega_m)]^2. \qquad \text{(V.13)}$$

Defining

$$\delta = \frac{150\pi^2}{7\zeta(3)}\left(\frac{T_c}{\varepsilon_F}\right)\int_0^{2k_F}\frac{dq}{2k_F}\left(\frac{\bar{I}}{(1 - \bar{I} + \bar{I}q^2/12k_F^2)}\right)^2, \qquad \text{(V.14)}$$

we can write

$$\Delta F^2 = -N(0)\left(\frac{T}{T_c}\right)\delta\left(\frac{7\zeta(3)}{150\pi^2}\right)\sum_{ij}\sum_{m}\int\frac{d\Omega_{\hat{q}}}{4\pi}[\delta\bar{\chi}_{ij}(\hat{\mathbf{q}}, i\omega_m)]^2. \quad \text{(V.15)}$$

Since $\delta\bar{\chi}$ depends only on T and Δ, we have separated off the spin-fluctuation parameters into a single variable δ. The particular choice of normalization of δ made above turns out to be convenient in discussing the terms in ΔF^2 that are fourth order in the free energy.

The stability of various phases just below T_c is determined by an examination of the fourth-order terms in Δ. As discussed in Section IV, from general invariance arguments[91, 35, 36] one can show that there are five fourth-order invariants. These can be written in terms of the $d_{\alpha i}$ previously introduced

$$F^4 = \frac{a_1}{2}\left|\sum_{\alpha i} d_{\alpha i}^2\right|^2 + \frac{a_2}{2}d^*_{\beta i}d^*_{\beta j}d_{\alpha i}d_{\alpha j} + \frac{a_3}{2}d^*_{\alpha i}d^*_{\beta i}d_{\alpha j}d_{\beta j}$$

$$+\frac{a_4}{2}\{\sum_{\alpha i}|d_{\alpha i}|^2\}^2 + \frac{a_5}{2}d^*_{\alpha i}d^*_{\beta j}d_{\alpha j}d_{\beta i}. \quad \text{(V.16)}$$

Using the weak-coupling theory presented in the previous section one can easily determine the weak-coupling values of the a's. We find that $a_2 = a_4 = a_5 = -a_3 = -2a_1$, and assuming that the $d_{\alpha i}$ are normalized so that $\Sigma_{\alpha i}|d_{\alpha i}|^2 = 3|\Delta|^2$, i.e., Δ is defined as the average energy gap then

$$a_1^{wc} = -\frac{N(0)}{(\pi T)^2}\left(\frac{7\zeta(3)}{120}\right). \quad \text{(V.17)}$$

The result for the spin-fluctuation contribution is

$$a_1^{SF} = a_1^{wc}\left(\frac{\delta}{10}\right)\left(\frac{T}{T_c}\right) \quad \text{(V.18)}$$

and the spin-fluctuation a's are such that $a_2 = 0{\cdot}5a_1$; $a_3 = 7{\cdot}0a_1$; $a_4 = -2{\cdot}0a_1$, and $a_5 = 5{\cdot}5a_1$. We can then write the total fourth-order free energy as (setting $T = T_c$)

$$F^4 = \frac{|a_1^{wc}|}{2}\{-(1{\cdot}0 + 0{\cdot}1\delta)|\sum_{\alpha i} d_{\alpha i}^2|^2 + (2{\cdot}0 - 0{\cdot}05\delta)d^*_{\alpha i}d^*_{\alpha j}d_{\beta i}d_{\beta j}$$

$$-(2{\cdot}0 + 0{\cdot}7\delta)d^*_{\alpha i}d^*_{\beta i}d_{\alpha j}d_{\beta j} + (2{\cdot}0 + 0{\cdot}2\delta)(\sum_{\alpha i}|d_{\alpha i}|^2)^2$$

$$+(2{\cdot}0 - 0{\cdot}55\delta)d^*_{\alpha i}d^*_{\beta j}d_{\alpha j}d_{\beta i}\}. \quad \text{(V.19)}$$

Using this expression along with the expression for the second-order free energy

$$F^2 = (1/3)N(0)\ln\left(\frac{T}{T_c}\right)|d_{\alpha i}|^2 \tag{V.20}$$

allows one to make a number of predictions regarding the thermodynamic properties near T_c. The most important prediction is that for δ greater than $\delta = 0{\cdot}47$ the Anderson–Morel state is the stable state just below T_c. This result is obtained by simply comparing the fourth-order terms, since the normalization of Δ is such that the second-order terms are equal:

$$F^4_{\text{BW}} = \frac{|a_1^{wc}|}{2} 9\Delta^4(\tfrac{5}{3} - 0{\cdot}33\delta) \tag{V.21}$$

$$F^4_{\text{AM}} = \frac{|a_1^{wc}|}{2} 9\Delta^4(2 - 1{\cdot}05\delta). \tag{V.22}$$

Equating these gives $\delta = 0{\cdot}47$.

One can calculate the discontinuity in the specific heat from the free energy. Using the fact that the weak-coupling value of the BW state is the same as the BCS value we find that

$$\frac{\Delta C_v}{C_v^>} = 1{\cdot}43(\tfrac{5}{3})/(2 - 1{\cdot}05\delta) \tag{V.23}$$

for the ABM state. Here $C_v^>$ is the value of the specific heat just above T_c. For the BW state we find

$$\frac{\Delta C_v}{C_v^>} = 1{\cdot}43/(1 - \tfrac{1}{5}\delta). \tag{V.24}$$

At the critical value of $\delta = 0{\cdot}47$ we find $\Delta C_v/C_v^> \simeq 1{\cdot}6$. This appears to be slightly smaller than the experimental value $\approx 1{\cdot}7$. If we assume that the change in $\Delta C_v/C_v^>$ with pressure is due to the change in δ, then we find that $\delta = 0{\cdot}63$ at the melting curve. One can also predict the behaviour of the splitting of the transition in the presence of a magnetic field,(28, 35, 36) as discussed in Section IV. Writing

$$\hat{\Delta}(\mathbf{k}) = (\hat{k}_x + i\hat{k}_y)\begin{pmatrix}\Delta_\uparrow & 0 \\ 0 & \Delta_\downarrow\end{pmatrix},$$

we note that the field-dependent term can be written as

$$F(H) = -N(0)\eta H(\Delta_\uparrow^2 - \Delta_\downarrow^2)$$

and determining the d's in terms of $\Delta_\uparrow$ and $\Delta_\downarrow$ one can calculate the transition temperatures for ordering of the up and down spin pairs respectively. The up spins always order at $T_c + \eta H$ while the down spins order at

$$T_c - \eta H\left(\frac{1-\delta/2}{1+0{\cdot}35\delta}\right).$$

Estimates of $\delta \approx 0{\cdot}6$ lead to a slope of $-0{\cdot}6\eta H$. This effect due entirely to strong coupling has recently been verified in NMR experiments by Osheroff and Anderson.[44] It is important to note that the reason the slope is not minus one is that the spin fluctuations introduce a coupling between the up spin pairs and the down spin pairs so that the down spin pairs order more easily after the up spins have ordered. The coupling induced by spin fluctuations between the up spin and down spin pairs also favours the parallel alignment of the orbital angular momentum of the up and down spins. Thus in weak-coupling theory the $\mathbf{l}$ of the up spins can be arbitrarily directed relative to the down spins. This led to confusion in the early papers and the label Anderson–Morel was used for states with $\mathbf{l}_\uparrow$ and $\mathbf{l}_\downarrow$ parallel or antiparallel with one another. We now regard AM states as the class of states with $\mathbf{l}_\uparrow$ pointing arbitrarily with respect to $\mathbf{l}_\downarrow$ and the A phase is regarded as a particular state of this class with $\mathbf{l}_\uparrow = \mathbf{l}_\downarrow$. This state minimizes the spin-fluctuation energy.

In order to obtain the complete phase diagram of He^3 versus pressure we assume that δ is an increasing function of pressure and first determine the ratio, T_{AB}/T_c, of the transition temperature between the ABM and BW states T_{AB} to T_c versus δ. We can then attempt to determine $\delta(P)$ from the effects discussed above and test the consistency of the theory. To do this, the expression for ΔF^2 in (V.15) is expanded to order $(\Delta)^6$ so that the slope of the T_{AB}/T_c versus δ curve can be calculated near T_c. It turns out that the largest contribution to the slope comes from the fact that the fourth-order spin-fluctuation terms are proportional to $(T\omega_s)^{-1}$ while the weak-coupling terms are proportional to T^{-2}. Therefore, as the temperature is lowered below T_c at the critical value of δ the weak-coupling terms become more important and the BW state becomes more stable. Thus we expect the phase diagram to be roughly what is seen experimentally, i.e., the A phase is suppressed at lower temperatures. By calculating the crossover point at zero temperature $\delta_c(T=0) = 0{\cdot}87$ and its slope as $T \to 0$, i.e.,

$$T_{AB}/T_c = \left(\frac{\delta - 0{\cdot}87}{8}\right)^{1/4},$$

we can plot the T_{AB}/T_c curve shown in Fig. 11.

There are several ways of obtaining experimental estimates of $\delta(P)$. The first is from the specific heat discontinuity on the melting curve which gives $\delta(P = 34\,\text{atm}) = 0{\cdot}63$. This value is probably somewhat large since state-independent strong-coupling corrections may affect the specific heat. Alternatively, one can determine the δ on the melting curve from the slope of the

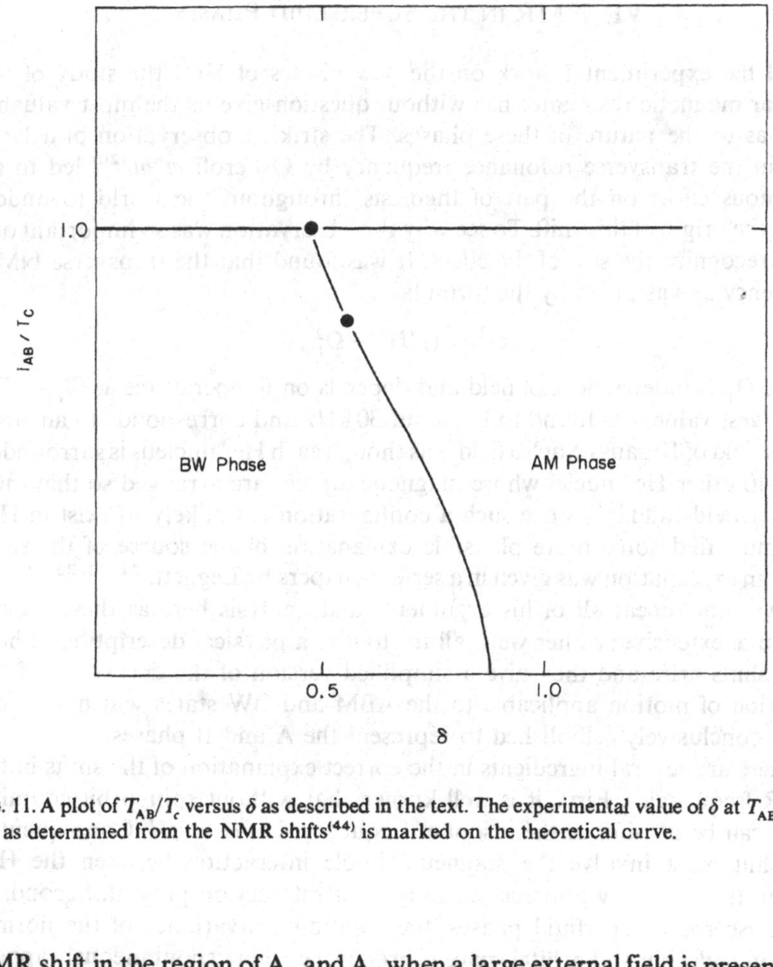

FIG. 11. A plot of T_{AB}/T_c versus δ as described in the text. The experimental value of δ at $T_{AB}/T_c = 0{\cdot}8$ as determined from the NMR shifts[44] is marked on the theoretical curve.

NMR shift in the region of A_1 and A_2 when a large external field is present.[44] This gives $\delta(34\ \text{atm}) = 0{\cdot}54$. This value is used to insert an experimental point for T_{AB}/T_c versus δ in Fig. 11. It is seen that the agreement between theory and experiment is quite good.

It should be noted that given the zero-field phase diagram the various aspects of the phase diagram in the presence of an external field, i.e. the splitting of T_c into two transitions when going into the A phase and the suppression of the B phase at all pressures by an external field, are automatic consequences of the free-energy expansions when the field-dependent terms are included. Therefore, it appears that the effect on the spin fluctuations due to the change in the excitation spectrum in superfluid states does indeed give a complete description of the thermodynamics of the superfluid phases.

VI. NMR in the Superfluid Phases

Of all the experimental work on the new phases of He^3, the study of the nuclear magnetic resonance has without question give us the most valuable clues as to the nature of these phases. The striking observation of a large shift in the transverse resonance frequency by Osheroff *et al.*[20] led to an enormous effort on the part of theorists throughout the world to understand the origin of this shift. To see why this observation was so important one must recognize the size of the effect. It was found that the transverse NMR frequency ω was given by the formula

$$\omega^2 = (\gamma H)^2 + \Omega_T^2 ,$$

where Ω_T is independent of field and depends on temperatures as $(T_c - T)^{\frac{1}{2}}$. Its largest value was found to be about 30 kHz and corresponds to an equivalent field of 10 gauss. Such a field is as though each He^3 nucleus is surrounded by 8–10 other He^3 nuclei whose magnetic dipoles are arranged so that their dipolar fields add.[23] Since such a configuration is unlikely to exist in He^3 one must find some more plausible explanation of the source of the shift. Such an explanation was given in a series of papers by Leggett.[21, 22, 23, 55, 112] We will not repeat all of his arguments and analysis here as these papers are quite extensive; rather we shall try to give a physical description of how such shifts arise and then give a simplified version of the derivation of the equation of motion applicable to the ABM and BW states which are now fairly conclusively established to represent the A and B phases.

There are several ingredients in the correct explanation of the shifts in the NMR frequencies. First, it is well known that without spin–orbit coupling there can be no shift in the resonant frequency, i.e., $\omega = \gamma H$. Consequently, the shift must involve the magnetic dipole interaction between the He^3 nuclei. It is the only appreciable spin–orbit interaction present. Second, in the anisotropic superfluid phases, the rotational invariance of the normal state in *both* spin and orbital space is broken, i.e., there spontaneously appear anisotropy axes such as the **l** and **w** vectors of the A phase. The importance of such broken spin–orbit symmetry was emphasized by Leggett.[23] In such states it is no longer true that the expectation value of the dipolar interaction

$$H_{\mathrm{D}} = \tfrac{1}{2}\gamma^2\hbar^2 \sum_{i \neq j} \left\{ \frac{\boldsymbol{\sigma}_i \cdot \boldsymbol{\sigma}_j}{|r_{ij}|^3} - 3\frac{\boldsymbol{\sigma}_i \cdot \mathbf{r}_{ij}\, \boldsymbol{\sigma}_j \cdot \mathbf{r}_{ij}}{|r_{ij}|^5} \right\}. \qquad \text{(VI.1)}$$

is zero. In fact, as we saw in Section IV, the expectation value of the dipolar interaction depends on the directions of the spin anisotropy axes relative to the orbital axes.

The system therefore establishes itself so that it minimizes the dipolar interaction in a fashion identical to the way a spin system orients along an easy axis. Finally, in spin resonance, the spin axes rotate relative to this minimum energy configuration and the dipolar interaction, therefore, acts as a restoring force and gives rise to a torque in the spin system. The spin system then acts like an oscillator whose "inertia" is the spin susceptibility and one obtains an additional contribution to the resonance frequency

$$\Omega_T^2 \propto \frac{1}{\chi}\langle H_D\rangle/\hbar^2. \tag{VI.2}$$

The expectation value of H_D is proportional to $(\gamma\hbar)/a^3$ times the number of superconducting pairs $(T_c/\varepsilon_F)^2$, so that

$$\Omega_T^2 \approx \frac{\gamma^2}{a^3}\frac{1}{\chi}\left(\frac{kT_c}{\varepsilon_F}\right)^2 \tag{VI.3}$$

which is the order of magnitude of the observed shift. The large value of the shift comes about because of the combination of the first-order dipolar energy being nonzero and the response of the spin system being given by the susceptibility. Since $\chi^{-1} \propto \varepsilon_F$ we have $\Omega_T^2 \propto \varepsilon_F\langle H_D\rangle$, so that the shift is large compared to the dipole energy itself. Obtaining a shift which is the product of a large "exchange" energy (ε_F) and a small anisotropy energy is well known in antiferromagnetic resonance. Indeed, the resonance in the A phase is very analogous to antiferromagnetic resonance[24, 25] as we shall see.

In order to understand the NMR in the various phases we must derive the equations of motion for the spin angular momentum and the anisotropy axes. To do this we first write the dipolar energy in a form which makes the resonance calculation more tractable. Fourier transforming (VI.1) and ignoring the surface-dependent terms we find that

$$H_D = \frac{2\pi}{3}(\gamma\hbar)^2 \sum_{\mathbf{q}} (\boldsymbol{\sigma}_{\alpha\beta}\cdot\boldsymbol{\sigma}_{\gamma\delta} - 3(\boldsymbol{\sigma}_{\alpha\beta}\cdot\hat{\mathbf{q}})(\boldsymbol{\sigma}_{\gamma\delta}\cdot\hat{\mathbf{q}}))a^\dagger_{\mathbf{k}+\mathbf{q}\alpha}a^\dagger_{\mathbf{k}'-\mathbf{q}\gamma}a_{\mathbf{k}'\delta}a_{\mathbf{k}\beta}, \tag{VI.4}$$

where $\hat{\mathbf{q}}$ is a unit vector in the direction of $\mathbf{q}$.

If we write this expression in terms of the anomalous expectation values

$$x_i(\mathbf{k}) \equiv 2\langle a_{-\mathbf{k}\alpha}a_{\mathbf{k}\beta}\rangle(\sigma_y\sigma_i)_{\alpha\beta} \tag{VI.5}$$

and

$$x_i^*(\mathbf{k}) \equiv 2\langle a^\dagger_{\mathbf{k}\alpha}a^\dagger_{-\mathbf{k}\beta}\rangle(\sigma_i\sigma_y)_{\alpha\beta} \tag{VI.6}$$

where the σ are the Pauli matrices we find that

$$H_D = +\frac{24\pi}{3}(\gamma\hbar)^2 \sum_{\mathbf{kk}'} (\mathbf{x}(\mathbf{k})\cdot\mathbf{x}^*(\mathbf{k}') - 3\mathbf{x}(\mathbf{k})\cdot\hat{\mathbf{q}}\mathbf{x}^*(\mathbf{k}')\cdot\hat{\mathbf{q}}) \tag{VI.7}$$

where $\hat{\mathbf{q}} = (\mathbf{k}-\mathbf{k}')/|\mathbf{k}-\mathbf{k}'|$. Now we can go one step further to write for

p-wave pairing

$$x_i(\mathbf{k}) = \Sigma x_{\alpha i}\hat{k}_\alpha, \tag{VI.8}$$

where $x_{\alpha i}$ are assumed to be roughly independent of the $|\mathbf{k}|$ in a region $v_F(k - k_F) = \pm\omega_c$ about the Fermi surface and zero elsewhere.

Doing the integrals over $\hat{\mathbf{k}}$ and $\hat{\mathbf{k}}'$ we find that the terms leading to spin–orbit coupling can be written as

$$\langle H_D \rangle = \Gamma_1(x_{ii}x_{jj}^* + x_{ij}x_{ji}^*) \tag{VI.9}$$

where Γ_1 is a constant of order $(\gamma\hbar)^2\rho^2(\omega_c/\varepsilon_F)^2$. So far we have written the dipolar energy in terms of $x_{\alpha i}$ and not in terms of the $d_{\alpha i}$ related to the gap function $\Delta_{\alpha\beta}(\mathbf{k})$. In weak coupling these two functions are related by a simple factor of $(2\lambda\omega_c)$. Since we are accustomed to thinking in terms of the $d_{\alpha i}$ we will work with these variables and write

$$\langle H_D \rangle = \Gamma(d_{ii}d_{jj}^* + d_{ij}d_{ji}^*) \tag{VI.9$'$}$$

where $\Gamma = \Gamma_1/(2\lambda\omega_c)^2$.

If we add the energy required to induce a net spin angular momentum

$$H_S = (\gamma^2/2)\mathbf{S}\cdot\overleftrightarrow{\chi}^{-1}\cdot\mathbf{S} - \gamma\mathbf{S}\cdot\mathbf{H}, \tag{VI.10}$$

where $\overleftrightarrow{\chi}^{-1}$ is the susceptibility tensor discussed in section IV, we have an energy function which when minimized with respect to the free variables in $d_{\alpha i}$ and the $\mathbf{S}$ gives us the equilibrium configuration for a given static field $\mathbf{H}$. As discussed in Section IV, for the A phase the equilibrium configuration is such that $\mathbf{w}$ is perpendicular to the external field. The dipole energy for this state reduces to

$$-3\Gamma\Delta^2((\mathbf{w}\cdot\mathbf{l})^2 - 1) \tag{VI.11}$$

where Δ^2 is the average of the square of the energy gap. Therefore $\mathbf{w}$ and $\mathbf{l}$ are parallel in equilibrium. They can lie anywhere in the plane perpendicular to $\mathbf{H}$ as shown in Fig. 12. In the BW state the situation is more complicated.

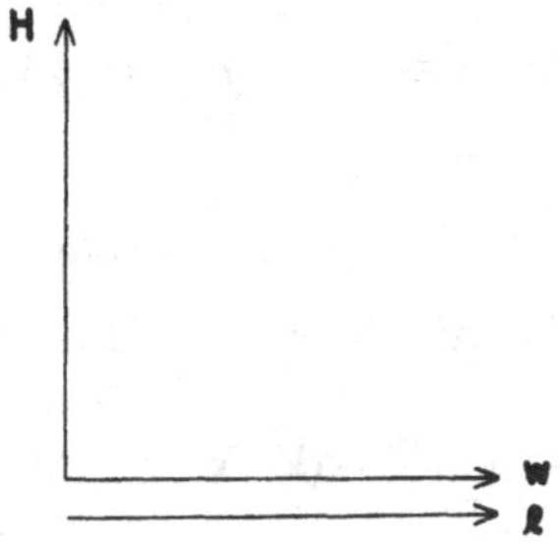

FIG. 12. The orientation of $\mathbf{l}$ and $\mathbf{w}$ in the presence of an external field $\mathbf{H}$ in the A phase.

The $d_{\alpha i}$ can be written as

$$d_{\alpha i} = \sqrt{3}\Delta\hat{O}(\mathbf{n}, \theta) \tag{VI.12}$$

where $\hat{O}(\mathbf{n}, \theta)$ is a rotation matrix specified by the axis of rotation $\mathbf{n}$ and the angle of rotation θ. The dipolar energy is then

$$(H_D)_{\mathrm{BW}} = 4\Gamma\Delta^2(\cos\theta + 2\cos^2\theta). \tag{VI.13}$$

It does not depend on the direction $\mathbf{n}$, but requires $\theta = \cos^{-1}(-\frac{1}{4})$.

Since the susceptibilities are isotropic in this phase there is nothing in (VI.10) to indicate how $\mathbf{n}$ should orient relative to the field. However, as Leggett[112] discussed (see also Engelsberg *et al.*[96]), there is a small term which is due to a combination of the depairing effects of the field acting back into the dipole energy which tends to make $\mathbf{n}$ parallel to $\mathbf{H}$. The equilibrium configuration for the B phase is therefore that shown in Fig. 13. The orbital

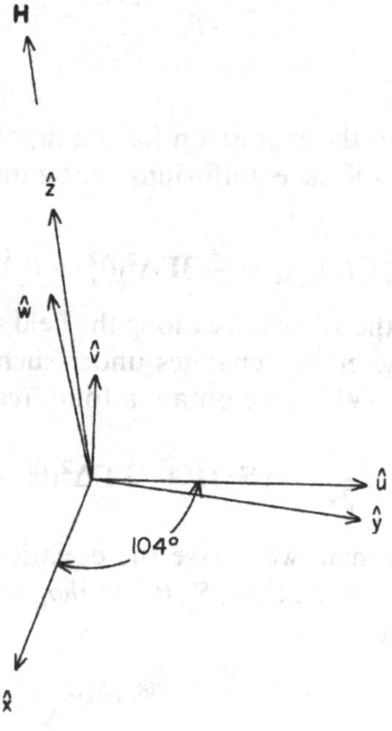

FIG. 13. The equilibrium orientation of the B phase. The spin axes (u, v, w) are rotated relative to the orbital axis (x, y, z) by 104°. The field is along the axis of rotation.

components (x, y, z) are to be paired with the spin components (u, v, w). The (u, v, w) coordinates are rotated by $\cos^{-1}(-\frac{1}{4})$ about the external field.

Now given these equilibrium configurations we desire to obtain the equations of motion determining NMR. Leggett in his work studied the commutation relation between **S** and the $x_i(\mathbf{k})$ defined without the expectation values as in (VI.5, VI.6). He wrote the $x_i(\mathbf{k})$ operator averaged over the magnitude of **k** as $T_i(\hat{k})$ and showed that

$$[S_i, T_j(\hat{k})] = i\varepsilon_{ijk} T_k(\hat{k}). \tag{VI.14}$$

This equation is simply the statement that the S_i act as generators of rotations in spin space. Therefore we derive the equations of motion using the S_i as conjugate variables to the angles describing the rotation of $d_{\alpha i}$ in spin space. For small oscillations, we can consider rotations of the $d_{\alpha i}$ about the three coordinate axes to be independent. Such rotations change $d_{\alpha i}$ to

$$d_{\alpha i}(\theta_x, \theta_y, \theta_z) = d_{\alpha j} R_{ji}(\theta_x, \theta_y, \theta_z),$$

where $\hat{R}(\theta_x, \theta_y, \theta_z)$ is the sum of the small rotations about the three axes

$$\hat{R}(\theta_x, \theta_y, \theta_z) = \begin{pmatrix} 1 - \frac{1}{2}(\theta_y^2 + \theta_z^2) & \theta_z & -\theta_y \\ -\theta_z & 1 - \frac{1}{2}(\theta_x^2 + \theta_z^2) & \theta_x \\ \theta_y & -\theta_x & 1 - \frac{1}{2}(\theta_x^2 + \theta_y^2) \end{pmatrix}. \tag{VI.15}$$

Inserting this result into the expression for the dipolar energy one finds that it is quadratic in the θ's if the equilibrium configuration is correct. In the A phase

$$(H_D)_{\text{ABM}} = +3\Gamma\Delta^2(\theta_x^2 + \theta_z^2), \tag{VI.16}$$

where we have chosen the z axis to be along the field and the y axis to be along **l**. Since only the dipole energy changes under such a rotation, adding this expression to equation (VI.10) we obtain a total free energy

$$E = \frac{\gamma^2 S^2}{2\chi} - \gamma \mathbf{S}\cdot\mathbf{H} + 3\,\Gamma\Delta^2(\theta_x^2 + \theta_z^2). \tag{VI.17}$$

Using this as a hamiltonian we derive the equation of motion of **S** and θ_i by using the commutation relation $[S_i, \theta_j] = i\hbar\delta_{ij}$ and the usual commutator for spin angular momenta

$$[S_i, S_j] = i\hbar\varepsilon_{ijk} S_k. \tag{VI.18}$$

The resulting equations are

$$\frac{\mathrm{d}\mathbf{S}}{\mathrm{d}t} = \gamma\mathbf{S}\times\mathbf{H} - \nu^{-2}\chi\hat{\Omega}^2\cdot\boldsymbol{\theta} \tag{VI.19}$$

and

$$\frac{\mathrm{d}\boldsymbol{\theta}}{\mathrm{d}t} = \gamma(\gamma\mathbf{S}/\chi - \mathbf{H}). \tag{VI.20}$$

Here $\hat{\Omega}^2$ is a dyadic. For the ABM state, it is

$$\left(\frac{6\gamma^2\Gamma\Delta^2}{\chi}\right)\begin{pmatrix}1 & 0 & 0\\ 0 & 0 & 0\\ 0 & 0 & 1\end{pmatrix} = \hat{\Omega}^2. \tag{VI.21}$$

Assuming that $\mathbf{S} = \mathbf{S}_0 + \mathbf{S}_1(t)$ and $\mathbf{H} = \mathbf{H}_0 + \mathbf{H}_{rf}(t)$ and S_1 and H_{rf} oscillate as $e^{i\omega t}$, we obtain the equation for $\mathbf{S}_1$

$$(\omega^2 - \hat{\Omega}^2\cdot)\mathbf{S}_1 - i\omega(\boldsymbol{\omega}_L \times \mathbf{S}_1) = -\gamma^{-1}\chi(\hat{\Omega}^2\cdot\mathbf{H}_{rf} + i\omega\boldsymbol{\omega}_L \times \mathbf{H}_{rf}) \tag{VI.22}$$

where $\boldsymbol{\omega}_L = \gamma\mathbf{H}_0$. This equation is identical to Eqn (5.7) of Leggett.[112] The relation between the matrix $\hat{\Omega}^2$ and the expansion of the dipolar energy in terms of the rotation angles is made transparent by our derivation. We have

$$\hat{\Omega}_{ij} = \frac{\gamma^2}{\chi}\frac{\partial^2 H_D}{\partial\theta_i\partial\theta_j}. \tag{VI.23}$$

Equation (VI.22) can now be used to calculate the dynamic susceptibilities defined as the solution to this equation

$$(S_1)_i = \gamma^{-1}\chi_{ij}H_{rfj}.$$

These susceptibilities have been discussed in detail by Leggett[112] and we will not discuss them here. Rather we concentrate on the frequencies of the normal modes and simply state their polarization. The modes are determined by the vanishing of the determinant of the coefficients of the S_1 on the left hand side of (VI.22). Returning to the particular form for the ABM state we find that the transverse resonance frequency is

$$\omega^2 = \omega_L^2 + \Omega_T^2 \tag{VI.24}$$

where $\Omega_T^2 = 6\gamma^2\Gamma\Delta^2/\chi$. Thus the shift in the transverse resonance Ω_T^2 varies as Δ^2. As discussed at the beginning of this section it is also of the right order of magnitude. The interesting feature of (VI.22) is that the equation for S_{1z} is

$$(\omega^2 - \Omega_T^2)S_{1z} = -\gamma^{-1}\chi\Omega_T^2 H_{rfz} \tag{VI.25}$$

so that there is a longitudinal resonance in this state whose frequency is Ω_T^2, i.e., exactly equal to the transverse resonance shift. The existence of this longitudinal resonance has recently been confirmed experimentally (Osheroff and Brinkman,[26] Bozler *et al.*,[113] and Webb *et al.*[97]).

Turning now to the B phase we need only calculate the $\hat{\Omega}^2$ tensor. With the equilibrium configuration discussed above we find that only the $\hat{\Omega}^2_{zz}$ component is nonzero and it is equal to

$$(6\gamma^2\Gamma\Delta^2/\chi_B)(5/2) = \hat{\Omega}^2_{zz} = \Omega_L^2. \tag{VI.26}$$

Substituting into (VI.22) we find that, in equilibrium, the B phase shows no shift in the transverse resonance, i.e., $\omega = \gamma H$, but that it should show a longitudinal resonance at Ω_L. The frequency of this resonance written in terms of the A phase resonance is

$$\hat{\Omega}^2_{zz} = (5/2)\frac{\chi_A}{\chi_B}\left(\frac{\Delta^2_B}{\Delta^2_A}\right)\Omega^2_T, \qquad \text{(VI.27)}$$

where χ_A/χ_B is the ratio of the susceptibilities in the A and B phases and (Δ^2_B/Δ^2_A) the ratio of the average of the square of their respective gaps. This relationship has been verified experimentally by Osheroff[27] to within 5%.

Work by Webb *et al.*[97, 98] does not appear to give this agreement. There are, however, difficulties in observing the resonance in the B phase because of texture effects which we will discuss later.

The texture effects in the B phase will involve the direction of the axis of rotation **n** relative to the external field. We can calculate the resonance frequencies by calculating the $\hat{\Omega}^2$ matrix under circumstances where **n** makes an angle ϕ with respect to the external field. The result of this calculation is that one obtains two resonance frequencies. These are plotted versus the magnitude of the external field, in Fig. 14. At small fields there are two modes, one at Ω_L and the other at $\gamma H \cos\phi$. These modes are longitudinal and transverse (with respect to **n**) respectively. Therefore in this situation there is a mode

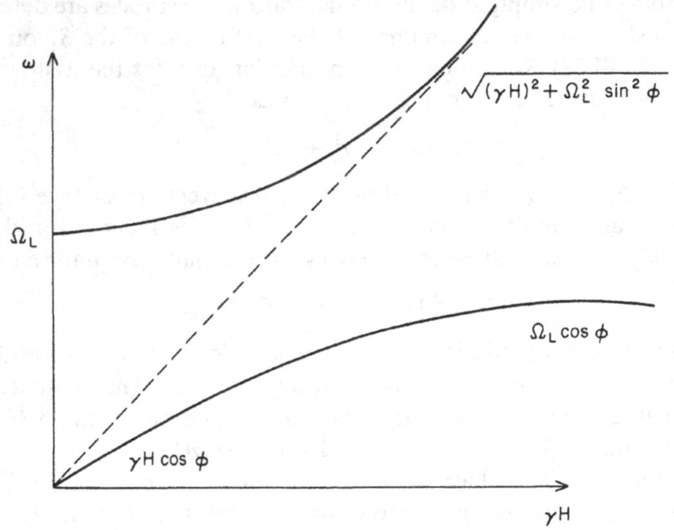

FIG. 14. A plot of the resonance frequencies as a function of field when the axis of rotation makes an angle ϕ with the external field.

with a component along the external field whose frequency is linear in H. At large fields $\gamma H \gg \Omega_L$ we have modes at

$$\omega = ((\gamma H)^2 + \Omega_L^2 \sin^2 \chi)^{\frac{1}{2}} \tag{VI.28}$$

and

$$\omega = \Omega_L \cos \chi, \tag{VI.29}$$

and these modes are polarized transversely and longitudinally with respect to the external field. The transverse-mode expression will be important in describing resonance line shape in situations where the texture of **n** is important.

Recently, Osheroff and Anderson[44] have investigated carefully the transverse shift in large fields where T_c splits into two transitions A_1 and A_2. They find that the shift is proportional to the square of the sum of the up-spin order parameter and the down-spin order parameter, $(\Delta_\uparrow + \Delta_\downarrow)^2$. As already mentioned, they were able to accurately predict the resonant frequency using the spin-fluctuation model to obtain the fourth-order free energy. Their results are shown in Fig. 15.

Another interesting result of the Leggett equations is the prediction of nonlinear ringing associated with the longitudinal resonance (Leggett[112], Maki and Tsuneto[60] and Webb *et al.*[98]). Consider the A phase and assume only θ_z and S_z to be nonzero. Then the total energy is

$$E = \frac{S_z^2}{2\chi} + \frac{\chi \Omega_T^2}{2} \sin^2 \theta_z. \tag{VI.30}$$

Since θ_z and S_z are conjugate variables, the energy expression is that of a pendulum with θ_z the angle the pendulum makes with respect to the vertical axis and S_z its angular momentum. The susceptibility acts like a moment of inertia. In the experiments to probe this effective hamiltonian one starts with an external field on, so that $S_z = \chi H$ and $\theta_z = 0$. One then suddenly turns off the field leaving the system with a "kinetic energy" $S_z^2/2\chi = \frac{1}{2}\chi H^2$. The system will then ring at a frequency determined by the amount of energy left in the system. Clearly, this frequency will become zero when the "kinetic energy" is such that the pendulum just swings into the upside down position. This occurs when $S_z^2/2\chi = \chi\Omega_T^2/2$ or $\gamma H = \Omega_T$. This was confirmed in the recent experiments by Webb *et al.*[97,98]

The B phase is more complicated. When **n** is along **H** the θ in (VI.13) is the same as θ_z, so that the energy is

$$\frac{S_z^2}{2\chi} + \chi\Omega_L^2(\tfrac{4}{15})(\cos\theta + 2\cos^2\theta). \tag{VI.31}$$

The equilibrium value is at $\theta = \cos^{-1}(-\frac{1}{4})$.

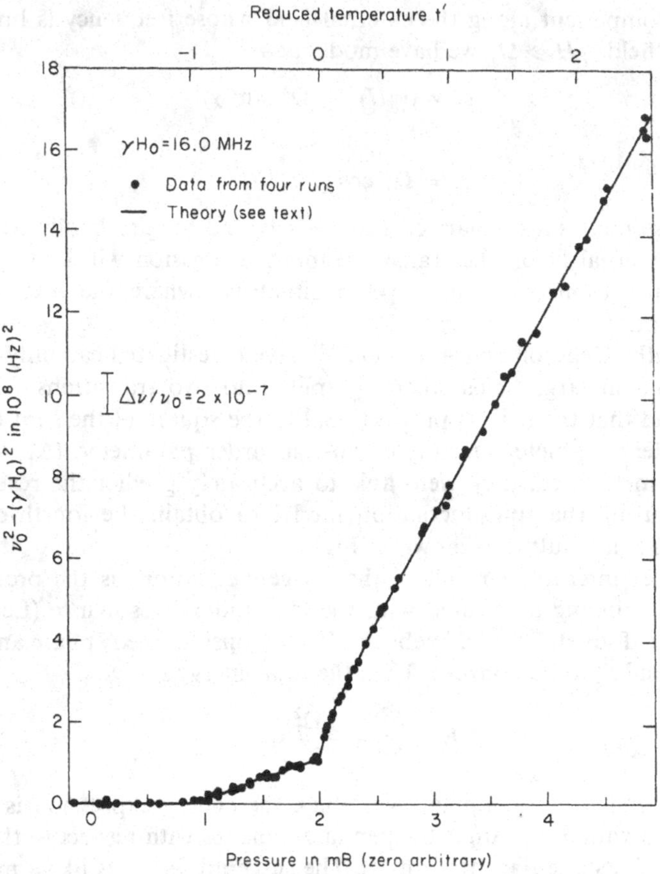

FIG. 15. The transverse resonance in the region of A_1 and A_2 as measured by Osheroff.[44]

The potential energy has two maxima in this case at $\theta = 0$ and π. These maxima should give rise to zeros in the ringing frequency at $(\gamma H/\Omega_L)^2 = \frac{3}{5}$ and $\frac{5}{3}$. It is not yet clear whether this prediction is correct or not.

The confirmation of all the above results leaves little question as to the validity of the theory of Leggett or that the two phases are any other than the ABM and BW states as first suggested by Anderson.[24] Several authors have tried to derive the expressions for the susceptibilities from microscopic theory (Maki and Ebisawa,[56, 114, 115] Takagi[116]) but have apparently enjoyed only partial success. We will consequently not attempt to review these approaches to the problem.

One aspect of the problem we have ignored in our discussion is the question of the role of the anisotropic susceptibilities in the calculation of the resonant frequencies. This question was discussed extensively by both Leggett[112] and Engelsberg *et al.*[96] The essential point in the discussion is that the S_i, as generators of rotation in spin space, rotate the susceptibility tensor as well as the magnetization and, consequently, the anisotropic susceptibilities do not appear in the equation of motion for **S**. They do appear in the equations for θ_i but do not affect the resonant frequencies.

It should be noted before we close this section that Combescot and Ebisawa[117] have given a rather successful theory of NMR line widths which gives a semiquantitative fit to Osheroff's longitudinal resonance data. Unfortunately we have no space to describe it here.

In concluding this section we should point out the relationship between the angles θ_i used in the present discussion and the Leggett discussion of the longitudinal resonance in terms of the Josephson effect. If one does a rotation of θ_z about the external field in the A phase the up-spin order parameter changes its phase by θ_z and the down-spin order parameter by $-\theta_z$. Therefore, the phase difference between the up-spin and down-spin superfluids is $2\theta_z$. In the longitudinal resonance the dipole interaction acts as a weak coupling between the two Fermi systems. The equation for S_z is then the equation for the current between the two systems. The equation for θ_z is the equation for the phase difference. In it the γH acts as a potential difference between the two spin systems and, in general, the right-hand side of (VI.20) is the difference in chemical potential. This description of the equations of motion is undoubtedly closer to what is actually occurring than the classical approach presented here. The classical approach offers the advantage of exhibiting the full symmetry of the problem and a rather simple visualization of the modes.

VII. Superflow and Spatial Variation: Generalized Ginzburg–Landau Approach

1. Generalized Ginzburg–Landau Theory

(a) *Ginzburg–Landau equation*

In Section IV we summarized the results of the BCS approach to superflow properties as carried through by Leggett[13, 14, 106] and Saslow.[45] The net result was expressions for ρ_n near equilibrium, which we expected to be exact and which turned out to be anisotropic in the ABM state. Assuming the existence of a uniquely defined v_s, Galilean invariance then defines $\rho_s = \rho - \rho_n$ for us, again as an anisotropic tensor quantity. All of this was subject to straightforward Fermi-liquid corrections. In a final section we carried the problem of superflow a bit farther, setting up a hydrodynamics of the normal fluid in sufficient detail to derive second sound according to Wölfle.[32] All of this was limited by the v_s assumption, as well as leaving out certain important relations which are not obvious in the perturbation approach.

In the Ginzburg–Landau theory of superconductivity the superflow properties are intimately related to the system's response to spatial variations, i.e. to an expansion of the free energy in terms of the gradient of the order parameter Ψ. One writes the free energy

$$F = \tfrac{1}{2}F_0(|\Psi|^2, T) + \tfrac{1}{2}F_{II}(|\Psi^2, T)(\nabla\Psi)^2 + \ldots. \tag{VII.1}$$

In the original Ginzburg–Landau theory,[118] as justified from BCS by Gor'kov,[119] F_0 was in turn expanded in a power series in $|\Psi|^2$, the first term being $a(T - T_c)|\Psi|^2$, while the coefficient F_{II} was a constant. This is useful only near T_c, and, ignoring F_{II}, is the expansion in $|\Delta|^2$ we repeatedly used in Section IV. Werthamer[120, 90] and Eilenberger[88] have shown that expansions in $\nabla\Psi$ such as (VII.1) have a general validity when gradients are small compared to $(\xi_0)^{-1}$; this is the "Generalized Ginzburg–Landau" approach.

The term F_{II} not only controls the spatial fluctuations of the order parameter, it is within a factor the superfluid density ρ_s as well. This is clear if we write

$$\mathbf{v}_s = \frac{\hbar}{2m}\nabla\varphi = \frac{\hbar}{2im}\frac{\nabla\Psi}{\Psi} \tag{VII.2}$$

and then clearly if

$$F_s = \frac{\rho_s v_s^2}{2}, \tag{VII.3}$$

$$\rho_s = F_{II}\frac{4m^2}{\hbar^2}|\Psi|^2. \tag{VII.4}$$

The basic point here is one of gauge invariance. If we introduce a particle charge q and a vector potential, the true one if we were speaking of electrons or a fictitious "baryon field" for neutral He atoms, gauge invariance tells us that F_{II} must multiply $|\nabla - (2q/i\hbar_c)\mathbf{A}|^2$, not $|\nabla|^2$, while the current must be defined as

$$\mathbf{j} = \frac{c}{q}\frac{\delta F}{\delta \mathbf{A}}. \tag{VII.5}$$

Hence the identity (VII.4).

De Gennes[50, 51] first pointed out that in the case of the complicated multidimensional order parameter of anisotropic superfluidity, a great deal of new physics could be obtained from the Ginzburg–Landau gradient expansion. A slightly more formal theory spelling out more aspects of the physics followed from Ambegaokar, de Gennes and Rainer.[52] Certain formal aspects of de Gennes' approach seem to us to be questionable and to have been more convincingly handled near T_c by Wölfle,[53] while the most complete treatment is that of Blount, Anderson, and Brinkman[54, 25] which will be presented here. It is important to realize that in this case not only the supercurrent controls macroscopic (as opposed to fluctuation) phenomena; the angular elastic stiffness which follows from F_{II}-type terms is also of importance in many phenomena, especially NMR.[61]

We will specialize from the start to the $l = 1$ case, although most of the computation could be carried out in general. Here we have the gap bivector $d_{\alpha i}$ defined by

$$\Delta_{\sigma\sigma'}(\mathbf{k}) = \sum_{\alpha i} \hat{k}_\alpha d_{\alpha i}(\sigma_2 \sigma_i)_{\sigma\sigma'}. \qquad \text{(VII.6)}$$

As we have already explained, the major terms in the uniform free energy F_0 are invariant to rotations of both spin and orbit variables, so we define "states" in which $d_{\alpha i}$ can be reduced by rotation to certain canonical forms: the important ones being the BW

$$\text{BW:} \quad d_{\alpha i} = \begin{pmatrix} d & 0 & 0 \\ 0 & d & 0 \\ 0 & 0 & d \end{pmatrix} \qquad \text{(VII.7)}$$

and the ABM

$$\text{ABM:} \quad d_{\alpha i} = \begin{pmatrix} 0 & 0 & d \\ 0 & 0 & id \\ 0 & 0 & 0 \end{pmatrix}. \qquad \text{(VII.8)}$$

In order to minimize dipolar energy, the relative spin and orbit orientation must be fixed in such a way that (VII.8) remains as it is while (VII.7) becomes

$$d\begin{pmatrix} \frac{1}{4} & -\sqrt{\frac{15}{4}} & 0 \\ \sqrt{\frac{15}{4}} & \frac{1}{4} & 0 \\ 0 & 0 & 1 \end{pmatrix}. \qquad \text{(VII.9)}$$

Let us first, following de Gennes[50, 51] and Wölfle,[53] make the standard Ginzburg–Landau expansion in both d and ∇d. In this case the F_{II} term is quadratic in d and its form is severely restricted by symmetry. The relevant energy terms are, like those responsible for F_0, separately spin and orbit invariant, and thus there are only three possible terms:

$$F_1 = \sum_{\alpha\beta} \sum_i |\nabla_\alpha d_{\beta i}|^2 \qquad \text{(VII.10)}$$

$$F_2 = \sum_i |\nabla_\alpha d_{\alpha i}|^2 = |\nabla \cdot \mathbf{d}|^2 \tag{VII.11}$$

$$F_3 = \sum_{\alpha\beta} \sum_i (\nabla_\alpha d^*_{\beta i})(\nabla_\beta d_{\alpha i}), \tag{VII.12}$$

and one could write $(F)^{\text{II}} = K_1F_1 + K_2F_2 + K_3F_3$. Note that in each case the form is a simple dot product on the spin index; as in (VII.11) we will often omit the spin index which is always simply summed over, and treat $\hat{d}$ as a vector in its first orbital index α.

De Gennes[51] and Ambegaokar, de Gennes and Rainer[52] follow liquid-crystal practice in using instead of (VII.12) the combination

$$(F_3)_{\text{DG}} = |\nabla \times \mathbf{d}|^2 = F_1 - F_3 \tag{VII.13}$$

as the third invariant. In principle this is equivalent but in practice leads to confusion.

From (VII.10) to (VII.12) one can construct one total divergence

$$F_3 - F_2 = \nabla \cdot [\mathbf{d}^* \cdot \nabla \mathbf{d} - \mathbf{d}^* \nabla \cdot \mathbf{d}]. \tag{VII.14}$$

In the total free energy such a term can lead only to a surface free energy, and since there are other, rather large sources of surface energies (VII.14) might as well be omitted from the free energy expression.

In particular, from the free energy there follows the Ginzburg–Landau differential equation for the spatial variation of the gap parameter, which is all-important in problems of textures, singularities, etc., as we shall see, and in this equation the form (VII.14) drops out upon partial integration. Thus for these purposes there are only two independent coefficients and one may use the form given by de Gennes

$$(F_{\text{DG}})^{\text{II}} = \tfrac{1}{2}K_L F_2 + \tfrac{1}{2}K_T(F_1 - F_3) \tag{VII.15}$$

instead of the form which is natural upon eliminating (VII.14) from (VII.10) to (VII.12)

$$(F)^{\text{II}} = \tfrac{1}{2}K_0 F_1 + \tfrac{1}{2}K_{\text{A}}(F_2 + F_3). \tag{VII.16}$$

Since K_0 and K_{A} are more directly related to the supercurrents (as we shall see) we will use (VII.16). From (VII.16) we may derive the equilibrium "Ginzburg–Landau" equation for the gap bivector, $(\delta F/\delta d_{\alpha i})\delta d_{\alpha i} = 0$. We introduce $F_0(|d|^2)$ which is the free energy (IV.50) as a function of a uniform gap, and obtain for the coefficient of $\delta d_{\alpha i}$

$$\frac{\delta F_0}{\delta d_{\alpha i}} - K_0 \nabla^2 d^*_{\alpha i} - 2K_{\text{A}} \nabla_\alpha (\nabla \cdot \mathbf{d}^*_i) = 0 \tag{VII.17}$$

[since the form $\nabla(\nabla \cdot \mathbf{d})$ results from either F_2 or F_3, de Gennes' form (VII.15) would lead to the same equation with $K_A = K_L - K_T$, $K_0 = K_T$, but we repeat that K_L and K_T are not correct coefficients for the currents].

(b) "*Ordinary*" *supercurrents*

From (VII.10) to (VII.12) one can derive three formal expressions for supercurrents by introducing the gauge field $\mathbf{A}_B$ and setting

$$\nabla \to \nabla - \frac{2q}{i\hbar c}\mathbf{A}_B$$

and finally using (VII.4). (Equally, one could do this by doing a uniform Galilean transformation, setting

$$d \to d \exp\left(i\frac{2m\mathbf{v}_s}{\hbar}\cdot\mathbf{r}\right)$$

and $\mathbf{j}_s = \delta F/\delta \mathbf{v}_s$, i.e. the terms multiplying $\nabla\varphi$ where φ is the phase of d.) We obtain thence

$$\mathbf{J}_1 = \frac{2i}{\hbar}[(d^*_{\alpha i}\nabla d_{\alpha i}) - \mathrm{C.C.}] \tag{VII.18}$$

$$\mathbf{J}_2 = \frac{2i}{\hbar}[\mathbf{d}^*(\nabla\cdot\mathbf{d}) - \mathrm{C.C.}] \tag{VII.19}$$

$$\mathbf{J}_3 = \frac{2i}{\hbar}[(\mathbf{d}^*\cdot\nabla)\mathbf{d} - \mathrm{C.C.}] \tag{VII.20}$$

As Blount pointed out, the fact that $F_2 - F_3$ is a divergence does not prevent us from using the general *formal* expression for F, nor can we see any unique way to choose a free energy "orthogonal" to $F_2 - F_3$. Thus *a priori* any three coefficients for (VII.18) to (VII.20) are compatible with symmetry, and we could write

$$\mathbf{J} = j_1\mathbf{J}_1 + j_2\mathbf{J}_2 + j_3\mathbf{J}_3$$

where (VII.16) merely enforces the conditions

$$j_1 = \tfrac{1}{2}K_0$$

$$j_2 + j_3 = \tfrac{1}{2}K_A. \tag{VII.21}$$

Nonetheless we (as well as Wölfle[53]) find from microscopic calculations, in accord with the intuitive (VII.16), that the major portion of the supercurrent is given by

$$\mathbf{J} = \tfrac{1}{2}[K_0\mathbf{J}_1 + K_A(\mathbf{J}_2 + \mathbf{J}_3)]. \tag{VII.22}$$

Given that the form of the supercurrent is (VII.22), the coefficients K_0 and K_A could be determined simply from Saslow's calculation[45] of the anisotropy of ρ_s in the ABM state, since that determines two independent coefficients. Wölfle[53] in this limit, and Blount for general T, have calculated the free energy and current coefficients (without Fermi-liquid corrections, which we will discuss shortly). Blount's method was a generalization of the Eilenberger[88] scheme which we discussed in Section IV and which was used by him to derive the generalized Ginzburg–Landau theory. His result is

$$J_\alpha = \frac{\hbar\rho}{2m} T_{\alpha\beta\gamma\delta} \frac{1}{2i} (d^*_{\beta i} \nabla_\gamma d_{\delta i} - \text{C.C.}) \tag{VII.23}$$

where the coefficient $T_{\alpha\beta\gamma\delta}$ is given, to lowest order in T_c/E_F, by

$$T_{\alpha\beta\gamma\delta} = \frac{3}{4\pi} \int \mathrm{d}\Omega \frac{f_s(\hat{\mathbf{k}})}{|\Delta|^2} \hat{k}_\alpha \hat{k}_\beta \hat{k}_\gamma \hat{k}_\delta. \tag{VII.24}$$

where

$$f_s(\hat{\mathbf{k}}) = \Delta^2(\hat{\mathbf{k}}) \int \frac{\mathrm{d}\varepsilon_{\mathbf{k}}}{E(\mathbf{k})} \frac{\mathrm{d}}{\mathrm{d}E_{\mathbf{k}}} \left(\frac{\tanh(\beta E_{\mathbf{k}}/2)}{E_{\mathbf{k}}} \right). \tag{VII.25}$$

In the Ginzburg–Landau region proper, $T \simeq T_c$,

$$T_{\alpha\beta\gamma\delta} \simeq \frac{7\zeta(3)}{20\pi^2 T_c^2} (\delta_{\alpha\beta}\delta_{\gamma\delta} + \delta_{\alpha\gamma}\delta_{\beta\delta} + \delta_{\alpha\delta}\delta_{\gamma\beta}). \tag{VII.26}$$

Comparing (VII.26) with (VII.23) and (VII.21), we find that

$$K_0 = K_A = \frac{\hbar^2 \rho}{m^2} \frac{7\zeta(3)}{80\pi^2 T_c^2}. \tag{VII.27}$$

To this order Blount finds that (VII.23) and (VII.24) hold at general temperatures, but of course in general (VII.24) is anisotropic. So long as we maintain axial symmetry, however, the general form of δ-function products in (VII.24) is still maintained if one calculates in the axis system of $\hat{d}$, and very similar current expressions are retained. For BW, where isotropy of Δ continues to hold, the expressions (VII.22), (VII.16) and $K_0 = K_A$ are correct for all T.

(c) *Balian–Werthamer: Complete expressions*

The BW case, because of its isotropy and because its gap function is essentially real, is a simple case. The general gap function is

$$d_{\alpha j} = d e^{i\phi} O_{ij}(\mathbf{n}, \theta) \delta_{\alpha i} \tag{VII.28}$$

where O_{ij} is the orthogonal matrix for rotation by θ about the axis $\mathbf{n}$.

The current terms $\mathbf{J}_1 - \mathbf{J}_3$ all lead to the same current:

$$\mathbf{J}_1 = 3\mathbf{J}_2 = 3\mathbf{J}_3 = \frac{4}{\hbar}\,\mathrm{Tr}\,(\hat{d}^2)\nabla\varphi \qquad \text{(VII.29)}$$

(d and O_{ij} are real, hence gradients of them cancel in $\mathbf{J}$; we also use the orthogonality of $\hat{O}$). Note that $\mathbf{J}_2 - \mathbf{J}_3 \equiv \mathbf{0}$ so that there the ambiguity of K's does not exist. We can then identify $K_0 = K_A = \frac{1}{5}\rho_s \times$ consts.. where ρ_s is the superfluid density already calculated in Section IV, since the current consists entirely of ordinary supercurrent with a well-defined $\mathbf{v}_s$ as assumed there.

Here the treatment of Fermi-liquid corrections requires some discussion. So far as we know, no rigorous treatment of Fermi-liquid corrections to the Ginzburg–Landau equation, as opposed to those to normal supercurrents, exists. From two rather qualitative but suggestive arguments we would suggest that K_0 and K_A for the free energy (VII.16) and for the current (VII.22) must remain identical even after Fermi-liquid corrections. First, the expressions F_1, F_2, F_3 were derived as rotational invariants and the Fermi-liquid corrections bring in no rotational asymmetries so should maintain their rotationally invariant structure; gauge invariance of course connects some of the F_i structure to the J_i. Second, the form of the Fermi-liquid correction is

$$\rho_s = \frac{\rho_s^0}{1 + (F_1/3)(\rho_n^0/Nm^*)} = \rho_s^0\left(1 - \frac{(F_1/3)N_n}{1 + (F_1/3)N_n}\right), \qquad \text{(VII.30)}$$

where ρ_s^0 is the superfluid density calculated in weak coupling and N_n is an effective relative density of normal quasiparticles. This suggests that we think heuristically of superfluid kinetic energies being reduced by a fraction for each normal particle: i.e. of a kind of "obstruction" effect by the normal fraction. Such observations should work equally well to reduce kinetic energies depending on ∇O_{ij} as those depending on $\nabla\varphi$. Thus in (VII.16) we will use the Fermi-liquid corrected K's, proportional to ρ_s as given in Section IV.3, recognizing however that this may be incorrect.

Brinkman *et al.*,[61] have given the form of F^{II} terms which results upon substitution of (VII.29) into (VII.16) (omitting even more terms which come from $\nabla\theta$ and ∇d)

$$\begin{aligned} F^{\mathrm{II}}_{\mathrm{BW}} = \frac{\hbar^2\rho_s}{8m^2}(\nabla\varphi)^2 + \frac{\hbar^2\rho_s}{20\,m^2}(1 - \cos\theta)[4(\mathbf{n}\times\nabla\times\mathbf{n})^2 \\ + (3 + \cos\theta)(\mathbf{n}\cdot\nabla\times\mathbf{n})^2 + (3 - \cos\theta)(\nabla\cdot\mathbf{n})^2 \\ - 2\sin\theta\,(\nabla\cdot\mathbf{n})(\mathbf{n}\cdot\nabla\times\mathbf{n})]. \qquad \text{(VII.31)} \end{aligned}$$

If we specialize as Brinkman *et al.* did to $\cos\theta = -\frac{1}{4}$, which will usually be fairly accurately fixed, we obtain the still complicated expression which will be used in BW texture problems:

$$F = \frac{\hbar^2 \rho_s}{8m^2}\left\{2(\nabla \times \mathbf{n})^2 + \tfrac{13}{4}(\nabla \cdot \mathbf{n})^2 - \frac{\sqrt{15}}{4}(\mathbf{n}\cdot\nabla\times\mathbf{n})(\nabla\cdot\mathbf{n}) - \tfrac{5}{8}(\mathbf{n}\cdot\nabla\times\mathbf{n})^2 \right.$$

$$\left. + 16\nabla\cdot((\mathbf{n}\cdot\nabla)\mathbf{n} - \mathbf{n}(\nabla\cdot\mathbf{n}))\right\}. \qquad \text{(VII.32)}$$

(d) *ABM: Orbital currents and orbital supercurrents*

Complicated as (VII.31) may appear, it pales beside the real complications of the ABM case. In that case, entirely aside from the difficulties of anisotropy we have those which result from the finite angular momentum of the pairs. The first difficulty is one of principle: the ambiguity of the $F_2 - F_3$ term. The fact that $F_2 - F_3$ is a divergence leads automatically to the conclusion that $\mathbf{J}_2 - \mathbf{J}_3$ is divergenceless,

$$\nabla\cdot(\mathbf{J}_2 - \mathbf{J}_3) = 0, \qquad \text{(VII.33)}$$

and in fact by actual expansion one can see that $J_2 - J_3$ is a curl:

$$\mathbf{J}_2 - \mathbf{J}_3 = \frac{i}{\hbar}\nabla\times(\mathbf{d}^*\times\mathbf{d})$$

$$= \frac{i}{\hbar}\nabla\times\mathbf{l} \qquad \text{(VII.34)}$$

where have introduced the vector $\mathbf{l}$ of "angular momentum of the pairs", which is of course *not* equal to $\langle L\rangle$.

Since $(F_2 - F_3)$ does not enter the Landau–Ginzburg equation nor, presumably, any hydrodynamic equations of motion, the corresponding current must be conserved automatically rather than as a consequence of the configuration; this is the physical meaning of (VII.33). This type of current is precisely analogous to the magnetization current $\mathbf{J}_M = \nabla\times\mathbf{M}$ of a ferromagnet: it transports nothing and follows simply as a consequence of the configuration. In particular, it can have no coupling with the conventional superflow transport processes, since transport implies a coupling with a source or sink, i.e. an energy term proportional to $\nabla\cdot\mathbf{J}$. In the ferromagnetic case there is an appropriate field with which it interacts; here, there is only its inertia, which is, however, all-important to the frequencies of orbit waves, as pointed out by Anderson[24, 25] and Wölfle,[53] just as the angular momentum of a ferromagnet is crucial to its spin waves.

Thus we must identify the coefficient of the current $(\mathbf{J}_2 - \mathbf{J}_3)$ with the "intrinsic" orbital angular momentum. We emphasize that this identification is neither rigorous nor unambiguous, since it is easy to invent configurations in which **d** varies in such a way as to give the sample as a whole a total angular momentum from the ($\mathbf{J}_2 + \mathbf{J}_3$) term or even the $\mathbf{J}_1$ one. (One of these may in fact be a common texture in a tube.) Nonetheless, if the concept of an $\langle L \rangle$ has a meaning, it must be this. As we have already seen ((VII.23–VII.25), to lowest order in Δ/E_F this coefficient vanishes; Anderson and Morel estimated that in fact the coefficient was $\propto \Delta/E_F$ (see Section IV) but more recent estimates by Blount and by Cross place it one order lower. These estimates were direct calculations of the coefficient K_l of $\mathbf{J}_2 - \mathbf{J}_3$ in

$$\mathbf{J} = \tfrac{1}{2}K_0\mathbf{J}_1 + \tfrac{1}{2}K_A(\mathbf{J}_2 + \mathbf{J}_3) + \tfrac{1}{2}K_l(\mathbf{J}_2 - \mathbf{J}_3)$$

$$K_l \simeq \frac{T_c^2}{E_F^2}K_0. \tag{VII.35}$$

But even omitting the effects of $\mathbf{J}_2 - \mathbf{J}_3$ as small, the K_A term has rather interesting properties in that unlike (VII.34) the current depends on gradients of direction angles as well as of φ. When F is expanded in terms of gradients of angles and $\nabla\varphi$, there are cross-terms and as a result there are what de Gennes calls "orbital supercurrents." As opposed to the orbital current ($\mathbf{J}_2 - \mathbf{J}_3$), these currents are just as big as normal supercurrents. Near T_c these have been correctly calculated by Wölfle and by Blount, and the structure of the current may be simply summarized as follows:

$$\mathbf{J} = \overleftrightarrow{\rho}_s\mathbf{v}_s + \overleftrightarrow{C}\frac{\hbar\nabla}{2m} \times \hat{\mathbf{l}} \tag{VII.36}$$

where

$$\frac{\overleftrightarrow{\rho}_s}{\rho} = \begin{pmatrix} 2 & 0 & 0 \\ 0 & 2 & 0 \\ 0 & 0 & 1 \end{pmatrix} \cdot \frac{7\zeta(3)}{10\,\pi^2}\frac{\langle\Delta^2\rangle}{T_c^2}\frac{m}{m^*}$$

and

$$\overleftrightarrow{C} = \tfrac{1}{2}\begin{pmatrix} 1 & 0 & 0 \\ 0 & 1 & 0 \\ 0 & 0 & -1 \end{pmatrix} \cdot \frac{7\zeta(3)}{10\,\pi^2}\frac{\langle\Delta^2\rangle}{T_c^2}\frac{m}{m^*}.$$

In each case the z-axis is taken parallel to **l**. The $\overleftrightarrow{C}$ term is the orbital supercurrent term; it never vanishes if $\nabla \times \mathbf{l}$ is finite but points in a different direction in general.

Blount has carried the current calculations to absolute zero within the weak-coupling approximation. Although, as we remarked in Section IV.2, at

$T = 0$ $\rho_s = \rho$, which is isotropic, the orbital supercurrent does not vanish but still retains exactly the same form:

$$\mathbf{J}_{T=0} = \frac{\rho\hbar}{2m}\left(\nabla\phi + \tfrac{1}{2}\begin{pmatrix}1 & 0 & 0\\ 0 & 1 & 0\\ 0 & 0 & -1\end{pmatrix}\nabla \times \mathbf{l}\right). \tag{VII.37}$$

The superfluid density isotropy is achieved not by vanishing of the terms corresponding to K_A but by virtue of the anisotropy of the K_0 coefficient [see (VII.38)].

The value of the second term in (VII.37) is again not absolutely independent of assumptions about Fermi-liquid corrections, but this is uncertain rather at intermediate temperatures—it is evident that at $T = 0$ the whole fluid must follow the equation of motion of the pairs. We shall not hazard a guess as to the correct temperature dependence of the Fermi-liquid coefficients, but will quote here Blount's general result for weak-coupling theory only:

$$\mathbf{J} = \frac{\hbar\rho}{2m}[4\alpha\nabla\varphi + (2\beta - 4\alpha)\hat{\mathbf{l}}(\hat{\mathbf{l}}\cdot\nabla\varphi) + 2\beta\nabla \times \mathbf{l} - 2\hat{\mathbf{l}}[\hat{\mathbf{l}}\cdot(\nabla \times \mathbf{l})]]. \tag{VII.38}$$

Here

$$\alpha = \frac{3}{32\pi}\int d\Omega\, f_s \sin^2\theta$$

$$\beta = \frac{3}{8\pi}\int d\Omega\, f_s \cos^2\theta; \tag{VII.39}$$

$f_s(\hat{\mathbf{k}})$ is as defined in (VII.25).

Blount has also worked out the general form of the Ginzburg–Landau free energy, again without Fermi-liquid corrections. We quote the rather complicated result:

$$\begin{aligned} F = \frac{\hbar^2\rho}{8m^2}\Big\{ & \alpha[4(\mathbf{l}\times\nabla\varphi)^2 + (\nabla\cdot\mathbf{l})^2] \\ & + \beta[2(\mathbf{l}\cdot\nabla\varphi)^2 + 2\nabla\varphi\cdot(\nabla\times\mathbf{l} - 2\mathbf{l}(\mathbf{l}\cdot\nabla\times\mathbf{l}))] \\ & + \left(\beta + \frac{2\gamma}{3}\right)(\mathbf{l}\times\nabla\times\mathbf{l})^2 + \left(\frac{\alpha + 2\beta}{3}\right)(\mathbf{l}\cdot\nabla\times\mathbf{l})^2\Big\}. \end{aligned} \tag{VII.40}$$

α and β are as in (VII.39) while

$$\gamma = \frac{3}{4\pi}\int d\Omega\, f_s(\hat{\mathbf{k}})\frac{\cos^4\theta}{\sin^2\theta}. \tag{VII.41}$$

Oddly, this coefficient diverges logarithmically as $T \to 0$, as a consequence of the zeros in the gap function. There appears to be an anomalous stiffness against the quantity $(\mathbf{l} \times \nabla \times \mathbf{l})$. Perhaps if the configuration were driven to have finite $(\mathbf{l} \times \nabla \times \mathbf{l})$, $\uparrow\downarrow$ pairing would be enforced in order to fill up the zeros.

Again, as for the current, we hesitate to speculate on Fermi-liquid effects. At $T = 0$ (VII.40) must be taken literally (here $\alpha = \frac{1}{4}$, $\beta = \frac{1}{2}$, $\gamma \to \ln T$). As $T \to T_c$,

$$\alpha = \beta = \frac{\gamma}{3} = \frac{d^2}{5} \frac{\rho_s}{\langle \Delta^2 \rangle} = \frac{7}{20\pi^2} \zeta(3) \frac{d^2}{T_c^2} \qquad \text{(VII.42)}$$

(d^2 is the coefficient of $\sin^2\theta$ in $\Delta^2 = d^2 \sin^2\theta$)

and, as for the current, in this limit Fermi-liquid effects amount merely to multiplying all ρ's by m/m^* after calculating in weak coupling. Given the two limits, not much leeway is left for estimating the variation between them.

It is too early to decide which of the many striking properties of the orbital supercurrents will be of most interest. We note that they are by no means either conserved or irrotational by nature. The former means that they will play a role in transport processes, and indeed we can expect that externally applied currents will cause twists in the orientation over and above the free-energy minimizing effects, especially for heat currents, discussed by de Gennes and Rainer[121] [who seem to have ignored the cross terms of (VII.40)]; equally, any means of twisting the orientation will drive currents. The latter means (as pointed out by Ambegaokar, de Gennes and Rainer[52]) that the superflow is by no means a potential irrotational flow, which may lead to interesting differences. The whole problem of Josephson-type interference effects raised by those authors becomes far more complicated because of the non-Abelian character of the group of transformations of the order parameter which can affect its phase. Since it is not out of the question that a reasonable sized orifice (≈ 1 micron) could have Josephson properties, this is an experimentally interesting question.

2. COLLECTIVE MODES IN ANISOTROPIC SUPERFLUID He^3

In Section IV.4 we discussed fairly completely the subject of soundlike modes in He^3. Those modes could be done reasonably completely using only conventionally defined ρ_s and ρ_n values. Here we wish to give a brief survey of those few calculations which have used more complete and rigorous theories, such as the Ginzburg–Landau techniques above, to discuss the various collective excitations involving the other (non-phase) degrees of freedom of d. The reader will recall that zero and first sound were at very high frequencies ($c \gg v_F$) and that second sound is confined to a very small frequency and temperature regime, and has a very slow velocity (< 1 m sec^{-1}). Therefore the other modes we discuss will be separated from these easily.

Brinkman and Smith,(59) Maki,(122) Maki and Ebisawa,(114, 115) Combescot,(57, 58) and Wölfle(53) have all given treatments of one aspect or another of the spin-wave problem, which on the whole are in good agreement. Maki and Tsuneto(60) have discussed very simply some aspects of the ABM phase which agrees with Combescot's more complete treatment of the hydrodynamic regime $\omega\tau_D \ll 1$ (where τ_D is the collision time between quasiparticles, which near T_c is of order 10^{-6}–10^{-7} sec). Brinkman and Smith, as well as Combescot, have discussed the BW case. The former treatment is particularly simple and we will reproduce it here to derive the spin waves for both phases.

For a rigorous theory including all Fermi-liquid effects, and especially for the collisionless regime at high frequencies, it is essential to go through one version or another of a full generalized RPA calculation, and this has been done by Maki and Ebisawa(56) and, particularly, by Combescot.(57) These theories are quite complicated and we will not discuss them in detail here, but note that they agree—with a suitable, but not unique, interpretation of Fermi-liquid corrections—with the heuristic theories at $T = 0$ and thus qualitatively in terms of number of modes, etc., they contribute nothing new.

In the theory of Brinkman and Smith the conjugate relationship between the spin angular momentum and the angles of rotation and spin space are further exploited to derive the spin waves. As already discussed in Section VI, the rotation angles are related to the phase difference between the up and down spin gap functions, so that this approach is identical to that of Maki and Tsuneto and may alternatively be thought of in terms of a Josephson effect between the up and down spin fermion reservoir. The advantage in thinking in terms of rotation angles is that the full symmetry of the problem is explicit.

Thus we introduce the same set of angles $(\theta_u, \theta_v, \theta_w)$ representing small rotations about the three axes. These angles now depend on position in space as well as time. The order parameter is

$$d_{\alpha i} = d^0_{\alpha j} O_{ji}(\theta_u, \theta_v, \theta_w), \tag{VII.43}$$

where $d^0_{\alpha j}$ is the equilibrium order parameter assumed to be uniform. The equations of motion are obtained from the local energy density F_T which includes the same terms as the NMR calculations in Section VI (VI.17),

$$E = \frac{\gamma^2 S^2}{2\chi} - \gamma \mathbf{S} \cdot \mathbf{H} + H_0(\theta_u, \theta_v, \theta_w) \tag{VII.44}$$

along with the stiffness energy described by (VII.16)

$$F^{\text{II}} \propto \nabla_\alpha d^*_{\beta i} \nabla_\alpha d_{\beta i} + \nabla_\alpha d^*_{\alpha i} \nabla_\beta d_{\beta i} + \nabla_\alpha d^*_{\beta i} \nabla_\beta d_{\alpha i}. \tag{VII.16}$$

Then, using the commutation relations

$$[S_i, \theta_j] = i\hbar\delta_{ij} \tag{VII.45}$$

and F_T as a hamiltonian, we obtain the equations of motion

$$\frac{\partial}{\partial t} S_i = \gamma(\mathbf{S} \times \mathbf{H})_i - \frac{\partial}{\partial \theta_i}(H_0 + F^{\text{II}}) \tag{VII.46}$$

$$\frac{\partial}{\partial t}\theta_i = \frac{\partial F_T}{\partial S_i} = \gamma\left(\frac{\gamma S_i}{\chi} - H_i\right). \tag{VII.47}$$

Let us now restrict ourselves to the ABM state. Then, as given by (VI.16)

$$H_0 = 3\Gamma\Delta^2(\theta_u^2 + \theta_v^2) \tag{VII.48}$$

and after some algebra

$$F^{\text{II}} = \frac{\hbar^2}{8m^2}(\nabla\theta_u \cdot \vec{\rho}_s \cdot \nabla\theta_u + \nabla\theta_v \cdot \vec{\rho}_s \cdot \nabla\theta_v), \tag{VII.49}$$

where ρ_s is the superfluid density discussed in Section IV and we have again taken w to be the axis along which $m_s = 0$. Then

$$\frac{\partial}{\partial t} S_u = \gamma(\mathbf{S} \times \mathbf{H})_u - 6\Gamma\Delta^2(\theta_u) - \frac{\hbar^2}{4(m)^2}\mathbf{q} \cdot \vec{\rho}_s \cdot \mathbf{q}\theta_u. \tag{VII.50}$$

Solving these equations we find two solutions just as in the uniform resonance case, a longitudinal mode in which **S** is parallel to **H** with a frequency

$$\omega_q^2 = \Omega_L^2 + \frac{\gamma^2}{\chi}\mathbf{q} \cdot \vec{\rho}_s \cdot \mathbf{q} \tag{VII.51}$$

and the transverse mode with frequency

$$\omega_q^{T^2} = \omega_q^2 + (\gamma H)^2. \tag{VII.52}$$

These results are identical to those of Maki and Tsuneto[60] for the wave associated with the longitudinal resonance. As we know, the expression used for F^{II} is only valid in the Landau–Ginzburg regime. However, the work of Combescot[57, 58] which starts from a less phenomenological approach gives an indication that the correct expression for the spin stiffness is (VII.49) with ρ_s given by the weak-coupling form divided by $(1 + F_1^s/3)$, provided one ignores the higher antisymmetric Fermi-liquid corrections F_1^a, etc. To see this we note that (VII.50) is simply the continuity equation for the spin density and therefore the last term must be identified with $\nabla \cdot \mathbf{j}_s$ where $\mathbf{j}_s$ is the spin current. Combescot has shown that the superfluid density occurring in the spin current is not given by the same expression as the superfluid mass density because spin currents are not conserved quantities. He shows that, if you ignore higher-order antisymmetric Fermi-liquid factors other than F_0^a, the spin density occurring in the current is just $\rho_s^{wc}/(1 + F_1^s/3)$, where ρ_s^{wc} is calculated without Fermi-liquid effects ($\rho_s^{wc}/\rho(1-d)$ in his notation). Therefore, this superfluid "spin density" does not go to one as T goes to zero but to $1/(1 + F_1^s/3)$.

Turning now to the collective spin waves in the B phase we remember that the $d_{\alpha i}^*$ is a rotation matrix with the axis of rotation along the field and the angle of rotation $\theta = \cos^{-1}(-\frac{1}{4})$. The spin-wave calculation is simplified if we measure our spin variables in a coordinate system rotated by this same rotation. Thus we have **u**, **v** axes that are rotated by θ about the z axis. Then the $d_{\alpha i}^*$ matrix is the unit matrix and the evaluation of F^{II} is quite simple:

$$F^{\mathrm{II}} = \rho_s \frac{\hbar^2}{20\,m^2}(4(\nabla_\alpha \theta_\beta)^2 - 2\nabla_\alpha\theta_\beta\nabla_\beta\theta_\alpha). \tag{VII.53}$$

Ignoring for the moment the dipolar and external field terms in E_T we obtain the following equations:

$$\frac{\partial}{\partial t}\mathbf{S} = -\rho_s \frac{\hbar^2}{4m^2}(\tfrac{2}{5})(4q^2\boldsymbol{\theta} - 2\mathbf{q}(\mathbf{q}\boldsymbol{\theta})) \tag{VII.54}$$

$$\frac{\partial}{\partial t}\boldsymbol{\theta} = \mathbf{S}/\chi. \tag{VII.55}$$

These equations give three modes, two modes polarized transverse to **q** and one longitudinal, with velocities given by

$$v_L = \left(\frac{\hbar^2 \rho_s}{5\,m^2 \chi}\right)^{\frac{1}{2}} \tag{VII.56}$$

and

$$v_T = \sqrt{2}\,v_L. \tag{VII.57}$$

It must be remembered that the "polarization" is measured in the rotated coordinate system. As pointed out by Combescot, the polarization found for such waves would determine the form for $d_{\alpha i}^*$.

The analysis of inclusion of the dipolar energy leads to the conclusion that one transverse mode is unaffected, while the longitudinal and the other transverse mode mix in a $\hat{\mathbf{q}}$-dependent way and one of them becomes the longitudinal resonance as $q \to 0$. For details see Brinkman and Smith.[59] Finally, for the situation in the presence of a magnetic field the analysis becomes even more complicated.

As we have already remarked above: (a) the ρ_s used in these equations is undoubtedly close to that in the absence of Fermi-liquid effects, so that the velocities are of order v_F divided by (enhancement factor)$^{\frac{1}{2}}$; (b) the equations are valid in the hydrodynamic regime or near $T = 0$ or, with considerable quantitative change, in the collisionless regime $\omega \gg 10^7$. Where $\omega\tau_D \approx 1$ there will be severe damping by diffusion of the normal component. This assumes that $\omega \approx v_F q$; for modes of very long wavelength the behaviour depends strongly on magnetic field.) The best regime for observation should be $\omega \approx 10^5$–10^6, $\lambda \approx 10^{-1}$–10^{-2} cm.

Orbit waves have been discussed only very briefly by Wölfle,[53] Anderson[24, 25] and de Gennes.[50, 51] The spin waves and the magnetic resonances we have discussed in the ABM state so far are all in essence oscillations of the spin portion of the gap bivector in the presence of an assumed stationary structure of the orbital portion. (This is not the case in the B phase, where only one rotational structure exists. As we have emphasized, *total* rotation in that case not only leaves the energy invariant, it leads to an identical state.) In actual fact, of course, the torque imparted by the spin system to the orbits must lead to (d/dt) $(\mathbf{L} + \mathbf{S}) \equiv \mathbf{0}$, in the absence of spin currents, but because there is a finite orbital angular momentum the net rotation of d_α is very small.

At some considerably lower frequency, one presumes, there must be a sequence of modes in which the orbit system alone precesses, or it carries the spin system with it, and the restoring forces are provided only by the intrinsic Landau–Ginzburg stiffness of the gap system and the pinning of the vector $\mathbf{l}$ to the walls of the container.

In the simplest case there is no magnetic field and thus literally there is no external orienting force. As Wölfle remarks, in this case the direction $\mathbf{l}$ of the orbital angular momentum $\mathbf{L}$, and $\mathbf{L}$ itself, are conjugate dynamical variables, so that

$$\frac{\partial}{\partial t}\delta L_x = -\frac{\delta F}{\delta l_y} = \frac{\rho_s}{8m}(-2\nabla_x\nabla_z\,\delta\varphi + (2\nabla_z^2 + \nabla^2)\delta l_y) \qquad \text{(VII.58)}$$

$$\frac{\partial}{\partial t}\delta L_y = -\frac{\rho_s}{8m}(2\nabla_y\nabla_z\delta\varphi + (2\nabla_z^2 + \nabla^2)\delta l_x) \qquad \text{(VII.59)}$$

where we have put $\mathbf{l}$ along z. Note the coupling of $\mathbf{L}$ to $\nabla\varphi = \mathbf{v}_s$, which means that orbital oscillations can be driven by superfluid flow.

Now the further progress depends on the ratio $\langle L\rangle = \mathbf{L}/\hat{\mathbf{l}}$. In AM this was estimated by the assumption

$$\langle L\rangle = \int d^3r\; r\langle \mathbf{j}(\mathbf{r})\rho(0)\rangle_\varphi, \qquad \text{(VII.60)}$$

which Wölfle evaluates as

$$\langle L \rangle_W = \frac{5m}{16m^*} \frac{T_c}{E_F} \frac{\rho_s}{\rho}. \tag{VII.61}$$

Cross and Blount, however, find no rigorous justification for (VII.60), and in particular suggest that if particle–particle and hole–hole contributions to $\langle n\mathbf{j} \rangle$ are nearly equal, the net coefficient K_l which we find in the preceding section to represent $\langle L \rangle$ is an order smaller than (VII.61). Leaving this coefficient indeterminate, Wölfle's orbit-wave frequencies become

$$\omega(q) = \left(\frac{\langle L \rangle_W}{\langle L \rangle} \right) \frac{2\hbar}{5m^*} \frac{E_F}{kT_c} (3q_\parallel^2 + q_\perp^2)^{\frac{1}{2}} \left[3q_\parallel^2 + q_\perp^2 - 2\frac{q_\parallel^2 q_\perp^2}{2q_\perp^2 + q_\parallel^2} \right]^{1/2}.$$

For $q \simeq 1\,\text{cm}^{-1}$, this amounts to $\omega \approx 10^{-2}(\langle L \rangle_W / \langle L \rangle)$. As Anderson observed, this range of frequency matches, if the L ratio were unity, rather nicely to the observed low-frequency fluctuations of NMR and zero-sound amplitudes in the A phase,[105, 123] if one assumes that the few lowest-frequency L-wave modes are involved. On the other hand, the Cross and Blount estimate leaves us with about 100 times higher frequencies, or $\omega \approx$ 1 Hz, which is altogether too high relative to the experimental results. Other aspects of these results, such as their magnetic-field dependence, are also rather questionable under this interpretation. Any simple interpretation suggests that **H** should raise the frequency by pinning **L** in a plane.[24, 25] That L-waves may be driven by the cross-coupling of $\hat{\mathbf{l}}$ and $\mathbf{v}_s$ is of course in good agreement both with these results (where random heat currents are likely to be present) and with Wheatley's heat transport instabilities.

One possible way out of this dilemma goes as follows. This is that often the A-phase texture may involve a curvature of $\hat{\mathbf{l}}$ near the vessel wall which would imply a rather large orbital supercurrent near the vessel wall. If $\hat{\mathbf{l}}$, for instance, must curve from perpendicular to the wall to axial along a tube (see Fig. 16c), the total surface current flowing is $\propto \int \nabla \times \hat{\mathbf{l}} \approx 1$ relative to Δ^2/E_F^2, as for the weak orbital currents. The inertia of this circulation will be 10^2–10^4 times that of the orbital moment, and thus macroscopic oscillations of $\hat{\mathbf{l}}$ will again revert to the 10^{-2} Hz frequency range. This, then, would suggest that not all, but a subset of orbit-wave modes will be found in the low-frequency range: a situation rather like the "Walker modes" of ferromagnetic insulators.[124]

Clearly this question is one of those most deserving of further study, both experimentally and theoretically, in the He^3 field. It seems possible that this is one of the few areas where the fundamentals of the theory are not yet established.

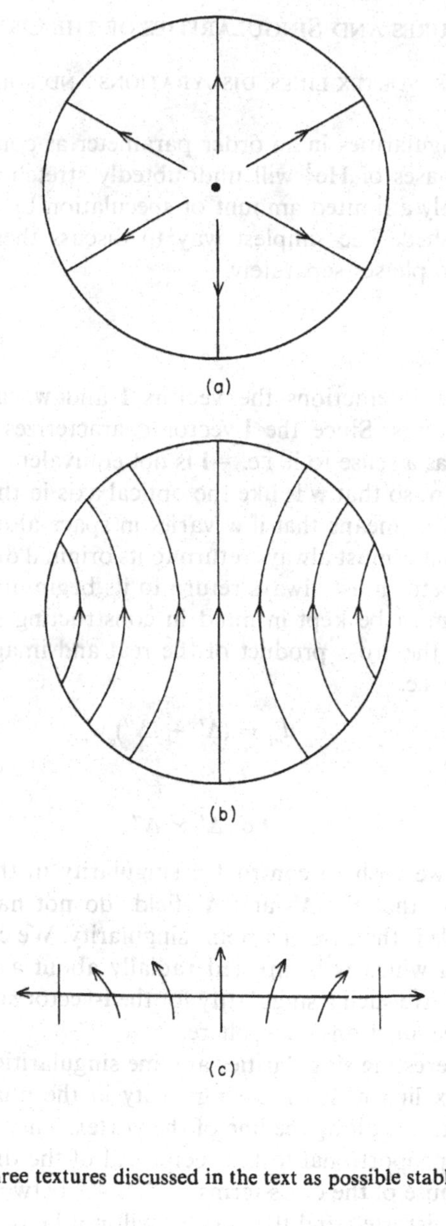

FIG. 16. The three textures discussed in the text as possible stable textures.

VIII. Textures and Singularities of the Order Parameter

1. Singularities, vortex lines, disgyrations and point defects

The possible singularities in an order parameter as complicated as those of the A and B phases of He^3 will undoubtedly stretch our imagination. So far, however, only a limited amount of speculation by de Gennes[50, 51, 121] has been published. The simplest way to discuss these singularities is to consider the two phases separately.

(i) "*A*" *phase*

Without dipolar interactions the vectors **l** and **w** can be considered as independent vectors. Since the **l**-vector characterizes the orbital angular momentum it has a sense to it, i.e. $-\mathbf{l}$ is not equivalent to **l**. By contrast $-\mathbf{w}$ is equivalent to **w**, so that **w** is like the optical axis in the nematic phase of a liquid crystal. This means that if **w** varies in space along a closed path it is not necessary that **w** must always return to its original direction; it can return as $-\mathbf{w}$. The **l**-vector must always return to its beginning direction. Another difference that must be kept in mind in constructing singularities in the **l**-field is that **l** is the cross product of the real and imaginary vectors of the order parameter i.e.

$$d_{\alpha i} = (\boldsymbol{\Delta}' + i\boldsymbol{\Delta}'')_\alpha w_i$$

and

$$\mathbf{l} \propto \boldsymbol{\Delta}' \times \boldsymbol{\Delta}''.$$

Therefore, if we wish to contruct a singularity in the **l**-field we must be careful to ensure that the $\boldsymbol{\Delta}'$- and $\boldsymbol{\Delta}''$-fields do not have singularities. The simplest example is the case of a point singularity. We can have a singularity in the **w**-field in which **w** is directed radially about a single point but it is not possible to have such a singularity for the **l**-vector since one cannot make $\boldsymbol{\Delta}'$ and $\boldsymbol{\Delta}''$ unique functions on a sphere.

The more interesting singularities are line singularities. One can construct the usual vortex line which is a singularity in the phase ϕ. However, it is only simple when **l** is along the line of the vortex. Then the current is simply tangential and proportional to the reciprocal of the distance from the line. Otherwise, because of the cross terms in (VII.40) between $(\nabla\phi)$ and $(\nabla \times \mathbf{l})$, **l** will be slightly distorted and the current will not be rotationally symmetric.

More interestingly, there can be line singularities in the **l**-field which de Gennes calls "disgyrations". These can be of two types; one called a tangential disgyration has **l** directed tangentially about the line axis ($\boldsymbol{\Delta}'$ along the line axis and Δ″ radial). It has a net current flowing along the line axis. Its magnitude can be obtained from (VII.36) to be

$$J = \frac{\hbar}{m^*}\left\{\frac{7\zeta(3)}{40\pi^2}\right\}\frac{\langle\Delta^2\rangle}{T_c^2}\frac{1}{r}.$$

The second, a radial disgyration, has **l** pointing radially outward from the line (Δ′ along the line axis and Δ″ tangential). There are no currents associated with a radial disgyration.

Just as it is possible for usual vortices to have more than one quantum of circulation, it is possible for **l** to rotate through multiples of 2π about the line axis. Such higher-order disgyrations would have a current along the line axis.

Similar behaviour can be obtained for the **w**-vector. However, as discussed in section VI the dipolar energy tends to keep **w** and **l** parallel. We can estimate a length ξ_D which determines the scale on which **w** will no longer follow **l** by equating the dipolar energy to the bending energy required to bend **w**. This gives

$$\xi_D = \xi\left(\frac{\Delta F}{E_D}\right)^{\frac{1}{2}} \approx 10\mu, \qquad \text{(VIII.1)}$$

where ΔF is the free energy gained in the superfluid state and E_D is the dipole energy. In regions where **w** or **l** bends rapidly on the scale of ξ_D they will not follow one another. For example, in small pores ($\leqslant 10\mu$) **l** will be fixed by the walls but **w** may actually be constant in space and be oriented by an external field.

(ii) "*B*" phase

In this phase one has possible singularities in the phase, the axis of rotation **n**, and possibly in the angle of rotation θ. Since θ is fixed by the dipolar energy the last can only occur if the scale of the distortion is small compared to the dipolar coherence length ξ_D. Since the direction of the vector **n** specifies the sense of rotation, **n** is like **l** in that its variation along a closed path must be such that it always returns to its original direction. Unlike **l**, however, it can exhibit a point singularity as described for the vector **k** since there are no difficulties with vectors perpendicular to it. In fact, in recent experiments reported by Osheroff a dip in the intensity of the NMR signal as a function of position along the axis of his cylindrical sample may be due to the presence of such a singularity. The point singularity could occur if two regions on the two ends of the cylinder grew together with **n** in opposite directions.

The vector **n** can also exhibit disgyration line-type singularities just as

l does. These may give rise to spin currents but they will not be discussed here.

An interesting feature of any singularities in the vector **n** is that within ξ_D of the singularity the angle of rotation θ may begin to rotate and reduce the kinetic energy due to the **n** distortion. At the centre of the singularity θ may be zero and the bending energy reduced to something of the order of the dipolar energy. A somewhat related suggestion was made previously by de Gennes for the A phase. He suggested that the core of the disgyration in that phase may be the BW state.

(iii) *Textures*

Having found that the order parameter in these new phases can exhibit singularities it is not too surprising that in the presence of boundaries the various vectors involved will exhibit textures similar to those found in liquid crystals. The possibility of such effects was first discussed by de Gennes[50, 51, 121] and in more detail by Ambegaokar, de Gennes and Rainer.[52] Recently, textures have been discussed for the B phase by Brinkman, Smith, Osheroff and Blount.[61]

The first question that must be answered is what are the boundary conditions for the order parameter. The answer to this question is a relatively simple one.[52] If one has spectral scattering of He^3 atoms then a wall only acts to depair that component of the triplet for which $m_l = 0$ normal to the wall. Said differently, when two particles forming a pair approach a wall they must circulate around one another in the plane of the wall. We will not go into the derivation of this result. It is obtained by studying the linearized gap equation and introducing the boundary conditions through the non-local kernel in the integral equation. A semiclassical approximation to the kernel allows one to obtain the solutions. Under circumstances when the scattering is diffuse it was found that all of the components of the order parameter are greatly reduced at the boundary but those with $m_l = 0$ still become zero.

Therefore, in terms of $d_{\alpha i}$ we must impose the condition that

$$s_\alpha d_{\alpha i} = 0 \tag{VIII.2}$$

at the surface, where **s** is a unit vector. In addition to these boundary conditions we must also make sure that the currents into the wall vanish. This requires that

$$(\mathbf{s} \cdot \nabla) d_{\alpha i} = 0 \tag{VIII.3}$$

for those components $d_{\alpha i}$ not going to zero by the first boundary condition. In the A phase the first condition is automatically satisfied if **l** is normal to the wall. Having **l** at any other direction than normal costs an energy of order $\Delta F\xi$; therefore, there is a strong force pinning **l** in this position.

Given this boundary condition we can examine various possible textures of **l** in restricted geometries. For a cylinder the following seems to be the situation. For small radii $r \leqslant 100$ the stable texture will be either one with **l** directed radially outward (or inward) from the axis of the cylinder with a disgyration down the centre or one with two tangential disgyrations running down opposite sides of the cylinder as illustrated in Fig. 16a, b. For larger radii, we expect to find a texture in which **l** simply rotates upward (or downward) as a function of radius and ends up pointing parallel to the cylindrical axis at the centre of the cylinder Fig. 16c. The form of the order parameter for such a texture can be obtained by starting with $\mathbf{\Delta}'$ and $\mathbf{\Delta}''$ in the plane perpendicular to the cylinder axis and then defining

$$\mathbf{\Delta}'(r) = \hat{O}(\hat{\boldsymbol{\phi}}, \theta(r))\mathbf{\Delta}' \qquad \text{(VIII.4)}$$

$$\mathbf{\Delta}(r) = \hat{O}(\hat{\boldsymbol{\phi}}, \theta(r))\mathbf{\Delta}'', \qquad \text{(VIII.5)}$$

where $\hat{O}$ is a rotation matrix and the axis of rotation is $\hat{\boldsymbol{\phi}}$, the tangential direction. The angle $\theta(r)$ is a function of position chosen so that $\theta(R) = \pi/2$ at the surface of the cylinder. This texture becomes more stable at large radius because it does not involve a disgyration whose energy becomes proportional to the $\ln(R/\xi)$ as the above textures do.

It should also be pointed out that such a texture has an angular momentum of order $\hbar\rho_s(\pi R^2)$ per unit length of the cylinder because it has a supercurrent flowing in the tangential direction.

When the radius of the cylinder is larger than $\approx 10\mu$, **w** and **l** will tend to bend together. Then for relatively small axial fields the last texture described (*c*) will be suppressed by the magnetic anisotropy associated with **w**. A crude estimate of when this will occur can be obtained by setting the magnetic anisotropy energy $\delta\chi H^2$ equal to the bending energy $\Delta F(\xi/R)^2$, giving

$$H_c \simeq \left(\frac{\xi}{R}\right)\left(\frac{\Delta F}{\delta\chi}\right)^{\frac{1}{2}} \approx \frac{\xi}{R}\left(\frac{\Delta}{\hbar\gamma}\right). \qquad \text{(VIII.6)}$$

This field is typically much less than one gauss.

Perhaps the best geometry for obtaining essentially uniform textures is that of two flat surfaces separated by a distance χ. The **l** and **w** vectors will be normal to the surfaces if the external field is applied parallel to the surface. If the field is applied perpendicular to the surface **w** and **l** will tend to bend. This situation has been studied recently by Rainer.[121]

In the same geometry de Gennes and Rainer[121] studied the effect of a current and showed that, because of anisotropy in ρ_s above a critical current given by

$$J_{\text{critical}} \propto \frac{\hbar}{m} \Delta^2 \left(\frac{\pi}{\chi}\right),$$

the **l** vector will tend to bend across the sample and at high enough currents it will be pointing essentially along the direction of flow. The prediction of a critical field appears to be incorrect because of the cross coupling between **l** and the gradient of ϕ in (VII.40).

A similar situation is expected for heat flow since the quasiparticles carry the entropy and their distribution is anisotropic. This can be seen in the definition of the entropy tensor discussed earlier (IV.128) and obtained from the work of Wölfle[31, 32]

$$\sigma_{ij} = -\frac{1}{T} \sum_{\mathbf{k}} \frac{k_i}{m} \frac{k_j}{m^*} \frac{\mathrm{d}n(E_{\mathbf{k}}/kT)}{\mathrm{d}E} \varepsilon_{\mathbf{k}}. \tag{VIII.7}$$

Here $\mathrm{d}n/\mathrm{d}E$ is the derivative of the Fermi factor. Clearly this expression is large along **l** where the gap is small. Indeed, the effect of a heat flow appears to have been seen in the recent data of Osheroff and Brinkman,[26] where both **l** and **w** appear to rotate parallel to the external field when the field is sufficiently weak.

Turning now to the B phase, since no matter what the values of **n** and θ are one has a component of the order parameter (such that $m_l = 0$ normal to the surface) which must go to zero, there are no large orientational effects due to a surface. However, because the component of $d_{\alpha i}$ normal to the surface is suppressed, i.e. $s_\alpha d_{\alpha i} = 0$, the susceptibility becomes anisotropic in this region and the dipolar energy changes. By constructing a trial function which properly satisfies this boundary condition Brinkman, Smith, Osheroff and Blount[61] calculated these terms. The term arising from the dipolar energy is of the form

$$F_D^s = -b \int_S \mathrm{d}^2 r[(\mathbf{s}\cdot\mathbf{n})^2 - \tfrac{5}{18}(\mathbf{s}\cdot\mathbf{n})^4], \tag{VIII.8}$$

where b is of order $E_D\xi$. This term favours **n** being perpendicular to the surface. From the susceptibility anisotropy they obtained the term

$$F_H^s = -d \int_S \mathrm{d}^2 r[\mathbf{s}\cdot \hat{O}(\mathbf{n},\theta)\cdot \mathbf{H}]^2 \tag{VIII.9}$$

where $\hat{O}(\mathbf{n},\theta)$ is the rotation matrix in (VII.28). The d is of order $\xi(\chi_N - \chi_S)$. This term favours a direction of **n** such that the rotation matrix takes **H** into $\pm\mathbf{s}$. With $\theta = \cos^{-1}(-\frac{1}{4}) = 104°$ one can always find a direction of **n** such that this condition is satisfied. Since this term does not tend to change θ and since a surface dipole energy, which in fact would like to have $\theta = 90°$ at the surface, cannot compete against the volume dipole energy, the textures of the B phase can be described completely in terms of **n** variation. In the absence of a magnetic field the characteristic length in the problem is

$$R_c \simeq \xi\left(\frac{\Delta F}{E_D}\right) \simeq 0{\cdot}1(1 - T/T_c)^{\frac{1}{2}}\text{cm}.$$

Again considering a cylindrical sample we find that the two textures b and c in Fig. 16 compete, the b texture being most stable for $R < 74R_c$.

In the presence of magnetic field along the cylinder we must include the anisotropy energy discussed in Section IV, eq. (IV.93). From it we obtain a field-dependent characteristic length $R_H = R_c H_B/H$, where $R_c H_B$ is estimated to be of order 10–20 cm gauss. As the field is turned on, the surface effects become less important and the texture c becomes more stable and the **n** vector becomes more and more parallel to the cylinder axis.

As we discussed in Section VI, the transverse resonance frequency shifts when **n** is not parallel to the external field, so the texture of **n** being on the scale of cm has a profound effect on the NMR resonance. The simplest effect is that $R_H \propto (1 - T/T_c)^{\frac{1}{2}}$, so that the characteristic length becomes small near T_c. If one is looking at either transverse or longitudinal resonance at fields that are small compared to those necessary to align **n** at low temperature one should see the signal recover as the temperature is raised near to T_c. Indeed, such behaviour of both the transverse and longitudinal resonances has been seen by Ahonen *et al.*(125) and by Webb *et al.*(97)

Operating at a given temperature one expects the amount of unshifted resonance signal to increase as a function of increasing field. This has been found in the recent experiments of Osheroff and Brinkman.(26)

We can actually make a quantitative theory of the transverse line shape in the high-field limit if we assume that each region of the sample resonates at its local frequency which is given by

$$\omega^2 = (\gamma H)^2 + \Omega_L^2 \sin^2 [\mathbf{n}(\mathbf{r}) \cdot \mathbf{H}]. \qquad \text{(VIII.10)}$$

In view of the fact that one expects collective modes as discussed in the previous section, the local approximation is not entirely justified. However, since we will be looking at small shifts of order a few hundred hertz which must be driven by the local shift, we assume it is a reasonable approximation.

A very good approximation to the variation of $[\mathbf{n}(\mathbf{r}) \cdot \mathbf{H}]$ can be obtained by using the simple formula

$$\mathbf{n}(\mathbf{r}) \cdot \mathbf{H} = A \sinh\left(\frac{rH}{R_c H_B}\right). \qquad \text{(VIII.11)}$$

The constant A is given by $A_0/\sinh(RH/R_c H_B)$, where A_0 is the angle of **n** relative to **H** at the surface and R is the radius of the sample. The calculated lineshape is

$$P(\omega) = 2\left(\frac{H_B R_c}{HR}\right)^2 \frac{1}{\Gamma} f\left[2\left(\frac{\omega - \gamma H}{\Gamma}\right)\right]; \quad \omega > \gamma H \qquad \text{(VIII.12)}$$

here Γ is the effective width,

$$\Gamma = (\Omega_L^2/\gamma H)A^2. \qquad \text{(VIII.13)}$$

Because of the sinh $(RH/R_c H_B)$ in this expression Γ can indeed be quite small. The function $f(x)$ is found to be

$$f(x) = \ln(x^{\frac{1}{2}} + \sqrt{x+1})/\sqrt{x+x^2} \qquad \text{(VIII.14)}$$

The integrated intensity between γH and $\gamma H + \Delta\omega$ can be calculated as a function of field:

$$N(\Delta\omega) = \int_{\gamma H}^{\gamma H + \Delta\omega} P(\omega)\,\mathrm{d}\omega$$

$$= \left\{1 - \frac{H_B R_c}{2HR} \ln\left(\frac{\Omega^2}{2\Delta\omega\gamma H} A_0^2\right)\right\}^2. \qquad \text{(VIII.15)}$$

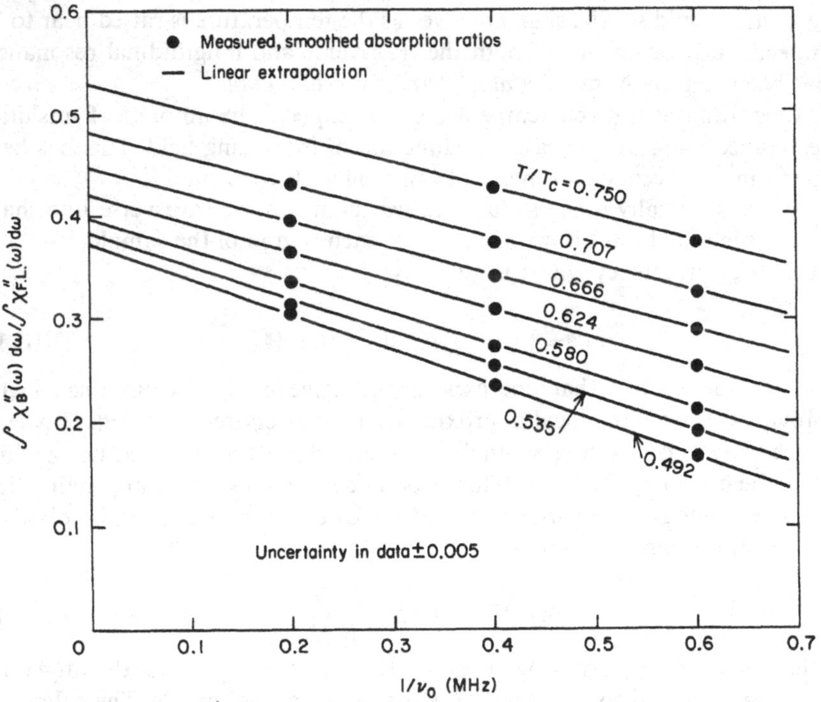

FIG. 17. A plot of the integrated absorption signal in the B phase as a function of 1/H. The measurements are those of Osheroff.[27]

This gives rise to an essentially $1/H$ saturation of the integrated intensity. Recent data of Osheroff[27] is shown in Fig. 17. The slopes of the various curves should be proportional to the coefficient of the bending energy in (VII.32) i.e. $\rho_s^{\frac{1}{s}}$. This expression does not, however, include Fermi-liquid effects. Recent calculations by Combescot[57, 58] and by Brinkman and Smith[59] suggest that this coefficient should be ρ_s^0, the superfluid density calculated in weak-coupling theory. The experimental curves vary on a scale quite close to ρ_s, which in the region of observation is varying more rapidly than ρ_s^0. This result could possibly be due to the neglect of dispersion already mentioned, since dispersion would presumably cause the signal to shift further than this calculation predicts. This effect may be stronger at lower temperatures where the dispersion is more important.

ACKNOWLEDGEMENTS

The authors would like to acknowledge contributions and discussions from E. I. Blount, D. D. Osheroff, H. Smith, M. C. Cross, S. Engelsberg and C. M. Varma. We also would like to thank Mrs N. J. Hines for her patience in preparing the manuscript.

REFERENCES

1. J. Bardeen, L. N. Cooper and J. R. Schrieffer. *Phys. Rev.* **108**, 1175 (1957).
2. D. J. Thouless. *Ann. Phys. (New York)* **10**, 553 (1960).
3. J. C. Fisher. Private communication (1958).
4. L. P. Pitaevskii. *Sov. Phys. JETP* **10**, 1267 (1960).
5. K. A. Brueckner, T. Soda, P. W. Anderson and P. Morel. *Phys. Rev.* **118**, 1442 (1960).
6. V. J. Emery and A. M. Sessler. *Phys. Rev.* **119**, 43 (1960).
7. L. N. Cooper, R. L. Mills and A. M. Sessler. *Phys. Rev.* **114**, 1377 (1959).
8. W. R. Abel, A. C. Anderson, W. C. Black and J. C. Wheatley. *Physics* **1**, 337 (1965).
9. P. W. Anderson and P. Morel. *Physica* **26**, 671 (1960).
10. P. W. Anderson and P. Morel. *Phys. Rev.* **123**, 1911 (1961).
11. K. Yosida. *Phys. Rev.* **110**, 769 (1958).
12. R. Balian and N. R. Werthamer. *Phys. Rev.* **131**, 1553 (1963).
13. A. J. Leggett. *Phys. Rev. Letts* **14**, 536 (1965).
14. A. J. Leggett. *Phys. Rev.* **140**, A1869 (1965).
15. V. J. Emery. *Ann. Phys. (New York)* **28**, 1 (1964).
16. N. F. Berk and J. R. Schrieffer. *Phys. Rev. Letts* **17**, 433 (1966).
17. W. F. Brinkman and S. Engelsberg. *Phys. Rev.* **169**, 417 (1968).
18. A. J. Layzer and D. Fay. Proc. *Int. Low Temp. Conf. St. Andrews* (LT-11) (1968).
19. D. D. Osheroff, R. C. Richardson and D. M. Lee. *Phys. Rev. Letts* **28**, 885 (1972).
20. D. D. Osheroff, W. J. Gully, R. C. Richardson and D. M. Lee. *Phys. Rev. Letts* **29**, 920 (1972).

21. A. J. Leggett. *Phys. Rev. Letts* **29**, 1227 (1972).
22. A. J. Leggett. *Phys. Rev. Letts* **31**, 352 (1973).
23. A. J. Leggett. *J. Phys. C* **6**, 3187 (1973).
24. P. W. Anderson. *Phys. Rev. Letts* **30**, 368 (1973).
25. P. W. Anderson. *Proc. Nobel Symp.* **24**, 103 (1973) (Uppsala, Nobel Foundation).
26. D. D. Osheroff and W. F. Brinkman. *Phys. Rev. Letts* **32**, 584 (1974).
27. D. D. Osheroff. *Phys. Rev. Letts* **33**, 1009 (1974).
28. V. Ambegaokar and N. D. Mermin. *Phys. Rev. Letts* **30**, 81 (1973).
29. C. M. Varma and N. R. Werthamer. *Bull. Am. Phys. Soc.* **18**, 8 (1973).
30. B. R. Patton. Unpublished preprint (1973).
31. P. Wölfle. *Phys. Rev. Letts* **30**, 1169 (1973).
32. P. Wölfle. *Phys. Rev. Letts* **31**, 1437 (1973).
33. R. C. Richardson and D. M. Lee. *Proc. Nobel Symp.* **24**, 84 (1973) (Uppsala, Nobel Foundation).
34. P. W. Anderson and W. F. Brinkman. *Phys. Rev. Letts* **30**, 1108 (1973).
35. W. F. Brinkman and P. W. Anderson. *Proc. Nobel Symp.* **24**, 116 (1973) (Uppsala, Nobel Foundation).
36. W. F. Brinkman and P. W. Anderson. *Phys. Rev.* **A8**, 2732 (1973).
37. W. F. Brinkman, J. Serene and P. W. Anderson. *Phys. Rev.* **A10**, 2386 (1974).
38. Y. Kuroda. *Prog. Theor. Phys.* **51**, 1269 (1974).
39. Y. Kuroda. *Prog. Theor. Phys.* **53**, 349 (1975).
40. R. A. Webb, T. J. Greytak, R. T. Johnson and J. C. Wheatley. *Phys. Rev. Letts* **30**, 210 (1973).
41. T. J. Greytak, R. T. Johnson, D. N. Paulson and J. C. Wheatley. *Phys. Rev. Letts* **31**, 452 (1973).
42. J. C. Wheatley. *Physica* **69**, 218 (1973).
43. D. N. Paulson, H. Kojima and J. C. Wheatley. *Phys. Rev. Letts* **32**, 1098 (1974).
44. D. D. Osheroff and P. W. Anderson. *Phys. Rev. Letts* **33**, 686 (1974).
45. W. M. Saslow. *Phys. Rev. Letts* **31**, 870 (1973).
46. T. A. Alvesalo, Yu. D. Anufriyev, H. K. Collan, O. V. Lounasmaa and P. Wennerström. *Phys. Rev. Letts* **30**, 962 (1973).
47. T. A. Alvesalo, Yu. D. Anufriyev, H. K. Collan, O. V. Lounasmaa and P. Wennerström. *Proc. Nobel Symp.* **24**, 91 (1973) (Uppsala, Nobel Foundation).
48. H. Kojima, D. N. Paulson and J. C. Wheatley. *Phys. Rev. Letts* **32**, 141 (1974).
49. A. W. Yanof and J. D. Reppy. *Phys. Rev. Letts* **33**, 631 (1974).
50. P. G. de Gennes. *Phys. Letts* **44A**, 271 (1973).
51. P. G. de Gennes. *Proc. Nobel Symp.* **24**, 112 (1973) (Uppsala, Nobel Foundation).
52. V. Ambegaokar, P. G. de Gennes and D. Rainier. *Phys. Rev.* **A9**, 2676 (1974).
53. P. Wölfle. *Phys. Letts* **47A**, 224 (1974).
54. E. I. Blount, P. W. Anderson and W. F. Brinkman. Unpublished notes incorporated into Section VII of this article (1974).
55. A. J. Leggett. *Proc. Nobel Symp.* **24**, 109 (1973) (Uppsala, Nobel Foundation).
56. K. Maki and H. Ebisawa. *Prog. Theor. Phys.* **50**, 1452 (1973).
57. R. Combescot. *Phys. Rev.* **A10**, 1700 (1974).
58. R. Combescot. *Phys. Rev. Letts* **33**, 946 (1974).
59. W. F. Brinkman and H. Smith. *Phys. Rev.* **A10**, 2325 (1974).
60. K. Maki and T. Tsuneto. *Prog. Theor. Phys.* **52**, 773 (1974).
61. W. F. Brinkman, H. Smith, D. D. Osheroff and E. I. Blount. *Phys. Rev. Letts* **33**, 624 (1974).
62. G. Baym and C. J. Pethick. To be published (1975).

63. J. C. Wheatley. *In* "Quantum Fluids" (D. F. Brewer, Ed.), Wiley, New York (1966).
64. S. Doniach and S. Engelsberg. *Phys. Rev. Letts* **17**, 750 (1966).
65. M. J. Rice. *Phys. Rev.* **159**, 153 (1967).
66. M. J. Rice. *Phys. Rev.* **162**, 189 (1967).
67. S. K. Ma, M. T. Beal-Monod and D. R. Fredkin. *Phys. Rev.* **174**, 227 (1968).
68. C. J. Pethick. *In* "Lectures in Theoretical Physics" (K. T. Mahunthappa, Ed.), Gordon and Breach, New York (1969).
69. V. J. Emery. *Phys. Rev.* **170**, 205 (1968).
70. J. Amit, J. W. Kane and H. Wagner. *Phys. Rev.* **175**, 313 (1968).
71. N. F. Mott. *Proc. Phys. Soc.* **62**, 416 (1949).
72. M. C. Gutzwiller. *Phys. Rev.* **137**, A1726 (1965).
73. W. F. Brinkman and T. M. Rice. *Phys. Rev.* **B2**, 4302 (1970).
74. W. F. Brinkman and T. M. Rice. *In* "Battelle Colloquium on Critical Phenomena" (1971), p. 593.
75. S. Babu and G. E. Brown. *Ann. Phys. (New York)* **78**, 1 (1973).
76. E. Feenberg. "Theory of Quantum Fluids", Academic Press, New York and London (1969).
77. K. A. Brueckner and J. L. Gammel. *Phys. Rev.* **109**, 1040 (1958).
78. E. Østgaard. *Phys. Rev.* **187**, 371 (1969).
79. L. P. Gor'kov and L. P. Pitaevskii. *Sov. Phys. Dokl.* **8**, 788 (1964).
80. H. Bookbinder. *Phys. Rev.* **A3**, 372 (1971).
81. A. J. Layzer and D. Fay. *Int. J. Mag.* **1**, 135 (1971).
82. S. Nakajima. *Prog. Theor. Phys.* **50**, 1101 (1973).
83. A. J. Layzer and D. Fay. *Solid State Commun.* **15**, 599 (1974).
84. J. R. Schrieffer. "Superconductivity," Benjamin, New York (1963), Ch. 5.
85. L. Tewordt. *J. Law Temp. Phys.* **15**, 349 (1974); *Zeit. f. Phys.* **268**, 207 (1974).
86. G. M. Eliashberg. *Sov. Phys. JETP* **11**, 696 (1960).
87. E. P. Wigner. "Theory of Groups in Quantum Mechanics", Springer, Berlin (1973).
88. G. Eilenberger. *Zeit. f. Phys.* **182**, 427 (1965).
89. J. M. Luttinger and J. C. Ward. *Phys. Rev.* **118**, 1417 (1960).
90. N. R. Werthamer. *In* "Superconductivity" (R. D. Parks, Ed.), Dekker, New York (1969), p. 321.
91. N. D. Mermin and G. Stare. *Phys. Rev. Letts* **30**, 1135 (1973).
92. I. S. Gradshteyn and I. M. Ryzhik. "Table of Integrals, Series and Products", Academic Press, New York and London (1965), 4.116.
93. J. M. Delrieu. Private communication (1973).
94. E. I. Blount and M. C. Cross. Private communication.
95. P. W. Anderson. *Rev. Mod. Phys.* **38**, 298 (1966).
96. S. Engelsberg, W. F. Brinkman and P. W. Anderson. *Phys. Rev.* **A9**, 2592 (1974).
97. R. A. Webb, R. L. Kleinberg and J. C. Wheatley. *Phys. Letts* **48A**, 421 (1974).
98. R. A. Webb, R. L. Kleinberg and J. C. Wheatley. *Phys. Rev. Letts* **33**, 145 (1974).
99. J. Bardeen and J. R. Schrieffer. *Prog. Low Temp. Phys.* **3**, 212 (1961).
100. M. C. Cross. Private communication.
101. A. W. Yanof, E. Smith, D. M. Lee, R. C. Richardson and J. D. Reppy. *Bull. Am. Phys. Soc.* **19**, 435 (1974).
102. O. Betbeder-Matibet and P. Nozières. *Ann. Phys. (New York)* **51**, 392 (1969).
103. D. N. Paulson, R. T. Johnson and J. C. Wheatley. *Phys. Rev. Letts* **31**, 746 (1973).
104. M. Vuorio. *J. Phys. C* **7**, L5 (1974).

105. D. T. Lawson, W. J. Gully, S. Goldstein, R. C. Richards and D. M. Lee. *Phys. Rev. Letts* **30**, 541 (1973).
106. A. J. Leggett. *Phys. Rev.* **147**, 146 (1966).
107. P. W. Anderson. *Phys. Rev.* **112**, 1900 (1958).
108. N. N. Bogoliubov, V. Tolmachev and D. V. Shirkov. "New Method in the Theory of Superconductors", Consultants Bureau, New York, (1959).
109. D. N. Paulson, R. T. Johnson and J. C. Wheatley. *Phys. Rev. Letts* **30**, 829 (1973).
110. P. W. Anderson and H. Suhl. *Phys. Rev.* **116**, 898 (1959).
111. J. Serene and D. Rainer. To be published (1975).
112. A. J. Leggett. *Ann. Phys.* (*New York*) **85**, 11 (1974).
113. H. M. Bozler, M. E. R. Bernier, W. J. Gully, R. C. Richardson and D. M. Lee. *Phys. Rev. Letts* **32**, 875 (1974).
114. K. Maki and H. Ebisawa. *J. Low Temp. Phys.* **15**, 212 (1974).
115. K. Maki and H. Ebisawa. *Phys. Rev. Letts* **32**, 520 (1974).
116. S. Takagi. *Prog. Theor. Phys.* **51**, 674 (1975).
117. R. Combescot and H. Ebisawa. *Phys. Rev. Letts* **33**, 810 (1974).
118. V. L. Ginzburg and L. D. Landau. *Zh. Eksp. Teor. Fiz.* **20**, 1064 (1950).
119. L. P. Gor'kov. *Sov. Phys. JETP* **9**, 1364 (1959).
120. N. R. Werthamer. *Phys. Rev.* **132**, 663 (1963).
121. P. G. de Gennes and D. Rainer. *Phys. Letts* **46A**, 429 (1974).
122. K. Maki. *Prog. Theor. Phys.* **52**, 745 (1974).
123. D. D. Osheroff, Thesis, Cornell University (1972).
124. L. R. Walker. *Phys. Rev.* **105**, 390 (1957).
125. A. I. Ahonen, M. T. Haikala and M. Kousins. *Phys. Letts* **47A**, 215 (1974).

PRINCIPLES OF A CLASSIFICATION OF DEFECTS IN ORDERED MEDIA

G. TOULOUSE and M. KLÉMAN

Laboratoire de Physique des Solides, Université Paris-Sud, 91405 Orsay, France

(*Reçu le 25 mars 1976, accepté le 15 avril 1976*)

Résumé. — Une classification des défauts dans les milieux ordonnés est présentée. Fondée sur des concepts purement topologiques, elle fournit une distinction entre défauts élémentaires (topologiquement stables) et défauts composés. Cette classification systématique contient des résultats dérivés antérieurement de manière empirique et permet de nouvelles prédictions. Elle révèle une liaison entre la nature des défauts dans une phase ordonnée et les phénomènes critiques à la transition de phase.

Abstract. — A classification of defects in ordered media is presented. Based on purely topological concepts, it provides a distinction between elementary (topologically stable) and compound defects. This systematic classification recovers some previous empirically derived results and allows new predictions. It exhibits a striking connection between the nature of defects in an ordered phase and the critical phenomena at the phase transition.

1. **Introduction.** — An important part of condensed matter and phase transitions physics is involved with the study of the defects which occur in the ordered phases and give rise to the variety of observed textures. It is a natural inclination of the physicist to try to find elementary objects, which can serve as building blocks to construct the others. What is needed, in other terms, is a classification of elementary defects and the set of rules governing their aggregation. In other words, one wants to apply to the defects in a given ordered medium the program which has been achieved for chemical objects, with a hierarchy of levels : atoms, molecules, condensed states.

Most of the past and considerable effort in the theory of defects has been concerned with energy calculations, although it has been recognized by many that topological concepts are important. It is shown here that it is possible to go far indeed with considerations based merely on continuity properties, that is with topology. For a given ordered medium of arbitrary space dimensionality, it is possible to give a systematic classification of the elementary defects of various dimensionalities (points, lines, walls, ...) and to attribute to them characteristic numbers, which govern the rules for their associations.

2. **The space of internal states.** — The nature of the ordering in an ordered medium can be characterized by an order parameter. Some examples may help to fix ideas ; for an ordered alloy, the order parameter is a real scalar ; for a superfluid, it is a complex scalar ; for an isotropic ferromagnet, it is a vector ; many more examples could be given. This order parameter is defined at each point of the ordered medium, and, by definition, it characterizes the internal state of the medium at that point. If there were no distortions, the internal state would be the same in each point of the sample. The presence of defects is accompanied by a variation of the internal state from point to point in the medium.

Now each internal state can be represented by a point in an abstract space, the space of internal states. We shall be interested in the subspace formed by all possible values of equal amplitude of the order parameter, and we shall call this subspace the manifold of internal states (we use the term manifold, instead of space, in order to avoid confusion with the real space in which the medium lies, and also, because it is a topological concept, to stress the topological properties).

Let us take again some examples ; for a real scalar order parameter, the manifold of internal states is two points ($\pm$ 1) ; for a complex scalar, it is a circle ; for a vector, it is a sphere ; etc... This manifold of internal states has important topological properties. First, its dimensionality. Second, its connectivity properties. In the theory of critical phenomena, that occur near a phase transition point, much emphasis has been put on the effect of the dimensionality. In this theory of defects, the emphasis will be put on the connectivity properties. Actually, the two topics are closely related.

3. ***The surrounding* of defects.** Let us begin with a known simple example, which will be generalized afterwards. Consider a line defect (vortex line) in a three dimensional sample of superfluid. To characterize this line defect, one surrounds it by a closed loop. The phase change $\Delta\varphi/2\pi$ of the complex order parameter as one completes a turn along the loop is a topological invariant : to one turn in real space, around the vortex line, is associated a certain closed path in the manifold of internal states. This closed path (more precisely, the class of equivalent paths into which this path can be continuously deformed within the manifold of internal states) then characterizes topologically our line defect. If the closed path can be continuously deformed into one point in the manifold, then the line defect is not topologically stable (it can be continuously reduced to no defect at all); if the closed path cannot be continuously deformed into one point in the manifold, then the line defect is topologically stable.

Let us generalize this construction to arbitrary space dimensionalities of the medium (d) and of the defect (d'). We wish to *surround* the defect by a subspace of dimensionality r such that :

$$d' + 1 + r = d .$$

The term 1 in the left hand side comes from the distance between the line defect and the subspace which surrounds it. In the preceding example, $d = 3$, $d' = 1$, and the surrounding subspace has dimensionality $r = 1$. Now it is seen that, in three-dimensional space, wall defects will be surrounded by two points (this is the 0-dimensional sphere S_0), line defects by a closed loop (this is the 1-dimensional sphere S_1), point defects by a sphere (this is the 2-dimensional sphere S_2).

In each point of the surrounding subspace S_r exists some internal state which is represented by a point in the manifold of internal states V. This defines a map of S_r into V. The possible maps of S_r into V can be classified into classes of equivalent maps (which can be continuously deformed into one another within V). The ensemble of these classes is called the rth homotopy group of V and is denoted $\pi_r(V)$.

Much is known in mathematics concerning the homotopy groups of many manifolds. In some cases, one simply recovers empirically known facts. For instance, in the preceding example of the three-dimensional superfluid, knowing that the manifold of internal states is $V = S_1$ and that

$$\pi_0(S_1) = 0 , \quad \pi_1(S_1) = Z , \quad \pi_2(S_1) = 0 ,$$

where 0 denotes the trivial group with only one element and Z the additive group of integers, one concludes that there are no stable walls, no stable points, but that there are stable vortex lines which can be characterized by an integer (positive or negative), the strength of the vortex. In three-dimensional isotropic ferromagnets, for which $V = S_2$, one finds stable points but no stable lines (this corresponds to the empirical phenomenon of *escape in third dimension*).

4. **Application to systems where the order parameter is an *n*-component vector (*n*-vector model).** — This includes a large category of systems which contains the examples previously given (real scalar order parameter : $n = 1$; complex scalar : $n = 2$; ordinary vector : $n = d$) and has been much studied in the context of critical phenomena.

For an n-vector order parameter, the manifold of internal states is $V = S_{n-1}$, since the amplitude is taken constant. Now it is known [1] that

$$\pi_r(S_m) = 0 \quad \text{for} \quad r < m ,$$

$$\pi_m(S_m) = Z .$$

Topologically stable defects have therefore the dimensionality

$$d' = d - n ,$$

which means that for $n > d$, there are no topologically stable defects, for $0 < n < d$ (this is the *triangle of defects* in the n, d plane) there is one kind of defect (points for $n = d$, lines for $n = d - 1$, walls for $n = d - 2$, ...; other defects may occur for $d > 4$, see note [1]), and finally for $n < 0$, there is again no topologically stable defect.

Note that the boundaries of the *triangle of defects* in the n, d plane are the diagonal $n = d$, which plays an important role in critical phenomena, and the line $n = 0$, which is known to describe disordered systems; it is interesting to notice that, as far as defects are concerned, the case $n = 0$ corresponds to stable defects having the dimensionality of real space, the whole system being in some sense the core of a defect, with no recognizable ordered domains.

5. **Application to some other systems.** — Let us consider uniaxial nematic liquid crystals, where the order parameter is a line element, that is a vector with no arrow. For an arbitrary number n of components of the order parameter, the manifold of internal states is $V = P_{n-1}$, which means real projective space of $(n - 1)$ dimensions. For usual nematics in three dimensional space, $V = P_2$, the projective plane; for two-dimensional nematics, $V = P_1 = S_1$.

It is then known that

$$\pi_1(P_m) = Z_2 ,$$

$$\pi_r(P_m) = \pi_r(S_m) \quad \text{for} \quad r > 1 ,$$

where Z_2 is the two-element group of the integers modulo 2.

As a consequence, for instance, the usual three-dimensional nematics will have, besides the point defects they share with the corresponding vector systems, topologically stable line defects which have the property of being their own antiparticle : two nematic line defects can disintegrate into points.

As a last example, let us consider the superfluid A phase of He_3 where the orbital order parameter is now estimated to be a frame of three orthogonal vectors (we neglect here the nuclear spin degrees of freedom, which amounts to considering only defects which do not break the dipolar energy). Then, the manifold of internal states V is $V = SO(3) = P_3$, so that the A phase appears as a kind of higher-dimensional nematic. One then predicts for a three-dimensional He_3 sample, no walls, no points and lines which are their own antiparticles (these lines can have a mixed vortex-disgyration [2] character). It is amusing to notice that if one tries to construct a point defect for one of the three orthogonal vectors, there is necessarily a *string* of singularities of the other two vectors, attached to the point; this situation is obviously reminiscent of the Dirac monopoles [3].

At this stage, it may be noticed that, through such a classification, the mere observation of the defects in a given phase may give a clue on the nature of the ordering; this poses a rather interesting inverse problem.

Obviously, this short exposition calls for development of both abstract and concrete aspects of the classification scheme; this will be presented in a more detailed publication [4].

6. **Conclusion.** — During the course of our study, we have discovered that topological concepts have been previously used in field theory by quite a few people, the emphasis being mainly on point singularities [5] or on global configurations of the whole space [6]. Actually, this similarity of concepts in the study of elementary particles and of defects in ordered media appears as a very promising feature. First, it brings some unity in physics. Second, it will probably lead to cross-fertilization. The field theorists (and the mathematicians) have an experience with rather complicated manifolds, the condensed matter physicists can exhibit many systems, with a lot of experimental control on them.

Acknowledgments. — It is a real pleasure for the authors to acknowledge much direct and indirect inspiration from R. Thom, and fruitful discussions with L. Michel and V. Poenaru.

References

[1] The $\pi_r(S_m)$, for $r > m$ and $m > 1$, exhibit a rich variety which is not discussed here because they do not enter into account for the physical dimensionalities $d < 4$.

[2] DE GENNES, P. G., in *Collective properties of physical systems*, Nobel Symposium (Academic Press) 1973.

[3] DIRAC, P. A. M., *Phys. Rev.* **74** (1948) 817.

[4] KLÉMAN, M., TOULOUSE, G., in preparation.

[5] HOOFT, G.'t, *Nucl. Phys. B* **79** (1974) 276; MONASTYRSKII, M. I., PERELOMOV, A. M., *J.E.T.P. Lett.* **21** (1975) 43.

[6] FINKELSTEIN, D., *J. Math. Phys.* **7** (1966) 1218.

Investigation of singularities in superfluid He^3 in liquid crystals by the homotopic topology methods

G. E. Volovik and V. P. Mineev

L. D. Landau Institute of Theoretical Physics, USSR Academy of Sciences; Institute of Solid State Physics, USSR Academy of Sciences
(Submitted December 27, 1976)
Zh. Eksp. Teor. Fiz. 72, 2256–2274 (June 1977)

Various singularities in the superfluid He^3 are considered: vortices, disgyrations, pointlike singularities, vortices with ends, singular surfaces, and particle-like states, as well as disclinations in cholesteric liquid crystals. A classification is presented of the topologically stable singularities. The methods of homotopic topology are used and are described with examples of well known systems such as superfluid He II, an isotropic ferromagnet, and a nematic liquid crystal. The possibility of applying these methods to ordinary crytals and to liquid crytals of the smectic type is discussed.

PACS numbers: 67.50.Fi, 61.30.–v

1. INTRODUCTION

We shall consider singularities in superfluid He^3 and in liquid crystals. These substances are typical examples of systems with spontaneously broken symmetry. Such systems are characterized by the fact that their equilibrium states, at given homogeneous external conditions (temperature, pressure, external fields) are degenerate with respect to one or several parameters. In other words, there are non-equivalent equilibrium states in which these parameters are different but the thermodynamic potential is the same. Thus, for He II the degeneracy parameter is the condensate phase shift Φ, for an isotropic ferromagnet it is the direction of the spontaneous magnetization $\mathbf{m}$ (the degeneracy parameters for liquid crystals and for superfluid He^3 will be given in the text).

In the case of an inhomogeneous state of the substance the degeneracy parameter is a function of the coordinates and of the time. Inhomogeneous states are possible in which, at a certain point or on a line in space, the degeneracy parameter is not defined, and this singular point or line cannot be eliminated without destroying at the same time the ordered state in a large volume of matter. This, for example, is the situation with the vortex in He II. This vortex constitutes a singular line and the degeneracy parameter (the phase Φ) changes by 2π after circling this line. On the line itself, the phase Φ is indeterminate. This singular line can be eliminated only by destroying the superfluid state in a large volume of liquid. It is easily seen that the existence of a vortex in He II is connected with the fact that the region in which the phase Φ varies is a circle of unit radius.

It is natural to expect the existence of singular lines and points of other ordered substances also to depend on the global properties of the region where the degeneracy parameter varies, i.e., on its topological structure. The purpose of the present article is to describe a regular method of classifying topologically stable singularities of the degeneracy parameter of an ordered system, by starting from the topological structure of the region where the latter varies. This method is based on the use of the so called homotopic group. It makes it possible to find all the types of topologically stable singularities, i.e., those which can be eliminated only by destroying the ordered state in a large volume, and also to set in correspondence with each singularity a homotopic-group element, by the same token making it possible to classify the types of singularities. In addition, by using this method, it is possible to identify the type of singularity that results from coalescence of singularities.

The homotopic topology method has not been used in the literature known to us on the investigation of different singularities of ordered systems. (For the use of this method in field theory, see the paper of Monastyrskiĭ and Perelomov.)[1] In a number of cases, therefore, the classifications obtained in the literature were either incomplete or contained singularities that could be eliminated topologically.

In Secs. 2–4 of the present paper the method of homotopic groups is described with He II, a ferromagnet, and a nematic liquid crystal as examples. In the exposition of this method, we cite elementary information on homotopic topology, details of which can be found, for example, in the book of Huismoller and Spanier.[2] In Sec. 5, the homotopic groups are used to classify the singularities in superfluid He^3 (some of the results were already published earlier).[3] In Sec. 6, this method is applied to the classification of singularities in liquid crystals of the cholesteric type, and the question of its application to ordinary crystals and liquid crystals of smectic type is also considered. The Appendix contains a brief description of the method of calculating the homotopic groups used in the article (see also[1,2]).

2. THE FUNDAMENTAL GROUP AND LINEAR SINGULARITIES OF He II

It is known that He II is characterized by a complex order parameter $\Psi = |\Psi| e^{i\Phi}$. At equilibrium the modulus of the order parameter assumes a fixed value $|\Psi| = C(T, P)$, which minimizes the condensation energy $F_c(|\Psi|)$, while the phase Φ can assume any value and is therefore the degeneracy parameter. In the nonequilibrium state, when $|\Psi|$ and Φ vary in space, an additional gradient energy F_{grad}, which depends on $\nabla|\Psi|$ and $\nabla\Phi$, is added to the condensation energy. We consider weakly inhomogeneous states, when the characteristic distances over which $|\Psi|$ and Φ vary are much larger than the coherence length $\xi(T, P)$. In this case $F_{grad} \ll F_c$, so that we can assume $|\Psi|$ to be practically constant, and only the degeneracy parameter Φ varies in space, and the gradient energy can be regarded as dependent only on $\nabla\Phi$, i.e., $F_{grad} = \rho_s v_s^2/2$, where $\mathbf{v}_s = \hbar\nabla\Phi/m_4$ is the superfluid velocity and m_4 is the mass of the He^4 atom. This no longer holds near the core of the vortex. Indeed, since the phase Φ is indeterminate on the vortex axis and $|\Psi|$ should accordingly vanish, the gradient energy near the core becomes comparable with condensation energy. However, if we are not interested in a region of dimension near the core, and move over to a distance $r \gg \xi$, then we have here $F_{grad} \ll F_c$ and we can again assume that $|\Psi| = C(T, P)$ and only Φ changes.

From the mathematical point of view, the complex order parameter $\Psi(\mathbf{r})$ determines the continuous mapping of the set of points $\mathbf{r}$ of the vessel on the complex plane. If we consider only weakly inhomogeneous states and neglect regions with dimensions $\sim \xi$ near the lines on which the phase is not determined, then we obtain $|\Psi| = \text{const}$, and $\Phi(\mathbf{r})$ maps continuously the set of points $\mathbf{r}$ of a vessel with a notched line L, on which the phase Φ is not defined, into the circle that constitutes the region of variation of the phase Φ (this circle will henceforth be designated S^1).

We must ascertain which of these singular lines can be eliminated by continuous deformation of the field $\Phi(\mathbf{r})$, and which cannot be eliminated by any continuous change of the field $\Phi(\mathbf{r})$. To this end, we surround the investigated singular line L by a simple closed contour γ (which passes, of course, at a distance $r \gg \xi$ from the line), which starts at a fixed point and goes in a fixed direction. This contour is mapped by the function $\Phi(\mathbf{r})$ on the circle S^1 also into a closed contour Γ with a fixed circuiting direction; the point $\mathbf{r}_0$ is mapped thereby into the point $A = \Phi(\mathbf{r}_0)$ on S^1.

Let us imagine that we can contract the contour Γ to a point A, by continuously deforming it on S^1. It is then easily seen that the investigated line is topologically not singular, since it can be eliminated by continuously transforming the field $\Phi(\mathbf{r})$ into a constant field $\Phi(\mathbf{r}) = \text{const}$. In the case when the investigated line is the core of a vortex, the contour surrounds the circle S^1 once or several times and is closed at the point A. In this case we cannot contract the contour Γ into a point by any deformation. Consequently, no continuous change of the field $\Phi(\mathbf{r})$ is capable of eliminating the singularity on the core of the vortex.

Thus, the line L is topologically singular whenever the contours γ which surround it are mapped on contours Γ that cannot be contracted into points in the region of variation of the degeneracy parameters. This is valid also for other ordered systems. To investigate singular lines of an ordered system it is therefore necessary to investigate the possible continuous deformations of the contours Γ in the region of variation of the degeneracy parameters (this region will henceforth be denoted R)

In topology, continuous deformation is called homotopy. Two closed contours emerging from a point A are called homotopic relative to each other or homotopically equivalent, or else belonging to a single homotopic class, if they can be deformed in continuous fashion into each other while leaving the point A immobile. The contours that contract to a point are referred to as homotopic to zero. Obviously, contours with different numbers of circuits along a circle are homotopically not equivalent. The product of two contours Γ_1 and Γ_2 is defined as a contour $\Gamma_2\Gamma_1$ such that the mapping of a point that runs along the contour γ, after leaving the point A, first goes around contour Γ_1 and then Γ_2. The element reciprocal to Γ is defined as the contour Γ^{-1} with opposite circuiting direction. The set of contours Γ belonging to one homotopic class must be regarded as a single contour within the homotopy accuracy. Just as for the contours, we can introduce definitions of multiplication of classes of contours, of a class that is the inverse of a given class, and also a unit class of contours that are homotopic to zero. It can be verified that multiplication of classes is associative.

Thus, the aggregate classes of homotopic contours forms a group—the so called fundamental group of space R, which we designate $\pi_1(A, R)$. This group is generally speaking non-commutative. An example of such a non-commutative group is the fundamental of space R for a cholesteric liquid crystal. Deferring the discussion of non-commutative fundamental groups until we reach this case, we consider for the time being Abelian fundamental groups. The elements of the latter do not depend on the choice of the point A, so that the fundamental Abelian group will henceforth be designated $\pi_1(R)$.

In the case of He II, each vortex corresponds to a class of contours Γ—the transforms of the contours γ that surround the vortex. Therefore each vortex can be set in correspondence to an element of the fundamental group $\pi_1(S^1)$, which is called the fundamental group of the circle. The latter is isomorphic to the group of integers (which will henceforth be designated Z), since each class of contours Γ can be set in correspondence with an integer N—the number of circuits of S^1 in the positive direction. Consequently, each vortex can be characterized by a whole-number index, N, which is equal in this case to the number of circulation quanta of the superfluid velocity $\mathbf{v}_s$ around the core of the vortex

$$N = \frac{m_4}{2\pi\hbar} \oint \mathbf{v}_s \, d\mathbf{l}.$$

In our case, too, each linear singularity of the degeneracy parameters can be set in correspondence with an element of the fundamental group $\pi_1(R)$. A nonsingular configuration of the degeneracy-parameter field corresponds to a unit element of this group, and coalescence of the singularities corresponds to multiplication of elements of these groups. Since we are interested only in Abelian groups with a finite number of generators, each singularity can be characterized by a set of whole-number indices $\{N_i\}$. When singularities coalesce, the indices N_i add up in modulo p_i, where p_i is the order of the i-th generator (for example, if $p = 2$, then $1 + 1 = 0$).

Degeneracy-parameter field configurations characterized by identical indices N_i can be continuously transformed into one another. Nevertheless, they differ in their energy and therefore, a potential barrier is possible when they are transformed into one another, i.e., different locally stable singularities with identical indices are possible. In the investigation of the stability it is important to know the heights of the barriers that hinder transitions into configuration with lower energy. Thus, for example, in He II a barrier is possible when a vortex with two circulation quanta ($N = 2$) decays (a process with decreasing energy) into vortices with one circulation quantum each ($N = 1 + 1 = 2$). An estimate shows that if such a barrier exists then its value is $\sim F_c \xi^3$. However, if we want to transform a vortex with $N = 2$ into a vortex with $N = 1$ or to annihilate a vortex (processes with decrease of energy), then no continuous change of the phase Φ can accomplish this. To reach this goal it is necessary to go over in the intermediate states from the circle S^1 into a complex plane on which the entire order parameter Ψ changes. It is easy to see that the path with minimal barrier on which one vortex is transformed into another, is the one in which $|\Psi|$ vanishes in the intermediate state on a certain surface S that borders on the vortex line. The phase Φ is not defined on this surface. The height of the barrier turns out to be $\sim F_c \xi S \gg F_c \xi^3$. In view of the tremendous sizes of the potential barriers, the probability of processes with change of indices is vanishingly small. We can therefore assume that the dynamics of an ordered system is such that the summary indices of the singularities situated in a given volume of the system are conserved, i.e., the motion has the following invariants N_i:

$$\sum_l N_i^{(l)} \pmod{p_i} = N_i, \tag{2.1}$$

where the sum is taken over all the singularities in the given volume. The invariants N_i can change only when a singularity enters or leaves the volume. Homotopic topology makes it therefore possible to single out classes of singularities having identical indices with macroscopically large barriers to transitions from one class to another.

Thus, the procedure for classifying the linear singularities must be the following: A homotopic classification is first carried out to subdivide the singularities into classes with large topology-governed barriers to transitions from one class to another. For this purpose it is necessary to find the fundamental group of the space R. If this group is commutative, then each linear singularity can be set in correspondence with an element of this group (or with a set of whole-number indices), by identifying the class of contours of the space R into which the contour surrounding the singular line is mapped. By the same token, the singularities are subdivided into classes characterized by identical indices with large barriers to transitions from class to class. The subsequent analysis, which is no longer homotopic, should consist of finding the locally stable singularities within each class by minimizing the functional of the energy.

3. THE HOMOTOPIC GROUP $\pi_2(R)$ AND POINT SINGULARITIES OF A FERROMAGNET

We proceed to the investigation of the singular points. By way of example we consider an isotropic ferromagnet. The order parameter of an isotropic ferromagnet is the magnetization vector $\mathbf{M}$. The states of the ferromagnet are degenerate with respect to the directions of this vector, and therefore the degeneracy parameter is the unit vector $\mathbf{m} = \mathbf{M}/|\mathbf{M}|$, while the space R coincides with the two-dimensional sphere S^2. We note that an isotropic ferromagnet has no topologically stable linear singularities, since the fundamental group of the space S^2 is trivial ($\pi_1(S^2) = 0$). Indeed, any contour on a sphere can be contracted into a point. This, of course, does not exclude the possible existence of locally stable singular lines of the vector $\mathbf{m}$ with a small barrier $\sim F_c \xi^3$ of non-topological character on going into a nonsingular configuration, where F_c is the ferromagnetic-ordering energy.

To investigate the singular points, we surround in the ferromagnet a point in which the vector $\mathbf{m}$ is not defined by a sphere of radius much larger than ξ. The function $\mathbf{m}(\mathbf{r})$ specifies the mapping of this sphere on a certain closed surface on the sphere S^2. If the sphere σ goes over in this case to a surface that is homotopic to zero, i.e., that can be contracted into a point on S^2, then the investigated singular point is topologically unstable, since the field $\mathbf{m}(\mathbf{r})$ can be made homogeneous by means of continuous deformation. On the other hand, if we consider the so called "hedgehog," i.e., a field of the type $\mathbf{m}(\mathbf{r}) = \hat{\mathbf{r}}$ with a singular point at the origin (where $\hat{\mathbf{r}}$, $\hat{\theta}$, and $\hat{\varphi}$ are the unit vectors of a spherical coordinate system in ordinary space), then the sphere σ surrounding the singular point is mapped by the function $\mathbf{m}(\mathbf{r})$ on the entire sphere S^2. This surface cannot be contracted into a point and remain on the sphere S^2, since the singular point cannot be eliminated by any continuous transformation of the field $\mathbf{m}(\mathbf{r})$.

Thus, to classify the topologically stable singular points it is necessary to find all the classes of the surfaces on the sphere S^2 (in our case, in the space R) that are not homotopic to zero, into which the sphere σ can be mapped. These classes, together with the class of surfaces homotopic to zero, are the elements of a homotopic group of dimensionality 2, designated $\pi_2(R)$. In the case of He II the group $\pi_2(S^1) = 0$ and there are no point singularities in He II.

In the case of an isotropic ferromagnet, the group $\pi_2(S^2)$ is isomorphic to the group of integers Z. Indeed, to each class of mappings of σ on S^2 one can set in correspondence an integer N that shows how many times the vector $\mathbf{m}$ runs over the sphere S^2, with allowance for the orientation, whenever $\mathbf{r}$ runs over σ. This number is called the degree of mapping and it can be expressed in terms of a surface integral of the field $\mathbf{m}(\mathbf{r})$, which coincides with the integral of the Gaussian curvature of the surface to which the vector $\mathbf{m}$ is the normal vector:

$$N = \frac{1}{4\pi} \int_\sigma d\theta\, d\varphi\, \mathbf{m} \left[\frac{\partial \mathbf{m}}{\partial \theta} \times \frac{\partial \mathbf{m}}{\partial \varphi} \right]. \tag{3.1}$$

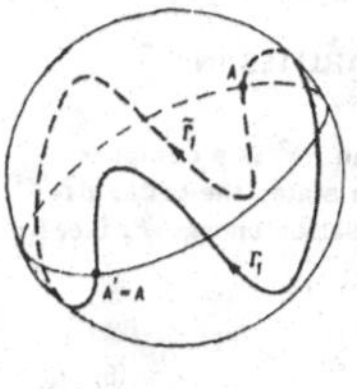

FIG. 1. Closed paths that are not homotopic to zero in the space S^2/Z_2.

Thus, each point singularity in the field of the unit vector $\mathbf{m}(\mathbf{r})$ is characterized by a whole-number index N that runs through the values from $-\infty$ to ∞. In the case of a "hedgehog" in the form $\mathbf{m}(\mathbf{r}) = \hat{\mathbf{r}}$, we have $N = 1$, and for the "hedgehog" of the type $\mathbf{m}(\mathbf{r}) = -\hat{\mathbf{r}}$ the number is $N = -1$. When the singularities coalesce, the indices N add up. Of course, within a singularity class characterized by a single index N, there can be different locally stable singularities. For one such singularity to go over into another with a lower energy it is necessary to overcome barriers on the order of $F_c\xi^3$. But to transform a singularity of one class into a singularity of the other class it is necessary, in the intermediate state, to make $\mathbf{m}$ indeterminate on the entire line L that emerges from the singular point. On this line, the vector $\mathbf{M}$ vanishes, and therefore the barrier is macroscopically large: $\sim F_c\xi^2 L \gg F_c\xi^3$.

When generalizing the foregoing arguments to the case of an arbitrary space R, it should be noted that we must stipulate in the definition of the group $\pi_2(R)$ that a certain point $\mathbf{r}_0 \in \sigma$ always goes over as a result of the mapping into the same point $A \in R$. If the fundamental group $\pi_1(R)$ is nontrivial, then it may turn out that when the point A is moved along a close contour that is not homotopic to zero the element of the group $\pi_2(R)$ goes over into another element of this group. This influence of the fundamental group $\pi_1(R)$ on the group $\pi_2(R)$ will be considered in the next section, using nematic liquid crystals as an example. On the other hand, if this phenomenon does not take place (for example, when $\pi_1(R) = 0$), then the point singularities of ordered systems are classified in the same way as linear singularities in substances with an Abelian fundamental group of the sapce R. Namely, one finds the group $\pi_2(R)$, and then each point singularity is set in correspondence with an element of this group. When the singularities coalesce, multiplication of the elements of the group takes place. It is known that all the groups $\pi_2(R)$ are Abelian, and therefore the pointlike singularities can be classified by means of whole-number integers. We then obtain singularity classes having identical indices, with large barriers for the transition between classes. The succeeding, no longer homotopic, investigation of the energy functional should reveal the locally stable singularities within each class.

4. SINGULARITIES IN NEMATIC LIQUID CRYSTALS

A nematic liquid crystal (NLC) is characterized by the unit vector $\mathbf{d}$ of the director, the states with $\mathbf{d}$ and $-\mathbf{d}$ being indistinguishable. Therefore the region of variation of the degeneracy parameter R is the sphere S^2, in which two diametrically opposed points are equivalent. This space is written in the form $R = S^2/Z_2$ (i.e., the sphere S^2 is factorized with respect to a group of two elements Z_2).

In contrast to a ferromagnet, a nematic liquid crystal has a nontrivial fundamental group $\pi_1(R) = Z_2$ (Z_2 is also called the group of residues in modulo 2). Indeed, consider a contour Γ_1 that joins two diametrically opposite A and A' on the sphere S^2 (see Fig. 1). This contour is closed because $A' = A$ and, in addition, cannot be contracted into a point. Therefore Γ_1 belongs to one of the classes of contours that are non-homotopic to zero. There is only one such class. Indeed, consider the contour Γ_1^{-1} that goes in the opposite direction. This path is equivalent to the path $\tilde{\Gamma}_1$ consisting of the diametrically opposite points. But the path $\tilde{\Gamma}_1$ can be deformed into Γ_1 over the surface of the sphere, therefore $\Gamma_1^{-1} = \Gamma_1$ and consequently $\Gamma_1 \cdot \Gamma_1 = \Gamma_0$, where Γ_0 is the class of paths that are homotopic to zero. Thus, all the contours can be homotopic either to Γ_1 or to zero. Consequently the group $\pi_1(S^2/Z_2)$ consists of two elements (see also the Appendix). The singular lines of a nematic liquid crystal are characterized by a whole-number index N that assumes only two values, 0 and 1, and the addition of the indices occurs on modulo 2 (i.e., $1 + 1 = 0$).

The linear singularities in NLC (disclinations) are customarily characterized by the Frank index m. The contour γ that surrounds the disclinations with Frank index m is mapped into the contour Γ_1^m which is homotopic to zero for even m and homotopic to Γ_1 for odd m. Therefore disclinations with an even Frank index $m = 2k$ belong to the class of singularities with index $N = 0$, and consequently any possible barrier to the transformation of these disclinations into a nonsingular configuration is small, $\sim F_c\xi^3$ (here ξ is of the order of the interatomic distances, and $F_c \sim K/\xi^2$, where K is the elastic modulus of the liquid crystal). Let us examine for example the elimination of the disclination $\mathbf{d} = \hat{\rho}$ with $m = 2$ ($\hat{\rho}$, $\hat{\mathbf{z}}$, and $\hat{\varphi}$ are the unit vectors of a cylindrical coordinate system). The field $\mathbf{d} = \hat{\rho}$ can be continuously deformed into a homogeneous field $\mathbf{d} = \hat{\mathbf{z}}$ by means of the continuous transition

$$\mathbf{d} = \hat{\rho}\cos(\pi t/2) + \hat{\mathbf{z}}\sin(\pi t/2),$$

by varying t from zero to unity (see Figs. 2a–2c). On the other hand, if we go through a potential barrier when t is varied (see, e.g., the paper of Anisimov and Dzyaloshinskiĭ),[4] then it is more convenient to eliminate the singularity by breaking the filament (see Fig. 2d). We then change the field in a volume of the order of ξ^3, and possibly surmount a barrier

$$\sim \frac{K}{\xi^2}\xi^3 \sim F_c\xi^3.$$

Disclinations with odd Frank indices $m = 2k + 1$ belong to the homotopic class $N = 1$. To eliminate them it is therefore necessary to surmount a barrier $\sim F_c\xi S$, where S is the area of the surface that bears on the disclination line. An investigation of the locally stable disclinations inside each homotopic class can be found in[4].

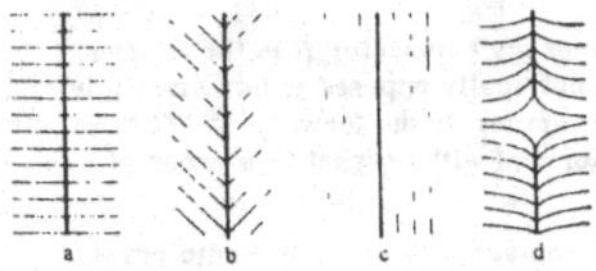

FIG. 2. Conversion of a disclination with $m = 2$ into a nonsingular configuration: a)–c) by continuous deformation, d) by breaking the filament. Thick line—disclination filament. Thin lines—lines of the director field in a plane passing through the filament.

We proceed now to point singularities in NLC. The group is $\pi_2(S^2/Z_2) = Z$ (see the Appendix), i.e., there is a group of homotopic mappings of the sphere σ on the space $R = S^2/Z_2$; this group is homotopic to the group of integers. These mappings can be easily obtained from the mappings of on the sphere S^2 followed by mapping S^2 on S^2/Z_2. Therefore the elements of $\pi_2(R)$ are characterized by a whole-number invariant N of the type (3.1), where **m** must be replaced by **d**. It is easy to see, however, that each singularity is characterized by two numbers, N and $-N$. Indeed, replacement of **d** by $-$**d** does not change the states, whereas in (3.1) N is replaced by $-N$. This is the consequence of the influence of the fundamental group $\pi_1(R)$ on the group $\pi_2(A, R)$. Let us examine this in greater detail.

Assume that we have a mapping of degree N of the sphere σ on R, and the point $\mathbf{r}_0$ goes over into a certain point $\mathbf{d}(\mathbf{r}_0) = A$. If we move the point $\mathbf{r}_0$ over a closed contour γ, then the point A will move over a closed contour Γ in the space R. If the contour γ does not enclose a singular line, then the vector $\mathbf{d}(\mathbf{r}_0)$ does not reverse sign after the circuit and N remains likewise unchanged. In the opposite case, when γ encloses a singular line, then $\mathbf{d}(\mathbf{r}_0)$ goes over into $-\mathbf{d}(\mathbf{r}_0)$ after the circuit, and consequently N reverses sign. The point A then runs over the contour Γ_1, which is not homotopic to zero. Thus, the influence of the fundamental group $\pi_1(R)$ on $\pi_2(R)$ consists in the fact that when the point A moves along a certain contour that is not homotopic to zero, the elements of the group $\pi_2(R)$ can go over into another element of this group. In this case the element N goes over into $-N$. In the general case of an arbitrary space R, we can state that each point singularity corresponds to an entire class of the group $\pi_2(R)$, whose elements are obtained from one another by motion of the point A over the contours of the group $\pi_1(R)$. Thus in this case each singularity corresponds to a class of a group of whole numbers with identical moduli, i.e., the singularities are characterized by the index $|N|$. When the singularities coalesce, multiplication of the classes takes place. In a nematic liquid crystal, the coalescence of singularities with $|N_1|$ and $|N_2|$ can result in a singularity having either the index $|N_1| + |N_2|$ or the index $||N_1| - |N_2|$ The particular singularity obtained depends on the manner of the coalescence. In order to clarify this, it is necessary to introduce continuously in the vicinity of the coalescence path, in place of the director field **d**, the field of the true director **d** and add the indices of the singular points of the field **d** (see Fig. 3).

5. CLASSIFICATION OF SINGULARITIES IN SUPERFLUID He^3

The order parameter in superfluid He^3 is a complex 3×3 matrix A_{ik}. In the equilibrium state, the order parameter that minimizes the condensation energy F_c (see Leggett's review),[5]

$$F_c = -\alpha A_{ik}A_{ik}^* + \beta_1 |A_{ik}A_{ik}|^2 + \beta_2 (A_{ik}A_{ik}^*)^2 + \beta_3 A_{ik}A_{il}A_{mk}^*A_{ml}^* + \beta_4 A_{ik}A_{il}^*A_{mk}A_{ml}^* + \beta_5 A_{ik}A_{il}^*A_{mk}^*A_{ml}, \tag{5.1}$$

takes the following form for the A phase and the B phase of He^3, respectively

$$A_{ik} = \mathrm{const} \cdot V_i (\Delta_k' + i\Delta_k''), \quad A_{ik} = \mathrm{const} \cdot e^{i\Phi} R_{ik}(\omega), \tag{5.2}$$

where **V**, $\mathbf{\Delta}'$, and $\mathbf{\Delta}''$ are arbitrary unit vectors connected by the relation $\mathbf{\Delta}' \perp \mathbf{\Delta}''$ (the vector product $\mathbf{\Delta}' \times \mathbf{\Delta}''$ specifies the direction **l** of the orbital angular momentum of the Cooper pair); Φ is the phase of the condensate, $R_{ik}(\omega)$ is the matrix of the rotation through the angle $|\omega| \leq \pi$ around the ω axis, and const $\approx (\alpha/\beta)^{1/2}$. In the B phase of He^3 the equilibrium states are degenerate with respect to the phase Φ and the rotation matrix R_{ik}, therefore the space R for the B phase is the product of S^1 (the region of variation of the phase Φ) by the space of the three-dimensional rotations SO_3, i.e.,

$$R_B = S^1 \times SO_3. \tag{5.3}$$

In the A phase the space R is the product of S^2 (the region of variation of the vector **V**) by SO_3 (rotations that specify the orientation of the triplet of vectors $\mathbf{\Delta}'$, $\mathbf{\Delta}''$, **l**). It must be recognized here that the states with **V**,

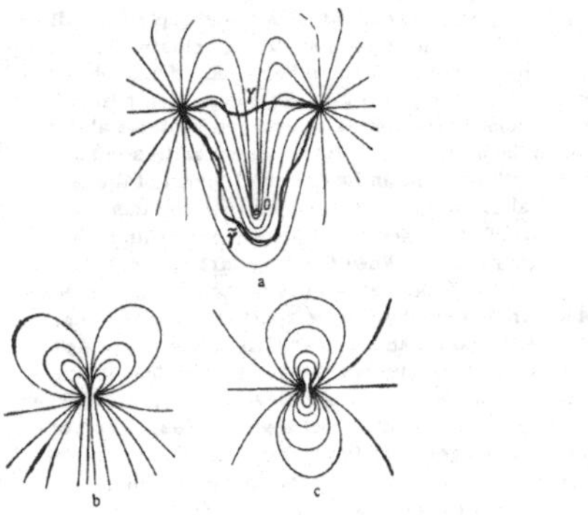

FIG. 3. Coalescence of point singularities with indices $|N_1| = |N_2| = 1$ in a nematic liquid crystal: a) coalescence along the paths γ and $\tilde{\gamma}$ passing on opposite sides of the disclination line (the point O) perpendicular to the plane of the figure yields respectively b) a point singularity with $|N| = 2$ and c) a point singularity with $|N| = 0$.

Δ', Δ'' and $-\mathbf{V}$, $-\Delta'$, $-\Delta''$ are indistinguishable, so that the space $S^2\times SO_3$ must be factorized with respect to the group Z_2, i.e.,

$$R_A=(S^2\times SO_3)/Z_2 \tag{5.4}$$

(Chechetkin[6] has incorrectly defined the spaces R_A and R_B, and this resulted in an incorrect classification of the singularities in the A and B phases of He^3).

The degeneracy in the A and B phases is partially lifted on account of the weak spin-orbit interaction F_{sl}:

$$F_{sl}=\lambda(|A_{ii}|^2+A_{ik}A_{ki}^*),\quad \lambda\ll\alpha. \tag{5.5}$$

In the A phase, the vector $\mathbf{V}$ is fixed in either the direction $\mathbf{l}$ or $-\mathbf{l}$, while in the B phase the degeneracy remains with respect only to those matrixes R_{ik} which describe rotation through a fixed angle $|\omega| = \theta_0 = \arccos(-\frac{1}{4}) \approx 104°$ relative to an arbitrary axis ω. The regions R_A and R_B go over in this case into the following:

$$R_A=U_2,\quad R_B=S^1\times S^2 \tag{5.6}$$

where the sphere S^2 is the region of variation of the unit vector ω/θ_0.

In the inhomogeneous state, the gradient energy

$$F_{grad}=\gamma\left(2\frac{\partial A_{\alpha i}}{\partial x_k}\frac{\partial A^*_{\alpha k}}{\partial x_i}-\frac{\partial A_{\alpha k}}{\partial x_i}\frac{\partial A^*_{\alpha k}}{\partial x_i}\right) \tag{5.7}$$

is added to F_c and F_{sl}, and the result is two length scales $\xi\sim(\gamma/\alpha)^{1/2}$ and $R_c\sim(\gamma/\lambda)^{1/2}$, with $R_c\sim(10^2-10^3)\xi \gg \xi$. If the characteristic distances over which A_{ik} varies are $r\gg R_c$, then the gradient energy $F_{grad}\ll F_{sl}F_c$, $F_{grad}\ll F_c$ does not change the structure of the order parameter and the region of variation of R takes the form (5.6). On the other hand, if the characteristic distances are $\xi\ll r\ll R_c$, then $F_{sl}\ll F_{grad}\ll F_c$, and therefore the spin-orbit interaction can be neglected, and does not influence the form of the order parameter, and consequently the region R is given by formulas (5.3) and (5.4). Therefore the classification of the singularities in superfluid He^3 depends essentially on the dimensions of the investigated region of the liquid. We shall consider only the A phase, since the classification of the linear and point singularities of the B phase is given quite completely in the preceding paper.[3] Exceptions are the singular surfaces (see below).

We consider the first when one of the characteristic dimensions of the region is $\xi\ll r\ll R_c$. In this case the space R is given by formula (5.4). The homotopic groups of this space are

$$\pi_1((S^2\times SO_3)/Z_2)=Z_4,\quad \pi_2((S^2\times SO_3)/Z_2)=Z$$

(see the Appendix). Here Z_4 is a group of residues in modulo 4, therefore the linear singularities are characterized by a whole-number index N that takes on values 0, 1, 2, and 3. Coalescence of the singularities leads to addition of the indices in modulo 4, for example, 3 + 3 = 2. The barrier for the transition of a singularity from one class to another is $\sim F_c\xi S$.

To find locally stable singularities within each class it is necessary to minimize the gradient-energy functional $\int d^3r F_{grad}$ with respect to the degeneracy parameters. Upon variation of this functional, equations are obtained for $\mathbf{V}$, Δ', and Δ''. We write out several solutions for these equations for linear singularities of various classes ($\hat{\mathbf{z}}$, $\hat{\boldsymbol{\rho}}$, $\hat{\boldsymbol{\varphi}}$ and $\hat{\mathbf{z}}$, $\hat{\mathbf{x}}$, $\hat{\mathbf{y}}$ are unit vectors of cylindrical and Cartesian coordinate systems, respectively, with the $\hat{\mathbf{z}}$ axis along the singular line):

$N=1$:

$$\Delta'+i\Delta''=e^{i\varphi/2}(\hat{\mathbf{x}}+i\hat{\mathbf{y}}),\quad \mathbf{V}=\hat{\mathbf{x}}\cos\frac{\varphi}{2}-\hat{\mathbf{y}}\sin\frac{\varphi}{2},\quad \mathbf{v}_s=\frac{\hat{\boldsymbol{\varphi}}}{4m_3\rho} \tag{5.8}$$

$N=2$:

$$\Delta'=\hat{\boldsymbol{\varphi}},\quad \Delta''=\hat{\mathbf{z}},\quad \mathbf{l}=\hat{\boldsymbol{\rho}},\quad \mathbf{V}=\text{const},\quad \mathbf{v}_s=0; \tag{5.9a}$$

$$\Delta'+i\Delta''=e^{i\varphi}(\hat{\mathbf{x}}+i\hat{\mathbf{y}}),\quad \mathbf{V}=\text{const},\quad \mathbf{v}_s=\hat{\boldsymbol{\varphi}}/2m_3\rho; \tag{5.9b}$$

$$\Delta'=\hat{\boldsymbol{\varphi}},\quad \Delta''=-\hat{\mathbf{z}},\quad \mathbf{l}=-\hat{\boldsymbol{\rho}},\quad \mathbf{V}=\text{const},\quad \mathbf{v}_s=0; \tag{5.9c}$$

$$\Delta'=\hat{\boldsymbol{\varphi}},\quad \Delta''=\hat{\mathbf{r}},\quad \mathbf{l}=\hat{\boldsymbol{\theta}},\quad \mathbf{V}=\text{const},\quad \mathbf{v}_s=\hat{\boldsymbol{\varphi}}/2m_3r; \tag{5.9d}$$

$N=3$:

$$\Delta'+i\Delta''=e^{-i\varphi/2}(\hat{\mathbf{x}}+i\hat{\mathbf{y}}),\quad \mathbf{V}=\hat{\mathbf{x}}\cos\frac{\varphi}{2}+\hat{\mathbf{y}}\sin\frac{\varphi}{2},\quad \mathbf{v}_s=-\frac{\hat{\boldsymbol{\varphi}}}{4m_3\rho}; \tag{5.10}$$

$N=0$:

$$\Delta'=\hat{\mathbf{x}},\quad \Delta''=\hat{\mathbf{y}},\quad \mathbf{l}=\hat{\mathbf{z}},\quad \mathbf{V}=\text{const},\quad \mathbf{v}_s=0; \tag{5.11}$$

$$\Delta'+i\Delta''=e^{2i\varphi}(\hat{\mathbf{x}}+i\hat{\mathbf{y}}),\quad \mathbf{V}=\text{const},\quad \mathbf{v}_s=\hat{\boldsymbol{\varphi}}/m_3\rho; \tag{5.11b}$$

$$\Delta'=\hat{\mathbf{x}},\quad \Delta''=\hat{\mathbf{y}},\quad \mathbf{l}=\hat{\mathbf{z}},\quad \mathbf{V}=\hat{\boldsymbol{\rho}},\quad \mathbf{v}_s=0. \tag{5.11c}$$

Here m_3 is the mass of the He^3 atom and $v_s^i=\Delta'\nabla^i\Delta''/2m_3$.

Singularities with $N=1$ and $N=3$ are vortices in which the circulation quantum of the superfluid velocity $\mathbf{v}_s$ is equal to $\frac{1}{2}$, superimposed on these vortices are disclinations of the vector $\mathbf{V}$ with a Frank index $m=1$ (the vector $\mathbf{V}$ turns out to be analogous to the vector $\mathbf{d}$ in the NLC). The written-out singularities with $N=2$ are the vortex (5.9b) with one circulation quantum and the disgyrations (see de Gennes' paper[7]) (5.9a) and (5.9c). An analysis of the energy functional $\int d^3r F_{grad}$ shows that both disgyrations are locally stable (see[8]) and have identical and apparently the lowest energies from among all singularities of the class $N=2$. The vortex (5.9b) is locally unstable and should go over without a barrier into one of the stable disgyrations. The locally stable solution (5.9d) is a junction of two disgyrations (5.9a) and (5.9c) at the point $\mathbf{r}=0$. Indeed, at $z>0$, $\rho\to 0$ we have $\mathbf{l}\to\hat{\boldsymbol{\rho}}$, $\Delta''\to\hat{\mathbf{z}}$, i.e., the solution takes the form (5.9a), and at $z<0$, $\rho\to 0$ we have $\mathbf{l}\to-\hat{\boldsymbol{\rho}}$; $\Delta''\to-\hat{\mathbf{z}}$ and the solution takes the form (5.9c).

Singularities with $N=0$—a vortex with two circulation quanta (5.11b) and a disclination in the field of the vector $\mathbf{V}$ (5.11c) with a Frank index $m=2$ can relax without a barrier into the homogeneous state (5.11a). A barrier $\sim F_c\xi^3$ can appear when account is taken of the structure of the cores of these singularities.

The point singularities of the A phase in the region $\xi\ll r\ll R_c$ are singularities in the vector field $\mathbf{V}$. They are analogous to the singularities of the vector $\mathbf{d}$ in a nematic liquid crystal and are characterized by the index $|N|$ in accordance with formula (3.1), in which $\mathbf{m}$ must be replaced by $\mathbf{V}$. The corresponding solutions of the equations for $\mathbf{V}$, Δ', and Δ'' are not simple in form even for $|N|=1$. The "hedgehog" of the form $\mathbf{V}=\mathbf{r}$, $\Delta'+i\Delta''=\text{const}$ is not a solution of these equations and relaxes to a stable singularity with $|N|=1$ (we note that

FIG. 4. Qualitative distribution of the vector fields in a plane perpendicular to the axis of a stable vortex with $N=2$ at distances $\rho>R_c$ from the axis. Thin lines—lines of vector field **l**. Thick lines—streamlines of superfluid velocity $\mathbf{v}_s$.

point singularities in the field of the vector **l**, in contrast to the statement made by Blaha,[9] are topologically removable.)

Let us see now what happens if we extend the region of the liquid to distances $r \gg R_c$, where the range of variation of the degeneracy parameters R_A narrows down to $\bar{R}_A$ of (5.5). This gives rise to two groups of singularities of different origin. The first group includes the singularities characterized by elements of homotopic groups of space $\bar{R}_A$. Let us examine first these singularities.

The homotopic groups are $\pi_1(\bar{R}_A)=Z_2$ and $\pi_2(\bar{R}_A)=0$, i.e., there are no point singularities, and there are two classes of linear singularities. It is convenient to characterize them by the same index N as the linear singularities in the region $\xi \ll r \ll R_c$, but now N can assume only two values, 0 and 2, with addition in modulo 4.

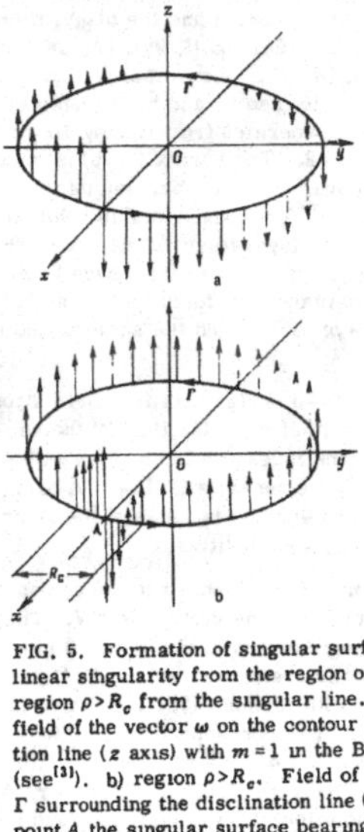

FIG. 5. Formation of singular surface in the propagation of a linear singularity from the region of distances $\rho<R_c$ into the region $\rho>R_c$ from the singular line. a) region $\rho<R_c$. The field of the vector ω on the contour Γ surrounding the disclination line (z axis) with $m=1$ in the B phase of He^3: $\omega=\hat{z}(\pi-\varphi)$ (see[11]). b) region $\rho>R_c$. Field of the vector ω on the contour Γ surrounding the disclination line (z axis) and crossing at the point A the singular surface bearing against the disclination line. In the region of distances larger than R_c from the singular surface we have $|\omega|\approx 104°$.

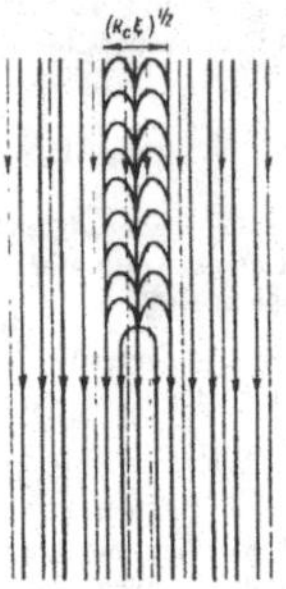

FIG. 6. Formation of singular lines in the propagation of the point singularity of the field of the vector **V** with $|N|=1$ from the region of distances $r<(R_c\xi)^{1/2}$ into the region $r>(R_c\xi)^{1/2}$ from the singular point. The qualitative distribution of the lines of the vector field **l** is shown by thin lines, and those of the field **V** by thick lines.

Among the singularities of the class $N=2$, the lowest energy is possessed by a vortex with one circulation quantum and vectors **l** and **v** perpendicular to the vortex axis. The corresponding solution has no simple form. The lines of the fields **l** and $\mathbf{v}_s$ in a plane perpendicular to the vortex axis are shown in Fig. 4. At distances $r \lesssim R_c$, this solution merges with one of the locally stable solutions of the class $N=2$ (see (5.9)). We note that in addition to linear singularities, the state $\bar{R}_A$ admits of the existence of singular surfaces.[1)] The reason is that the space O_3 is doubly connected (in other words $\pi_0(O_3)=Z_2$, where the homotopic group $\pi_0(R)$ determines the connectivity of the space R). There is therefore one class of singular surfaces joining the region with $\mathbf{V}=\mathbf{l}$ and the region with $\mathbf{V}=-\mathbf{l}$. The width of the surface is $\sim R_c$. These surfaces recall the domain walls in magnets.

The second group of singularities occurs when linear singularities with $N=1$ and 3 and point singularities propagated from the region $\xi \ll r \ll R_c$ into the region $r \gg R_c$. In this case, inasmuch as we cannot satisfy the condition $\mathbf{V} \parallel \mathbf{l}$ in the region $r \gg R_c$, the energy of these singularities becomes proportional to $F_{sl}V$, where V is the volume of space in which $\mathbf{V} \neq \pm\mathbf{l}$. In the case of linear singularities, this volume is minimal if $\mathbf{V} \neq \pm\mathbf{l}$ on a certain surface of thickness R_c, bearing on the singular line. Thus, the singular lines with $N=1$ and 3 go over into singular surfaces that border on these lines. A similar surface exists also in the B phase (see Fig. 5).

Analogously, the point singularity of the field **V** with index $|N|$ goes over into a singular line of thickness $(R_c\xi)^{1/2}$, which starts out from a singular point (see Fig. 6). In the same homotopic class there exists another singular line, albeit less convenient because of the larger gradient energy, but apparently locally stable. This singular line is a vortex with an end (vorton), the possible existence of which was noted by Blaha and by us[9,10]. The exact solution for this singularity does not have a simple form, and we therefore present one of the possible configurations of the fields **V** and **l**, which are characterized by the index $|N|=1$ (see Fig. 7):

$$\mathbf{V}=\hat{\mathbf{r}}, \quad \mathbf{v}_s=\frac{1-\hat{\mathbf{z}}\mathbf{l}}{2m_3\rho}\,\boldsymbol{\varphi}, \quad \mathbf{l}=\begin{cases}\hat{\mathbf{z}}, & z<0,\ \rho\leqslant R_c\\ \hat{\mathbf{r}} & \text{in the rest of space}\end{cases}$$

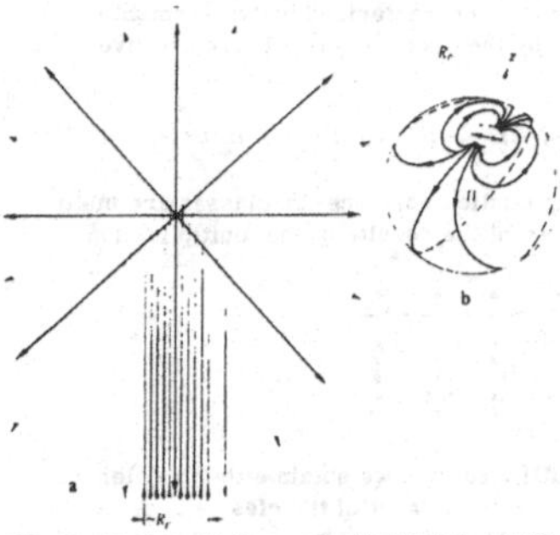

FIG. 7. Vortex with free and with $|N|=1$. a) The qualitative distribution of the lines of the vector field **l** is shown by thin lines, and those of the field **V** by thick lines. b) Distribution of the lines of the vector field Δ' on a sphere at distances $\rho > R_c$ from the vortex line coinciding with the semiaxis $z < 0$.

At a distance $\rho > R_c$ from the lower semiaxis the field of the superfluid velocity $\mathbf{v}_s$ and the field **l** take the form

$$\mathbf{v} = \frac{\hat{\varphi}}{2m_3\rho}(1-\cos\theta), \quad \mathbf{l}=\hat{\mathbf{r}},$$

which corresponds to a vortex with two circulation quanta that terminate at the center of the "hedgehog" (see[10]). The width of the core of the vortex is $\sim R_c$. In the region $r < R_c$ there are no singularities in the fields $\mathbf{v}_s$ and **l**. Such a vorton cannot vanish because of the topologically stable point singularity in the field **V** with index $|N| = 1$ at the origin. However, it is separated by an energy barrier $\sim F_{sl}R_c^3$ from the singular line with the lowest energy, shown in Fig. 6. On the other hand, if there is no singularity in the field of the vector **V**, then such a vortex relaxes into a nonsingular configuration.

We have considered all types of singular lines, points, and surfaces in the *A* phase in the absence of external fields. The presence of a magnetic field complicates the situation, since a third length scale appears, namely the magnetic lengths $R_H \sim (\gamma\alpha/\beta\chi H^2)^{1/2}$, where χ is the magnetic susceptibility of He^3. The classification depends on the relations between ξ, R_c, and R_H and on the length interval in which the characteristic dimensions of the investigated regions of the liquid are situated. For lack of space, we shall not describe this classification. It is obtained by the same method as in the case of two lengths.

We note in conclusion that the topology admits of the existence in the *A* phase of He^3 of particle-like solutions that have no singularities. By particle-like solutions we mean solutions characterized by a topological invariant that does not perturb the field of the degeneracy parameters at large distances from the particle, so that at infinity the field of the degeneracy parameters is homogeneous, i.e., all of infinity is mapped on a single point of the space *R*. The usual three-dimensional space R^3, all the points of which are equivalent at infinity, has the same topology as a three-dimensional sphere S^3 in four-dimensional space R^4 (exactly just as the plane R^2, all of the infinitely remote points of which are mapped on a single point via, e.g., a stereographic projection, is equivalent to a two-dimensional sphere S^2). Thus, the field of the degeneracy parameter specifies the mapping of S^3 in *R*. The homotopic classes of these mappings form the group $\pi_3(R)$. For the *A* phase, for example, in the region $r > R_c$ we have $\pi_3(\tilde{R}_A) = Z$ (see the Appendix). Therefore particle-like solutions are characterized by a whole-number invariant, which in this case can be written in the form

$$N = (m_3/2\pi\hbar)^2 \int d^3r\, \mathbf{v}_s \,\mathrm{rot}\, \mathbf{v}_s. \qquad (5.13)$$

From dimensionality considerations, the momentum and energy of such particles are of the order of

$$p \sim \rho_s \frac{\hbar r_0^3}{m_3}, \quad E \sim \rho_s \frac{\hbar^2}{m_3^2} r_0, \quad E \sim p^{1/3}, \qquad (5.14)$$

where ρ_s is the density of the superfluid component, r_0 is the characteristic dimension of the region of space where the fields **V**, Δ', and Δ'' are inhomogeneous. The spectrum of such particles $E \sim p^{1/2}$ is reminiscent of the spectrum of vortex rings in He II, but the field of the vectors **V**, Δ', and Δ'' has no singularities anywhere; in addition, the spectrum can be anisotropic.

The particles can also have dimensions smaller than R_c, since $\pi_3(R_A) = Z + Z$ (see the Appendix). The smallest dimension of these formations is $\sim \xi$. If $r_0 \lesssim \xi$, then $F_{grad} \sim F_c$ in this region, and consequently the order parameter A_{ik} changes already not in the vicinity of R_A, but in the entire linear space R^{18}. Since π_3 is trivial for any linear space *R*, it follows that the topological invariant *N* ceases to exist. Thus, if the particle momentum decreases to $\rho_s \hbar\xi^2/m_3$, then the particle can go over continuously into a homogeneous state.

6. CHOLESTERIC LIQUID CRYSTALS

In cholesteric liquid crystals (CLC) the spatial distribution of the director **d** about an arbitrary point $\mathbf{r}_0$ takes the following form (see the review of Stephen and Straley[11]):

$$\mathbf{d}(\mathbf{r}) = \mathbf{d}(\mathbf{r}_0)\cos\left\{\frac{2\pi}{L}\mathbf{t}(\mathbf{r}_0)(\mathbf{r}-\mathbf{r}_0)\right\} + [\mathbf{t}(\mathbf{r}_0)\mathbf{d}(\mathbf{r}_0)]\sin\left\{\frac{2\pi}{L}\mathbf{t}(\mathbf{r}_0)(\mathbf{r}-\mathbf{r}_0)\right\}, \qquad (6.1)$$

i.e., at each point $\mathbf{r}_0$ of space there is specified an orthonormal basis **t**, **d**, $\mathbf{t}\times\mathbf{d}$ (analogous to **l**, Δ', Δ'' in the *A* phase of He^3); **t** is a unit vector of the helix axis, and *L* is the pitch of the helix. Let us find the space *R* for CLC. We note for this purpose that the region of variation of the triplet of vectors **t**, **d**, $\mathbf{t}\times\mathbf{d}$ is the three-dimensional group SO_3 of rotations of this triplet relative, e.g., the bases $\hat{\mathbf{x}}, \hat{\mathbf{y}}, \hat{\mathbf{z}}$. It is known that each three-dimensional rotation corresponds to two 2×2 complex unimodular matrices $\hat{U}$ and $-\hat{U}$. If we express the matrix $\hat{U}$ in the form

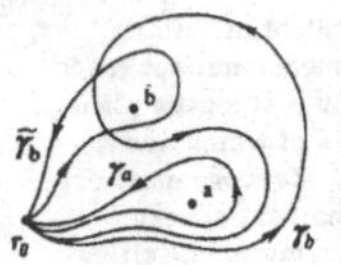

FIG. 8. The paths $\gamma_a, \gamma_b, \tilde{\gamma}_b = \gamma_a^{-1}\gamma_b\gamma_a$ surrounding the singular lines marked by the points a and b.

$$\hat{U} = \begin{pmatrix} x_1+ix_2 & x_3+ix_4 \\ -x_3+ix_4 & x_1-ix_2 \end{pmatrix}, \qquad (6.2)$$

then, by virtue of unimodularity we have $x_1^2+x_2^2+x_3^2+x_4^2 = 1$, and consequently the region of variation of the matrix $\hat{U}$ is the three-dimensional sphere S^3 in four-dimensional space. Each three-dimensional rotation corresponds to two diametrically opposite points on the sphere S^3. Consequently $SO_3 = S^3/Z_2$.

In CLC, however, the factoring of S^3 is not confined to the inversion group in four-dimensional space. Indeed, the states obtained from the initial one by making in (6.1) the substitutions

$$\mathbf{d}(\mathbf{r}_0) \to -\mathbf{d}(\mathbf{r}_0), \quad \mathbf{t}(\mathbf{r}_0) \to -\mathbf{t}(\mathbf{r}_0), \qquad (6.3)$$

are equivalent to the initial state. Therefore each point on the sphere S^3 has already 7 equivalent points that are obtained by inversion and rotations in three-dimensional space through an angle π about the axes $\mathbf{t}$, $\mathbf{d}$, $\mathbf{t}\times\mathbf{d}$. Thus, for example, the point corresponding to the unit matrix $\hat{U} = \hat{\sigma}_0$ on the sphere S^3 has equivalent points corresponding to the matrices

$$-\hat{\sigma}_0, \quad \pm i\hat{\sigma}_x, \quad \pm i\hat{\sigma}_y, \quad \pm i\hat{\sigma}_z, \qquad (6.4)$$

where $\hat{\sigma}_x$, $\hat{\sigma}_y$, $\hat{\sigma}_z$ are Pauli matrices. The matrices (6.4) together with the unit matrix form a group that is isomorphic to the group Q of the quaternion units, which consists of 8 elements $(1, -1, i, -i, j, -j, k, -k)$, such that $ij = -ji = k$, $jk = -kj = i$, $ki = -ik = j$, $ii = kk = jj = -1$. It can be shown that any 8 equivalent points on S^3 also form a group isomorphic to Q. Therefore the space R for CLC is $R = S^3/Q$.

The homotopic groups of this space are $\pi_2(R) = 0$ and $\pi_1(A, R) = Q$ (see the Appendix). CLC have no pointlike topologically stable singularities. The fundamental group $\pi_1(A, R) = Q$ is noncommutative. As a result, each linear singularity in CLC is characterized not by an element of the fundamental group, but by a class of conjugate elements of this group. Let us consider the two singular lines marked in Fig. 8 by the points a and b. We take the point $\mathbf{r}_0$ and surround a and b by contours γ_a, γ_b, and $\tilde{\gamma}_b$ that start out from this point. It can be verified that the contour $\tilde{\gamma}_b$ can be represented in the form

$$\tilde{\gamma}_b = \gamma_a^{-1}\gamma_b\gamma_a.$$

If γ_a is mapped into element a of the group $\pi_1(R)$ and γ_b into element b, then $\tilde{\gamma}_b$ is mapped into the element $\tilde{b} = a^{-1}ba$. Let $a = i$ and $b = j$; then $\tilde{b} = -j$. This means that the singularity b is characterized by two elements, j and $-j$, i.e., by the class $\{j, -j\}$. There are five such classes:

$$e = \{1\}, \quad e_1 = \{-1\}, \quad a = \{i, -i\}, \quad b = \{j, -j\}, \quad c = \{k, -k\}.$$

When the singularities coalesce the classes are multiplied. We present the results of the multiplication

	e	e_1	a	b	c
e	e	e_1	a	b	c
e_1	e_1	e	a	b	c
a	a	a	e, e_1	c	b
b	b	b	c	e, e_1	a
c	c	c	b	a	e, e_1

When a is multiplied by a we obtain either an element of the class e or an element of the class e_1. An analogous result is obtained for bb and cc. This means that in the foregoing three cases, and only in these cases, the type of the resultant singularity depends on the coalescence path, in analogy with coalescence of singular points in NLC (see Fig. 9).

Kleman and Friedel[12] (see also[11]) have presented a classification of the disclinations in CLC. According to this classification there are three types of disclinations: $\chi^{(m)}$, $\lambda^{(m)}$, $\tau^{(m)}$, where m is the Frank index of the disclination. We write out the distribution of these singularities among the classes of the elements of group Q:

e: $\chi^{(4n)}$, $\lambda^{(4n)}$, $\tau^{(4n)}$,
e_1: $\chi^{(4n+2)}$, $\lambda^{(4n+2)}$, $\tau^{(4n+2)}$;
a: $\chi^{(2n+1)}$,
b: $\lambda^{(2n+1)}$,
c: $\tau^{(2n+1)}$.

We recall that within each class the disclinations can continuously go over into one another. The question of the local stability of the disclinations within each class is no longer homotopic and calls for an examination of the functional of the energy.

A few words now concern smectic liquid crystals and ordinary crystals. The states of an ordinary crystal are degenerate with respect to translation through an arbitrary vector $\mathbf{u}$ that runs through the linear space R^3, and with respect to the three-dimensional rotations defined in the space SO_3. In the space $R^3 \times SO_3$, each point corresponds to a set of equivalent points that transform the crystal into a state identical with the initial one. This set forms a subgroup G of the a crystal space group whose elements contain no inversion. Therefore for a crystal the space is $R = (R^3 \times SO_3)/G$. The homotopic group is $\pi_2(R) = 0$, so that there are no point singular-

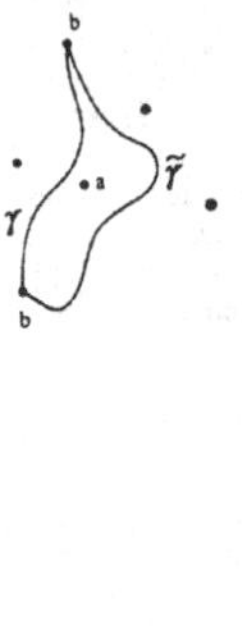

FIG. 9. Coalescence of singular lines in a cholesteric liquid crystal. The results of the coalescence is $b \cdot b = e_1$ along the path γ and $b \cdot b = e$ along the path $\tilde{\gamma}$.

ities of topological character in the crystals. The fundamental group $\pi_1(R)$ depends on the form of G and admits, besides dislocations, of the existence of disclinations. These disclinations, however, have an energy proportional to the volume, which is the result of the requirement that the distances between the crystal planes be equal; they can exist only in very soft crystals. The homotopic classification of the dislocations yields nothing new in comparison with the known classification in accordance with the Burgers vector. From the point of view of homotopic topology, edge and screw dislocations, characterized by an identical Burgers vectors belong to one homotopic class.

A smectic liquid crystal constitutes a system of equidistant surfaces separated from one another by atomic distances. These surfaces can bend, so that in smectic liquid crystals there are no dislocations. However, the requirement that the distances between layers be constant makes impossible many continuous deformations and therefore limits the applicability of the homotopic classification. In CLC the requirement that the pitch L of the helix be constant is not so stringent, since L is much larger than the dimension of the molecules and can vary slowly from point to point.

In conclusion, the authors thank S. P. Novikov, O. I. Bogoyavlenskiĭ, M. I. Monastyrskiĭ, and V. L. Golo for valuable consultations on topology, and also I. E. Dzyaloshinskiĭ for interesting discussions of the questions touched upon in the paper.

APPENDIX

The spaces R used in this paper have the general form P/G, where G is a discrete group, for example Z_2 or the group of quaternion units Q; $P=R_1\times R_2$, where R_1 and R_2 are spaces with known homotopic groups. In the calculation of the homotopic groups of the space R of this type it is necessary to use certain very simple rules which make it possible to express $\pi_n(R)$ in terms of the homotopic groups R_1 and R_2. First,

$$\pi_n(P)=\pi_n(R_1)+\pi_n(R_2). \quad \text{(A. 1)}$$

Second, the sequence of homomorphisms of the groups

$$\to\pi_n(G)\to\pi_n(P)\to\pi_n(P/G)\to\pi_{n-1}(G)\to\ldots \quad \text{(A. 2)}$$

is exact, i.e., for each triplet of successive groups from this sequence $G_1\to G_2\to G_3$ the transform of the homomorphism $G_1\to G_2$ is the kernel of the homomorphism $G_2\to G_3$. Third, π_0 from a connected space is trivial, and

$$\pi_0(G)=G. \quad \text{(A. 3)}$$

1. Nematic liquid crystal

The spaces $R=S^2/Z_2$. It is known that $\pi_2(S^2)=Z$ and $\pi_1(S^2)=\pi_0(S^2)=0$. To find $\pi_1(R)$ and $\pi_2(R)$ we write down the exact sequence (A. 2):

$$\pi_2(Z_2)\to\pi_2(S^2)\to\pi_2(S^2/Z_2)\to\pi_1(Z_2)\to$$
$$\to\pi_1(S^2)\to\pi_1(S^2/Z_2)\to\pi_0(Z_2)\to\pi_0(S^2),$$

or, substituting the known homotopic group, we obtain

$$0\to Z\to\pi_2(S^2/Z_2)\to 0\to 0\to\pi_1(S^2/Z_2)\to Z_2\to 0.$$

Using the definition of the exact sequence, we obtain

$$\pi_1(S^2/Z_2)=Z_2,\quad \pi_2(S^2/Z_2)=Z. \quad \text{(A. 4)}$$

2. Cholesteric liquid crystals

The space is $R=S^3/Q$. It is known that $\pi_0(S^3)=\pi_1(S^3)=\pi_2(S^3)=0$ and $\pi_3(S^3)=Z$; using the exact sequence, we obtain

$$\pi_1(S^3/Q)=Q,\quad \pi_2=(S^3/Q)=0,\quad \pi_3(S^3/Q)=Z. \quad \text{(A. 5)}$$

3. The *B* phase of He[3]

The space is $R_B=S^1\times SO_3$. We first obtain $\pi_n(SO_3)$. For this purpose we use the fact that $SO_3=S^3/Z_2$ (see Sec. 6) and then we have in analogy with (A. 5) $\pi_1(SO_3)=Z_2$, $\pi_2(SO_3)=0$, $\pi_3(SO_3)=Z$. Using (A. 1) we obtain

$$\pi_2(R_B)=\pi_2(S^1)+\pi_2(SO_3)=0,$$
$$\pi_1(R_B)=\pi_1(S^1)+\pi_1(SO_3)=Z+Z_2,\quad \pi_3(R_B)=Z. \quad \text{(A. 6)}$$

4. The *A* phase of He[3]

The space is $R_A=(S^2\times SO_3)/Z_2$. We have

$$\pi_2((S^2\times SO_3)/Z_2)=\pi_2(S^2\times SO_3)=\pi_2(S^2)+\pi_2(SO_3)=Z,$$
$$\pi_3(R_A)=\pi_3(S^2)+\pi_3(SO_3)=Z+Z,\quad \text{since}\quad \pi_3(S^2)=Z. \quad \text{(A. 7)}$$

To calculate $\pi_1(R_A)$ we make the exact sequence

$$\pi_1(Z_2)\to\pi_1(S^2\times SO_3)\to\pi_1(R_A)\to\pi_0(Z_2)\to\pi_0(S^2\times SO_3)$$

or, substituting in the known homotopic groups, we obtain

$$0\to Z_2\to\pi_1(R_A)\to Z_2\to 0.$$

It follows therefore that either $\pi_1(R_A)=Z_4$ or $\pi_1(R_A)=Z_2+Z_2$. It can be shown that for the A phase the first possibility is realized.

The space is $\bar{R}_A=O_3=SO_3\times Z_2$. We have

$$\pi_0(\bar{R}_A)=Z_2,\quad \pi_1(\bar{R}_A)=Z_2,\quad \pi_2(\bar{R}_A)=0,\quad \pi_3(\bar{R}_A)=Z. \quad \text{(A. 8)}$$

[1)]This circumstance was called to our attention by I. E. Dzyaloshinskiĭ.

[1]M. I. Monastyrskiĭ and A. M. Perelomov, Pis'ma Zh. Eksp. Teor. Fiz. 21, 94 (1975) [JETP Lett. 21, 43 (1975)]; M. I. Monastyrsky and A. M. Perelomov, Preprint, Institute for Theoretical and Experimental Physics, 56, Moscow, 1974.
[2]D. Huismoller, Stratified Spaces (Russ. Transl.) Mir, 1970. E. Spanier, Algebraic Topology, McGraw, 1966 (Russ. transl. Mir, 1971).
[3]G. E. Volovik and V. P. Mineev, Pis'ma Zh. Eksp. Teor. Fiz. 24, 605 (1976) [JETP Lett. 24, 595 (1976)].
[4]S. I. Anisimov and I. E. Dzyaloshinskiĭ, Zh. Eksp. Teor. Fiz. 63, 1460 (1972) [Sov. Phys. JETP 36, 774 (1973)].
[5]A. J. Leggett, Rev. Mod. Phys. 47, 331 (1975).
[6]V. R. Chechetkin, Zh. Eksp. Teor. Fiz. 71, 1463 (1976)

[Sov. Phys. JETP 44, 766 (1977)].
[7]P. G. de Gennes, Phys. Lett. 44A, 271 (1973).
[8]F. Fishman and I. A. Privorotskii, J. Low Temp. Phys. 25, 225 (1976).
[9]S. Blaha, Phys. Rev. Lett. 36, 874 (1976).
[10]G. E. Volovik and V. P. Mineev, Pis'ma Zh. Eksp. Teor. Fiz. 23, 647 (1976) [JETP Lett. 23, 593 (1976)].
[11]M. J. Stephen and J. P. Straley, Rev. Mod. Phys. 46, 617 (1974).
[12]M. Kleman and J. Friedel, J. Phys. (Paris) 30, Suppl. C4, 43 (1969).

Translated by J. G. Adashko

Phase Slippage without Vortex Cores: Vortex Textures in Superfluid ^{3}He

P. W. Anderson
Bell Laboratories, Murray Hill, New Jersey 07974

and

G. Toulouse
Université de Paris-Sud, Laboratoire de Physique des Solides, 91405 Orsay, France
(Received 14 September 1976)

The characteristic dissipation process for conventional superfluid flow is phase slippage: motion of quantized vortices in response to the Magnus force, which allows finite chemical potential differences to occur. Topological considerations and actual construction are used to show that in liquid ^{3}He-A, textures with vorticity but no vortex core can easily be constructed, so that dissipation of superfluid flow can occur by motion of textures alone without true vortex lines, dissipation occurring via the Cross viscosity for motions of $\hat{l}$.

Usually dissipative relaxation of the order parameter of a broken-symmetry condensed system occurs by motion of order-parameter singularities. For instance, magnetic hysteresis involves the motion of domain walls, slip of solids that of dislocations, and self-diffusion that of vacancies or interstitials. These are 2-, 1-, and 0-dimensional "order-parameter singularities." All are characterized by a "core" of atomic dimension where the order parameter departs substantially from its equilibrium value.

One of the clearest examples of this general rule is phase slippage in superconductors (flux flow and creep) and in liquid helium II: The only way in which these superfluides in bulk form can sustain a gradient of chemical potential, and thus flow dissipatively, is by the continual motion of quantized vortex lines transverse to that gradient. The controlling equation is[1]

$$\langle \mu_1 - \mu_2 \rangle = \left\langle \frac{\hbar\, d(\varphi_1 - \varphi_2)}{dt} \right\rangle = h \frac{dn_{\text{vortices}}}{dt}, \tag{1}$$

where $\langle\mu_1-\mu_2\rangle$ is the time-averaged chemical potential difference between two points, $\varphi_1-\varphi_2$ is the phase difference of the mean particle (or pair for superconductors) field, and dn_{vortices}/dt is the rate of passage of quantized vortices across the line joining 1 and 2.

Quantized vortex lines can exist[2] in the anisotropic superfluid ^{3}He-A, but as one of us has shown,[3] the circulation around such a vortex is not a topological invariant, as it is in the simple superfluids; it is, in those, equivalent to the so-called "winding number." In ^{3}He-A, on the other hand, by making a rotation of the order parameter which is continuous everywhere, i.e., by superposing a "texture," one may convert a vortex line of either sign into the opposite one or into a de Gennes disgyration; and two vortex lines of the same sign can in principle annihilate each other. Those topological results demonstrate that quantization of vorticity and vortex line motion are not the keys to dissipative processes that they are in the conventional cases, since they destroy the second equality of Eq. (1). Nonetheless the first equality of Eq. (1) shows that phase slippage by one mechanism or another is necessary in order to feed energy from superfluid flow into dissipation processes (since the energy dissipated is $\Delta\mu\, dN/dt = \Delta\mu\, J_{\text{tot}}$). We describe here some likely mechanisms by which dissipation can occur by phase slippage *without* order-parameter singularities, by the motion of textures alone.

Briefly the geometry of the order parameter in ^{3}He-A is that of the orthogonal triad of vectors $\hat{l}$, $\vec{\Delta}_1$, and $i\vec{\Delta}_2$, where $\hat{l}$ is the orbital angular momentum of the pairs, and $\vec{\Delta}_1$ and $i\vec{\Delta}_2$ are real and imaginary vectors representing the two components of the anisotropic energy gap. Rotations of $\vec{\Delta}_1$ and $i\vec{\Delta}_2$ about $\hat{l}$ are phase changes of the order parameter, while rotations of $\hat{l}$ rotate the anisotropy axis of the system. Rotations of a rigid frame generate the group SO(3) which has the topology of projective space P_3, the three-dimensional sphere S_3 with diametrically opposite points on the surface identified. The rule of Ref. 3 is that one maps paths around the singularities in real space into this gap-parameter space, and if the path cannot be deformed continuously into a point, the singularity is stable. 360° rotations are equivalent to paths between the two identified poles on the sphere and are topologically nontrivial, but a second 360° rotation in any direction returns one to the starting point and the resultant is equivalent topologically to no rotation at all.

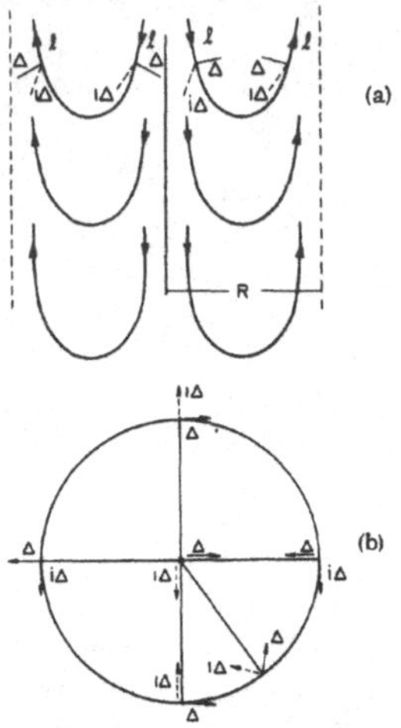

FIG. 1. Sketch of the 4π vortex texture. Lines are stream of $\hat{l}$, and $\hat{\Delta}$ and $i\hat{\Delta}$ rotate as sketched. (a) Side view; (b) view along the length.

Thus, in principle a 720° vortex requires no core.

These ideas led us to search for a texture which can play the role of a double vortex line, of which an example is the following (see Fig. 1). On the axis of a circular cylinder $\hat{l}$ is in the $-\hat{z}$ direction, $\vec{\Delta}_1$ and $i\vec{\Delta}_2$ in say the $\hat{x}$ and $\hat{y}$ directions, respectively. Along radii of this cylinder $\hat{l}$ rotates about the $\hat{\varphi}$ direction through an angle $f(r)$, with $f(R)=\pi$, $f(0)=0$. At the cylinder R, we will find $\hat{l}$ pointing in the $+\hat{z}$ direction and the phase rotating by 4π as we circumnavigate the cylinder. (This is a simple modification of the Brinkman-Osheroff texture.[4]) A generalization of this to a texture equivalent to a vortex sheet can easily be made.

Another peculiar property of this texture is that it can terminate in a "hedgehog" or pointlike object at which the lines of $\hat{l}$ splay out in all directions. The phase rotates by 360° while $\hat{l}$ rotates by 360° around the circumference of the hedgehog (see Fig. 2).

These vortex textures have two properties which suggest that they may play a role in dynamical processes. First, having no core, the energy will be less than that of two normal 2π vortices by a factor of order $\frac{1}{2}\ln(R/\xi)$ (2 for the double quantum) which will usually make them energetically cheaper. Second, they have no localized

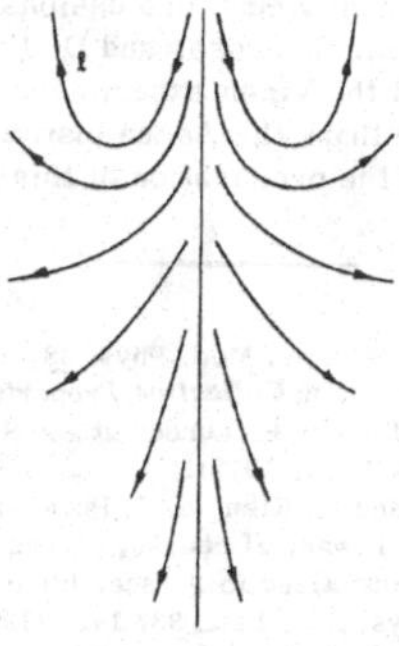

FIG. 2. Termination of 4π vortex texture in a "hedgehog," which is the texture equivalent of a "monopole" attached to the end of a vortex line.

pattern of high superfluid velocity and thus the characteristic and puzzling problem of vortex nucleation will not be as serious.

On the other hand, the phase slippage theorem [the first equality of Eq. (1)] is still valid between any two points in the fluid where the orientation of $\hat{l}$ does not change. This is because the usual argument from the number-phase commutation relation or from gauge invariance[1] that $d\varphi/dt$ equals $\partial E/\partial N$ is universal to all superfluids, but, if $\hat{l}$ is moving too rapidly, it is not safe to equate $\langle \partial E/\partial N\rangle$ with μ: hence the proviso on motions of $\hat{l}$. φ is, of course, not a velocity potential but is uniquely definable when $\hat{l}$ is fixed, as it will be effectively, in most physical situations, at boundaries by the boundary condition on $\hat{l}$, and in many other situations in the bulk of the system because of orienting flows and fields or the Cross "normal pinning" effect.[5] Equation (1), not its character as a velocity potential, is the key to the importance of the phase φ in dissipative processes. For instance, flow between two orifices at which $\hat{l}$ can be expected to be pinned (see, for example, Wheatley[6]) will see a chemical potential difference obeying the phase slippage equation, and dissipation can occur by motions of vortex textures across the path between the orifices. The importance of vortex textures in this context arises from the fact that $\hat{l}$ is fixed in direction in the exterior region, and with $\hat{l}$ fixed in the exterior region *changes* in vorticity are still quantized. The problem of boundary conditions will be discussed in more detail in a future publication.

In this situation at least, the textures will experience the characteristic Magnus force in the presence of a background superfluid flow field v_s,

$$F_M = \rho \vec{v}_s \times \vec{\Omega}, \tag{2}$$

where Ω is the circulation, in this case equal to h/m in magnitude. But much more general situations can be understood if we simply retain the idea of a locally definable phase, except possibly where $\hat{l}$ is moving too fast, and rely on the second fundamental equation

$$-\frac{dN_{\text{pairs}}}{dt} = \frac{1}{\hbar}\frac{\partial\langle E\rangle}{\partial\varphi}. \tag{3}$$

Equation (3) implies the more general remark[7] that a current source may always be inserted as a phase-dependent term in the Hamiltonian:

$$H_{\text{source}} = -dN/dt_{\text{ext}}\,\varphi$$

Hence, a current source exerts an appropriate force on any texture whose motion can allow phase slippage.

The final physical fact is that textures move relatively slowly and dissipatively in ^{3}He because of the Cross "normal pinning" effect,[5] so that vortex textures can only affect very low-frequency phenomena. We envisage three flow regimes.

(1) For high-frequency phenomena such as fourth sound and vibrating wires, or for large chemical potential differences, ^{3}He will behave like a conventional but anisotropic superfluid, since orbital motion will be pinned and phase slippage can only occur by motion of conventional vortices.

(2) In the absence of a magnetic field and for moderately low frequencies, $\hat{l}$ will be free to orient at will and vortex textures will play a great role in causing dissipation. The critical velocity for nucleating textures should be extremely low, of order h/mR where R is an apparatus size, or $\sim 10^{-2}$ cm/sec. This is even lower than what is observed in heat flow experiments by Wheatley's group, but in the right range. It seems possible that in the absence of a magnetic field almost any flow will fill the sample with enough vorticity to damp out fluctuations and the relatively quiet behavior at low fields could be a turbulent regime.

(3) In a magnetic field $\hat{l}$ is oriented by the dipolar energy for structures larger than the length R_S of order ~ 100 μm[4] determined by the ratio of dipolar to current energy. The vortex structures must be of this order and thus contain velocities of order $h/mR_S \sim 10^{-1}$ cm/sec. This is again the right order of magnitude, but a bit fast relative to some critical velocities measured in resonance

and other experiments. In both of these cases, it is easily possible that a regular motion of vortex structures can be set up under appropriate flow conditions, which we speculate may be related to certain observations of regular and irregular orbital fluctuations.[6]

It is interesting to speculate on the outcome of a measurement of quantized vorticity in ^{3}He-A by the Vinen[8] vibrating-wire experiment or otherwise. At a Vinen wire the boundary condition will require $\hat{l}$ to be radial and the phase may rotate by any integer number of units 2π; a texture in, for instance, a cylinder can simply add or subtract 4π to this, so that any integer amount of vorticity is possible. However, the results might be chaotic in the absence of a field because of vortex textures throughout the liquid. With a field the wire can again have integer vorticity but in the surrounding liquid the 4π double vorticity is the most stable vortex line, consisting of a "core" which is a Fig. 1 texture of size $\sim R_s$, and a conventional outer region. Thus, one may tend to add or subtract *double* units.

In summary, the most important point to be made is that dissipation in this superfluid is qualitatively different from that in other superfluids and in most broken-symmetry systems, in that it can occur by the motion of textures (rather like "topological solitons") and not only by singularities of the order parameter. Thus, the property of superfluidity takes a very novel form in this case.

We wish to acknowledge discussions with W. F. Brinkman and M. C. Cross, and D. J. Thouless's suggestion that the Vinen experiment should be examined. We thank the Aspen Institute for hospitality during the preparation of this Letter.

[1]P. W. Anderson, Rev. Mod. Phys. 38, 298 (1966).

[2]P.-G. de Gennes, in *Collective Properties of Physical Systems*, edited by B. Lundqvist and S. Lundqvist (Academic, New York, 1973).

[3]G. Toulouse and M. Kléman, J. Phys. (Paris), Lett. 37, 149 (1976). It was, of course, already proven that vorticity is not quantized in A [see, for example, R. Graham, Phys. Rev. Lett. 33, 1431 (1974)], but the topological argument gives a new and very fundamental point of view.

[4]W. F. Brinkman and D. D. Osheroff, Phys. Rev. Lett. 32, 584 (1974). See also P. W. Anderson and W. F. Brinkman, in *The Helium Liquids*, edited by J. G. M. Armitage and I. E. Farquhar (Academic, New York, 1975), p. 315.

[5]M. C. Cross and P. W. Anderson, in *Proceedings of the Fourteenth International Conference on Low Temperature Physics, Otaniemi, Finland, 1975* edited by M. Krusius and M. Vuorio (North-Holland, Amsterdam, Vol. 1, p. 29

[6]See J. C. Wheatley, Rev. Mod. Phys. 47, 415 (1975), especially Sect. VII; also D. N. Paulson, M. Krusius, and J. C. Wheatley, Phys. Rev. Lett. 37, 599 (1976).

[7]P. W. Anderson, in *The Many Body Problem*, edited by E. R. Caianello (Academic, New York, 1964).

[8]W. F. Vinen, Proc. Roy Soc. London, Ser. A 260, 218 (1961).

CXXXVI. *On the Quantum Mechanics of Helium II.*

By O. Penrose,
Royal Society Mond Laboratory, Cambridge*.

[Received June 2, 1951; revised June 29, 1951.]

Abstract.

The quantum mechanics of a system of identical interacting particles must lead to the classical hydrodynamic equations of motion at high temperatures, because of the correspondence principle. On the other hand, the behaviour of helium II shows that this is not always the case at low temperatures. In this paper it is shown that in certain cases the quantum description requires an extra parameter, which is the potential of a new velocity field superimposed on the classical motion. Expressed in semi-classical terms, the condition for the existence of this new parameter is that the probability, in the equilibrium state, of a particle having a very large de Broglie wavelength (that is, a negligible momentum) is finite. This condition is satisfied in one model of a superfluid system, a condensed Bose–Einstein gas, but not in a crystal. A tentative theoretical interpretation of two basic equations of the empirical two-fluid theory of helium II is given, in which this new parameter determines the velocity of the superfluid.

---

§ 1. Introduction.

Born and Green (1947, 1948) described a method of studying transport processes in a system of identical interacting particles obeying quantum mechanics by using the analogy with the corresponding classical system. Gurov (1948, 1950) carried out a similar calculation. In both cases the conclusion was that the equations of hydrodynamics and heat transfer were identical in form with the classical equations. These treatments assumed that the three quantities describing a " normal " state of the system (Born and Green 1947, Chapman and Cowling 1939) were the same as for a classical system : the density, velocity and temperature fields. In other words, if at one instant the system is in a state described by given fields of these three parameters, its state at a later time is sufficiently described by the new fields of the parameters resulting from the thermo-hydrodynamical equations of motion. From the correspondence principle it is clear that the three parameters of classical mechanics are sufficient in quantum mechanics at high temperatures, but, as we shall see, this is not always true at low temperatures.

---

* Communicated by the Author.

§ 2. Properties of the Reduced Density Matrices.

Suppose the system consists of N identical particles of mass m interacting by central forces whose potential is U, and let the Schrödinger representative of the density matrix of the corresponding Gibbs ensemble (Dirac 1947) be

$$D(\mathbf{q}_1, \ldots \mathbf{q}_N ; \mathbf{q}'_1, \ldots \mathbf{q}'_N),$$

where $\mathbf{q}_1, \ldots . \mathbf{q}_N$ are the position vectors of the N particles, and D is Hermitian. The reduced density matrices, which are also Hermitian, are defined by

$$R_k(\mathbf{q}_1, \ldots \mathbf{q}_k ; \mathbf{q}'_1, \ldots \mathbf{q}'_k) = [N!/(N-k)!] \times \int \ldots \int D(\mathbf{q}_1 \ldots \mathbf{q}_N ; \mathbf{q}'_1 \ldots \mathbf{q}'_k, \mathbf{q}_{k+1} \ldots \mathbf{q}_N)\, d\mathbf{q}_{k+1} \ldots d\mathbf{q}_N \quad (1)$$

The equation of motion for D is

$$i\hbar\, \partial D/\partial t = HD - DH, \quad (2)$$

where H is the Hamiltonian operator of the system. After setting $\mathbf{q}_2 = \mathbf{q}'_2, \ldots \mathbf{q}_N = \mathbf{q}'_N$, this may be integrated over $\mathbf{q}_2, \ldots \mathbf{q}_N$, giving

$$i\hbar \partial R_1(\mathbf{q}_1 ; \mathbf{q}'_1)/\partial t = -(\hbar^2/2m)(\nabla_1^2 - \nabla_1'^2) R_1(\mathbf{q}_1 ; \mathbf{q}'_1) + \int [U(\mathbf{q}_1 - \mathbf{q}_2) - U(\mathbf{q}'_1 - \mathbf{q}_2)] R_2(\mathbf{q}_1, \mathbf{q}_2 ; \mathbf{q}'_1, \mathbf{q}_2)\, d\mathbf{q}_2, \quad (3)$$

where $\nabla_j = \partial/\partial \mathbf{q}_j$, etc.

By using Wigner's result (Wigner 1932, Moyal 1949) that a density matrix $R_1(\mathbf{q}_1 ; \mathbf{q}'_1)$ corresponds to a classical probability density $F(\mathbf{x}, \mathbf{p})$ for the position $\mathbf{x}$ and momentum $\mathbf{p}$ of one particle such that

$$R_1(\mathbf{q}_1 ; \mathbf{q}'_1) \sim \int d\mathbf{p} F(\tfrac{1}{2}[\mathbf{q}_1 + \mathbf{q}'_1], \mathbf{p}) \exp [i\mathbf{p} \cdot (\mathbf{q}_1 - \mathbf{q}'_1)/\hbar], \quad (4)$$

combined with the classical result

$$F(\mathbf{x}, \mathbf{p}) \propto \exp [-\mathbf{p}^2/2mkT], \quad (5)$$

or alternatively by approximately solving the Bloch equation $\partial D/\partial (kT)^{-1} = -HD$ (Husimi 1940), we obtain a high temperature approximation

$$R_1(\mathbf{q}_1 ; \mathbf{q}'_1) \propto \exp [-(\mathbf{q}_1 - \mathbf{q}'_1)^2 mkT/2\hbar^2], \quad (6)$$

for equilibrium at the temperature T. Therefore

$$\lim_{|\mathbf{q}_1 - \mathbf{q}_1'| \to \infty} R_1(\mathbf{q}_1 ; \mathbf{q}'_1) = 0. \quad (7)$$

At low temperatures (7) may not be satisfied for every type of system. To see the physical significance of the failure of (7) let us assume for the moment that $R_1(\mathbf{q}_1 ; \mathbf{q}'_1)$ depends only on $|\mathbf{q}_1 - \mathbf{q}'_1|$ and approaches a constant value L when $|\mathbf{q}_1 - \mathbf{q}'_1| \to \infty$. Then the Fourier inverse of (4) yields a divergent integral which may be interpreted as

$$F(\mathbf{x}, \mathbf{p}) = f(\mathbf{p}) + L\delta(\mathbf{p}), \quad (8)$$

where $f(\mathbf{p})$ is a regular function. That is to say, the probability of one particle having zero momentum is the finite fraction $L/R_1(\mathbf{x} ; \mathbf{x})$. Since (8) is in fact an expression for the mean occupation numbers of single-particle levels, L must vanish (and (7) therefore be satisfied) in a Fermi-Dirac system, where the exclusion principle limits the occupation numbers.

§ 3. Asymptotic Behaviour of the Reduced Density Matrix for One Particle.

In the general case when (7) does not hold we shall consider the behaviour of $R_1(\mathbf{q}_1\,;\,\mathbf{q}_1')$ for values of $|\mathbf{q}_1-\mathbf{q}_1'|$ large compared with both $\hbar(mkT)^{-1/2}$ and the range of U. The integral in (3) may be written, using the Hermitian property of R_2,

$$\int U(\mathbf{r})[R_2(\mathbf{q}_1,\mathbf{q}_1+\mathbf{r}\,;\,\mathbf{q}_1',\mathbf{q}_1+\mathbf{r})-R_2^*(\mathbf{q}_1',\mathbf{q}_1'+\mathbf{r}\,;\,\mathbf{q}_1,\mathbf{q}_1'+\mathbf{r})]\,d\mathbf{r}. \quad . \quad (9)$$

Suppose that the asymptotic form of $R_2(\mathbf{q}_1,\mathbf{q}_1+\mathbf{r}\,;\,\mathbf{q}_1',\mathbf{q}_1+\mathbf{r})$ for large $|\mathbf{q}_1-\mathbf{q}_1'|$ is

$$R_2(\mathbf{q}_1,\mathbf{q}_1+\mathbf{r}\,;\,\mathbf{q}_1',\mathbf{q}_1+\mathbf{r})\sim A(\mathbf{q}_1,\mathbf{r})R_1(\mathbf{q}_1\,;\,\mathbf{q}_1'), \quad . \quad . \quad . \quad (10)$$

just as the neighbour distribution function of a liquid obeys

$$R_2(\mathbf{q}_1,\mathbf{q}_2\,;\,\mathbf{q}_1,\mathbf{q}_2)\sim R_1(\mathbf{q}_1\,;\,\mathbf{q}_1)R_1(\mathbf{q}_2\,;\,\mathbf{q}_2) \quad . \quad . \quad . \quad (11)$$

for large $|\mathbf{q}_1-\mathbf{q}_2|$. For large $|\mathbf{q}_1-\mathbf{q}_1'|$ (3) becomes

$$i\hbar\,\partial R_1(\mathbf{q}_1\,;\,\mathbf{q}_1')/\partial t\sim-(\hbar^2/2m)(\nabla_1^2-\nabla_1'^2)R_1(\mathbf{q}_1\,;\,\mathbf{q}_1')$$
$$+[X(\mathbf{q}_1)-X^*(\mathbf{q}_1')]R_1(\mathbf{q}_1\,;\,\mathbf{q}_1'), \quad . \quad . \quad (12)$$

where

$$X(\mathbf{q})=\int A(\mathbf{q},\mathbf{r})U(\mathbf{r})\,d\mathbf{r}=V(\mathbf{q})+iW(\mathbf{q}). \quad . \quad . \quad . \quad . \quad (13)$$

Since the variables in (12) are separable the general solution is a linear combination of expressions of the form

$$\Psi(\mathbf{q}_1)\Psi^*(\mathbf{q}_1'), \quad . \quad . \quad . \quad . \quad . \quad . \quad . \quad (14)$$

where

$$i\hbar\partial\Psi/\partial t=-(\hbar^2/2m)\nabla^2\Psi+X\Psi. \quad . \quad . \quad . \quad . \quad (15)$$

At equilibrium $D=e^{-\beta H}$, and since the Schrödinger representative of H is real, those of D and the R's are also real. The state of the liquid is uniform, and, therefore, for large enough $|\mathbf{q}_1-\mathbf{q}_1'|$ it may be assumed that

$$R_1(\mathbf{q}_1\,;\,\mathbf{q}_1')\sim\text{const.}$$

This is of the form (14) with $\Psi=\text{const.}$ If the density matrix of a non-equilibrium state is sufficiently similar to $e^{-\beta H}$ the required solution of (12) will still be a single expression of the form (14):

$$R_1(\mathbf{q}_1\,;\,\mathbf{q}_1')\sim\Psi(\mathbf{q}_1)\Psi^*(\mathbf{q}_1'). \quad . \quad . \quad . \quad . \quad (16)$$

The example of a condensed Bose gas may clarify the argument of this section. Equation (16) is true, with Ψ the wave-function of the single-particle state into which condensation has taken place. Husimi's formula (1940)

$$R_2(\mathbf{q}_1,\mathbf{q}_2\,;\,\mathbf{q}_1',\mathbf{q}_2')=R_1(\mathbf{q}_1\,;\,\mathbf{q}_1')R_1(\mathbf{q}_2\,;\,\mathbf{q}_2')+R_1(\mathbf{q}_1\,;\,\mathbf{q}_2')R_1(\mathbf{q}_2\,;\,\mathbf{q}_1')$$

is not valid for a condensed Bose gas; for example, at absolute zero, when all the particles are in the ground state,

$$R_1(\mathbf{q}_1\,;\,\mathbf{q}_1')=\Psi(\mathbf{q}_1)\Psi^*(\mathbf{q}_1'),$$
$$R_2(\mathbf{q}_1,\mathbf{q}_2\,;\,\mathbf{q}_1',\mathbf{q}_2')=\Psi(\mathbf{q}_1)\Psi(\mathbf{q}_2)\Psi^*(\mathbf{q}_1')\Psi^*(\mathbf{q}_2').$$

The correct formula is

$$R_2(\mathbf{q}_1,\mathbf{q}_2\,;\,\mathbf{q}_1',\mathbf{q}_2')=R_1(\mathbf{q}_1\,;\,\mathbf{q}_1')R_1(\mathbf{q}_2\,;\,\mathbf{q}_2')+R_1(\mathbf{q}_1\,;\,\mathbf{q}_2')R_1(\mathbf{q}_2\,;\,\mathbf{q}_1')$$
$$-\Psi(\mathbf{q}_1)\Psi(\mathbf{q}_2)\Psi^*(\mathbf{q}_1')\Psi^*(\mathbf{q}_2') \quad . \quad . \quad . \quad . \quad . \quad (17)$$

(compare London 1943). Assuming that this formula holds for non-equilibrium states, (10) and (11) may be verified, with

$$A(\mathbf{q}, \mathbf{r}) = R_1(\mathbf{q}+\mathbf{r}\,;\ \mathbf{q}+\mathbf{r}) + R(\mathbf{q}\,;\ \mathbf{q}+\mathbf{r})\Psi(\mathbf{q}+\mathbf{r})/\Psi(\mathbf{q}) - \Psi(\mathbf{q}+\mathbf{r})\Psi^*(\mathbf{q}+\mathbf{r}).$$

Madelung's transformation (1927) brings equations (15) and (16) into a form similar to the hydrodynamic equations of motion with velocity potential ϕ, density $|\Psi|^2$, external forces of potential V and a source distribution of strength $2|\Psi|^2 W/\hbar$:

$$R_1(\mathbf{q}_1\,;\ \mathbf{q}_1') \sim Q_\infty(\mathbf{q}_1\,;\ \mathbf{q}_1') \exp\{im[\phi(\mathbf{q}_1)-\phi(\mathbf{q}_1')]/\hbar\}, \quad . \ . \quad (18)$$

where

$$\left.\begin{aligned} Q_\infty(\mathbf{q}_1\,;\ \mathbf{q}_1') &= |\Psi(\mathbf{q}_1)|\,|\Psi(\mathbf{q}_1')|,\\ \partial|\Psi|^2/\partial t + \nabla\cdot(|\Psi|^2\nabla\phi) &= 2|\Psi|^2 W/\hbar,\\ \partial\phi/\partial t + \tfrac{1}{2}(\nabla\phi)^2 + V/m &= \hbar^2\nabla^2|\Psi|/2m^2|\Psi| \simeq 0. \end{aligned}\right\} \quad . \ . \ . \quad (19)$$

At equilibrium Q_∞ depends only on the density and temperature, so that even for non-equilibrium states it is a function of the parameters of the classical description. On the other hand, the velocity $\nabla\phi$ corresponds to no quantity in the classical description; it differs from the classical velocity in its laws of change and in being irrotational.

§ 4. The Two-Fluid Model of Helium II.

The two-fluid model may be interpreted tentatively as follows : Assume that real functions $Q_n(\mathbf{q}_1\,;\ \mathbf{q}'_1)$ and $Q_s(\mathbf{q}_1\,;\ \mathbf{q}_1')$ can be chosen, with $Q_n \to 0$ as $|\mathbf{q}_1 - \mathbf{q}_1'| \to \infty$, such that if at one instant

$$\begin{aligned} R_1(\mathbf{q}_1\,;\ \mathbf{q}_1') = {} & Q_n(\mathbf{q}_1\,;\ \mathbf{q}_1') \exp\{im(\mathbf{q}_1-\mathbf{q}_1')\cdot\mathbf{u}(\tfrac{1}{2}[\mathbf{q}_1+\mathbf{q}_1'])/\hbar\}\\ & + Q_s(\mathbf{q}_1\,;\ \mathbf{q}_1') \exp\{im[\phi(\mathbf{q}_1)-\phi(\mathbf{q}_1')]/\hbar\}\\ & + O(\nabla\mathbf{u}) + O(\nabla\nabla\phi). \quad . \ . \ . \ . \ . \ . \ . \ . \ . \ . \quad (20) \end{aligned}$$

(where the velocities $\mathbf{u}$ and $\nabla\phi$ do not alter appreciably in a distance $\hbar(mkT)^{-1/2}$), then at later instants it is of the same form, with new real functions Q_n, Q_s, $\mathbf{u}$, and ϕ. If $Q_s = 0$ this would describe a system with classical velocity field $\mathbf{u}$ (Gurov 1948). The density is

$$\begin{aligned} \rho(\mathbf{x}) &= mR(\mathbf{x}\,;\ \mathbf{x}) = mQ_n(\mathbf{x}\,;\ \mathbf{x}) + mQ_s(\mathbf{x}\,;\ \mathbf{x})\\ &= \rho_n(\mathbf{x}) + \rho_s(\mathbf{x}), \quad . \ . \ . \ . \ . \ . \ . \ . \ . \ . \ . \quad (21) \end{aligned}$$

and the current density (Born and Green 1948)

$$\begin{aligned} \mathbf{J}(\mathbf{x}) &= -(i\hbar/2)(\nabla_1 - \nabla_1')R_1(\mathbf{q}_1\,;\ \mathbf{q}_1') \qquad [\mathbf{q}_1 = \mathbf{q}_1' = \mathbf{x}]\\ &= \rho_n(\mathbf{x})\mathbf{u}(\mathbf{x}) + \rho_s(\mathbf{x})\nabla\phi(\mathbf{x}). \quad . \ . \ . \ . \ . \ . \ . \ . \ . \quad (22) \end{aligned}$$

These two equations are the basis of the two-fluid theory (Landau 1941, Tisza 1947) ; in Landau's version the superfluid velocity is irrotational.

§ 5. Examples.

Two examples will illustrate the argument. In a crystal the single-molecule wave-functions are confined to small regions, so that (7) is true and classical ideas should explain its mechanical properties. In a condensed

Bose–Einstein gas with negligible repulsive forces and periodic boundary conditions (Bogoliubov 1947)

$$\left.\begin{aligned} R_I(\mathbf{q}_I\,;\ \mathbf{q}_I') &= B^{-1}\sum_{\mathbf{l}} n_{\mathbf{l}} \exp i\mathbf{l}\cdot(\mathbf{q}_I-\mathbf{q}_I'), \\ Q_\infty(\mathbf{q}_I\,;\ \mathbf{q}_I') &= n_0/B \neq 0, \end{aligned}\right\} \quad . \quad . \quad . \quad (23)$$

where B is the volume of the container and $n_{\mathbf{l}}$ is the mean occupation number of the single-particle state whose wave-number is **l**, given by

$$\left.\begin{aligned} n_{\mathbf{l}} &= [\exp(\mathbf{l}^2\hbar^2/2mkT)-1]^{-1}, \qquad [\mathbf{l}\neq 0], \\ \sum_{\mathbf{l}} n_{\mathbf{l}} &= N. \end{aligned}\right\} \quad . \quad . \quad . \quad (24)$$

Bogoliubov claimed that such a system shows superfluidity.

We conclude that (7) is the condition for the classical equations of heat transfer and hydrodynamics to apply, and that when (7) is not satisfied a new quantity enters the equations. This quantity is the potential of a new velocity field superimposed on the classical motion.

ACKNOWLEDGMENTS

I am very much indebted to Mr. H.N.V. Temperley for supervising this work, to Mr. P.J. Price for discussions, to many of my colleagues for criticizing the manuscript, and to Kings College, Cambridge, for awarding me the R.J. Smith Memorial Student.

REFERENCES

BOGOLIUBOV, N.N., 1947, J. Phys. U.S.S.R., **11**, 23.

BORN, M., and GREEN, H.S., 1947, Proc. Roy. Soc. A., **190**, 455; 1948, Ibid., **191**, 168

CHAPMAN, S., and COWLING, T.G., 1939, The Mathematical Theory of Non-Uniform Gases, (Cambridge: University Press) p. 117.

DIRAC, P.A.M., 1947, The Principles of Quantum Mechanics, 3rd edition, (Oxford: University Press) pp. 131-135.

GUROV, K.P. 1948, Jour. Exp. and Theor. Phys. U.S.S.R., **18**, 110; 1950, Ibid., **20**, 279.

HUSIMI, K. 1940, Proc.Physico-Math., Soc. Japan, (3)**22**, 264.

LANDAU, L., 1941, J. Phys. U.S.S.R., **5**, 71.

LONDON, F., 1943, J. Chem. Phys., **11**, 203.

MADELUNG, E., 1927, Zeit. fur Phys., **40**, 322.

MOYAL, J.,1949, Proc. Camb. Phil. Soc., **45**, 99.

TISZA, L., 1947, Phys. Rev., **72**, 838.

WIGNER, E.P., 1932, Phys. Rev., **40**, 749.

Bose-Einstein Condensation and Liquid Helium

OLIVER PENROSE* AND LARS ONSAGER
Sterling Chemistry Laboratory, Yale University, New Haven, Connecticut*

(Received July 30, 1956)

The mathematical description of B.E. (Bose-Einstein) condensation is generalized so as to be applicable to a system of interacting particles. B.E. condensation is said to be present whenever the largest eigenvalue of the one-particle reduced density matrix is an extensive rather than an intensive quantity. Some transformations facilitating the practical use of this definition are given.

An argument based on first principles is given, indicating that liquid helium II in equilibrium shows B.E. condensation. For absolute zero, the argument is based on properties of the ground-state wave function derived from the assumption that there is no "long-range configurational order." A crude estimate indicates that roughly 8% of the atoms are "condensed" (note that the fraction of condensed particles need not be identified with ρ_s/ρ). Conversely, it is shown why one would not expect B.E. condensation in a solid. For finite temperatures Feynman's theory of the lambda-transition is applied: Feynman's approximations are shown to imply that our criterion of B.E. condensation is satisfied below the lambda-transition but not above it.

1. INTRODUCTION

THE analogy between liquid He^4 and an ideal Bose-Einstein gas was first recognized by London.[1,2] He suggested that the lambda-transition in liquid helium could be understood as the analog for a liquid of the transition[3,4] which occurs in an ideal B.E. (Bose-Einstein) gas at low temperatures. The fact[5] that no lambda-transition has been found in He^3 supports London's viewpoint. Further support comes from recent theoretical work[6–8] which shows in more detail how a system of interacting particles can exhibit a transition corresponding to the ideal-gas transition.

Tisza[2,9] showed that the analogy between liquid He^4 and an ideal B.E. gas is also useful in understanding the transport properties of He II. Below its transition temperature a B.E. gas in equilibrium has a characteristic property: a finite fraction of the particles occupy the lowest energy level. Tisza reasoned that the presence of these "condensed" particles would make necessary a special two-fluid hydrodynamical description for such a gas. His idea that this two-fluid description applies also to He II has been strikingly verified by many experiments.[2,10,11]

In theoretical treatments where the forces between helium atoms are taken into account, the ideal-gas analogy takes on forms differing widely from one treatment to another. For example, Matsubara[6] and Feynman[7] account for the lambda-transition by writing the partition function for liquid helium in a form similar to the corresponding expression for an ideal gas. In Bogolyubov's theory[12–14] of the superfluidity of a system of weakly repelling B.E. particles, it is the distribution of the momenta of the particles which resembles that of an ideal gas. Yet another form for the analogy has been suggested by Penrose[15]; this work will be discussed in more detail in Sec. 4 below. A further complication is that the excitation theory of superfluidity[16,17] is apparently independent of the ideal-gas analogy (though Bogolyubov's work[12] suggests that there actually is a connection).

The object of the present paper is, first, to unify the varied forms of the ideal-gas analogy mentioned above by showing how they are all closely related to a single *criterion for B.E. condensation*, applicable in either a liquid or a gas, and, secondly, to give an argument based on first principles indicating that this criterion actually is satisfied in He II. The relation between B.E. condensation and the excitation theory of superfluidity will be discussed in a later paper.

2. PRELIMINARY DEFINITIONS

We make the usual approximation of representing liquid He^4 by a system of N interacting spinless B.E. particles, each of mass m, with position and momentum vectors $\mathbf{q}_1\cdots\mathbf{q}_N$ and $\mathbf{p}_1\cdots\mathbf{p}_N$, respectively. The Hamiltonian is taken to be

$$H\equiv\sum_j p_j^2/2m+\sum_{i<j} U_{ij}. \quad (1)$$

Here U_{ij} stands for $U(|\mathbf{q}_i-\mathbf{q}_j|)$, where $U(r)$ is the interaction energy of two He^4 atoms separated by a

* Present address: Imperial College, London, England.
[1] F. London, Nature 141, 643 (1938); Phys. Rev. 54, 947 (1938).
[2] F. London, *Superfluids* (John Wiley and Sons, Inc., New York, 1954), Vol. 2, especially pp. 40–58, 199–201.
[3] A. Einstein, Sitzber. preuss. Akad. Wiss. physik-math. Kl. 1924, 261; *ibid.* 1925, 3, 18.
[4] For bibliography, see P. T. Landsberg, Proc. Cambridge Phil. Soc. 50, 65 (1954).
[5] E. F. Hammel in *Progress in Low-Temperature Physics*, edited by C. J. Gorter (North Holland Publishing Company, Amsterdam, 1955), Vol. 1, pp. 78–107.
[6] T. Matsubara, Progr. Theoret. Phys. Japan 6, 714 (1951).
[7] R. P. Feynman, Phys. Rev. 91, 1291 (1953).
[8] G. V. Chester, Phys. Rev. 100, 455 (1955).
[9] L. Tisza, Nature 141, 913 (1938).
[10] V. Peshkov, J. Phys. U.S.S.R. 8, 381 (1944); 10, 389 (1946).
[11] E. Andronikashvili, J. Phys. U.S.S.R. 10, 201 (1946).
[12] N. N. Bogolyubov, J. Phys. U.S.S.R. 11, 23 (1947).
[13] N. N. Bogolyubov and D. N. Zubarev, Zhur. Eksptl. i Teort. Fiz. 28, 129 (1955); English translation in Soviet Phys. 1, 83 (1955).
[14] D. N. Zubarev, Zhur. Eksptl. i Teort. Fiz. 29, 881 (1955).
[15] O. Penrose, Phil. Mag. 42, 1373 (1951).
[16] L. D. Landau, J. Phys. U.S.S.R. 5, 71 (1941).
[17] R. P. Feynman, Phys. Rev. 94, 262 (1954).

distance r, and $|\mathbf{q}_i-\mathbf{q}_j|$ means the length of the vector $\mathbf{q}_i-\mathbf{q}_j$ (except when the artifice of periodic boundary conditions is used, in which case $|\mathbf{q}_i-\mathbf{q}_j|$ means the length of the shortest vector congruent to $\mathbf{q}_i-\mathbf{q}_j$). Many-body interactions are omitted from (1), but including them would make no essential difference. The interaction between the He^4 atoms and those of the container is also omitted from (1); these could be included, but it is simpler to represent the container by a closed geometrical surface, considering only configurations for which all particles are within or on this surface, and imposing a suitable boundary condition on the wave function when any particle is on the surface. This boundary condition must be chosen to make H Hermitian. We denote the volume inside the container by V, and integrations over V by $\int_V \cdots d^3\mathbf{x}$ or $\int \cdots d^3\mathbf{x}$ or $\int \cdots d\mathbf{x}$.

As always in statistical mechanics, we are concerned here with very large values of N. Therefore we can often neglect quantities (for example $N^{-\frac{1}{3}}$) which are small when N is very large. A relation holding approximately by virtue of N being very large will be written in one of the forms $A\cong B$ or $A=B+o(1)$. These mean respectively that A/B is approximately 1 and that $A-B$ is negligible compared with 1, when N is large enough. We shall also use the notation $A=e^{O(1)}$ to mean that positive upper and lower bounds are known for A, but that a relation of the form $A\cong\text{const}$ has not been established. Evidently $A\cong\text{const}>0$ implies $A=e^{O(1)}$, but the example $A=2+\sin N$ shows that the converse does not hold in general. We shall use the phrase "A is finite" to mean $A=e^{O(1)}$.

We can give more precise meanings to the symbols $\cong$, etc., by considering not a single system but an infinite sequence of systems with different values of N. The boundary conditions for the different members of the sequence should be the same, and should be specified on boundary surfaces of the same shape but of sizes such that N/V is independent of N. Then, if the quantities A, B, etc., are defined for each member of the sequence, $A\cong B$ means $\lim_{N\to\infty}(A/B)=1$, $A=B+o(1)$ means $\lim_{N\to\infty}(A-B)=0$, and $A=e^{O(1)}$ means that positive constants a_1, a_2, and N_1 exist such that $N>N_1$ implies that $a_1<A<a_2$.

We shall use Dirac's notation[18] for matrix elements and for eigenvalues of operators.

3. A GENERALIZED CRITERION OF B.E. CONDENSATION

It is characteristic of an ideal B.E. gas in equilibrium below its transition temperature that a finite fraction of the particles occupies the lowest single-particle energy level. Using the notation of Sec. 2, we can therefore give the following criterion of B.E. condensation[4] for an ideal gas in equilibrium:

$$\begin{aligned}\langle n_0\rangle_{\text{Av}}/N&=e^{O(1)}\leftrightarrow\text{B.E. condensation},^{19}\\ \langle n_0\rangle_{\text{Av}}/N&=o(1)\leftrightarrow\text{no B.E. condensation},\end{aligned}\tag{2}$$

where $\langle n_0\rangle_{\text{Av}}$ is the average number of particles in the lowest single-particle level and the sign $\leftrightarrow$ denotes logical equivalence. This criterion has meaning for noninteracting particles only, because single-particle energy levels are not defined for interacting particles.

To generalize the criterion (2), we rewrite it in a form which has meaning even when there are interactions. This can be done by using von Neumann's statistical operator,[20] σ, whose position representative[21] $\langle q_1'\cdots q_N'|\sigma|q_1''\cdots q_N''\rangle$ is known as the density matrix.[22] We define a reduced statistical operator, σ_1, as follows[23,24]:

$$\sigma_1=N\,\mathrm{tr}_{2\cdots N}(\sigma),\tag{3}$$

where $\mathrm{tr}_{2\ldots N}(\sigma)$ means the trace of σ taken with respect to particles $2\cdots N$ but not particle 1. For an ideal gas in equilibrium, the eigenstates of σ_1 are the single-particle stationary states, and the corresponding eigen values are the average numbers of particles in these stationary states.[25] Consequently, (2) may be rewritten as follows:

$$\begin{aligned}n_M/N&=e^{O(1)}\leftrightarrow\text{B.E. condensation},\\ n_M/N&=o(1)\leftrightarrow\text{no B.E. condensation},\end{aligned}\tag{4}$$

where n_M denotes the largest eigenvalue of σ_1. In this form the criterion has meaning for interacting as well as for noninteracting particles, since σ_1 is defined in either case; thus (4) provides a suitable generalization of the ideal-gas criterion (2).

According to our criterion (4), B.E. condensation cannot occur in a Fermi system, because[26] the exclusion principle implies that $0\leqslant n_M\leqslant 1$. For Bose systems, however, the only general restriction on n_M is $0\leqslant n_M\leqslant N$, a consequence of the identity $\mathrm{tr}(\sigma_1)=N$ and the fact that σ_1 is positive semidefinite.

The application of (4) is most direct when the system satisfies periodic boundary conditions and is spatially uniform (a homogeneous phase in the thermodynamic sense). For, in this case, the reduced density matrix $\langle\mathbf{q}'|\sigma_1|\mathbf{q}''\rangle$ is a function of $\mathbf{q}'-\mathbf{q}''$ only, and specifying this function is equivalent to specifying the single-particle momentum distribution. In fact, the momentum

[18] P. A. M. Dirac, *The Principles of Quantum Mechanics* (Oxford University Press, London, 1947).

[19] A possible alternative to this equation is $\langle n_0\rangle_{\text{Av}}/N\cong\text{const}>0$, but the weaker form used in (2) is easier to apply.

[20] J. von Neumann, *Mathematical Foundations of Quantum Mechanics* (Princeton University Press, Princeton, 1955), Chap. 4.

[21] P. A. M. Dirac, reference 18, Chap. 3.

[22] P. A. M. Dirac, reference, pp. 130–135.

[23] K. Husimi, Proc. Phys. Math. Soc. Japan **22**, 264 (1940).

[24] J. de Boer, Repts. Progr. Phys. **12**, 313–316 (1949).

[25] K. Husimi, reference 23, Eq. (10.6).

[26] See, for example, P-O. Löwdin, Phys. Rev. **97**, 1474 (1955).

representative[21] of σ_1, given by

$$\langle \mathbf{p}'|\sigma_1|\mathbf{p}''\rangle \equiv \int\int \langle \mathbf{p}'|\mathbf{q}'\rangle d\mathbf{q}'\langle \mathbf{q}'|\sigma_1|\mathbf{q}''\rangle d\mathbf{q}''\langle \mathbf{q}''|\mathbf{p}''\rangle$$

$$= V^{-1}\int\int \exp[i(\mathbf{p}''\cdot\mathbf{q}''-\mathbf{p}'\cdot\mathbf{q}')/\hbar] \times \langle \mathbf{q}'|\sigma_1|\mathbf{q}''\rangle d\mathbf{q}'d\mathbf{q}''$$

is a diagonal matrix, so that n_M is the largest diagonal element of this matrix. Now, since

$$\langle \mathbf{p}_1'\cdots\mathbf{p}_N'|\sigma|\mathbf{p}_1'\cdots\mathbf{p}_N'\rangle$$

is the probability distribution in the (discrete) momentum space of N particles,

$$\langle \mathbf{p}_1'|\sigma_1|\mathbf{p}_1'\rangle = N\sum_{\mathbf{p}_2'}\cdots\sum_{\mathbf{p}_N'}\langle \mathbf{p}_1'\cdots\mathbf{p}_N'|\sigma|\mathbf{p}_1'\cdots\mathbf{p}_N'\rangle$$

must be the average number of particles with momentum $\mathbf{p}_1'$. [To confirm this interpretation, note that[14]

$$\langle \textstyle\sum_j f(\mathbf{p}_j)\rangle_{\text{Av}} = \sum_{\mathbf{p}'}\langle \mathbf{p}'|\sigma_1|\mathbf{p}'\rangle f(\mathbf{p}')$$

for arbitrary $f(\mathbf{p})$.] Therefore, according to (4), B.E. condensation is present for a spatially uniform system with periodic boundary conditions whenever a finite fraction of the particles have identical momenta. The work of Bogolybov[12] shows that this form of the criterion is satisfied in a system of weakly interacting B.E. particles at very low temperatures.

4. ALTERNATIVE FORMS OF THE CRITERION

When the system is not spatially uniform, it is more difficult to diagonalize σ_1, and some transformations of the criterion (4) are useful. The simplest of these depends on the following inequality:

$$n_M^2 \leqslant \sum_a n_a^2 \leqslant n_M \sum_a n_a = n_M N, \tag{5}$$

where the n_a's are the eigenvalues of σ_1; the fact that $\sum_a n_a = \text{tr}(\sigma_1) = N$ follows from (3). We define

$$A_2 \equiv N^{-2}\int_V\int_V |\langle \mathbf{q}'|\sigma_1|\mathbf{q}''\rangle|^2 d^3\mathbf{q}'d^3\mathbf{q}''. \tag{6}$$

It is clear that $A_2 = N^{-2}\,\text{tr}(\sigma_1^2) = N^{-2}\sum n_a^2$, and hence, by (5), that $(n_M/N)^2 \leqslant A_2 \leqslant n_M/N$. It follows that

$A_2 = e^{O(1)} \leftrightarrow n_M/N = e^{O(1)}$, while $A_2 = o(1) \leftrightarrow n_M/N = o(1)$.

The following criterion is therefore equivalent to (4):

$$\begin{aligned} A_2 &= e^{O(1)} \leftrightarrow \text{B.E. condensation,} \\ A_2 &= o(1) \leftrightarrow \text{no B.E. condensation.} \end{aligned} \tag{7}$$

Another form of the criterion depends on inequalities satisfied by

$$A_1 \equiv (NV)^{-1}\int_V\int |\langle \mathbf{q}'|\sigma_1|\mathbf{q}''\rangle| d^3\mathbf{q}'d^3\mathbf{q}''. \tag{8}$$

Unlike A_2, this quantity has no simple interpretation in terms of the eigenvalues of σ_1/N, but it is easier to use. An upper bound for A_1 comes from the fact that $(A_1N/V)^2$, the square of the mean value of the function $|\langle \mathbf{q}'|\sigma_1|\mathbf{q}''\rangle|$, cannot exceed $A_2(N/V)^2$, the mean value of the square of this function; therefore we have

$$A_1^2 \leqslant A_2. \tag{9}$$

To find a lower bound for A_1, we use the fact that σ_1 is positive semidefinite (this follows intuitively from the probability interpretation of the eigenvalues of σ_1/N; alternatively it can be proved rigorously from (3) and the fact that σ is positive semidefinite). Since σ_1 has no negative eigenvalues, its square root is Hermitian. Applying the Schwartz inequality to the two state vectors $\sigma_1^{\frac{1}{2}}|\mathbf{q}'\rangle$ and $\sigma_1^{\frac{1}{2}}|\mathbf{q}''\rangle$, we obtain

$$|\langle \mathbf{q}'|\sigma_1|\mathbf{q}''\rangle| \leqslant [\langle \mathbf{q}'|\sigma_1|\mathbf{q}'\rangle\langle \mathbf{q}''|\sigma_1|\mathbf{q}''\rangle]^{\frac{1}{2}} \leqslant \alpha N/V, \tag{10}$$

where $\alpha N/V$ is any upper bound of $\langle \mathbf{q}'|\sigma_1|\mathbf{q}'\rangle$. Combining (6), (8), and (10), we find

$$A_2 \leqslant \alpha A_1. \tag{11}$$

For any physical system, α can be chosen independent of N; for $\langle \mathbf{q}'|\sigma_1|\mathbf{q}'\rangle$ is the average number density at the point $\mathbf{q}'$ and cannot become indefinitely large. For example, in a liquid at thermal equilibrium, $\langle \mathbf{q}'|\sigma_1|\mathbf{q}'\rangle$ is approximately N/V except near the boundary, so that α can be chosen just greater than 1. Treating α as finite, and combining (7) with (9) and (11), we obtain

$$\begin{aligned} A_1 &= e^{O(1)} \leftrightarrow \text{B.E. condensation,} \\ A_1 &= o(1) \leftrightarrow \text{no B.E. condensation.} \end{aligned} \tag{12}$$

A third form of the criterion is valuable when the reduced density matrix $\langle \mathbf{q}'|\sigma_1|\mathbf{q}''\rangle$ has the asymptotic form $\Psi(\mathbf{q}')\Psi^*(\mathbf{q}'')$ for large $|\mathbf{q}'-\mathbf{q}''|$ (Ψ^* is the complex conjugate of Ψ). Some consequences of assuming this asymptotic relation for He II were discussed by Penrose.[15] Here we formulate the assumption as follows:

$$|\langle \mathbf{q}'|\sigma_1|\mathbf{q}''\rangle - \Psi(\mathbf{q}')\Psi^*(\mathbf{q}'')| \leqslant (N/V)\gamma(|\mathbf{q}'-\mathbf{q}''|), \tag{13}$$

where the (non-negative) function $\gamma(r)$ is independent of N and satisfies

$$\lim_{r\to\infty}\gamma(r) = 0. \tag{14}$$

To use (13), we need the following lemma:

$$\Gamma(\mathbf{x}) \equiv V^{-1}\int \gamma(|\mathbf{x}'-\mathbf{x}|)d^3\mathbf{x}' = o(1). \tag{15}$$

Proof.—Let ϵ be an arbitrary positive number. Then, by (14), there exists a number R (depending on ϵ) such that

$$0 \leqslant \gamma(r) < \tfrac{1}{2}\epsilon \quad \text{if} \quad r > R. \tag{16}$$

We also have

$$0 \leqslant \gamma(r) < \gamma_M + \tfrac{1}{2}\epsilon \quad \text{if} \quad 0 \leqslant r \leqslant R,$$

where γ_M is the maximum of the function $\gamma(r)$. Using (16) in (15) we obtain $0\leqslant\Gamma(\mathbf{x})<\frac{1}{2}\epsilon+V^{-1}\gamma_M V_R(\mathbf{x})$, where $V_R(\mathbf{x})$ is the volume of $\mathbf{x}'$-space for which $|\mathbf{x}'-\mathbf{x}|\leqslant R$. Since $V_R(\mathbf{x})\leqslant 4\pi R^3/3$, we can ensure that $0\leqslant\Gamma<\epsilon$ by choosing $V>8\pi\gamma_M R^3/3\epsilon$. By the definition of a limit, this implies $\lim_{V\to\infty}\Gamma=0$; that is, (15) is true.

We can now obtain a criterion of B.E. condensation, using the relation

$$(NV)^{-1}\int\int|\langle\mathbf{q}'|\sigma_1|\mathbf{q}''\rangle-\Psi(\mathbf{q}')\Psi^*(\mathbf{q}'')|d\mathbf{q}'d\mathbf{q}''=o(1), \tag{17}$$

which follows from (13) and (15). Combining this with (8) with the help of the elementary inequality $-|u-v|\leqslant|u|-|v|\leqslant|u-v|$ gives

$$A_1=(NV)^{-1}\left[\int|\Psi(\mathbf{x})|d^3\mathbf{x}\right]^2+o(1). \tag{18}$$

Using (12), we obtain the criterion, valid whenever (13) holds:

$$V^{-1}\int|\Psi|d^3\mathbf{x}=e^{O(1)}\leftrightarrow\text{B.E. condensation,}$$
$$V^{-1}\int|\Psi|d^3\mathbf{x}=o(1)\leftrightarrow\text{no B.E. condensation.} \tag{19}$$

The function Ψ has a simple interpretation when B.E. condensation is present: we can show that $\Psi(\mathbf{x})$ is a good approximation to the eigenfunction of the matrix $\langle\mathbf{q}'|\sigma_1|\mathbf{q}''\rangle$ corresponding to the eigenvalue n_M, and also that its normalization is

$$n_\Psi\equiv\int|\Psi(\mathbf{x})|^2d^3\mathbf{x}\cong n_M. \tag{20}$$

We note first that all eigenvalues of the matrix

$$N^{-1}\langle\mathbf{q}'|\tau|\mathbf{q}''\rangle\equiv N^{-1}[\langle\mathbf{q}'|\sigma_1|\mathbf{q}''\rangle-\Psi(\mathbf{q}')\Psi^*(\mathbf{q}'')]$$

are $o(1)$, since, by (17), (8), and (12), a system whose reduced density matrix was τ would not show B.E. condensation. It follows that

$$f\{\varphi(\mathbf{x})\}\equiv N^{-1}\int\int\varphi^*(\mathbf{q}')\langle\mathbf{q}'|\sigma_1|\mathbf{q}''\rangle\varphi(\mathbf{q}'')d\mathbf{q}'d\mathbf{q}''$$
$$=|(\varphi,\Psi)|^2/N+o(1), \tag{21}$$

where φ is an arbitrary normalized function and

$$(\varphi,\Psi)\equiv\int\varphi^*(\mathbf{x})\Psi(\mathbf{x})d^3\mathbf{x}.$$

The arbitrary function $\varphi(\mathbf{x})$ in (21) can be written in the form $\varphi(\mathbf{x})=n_\Psi^{-\frac{1}{2}}[a\Psi(\mathbf{x})+b\Phi(\mathbf{x})]$, where Φ is chosen to make $(\Psi,\Phi)=0$ and $(\Phi,\Phi)=n_\Psi$, and where (since φ is normalized) $|a|^2+|b|^2=1$. Inserting this expression for φ into (21) gives

$$f\{\varphi\}=|a|^2n_\Psi/N+o(1)=(1-|b|^2)n_\Psi/N+o(1).$$

Now, the maximum value of $f\{\varphi\}$ is n_M/N, and it is attained when φ equals φ_M, the normalized eigenfunction of $\langle\mathbf{q}'|\sigma_1|\mathbf{q}''\rangle$ corresponding to the eigenvalue n_M. The last expression for $f\{\varphi\}$ shows that this maximum is $n_\Psi/N+o(1)$ and is attained with $|b|=o(1)$. It follows that $n_M\cong n_\Psi$ in agreement with (20), and also that

$$\int|an_\Psi^{-\frac{1}{2}}\Psi(\mathbf{x})-\varphi_M(\mathbf{x})|^2d^3\mathbf{x}=|b|^2=o(1). \tag{22}$$

This equation tells us that $\Psi(\mathbf{x})$ is to a good approximation proportional to $\varphi_M(\mathbf{x})$. In view of these results, we may call Ψ the *wave function of the condensed particles*, and n_Ψ/N the *fraction of condensed particles*.

5. GROUND STATE OF A B.E. FLUID

In this section we derive some general properties of the ground-state wave function. These will be needed in Sec. 6.

Let us define the ground-state wave function $\psi(\mathbf{x}_1\cdots\mathbf{x}_N)$ to be the real symmetric function which minimizes the expression

$$\int\cdots\int[\hbar^2\sum_i(\nabla_i\psi)^2/2m+\sum_{i<j}U_{ij}\psi^2]d\mathbf{x}_1\cdots d\mathbf{x}_N, \tag{23}$$

while at the same time satisfying the boundary conditions and the normalization condition. The Euler equation of this variation problem shows that ψ satisfies Schrödinger's equation for the Hamiltonian (1). Now, the function $|\psi|$ also conforms to the above definition, and so it too satisfies Schrödinger's equation. The first derivative of $|\psi|$ must therefore be continuous wherever the potential energy is finite. This is possible only if ψ does not change sign. We may therefore take ψ to be non-negative. Suppose now that ψ_1 and ψ_2 are two different non-negative functions conforming to the above definition. Then, since Schrödinger's equation is linear, $\psi_1-\psi_2$ also conforms to the definition, and (by the result just proved) does not change sign; but this contradicts the original assumption that both ψ_1 and ψ_2 are normalized. Hence the above definition yields a unique, non-negative function[27] ψ.

For a fluid phase, we can obtain further information if we assume [28] that there is *no long-range configurational*

[27] These properties of the ground-state wave function are fairly well known [see, for example, R. P. Feynman, Phys. Rev. **91**, 1301 (1953)], but the authors have seen no proof in the literature. A proof for the special case $N=1$ (to which no symmetry requirements apply) is given by R. Courant and D. Hilbert, *Methoden der Mathematischen Physik* (Verlag Julius Springer, Berlin, 1931), Vol. 1, Chap. 6, Secs. 6, 7.

[28] A similar principle is often used in the classical theory of liquids—for example by J. E. Mayer and E. W. Montroll, J. Chem. Phys. **9**, 2 (1941). Its use here amounts to assuming that

order. By this we mean that there is a finite "range of order" R with the following property: for any two concentric spheres S_1 and S_2 with radii R_1 and R_1+R, respectively, the relative probabilities of the various possible configurations of particles inside S_1 are approximately[29] independent of the situation outside S_2. By the "situation" outside S_2, we mean here the number of particles ouside S_2, their positions, and the position of the part of the boundary surface outside S_2.

This assumption implies that, if the configuration of the particles inside S_1 is altered while everything else remains the same, then the probability density in configuration space changes by a factor approximately independent of the situation outside S_2. Hence, if the point $\mathbf{x}_i$ is inside S_1, then $\nabla_i \log\psi$ is approximately independent of the situation outside S_2. This is true for any choice of S_1 provided S_1 encloses $\mathbf{x}_i$, and in particular it is true when R_1 is vanishingly small. Therefore, $\nabla_i \log\psi$ is independent of the situation outside a sphere $S(\mathbf{x}_i)$ with center $\mathbf{x}_i$ and radius R.

Setting $i=1$ in this result and integrating shows that ψ can be written in the form

$$\psi(\mathbf{x}_1\cdots\mathbf{x}_N)=\theta(\mathbf{x}_2\cdots\mathbf{x}_N)\chi(\mathbf{x}_1;\mathbf{x}_2\cdots\mathbf{x}_N), \quad (24)$$

where the functions θ and χ are symmetric in $\mathbf{x}_2\cdots\mathbf{x}_N$, and χ is approximately[30] independent of the situation outside $S(\mathbf{x}_1)$.

The function θ in (24) has a simple physical meaning. To find this, we write the Schrödinger equation satisfied by ψ in the form

$$-(\hbar^2/2m)\sum_j[\nabla_j^2\log\psi+(\nabla_j\log\psi)^2]+\sum_{i<j}U_{ij}=\text{const.} \quad (25)$$

Taking the gradient with respect to x_i $(i\neq 1)$ and substituting from (24), we obtain after some rearrangement

$$\begin{aligned}\nabla_i\{-(\hbar^2/2m)\textstyle\sum_j'[\nabla_j^2\log\theta+(\nabla_j\log\theta)^2]+\sum_j''U_{ij}\}\\ =(\hbar^2/2m)\textstyle\sum_j[\nabla_j^2+2(\nabla_j\log\theta)\cdot\nabla_j]\nabla_i\log\chi\\ +(\hbar^2/m)\textstyle\sum_j(\nabla_j\log\chi)\cdot\nabla_j\nabla_i\log\psi-\nabla_iU_{1i}, \quad (26)\end{aligned}$$

where $\sum_j'$ means a sum with the $j=1$ term omitted, and $\sum_j''$ means a sum with the $j=1$ and $j=i$ terms omitted. Since the left member of (26) does not contain $\mathbf{x}_1$, the right member must be independent of $\mathbf{x}_1$. To evaluate the right member, we may therefore choose $\mathbf{x}_1$ to make $|\mathbf{x}_i-\mathbf{x}_1|>2R$. The properties of χ then imply that $\nabla_i\log\chi\simeq 0$, so that the first sum vanishes approximately; they also imply that the summand in the second sum is negligible unless $|\mathbf{x}_j-\mathbf{x}_1|\leqslant R$. The argument preceding (24) shows, however, that $\nabla_i\log\psi$ is independent of $\mathbf{x}_j$ unless $|\mathbf{x}_j-\mathbf{x}_i|\leqslant R$. Since $|\mathbf{x}_j-\mathbf{x}_1|$ and $|\mathbf{x}_j-\mathbf{x}_i|$ cannot both be less than R (because $|\mathbf{x}_1-\mathbf{x}_i|>2R$), the summand in the second sum is always negligible. The term ∇_iU_{1i} also vanishes, because the interaction has a short range. Thus the entire right member of (26) vanishes approximately.[31] The expression in curly brackets is therefore approximately independent of $\mathbf{x}_i$ for $i=2\cdots N$. This means that $\theta(\mathbf{x}_2\cdots\mathbf{x}_N)$ approximately satisfies an equation, analogous to (25), which is equivalent to Schrödinger's equation for a system of $N-1$ particles. Since θ is non-negative, it must therefore have the form

$$\theta(\mathbf{x}_2\cdots\mathbf{x}_N)\simeq c\vartheta(\mathbf{x}_2\cdots\mathbf{x}_N), \quad (27)$$

where c is a constant and ϑ is the normalized ground-state wave function for $N-1$ particles.

A simple illustration of (24) is provided by a type of approximation to $\psi(\mathbf{x}_1\cdots\mathbf{x}_N)$ used by various authors[32,33,18]:

$$\psi(\mathbf{x}_1\cdots\mathbf{x}_N)\propto[\prod_j\mu(\mathbf{x}_j)]\prod_{i<j}\omega(|\mathbf{x}_i-\mathbf{x}_j|), \quad (28)$$

where $\omega(r)\rightarrow 1$ when r is large. In this approximation, (24) can be satisfied by taking

$$\theta(\mathbf{x}_2\cdots\mathbf{x}_N)\propto[\prod_i'\mu(\mathbf{x}_i)]\prod_{i<j}'\omega(|\mathbf{x}_i-\mathbf{x}_j|), \quad (29)$$

$$\chi(\mathbf{x}_1,\mathbf{x}_2\cdots\mathbf{x}_N)=\mu(\mathbf{x}_1)\prod_i'\omega(|\mathbf{x}_i-\mathbf{x}_1|), \quad (30)$$

where Π' means a product with all $i=1$f actors omitted. It is clear that (29) is consistent with (27), and that, if R is large enough, χ as defined in (30) is approximately independent of the positions of the particles outside $S(\mathbf{x}_1)$.

6. LIQUID HELIUM-4 AT ABSOLUTE ZERO

At absolute zero, the density matrix is given by

$$\langle\mathbf{q}_1'\cdots\mathbf{q}_N'|\sigma|\mathbf{q}_1''\cdots\mathbf{q}_N''\rangle=\psi(\mathbf{q}_1'\cdots\mathbf{q}_N')\psi(\mathbf{q}_1''\cdots\mathbf{q}_N''),$$

since the ground-state wave function ψ is real and normalized. The reduced density matrix is therefore

$$\langle\mathbf{q}'|\sigma_1|\mathbf{q}''\rangle=N\int\cdots\int\psi(\mathbf{q}',\mathfrak{x})\psi(\mathbf{q}'',\mathfrak{x})d\mathfrak{x}, \quad (31)$$

---

the probability density in configuration space is qualitatively similar to the corresponding probability density for a classical liquid. The importance of this principle for the ground state of a quantum liquid was noted by A. Bijl, Physica 7, 869 (1940).

[29] The meaning of the word "approximately" is purposely left vague, since it would complicate the discussion too much to attempt a rigorous formulation. As we see it, a rigorous formulation would have to depend on a limit operation $R\rightarrow\infty$: that is, it would assume that the approximation of statistical independence could be made arbitrarily good by choosing R large enough.

[30] If the theory were formulated more rigorously (see reference 29), the corresponding property of χ might be $\nabla_j\chi(\mathbf{x}_1;\mathbf{x}_2\cdots\mathbf{x}_N)\leqslant K(|\mathbf{x}_j-\mathbf{x}_1|)$ where $K(r)\rightarrow 0$ in a suitable way as $r\rightarrow\infty$.

[31] Only a rigorous treatment can completely justify the implicit assumption that the sum of N negligible terms is itself negligible. The present methods can, however, be used to show that the contribution of a given j value to the sums in the right member of (26) is negligible compared with its contribution to the sums in the left (with a finite number of exceptions, for which $|\mathbf{x}_j-\mathbf{x}_1|\leqslant R$ and the contributions on both sides are negligible).

[32] A. Bijl, reference 28.

[33] R. B. Dingle, Phil. Mag. 40, 573 (1949).

where $\mathfrak{x}$ and $d\mathfrak{x}$ are abbreviations for $\mathbf{x}_2\cdots\mathbf{x}_N$ and $d^3\mathbf{x}_2\cdots d^3\mathbf{x}_N$, respectively. For a preliminary discussion of (31), we use a crude approximation to ψ suggested by Feynman[7]:

$$\psi(\mathbf{x}_1\cdots\mathbf{x}_N)\simeq(\Omega_N)^{-\frac{1}{2}}F_N(\mathbf{x}_1\cdots\mathbf{x}_N), \tag{32}$$

where Ω_N is a normalizing constant, and $F_N(\mathbf{x}_1\cdots\mathbf{x}_N)$ by definition takes the value 1 whenever $\mathbf{x}_1\cdots\mathbf{x}_N$ is a possible configuration for the centers of N hard spheres of diameter d and the value 0 for all other configurations. Here $d\simeq2.6$A is the diameter of a He^4 atom. The approximation amounts to using (28) with $\mu=1$ and with $\omega(r)=0$ for $r<d$ and $\omega=1$ for $r\geqslant d$.

The normalization integral corresponding to (32) shows that $\Omega_N/N!$ is the configurational partition function for a classical system of N noninteracting hard spheres. Moreover, the integral in (31) is now closely related to the pair distribution function for $N+1$ hard spheres, defined as follows[24]:

$$n_2(\mathbf{q}',\mathbf{q}'')\equiv(N+1)N\int\cdots\int F_{N+1}(\mathbf{q}',\mathbf{q}'',\mathfrak{x})d\mathfrak{x}/\Omega_{N+1}. \tag{33}$$

Under the approximation (32), the integrand in (31) is $1/\Omega_N$ times that in (33) when $|\mathbf{q}'-\mathbf{q}''|\geqslant d$, so that

$$\langle\mathbf{q}'|\sigma_1|\mathbf{q}''\rangle=z^{-1}n_2(\mathbf{q}',\mathbf{q}'')\quad\text{if}\quad|\mathbf{q}'-\mathbf{q}''|\geqslant d. \tag{34}$$

Here $z\equiv(N+1)\Omega_N/\Omega_{N+1}$ is the activity of the hard-sphere system. The physical meaning of n_2 shows that, for large $|\mathbf{q}'-\mathbf{q}''|$, n_2 tends to $(N/V)^2$. Hence (13) can be satisfied by taking $\Psi\cong\text{const}\cong z^{-\frac{1}{2}}N/V$ (except, possibly, near the boundary). Hence, by (19), B.E. condensation is present; moreover, by (20) and the discussion following (20), the fraction of condensed particles is

$$n_M/N\cong n_\Psi/N\cong\Psi^2V/N\cong N/Vz. \tag{35}$$

The right member of (35) can be calculated from the virial series for hard spheres.[24] Taking the density of He II to be 0.28 times the density at closest packing, we obtain the result 0.08. Thus, Feynman's approximation (32) implies that B.E. condensation is present in He II at absolute zero and that the fraction of condensed particles is about 8%.

The above discussion makes it plausible that a treatment based on the true wave function will also indicate the presence of B. E. condensation. To supply such a treatment, we first substitute from (24) and (27) into (31). This yields

$$\langle\mathbf{q}'|\sigma_1|\mathbf{q}''\rangle=c^2N\langle\chi(\mathbf{q}';\mathfrak{x})\chi(\mathbf{q}'';\mathfrak{x})\rangle_\vartheta, \tag{36}$$

where, for any function $f(\mathfrak{x})$

$$\langle f\rangle_\vartheta\equiv\langle f(\mathfrak{x})\rangle_\vartheta\equiv\int\cdots\int f(\mathfrak{x})\vartheta^2(\mathfrak{x})d\mathfrak{x} \tag{37}$$

is the expectation value of $f(\mathfrak{x})$ in the ground state of a liquid of $N-1$ particles whose configuration is $\mathfrak{x}\equiv\mathbf{x}_2\cdots\mathbf{x}_N$.

In studying (36) it will be convenient to look on $\mathbf{q}'$ and $\mathbf{q}''$ as parameters and to treat $\chi'\equiv\chi(\mathbf{q}';\mathfrak{x})$ and $\chi''\equiv\chi(\mathbf{q}'';\mathfrak{x})$ as variables depending on the configuration $\mathfrak{x}$ of a liquid of $N-1$ particles. The correlation coefficient[25] of χ' and χ'' is defined by

$$\rho(\mathbf{q}',\mathbf{q}'')\equiv\frac{\langle\chi'\chi''\rangle_\vartheta-\langle\chi'\rangle_\vartheta\langle\chi''\rangle_\vartheta}{[\langle\chi'^2\rangle_\vartheta-\langle\chi'\rangle_\vartheta^2]^{\frac{1}{2}}[\langle\chi''^2\rangle_\vartheta-\langle\chi''\rangle_\vartheta^2]^{\frac{1}{2}}}. \tag{38}$$

Now, it was shown in Sec. 5 that χ' is independent of the "situation" outside $S(\mathbf{q}')$ and that χ'' is independent of the situation outside $S(\mathbf{q}'')$. By applying the principle of no long-range configurational order, given in Sec. 5, to the ground state of a liquid of $N-1$ particles, with the sphere S_1 chosen large enough to enclose both $S(\mathbf{q}')$ and $S(\mathbf{q}'')$, we find that $\rho(\mathbf{q}',\mathbf{q}'')$ is independent of V for large enough V. By applying the same principle with S_1 this time taken to coincide with $S(\mathbf{q}')$, we find that χ' and χ'' are approximately statistically independent if $S(\mathbf{q}'')$ is entirely outside S_2; that is, $\rho(\mathbf{q}',\mathbf{q}'')$ approximately vanishes if $|\mathbf{q}'-\mathbf{q}''|>3R$.

We can now show that (13) holds, with

$$\Psi(\mathbf{q}')=cN^{\frac{1}{2}}\langle\chi'\rangle_\vartheta. \tag{39}$$

For, substituting (36) and (39) into the left member of (13) gives $c^2N[\langle\chi'\chi''\rangle_\vartheta-\langle\chi'\rangle_\vartheta\langle\chi''\rangle_\vartheta]$, which, by (38), is less than $c^2N\rho(\mathbf{q}',\mathbf{q}'')[\langle\chi'^2\rangle_\vartheta\langle\chi''^2\rangle_\vartheta]^{\frac{1}{2}}$. Setting $\mathbf{q}'=\mathbf{q}''$ in (36) shows that this last expression equals $\rho(\mathbf{q}',\mathbf{q}'')\times[\langle\mathbf{q}'|\sigma_1|\mathbf{q}'\rangle\langle\mathbf{q}''|\sigma_1|\mathbf{q}''\rangle]^{\frac{1}{2}}$. Therefore, by (10) and the properties of $\rho(\mathbf{q}',\mathbf{q}'')$ given above, (13) can be satisfied by making $\gamma(|\mathbf{q}'-\mathbf{q}''|)\geqslant\alpha\rho(\mathbf{q}',\mathbf{q}'')$ for every $\mathbf{q}'$ and $\mathbf{q}''$.

If the distance from $\mathbf{q}'$ to the boundary exceeds $2R$, then $\langle\chi'\rangle_\vartheta$ and $\langle\chi'^2\rangle_\vartheta$ are (approximately) positive constants independent of N and $\mathbf{q}'$. For we may take the sphere S_1 defined in Sec. 5 to be $S(\mathbf{q}')$; then the relative probabilities for the various configurations of particles inside $S(\mathbf{q}')$—on which alone χ' depends—are independent of N and the relative positions of S_2 and the boundary. It follows that $V^{-1}\int_V\langle\chi'\rangle_\vartheta d^3\mathbf{q}'\cong\text{const}>0$ and also, by (36), that $c^2N\cong\text{const}>0$ since $\langle\mathbf{q}'|\sigma_1|\mathbf{q}'\rangle\cong N/V$ if $\mathbf{q}'$ is far from the boundary. Applying the criterion (19) to the Ψ defined in (39), we conclude that B.E. condensation is present in liquid He^4 at absolute zero.

The above discussion would not lead one to expect B.E. condensation in a solid, because the assumption of no long-range configurational order is valid for a fluid phase only. In fact, it can be argued that a solid does *not* show B.E. condensation, at least for $T=0$°K. We assume that a solid at $T=0$°K is a perfect crystal

[24] M. N. Roxenbluth and A. W. Rosenbluth, J. Chem. Phys. 22, 881 (1954).

[25] H. Cramer, *Mathematical Methods of Statistics* (Princeton University Press, Princeton, 1946), p. 277.

—i.e., that there exists a set of lattice sites such that ψ is small unless one particle is near each lattice site.[36] In the expression (31) for the reduced density matrix, therefore, the integrand will be appreciable only if every one of the points $\mathbf{x}_2\cdots\mathbf{x}_N$ is near a separate lattice site, while both $\mathbf{q}'$ and $\mathbf{q}''$ are near the remaining site. When $|\mathbf{q}'-\mathbf{q}''|$ is large, this last condition cannot be fulfilled, so that $\langle\mathbf{q}'|\sigma_1|\mathbf{q}''\rangle$ will tend to 0 for large $|\mathbf{q}'-\mathbf{q}''|$. This indicates that the function Ψ of (13) will be 0, so that, by (20), there is no B.E. condensation in a solid at absolute zero.

Our result that B.E. condensation occurs in liquid He^4 at $T=0°K$ must now be extended to nonzero temperatures. (The need for such an extension is illustrated by the example of a two-dimensional B.E. gas, which[4] shows B.E. condensation at $T=0°K$ but not for $T\neq 0°K$.) This will be done in the next section.

7. B.E. CONDENSATION AND THE LAMBDA-TRANSITION

Feynman,[7] and also Matsubara,[6] have studied the lambda-transition in liquid helium by expressing the partition function in the form

$$Z=\sum_{\{m_l\}}\Big[\prod_l (m_l!l^{m_l})\Big]^{-1}\operatorname{tr}(Pe^{-\beta H}), \qquad (40)$$

where the sum is over all partitions of the number N (that is, over all sets $\{m_l\}$ of non-negative integers satisfying $\sum_l m_l=N$), P is any permutation containing m_l cycles[37] of length l $(l=1\cdots N)$, and $\beta\equiv 1/kT$ with $k\equiv$ Boltzmann's constant. Evaluating (40) with the help of approximations for $\operatorname{tr}(Pe^{-\beta H})$, they showed how it could exhibit a transition, which they identified with the lambda-transition. In the present section, we shall show that Feynman's approximations also imply that the criterion (4) of B.E. condensation is satisfied for He II in equilibrium.

The statistical operator for thermal equilibrium is

$$\sigma=(N!Z)^{-1}\sum_P Pe^{-\beta H}, \qquad (41)$$

where the sum is over all permutations P of the N particles. Feynman's path integral[7] for the density matrix shows that the position representative of (41) is non-negative. Therefore the corresponding reduced density matrix, calculated according to (3), is also non-negative, so that the quantity defined in (8) is now given by

$$A_1=(N!ZV)^{-1}\sum_P\int\int d\mathbf{q}'_1 d\mathbf{q}_1''\langle\mathbf{q}_1'|\operatorname{tr}_{2\ldots N}Pe^{-\beta H}|\mathbf{q}_1''\rangle. \qquad (42)$$

All permutations corresponding to a given partition $\{m_l\}$ and also having particle No. 1 in a cycle of given length L contribute equal terms to the above sum, since a suitable relabeling of the particles $2\cdots N$ will turn any one such term into any other. Collecting together, for each $\{m_l\}$ and L, the $(L/N)N!/\prod_l(m_l!l^{m_l})$ equal terms, we can write A_1 as a sum over L and $\{m_l\}$, obtaining

$$A_1=N^{-1}\sum_L L\langle m_L A_{1,L}\{m_l\}\rangle. \qquad (43)$$

Here we have defined, for any function $f\equiv f\{m_l\}$ depending on the set of numbers $\{m_l\}$, a quantity

$$\langle f\rangle\equiv Z^{-1}\sum_{\{m_l\}}\Big[\prod_l(m_l!l^{m_l})\Big]^{-1}f\{m_l\}\operatorname{tr}(Pe^{-\beta H}), \qquad (44)$$

where P is any permutation corresponding to the partition $\{m_l\}$. We have also defined

$$A_{1,L}\{m_l\}\equiv\frac{\int\int d\mathbf{q}_1'd\mathbf{q}_1''\langle\mathbf{q}_1'|\operatorname{tr}_{2\ldots N}Pe^{-\beta H}|\mathbf{q}_1''\rangle}{V\operatorname{tr}(Pe^{-\beta H})}, \qquad (45)$$

where P is any permutation which corresponds to the partition $\{m_l\}$ and also has particle No. 1 in a cycle of length L.

To use (43), we introduce two approximations due to Feynman.[7,38] The first is

$$\langle\mathbf{q}_1'\cdots\mathbf{q}_N'|Pe^{-\beta H}|\mathbf{q}_1''\cdots\mathbf{q}_N''\rangle \simeq K\lambda^{-3N}\phi(\mathbf{q}_1'\cdots\mathbf{q}_N')\phi(\mathbf{q}_1''\cdots\mathbf{q}_N'')\times\exp[-(\pi/\lambda^2)\textstyle\sum_j(\mathbf{q}_j'-\mathbf{q}_{Pj}'')^2], \qquad (46)$$

where K is a constant, λ means $h(2\pi m'kT)^{-\frac{1}{2}}$ with m' an effective mass, and $\phi(\mathbf{x}_1\cdots\mathbf{x}_N)$ is a normalized non-negative symmetric function which reduces to the ground-state wave function when $T\to 0°K$. (We deviate slightly from Feynman's usage: he does not take ϕ to be normalized.) Feynman's other approximation is used in evaluating integrals over configuration space involving (46); it is to replace the factor containing ϕ by its value averaged over the region of integration and to replace each factor $\exp[-\pi(\mathbf{x}_i-\mathbf{x}_j)^2/\lambda^2]$ by

$$G(\mathbf{x}_i-\mathbf{x}_j)\equiv p(|\mathbf{x}_i-\mathbf{x}_j|)\exp[-\pi(\mathbf{x}_i-\mathbf{x}_j)^2/\lambda^2], \qquad (47)$$

where $p(0)\equiv 1$ and $p(r)$ for $r>0$ is the radial distribution function, tending to 1 as $r\to\infty$.

Using these approximations in (45), we obtain

$$A_{1,L}\{m_L\}\simeq A_{1,\infty}\delta_L/f_L, \qquad (48)$$

[36] For equilibrium at a temperature $T\neq 0°K$, a few atoms will be in interstitial positions far from their proper lattice sites. The fraction of interstitial atoms will be $e^{-W/kT}$, where W is the energy required to excite one atom from a lattice to an interstitial site. This fraction tends to 0 as T tends to 0°K.

[37] For the definition of a cycle, see R. P. Feynman, reference 7, or H. Margenau and G. M. Murphy, *The Mathematics of Physics and Chemistry* (D. Van Nostrand Company, Inc., New York, 1943), p. 538.

[38] For a critical discussion of these approximations, see G. V. Chester, Phys. Rev. **93**, 1412 (1954).

where (with $\mathfrak{x}$ standing, as before, for $\mathbf{x}_2\cdots\mathbf{x}_N$)

$$A_{1,\infty}\equiv V^{-1}\int\cdots\int\phi(\mathbf{q}',\mathfrak{x})\phi(\mathbf{q}'',\mathfrak{x})d\mathbf{q}'d\mathbf{q}''d\mathfrak{x}, \quad (49)$$

$$\delta_L\equiv V^{-1}\int\cdots\int\prod_{j=1}^{L}G(\mathbf{x}_{j+1}-\mathbf{x}_j)d\mathbf{x}_1\cdots d\mathbf{x}_{L+1}, \quad (50)$$

$$f_L\equiv\int\cdots\int G(\mathbf{x}_1-\mathbf{x}_L)\prod_{j=2}^{L}G(\mathbf{x}_j-\mathbf{x}_{j-1})d\mathbf{x}_1\cdots d\mathbf{x}_L. \quad (51)$$

[For $L=1$ we interpret (51) as $f_1\equiv V$.]

To find the order of magnitude of $A_{1,\infty}$, we replace ϕ in (49) by the ground-state wave function ψ (since ϕ is qualitatively similar to ψ and both are normalized). Then, by (31) and (8), $A_{1,\infty}$ roughly equals the value of A_1 for $T=0°K$; this is finite, by (12) and the result of Sec. 6. Feynman[7] has suggested using the approximation (32) for ϕ as well as for ψ; this leads, by (18) and (35), to the rough estimate $A_{1,\infty}\simeq 0.08$.

We estimate δ_L by replacing the integrals over $d\mathbf{x}_2\cdots d\mathbf{x}_{L+1}$ in (50) by the corresponding infinite integrals. This gives $\delta_L\simeq\delta_1{}^L$, which, when combined with (48) and (43), yields

$$A_1/A_{1,\infty}\simeq N^{-1}\sum_L L\langle m_L\rangle\delta_1{}^L/f_L. \quad (52)$$

To study (52), we note that Feynman's approximations (46) and (47) also imply[7] $\operatorname{tr}(Pe^{-\beta H})\simeq K\lambda^{-3N}\times\prod_l(f_l{}^{m_l})$. Substituting this into (40) and (44) yields

$$\langle Lm_L/f_L\rangle=Q_{N-L}/Q_N, \quad (53)$$

where

$$Q_M\equiv\sum_{\{m_l\}}\prod_l(f_l/l)^{m_l}/m_l!, \quad (54)$$

the sum being over all partitions of the arbitrary integer M. Equation (54) is just Mayer's expression[39] for the configurational partition function of an imperfect gas of M particles with cluster integrals $b_l=f_l/lV$. Therefore if $z\equiv Q_{N-1}/Q_N\cong\text{const}$ is the activity of this imperfect gas when it contains N particles, the approximation $Q_{N-L}/Q_N\cong z^L$ will hold provided $L\ll N$.

Using this approximation with (53) and (52), we obtain

$$A_1/A_{1,\infty}\simeq N^{-1}\sum_{L=1}^{N}(z\delta_1)^L\simeq N^{-1}(1-z\delta_1)^{-1}=o(1) \quad (55)$$

provided that $z\delta_1\cong\text{const}<1$. Feynman's work[7] shows that this condition holds above the transition temperature; therefore, since $A_{1,\infty}=e^{O(1)}$ and (12) holds, there is no B.E. condensation in HeI.

This argument fails below the transition temperature, where $z\delta_1\cong 1$. To study this case, we combine (52) with the identity $1=N^{-1}\sum_L L\langle m_L\rangle$ [which follows from (40) and (44)] and use (53); this gives

$$1-A_1/A_{1,\infty}\simeq N^{-1}\sum_L(f_L-\delta_1{}^L)Q_{N-L}/Q_N. \quad (56)$$

Feynman[7] estimates that, unless L is a small integer,

$$f_L\simeq(L^{-\frac{3}{2}}V\Delta+1)\delta_1{}^L, \quad (57)$$

where

$$\Delta\equiv\left[3\delta_1\Big/8\pi^2\int_0^\infty G(r)r^4dr\right]^{\frac{3}{2}}$$

Therefore, although the approximation $Q_{N-L}/Q_N\simeq z^L$ is no longer legitimate in (52), it is still legitimate in (56), according to (57), a convergent series results even though $z\delta_1\cong 1$.

$$1-A_1/A_{1,\infty}\cong\text{const}+N^{-1}\sum_{L=1}^{\infty}V\Delta L^{-\frac{3}{2}}\cong\text{const}, \quad (58)$$

where the first "const" takes care of the error due to the failure of (57) for small L. Feynman's work[7] shows that the right-hand side of (58) is less than 1 below the transition. Hence $A_1/A_{1,\infty}\cong\text{const}>0$, and, by (12), B.E. condensation does occur in He II.

The deductions we have made from Feynman's approximations can be paraphrased as follows: the quantity $\langle A_{1,L}\{m_l\}\rangle$ is very small if $L\ll(V\Delta)^{\frac{2}{3}}$ (where Δ is finite), and equals the finite quantity $A_{1,\infty}$ if $L\gg(V\Delta)^{\frac{2}{3}}$. Hence, by (43), $A_1/A_{1,\infty}$ equals the contribution of large L values to the sum $N^{-1}\sum L\langle m_L\rangle$; that is, it equals the fraction of particles in large cycles. Above the lambda-transition this fraction is negligible, so that, by (12), there is no B.E. condensation; below the transition this fraction is finite, so that B.E. condensation is present.

8. DISCUSSION

Equation (4) provides a mathematical definition of B.E. condensation, applicable for a system of interacting particles as well as for an ideal gas. Physically, the definition means that B.E. condensation is present whenever a finite fraction—n_M/N— of the particles occupies one single-particle quantum state, φ_M. The definitions of n_M and φ_M are given in Sec. 3 and Sec. 4, respectively. Even for an ideal gas, our definition is more general than the usual one, since here φ_M is not necessarily the lowest single-particle energy level. The close relation between our definition of B.E. condensation and London's suggested[1,2] "condensation in momentum space" is illustrated in the last paragraph of Sec. 3 above, where it is shown that under suitable conditions φ_M actually is an eigenstate of momentum.

The reasoning of Secs. 5, 6, and 7 indicates that liquid helium II satisfies our criterion of B.E. condensation. For $T=0°K$ the only physical assumption used is that a quantum liquid—as distinct from a solid—lacks long-range configurational order (though the mathematical treatment of this assumption is not yet

[39] J. E. Mayer and M. G. Mayer, *Statistical Mechanics* (John Wiley and Sons, Inc., New York, 1940), pp. 277–282.

completely rigorous). For $T > 0°K$, some fairly crude approximations, taken from Feynman's theory of the lambda-transition, have to be introduced. This part of the theory is therefore open to improvement—possibly in the form of a more rigorous proof that Feynman's implied criterion for B.E. condensation [the importance of long cycles in the sum (40) for the partition function] is equivalent to our criterion (4) at thermal equilibrium. Despite these imperfections, however, our analysis would appear to strengthen materially the case put forward previously by London[1,2] and Tisza[9] for the importance of B.E. condensation in the theory of liquid helium.

We have not considered here how B.E. condensation is related to superfluidity and to the excitation theory[16,17] of liquid helium. This will be done in another paper, where some of the results already obtained by Bogolyubov[12] for weakly repelling B.E. particles will be extended[40] to the case of interacting He^4 atoms.

9. ACKNOWLEDGMENTS

The authors are indebted to the National Science Foundation and to the United States Educational Commission in the United Kingdom for financial support. They would also like to thank Dr. G. V. Chester and Dr. D. W. Sciama for helpful discussions.

[40] A brief account of this work was given at the National Science Foundation Conference on Low-Temperature Physics and Chemistry, Baton Rouge, Louisiana, December, 1955 (unpublished).

APPLICATION OF THE METHODS OF QUANTUM FIELD THEORY TO A SYSTEM OF BOSONS

S. T. BELIAEV

Academy of Sciences, U.S.S.R.

Submitted to JETP editor August 2, 1957

J. Exptl. Theoret. Phys. (U.S.S.R.), 34, 417-432 (February, 1958)

It is shown that the techniques of quantum field theory can be applied to a system of many bosons. The Dyson equation for the one-particle Green's function is derived. Properties of the condensed phase in a system of interacting bosons are investigated.

1. INTRODUCTION

In recent years Green's functions have been widely used[1] in quantum field theory, and in particular in quantum electrodynamics. This has made possible the development of methods[2] which escape from ordinary perturbation theory. The method of Green's functions has also been shown* to be applicable to many-body problems. In such problems the one-particle Green's function determines the essential characteristics of the system, the energy spectrum, the momentum distribution of particles in the ground state, etc.[3]

The present paper develops the method of Green's functions for a system consisting of a large number N of interacting bosons. The special feature of this system is the presence in the ground state of a large number of particles with momentum $p = 0$ (condensed phase), which prevent the usual methods of quantum field theory from being applied. We find that for large N the usual technique of Feynman graphs can be used for the particles with $p \neq 0$, while the condensed phase (we show that it does not disappear when interactions are introduced) can be considered as a kind of external field.

The Green's function is expressed in terms of three effective potentials Σ_{ik}, describing pair-production, pair-annihilation and scattering, and in terms of a chemical potential μ. This is the analog of Dyson's equation in electrodynamics.[4,1] Some approximation must be made in the calculation of Σ_{ik} and μ. If these quantities are computed by perturbation theory, the quasi-particle spectrum of Bogoliubov[5] is obtained. In the following paper[6] we evaluate Σ_{ik} and μ in the limit of low density.

*Private communication from A. B. Migdal.

2. STATEMENT OF THE PROBLEM. FEYNMAN GRAPHS

We consider a system of N spinless bosons with mass $m = 1$, enclosed in a volume V. We suppose N and V become infinite, the density $N/V = n$ remaining finite. A summation over discrete momenta is then replaced by an integral according to the rule

$$\sum_p \to (2\pi)^{-3} V \int dp.$$

The Hamiltonian of the system is $H = H_0 + H_1$, where

$$H_0 = \frac{1}{2}\int \nabla\Psi^+(x)\,\nabla\Psi(x)\,dx = \sum_p \epsilon_p^0 a_p^+ a_p, \quad \epsilon_p^0 = \frac{p^2}{2}. \tag{2.1}$$

$$H_1 = \frac{1}{2}\int \Psi^+(x)\,\Psi^+(x')\,U(x-x')\,\Psi(x')\,\Psi(x)\,dx\,dx' = \frac{1}{2V}\sum_{pp'q} U_q a_p^+ a_{p'}^+ a_{p'-q} a_{p+q}. \tag{2.2}$$

The units are chosen so that $\hbar = 1$. $U(\mathbf{x}-\mathbf{x}')$ is the interaction between a pair of particles, $U_q = \int e^{-iqx}\, U(x)\, dx$ is its Fourier transform, and

$$\Psi = V^{-1/2} \sum_p e^{ipx} a_p, \quad \Psi^+ = V^{-1/2} \sum_p e^{-ipx} a_p^+,$$

where a_p and a_p^+ are the usual boson operators with the commutation law $[a_p, a_{p'}^+] = \delta_{pp'}$.

The one-particle Green's function may be defined in two equivalent ways. In terms of Heisenberg-representation operators we may write

$$iG(x-x') = \langle \Phi_0^N,\ T\{\Psi(x)\Psi^+(x')\}\,\Phi_0^N\rangle, \tag{2.3}$$

with the expectation value taken in the ground-state of the N interacting particles. In terms of interaction-representation operators we may write

$$iG(x-x') = \langle T\{\Psi(x)\Psi^+(x')S\}\rangle / \langle S\rangle, \tag{2.4}$$

with the expectation value taken in the ground-state of the non-interacting particles, which has all the particles in the condensed phase so that $N_{p\neq 0} = 0$, $N_0 = N$. The S-matrix for this system has the form

$$S = T\Big\{\exp\Big(-\frac{i}{2}\int d^4x_1 d^4x_2 U (1-2)\Psi^+(1)\Psi^+(2)\Psi(2)\Psi(1)\Big)\Big\}, \tag{2.5}$$

where we have written for convenience $U(1-2) = U(\mathbf{x}_1-\mathbf{x}_2)\delta(t_1-t_2)$. Here and henceforth $x, \ldots, p$ are four-vectors, and $px = \mathbf{px} - p_0x_0$. The definition (2.3) is convenient for relating G to physical quantities, while Eq. (2.4) is convenient for calculations.

In the numerator of Eq. (2.4) we expand the S-matrix in a series, each term of which is a T-product of a certain number of factors Ψ and Ψ^+. A T-product can be expressed by standard methods[7] as a sum of normal products in which some of the factors Ψ and Ψ^+ have been paired. In quantum electrodynamics the vacuum expectation value of every term which contains an unpaired annihilation operator vanishes from this sum. The surviving terms, which contain only pairs of Ψ and Ψ^+, are represented by certain Feynman graphs. In our case the expectation value is taken in a state containing N particles with momentum $p = 0$. The expectation value of an N-product containing a_0 does not vanish, and the usual method of constructing graphs is not applicable.

Because of the special role of the state with $p = 0$, it is convenient to separate the operators a_0 and a_0^+ from Ψ and Ψ^+. Thus we write

$$\Psi = \Psi' + a_0/\sqrt{V}; \quad \Psi^+ = \Psi'^+ + a_0^+/\sqrt{V}. \tag{2.6}$$

The Green's function (2.4) is also divided into two parts. The uncondensed particles give

$$iG'(x-x') = \langle T\{\Psi'(x)\Psi'^+(x')S\}\rangle / \langle S\rangle \tag{2.7}$$

while the Green's function of the condensed phase, a function of $(t-t')$ only, is

$$iG_0(t-t') = \langle T\{a_0(t)a_0^+(t')S\}\rangle / V\langle S\rangle. \tag{2.8}$$

The two functions are not independently determined, since the S-matrix appears in the definition of both and itself contains both Ψ' and a_0 operators. We shall prove later that when N is large the usual method of Feynman graphs can be adapted to the calculation of G', the condensed phase behaving just like an external field.

We divide the operations T and $\langle\ldots\rangle$ into two successive operations, the first acting only upon Ψ' and Ψ'^+, the second acting only upon a_0 and a_0^+. Thus

$$T = T^0 T', \quad \langle\ \ \rangle = \langle\langle\ \ \rangle'\rangle^0,$$

where T^0 and $\langle\ldots\rangle^0$ act on a_0 and a_0^+.

We now drop the prime from G' and write Eq. (2.7) in the form

$$iG(x-x') = \langle T^0\{\mathfrak{G}(x-x')\}\rangle^0 / \langle S\rangle, \tag{2.9}$$

with

$$\mathfrak{G}(x-x') = \langle T'\{\Psi'(x)\Psi'^+(x')S\}\rangle'. \tag{2.10}$$

Eq. (2.10) has the same structure as the numerator of Eq. (2.7), but the operators a_0, a_0^+ occurring in S are now to be treated as parameters. The expectation value in Eq. (2.10) is taken in the ground state of the operators Ψ', Ψ'^+. This equivalent to a vacuum expectation value, and so the usual formalism of Feynman graphs can be used for calculating $\mathfrak{G}$.

We represent the potential $-iU(1-2)$ by a dotted line joining the points 1 and 2. The pair of operators $\dot{\Psi}'(1)\ \dot{\Psi}'^+(2) = iG^{(0)}(1-2)$ is represented by a continuous line directed from 2 to 1. From the form of the interaction Hamiltonian (2.2) it follows that every graph contributing to Eq. (2.10) is a combination of the eight elementary graphs shown in Fig. 1. These correspond to the various terms which appear in Eq. (2.2) after the substitution (2.6). A missing continuous line (incomplete vertex) corresponds to a factor $(a_0/\sqrt{V})$ or $(a_0^+/\sqrt{V})$. Fig. 2 shows an example of one graph which appears in $\mathfrak{G}(x_1-x_2)$, corresponding to the integral

$$\mathfrak{M}_2(x_1;\ x_2) = i^2\int G^{(0)}(1-3)\,U(3-4)\,G^{(0)}(3-5)\,G^{(0)}(4-6)\,U(5-6) \times G^{(0)}(6-2)\,V^{-1}a_0^+(t_4)\,a_0(t_5)\,d^4x_3d^4x_4d^4x_5d^4x_6. \tag{2.11}$$

Let $\mathfrak{M}(x; x')$ be any graph contributing to Eq. (2.10) and not containing disconnected parts or vacuum loops. Together with $\mathfrak{M}$ we may consider all graphs differing from $\mathfrak{M}$ by the addition of vacuum loops. The totality of such graphs gives $\mathfrak{M}$ multiplied by a factor which is just the vacuum expectation value of the S matrix, namely $\langle S \rangle'$ in this case, since we are taking matrix elements only of Ψ' and Ψ'^+ Thus the inclusion of vacuum loops changes $\mathfrak{M}$ into

$$\mathfrak{M}(x; x')\langle S\rangle' \qquad (2.12)$$

In quantum electrodynamics the factor $\langle S \rangle$ cancels the denominator of Eq. (2.4), so that we can ignore the vacuum loops and merely omit this denominator. In our case, as we shall see later, the factor $\langle S \rangle'$ has a real significance.

Eq. (2.12) substituted into Eq. (2.9) gives

$$\langle T^0 \{\mathfrak{M}(x; x')\langle S\rangle'\}\rangle^0 / \langle S\rangle, \qquad (2.13)$$

where the operation T^0 acts on the factors a_0, a_0^+ occurring in $\mathfrak{M}$ and in $\langle S \rangle'$. Suppose that $\mathfrak{M}$ contains m pairs of operators a_0, a_0^+. Then

$$\mathfrak{M}(x; x') = V^{-m}\int M(x; x', t_1 \ldots t_m, t_1' \ldots t_m')\, a_0(t_1) \ldots a_0(t_m)\, a_0^+(t_1') \ldots a_0^+(t_m')\,(dt)\,(dt'),$$

and Eq. (2.13) becomes

$$\int MiG_0(t_1 \ldots t_m, t_1' \ldots t_m')\,(dt)\,(dt'), \qquad (2.14)$$

where

$$iG_0(t_1 \ldots t_m; t_1' \ldots t_m') = \langle T\{a_0(t_1) \ldots a_0^+(t_m')\, S\}\rangle / V^m \langle S\rangle \qquad (2.15)$$

is the m-particle Green's function of the condensed phase, Eq. (2.8) being the special case m = 1.

The graphs for the Green's function (2.9) thus coincide with the graphs for $\mathfrak{G}$, only the factors $(a_0 a_0^+/V)$ in the integrals are replaced by the corresponding Green's function of the condensed phase. For example, in the integral (2.11), the factor $(a_0^+(t_4)\, a_0(t_5))/V)$ is replaced by $iG_0(t_5 - t_4)$. We need not consider graphs with disconnected parts, since these are already included in G_0. The problem is therefore reduced to the determination of the Green's functions G_0 of the condensed phase.

3. THE GREEN'S FUNCTIONS OF THE CONDENSED PHASE

We write the m-particle Green's function (2.15) of the condensed phase in the form

$$iG_0(t_1 \ldots t_m, t_1' \ldots t_m') = \frac{1}{V^m \langle S\rangle}\langle T^0\{a_0(t_1) \ldots a_0(t_m)\, a_0^+(t_1') \ldots a_0^+(t_m')\langle S\rangle'\}\rangle^0 \qquad (3.1)$$

The quantity $\langle S \rangle'$ is the sum of contributions from all vacuum loops. If λ is the sum of contributions from all connected vacuum loops, then[8] the sum of contributions from all pairs of connected loops is $(\lambda^2/2!)$, the sum of contributions from all triples is $(\lambda^3/3!)$, and so on. Therefore $\langle S \rangle' = e^{\lambda}$. In our case λ is a functional of $a_0 a_0^+$, and is proportional to the volume V if we take $(a_0 a_0^+/V)$ to be finite. This can be seen by considering any vacuum loop as obtained from a graph with two free ends, carrying momenta $\mathbf{p}$ and $\mathbf{p}'$, by setting $\mathbf{p} = \mathbf{p}' = 0$. The graph with free ends gives a contribution proportional to $\delta(\mathbf{p} - \mathbf{p}') \sim (2\pi)^{-3} V \delta_{\mathbf{pp}'}$, and so the vacuum loop becomes proportional to V. Therefore $\lambda = V\sigma$, where σ is a finite functional of $(a_0 a_0^+/V)$, and

$$\langle S\rangle' = e^{V\sigma}. \qquad (3.2)$$

The commutator of a_0 and a_0^+ is unity, and is small compared with their product which is of order N. At first glance it would seem that the order of factors a_0, a_0^+ was unimportant, and that the T-product in Eq. (3.1) could be omitted. But one must remember that the T^0 in Eq. (3.1) links the product $a_0 \ldots a_0^+$ with the quantity $e^{V\sigma}$, which contains all powers of the volume and hence may compensate for the smallness of the commutator of a_0 with a_0^+. Only after disentangling $a_0 \ldots a_0^+$ from the T-product may we neglect the commutators. We observe that a_0 and a_0^+ commute with H_0 given by Eq. (2.1), so these operators are independent of time in the interaction representation. The arguments of the $a_0(t)$ and $a_0^+(t')$ in Eq. (3.1) are only ordering symbols for the operation of T^0. After carrying out the T-ordering we may consider a_0 and a_0^+ as time-independent.

The disentangling of $a_0 \ldots a_0^+$ from the T-product is done by means of the following theorem. Let $B(a_0 a_0^+/V)$ and $\sigma(a_0 a_0^+/V)$ be any functionals of $(a_0 a_0^+/V)$, which is considered as a finite quantity. The "disentangling rule"

$$T^0\{B(a_0 a_0^+/V)\, e^{V\sigma}\} = B(AA^+)\, T^0\{e^{V\sigma}\}. \qquad (3.3)$$

holds with an error of order $(1/V)$. The quantities A and A^+ are defined by the integral equations

$$A(t) = C(AA^+) + \int dt'\, \theta(t - t')\, \delta\sigma(AA^+)/\delta A^+(t'),$$
$$A^+(t) = C^+(AA^+) + \int dt'\, \theta(t' - t)\, \delta\sigma(AA^+)/\delta A(t'), \qquad (3.4)$$

where $\theta(t-t')$ is the contribution from a factor-pair (a_0, a_0^+),

$$\theta(t-t') = \dot{a}_0(t)\dot{a}_0^+(t') = \begin{cases} 1 & \text{for } t > t' \\ 0 & \text{for } t < t', \end{cases} \tag{3.5}$$

and C, C^+ are time-independent functionals defined by the quadratic equations

$$C^2 + C\int \frac{\delta\sigma(AA^+)}{\delta A^+(t)}\,dt = \frac{a_0^2}{V},$$

$$C^{+2} + C^+\int \frac{\delta\sigma(AA^+)}{\delta A(t)}\,dt = \frac{a_0^{+2}}{V}. \tag{3.6}$$

A proof of this theorem is given in the Appendix.

Applying Eq. (3.3) to (3.1), we obtain

$$iG_0(t_1 \ldots, \quad t_m) = \langle A(t_1)\ldots A^+(t'_m)\rangle^0.$$

The denominator of Eq. (3.1) cancels against $\langle T^0 \langle S\rangle' \rangle^0 = \langle S\rangle$. When the expectation value of the product $(A \ldots A^+)$ is taken, we may with an error of order $(1/V)$ replace all factors a_0, a_0^+ by $\sqrt{N}$. Let K and K^+ denote the result of making this replacement in A and A^+. Then

$$iG_0(t_1 \ldots t_m; t'_1 \ldots t'_m)$$
$$= K(t_1)\ldots K(t_m)K^+(t'_1)\ldots K(t'_m), \tag{3.7}$$

holds, with K and K' given according to Eq. (3.4) by the integral equations

$$K(t) = \bar{C} + \int dt'\theta(t-t')\frac{\delta\sigma(KK^+)}{\delta K^+(t')},$$

$$K^+(t) = \bar{C}^+ + \int dt'\theta(t'-t)\frac{\delta\sigma(KK^+)}{\delta K(t')}, \tag{3.8}$$

and with $\bar{C}$ and $\bar{C}^+$ defined by

$$\bar{C}^2 + \bar{C}\int \frac{\delta\sigma(KK^+)}{\delta K^+(t)}\,dt = \frac{N}{V};$$

$$\bar{C}^{+2} + \bar{C}^+\int \frac{\delta\sigma(KK^+)}{\delta K(t)}\,dt = \frac{N}{V}. \tag{3.9}$$

Eq. (3.7) shows that the Green's functions of the condensed phase are products of factors, each factor being a function of one time variable. The physical meaning of this result may be clarified by the following qualitative argument. For simplicity we consider the one-particle function for non-interacting particles $iG_0^{(0)}(t-t') = \langle a_0(t)\, a_0^+(t')\rangle/V$. It describes the propagation of a particle from t' to t. If there was originally a vacuum, then this process can proceed only by creating a particle at time t' and annihilating it at the later time t, which is represented by the factor-pairing $\dot{a}_0\dot{a}_0^+ = \theta$. In this case $G^{(0)}$ coincides with the factor-pairing, as is the case in electrodynamics. But if the process occurs in the presence of N particles of the same type, the created and absorbed particles may be different. In this case the propagation of a particle from t' to t is composed of two processes, the creation of an extra particle in the condensed phase at time t', and the absorption of one particle from the condensed phase at time t. The time sequence of these two events is immaterial, to order N^{-1}, if N is large. The processes are therefore independent. These arguments are valid also for the exact function G_0. It is also a product of two factors $K(t)$ and $K^+(t')$, describing the two independent processes of emission and absorption of a particle at the two corresponding times.

We consider in greater detail the one-particle function of the condensed phase

$$iG_0(t-t') = K(t)K^+(t'). \tag{3.10}$$

The left side is a function of the difference $(t-t')$. The right side is a product of functions of t and t'. Therefore $K(t)$ and $K^+(t')$ must be exponentials

$$K(t) = \sqrt{n_0}\,e^{-i\mu t}; \quad K^+(t) = \sqrt{n_0}\,e^{i\mu t} \tag{3.11}$$

so that

$$iG_0(t-t') = n_0 e^{-i\mu(t-t')}. \tag{3.12}$$

To understand the physical meaning of the quantities n_0 and μ, we go back to the definition (2.3) of G_0

$$iG_0(t-t') = \langle \Phi_0^N, T\{a_0(t)\, a_0^+(t')\}\,\Phi_0^N\rangle / V \tag{3.13}$$

Putting $t' = t$ in Eq. (3.13) we find

$$iG_0(0) = \langle \Phi_0^N, a_0^+ a_0 \Phi_0^N\rangle / V = \bar{N}_0/V. \tag{3.14}$$

Comparing Eq. (3.14) with (3.12), we see that $n_0 = (\bar{N}_0/V)$ is the mean density of particles in the condensed phase.

Next, suppose for definiteness $t > t'$, and write Eq. (3.13) in the form

$$iG_0(t-t') = \frac{1}{V}\langle \Phi_0^N a_0(t)\,\Phi_0^{N+1}\rangle\langle \Phi_0^{N+1} a_0^+(t')\,\Phi_0^N\rangle$$
$$+ \frac{1}{V}\sum_{s\neq 0}\langle \Phi_0^N a_0(t)\,\Phi_s^{N+1}\rangle\langle \Phi_s^{N+1} a_0^+(t')\,\Phi_0^N\rangle,$$

Separating out the time dependence of the Heisenberg operators, this expression becomes

$$iG_0(t-t') = \frac{1}{V} \exp\{-i(E_0^{N+1} - E_0^N)(t-t')\} \langle \Phi_0^N a_0 \Phi_0^{N+1} \rangle \langle \Phi_0^{N+1} a_0^+ \Phi_0^N \rangle$$
$$+ \frac{1}{V} \sum_{s \neq 0} \exp\{-i(E_s^{N+1} - E_0^N)(t-t')\} \langle \Phi_0^N a_0 \Phi_s^{N+1} \rangle \langle \Phi_s^{N+1} a_0^+ \Phi_0^N \rangle. \quad (3.15)$$

We compare the exact Eq. (3.15) with the approximation (3.12) which is valid as $N \to \infty$, and conclude that the second term in Eq. (3.15) must vanish as $N \to \infty$. Comparison of the time dependence of the first term in Eq. (3.15) with that of Eq. (3.12) then shows that μ is the chemical potential of the system,

$$\mu = E_0^{N+1} - E_0^N \approx \partial E_0^N / \partial N. \quad (3.16)$$

The parameters n_0 and μ which appear in K and K^+ can be calculated in principle by solving Eq. (3.8). In practice this is very difficult. The trouble is that, in calculating the vacuum loops which contribute to σ, one has first to integrate over a finite time interval $(-T, T)$, so that the parameter T appears in Eq. (3.8). One may pass to the limit $T \to \infty$ in the solutions, but not in the equations. Thus it is incorrect to use in $\sigma(KK^+)$ the limiting expressions (3.11) for K and K^+. One has instead to solve the nonlinear equations (3.8) directly.

We can obtain from Eq. (3.8) one relation between the quantities n_0 and μ. Differentiating Eq. (3.8) with respect to t, and remembering that $d\theta(t-t')/dt = \delta(t-t')$, we obtain the differential equations

$$dK/dt = \delta\sigma(KK^+)/\delta K^+(t);$$
$$dK^+/dt = -\delta\sigma(KK^+)/\delta K(t). \quad (3.17)$$

Let σ be expanded in a series

$$\sigma(KK^+) = -i \sum_{(W)} \frac{1}{m} \int W_m(t_1' \ldots t_m'; t_1 \ldots t_m) K^+(t_1') \ldots$$
$$\ldots K^+(t_m') K(t_1) \ldots K(t_m)\,(dt)\,(dt'), \quad (3.18)$$

in which each term corresponds to a certain vacuum loop with m pairs of incomplete vertices, and the sum is taken over all such loops. The "vacuum amplitudes" W_m are functions of only $(2m-1)$ variables (time differences), so that the limiting values of K and K^+ would give an infinite result when substituted into Eq. (3.18). If Eq. (3.18) is varied with respect to K^+, one integration disappears, and the result becomes finite. The Fourier transform of $W_m(t'; t)$ may be written

$$W_m(\omega_1' \ldots; \ldots \omega_m)\, \delta\left(\sum \omega - \sum \omega'\right)$$
$$= \int W_m(t'; t)\, e^{i(\omega' t') - i(\omega t)}\, (dt)\,(dt'),$$

Then Eq. (3.11) and (3.18) give

$$\frac{\delta\sigma(KK^+)}{\delta K^+(t)}$$
$$= -i\sqrt{n_0} \sum_{(W)} n_0^{m-1} W_m(\mu \ldots, \ldots \mu)\, e^{-i\mu t}. \quad (3.19)$$

Substituting Eq. (3.19) into (3.17) and using Eq. (3.11), we obtain the desired relation

$$\mu = \sum_{(W)} n_0^{m-1} W_m(\mu \ldots \mu; \mu \ldots \mu). \quad (3.20)$$

The summation here extends over the various vacuum loops which contribute to the quantity $(\delta\sigma/\delta K^+(t))$ according to Eq. (3.19). Each such loop is to be taken with unit weight, remembering that there is one special incomplete vertex t, at which the variation with respect to K^+ was taken. Two loops are to be counted as different if they have the same geometrical structure and differ only in the position of the special vertex.

Equation (3.20) may be considered as an equation for $\mu(n_0)$. There is one free parameter in the problem, the total particle number N or the density n. Thus μ and n_0 ought to be expressible in terms of n. However n does not appear explicitly in the equation. It is thus convenient to consider n_0 instead of n as the free parameter, and to express all other quantities as functions of n_0. The connection between n_0 and n can be found after the problem is solved. From this standpoint, Eq. (3.20) completely determines K and K^+ and consequently all the Green's functions of the condensed phase. We might also solve the problem with two free parameters μ and n_0, and only consider the connection between them in the final result.[6] But this procedure would considerably increase the mathematical difficulties.

4. PROPERTIES OF THE CONDENSED PHASE

The form of the functions G_0 of the condensed phase leads to some deductions concerning the properties of the condensed phase in a system of interacting particles.

In the absence of interaction, the momentum distribution of particles in the ground state is $\delta(\mathbf{p}) = (2\pi)^{-3} V \delta_{\mathbf{p}0}$. When interactions are introduced the distribution is smeared out. In principle two possibilities are open. Either the term in $\delta(\mathbf{p})$ completely disappears and the distribution becomes continuous (there is no condensed phase), or a term in $\delta(\mathbf{p})$ remains and the state $\mathbf{p} = 0$ is still exceptional (there is a condensed phase). In the first case all average occupation numbers $\overline{N}_{\mathbf{p}}$ are finite, and $\overline{N}_{\mathbf{p}} \to \overline{N}_0$ as $\mathbf{p} \to 0$. In the second case $\overline{N}_{\mathbf{p}\neq 0}$ is finite but $\overline{N}_0 \sim V$.

The neglect of the second term in Eq. (3.15) is equivalent to the assumption that a_0, operating on the ground state Φ_0^N, does not excite the system, or in symbols

$$a_0\Phi_0^N \approx (N_0)^{1/2}\Phi_0^{N-1} \tag{4.1}$$

This assumption seems at first glance strange. A change in the number of particles with $\mathbf{p} = 0$, disturbing the stationary relation between the occupation numbers, must excite the system. If N_0 were finite, a change of it by one unit would change the state appreciably, but if $N_0 \sim V$ this change will practically not disturb the ground state. Equation (4.1) supports the second alternative. Therefore the introduction of interactions never causes the condensed phase to disappear entirely.

We next examine the problem of the fluctuation of the number of particles in the condensed phase. The quantity N_0 does not have an exact value in the state Φ_0^N. We expand Φ_0^N into eigenstates of the operator N_0. The expansion may be written

$$\Phi_0^N = \sum_{m=0}^{N} C_{N-m}^N \varphi_0^{N-m} \chi_m^N, \tag{4.2}$$

where $\varphi_0^{N_0}$ is a function only of the occupation number of the condensed phase, while χ_m^N depends on the other variables. χ_m^N describes a state of m particles with momenta distributed over all values $\mathbf{p} \neq 0$. It is a superposition of states with definite occupation numbers for the momenta $\mathbf{p} \neq 0$. The coefficients in this superposition depend on the upper index N. The normalization of χ_m^N is given by $<\chi_m^N\chi_{m'}^N> \delta_{mm'}$.

Equation (4.1), with the orthogonality of $\varphi_0^{N_0}$ and $\varphi_0^{N_0'}$ for $N_0 \neq N_0'$, now gives the result

$$\langle\Phi_0^{N-1}a_0\Phi_0^N\rangle = \sum_{m=0}^{N-1}\sqrt{N-m}\,(C_{N-m-1}^{N-1})^* C_{N-m}^N\langle\chi_m^{N-1}\chi_m^N\rangle \tag{4.3}$$

We assume that C_{N-m}^N and χ_m^N are smooth functions of N, so that

$$C_{N-m-1}^{N-1} \approx C_{N-m}^N - \partial C_{N-m}^N/\partial N = C_{N-m}^N\{1 + O(N^{-1})\};$$
$$\chi_m^{N-1} \approx \chi_m^N - \partial\chi_m^N/\partial N = \chi_m^N\{1 + O(N^{-1})\}.$$

Then Eq. (4.3) becomes

$$\langle\Phi_0^{N-1}a_0\Phi_0^N\rangle = \sum_{m=0}^{N-1}\sqrt{N-m}\,|C_{N-m}^N|^2\{1 + O(N^{-1})\}.$$

The sum here is simply $N_0^{1/2}$, so that

$$\langle\Phi_0^{N-1}a_0\Phi_0^N\rangle = \overline{N_0^{1/2}}\{1 + O(N^{-1})\}. \tag{4.4}$$

A similar expression naturally holds also for $<\Phi_0^{N+1}a_0^+\Phi_0^N>$.

We estimate the sum in Eq. (3.15) after setting $t = t'$. Using Eq. (4.4) and (3.14), we find

$$\sum_{s\neq 0}\langle\Phi_0^N a_0\Phi_s^{N+1}\rangle\langle\Phi_s^{N+1}a_0^+\Phi_0^N\rangle \approx \overline{N}_0 - (\overline{N_0^{1/2}})^2, \tag{4.5}$$

from which it is clear that the sum is connected with the magnitude of the fluctuations in the number of particles in the condensed phase. From the fact that the sum is negligible as $N \to \infty$, we conclude

$$[\overline{N}_0 - (\overline{N_0^{1/2}})^2]/\overline{N}_0 \to \infty \text{ for } N \to \infty, \tag{4.6}$$

Thus the fluctuations in the number of particles in the condensed phase are relatively small.

5. GREEN'S FUNCTION FOR A PARTICLE WITH $\mathbf{p} \neq 0$

The expressions obtained in Sec. 3 for the functions of the condensed phase allow us to reformulate the rules which were described in Sec. 2 for the construction of graphs.

Every graph is a combination of eight elementary graphs (Fig. 1). Every incomplete vertex carries a factor $K(t) = \sqrt{n_0}\,e^{-i\mu t}$ corresponding to a missing incoming continuous line, or a factor $K^+(t) = \sqrt{n_0}\,e^{i\mu t}$ corresponding to a missing outgoing line. These factors mean that the interaction

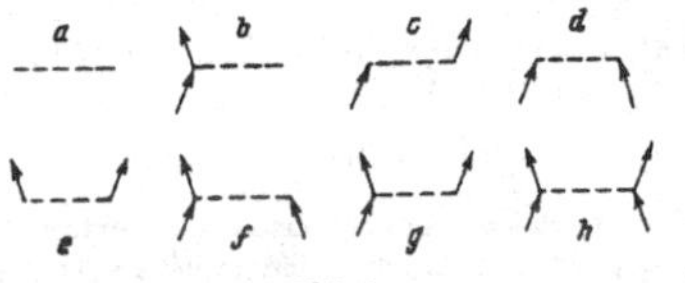

FIG. 1

involves the absorption or the emission of a particle of energy μ in the condensed phase. We may draw a wavy line corresponding to every incoming or outgoing particle of the condensed phase. All such lines have free ends. In analogy with quantum electrodynamics, we may say that the condensed phase behaves like an external field with frequency μ.

Consider the general structure of a graph which contributes to the Green's function (2.7). Every

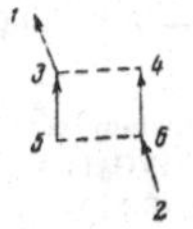

FIG. 2

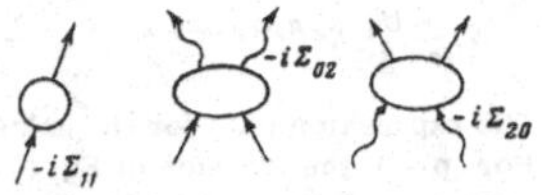

FIG. 3

graph contributing to G has the form of a chain consisting of separate irreducible parts connected to each other by only one continuous line. There are only three types of irreducible parts (i.e., parts which cannot be separated into pieces joined by only one continuous line). The three types differ in the number of outgoing and incoming continuous lines (Fig. 3). The sums of the contributions from all irreducible parts of each type we call respectively $-i\Sigma_{11}$, $-i\Sigma_{02}$, $-i\Sigma_{20}$. Σ_{11} describes processes in which the number of particles out of the condensed phase is conserved. Σ_{02} and Σ_{20} describe the absorption and emission of two particles out of the condensed phase; in these processes two particles in the condensed phase must be simultaneously emitted or absorbed, and their energy 2μ must be taken into account. In the momentum representation, $\Sigma_{11}(p_1; p_2)$ contains a factor $\delta(p_1 - p_2)$, while $\Sigma_{02}(p_1 p_2)$ and $\Sigma_{20}(p_1 p_2)$ contain* $\delta(p_1 + p_2 - 2\mu)$. Henceforth we shall assume momentum conservation in the arguments of the functions Σ_{ik}, representing the quantities which multiply the δ-functions by the notations

$$\Sigma_{11}(p;\ p) \equiv \Sigma_{11}(p);\ \Sigma_{02}(p+\mu, -p+\mu) \equiv \Sigma_{02}(p+\mu);$$
$$\Sigma_{20}(p+\mu, -p+\mu) \equiv \Sigma_{20}(p+\mu). \qquad (5.1)$$

The functions Σ_{ik} are characteristic of the particle interactions, and we may call them the effective potentials of the pair interaction.

Besides the Green's function G we introduce an auxiliary quantity $\hat{G}$, consisting of the sum of contributions from graphs with two ingoing lines. Graphs contributing to G have one ingoing and one outgoing. Figure 4 shows some of the graphs which contribute to $\hat{G}$. The quantity $\hat{G}$ describes the transition of two particles into the condensed phase. In momentum representation we write $\hat{G}(p+\mu)$ when the ingoing lines carry momenta $(p+\mu)$ and $(-p+\mu)$.

There are two equations, analogous to the Dyson equation in electrodynamics,[5,1] for the functions G and $\hat{G}$,

$$G(p+\mu) = G^{(0)}(p+\mu) + G^{(0)}(p+\mu)\Sigma_{11}(p+\mu)G(p+\mu)$$
$$+ G^{(0)}(p+\mu)\Sigma_{20}(p+\mu)\hat{G}(p+\mu),$$
$$\hat{G}(p+\mu) = G^{(0)}(-p+\mu)\Sigma_{11}(-p+\mu)\hat{G}(p+\mu)$$
$$+ G^{(0)}(-p-\mu)\Sigma_{02}(p+\mu)G(p+\mu). \qquad (5.2)$$

The structure of these equations is illustrated graphically by Fig. 5 and does not need any further explanation. Solving the system (5.2) for G and $\hat{G}$, we find

$$G(p+\mu)$$
$$= (G^{(0)^{-1}} - \Sigma_{11})^- \{(G^{(0)^{-1}} - \Sigma_{11})^+ (G^{(0)^{-1}} - \Sigma_{11})^- - \Sigma_{20}\Sigma_{02}\}^{-1},$$
$$\hat{G}(p+\mu) = \Sigma_{02}\{(G^{(0)^{-1}} - \Sigma_{11})^+ (G^{(0)^{-1}} - \Sigma_{11})^- - \Sigma_{20}\Sigma_{02}\}^{-1}, \qquad (5.3)$$

where the suffixes $\pm$ indicate the values $(\pm p + \mu)$ of the arguments. Equation (5.3) for G may be written in the usual form of a Dyson equation

$$G^{-1} = G^{(0)^{-1}} - \Sigma,$$
$$\Sigma(p+\mu) = \Sigma_{11}(p+\mu)$$
$$+ \Sigma_{20}\Sigma_{02} / [G^{(0)^{-1}}(-p+\mu) - \Sigma_{11}(-p+\mu)]. \qquad (5.4)$$

Into Eq. (5.3) we substitute the explicit form of the free-particle Green's function,

$$G^{(0)^{-1}}(p) = p^0 - \varepsilon_p^0 + i\delta; \qquad (\varepsilon_p^0 = p^2/2;\quad \delta \to +0), \qquad (5.5)$$

and obtain for G and $\hat{G}$ the expressions

$$G(p+\mu)$$
$$= \frac{p^0 + \varepsilon_p^0 + \Sigma_{11}^- - \mu}{[p^0 - (\Sigma_{11}^+ - \Sigma_{11}^-)/2]^2 - [\varepsilon_p^0 + (\Sigma_{11}^+ + \Sigma_{11}^-)/2 - \mu]^2 + \Sigma_{20}\Sigma_{02}}, \qquad (5.6)$$

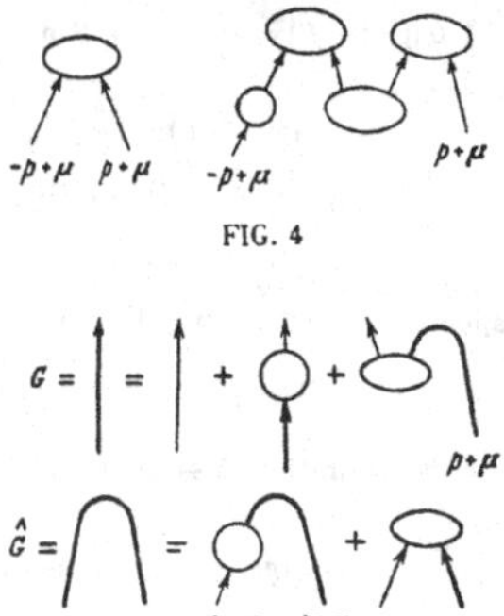

FIG. 4

FIG. 5

*Here μ represents a 4-vector having only its fourth component non-zero.

$$= \frac{\hat{G}(p+\mu) \\ -\Sigma_{02}}{[p^0 - (\Sigma_{11}^{+} - \Sigma_{11}^{-})/2]^2 - [\varepsilon_p^0 + (\Sigma_{11}^{+} + \Sigma_{11}^{-})/2 - \mu]^2 + \Sigma_{20}\Sigma_{02}} . \tag{5.7}$$

Equation (5.6) determines the Green's function in terms of the effective potentials Σ_{ik} and the chemical potential μ of the system. Equations for Σ_{ik} and μ cannot be obtained in so general a form. To calculate these quantities we have to use approximate methods to sum over series of graphs. In Sec. 7 we shall calculate Σ_{ik} and μ by perturbation theory. In the following paper[6] we develop an approximation in which the density is considered as a small parameter.

6. CONNECTION BETWEEN THE GREEN'S FUNCTION AND PROPERTIES OF THE SYSTEM

The energy E_0 of the ground state is the expectation value of the Hamiltonian (2.1), (2.2) in the state Φ_0^N,

$$E_0 = \langle \Phi_0^N H \Phi_0^N \rangle = \sum_p \varepsilon_p^0 \langle a_p^+ a_p \rangle + \frac{1}{2V} \sum_{pp'q} U_q \langle a_p^+ a_{p'}^+ a_{p'-q} a_{p+q} \rangle . \tag{6.1}$$

The last term in Eq. (6.1) is connected with the Green's function G. Consider G in the (p, t) representation, i.e., in the momentum representation for the space-components only. Taking the expectation value in the state Φ_0^N, Eq. (2.3) becomes

$$iG(\mathbf{p};\, t-t') = \langle T\{a_p(t)\, a_p^+(t')\}\rangle, \qquad (\mathbf{p} \neq 0), \tag{6.2}$$

from which it is easy to deduce

$$(i\partial/\partial t - \varepsilon_p^0)\, G(\mathbf{p};\, t-t') = \delta(t-t') + R(\mathbf{p};\, t-t'), \tag{6.3}$$

with

$$R(\mathbf{p};\, t-t') = -\frac{i}{V} \sum_{p'q} U_q \langle T\{a_{p'}^+(t)\, a_{p'-q}(t)\, a_{p+q}(t) a_p^+(t')\}\rangle . \tag{6.4}$$

We multiply Eq. (6.3) by $e^{ip^0(t-t')}$ and integrate with respect to t. Then using Eq. (5.5) we obtain

$$G^{(0)^{-1}}(p)\, G(p) = 1 + R(p),$$

This, with the definition (5.4) of Σ, gives immediately

$$R(p) = \Sigma(p)\, G(p), \qquad (\mathbf{p} \neq 0). \tag{6.5}$$

On the other hand, Eq. (6.4) shows that $R(\mathbf{p}; -0)$ is related to the last sum in Eq. (6.1), namely

$$\frac{1}{2V} \sum_{pp'q} U_q \langle a_p^+ a_{p'}^+ a_{p'-q} a_{p+q} \rangle = \frac{i}{2} \sum_p R(\mathbf{p};\, -0). \tag{6.6}$$

The expression (6.5) for R holds only when $\mathbf{p} \neq 0$. For $\mathbf{p} = 0$ the left side of Eq. (6.2) is $iVG_0(t-t')$ according to Eq. (3.13). Instead of Eq. (6.3) we have in this case

$$i \frac{\partial}{\partial t} G_0(t-t') = \frac{1}{V} \delta(t-t') + \frac{1}{V} R(0;\, t-t'). \tag{6.7}$$

We neglect the δ-function since it is of order V^{-1}, and use Eq. (3.12) for G_0. This gives

$$R(0;\, -0) = i\left[\frac{\partial}{\partial \tau} G_0(\tau)\right]_{\tau=-0} = -i\mu n_0. \tag{6.8}$$

Equations (6.6) and (6.8) bring the expression (6.1) for E_0 into the form

$$E_0 = \sum_p \varepsilon_p^0 \langle a_p^+ a_p \rangle + \frac{i}{2} \sum_{p \neq 0} R(\mathbf{p};\, -0) + \frac{1}{2} \mu n_0 . \tag{6.9}$$

From Eq. (6.2) we find

$$\overline{N}_p = \langle a_p^+ a_p \rangle = iG(\mathbf{p};\, -0) = i\int G(p)\, dp^0 / 2\pi \tag{6.10}$$

Using Eq. (6.5) and (6.10), and passing from summation to integration in Eq. (6.9), we obtain the following expression* for the ground-state energy E_0,

$$E_0/V = i\int [\varepsilon_p^0 + {}^1/_2 \Sigma(p)]\, G(p)\, d^4p/(2\pi)^4 + \mu n_0/2 \tag{6.11}$$

The p_0-integration is to be taken with a small detour into the upper half-plane.

The quantities Σ and G depend parametrically upon μ and n_0, supposing that Eq. (3.20) has not been used in order to eliminate one of these parameters. Therefore Eq. (6.11) gives a relation between (E_0/V), μ, and n_0. There are two further relations between these quantities. First there is the definition of the chemical potential μ,

$$\mu = \frac{\partial E_0}{\partial N} = \frac{\partial}{\partial n}\left(\frac{E_0}{V}\right), \tag{6.12}$$

and second there is the condition that the total number of particles is conserved, which by Eq. (6.10) can be written in the form

$$n = n_0 + i\int G(p)\, d^4p/(2\pi)^4 \tag{6.13}$$

Equations (6.11), (6.12), and (6.13) determine (E_0/V), μ, and n_0 in terms of the density n, or determine any three of these quantities in terms of the fourth. The relation (3.20) which we found earlier does not give any new information; it is

*V. M. Galitskii informed me that a similar relation exists for Fermi systems.

satisfied identically when Eq. (6.11), (6.12), and (6.13) hold.

7. PERTURBATION THEORY APPROXIMATION TO Σ_{ik} AND μ

In the first order of perturbation theory, the graphs which contribute to Σ_{ik} are the elementary graphs shown in Fig. 1. Graphs b and c refer to Σ_{11}, d to Σ_{02} and e to Σ_{20}. These graphs give the contributions

$$\Sigma_{02} = \Sigma_{20} = n_0 U_p, \qquad \Sigma_{11}^{+} = n_0 (U_0 + U_p). \tag{7.1}$$

In first approximation the only vacuum loop is the elementary graph a of Fig. 1. Thus Eq. (3.20) gives for μ the value

$$\mu = n_0 U_0. \tag{7.2}$$

Inserting Eq. (7.1) and (7.2) into the expression (5.6) for the Green's function, we find

$$G(p + \mu) = p^0 + \varepsilon_p^0 + n_0 U_p / (p^{0\,2} - \varepsilon_p^{0\,2} - 2 n_0 U_p \varepsilon_p^0 + i\delta). \tag{7.3}$$

The Green's function $G(p + \mu)$ has a pole at a value $p_0(\mathbf{p})$ which defines the energy of an elementary excitation of the system[3] (quasi-particle). Equation (7.3) gives for the quasi-particle energy

$$\varepsilon_p = \sqrt{\varepsilon_p^{0\,2} + 2 n_0 U_p \varepsilon_p^0}. \tag{7.4}$$

Substituting Eq. (7.3) into (6.10), we obtain the mean occupation number of the ground state,

$$\begin{aligned} \overline{N}_p &= (-\varepsilon_p + \varepsilon_p^0 + n_0 U_p)/2\varepsilon_p \\ &= (n_0 U_p)^2 / 2\varepsilon_p (\varepsilon_p + \varepsilon_p^0 + n_0 U_p). \end{aligned} \tag{7.5}$$

Equations (7.4) and (7.5) coincide with the results of the well-known work of Bogoliubov.[5]

APPENDIX. PROOF OF THEOREM (3.3)

We use the method of Wick[9] to transform the T-product in Eq. (3.3), which may be written symbolically

$$T\{Be^{V\sigma}\} = N\{e^{\Delta} B e^{V\sigma}\}, \tag{A.1}$$

where Δ is an operator which changes a pair $a_0 a_0^+$ into its replacement (3.5),

$$\begin{aligned} \Delta &= \frac{1}{V} \int dt\,dt'\theta(t - t') \frac{\delta^2}{\delta\alpha(t)\,\delta\alpha^+(t')}; \\ (\alpha &= a_0/\sqrt{V};\ \ \alpha^+ = a_0^+/\sqrt{V}). \end{aligned} \tag{A.2}$$

The proof of the theorem proceeds in two stages: (1) pulling B out across the operator e^{Δ}, and (2) a final disentangling of the N-product.

(1) The result of the first stage can be formulated as follows. With an error of order V^{-1} we have

$$e^{\Delta}\{B(\alpha;\ \alpha^+) e^{V\sigma}\} = B(\beta;\ \beta^+) e^{\Delta}\{e^{V\sigma}\}, \tag{A.3}$$

where β and β^+ are defined by the equations

$$\begin{aligned} \beta(t) &= \alpha + \int dt'\theta(t - t')\,\delta\sigma(\beta\beta^+)/\delta\beta^+(t'); \\ \beta^+(t) &= \alpha^+ + \int dt'\theta(t' - t)\,\delta\sigma(\beta\beta^+)/\delta\beta(t'). \end{aligned} \tag{A.4}$$

<u>Proof</u>. The factor V^{-1} in Δ can be compensated on the left side of Eq. (A.3) only if $e^{V\sigma}$ is involved in at least one operation of Δ. We write $\Delta = \Delta_{\sigma\sigma} + \Delta_{B\sigma} + \Delta_{\sigma B}$, where the first suffix indicates the object upon which the variation with respect to α operates, and the second suffix refers to the variation with respect to α^+. The result of operating with $e^{\Delta\sigma\sigma}$ can be written

$$e^{\Delta_{\sigma\sigma}}\{e^{V\sigma}\} = e^{V\sigma'} \tag{A.5}$$

It will be shown later that σ' is independent of V. Equation (A.5) gives

$$e^{\Delta}\{Be^{V\sigma}\} = \exp(\Delta_{B\sigma} + \Delta_{\sigma B})\{Be^{V\sigma'}\}. \tag{A.6}$$

We let $e^{\Delta_{B\sigma}}$ operate first on $e^{V\sigma'}$. From Eq. (A.2) we obtain

$$\Delta_{B\sigma} e^{V\sigma'} = e^{V\sigma'} \int dt\,dt'\theta(t - t') \frac{\delta\sigma'}{\delta\alpha^+(t')} \left[\frac{\delta}{\delta\alpha(t)}\right]_B \equiv e^{V\sigma'} D_B. \tag{A.7}$$

Successive application of Eq. (A.7) gives $(\Delta_{B\sigma})^k \times e^{V\sigma'} = e^{V\sigma'}(D_B)^k$, since the operators $\Delta_{B\sigma}$ produce variations only in the exponential. Therefore

$$e^{\Delta_{B\sigma}} e^{V\sigma'} = e^{V\sigma'} e^{D_B}. \tag{A.8}$$

From (A.7) it is clear that e^{D_B} is a displacement operator, displacing $\alpha(t)$ by the quantity

$$\alpha_1(t) = \int dt'\theta(t - t')\,\delta\sigma'/\delta\alpha^+(t').$$

Therefore

$$\begin{aligned} e^{\Delta_{B\sigma}}\{B(\alpha;\ \alpha^+) e^{V\sigma'}\} &= e^{V\sigma'} e^{D_B} B(\alpha;\ \alpha^+) \\ &= e^{V\sigma'} B(\alpha + \alpha_1;\ \alpha^+). \end{aligned} \tag{A.9}$$

An analogous result holds for the operator $e^{\Delta_{\sigma B}}$, which displaces $\alpha^+(t)$. We obtain finally from Eq. (A.6)

$$e^{\Delta}\{B(\alpha;\ \alpha^+) e^{V\sigma}\} = B(\beta;\ \beta^+) e^{V\sigma'} = B(\beta;\ \beta^+) e^{\Delta}\{e^{V\sigma}\}, \tag{A.10}$$

where

$$\begin{aligned} \beta(t) &= \alpha + \int dt'\theta(t - t')\,\delta\sigma'/\delta\alpha^+(t');\ \ \beta^+(t) \\ &= \alpha^+ + \int dt'\theta(t' - t)\,\delta\sigma'/\delta\alpha(t'); \end{aligned} \tag{A.11}$$

Equation (A.10) is identical with (A.3). It remains to show that Eq. (A.11) and (A.4) are identical. Varying both sides of Eq. (A.5) with respect to

$\alpha(t)$ and $\alpha^+(t)$, and using Eq. (A.10), we obtain

$$\begin{aligned} \delta\sigma'(\alpha\alpha^+)/\delta\alpha(t) &= \delta\sigma(\beta\beta^+)/\delta\beta(t); \\ \delta\sigma'(\alpha\alpha^+)/\delta\alpha^+(t) &= \delta\sigma(\beta\beta^+)/\delta\beta^+(t); \end{aligned} \tag{A.12}$$

Equations (A.12) and (A.11) define σ' When Eq. (A.12) is substituted into (A.11), the result is Eq. (A.4).

(2) In the second stage of the proof we may consider α and α^+ to be constant operators (see the beginning of Sec. 3). Then σ' and $B(\beta\beta^+) = B'(\alpha\alpha^+)$ are functions of α, α^+ instead of functionals. By integrating Eq. (A.12) with respect to time we obtain the connection between the functions $\sigma'(\alpha\alpha^+)$ and σ,

$$\int \frac{\delta\sigma(\beta\beta^+)}{\delta\beta(t)}\,dt = \int \frac{\delta\sigma'(\alpha\alpha^+)}{\delta\alpha(t)}\,dt = \frac{\partial\sigma'}{\partial\alpha}\cdot \int \frac{\delta\sigma(\beta\beta^+)}{\delta\beta^+(t)}\,dt = \frac{\partial\sigma'}{\partial\alpha^+} \tag{A.13}$$

For the following argument it is important that σ' and β' depend only on $\nu = \alpha^+\alpha$. This being so, the disentangling proceeds according to the rule

$$N\{B'(\nu)e^{V\sigma'(\nu)}\} = B'(\bar{\nu})\,N\{e^{V\sigma'}\}, \tag{A.14}$$

with $\bar{\nu}$ obtained from ν by the relation

$$\nu = \bar{\nu}\,[1 + \partial\sigma'(\bar{\nu})/\partial\bar{\nu}] \equiv \bar{\nu}X^2 \tag{A.15}$$

We defer the proof of Eq. (A.14), and show first that Eq. (A.3) and (A.14) imply the truth of the theorem (3.3). After the infinite factor $e^{V\sigma'}$ is removed, the lack of commutativity of α and α^+ can be neglected. The substitution $\nu \to \bar{\nu}$ which appears in Eq. (A.14), can therefore be divided into the two substitutions $\alpha \to \bar{\alpha}$, $\alpha^+ \to \bar{\alpha}^+$, where $\alpha = \bar{\alpha}X$ and $\alpha^+ = \bar{\alpha}^+X$ according to Eq. (A.15). After some algebra we find

$$\alpha^2 = \bar{\alpha}^2 + \bar{\alpha}\,\partial\sigma'(\bar{\nu})/\partial\bar{\alpha}^+; \quad \alpha^{+2} = \bar{\alpha}^{+2} + \bar{\alpha}^+\partial\sigma'(\bar{\nu})/\partial\bar{\alpha}. \tag{A.16}$$

We denote by A, A^+ the quantities into which β, β^+ are transformed under the substitution $\alpha, \alpha^+ \to \bar{\alpha}, \bar{\alpha}^+$. The equations for A and A^+ are obtained from Eq. (A.4) by changing the terms outside the integrals into $\bar{\alpha}$ and $\bar{\alpha}^+$, thus

$$\begin{aligned} A(t) &= \bar{\alpha} + \int dt'\theta(t-t')\,\delta\sigma(AA^+)/\delta A^+(t'); \\ A^+(t) &= \bar{\alpha}^+ + \int dt'\theta(t'-t)\,\delta\sigma(AA^+)/\delta A(t'). \end{aligned} \tag{A.17}$$

By the definition of $B'(\nu)$, $B'(\bar{\nu}) = B(AA^+)$, and so Eq. (A.3) and (A.14) imply (3.3). To complete the proof of the theorem it remains to show the equivalence of Eq. (A.17) and (3.4). To do this, we substitute $\alpha, \alpha^+ \to \bar{\alpha}, \bar{\alpha}^+$ in Eq. (A.13), and find the result

$$\frac{\partial\sigma'(\bar{\nu})}{\partial\bar{\alpha}} = \int \frac{\delta\sigma(AA^+)}{\delta A(t)}\,dt; \quad \frac{\partial\sigma'(\bar{\nu})}{\partial\bar{\alpha}^+} = \int \frac{\delta\sigma(AA^+)}{\delta A^+(t)}\,dt. \tag{A.18}$$

Equation (A.16) for $\bar{\alpha}$ and $\bar{\alpha}^+$ are identical with Eq. (3.6) by virtue of Eq. (A.18). Therefore $\bar{\alpha}$, $\bar{\alpha}^+$ are identical with C, C^+, and the theorem is proved.

We now return to the proof of Eq. (A.14). Let L be a quantity related to σ' by the equation

$$N\{e^{V\sigma'}\} = e^{VL} \tag{A.19}$$

We shall later express L explicitly in terms of σ' and shall verify that L is independent of V. Suppose that $B'(\nu)$ has the form

$$B'(\nu) = \sum_k b_k \nu^k. \tag{A.20}$$

Then Eq. (A.19) implies

$$N\{B'e^{V\sigma'}\} = \sum_k b_k \alpha^{+k} N\{e^{V\sigma'}\}\alpha^k = \sum_k b_k \alpha^{+k} e^{VL}\alpha^k. \tag{A.21}$$

The commutation relation $\alpha y(\nu) = (y + \frac{1}{V}\frac{\partial y}{\partial\nu})\alpha$ holds for any function $y(\nu)$. Applying it repeatedly, we find

$$\alpha^k e^y = \exp\left\{\left(1 + \frac{1}{V}\frac{\partial}{\partial\nu}\right)^k y\right\}\alpha^k \tag{A.22}$$

We choose y to satisfy $(1 + \frac{1}{V}\frac{\partial}{\partial\nu})^k y = VL$. Then Eq. (A.22) gives the rule for pulling α^k through e^{VL},

$$\begin{aligned} \exp\{VL\}\alpha^k &= \alpha^k \exp\left\{\left(1 + \frac{1}{V}\frac{\partial}{\partial\nu}\right)^{-k} VL\right\} \\ &\approx \alpha^k \exp\left(-k\frac{\partial L}{\partial\nu}\right)\exp\{VL\}. \end{aligned} \tag{A.23}$$

Applying Eq. (A.23) to Eq. (A.21), we obtain

$$N\{B'e^{V\sigma'}\} = \sum_k b_k (\nu e^{-\partial L/\partial\nu})^k e^{VL} = B'(\bar{\nu})\,e^{VL}, \tag{A.24}$$

with

$$\bar{\nu} = \nu e^{-\partial L/\partial\nu}. \tag{A.25}$$

L can be determined by differentiating both sides of Eq. (A.19) with respect to α. On the left side we find

$$\frac{1}{V}\frac{\partial}{\partial\alpha} N\{e^{V\sigma'}\} = N\left\{\frac{\partial\sigma'}{\partial\alpha}e^{V\sigma'}\right\} = \alpha^+ N\left\{\frac{\partial\sigma'}{\partial\nu}e^{V\sigma'}\right\}$$

which with Eq. (A.24) and (A.19) gives

$$\frac{1}{V}\frac{\partial}{\partial\alpha} N\{e^{V\sigma'}\} = \alpha^+ \frac{\partial\sigma'(\bar{\nu})}{\partial\bar{\nu}} e^{VL} \tag{A.26}$$

In differentiating the right side of Eq. (A.19) with respect to α, we must remember that $(\partial L/\partial\alpha)$ and L do not commute. Since $(\partial L/\partial\alpha) = \alpha^+(\partial L/\partial\nu)$, the commutation rule $L\alpha^+ = \alpha^+(L + \frac{1}{V}\frac{\partial L}{\partial\nu})$ implies

$$\frac{1}{V}\frac{\partial}{\partial\alpha} L^k = \alpha^+\left\{\left(L + \frac{1}{V}\frac{\partial L}{\partial\nu}\right)^k - L^k\right\}, \tag{A.27}$$

and hence

$$\frac{1}{V}\frac{\partial}{\partial\alpha}e^{VL} = \alpha^{+}(e^{\partial L/\partial\nu} - 1)e^{VL} \qquad (A.28)$$

By comparing Eq. (A.28) with (A.26), we obtain the desired relation between L and σ',

$$\partial\sigma'(\bar{\nu}) / \partial\bar{\nu} = e^{\partial L/\partial\nu} - 1. \qquad (A.29)$$

Equations (A.29) and (A.25) imply Eq. (A.15), and by Eq. (A.24) this proves Eq. (A.14).

[1] V. B. Berestetskii and A. D. Galanin, Проблемы современной физики (Problems of Modern Physics), No. 3 (1955), (introductory paper).

[2] Abrikosov, Landau, and Khalatnikov, Dokl. Akad. Akad. Nauk SSSR 95, 497, 773 and 1177; 96, 261 (1954).

[3] V. M. Galitskii and A. B. Migdal, J. Exptl. Theoret. Phys. (U.S.S.R.) 34, 139 (1958); Soviet Phys. JETP 7, 96 (1958).

[4] F. J. Dyson, Phys. Rev. 75, 1736 (1949).

[5] N. N. Bogoliubov, Izv. Akad. Nauk SSSR, ser. fiz. 11, 77 (1947).

[6] S. T. Beliaev, J. Exptl. Theoret. Phys. (U.S.S.R.) 34, 433 (1958); Soviet Phys. JETP 7, 289 (1958) (this issue).

[7] G. C. Wick, Phys. Rev. 80, 268 (1950).

[8] R. P. Feynman, Phys. Rev. 76, 749 (1949).

[9] S. Hori, Prog. Theoret. Phys. 7, 578 (1952).

Translated by F. J. Dyson
77

ENERGY-SPECTRUM OF A NON-IDEAL BOSE GAS

S. T. BELIAEV

Academy of Sciences, U.S.S.R.

Submitted to JETP editor August 2, 1957

J. Exptl. Theoret. Phys. (U.S.S.R.) 34, 433-446 (February, 1958)

The one-particle Green's function is calculated in a low-density approximation for a system of interacting bosons. The energy spectrum of states near to the ground state (quasi-particle spectrum is derived.

1. INTRODUCTION

In the preceding paper[1] the method of Green's functions was developed for a system consisting of a large number of bosons. The one-particle Green's function was expressed in terms of the effective potentials Σ_{ik} of pair interactions and the chemical potential μ of the system. Approximate methods must be used to determine Σ_{ik} and μ. In the present paper we study a "gaseous" approximation, in which the density n, or the ratio between the volume occupied by particles and the total volume, is treated as a small parameter. The interaction between particles is assumed to be central and short-range, but not necessarily weak. The first two orders of approximation involve only the scattering amplitude f of a two-particle system. But in the next order (proportional to $(\sqrt{nf^3})^2$) the effects of three-particle interaction amplitudes appear, which means that practical calculations to this order are hardly possible.

From the Green's function which we calculate, we derive the energy spectrum of excitations or quasi-particles, the energy of the ground state, and also the momentum distribution of particles in the ground state.

2. ESTIMATE OF THE GRAPHS CONTRIBUTING TO THE EFFECTIVE POTENTIALS

The definition of the potentials Σ_{ik}, and the rules for constructing Feynman graphs, were described in our earlier paper,[1] which we shall call I.

We shall estimate by perturbation theory the various graphs contributing to Σ_{ik} and μ. For the Fourier transform of the potential $U(\mathbf{p}) = U_{\mathbf{p}}$, we assume for simplicity* $U_p = U_0$ for $p < 1/a$, and $U_p = 0$ for $p > 1/a$. Then a is of the order of magnitude of the particle radius.

For definiteness we examine Σ_{20}. The graphs for Σ_{02}, Σ_{11} and μ are essentially similar. The first order of perturbation theory, as we saw in Sec. (I, 7), gives $\Sigma_{20}^{(1)} = n_0 U_p$; $\mu = n_0 U_0$.

In the estimate of any graph there may appear three parameters — U_0 and a, characterizing the interaction, and n_0, characterizing the density of particles in the condensed phase. The three parameters can be combined into two dimensionless ratios,

$$\xi = U_0/a; \qquad \beta = \sqrt{n_0 a^3}. \tag{2.1}$$

The quantity ξ is the usual parameter which appears in perturbation theory (in ordinary units $\xi \sim mU(r)a^2/\hbar^2$), while β is a parameter of gas-density.

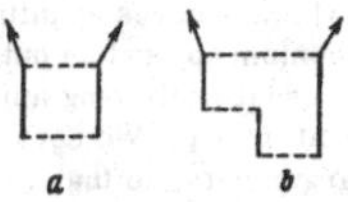
FIG. 1

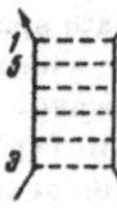
FIG. 2

The only non-vanishing graph in second order is the one shown in Fig. 1a. This gives a contribution

$$M_a \sim n_0 \int G^0(q+\mu)\, G^0(-q+\mu)\, U_0 U_{p+q} d^4q$$

Substituting for G^0 from

$$G^0(p) = (p^0 - \varepsilon_p^0 + i\delta)^{-1}, \qquad \varepsilon_p^0 = p^2/2, \qquad \delta \to +0 \tag{2.2}$$

and carrying out the q^0-integration, we find

$$M_a \sim n_0 U_0^2 \int_{qa<1} d\mathbf{q} \int \frac{dq^0}{(q^0+\mu-\varepsilon_q^0+i\delta)(-q^0+\mu-\varepsilon_q^0+i\delta)}$$

$$\sim n_0 U_0^2 \int_{qa<1} d\mathbf{q} \frac{1}{\mu-\varepsilon_q^0+i\delta}.$$

In the last integral the main contribution comes from $q \sim 1/a$, where $\mu/\varepsilon^0 \sim n_0 U_0 a^2 = \xi\beta^2 \ll 1$. Therefore

$$M_a \sim n_0 U_0^2/a = \Sigma_{20}^{(1)}\xi. \tag{2.3}$$

We consider next the third-order graph (1b). This gives a contribution

$$M_b \sim n_0^2 \int G^0(q+\mu)\,[G^0(-q+\mu)]^2 U_0 U_q U_{p-q} d^4q$$
$$\sim n_0^2 U_0^3 \int d\mathbf{q}/(\mu - \varepsilon_q^0 + i\delta)^2$$

The last integral, unlike the previous one, converges at the upper limit, and the main contribution now comes from the range $q \sim \sqrt{\mu} = \sqrt{n_0 U_0}$. Therefore

$$M_b \sim n_0^2 U_0^3/\sqrt{\mu} = \Sigma_{20}^{(1)}\xi^{1/2}\beta. \tag{2.4}$$

From Eqs. (2.3) and (2.4) we see that $M_b/M_a \sim \xi^{1/2}\beta$. This is a consequence of the fact that M_a contains an integral of a product of two factors G^0, formally diverging at the upper limit, while M_b contains an integral of a product of three factors G^0 and converges without any cut-off. In the graphs this difference is indicated by the number of continuous lines in the closed circuit formed by the continuous and dotted lines. The same result holds when the circuits form part of a more complicated graph.

Thus every circuit containing more than two continuous lines introduces the small parameter β, while circuits with two continuous lines do not involve β. In the lowest order we need consider only graphs whose circuits are all of the two-line type. All such graphs are of the "ladder" construction shown in Fig. 2. We denote by $-i\Gamma(12;34)$ the total contribution from all such graphs. The first-order approximation in β then differs from the first-order approximation in perturbation theory by changing the potential U (arising from a ladder with one rung) into Γ (arising from ladders of all lengths). Similar conclusions hold also for the higher approximations. A summation over a set of graphs, differing only by the insertion of ladder circuits into a fixed skeleton, produces a change of U into Γ. If we represent Γ by a rectangle, all graphs can be constructed by means of rectangles and continuous lines only. In this way the potential U is eliminated from the problem. The effective potential is Γ.

3.* EQUATION FOR THE EFFECTIVE POTENTIAL Γ

We can write down an integral equation

$$\Gamma(12;34) = U(1-2)\,\delta(1-3)\,\delta(2-4)$$
$$+ i \int U(1-2)\, G^0(1-5)\, G^0(2-6)\, \Gamma(56;34)\, d^4x_5 d^4x_6. \tag{3.1}$$

for the sum of contributions from all graphs of the ladder type (see Fig. 2). The notations are the

*The letters p, q, ... are used to denote the lengths of 3-vectors, or to denote 4-vectors. There can be no confusion, because they denote 4-vectors only when they appear as arguments in G(p), Σ(p), etc.

*The problems connected with Γ were solved in collaboration with V. M. Galitskii, who was working simultaneously on the analogous problems in Fermion systems.

same as in I. We next transform Eq. (3.1) into momentum representation. In order to relieve the equations of factors of 2π, we shall use the conventions

$$d^4p = (2\pi)^{-4}dp^1dp^2dp^3dp^0;$$
$$\delta(p) = (2\pi)^4\,\delta(p^1)\,\delta(p^2)\,\delta(p^3)\,\delta(p^0),$$

and similarly we understand $\mathbf{dp}$ and $\delta(\mathbf{p})$ to carry factors of $(2\pi)^3$. We write

$$\Gamma(p_1p_2;\,p_3p_4)\,\delta(p_1+p_2-p_3-p_4) = \int \exp\{-ip_1x_1 - ip_2x_2 + ip_3x_3 + ip_4x_4\}\,\Gamma(12;34)\,d^4x_1d^4x_2d^4x_3d^4x_4, \tag{3.2}$$

and introduce the relative and total momenta by

$$p_1 + p_2 = P';\quad p_3 + p_4 = P;$$
$$p_1 - p_2 = 2p',\quad p_3 - p_4 = 2p, \tag{3.3}$$

Then, by Eq. (3.1), $\Gamma(p';\,p;\,P) \equiv \Gamma(p_1p_2;\,p_3p_4)$ satisfies the equation

$$\Gamma(p';p;P) = U(p'-p) + i\int d^4qU(p'-q)\,G^0(P/2+q) \times G^0(P/2-q)\,\Gamma(q;p;P). \tag{3.4}$$

Since the interaction U is instantaneous, $U(1-2) = U(\mathbf{x}_1 - \mathbf{x}_2)\,\delta(t_1 - t_2)$, and therefore the points 1, 2 and 3, 4 in $\Gamma(12;34)$ must be simultaneous. In momentum representation this means that $\Gamma(p_1p_2;\,p_3p_4)$ depends on the fourth components only in the combination $p_1^0 + p_2^0 = p_3^0 + p_4^0 = P^0$. Therefore $\Gamma(p';\,p;\,P)$ is independent of the fourth components of its first two arguments (the relative momenta). The q^0-integration in Eq. (3.4) can thus be carried out, giving

$$\int dq^0G^0\Big(\frac{1}{2}P+q\Big)G^0\Big(\frac{1}{2}P-q\Big) = -i\Big(P^0 - \frac{1}{4}\mathbf{P}^2 - \mathbf{q}^2 + i\delta\Big)^{-1}, \tag{3.5}$$

and then Eq. (3.4) takes the form

$$\Gamma(\mathbf{p}';\mathbf{p};P) = U(\mathbf{p}'-\mathbf{p}) + \int d\mathbf{q}\,\frac{U(\mathbf{p}'-\mathbf{q})\,\Gamma(\mathbf{q};\mathbf{p};P)}{k_0^2 - q^2 + i\delta};$$
$$k_0^2 = P^0 - \frac{1}{4}\mathbf{P}^2. \tag{3.6}$$

Equation (3.6) cannot be solved explicitly, but its solution can be expressed in terms of the scattering amplitude of two particles in a vacuum. We write $\chi(\mathbf{q}) = (k_0^2 - q^2 + i\delta)^{-1}\,\Gamma(\mathbf{q};\,\mathbf{p};\,P)$. Then Eq. (3.6) becomes

$$(k_0^2 - p'^2)\,\chi(\mathbf{p}') - \int U(\mathbf{p}'-\mathbf{q})\,\chi(\mathbf{q})\,d\mathbf{q} = U(\mathbf{p}'-\mathbf{p}). \tag{3.7}$$

Let $\Psi_{\mathbf{k}}(\mathbf{p}')$ be the normalized wave-function which satisfies the equation

$$(k^2 - p'^2)\,\Psi_{\mathbf{k}}(\mathbf{p}') - \int U(\mathbf{p}'-\mathbf{q})\,\Psi_{\mathbf{k}}(\mathbf{q})\,d\mathbf{q} = 0, \tag{3.8}$$

Then the solution of Eq. (3.7) may be written

$$\chi(\mathbf{p}') = \int \frac{\Psi_{\mathbf{k}}(\mathbf{p}')\,\Psi_{\mathbf{k}}^*(\mathbf{q})}{k_0^2 - k^2 + i\delta}\,U(\mathbf{q}-\mathbf{p})\,d\mathbf{q},$$

and so $\Gamma(\mathbf{p}';\,\mathbf{p};\,P)$ becomes

$$\Gamma(\mathbf{p}';\mathbf{p};P) = (k_0^2 - p'^2)\int \frac{\Psi_{\mathbf{k}}(\mathbf{p}')\,\Psi_{\mathbf{k}}^*(\mathbf{q})}{k_0^2 - k^2 + i\delta}\,U(\mathbf{q}-\mathbf{p})\,d\mathbf{q}. \tag{3.9}$$

We observe now that Eq. (3.8) is the Schrödinger equation in momentum representation. Thus $\Psi_{\mathbf{k}}(\mathbf{p})$ is the wave-function for a scattering problem with potential U. The scattering amplitude* $f(\mathbf{p}'\,\mathbf{p})$ is related to the Ψ-function by

$$f(\mathbf{p}'\mathbf{p}) = \int e^{-i\mathbf{p}'\mathbf{r}}\,U(\mathbf{r})\,\Psi_{\mathbf{p}}(\mathbf{r})\,d\mathbf{r} = \int U(\mathbf{p}'-\mathbf{q})\,\Psi_{\mathbf{p}}(\mathbf{q})\,d\mathbf{q}, \tag{3.10}$$

or by

$$\Psi_{\mathbf{p}}(\mathbf{p}') = \delta(\mathbf{p}-\mathbf{p}') + f(\mathbf{p}'\mathbf{p})/(p^2 - p'^2 + i\delta). \tag{3.11}$$

In the first Eq. (3.10), $\Psi_{\mathbf{p}}(\mathbf{r})$ is the wave-function in coordinate space which behaves at infinity like a plane wave with momentum $\mathbf{p}$ and an outgoing spherical wave. The usual scattering amplitude is the value of $f(\mathbf{p}'\,\mathbf{p})$ at $p' = p$. We consider arbitrary values of the arguments, so that $f(\mathbf{p}'\,\mathbf{p})$ is in general defined by Eq. (3.10).

Because $\Psi_{\mathbf{p}}(\mathbf{p}')$ satisfies orthogonality conditions in both its arguments, $f(\mathbf{p}'\,\mathbf{p})$ satisfies the unitarity conditions

$$f(\mathbf{p}'\mathbf{p}) - f^*(\mathbf{p}'\mathbf{p}) = \int d\mathbf{q}f(\mathbf{p}'\mathbf{q})\,f^*(\mathbf{p}\mathbf{q})\Big[\frac{1}{q^2 - p'^2 + i\delta} - \frac{1}{q^2 - p^2 - i\delta}\Big] = \int d\mathbf{q}f^*(\mathbf{q}\mathbf{p}')\,f(\mathbf{q}\mathbf{p})\Big[\frac{1}{q^2 - p'^2 + i\delta} - \frac{1}{q^2 - p^2 - i\delta}\Big] \tag{3.12}$$

When $\mathbf{p}' = \pm\mathbf{p}$, Eq. (3.12) gives the imaginary part of the forward and backward scattering amplitudes. Since $f(-\mathbf{p}' - \mathbf{p}) = f(\mathbf{p}'\,\mathbf{p})$, Eq. (3.12) implies

$$\operatorname{Im} f(\pm\mathbf{p}\mathbf{p}) = -i\pi\int d\mathbf{q}f(\mathbf{p}\mathbf{q})\,f^*(\pm\mathbf{p}\mathbf{q})\,\delta(q^2 - p^2). \tag{3.13}$$

For the forward scattering amplitude, Eq. (3.13) gives just the well-known relation between the imaginary part of the amplitude and the total cross-section σ, $\operatorname{Im} f(\mathbf{p}\,\mathbf{p}) = -ip\sigma$.

We substitute Eq. (3.11) into (3.9) and use Eq. (3.12). This gives two equivalent expressions for $\Gamma(\mathbf{p}';\,\mathbf{p};\,P)$,

$$\Gamma(\mathbf{p}';\mathbf{p};P) = f(\mathbf{p}'\mathbf{p}) + \int d\mathbf{q}f(\mathbf{p}'\mathbf{q})\,f^*(\mathbf{p}\mathbf{q})\Big[\frac{1}{k_0^2 - q^2 + i\delta} + \frac{1}{q^2 - p^2 - i\delta}\Big] \tag{3.14}$$
$$= f^*(\mathbf{p}\mathbf{p}') + \int d\mathbf{q}f(\mathbf{p}'\mathbf{q})\,f^*(\mathbf{p}\mathbf{q})\Big[\frac{1}{k_0^2 - q^2 + i\delta} + \frac{1}{q^2 - p'^2 + i\delta}\Big].$$

*The quantity $f(\mathbf{p}'\,\mathbf{p})$ differs by a numerical factor from the usual amplitude $a(\mathbf{p}'\,\mathbf{p})$, in fact $f = -4\pi a$.

expressing the effective potential $\Gamma(\mathbf{p}'; \mathbf{p}; P)$ in terms of the scattering amplitudes of a two-particle system.

4. FIRST-ORDER GREEN'S FUNCTION

The effective potentials Σ_{ik} are determined by special values which Γ takes when two out of the four particles involved in a process belong to the condensed phase. Thus two of the four particles must have $\mathbf{p} = 0$, $p^0 = \mu$. Each particle of the condensed phase also carries a factor $\sqrt{n_0}$. Therefore we find

$$\Sigma_{20}(p+\mu) = n_0\Gamma(\mathbf{p}; 0; 2\mu);\ \Sigma_{02}(p+\mu) = n_0\Gamma(0; \mathbf{p}; 2\mu),$$

$$\Sigma_{11}(p+\mu) = n_0\Gamma(\mathbf{p}/2; \mathbf{p}/2; p + 2\mu) + n_0\Gamma(-\mathbf{p}/2; \mathbf{p}/2; p+2\mu). \tag{4.1}$$

To obtain the chemical potential we must let all four particles in Γ belong to the condensed phase, and divide by one power of n_0 [see Eq. (I, 3.20)]. We then have

$$\mu = n_0\Gamma(0; 0; 2\mu). \tag{4.2}$$

Substituting into Eqs. (4.1) and (4.2) the value of Γ from Eq. (3.14), we find

$$\mu = n_0 f(00) + n_0\int d\mathbf{q}\,|f(0\mathbf{q})|^2\left[\frac{1}{2\mu - q^2 + i\delta} + \frac{1}{q^2}\right],$$

$$\Sigma_{20}(p+\mu) = n_0 f(\mathbf{p}0) + n_0\int d\mathbf{q}\, f(\mathbf{pq})\, f^*(0\mathbf{q})\left[\frac{1}{2\mu - q^2 + i\delta} + \frac{1}{q^2}\right],$$

$$\Sigma_{02}(p+\mu) = n_0 f^*(\mathbf{p}0) + n_0\int d\mathbf{q}\, f(0\mathbf{q})\, f^*(\mathbf{pq})\left[\frac{1}{2\mu - q^2 + i\delta} + \frac{1}{q^2}\right],$$

$$\Sigma_{11}(p+\mu) = 2n_0 f_s\left(\frac{\mathbf{p}}{2}\,\frac{\mathbf{p}}{2}\right) + 2n_0\int d\mathbf{q}\left|f_s\left(\frac{\mathbf{p}}{2}\,\mathbf{q}\right)\right|^2\left[\frac{1}{p^0 + 2\mu - \mathbf{p}^2/4 - q^2 + i\delta} + \frac{1}{q^2 - \mathbf{p}^2/4 - i\delta}\right]. \tag{4.3}$$

In the last equation we have introduced the symmetrized amplitude

$$f_s(\mathbf{p}'\mathbf{p}) = [f(\mathbf{p}'\mathbf{p}) + f(-\mathbf{p}'\mathbf{p})]/2.$$

All the integrals in Eq. (4.3) converge at high momentum, even if the amplitudes are taken to be constant. For dimensional reasons these terms are of order $n_0 f^2\sqrt{\mu}$. Compared with the first terms in Eq. (4.3), these terms contain an extra factor $\sqrt{n_0 f^3}$, which is just the gas-density parameter (2.1) obtained by substituting the amplitude f for the particle radius. In first approximation we neglect the integral terms in Eq. (4.3) and obtain

$$\mu = n_0 f(00);\ \Sigma_{20}(p+\mu) = \Sigma_{02}^*(p+\mu) = n_0 f(\mathbf{p}0);$$
$$\Sigma_{11}^{\pm} \equiv \Sigma_{11}(\pm p + \mu) = 2n_0 f_s\left(\frac{\mathbf{p}}{2}\,\frac{\mathbf{p}}{2}\right). \tag{4.4}$$

The Green's function G is given by Eq. (I, 5.6),

$$G(p+\mu) = \frac{p^0 + \epsilon_0^0 + \Sigma_{11}^- - \mu}{[p^0 - (\Sigma_{11}^+ - \Sigma_{11}^-)/2]^2 - [\epsilon_p^0 + (\Sigma_{11}^+ + \Sigma_{11}^-)/2 - \mu]^2 + \Sigma_{20}\Sigma_{02} + i\delta} \tag{4.5}$$

and after substituting from Eq. (4.4) this becomes

$$G(p+\mu) = \frac{p^0 + \epsilon^0 + 2n_0 f_s\left(\frac{\mathbf{p}}{2}\,\frac{\mathbf{p}}{2}\right) - n_0 f(00)}{p^{02} - \epsilon_p^2 + i\delta}, \tag{4.6}$$

with

$$\epsilon_p = \sqrt{\left[\epsilon_p^0 + 2n_0 f_s\left(\frac{\mathbf{p}}{2}\,\frac{\mathbf{p}}{2}\right) - n_0 f(00)\right]^2 - n_0^2|f(\mathbf{p}0)|^2} \tag{4.7}$$

The point $p_0(\mathbf{p})$, at which the Green's function $G(p+\mu)$ has a pole, determines the energy $\epsilon_\mathbf{p}$ of elementary excitations or quasi-particles[2] carrying momentum $\mathbf{p}$. To calculate $\epsilon_\mathbf{p}$ we must know three distinct amplitudes. $f_s(\mathbf{p}/2\ \mathbf{p}/2)$ is the ordinary symmetrized amplitude for forward scattering, and $f(0\ 0)$ is a special value of the same amplitude. However, $f(\mathbf{p}\ 0)$ does not have any obvious meaning in the two-particle problem, since it refers to a process which is forbidden for two particles in a vacuum.

At small momenta, we may neglect the momentum dependence of $f_s(\mathbf{p}/2\ \mathbf{p}/2)$ and of $f(\mathbf{p}\ 0)$, setting $f_s(\mathbf{p}/2\ \mathbf{p}/2) \approx f(\mathbf{p}\ 0) \approx f(0\ 0) \equiv f_0$. This approximation is allowed when the wavelength is long compared with the characteristic size of the interaction region, which has an order of magnitude given by the scattering amplitude f_0. Therefore when $p < f_0^{-1}$ we may consider all the amplitudes in Eq. (4.6) and (4.7) to be constant. For higher excitations with $p \gtrsim f_0^{-1}$, the momentum dependence of the amplitude becomes important, and the problem cannot be treated in full generality. We shall examine the higher excitations (in Sec. 8) for the special example of a hard-sphere gas.

Confining ourselves to the case $pf_0 < 1$, we deduce from Eq. (4.6) and (4.7)

$$G(p+\mu) = (p^0 + \epsilon_p^0 + n_0 f_0)/(p^{02} - \epsilon_p^2 + i\delta), \tag{4.8}$$

with

$$\epsilon_p = \sqrt{\epsilon_p^{02} + 2n_0 f_0\epsilon_p^0}. \tag{4.9}$$

Equations (4.8) and (4.9) are formally identical with the results obtained from perturbation theory in Eq. (I, 7.3) and (I, 7.4). Only the scattering amplitude f_0 now appears instead of the Fourier transform $U_\mathbf{p}$ of the potential.

Equation (4.9) shows that quasi-particles with $p \ll \sqrt{n_0 f_0}$ have a sound-wave type of dispersion law $\epsilon_\mathbf{p} \approx p\sqrt{n_0 f_0}$. When $p \gg \sqrt{n_0 f_0}$ they go over into almost free particles with $\epsilon_\mathbf{p} \approx \epsilon_\mathbf{p}^0 + n_0 f_0$.

This sort of energy spectrum appears also when one considers particles moving in a continuous medium with a refractive index. The transition from phonon to free particle behavior occurs at $p \sim \sqrt{n_0 f_0} \ll 1/f_0$, so that the approximation of constant amplitudes is valid in both ranges.

The conditions $\sqrt{n_0 f_0^3} \ll 1$ and $pf_0 \ll 1$ are not independent. If we look at momenta p not greatly exceeding $\sqrt{n_0 f_0}$, then the second condition is a consequence of the first. If we are then neglecting quantities of order $\sqrt{n_0 f_0^3}$, we must also treat the amplitudes as constant.

In Sec. (I, 5) we introduced the quantity $\hat{G}(p+\mu)$, the analog of the Green's function G but constructed from graphs with two ingoing ends instead of one ingoing and one outgoing. The analogous quantity with two outgoing ends will be denoted by $\check{G}(p+\mu)$. It is obtained from $\hat{G}(p+\mu)$ when Σ_{02} is replaced by Σ_{20}. In the constant-amplitude approximation, Eq. (I, 5.7) and (4.4) give

$$\hat{G}(p+\mu) = \check{G}(p+\mu) = -n_0 f_0/(p^{02} - \varepsilon_p^2 + i\delta). \quad (4.10)$$

5. SECOND APPROXIMATION FOR THE GREEN'S FUNCTION

For the second approximation to Σ_{ik} and μ, we must retain quantities of order $\sqrt{n_0 f_0^3}$. As we saw at the end of the preceding section, we must then also retain terms of order pf_0 in the amplitudes. The real part of the amplitude involves only even powers of p, and the imaginary part only odd powers. Terms of order pf_0 arise only from the lowest approximation to the imaginary part of the amplitudes. The imaginary part of $f_s(\mathbf{p}/2\ \mathbf{p}/2)$ is given by Eq. (3.13), and from Eq. (3.10) we see that the amplitude $f(\mathbf{p}\ 0)$ is real [and anyway in this approximation we need only the square of the modulus of $f(\mathbf{p}\ 0)$].

The graphs of the first approximation give terms of order $\sqrt{n_0 f_0^3}$, namely the integral terms in Eq. (4.3). In these terms, as in Eq. (3.13), we may take the amplitudes to be constant. We have seen in Sec. 2 that graphs containing one circuit with three or more continuous lines give contributions of the same order. The summation over sets of graphs, which differ only in the number of continuous lines in a circuit, is automatically performed if one replaces the zero-order Green's function G^0 by the first-order functions G, $\hat{G}$ and $\check{G}$. We therefore consider immediately the circuits which can be built out of G, $\hat{G}$, $\check{G}$ and Γ. There are altogether ten essentially different circuits (see Fig. 3). A rectangle with a cross denotes a sum of two rectangles, one being a direct interaction and the other an exchange interaction. The two differ only by an

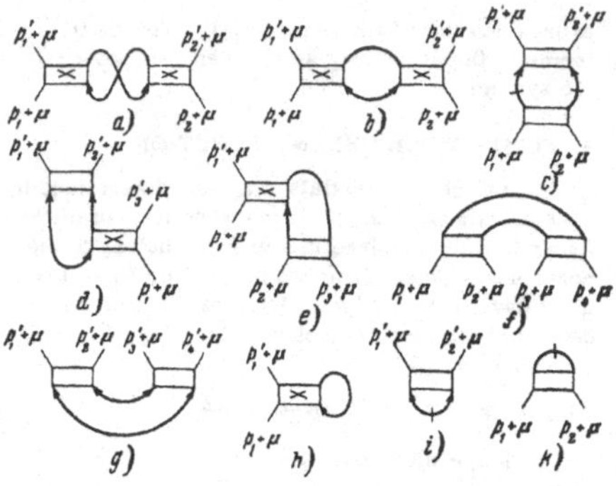

FIG. 3

interchange of the upper or the lower ends. The sum of the two rectangles introduces a factor $-i[\Gamma(12;34) + \Gamma(12;43)]$, or in momentum representation $-i[\Gamma(\mathbf{p}';\mathbf{p};P) + \Gamma(-\mathbf{p}';\mathbf{p};P)]$. If G, $\hat{G}$ and $\check{G}$ are expanded in powers of the effective potential Γ, then in the lowest approximation the graphs (3c, 3i, 3k) become circuits with two continuous lines. But all such circuits are already included in Γ and must therefore be omitted. This omission is represented in Fig. 3 by the strokes across the continuous lines. Let $-iF_{a,b\ldots}(p_1', \ldots, p_1 \ldots)$ denote the contributions from the graphs of Fig. 3. In the constant-amplitude approximation these contributions are:

$$\begin{aligned}
F_a(p_1' p_2', p_1 p_2) &= i4f_0^2 \int G(q+\mu)\, G(p_1 - p_1' + q + \mu)\, d^4q; \\
F_b &= i4f_0^2 \int \hat{G}(q+\mu)\, \check{G}(p_1 - p_1' + q + \mu)\, d^4q; \\
F_c &= if_0^2 \int \{G(q+\mu)\, G(p_1 + p_2 - q + \mu) \\
&\quad - G^0(q+\mu)\, G^0(p_1 + p_2 - q + \mu)\}\, d^4q; \\
F_d &= i2f_0^2 \int \hat{G}(q+\mu)\, G(p_2' + p_3' - q + \mu)\, d^4q, \\
F_e &= i2f_0^2 \int \check{G}(q+\mu)\, G(p_1' + p_2' - q + \mu)\, d^4q; \\
F_f &= if_0^2 \int \hat{G}(q+\mu)\, \hat{G}(p_1 - p_2 - q + \mu)\, d^4q; \\
F_k &= if_0^2 \int \check{G}(q+\mu)\, \check{G}(p_1' + p_2' - q + \mu)\, d^4q; \\
F_h &= i2f_0 \int G(q+\mu)\, d^4q; \\
F_i &= if_0 \int \{\check{G}(q+\mu) - n_0 f_0 G^0(q+\mu)\, G^0(-q+\mu)\}\, d^4q; \\
F_k &= if_0 \int \{\hat{G}(q+\mu) - n_0 f_0 G^0(q+\mu)\, G^0(-q+\mu)\}\, d^4q.
\end{aligned} \quad (5.1)$$

Momentum conservation $\Sigma p = \Sigma p'$ is assumed to hold everywhere. The q^0-integration in F_h is performed with a detour into the upper half-plane, since this contribution must vanish as $G \to G^0$. Everywhere on the right of Eq. (5.1) the first-approximation value $\mu^{(1)} = n_0 f_0$ should be substituted for μ.

The Σ_{ik} involve special values of the F, together with a factor $\sqrt{n_0}$ for each particle of the condensed phase:

$$\begin{aligned}
\Sigma'_{20}(p+\mu) &= n_0 F_a(p-p;\ 00) \\
&+ n_0 F_b(p-p;\ 00) + n_0 F_e(p0-p;\ 0) \\
&+ n_0 F_e(0p-p;\ 0) + n_0 F_e(-p0p;\ 0) \\
&+ n_0 F_e(0-pp;\ 0) + n_0 F_g(p0-p0;) \\
&+ n_0 F_g(p00-p;) + F_i(p-p;); \\
\Sigma'_{02}(p+\mu) &= n_0 F_a(00;\ p-p) \\
&+ n_0 F_b(00;\ p-p) + n_0 F_d(0;\ p-p0) \\
&+ n_0 F_d(0;\ p0-p) \\
&+ n_0 F_d(0;\ -pp0) + n_0 F_d(0;\ -p0p) \\
&+ n_0 F_f(;\ p0-p0) \\
&+ n_0 F_f(;\ p00-p) + F_k(;\ p-p);
\end{aligned} \tag{5.2}$$

$$\begin{aligned}
\Sigma'_{11}(p+\mu) &= n_0 F_a(p0;\ 0p) + n_0 F_b(p0;\ 0p) \\
&+ n_0 F_c(p0;\ p0) + n_0 F_c(p0;\ 0p) \\
&+ n_0 F_d(p;\ 0p0) + n_0 F_d(p;\ 00p) \\
&+ n_0 F_e(p00;\ p) + n_0 F_e(0p0;\ p) + F_h(p;\ p).
\end{aligned}$$

To enumerate the vacuum loops which contribute to μ, we must first distinguish one incoming or outgoing particle of the condensed phase (see Section (I, 4)). After this we must sum the loops, counting separately all possible geometric structures and all possible positions of the distinguished particle. The vacuum loops include three types of rectangle, differing in the numbers of incoming and outgoing continuous lines, and corresponding to factors $\Sigma_{11}^{(1)}$, $\Sigma_{02}^{(1)}$ and $\Sigma_{20}^{(1)}$. The distinguished particle of the condensed phase may come out from $\Sigma_{11}^{(1)}$ or from $\Sigma_{02}^{(1)}$. The sums of contributions from graphs of these two types are respectively $-iF_h(0;0)$ and $-iF_i(0\,0;)$. The term in μ arising from all these vacuum loops is thus

$$\mu' = F_h(0;\ 0) + F_i(00;). \tag{5.3}$$

To carry out the q^0-integration in Eq. (5.1), it is convenient to represent G and $\hat{G} = \check{G}$ in the following form,

$$G(q+\mu) = \frac{A_{\mathbf{q}}}{q^0 - \epsilon_{\mathbf{q}} + i\delta} - \frac{B_{\mathbf{q}}}{q^0 + \epsilon_{\mathbf{q}} - i\delta};$$

$$\hat{G} = \check{G}(q+\mu)$$

$$= -C_{\mathbf{q}}\left[\frac{1}{q^0 - \epsilon_{\mathbf{q}} + i\delta} - \frac{1}{q^0 + \epsilon_{\mathbf{q}} - i\delta}\right], \tag{5.4}$$

with

$$\begin{aligned}
A_{\mathbf{q}} &= (\epsilon_{\mathbf{q}} + \epsilon_{\mathbf{q}}^0 + n_0 f_0) / 2\epsilon_{\mathbf{q}}, \\
B_{\mathbf{q}} &= (-\epsilon_{\mathbf{q}} + \epsilon_{\mathbf{q}}^0 + n_0 f_0) / 2\epsilon_{\mathbf{q}} \\
&= n_0^2 f_0^2 / 2\epsilon_{\mathbf{q}}(\epsilon_{\mathbf{q}} + \epsilon_{\mathbf{q}}^0 + n_0 f_0);\ C_{\mathbf{q}} = n_0 f_0 / 2\epsilon_{\mathbf{q}}
\end{aligned} \tag{5.5}$$

depending only on $|\mathbf{q}|$. The q^0-integrations are now performed and the results substituted into Eq. (5.2) and (5.3). After some manipulations we obtain

$$\begin{aligned}
\Sigma'_{02\,(20)}(p+\mu) &= 2n_0 f_0^2 \int d\mathbf{q}\, [(A_{\mathbf{q}}, B_{\mathbf{k}}) - (A_{\mathbf{q}} + B_{\mathbf{q}}, C_{\mathbf{k}}) + 3C_{\mathbf{q}}C_{\mathbf{k}}] \\
&\times \left(\frac{1}{p^0 - \epsilon_{\mathbf{q}} - \epsilon_{\mathbf{k}} + i\delta} - \frac{1}{p^0 + \epsilon_{\mathbf{q}} + \epsilon_{\mathbf{k}} - i\delta}\right) \\
&- f_0 \int d\mathbf{q} \left\{C_{\mathbf{q}} + \frac{n_0 f_0}{2n_0 f_0 - 2\epsilon_{\mathbf{q}}^0 + i\delta}\right\},
\end{aligned}$$

$$\begin{aligned}
\Sigma'_{11}(p+\mu) &= 2n_0 f_0^2 \int d\mathbf{q} \left\{\frac{(A_{\mathbf{q}}; B_{\mathbf{k}}) + 2C_{\mathbf{q}}C_{\mathbf{k}} + A_{\mathbf{q}}A_{\mathbf{k}} - 2(A_{\mathbf{q}}; C_{\mathbf{k}})}{p^0 - \epsilon_{\mathbf{q}} - \epsilon_{\mathbf{k}} + i\delta}\right. \\
&- \frac{(A_{\mathbf{q}}, B_{\mathbf{k}}) + 2C_{\mathbf{q}}C_{\mathbf{k}} + B_{\mathbf{q}}B_{\mathbf{k}} - 2(B_{\mathbf{q}}, C_{\mathbf{k}})}{p^0 + \epsilon_{\mathbf{q}} + \epsilon_{\mathbf{k}} - i\delta} \\
&\left. - \frac{1}{p^0 + 2n_0 f_0 - \epsilon_{\mathbf{q}}^0 - \epsilon_{\mathbf{k}}^0 + i\delta}\right\} + 2f_0 \int d\mathbf{q}\, B_{\mathbf{q}},
\end{aligned} \tag{5.6}$$

$$\mu' = 2f_0 \int d\mathbf{q}\, B_{\mathbf{q}} - f_0 \int d\mathbf{q} \left\{C_{\mathbf{q}} + \frac{n_0 f_0}{2n_0 f_0 - 2\epsilon_{\mathbf{q}}^0 + i\delta}\right\},$$

Here $\mathbf{k} = \mathbf{p} - \mathbf{q}$, and the symbol (;) denotes a symmetrized product, $(A_{\mathbf{q}}; B_{\mathbf{k}}) = A_{\mathbf{q}}B_{\mathbf{k}} + B_{\mathbf{q}}A_{\mathbf{k}}$. The integrands are all symmetrical in $\mathbf{q}$ and $\mathbf{k}$.

Before we add to Eq. (5.6) the second-order terms from Eq. (4.2), we transform the expression (4.3) for Σ_{11}. Remembering that

$$q^2 + p^2/4 = \epsilon_{\mathbf{p}/2+\mathbf{q}}^0 + \epsilon_{\mathbf{p}/2-\mathbf{q}}^0$$

and introducing the new integration variable $\mathbf{q}' = \mathbf{q} + \mathbf{p}/2$, we find that Eq. (4.3) gives to the required approximation

$$\begin{aligned}
\Sigma_{11}(p+\mu) &= 2n_0 f_0 + 2n_0 \operatorname{Im} f_s\left(\frac{\mathbf{p}}{2}\frac{\mathbf{p}}{2}\right) \\
&+ 2n_0 f_0^2 \int d\mathbf{q} \left[\frac{1}{p^0 + 2n_0 f_0 - \epsilon_{\mathbf{q}}^0 - \epsilon_{\mathbf{k}}^0 + i\delta}\right. \\
&\left. - \frac{1}{\epsilon_p^0 - \epsilon_{\mathbf{q}}^0 - \epsilon_{\mathbf{k}}^0 + i\delta}\right].
\end{aligned} \tag{5.7}$$

The total of all second-order terms in now obtained from Eq. (5.6), (4.3), and (5.7), and after some algebra becomes

$$\mu^{(2)} = 2f_0 \int d\mathbf{q}\, B_\mathbf{q} + \frac{1}{2} n_0 f_0^2 \int d\mathbf{q} \left(\frac{1}{\varepsilon_\mathbf{q}^0} - \frac{1}{\varepsilon_\mathbf{q}} \right),$$

$$\begin{aligned}\Sigma_{20(02)}^{(2)}(p+\mu) = {} & 2n_0 f_0^2 \int d\mathbf{q}\, [(A_\mathbf{q}, B_\mathbf{k}) \\ & - (A_\mathbf{q} + B_\mathbf{q};\, C_\mathbf{k}) + 3C_\mathbf{q} C_\mathbf{k}] \\ & \times \left(\frac{1}{p^0 - \varepsilon_\mathbf{q} - \varepsilon_\mathbf{k} + i\delta} - \frac{1}{p^0 + \varepsilon_\mathbf{q} + \varepsilon_\mathbf{k} - i\delta} \right) \\ & + \frac{1}{2} n_0 f_0^2 \int d\mathbf{q} \left(\frac{1}{\varepsilon_\mathbf{q}^0} - \frac{1}{\varepsilon_\mathbf{q}} \right),\end{aligned}$$

$$\begin{aligned}\Sigma_{11}^{(2)}(p+\mu) \\ = 2n_0 f_0^2 \int d\mathbf{q} \Big\{ & \frac{(A_\mathbf{q}, B_\mathbf{k}) + 2C_\mathbf{q} C_\mathbf{k} + A_\mathbf{q} A_\mathbf{k} - 2(A_\mathbf{q};\, C_\mathbf{k})}{p^0 - \varepsilon_\mathbf{q} - \varepsilon_\mathbf{k} + i\delta} \\ & - \frac{(A_\mathbf{q}, B_\mathbf{k}) + 2C_\mathbf{q} C_\mathbf{k} + B_\mathbf{q} B_\mathbf{k} - 2(B_\mathbf{q}, C_\mathbf{k})}{p^0 + \varepsilon_\mathbf{q} + \varepsilon_\mathbf{k} - i\delta} \\ & + \frac{1}{4} \left(\frac{1}{\varepsilon_\mathbf{q}} + \frac{1}{\varepsilon_\mathbf{k}} \right) \Big\} + 2f_0 \int d\mathbf{q}\, B_\mathbf{q} \\ & + 2n_0 \operatorname{Im} f_s\left(\frac{\mathbf{p}}{2} \frac{\mathbf{p}}{2} \right) \\ & - 2n_0 f_0^2 \int d\mathbf{q} \left[\frac{1}{\varepsilon_p^0 - \varepsilon_\mathbf{q}^0 - \varepsilon_\mathbf{k}^0 + i\delta} + \frac{1}{4\varepsilon_\mathbf{q}} + \frac{1}{4\varepsilon_\mathbf{k}} \right].\end{aligned} \tag{5.8}$$

The value of $\operatorname{Im} f_s(\mathbf{p}/2\ \mathbf{p}/2)$ can be obtained from Eq. (3.13), and the integrals not involving p^0 can be carried out exactly. In this way Eq. (5.8) becomes

$$\begin{aligned}\Sigma_{20(02)}^{(2)}(p+\mu) = {} & \frac{1}{2} n_0 f_0^2 \int \frac{d\mathbf{q}}{\varepsilon_\mathbf{q} \varepsilon_\mathbf{k}} R(qk) \Big[\frac{1}{p^0 - \varepsilon_\mathbf{q} - \varepsilon_\mathbf{k} + i\delta} - \\ & - \frac{1}{p^0 + \varepsilon_\mathbf{q} + \varepsilon_\mathbf{k} - i\delta} \Big] + \frac{1}{\pi^2} \sqrt{n_0 f_0^3}\, n_0 f_0,\end{aligned} \tag{5.9}$$

$$\begin{aligned}\Sigma_{11}^{(2)}(p+\mu) = {} & \frac{1}{2} n_0 f_0^2 \int \frac{d\mathbf{q}}{\varepsilon_\mathbf{q} \varepsilon_\mathbf{k}} \Big[\frac{Q^-(qk)}{p^0 - \varepsilon_\mathbf{q} - \varepsilon_\mathbf{k} + i\delta} - \\ & - \frac{Q^+(qk)}{p^0 + \varepsilon_\mathbf{q} + \varepsilon_\mathbf{k} - i\delta} + \varepsilon_\mathbf{q} + \varepsilon_\mathbf{k} \Big] + \frac{8}{3\pi^2} \sqrt{n_0 f_0^3}\, n_0 f_0,\end{aligned} \tag{5.10}$$

$$\mu^{(2)} = (5/3\pi^2) \sqrt{n_0 f_0^3} \cdot n_0 f_0, \tag{5.11}$$

with

$$\begin{aligned} R(qk) &= 2\varepsilon_\mathbf{q}^0 \varepsilon_\mathbf{k}^0 - 2\varepsilon_\mathbf{q} \varepsilon_\mathbf{k} + n_0^2 f_0^2, \\ Q^\mp(qk) &= 3\varepsilon_\mathbf{q}^0 \varepsilon_\mathbf{k}^0 - \varepsilon_\mathbf{q} \varepsilon_\mathbf{k} + n_0 f_0 (\varepsilon_\mathbf{q}^0 + \varepsilon_\mathbf{k}^0) + \\ & + n_0^2 f_0^2 \mp [n_0 f_0 (\varepsilon_\mathbf{q} + \varepsilon_\mathbf{k}) - \varepsilon_\mathbf{q} \varepsilon_\mathbf{k}^0 - \varepsilon_\mathbf{k} \varepsilon_\mathbf{q}^0]. \end{aligned} \tag{5.12}$$

It is convenient to express the Green's function (4.5) in a form analogous to Eq. (5.4). In this approximation we find

$$G(p+\mu) = \frac{A_p + \alpha_p}{p^0 - \varepsilon_p - \Lambda_p^-} - \frac{B_p + \alpha_p}{p^0 + \varepsilon_p + \Lambda_p^+}, \tag{5.13}$$

where α_p and $\Lambda_p^\mp$ are the second-order corrections

$$\alpha_p = \frac{n_0 f_0}{4\varepsilon_\mathbf{p}^3} \{ 2\varepsilon_\mathbf{p}^0 \Sigma_{20}^{(2)} - n_0 f_0 (\Sigma_{11}^+ + \Sigma_{11}^- - 2\mu - 2\Sigma_{20})^{(2)} \};$$

$$\begin{aligned}\Lambda_p^\mp = {} & \frac{\varepsilon_\mathbf{p}^0}{2\varepsilon_\mathbf{p}} (\Sigma_{11}^+ + \Sigma_{11}^- - 2\mu)^{(2)} + \\ & + \frac{n_0 f_0}{2\varepsilon_p} (\Sigma_{11}^+ + \Sigma_{11}^- - 2\mu - 2\Sigma_{20})^{(2)} \pm \frac{1}{2} (\Sigma_{11}^+ - \Sigma_{11}^-)^{(2)}\end{aligned} \tag{5.14}$$

These α_p and $\Lambda_p^\mp$ are combinations of the integrals (5.9) and (5.10). In the limits of small and large momentum (compared with $\sqrt{n_0 f_0}$), explicit expressions can be obtained for the functions $\alpha_p = \alpha(p^0; \mathbf{p})$ and $\Lambda_p^\mp = \Lambda^\mp(p^0; \mathbf{p})$. When these are examined it is found that there are no new poles of the Green's function. We here exhibit the behavior of the Green's function near to the poles $p_0 \approx \pm\epsilon_\mathbf{p}$. In this region we may write $|p^0| = \epsilon_\mathbf{p}$ in α_p, and we need retain only terms of first order in the difference $(\epsilon_\mathbf{p} \mp p^0)$ in $\Lambda_p^\mp$. For small momenta $(p \ll \sqrt{n_0 f_0})$ we then find

$$\alpha_p = \sqrt{n_0 f_0^3} \left(\frac{2}{3\pi^2} \frac{n_0 f_0}{\varepsilon_p} + i \frac{1}{64\pi} \frac{\varepsilon_p}{n_0 f_0} \right)$$

$$(p \ll \sqrt{n_0 f_0},\ \varepsilon_p \approx p\sqrt{n_0 f_0}),$$

$$\begin{aligned}\Lambda_p^\mp \equiv \Omega_p + \lambda_p (\varepsilon_p \mp p^0) = {} & \sqrt{n_0 f_0^3} \left(\frac{7}{6\pi^2} \varepsilon_\mathbf{p} - i \frac{3}{640\pi} \frac{\varepsilon_p^5}{n_0^4 f_0^4} \right) \\ & + (\varepsilon_\mathbf{p} \mp p^0) \sqrt{n_0 f_0^3} \left(\frac{1}{2\pi^2} + i \frac{1}{32\pi} \frac{\varepsilon_\mathbf{p}^2}{n_0^2 f_0^2} \right).\end{aligned} \tag{5.15}$$

For large momenta only the imaginary part of $\Lambda_p^\mp$ is important,

$$\Lambda_p^\mp = \Omega_p = -\frac{i}{4\pi} p f_0 n_0 f_0 \quad (p \gg \sqrt{n_0 f_0}). \tag{5.16}$$

For small momenta, in virtue of Eq. (5.13) and (5.15), the Green's function near to the poles may be written in the form

$$G(p+\mu) = (1 - \lambda_\mathbf{p}) \left[\frac{A_\mathbf{p} + \alpha_p}{p^0 - \varepsilon_\mathbf{p} - \Omega_p} - \frac{B_\mathbf{p} + \alpha_p}{p^0 + \varepsilon_p + \Omega_p} \right]. \tag{5.17}$$

For large momenta, α_p and λ_p may be neglected in Eq. (5.17).

6. QUASI-PARTICLE SPECTRUM AND GROUND-STATE ENERGY

We have already mentioned that the energy of a quasi particle is determined by the value of $p^0(\mathbf{p})$ at a pole of $G(p+\mu)$. Only those poles are to be considered for which the imaginary part of the energy is negative, so that the damping is positive. In the range $p \ll \sqrt{n_0 f_0}$, Eq. (5.15) and (5.17) give

$$\begin{aligned}\varepsilon = {} & p\sqrt{n_0 f_0} \left(1 + \frac{7}{6\pi^2} \sqrt{n_0 f_0^3} \right) \\ & - i \frac{3}{640\pi} \sqrt{n_0 f_0^3} \frac{p^5}{(n_0 f_0)^{3/2}} \quad (p \ll \sqrt{n_0 f_0}),\end{aligned} \tag{6.1}$$

In the high-momentum range, according to Eq. (5.16), we have

$$\varepsilon = \varepsilon_p^0 + n_0 f_0\left(1 - \frac{i}{4\pi} p f_0\right) \approx \varepsilon_p^0 + n_0 f(\mathrm{pp}) \quad (p \gg \sqrt{n_0 f_0}). \tag{6.2}$$

Equation (6.1) shows that for small p the quasi particles are phonons. The second approximation gives a correction to the sound velocity, and a damping proportional to p^5 which is connected with a process of decay of one phonon into two. In the high-momentum range, the second approximation gives a damping which is related to the imaginary part of the forward scattering amplitude, and so to the total cross section.

In Sec. (I, 7) we found connections between the Green's function and various physical properties of the system. The mean number of particles $\overline{N}_p$ with a given momentum p in the ground state of the system is related to the residue of the Green's function at its upper pole,

$$\overline{N}_p = i\int G\, dp^0/2\pi = (B_p + \alpha_p)(1 - \lambda_p). \tag{6.3}$$

When $p \ll \sqrt{n_0 f_0}$, Eqs. (5.15) and (5.5) give

$$\overline{N}_p = \frac{n_0 f_0}{2\varepsilon_p}\left(1 + \frac{5}{6\pi^2}\sqrt{n_0 f_0^3}\right) \tag{6.4}$$

The imaginary parts of α_p and λ_p here cancel, as they should. To find the total number of particles with $p \neq 0$, we need to know $\overline{N}_p$ for all momenta. We therefore use only the first approximation formula for $\overline{N}_p$, namely $\overline{N}_p = B_p$. For the density of particles with $p \neq 0$ we find

$$n - n_0 = i\int G(p+\mu)\, d^4p = \int B_p\, d\mathrm{p} = \sqrt{n_0 f_0^3}\, n_0/3\pi^2. \tag{6.5}$$

Equation (6.5) gives the relation between the density n_0 of particles in the condensed phase, which appeared as a parameter in all our equations, and the total number of particles in the system.

We note here one important point. It can be seen from the way the calculations were done that the validity of the "gaseous" approximation requires that n_0 be small. It is not directly required that the total density n be small, since n does not appear explicitly in the problem. But Eq. (6.5) shows that when n_0 is small n is necessarily small, too. This means that it is not possible to decrease significantly the density of the condensed phase by increasing the interaction or the total density, so long as $n_0 \ll f_0^{-3}$. This result confirms and strengthens the assertion made in I that the condensed phase does not disappear when interactions are introduced.

We can calculate the ground-state energy from the chemical potential μ. By Eq. (4.4) and (5.11),

$$\mu = n_0 f_0\left(1 + \frac{5}{3\pi^2}\sqrt{n_0 f_0^3}\right). \tag{6.6}$$

Expressing n_0 in terms of n by means of Eq. (6.5), we have in the same approximation

$$\mu = n f_0\left(1 + \frac{4}{3\pi^2}\sqrt{n f_0^3}\right). \tag{6.7}$$

By definition we have $\mu = \frac{\partial}{\partial n}\left(\frac{E_0}{V}\right)$. Therefore, integrating Eq. (6.7) with respect to n, we obtain the ground-state energy

$$\frac{E_0}{V} = \frac{1}{2} n^2 f_0\left(1 + \frac{16}{15\pi^2}\sqrt{n f_0^3}\right). \tag{6.8}$$

This coincides with the result of Lee and Yang[3] for the hard-sphere gas, if we remember that in that case $f_0 = 4\pi a$.

The condition for the system to be thermodynamically stable is $\partial P/\partial V = -\partial^2 E/\partial V^2 < 0$. This condition reduces to $f_0 > 0$. Our results are only meaningful when this condition is satisfied.

7. POSSIBILITY OF HIGHER APPROXIMATIONS

In the first two approximations, all the results can be expressed in terms of the amplitudes f. Thus the problem of many interacting particles is reducible to the problem of two particles.

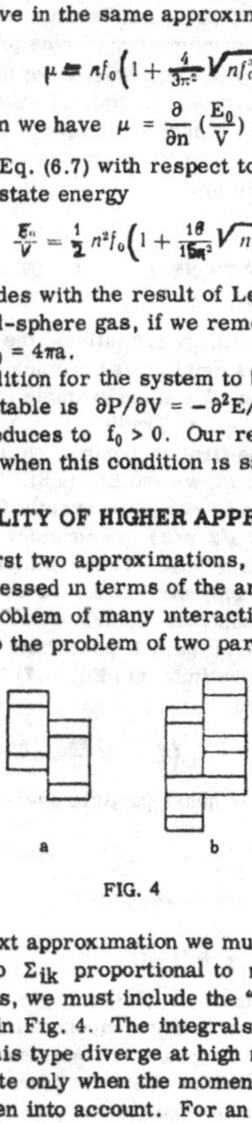

FIG. 4

In the next approximation we must consider contributions to Σ_{ik} proportional to $n_0 f_0^3$. Among other graphs, we must include the "triple ladders" illustrated in Fig. 4. The integrals arising from graphs of this type diverge at high momenta and become finite only when the momentum dependence of f is taken into account. For an estimate we may cut the integrals off at a momentum $p \sim f_0^{-1}$. We see then that an increase in the number of "rungs" does not change the order of magnitude of the integral. In fact, each rung adds a factor $f_0 G^2$ and an integration over one momentum 4-vector. For a rough estimate we take $q^0 \sim q^2$, $G \sim q^{-2}$, and find

$$\int f_0 G^2 d^4q \sim f_0 \int dq \sim 1.$$

Therefore we have to consider simultaneously all such graphs with any number of rungs. The totality of these triple ladders describes completely the interaction of three particles. Therefore the sum of contributions from such graphs can be expressed only by means of three-particle amplitudes.

In the third approximation (terms proportional to $n_0 f_0^3$) we thus require a solution of the three-particle problem (see also Ref. 4). Since the problem of three strongly interacting particles is in general insoluble, the higher approximations to the many-particle problem are physically meaningless.

8. HIGH EXCITATIONS ($pf_0 \sim 1$) IN A HARD-SPHERE GAS

For the high-energy excitations, the momentum dependence of the amplitudes becomes important. We therefore consider as an example the case of a gas of hard spheres of radius $(a/2)$. We also consider only the first approximation in the density expansion, i.e., we use Eq. (4.6). The amplitude $f(\mathrm{p}\,0)$ can be computed exactly from Eq. (3.10). For $f_s(\mathrm{p}/2\ \mathrm{p}/2)$ we consider only s-waves. The higher waves (the symmetrized amplitude involves only even values of l) add a numerically unimportant contribution. For example the d-waves at $pa \sim 1$ contribute about 10 per cent. We substitute into Eq. (4.7) the values of the amplitudes

$$f(\mathrm{p}0) = 4\pi \frac{\sin pa}{p}; \quad f_s\left(\frac{\mathrm{p}}{2}\,\frac{\mathrm{p}}{2}\right) = \frac{8\pi}{p} \sin\frac{pa}{2}\, e^{-ipa/2}, \qquad (8.1)$$

and obtain for the quasi-particle energy

$$\varepsilon = \left[\left(\frac{p^2}{2} + 8\pi n_0 \frac{\sin pa}{p} - 4\pi n_0 a\right)^2 - 16\pi^2 n_0^2 \frac{\sin^2 pa}{p^2}\right]^{1/2}. \qquad (8.2)$$

At high momenta this becomes

$$\varepsilon \approx \frac{p^2}{2} + 4\pi n_0 a \left(2\,\frac{\sin pa}{pa} - 1\right). \qquad (8.3)$$

The second term in Eq. (8.3) changes sign at $pa \approx 1.9$. An oscillating component is superimposed on the usual parabolic dependence. This oscillation will not be important since the magnitude of the term is small; when $pa \sim 1$ it is of relative order $n_0 a^3$. However, if one formally allows the parameter $n_0 a^3$ to become larger in Eqs. (8.3) or (8.2), the second term of Eq. (8.3) produces an increasing departure of the dispersion law from the parabolic form, until at sufficiently high densities there appears first a point of inflection and finally a maximum and a minimum in the curve. The spectrum then resembles qualitatively the spectrum postulated by L. D. Landau[5] to explain the properties of liquid helium II. This extrapolation is certainly unwarranted. But it allows one to suppose that the difference between liquid helium and a non-ideal Bose gas is only a quantitative one, and that no qualitatively new phenomena arise in the transition from gas to liquid.

9. CONCLUSION

We summarize the main features of the approximation which we have studied.

(1) The interaction between particles is specified not by a potential but by an exact scattering amplitude. This allows us to deal with strong interactions. After the potential has been replaced by the amplitude, it is possible to make a perturbation expansion in powers of the amplitude, or more precisely in powers of $\sqrt{n_0 f_0^3}$.

(2) We make a series expansion not of the quasi-particle energy (this appears as the denominator of the Green's function), but of the effective interaction potentials Σ_{ik} and the chemical potential μ. The formula giving the Green's function in terms of Σ_{ik} and μ is exact.

From Eq. (4.7) and (4.9) we see the ε_p can be expanded in powers of f only for high momentum excitations with $p \gg \sqrt{n_0 f_0}$. The low-lying excitations of the system are in principle impossible to obtain by perturbation theory. For this reason, the expression obtained by Huang and Yang[4] for the energy of the low excitations of a Bose hard-sphere gas is incorrect. They used perturbation theory with a "pseudopotential," and their result agrees with a formal expansion of Eq. (4.7) in powers of f_0.

In conclusion I wish to thank A. B. Migdal and especially V. M. Galitskii for fruitful discussions, and also L. D. Landau for criticism of the results.

[1] S. T. Beliaev, J. Exptl. Theoret. Phys. (U.S.S.R.) 34, 417 (1958); Soviet Phys. JETP 7, 289 (1958) (this issue).

[2] V. M. Galitskii and A. B. Migdal, J. Exptl. Theoret. Phys. (U.S.S.R.) **34**, 139 (1958); Soviet Phys. JETP **7**, 96 (1958).

[3] T. D. Lee and C. N. Yang, Phys. Rev. **105**, 1119 (1957).

[4] K. Huang and C. N. Yang, Phys. Rev. **105**, 767 (1957).

[5] E. M. Lifshitz, Usp. Fiz. Nauk **34**, 512 (1948).

Translated by F. J. Dyson
78

A "Fermi-Liquid" Description of the Kondo Problem at Low Temperatures

P. Nozières

Institut Laue–Langevin, Grenoble, France

(Received April 5, 1974)

We take as granted Anderson's statement that in the low-temperature limit the usual Kondo s–d *model evolves toward a fixed point in which the effective exchange coupling of the impurity with the conduction electrons is infinitely strong. The low-temperature properties* ($T \ll T_K$) *are then described phenomenologically in the same spirit as the usual Landau theory of Fermi liquids. The specific heat, spin susceptibility, and resistivity are expressed in terms of a small number of numerical parameters. In the strong coupling case the latter may be obtained via perturbation theory; in the opposite weak coupling limit they must be fitted to Wilson's recent numerical results.*

1. INTRODUCTION

We consider a single magnetic impurity (spin 1/2) in a metal, described by the well-known Kondo Hamiltonian

$$H = \sum_{\mathbf{k},\sigma} \varepsilon_k c^*_{\mathbf{k}\sigma} c_{\mathbf{k}\sigma} + \sum_{\substack{\mathbf{k}\sigma \\ \mathbf{k}'\sigma'}} (J/N)\mathbf{S} \cdot \mathbf{s}_{\sigma\sigma'} c^*_{\mathbf{k}\sigma} c_{\mathbf{k}'\sigma'} \tag{1}$$

In the case of antiferromagnetic coupling, $J > 0$, the existence of singular behavior near the Kondo temperature T_K is well known. What happens below T_K was established by Anderson *et al.*,* using a "scaling approach" which was nearest to being rigorous. The validity of Anderson's views was demonstrated recently by Wilson.[2] His method, inspired by renormalization group considerations, is based on a numerical study of the lowest energy levels of the many-body system, carried out on an appropriately simplified model. In this way he shows that when the energy scale goes to zero (or, equivalently, the size of the crystal goes to infinity) the distribution of eigenstates becomes similar to that which would prevail if J were

*For a recent review, see Anderson.[1]

going to $+\infty$. Thus, even if J was initially small, there is a smooth *crossover* from the weak coupling to the strong coupling case as T goes below T_K.

Such an approach contains two steps: (i) identification of the fixed point which one reaches when the energy is scaled down to zero, and (ii) description of the low-temperature properties, carried out by Wilson using a two-parameter pseudopotential fitted to his numerical results. This latter description is somewhat cumbersome, and not fully convincing: why two parameters only, why treat in first-order perturbation theory pseudopotentials which are clearly huge? The purpose of this note is to show that these difficulties may be bypassed entirely if one uses a phenomenological point of view similar to the Landau theory of Fermi liquids. Step (i) being taken as granted, one studies directly the degrees of freedom of the remaining low-temperature system. That can be described exactly in terms of a few numerical parameters that are obtained directly from Wilson's numerical result. It should be emphasized that such a "strong coupling" description, much as the weak coupling expansion which allows one to express T_K as a function of a small J, is a rather straightforward problem which could have been studied long ago. The really hard part is step (i), i.e., the numerical study of the transition region between the above two limits—in practice the numerical fitting of the high-T and low-T expansion parameters. That is where the real breakthrough lies, and why Wilson's work has such methodological importance. Here, we are much less ambitious, and we try only to simplify the low-T end of the story. As a bonus, we also calculate other physical properties, such as the conductivity, an easy task in our simple formulation. We discuss possible numerical relationships between these quantities, and display the physical processes of importance.

2. THE STRONG COUPLING LIMIT

Consider an exchange interaction localized on the impurity lattice site. For infinite, positive J, a spin-1/2 impurity traps a conduction electron, and is thereby locked into a rigid *singlet* state. Any attempt to break the singlet by transferring one electron in or out of the impurity site costs an infinite amount of energy. That site is thus out of the way. For the conduction electrons, it acts as a nonmagnetic, infinitely repulsive impurity.

If J is large, yet finite, real breakage of the singlet remains impossible at low temperature, since it requires a finite energy. On the other hand, virtual excitation of the singlet becomes possible, first into the $n = 0$ or $n = 2$ impurity cell occupation, eventually into the triplet $n = 1$ state: The impurity singlet has become polarizable. Such a polarizability gives rise to an *interaction* between the remaining conduction electrons: Electron 1 polarizes the singlet, and that affects a nearby electron 2. The situation is similar to the phonon-mediated attraction in ordinary metals: At temperatures well below

θ_D, real phonons are never excited, yet they provide an interaction mechanism between electrons.

An expansion in powers of $1/J$ is straightforward. We need only write the Hamiltonian (1) in real space, rather than with Bloch waves:

$$H = \sum_{ij} T_{ij} c_{i\sigma}^{*} c_{j\sigma} + J\mathbf{S} \cdot \mathbf{s}_{\sigma\sigma'} c_{0\sigma}^{*} c_{0\sigma'} \tag{2}$$

(i is an arbitrary lattice site, 0 the impurity site). We treat as a perturbation the hopping T_{0j} between the impurity and the rest of the crystal; we thus construct an effective Hamiltonian on the "singlet subspace." J occurs in the energy denominators, hence the $1/J$ expansion. The change in density of states, $\delta\nu$, occurs already to order 0 in J^{-1}; it is opposite in sign to the pure crystal density of states ν_0, since the impurity acts to remove one state out of the band.* The impurity-induced electron–electron interaction enters only in higher orders. It turns out that the first such term arises from excursions of the impurity cell into the triplet state (via the $n = 0$ or $n = 2$ states). In the simple case of nearest-neighbor hopping, the leading interaction is between electrons of *opposite spins* both lying next to the impurity. The interaction is found to be *repulsive*, of order D^4/J^3, where D is the bandwidth. If one pushes the calculation further, the interaction certainly extends beyond the first neighbors; it may involve electrons with parallel spins, possibly also more than two electrons.

In the weak or intermediate coupling case ($|J| \lesssim D$) the above expansion is of course useless. Yet, taking as granted Wilson's contention that below T_K the "effective coupling" goes to infinity, we may draw the following qualitative conclusions: (i) Well below T_K the impurity spin is frozen into a singlet, which will never be broken in real, energy-conserving transitions (i.e., the "spin-flip" scattering rate is assumed to be negligible). (ii) Because the singlet remains somewhat polarizable, an indirect interaction between conduction electrons appears. Put another way, we have replaced a magnetic impurity in a noninteracting electron gas by a nonmagnetic impurity together with a *localized interaction* (reminiscent of the local spin fluctuation model). The impurity site is in fact out of the way.

As we shall now see, such a qualitative statement, dictated by Wilson's results, is enough to describe the low-temperature properties in elementary fashion.

3. "FERMI LIQUID" EXPANSION OF PHASE SHIFTS

Since the impurity singlet is frozen, the only degrees of freedom that remain are those of a gas with $N - 1$ electrons in a lattice derived from the impurity site. In the spirit of the Landau theory of Fermi liquids, we assume

*$\delta\nu(J = \infty)$ is easily calculated for any specified band shape.

a one-to-one correspondence between the corresponding eigenstates and those found when $J \to \infty$.* The state of the system is described by the distribution function n_α for the *quasiparticle* scattering states, and all physical properties are functionals of n_α. In view of the singlet induced interaction, these functionals are nonlinear.

Since the impurity is local, each quasiparticle state α is characterized by phase shifts—here only one, δ_α, since we limit ourselves to s-wave scattering.† Usually, δ_α depends only on the energy ε_α of the state. Here, because the particles are interacting, we must also allow for a dependence of δ_α on the distribution n_β. Thus the phase shift for state α has the general form

$$\delta_\alpha[\varepsilon_\alpha, n_\beta] \tag{3}$$

Note that (3) is an *exact* description as long as we ignore the explicit fluctuations of the central singlet.

The next step is of course to expand (3) in powers of ε (measured from the chemical potential $\mu = 0$), and of the departure from the ground state

$$\delta n_\beta = n_\beta - n_{\beta 0}$$

$\delta n_\beta = \delta n_\sigma(\varepsilon_\beta)$ depends only on the spin and energy of the quasiparticle. We may thus expand (3) as

$$\begin{aligned} \delta_\sigma(\varepsilon) = \delta_0(\varepsilon) &+ \sum_{\varepsilon'\sigma'} \phi_{\sigma\sigma'}(\varepsilon, \varepsilon')\, \delta n_{\sigma'}(\varepsilon') \\ &+ \sum_{\substack{\varepsilon'\sigma' \\ \varepsilon''\sigma''}} \chi_{\sigma\sigma'\sigma''}(\varepsilon, \varepsilon', \varepsilon'')\, \delta n_{\sigma'}(\varepsilon')\, \delta n_{\sigma''}(\varepsilon'') + \cdots \end{aligned} \tag{4}$$

If we further assume that for strong coupling everything is analytic near the Fermi surface, we also write

$$\delta_0(\varepsilon) = \delta_0 + \alpha\varepsilon + \beta\varepsilon^2 + \cdots, \qquad \phi_{\sigma\sigma'}(\varepsilon, \varepsilon') = \phi_{\sigma\sigma'} + \psi_{\sigma\sigma'}(\varepsilon + \varepsilon') + \cdots \tag{5}$$

The quantities δ_0, α, $\phi_{\sigma\sigma'}, \ldots$, are now *numbers* that describe phenomenologically the low-temperature behavior.

At low T and H we need only δ up to first order in ε, T, H. Since δn is linear in T and H, only four quantities enter: δ_0, α, $\phi_{\sigma,\pm\sigma} = \phi^s \pm \phi^a$. The total number of electrons is constant and ϕ^s never shows up. If we set

$$n_\uparrow - n_\downarrow = m$$

*We thereby shrink the space of relevant states, a procedure which is familiar in many-body theory: It rests on the idea that the relevant frequencies, $\sim kT$, are much smaller than the excitation frequencies of the triplet, etc. The detailed dynamics of these rapid fluctuations is irrelevant, and accounts only for a renormalization of the reduced manifold of states. If T becomes $\gtrsim T_K$, we can no longer ignore the dynamics of triplet, ..., fluctuations, and a description of the full Hilbert space becomes necessary.

†The extension to other l values is straightforward, but of little interest as the number of parameters increases rapidly.

we obtain a "molecular field" expression of the phase shift

$$\delta_\sigma(\varepsilon) = \delta_0 + \alpha\varepsilon + \sigma\phi^a m \tag{6}$$

The new feature as compared to standard theories is the "molecular field" coefficient ϕ^a in (6).

4. SPECIFIC HEAT AND SPIN SUSCEPTIBILITY

From δ we find the new electron energy

$$\tilde{\varepsilon}_\sigma = \varepsilon - [\delta_\sigma(\varepsilon)/\pi\nu_0] \tag{7}$$

(ν_0 is the density of states of a pure system for *one* spin direction). Hence a change in that density of states at $H = 0$

$$\delta\nu = \nu_0[(d\varepsilon/d\tilde{\varepsilon}) - 1] = \alpha/\pi$$

leading to a specific heat correction

$$\delta C_v/C_v = \alpha/\pi\nu_0 \tag{8}$$

α/ν_0 is thus directly accessible from experiment.

At finite H and zero T the electron energies for up and down spins are

$$\tilde{\varepsilon}_\uparrow = \varepsilon - g\beta H - \frac{\alpha\varepsilon}{\pi\nu_0} - \frac{\phi^a}{\pi\nu_0}m, \qquad \tilde{\varepsilon}_\downarrow = \varepsilon + g\beta H - \frac{\alpha\varepsilon}{\pi\nu_0} + \frac{\phi^a}{\pi\nu_0}m$$

In equilibrium the Fermi levels for each spin direction are fixed by the conditions

$$\tilde{\varepsilon}_\uparrow = \tilde{\varepsilon}_\downarrow = \mu = 0$$

The magnetization, $g\beta m$, and the Pauli spin susceptibility follow at once:

$$\chi = \frac{g\beta m}{H} = 2\nu_0(g\beta)^2\left[1 + \frac{\alpha}{\pi\nu_0} + \frac{2\phi^a}{\pi}\right] \tag{9}$$

(remember that we are dealing here with *one* impurity effects, of order $1/N$). ϕ^a describes the "exchange enhancement" of χ due to interactions between electrons (the latter arising from fluctuations of the singlet). From (8) and (9) we infer that

$$(\delta\chi/\chi)/(\delta C_v/C_v) = 1 + (2\nu_0\phi^a/\alpha) \tag{10}$$

Many-body effects are contained in the dimensionless factor $2\nu_0\phi^a/\alpha$.

5. ORDER OF MAGNITUDE

In the strong coupling case ($J \gg D$), α is *negative*, of order $1/D$. The interaction energy is limited to opposite spin electrons on the first neighbors of the impurity, and is *repulsive*. From (4) and (7), we infer that

$$\phi^a + \phi^s \approx 0, \qquad 0 < 2\phi^a/\pi\nu_0 \sim D^4/N^2J^3$$

The enhancement factor is thus coupling dependent:

$$2\nu_0\phi^a/\alpha \sim -D^3/J^3 \tag{11}$$

In the opposite, weak coupling case below T_K, we must resort to Wilson's numerical result. First of all, it is obvious that the low-temperature thermodynamic quantities depend only on *two* parameters: α and ϕ^a. These can be mimicked by a pseudopotential, as Wilson does—but this step is unnecessary, and α and ϕ^a can be obtained *directly* from his numerical evaluation of the lowest eigenstate energies [using (4)]. The treatment to first order of a large pseudopotential is thus bypassed.

In practice, α depends only on the Kondo temperature,

$$\alpha = A/T_K \qquad (0 < A \sim 1) \tag{12}$$

We note the change of sign of α as compared to the other limit $J \to \infty$ (what was the impurity spin degree of freedom seems to *increase* the density of states). The value of A depends on how one defines T_K. If T_K is fitted to a small-J expansion above the crossover, the value of A depends on a full numerical treatment of the crossover region—this is just what Wilson does. Here we would rather use (12) as a definition of T_K, setting, for instance, $A = 1$.

In much the same way, we have

$$0 < \phi^a \sim 1/\nu_0 T_K$$

The ratios $2\nu_0\phi^a/\alpha$ and $2\nu_0\phi^s/\alpha$ are universal dimensionless constants, independent of T_K (i.e., of J). This is just a statement of the *universality of scaling*: If all energies are measured in units T_K, the behavior of the system in the critical region, $T \ll T_K$, is completely determined. It is clear that such a universal behavior holds only if $T_K \ll D$ (i.e., $J \ll 1$), a point which is clearly stressed by Wilson [if $J \gtrsim 1$, $2\nu_0\phi^a/\alpha$ becomes J dependent; see (11)]. In the universal region Wilson finds

$$2\nu_0\phi^a/\alpha \approx 1 \tag{13}$$

(i.e., $\Delta\chi$ is twice what a naive, noninteracting electron picture would predict). Wilson gives no estimate of $2\phi^s\nu_0/\alpha$.

Actually, these dimensionless ratios may be reached directly by means of simple arguments. First of all, it is clear that when $T_K \ll D$, the Kondo singularity *is tied to the Fermi level*: If both ε and the chemical potential μ are shifted by the same amount, the phase shift $\delta|\varepsilon, n_0(\mu)|$ is unchanged. In view of (4), that implies

$$\alpha + 2\nu_0\phi^s = 0 \tag{14}$$

Equation (14) is certainly *exact* in Wilson's case $J \ll 1$; it would break down for strong coupling, since the Kondo singularity then extends as far as the band edges.

One might also claim that, just as in the case $J \sim \infty$, the interaction couples only electrons with antiparallel spins, i.e.,

$$\phi^s + \phi^a = 0 \tag{15}$$

It is remarkable that Wilson's numerical result (13) follows at once from (14) and (15)—a fact which is maybe not surprising, since indeed Wilson's pseudopotential includes only antiparallel spin interactions. As it stands, there is no compelling reason to exclude parallel spin coupling when $J \ll 1$: Wilson's choice of a pseudopotential might therefore be too restrictive, and the agreement with numerical results somewhat accidental.

In any case, it is clear that once T_K is fixed (i.e., α), there is no adjustable parameter in the theory.

6. TRANSPORT PROPERTIES AT $T = 0$

There are two properties, C_v and χ, for two coefficients, α and ϕ^a; a third physical quantity would be helpful in providing a test of the theory—e.g., the conductivity $\sigma(T, H)$.

In the present case, where only *s*-wave scattering is included, every collision completely restores angular equilibrium. The collision integral $I(n)$ which enters the Boltzmann equation takes the simple form

$$I(n) = -n_{1\sigma}(\varepsilon)W_\sigma(\varepsilon) \tag{16}$$

where $n_1 = n - n_0(T)$ is the current carrying departure from thermal equilibrium, $W_\sigma(\varepsilon)$ being the *total relaxation rate* of $n_\sigma(\varepsilon)$, whether due to elastic or inelastic scattering. In terms of $W_\sigma(\varepsilon)$, the conductivity takes the usual form

$$\sigma = -\frac{\nu_0 e^2 v_F^2}{3} \sum_\sigma \int d\varepsilon \frac{df_0/d\varepsilon}{W_\sigma(\varepsilon)} \tag{17}$$

v_F is the Fermi velocity and f_0 is the thermal Fermi factor. Note that here σ is the spin of the *carrier*: we thus add *conductivities* for up and down spins.*

At zero temperature, only elastic scattering is possible. The corresponding relaxation rate is related to the phase shift according to

$$W_\sigma(\varepsilon) = (2n_i/\pi\nu_0)\sin^2\delta_\sigma(\varepsilon) \tag{18}$$

where n_i is the impurity density. In the absence of a magnetic field, $\delta_\sigma(\mu) = \delta_0$, and the conductivity is

$$\sigma_0 = \tfrac{1}{3}\pi\nu_0^2 v_F^2 e^2/(n_i \sin^2\delta_0)$$

If $\delta_0 = \pi/2$, σ_0 reaches the so-called *unitarity limit*.

For $H \neq 0$, $T = 0$, we introduce the phase shifts $\delta_\uparrow$ and $\delta_\downarrow$ at the Fermi level: The net magnetization is

$$m = m_0 + [(\delta_\uparrow - \delta_\downarrow)/\pi] = m_0 + \delta m \tag{19}$$

where m_0 is the magnetization of the pure matrix, δm the correction due to the impurity, given by the Friedel sum rule [which follows from (7)]. The conductivity is given by

$$\frac{\sigma}{\sigma_0} = \left(\frac{1}{\sin^2\delta_\uparrow} + \frac{1}{\sin^2\delta_\downarrow}\right)\frac{\sin^2\delta_0}{2} \tag{20}$$

If we find a way to determine $(\delta_\uparrow + \delta_\downarrow)$—for instance, the Friedel perfect screening sum rule—(19) and (20) provide an exact relationship between susceptibility and magnetoresistance $\sigma(0, H)$. Such a point of view is essentially that of Souletie *et al.*[3,4]

In the absence of such extra conditions, $\delta_\uparrow$ and $\delta_\downarrow$ must be found from (4). In order to obtain the leading H^2 correction to σ_0, we need δ_σ to that order; since the thermodynamic coefficients C_v and χ depend only on the *linear* terms of δ_σ, they are *not* systematically related to the magnetoresistance (the latter involving further "Landau" coefficients). This disappointing result holds unless $\delta_0 = \pi/2$, in which case the H^2 terms of $\sin^2\delta$ depend only on the *linear* correction to δ. Thus, *if we are close to the unitarity limit*, the magnetoresistance may be obtained from the susceptibility. Indeed, on

*A somewhat similar analysis was carried out by Souletie[3] and Genicon *et al.*[4] In these papers the phase shifts are supposed to be functions of T and H, and the Friedel sum rules are used to correlate the spin susceptibility and resistivity. However, the impurity spin is not assumed to be locked, and as a result, δ depends on the *relative* spins of the carrier and the scatterer. For a given carrier spin σ one must add the *resistivities* due to scatterers of the two spins, in contrast to what we are doing. The difference stems from our assumptions on the physical nature of the ground state: We assume the impurity spin to be locked in a unique state, while for Souletie it still can have two orientations at $T \ll T_K$.

linearizing (20), we find that

$$\sigma/\sigma_0 = 1 + (\pi\,\delta m/2)^2 \tag{21}$$

where $\delta m = H\,\delta\chi$ is the magnetization due to the impurity.

7. TRANSPORT PROPERTIES AT $T \neq 0$

We now consider the T^2 dependence of the conductivity in zero field, $\sigma(T, 0)$. In the general expression (17) for σ, the Fermi factor broadens over a range of width T. If we consider only *elastic* scattering, we may use (18): We then need corrections to $\sin^2\delta$ of order ε^2. The situation is similar to the one discussed above. If $\delta_0 \neq \pi/2$, corrections to σ involve higher coefficients of the expansions (4) and (5), and there is no simple relation between C_v and $\sigma(T)$. If, on the other hand, $\delta_0 = \pi/2$, we have

$$\sin^2\delta(\varepsilon) = 1 - \alpha^2\varepsilon^2 + O(\varepsilon^4)$$

and the T^2 part of the conductivity follows at once:

$$\sigma(T)/\sigma_0 = 1 + \tfrac{1}{3}\pi^2\alpha^2T^2 \tag{22}$$

From now on we assume that the unitarity limit is indeed reached at low temperatures. This is true for strong coupling if one has electron–hole symmetry; for weak coupling such a contention is supported by previous approximate methods;* it should be easy to check it on Wilson's numerical results, since δ_0 acts to shift all the eigenstates by a constant energy. Within that assumption, (22) relates $d\sigma/dT^2$ to δC_v.

Actually, (22) is wrong, since it ignores inelastic scattering. We discarded from the outset real processes that would break the singlet.† One may claim that the singlet binding energy is $\sim T_K$, in which case the probability to break it will decrease exponentially as $T \to 0$. Here we take this point of view as granted, and we ignore spin-flip collisions. Inelastic scattering arises only as a consequence of electron–electron interactions: One electron scatters on the impurity, and at the same time excites one or several electron–hole pairs. At low ε and T the exclusion principle, together with energy conservation, severely limits the available phase space. To order T^2, the only important process is that in which an electron $(\varepsilon\sigma)$ scatters into the state $(\varepsilon_1\sigma)$ while emitting another electron $(\varepsilon_2\sigma')$ and a hole $(\varepsilon_3\sigma')$.

*See, for instance, Suhl.[5]

†Such spin-flip scattering is not negligible in the dispersion theory formulation first proposed by Suhl:[5] The corresponding transition probability is of order T^2, and it is comparable to the effect we are looking for. It should be recognized, however, that Suhl's theory does not display scaling, and probably does not describe singlet locking; as such, it does not disprove our formulation.

The contribution of this new channel to the collision integral takes the form

$$I(n) = 2\pi n_i \nu_0^3 \int d\varepsilon_1\, d\varepsilon_2\, d\varepsilon_3$$

$$[|A_{\uparrow\downarrow}|^2 + \tfrac{1}{2}|A_{\uparrow\uparrow}|^2]\, \delta(\varepsilon + \varepsilon_3 - \varepsilon_1 - \varepsilon_2)$$
$$\times [(1 - n)(1 - n_{30})n_{10}n_{20} - nn_{30}(1 - n_{10})(1 - n_{20})] \tag{23}$$

where $A_{\sigma\sigma'}$ is the full scattering amplitude (to which we return later) and n_0 is the Fermi factor. The 1/2 in the brackets avoids double counting of final states with two parallel spin electrons. To order ε^2, T^2, we may neglect the energy dependence of $A_{\sigma\sigma'}(\varepsilon, \varepsilon_1, \varepsilon_2, \varepsilon_3)$: The integrations in (23) are then straightforward, and they yield a relaxation rate

$$W^{\text{in}}(\varepsilon) = \pi\nu_0^3 n_i(\pi^2 T^2 + \varepsilon^2)[|A_{\uparrow\downarrow}|^2 + \tfrac{1}{2}|A_{\uparrow\uparrow}|^2] \tag{24}$$

Equation (24) must be added to the former elastic relaxation rate.

At first sight, one is tempted to simply add (18) and (24): The total rate W would then be subject to two conflicting temperature corrections: the decrease of $\sin^2\delta$ and the added term W^{in}. The latter may predominate, and W would then exceed the unitarity limit. The mistake stems from the fact that the elastic scattering rate is no longer given by (18) when other channels are present, since the latter absorb part of the incoming beam. In order to find the correct answer, we must reconsider the concepts of unitarity and phase shift.

Consider a given one-particle excited state α (not a statistical ensemble). The S- and t-matrices are related in the usual way:

$$S_{\alpha\beta} = \delta_{\alpha\beta} - 2i\pi T_{\alpha\beta}\delta(\varepsilon_\alpha - \varepsilon_\beta)$$
$$= \delta_{\alpha\beta}[1 - 2i\pi\nu_0 T_{\alpha\alpha}] - 2i\pi T^{\text{in}}_{\alpha\beta}\delta(\varepsilon_\alpha - \varepsilon_\beta)$$

[remember that we consider s-scattering only: in the elastic channel we have $\delta(\varepsilon_\alpha - \varepsilon_\beta) = \nu_0\delta_{\alpha\beta}$]. $T^{\text{in}}_{\alpha\beta}$ is the inelastic t-matrix, yielding an inelastic *transition probability* U^{in}_α from state α:

$$U^{\text{in}}_\alpha = \sum_\beta 2\pi|T^{\text{in}}_{\alpha\beta}|^2\delta(\varepsilon_\alpha - \varepsilon_\beta)$$

Expressing the fact that S is unitary, we find that

$$1 - 2i\pi\nu_0 T_{\alpha\alpha} = (1 - 2\pi\nu_0 U^{\text{in}}_\alpha)^{1/2}e^{-2i\delta} \tag{25}$$

where δ is a real phase, indeed the correct definition of the elastic phase shift in the presence of inelastic scattering.* The *total* transition probability is

*The square root in (25) may be absorbed into the phase shift δ by assigning to the latter a small imaginary part δ'. We shall see later that $\delta' \sim T^2$. It is thus negligible as compared to the linear terms which control the thermodynamic properties, C_v and χ.

given by

$$U_\alpha = 2\,\mathrm{Im}\,T_{\alpha\alpha} = 2[(\sin^2\delta)/\pi\nu_0] + U_\alpha^{\mathrm{in}}\cos 2\delta \tag{26}$$

[we expanded the square root in (25)]. We note that U_α^{in} is multiplied by $\cos 2\delta$, i.e., by -1 near the unitarity limit; the total U_α is always *below* the unitarity limit, as it should be.*

A relation such as (26) holds for each state α, and thus also for the thermally averaged relaxation rates. The total rate is therefore equal to

$$W = 2[n_i(\sin^2\delta)/\pi\nu_0] + W^{\mathrm{in}}\cos 2\delta \tag{27}$$

where W^{in} is given by (24). We insert (27) into (16), and we again assume $\delta_0 = \pi/2$. In lowest order, we find

$$\sigma(T)/\sigma_0 = 1 + \tfrac{1}{3}\pi^2\alpha^2T^2 + \tfrac{2}{3}\pi^4\nu_0^4T^2[|A_{\uparrow\downarrow}|^2 + \tfrac{1}{2}|A_{\uparrow\uparrow}|^2] \tag{28}$$

In principle, the result (28) is exact to order T^2.

To conclude this analysis, we need an estimate of the scattering amplitude $A_{\sigma\sigma'}$. In the usual theory of Fermi liquids it is known that the *forward* scattering amplitude is related to the quasiparticle interaction energy $f_{kk'}$—they correspond to different limits of the same vertex operator. In the present case where all quantities are small, $\sim 1/N$, all these limits would be equal: One need not worry about such complications. Moreover, (i) there is no angular dependence, (ii) we ignore the energy dependence of $A_{\sigma\sigma'}$; the latter is simply equal to its forward value. It is thus tempting to transpose to our impurity problem the results of the Landau theory of homogeneous liquids, and to speculate that

$$A_{\sigma\sigma'} = -\phi_{\sigma\sigma'}/\pi\nu_0$$

Admittedly, this is not a proof—yet, it looks like a sensible guess. If we rely on it, (28) takes the explicit form

$$\sigma(T)/\sigma(0) = 1 + \tfrac{1}{3}\pi^2T^2[\alpha^2 + \nu_0^2(3\phi^{s2} + 3\phi^{a2} - 2\phi^s\phi^a)] \tag{29}$$

The temperature-dependent conductivity is expressed entirely in terms of thermodynamic quantities. If we use Wilson's result

$$\alpha = -2\nu_0\phi^s = 2\nu_0\phi^a$$

we obtain the simple relation

$$\sigma(T)/\sigma_0 = 1 + \pi^2T^2\alpha^2 \tag{30}$$

As expected, the correction to σ is of order T^2/T_K^2. If true, it should be possible to check (30) readily by experiment.

*If we write (26) in the form $U_\alpha = U_\alpha^{\mathrm{in}} + [2(\sin^2\delta)/\pi\nu_0][(1 - \pi\nu_0U_\alpha^{\mathrm{in}}])$, we clearly see the reduction of the elastic transition probability due to the other channels.

8. CONCLUSION

Even in the restricted framework of the Kondo model, this analysis remains largely conjectural. A number of questions arise:

1. To what extent can we apply the Landau phenomenological approach to an impurity problem, i.e., does the phase shift depend only on the *distribution* of quasiparticles, rather than on the complete density matrix? Is it fair to interpret $\phi_{\sigma\sigma'}/\pi v_0$ as an interaction energy? If this is acceptable, the thermodynamic properties follow.
2. Is our estimate of inelastic scattering amplitudes reliable? Such an estimate is a prerequisite to any comparison between $d\sigma/dT^2$ and C_v.
3. Can one really ignore singlet breaking scattering below T_K?
4. Does one always reach the unitarity limit at $T = H = 0$?

Within these uncertainties, such a "Fermi liquid" analysis of phase shifts is remarkably simple. As a matter of fact, it is not restricted to the Kondo problem and the *s–d* model. It should apply to any *impurity* problem at temperatures for which *the internal degree of freedom* is locked (i.e., fluctuating rapidly enough that it induces only electron renormalization *and interactions*). Such a concept should, for instance, be useful in the Anderson model.

ACKNOWLEDGMENTS

The author expresses his gratitude to J. Souletie and G. Toulouse for many enlightening discussions on this approach and on its relationship to other formulations of the Kondo problem.

REFERENCES

1. P. W. Anderson, *Comments in Solid State Phys.* **5**, 73 (1973).
2. K. G. Wilson, *Collective Properties of Physical Systems*, Nobel Symposium 24 (Academic Press, 1974).
3. J. Souletie, *J. Low Temp. Phys.* **7**, 141 (1972).
4. J. L. Genicon, F. Lapierre, and J. Souletie, to be published.
5. H. Suhl, *Physics* **2**, 39 (1965); **3**, 17 (1967).

CONDUCTIVITY FROM CHARGE OR SPIN DENSITY WAVES

P.A. Lee, T.M. Rice and P.W. Anderson*

Bell Laboratories Murray Hill, New Jersey 07974

(*Received* 24 *September* 1973 *by A.G. Chynoweth*)

Below a Peierls transition the coupled electron phonon collective mode plays an important role in the conductivity of one-dimensional metal models such as have been recently postulated for various organic compounds. Within the jellium model, or in an incommensurate situation, the mode frequency goes to zero for $q \to 0$ and is responsible for the infinite conductivity first proposed by Fröhlich. Impurities, lattice commensurability and three dimensional ordering introduce a gap into the mode spectrum. The low frequency conductivity and a large dielectric constant are predicted. Similar effects are predicted for a spin density wave.

RECENTLY there has been a lot of interest in one dimensional systems which undergo a Peierls distortion from a metallic to a semiconducting state.[1,2] In an earlier paper[3] we discussed the effects of fluctuations on the Peierls transition and found that while fluctuations are important over a wide temperature range, very long range correlation develops below about a quarter of the mean-field T_c. In this work we study electrical conductivity and dielectric response in the low temperature phase with particular emphasis on the role played by the collective modes.

As a model of the Peierls transition Fröhlich[4] considered the coupling of noninteracting electrons to phonons in a jellium model. We shall later examine the effects of the Coulomb interaction and the periodicity of the lattice, but we shall begin with Fröhlich's Hamiltonian.

$$\mathcal{H} = \sum_{k\sigma} \epsilon_k c^{+}_{k\sigma} c_{k\sigma} + \sum_q \omega_q (b^{+}_q b_q + b^{+}_{-q} b_{-q}) + \sum_q \left[\frac{ig}{\sqrt{N}} \sum_{k\sigma} c^{+}_{k+q\sigma} c_{k\sigma} (b_q + b^{+}_{-q}) + c.c. \right] \tag{1}$$

where $c^{+}_{k\sigma}$ and b^{+}_q are creation operators for a 1 – dim. Bloch electron and a longitudinal phonon q with energies ϵ_k and ω_q respectively and N is the number of atoms in the chain. In the mean field theory one singles out the interaction with the phonon at $Q = 2k_F$ and frequency ω_Q and describes the distortion of the lattice by the order parameter

$$\Delta = 2g \langle b_Q \rangle / \sqrt{N}. \tag{2}$$

Fröhlich showed that the distorted phase can carry a current by propagating the combined lattice and electronic charge distortion as a travelling wave. Since an energy gap 2Δ has been introduced in the electronic spectrum, Fröhlich argued that this wave will travel unattenuated and the conductivity will be infinite. The result can also be understood as a consequence of translational invariance of the charge density wave (CDW) relative to the laboratory frame in a jellium model. This invariance continues to hold if the lattice periodicity is incommensurate with the CDW periodicity Q. We shall show that the collective mode associated with the broken symmetry plays an important role in the conductivity.

To study the collective mode in the distorted phase it is convenient to introduce the phonon propagators in matrix form

$$D_{mn}(q, t) = -\frac{i}{\omega_Q} \left\langle T\left(b_{mQ+q}(t) b^{+}_{nQ+q}(0)\right) \right\rangle \tag{3}$$

where $m, n = \pm$. Similarly we introduce anomalous

---

* Also at Cavendish Laboratory, Cambridge, England.

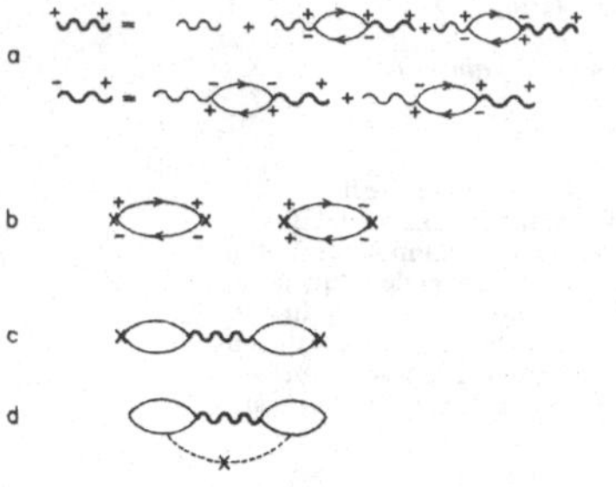

FIG. 1. (a) Dyson's equation for the collective mode. The modes D_{++} and D_{-+} are represented by heavy wavy lines with appropriate signs on each end and the bare phonon propagator D_0 is represented by a light wavy line. The electronic Green's functions G_{+-}, G_{++} are represented by solid lines. (b) Single particle contribution to the conductivity. (c) Collective mode contribution to the conductivity. (d) Self energy of the collective mode due to impurity scattering denoted by dashed lines.

electronic Green's function in analogy with the theory of superconductivity and of the excitonic insulator

$$G_{mn}(k,t) = i\left\langle T\left(c_{k+mQ/2}(t)c^{+}_{k+nQ/2}(0)\right)\right\rangle \quad (4)$$

where $m, n = \pm$ and $|k| < Q/2$. In equation (4) we have artificially separated the single band into + and − parts for positive and negative k and the formalism becomes identical to that of the excitonic insulator which couples an electron and a hole band.[5,6] The thermal Green's function is given by

$$G^{-1}(k,\omega_n) = \begin{bmatrix} i\omega_n - \epsilon_{k+Q/2} & -\Delta \\ -\Delta & i\omega_n - \epsilon_{k-Q/2} \end{bmatrix}. \quad (5)$$

We can write down the Dyson equation as shown in Fig. 1(a) for D, and it is easy to show that

$$D_{++}(q,\omega_n) \pm D_{+-}(q,\omega_n) =$$
$$D_0/[1 + 2g^2\omega_Q D_0 \sum_{k,\nu} G_{++}(k,\nu)G_{--}(k+q,\nu+\omega_n)$$
$$\pm G_{+-}(k,\nu)G_{-+}(k+q,\nu+\omega_n)] \quad (6)$$

where $D_0 = (\omega_n^2 + \omega_Q^2)^{-1}$. From equation (5) it follows that $G_{+-} = \Delta \det G$ from which we conclude that the denominator for the $D_{++} - D_{+-}$ mode becomes identical to the usual gap equation and the frequency of the mode goes to zero for $q = 0$. Let us define

$$A_{\pm}(q,\omega) = \tfrac{1}{2}(D_{++}(q,\omega) \pm D_{+-}(q,\omega))$$
$$= \frac{i}{2\omega_Q}\int dt\, e^{i\omega t}\left\langle T\left(b_{Q+q}(t) \pm b_{-Q+q}(t)\right)\right.$$
$$\left.\left(b^{+}_{Q+q}(0) \pm b^{+}_{-Q+q}(0)\right)\right\rangle \quad (7)$$

Explicit calculation[7] shows that for small q,

$$A_{\pm}(q,\omega) = (1 - m/m^*)/(\omega_{\pm}^2 - \omega^2) \quad (8)$$

where $\omega_+^2 = \lambda\omega_Q^2 + \frac{4}{3}(m/m^*)v_F^2 q^2$ and $\omega_-^2 = (m/m^*)v_F^2|q|^2$. The ratio of the effective mass m^* to the band mass m is

$$m^*/m = 1 + 4\Delta^2/\lambda\omega_Q^2. \quad (9)$$

The dimensionless electron phonon coupling constant $\lambda = \nu g^2/\omega_Q\epsilon_F$ where ν is the number of conduction electrons per atom. The dispersion of both modes contain the effective mass ratio m^*/m because of the slow response time of the phonon system compared with the electronic system. A similar expression for the effective mass has been derived by Fröhlich.[4] It is instructive to examine the modes in terms of small oscillations of the phonon field about the mean field $b_0 = \Delta/2g$. Writing $b_{\pm Q} = (b_0 + \delta b)e^{\pm i\Phi}$ we see that $b_Q - b_{-Q} \approx b_0 2i\Phi$ and $b_Q + b_{-Q} \approx 2b_0 + 2\delta b$ to lowest order in δb and Φ. The two modes can then be thought of as phase and amplitude fluctuation of the order parameter. From the definition of the displacement operator in terms of b it is clear that the phase Φ denotes the position of the CDW relative to the laboratory frame. The A_- mode is then the sliding mode discussed earlier and is expected to be current carrying. Furthermore an oscillation in Φ produces a dipole between the CDW and the background positive charge and the A_- mode is expected to be optically active.

To verify these assertions we have calculated the conductivity taking into account the collective modes. This is most conveniently done diagrammatically, following the formalism developed for the excitonic insulator problem.[5,6] In addition to the single particle contribution [Fig. 1(b)] it is important to include the collective contribution [Fig. 1(c)]. In the excitonic insulator problem Fig. 1(c) does not contribute because the average velocity of the electron and hole band separately vanishes. In the present problem we have split the energy band into + and − sides, each of which has a fixed velocity $\pm v_F$, and Fig. 1(c) is nonzero. Alternatively the excitonic insulator is formed from an electron and a hole band and has net charge zero. A translation of a neutral structure does not carry a current. The Fröhlich state, on the other hand, is current carrying because it is a charged structure formed from a single band.

The contribution of the single particle term [Fig. 1(b)] together with the diamagnetic term at zero temperature is

$$\sigma_0 = (ne^2/i\omega m)\left(f(\omega) - f(0)\right) \qquad (10)$$

where

$$f(\omega) = -\int d\xi \frac{2\Delta^2/E}{(\omega + i\eta)^2 - 4E^2}$$

$$= \frac{2\Delta^2}{\omega y}\left(\pi i + \ln\frac{1-y}{1+y}\right)$$

where $E^2 = \xi^2 + \Delta^2$, $\xi_k = \epsilon_k - \epsilon_F$ and $y = (1 - 4\Delta^2/\omega^2)^{1/2}$. This is the form appropriate to an insulator, describing band to band transition and has zero d.c. current. Our calculation shows that inclusion of the collective term Fig. 1 (c) yields a total conductivity

$$\sigma = \frac{ne^2}{i\omega m}\left(\frac{f(\omega)}{1 + (\lambda\omega_Q^2/4\Delta^2)f(\omega)} - f(0)\right). \qquad (11)$$

At $\omega = 0$, $\sigma(0) = (ne^2/i\omega m)(1 + (4\Delta^2/\lambda\omega_Q^2)f^{-1}(0))^{-1} = ne^2/i\omega m^*$. By the Kramers–Kronig relation, this implies a delta function at the origin for the real part of σ of relative weight m/m^*. This weight is taken out of the a.c. conductivity, and the square-root singularity at $\omega = 2\Delta$ given by equation (10) becomes a square root edge. This behavior is illustrated in Fig. 2 where the real part of σ is plotted vs ω for $m^*/m = 6$.

We would like to emphasize that the existence of a zero frequency mode is crucial to obtaining a d.c. conductivity. In Fig. 1(c), each bubble turns out to be proportional to ω, and it is the fact that $A_- \sim \omega^{-2}$ that results in a nonzero contribution in the limit of $\omega \to 0$. We shall see that various mechanisms will introduce a gap ω_T into the A_- mode. Instead of a d.c. conductivity there will be a large low frequency a.c. conductivity. This is most conveniently discussed in terms of the dielectric function $\epsilon(q, \omega)$. In a situation where chains are excited in phase, the long range three dimensional Coulomb energy is important. The dielectric function $\epsilon(q, \omega)$ can be calculated by considering the density response to a driving field $\langle\rho\rangle_{ext}$

$$\epsilon^{-1} = (\langle\rho\rangle + \langle\rho\rangle_{ext})/\langle\rho\rangle_{ext}. \qquad (12)$$

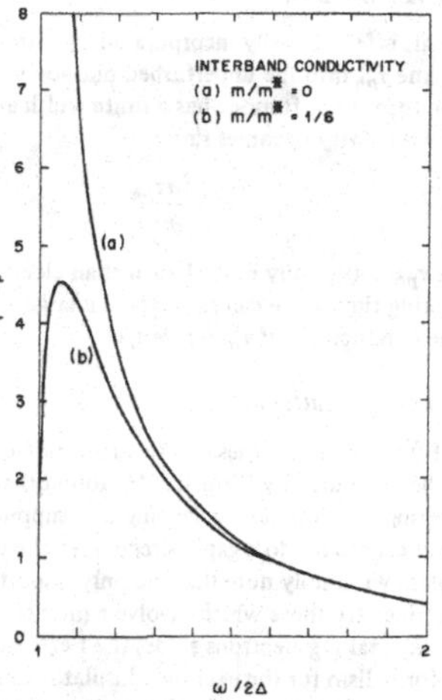

FIG. 2. The interband conductivity σ, as a function of frequency ω. (a) Infinite mass in the condensate. (b) Finite as. In case (a) all the weight in the conductivity sum rule is in the interband term while in case (b) a fraction m/m^* of the weight is in the δ-function at $\omega = 0$.

As pointed out earlier only the A_- mode is optically active. The A_+ mode contributes to ϵ only if particle hole symmetry is violated. This small coupling will be neglected here. The result is identical to that obtained from equation (11) using the relation $\epsilon - 1 = 4\pi i\sigma/\omega$. It is instructive to consider the case $\omega \ll \Delta$ when ϵ can be written in the form

$$\epsilon(0, \omega) = 1 + \frac{4\pi e^2 n/m}{6\Delta^2} + \frac{4\pi e^2 n/m^*}{\omega_T^2 - \omega^2}. \qquad (13)$$

This is the standard expression for an optical mode and the Fröhlich state corresponds to a situation when the restoring force $\omega_T = 0$. Coulomb interaction raises this mode to a finite frequency ω_{LO} given by the condition $\epsilon - 1 = 0$,

$$\omega_{LO} = (1.5\lambda)^{1/2}\omega_Q. \qquad (14)$$

Next we shall consider a number of effects that will destroy the infinite conductivity.

(i) Phonon lifetime.

This effect is easily incorporated by introducing a lifetime τ_{ph}^{-1} into the unperturbed phonon propagator D_0. In turn the soft mode has a finite width and the DC conductivity becomes finite

$$\sigma_{\text{d.c.}} = \frac{e^2 n \tau_{ph}}{m^*} \qquad (15)$$

Since τ_{ph} is typically much longer than electronic scattering time τ, we can still expect a large enhancement of the conductivity if $\tau_{ph}/m^* \gg \tau/m$.

(ii) Impurity scattering.

Effects of impurities on the excitonic insulator have been studied by Zittartz.[6] He found that impurity scattering is pair breaking, leading to a suppression of T_c and eventually to a gapless region. In the present problem we simply note that the only important scatterings are those which involve a momentum transver of $\sim 2k_F$, taking electrons across the Fermi surface. The formalism for the excitonic insulator can then be transcribed to this problem with only slight modifications. To calculate the conductivity we find that inclusion of scattering within an electronic bubble in Fig. 1(c) introduces a lifetime in d.c. conductivity which varies as $\exp(E_g/kT)$ where $2E_g$ is the gap in the single particle spectrum in the presence of impurities.[6] This effect is negligible at low temperatures. The important diagram is that shown in Fig. 1(d) which would introduce damping of the collective mode. The final state of this diagram is a collective mode which has given up momentum $\sim Q$ to the impurities. The collective mode is coupled to the impurities via the electronic bubble, which we find to be a constant at $\omega = q \doteqdot 0$. This is in contrast to the usual deformation coupling of acoustic modes. This is because the impurity couples directly to the phonon displacement via the modulation of the charge density. The damping is given by

$$\text{Im}\,\Sigma = n_{\text{imp}} \pi^2 V_Q^2 \lambda^2 \omega_Q^4 \Delta^{-2} \sum_q \text{Im}\, A_- \qquad (16)$$

where n_{imp} and V_Q are the impurity concentration and scattering potential. This is simply the golden rule. However in one dimension the density of states of a soft phonon $\sim \omega^{-1}$ and $\text{Im}\,\Sigma \sim \omega^{-1}$ is clearly an unacceptable result for small ω.[8] The only consistent solution is one in which a gap develops in A. We interpret this as implying that the collective mode is pinned by impurities. In the limit of dilute impurities one can imagine that the CDW will adjust its periodicity slightly so that the charge density sits in the potential minimum of each impurity. Translational invariance is lost and instead of d.c. conduvtivity one expects large low frequency conductivity. This behavior is very reminiscent of static friction.

When fluctuations are taken into account it becomes very hard even to speculate on this behavior. Presumably the CDW is no longer pinned at each impurity and a very impurity – and temperature-sensitive conductivity can be expected. One speculation worth suggesting is that, as in the similar mechanism of stick-slip friction, there will be nonlinearities in the conduction process.

(iii) Commensurability with lattice.

If lattice periodicity and the CDW periodicity form a rational fraction, translational invariance is lost and for some integer M, $\epsilon_{k+MQ} = \epsilon_k$. The effect of the lattice potential can be taken into account by including coupling of electronic states with momenta $k + nQ$, $n = 0$ to $M - 1$. Instead of a 2×2 matrix we have to consider the following $M \times M$ matrix for the eigenvalue E

$$\det \begin{bmatrix} \epsilon_k - E & \Delta & 0 & & \Delta^* \\ \Delta^* & \epsilon_{k+Q} - E & \Delta & & \\ & & \ddots & & \\ & & \Delta^* & \epsilon_{k+(M-2)Q} - E & \Delta \\ \Delta & & 0 & \Delta^* & \epsilon_{k+(M-1)Q} - E \end{bmatrix} = 0 \qquad (17)$$

We take Δ to be complex and its phase Φ determines the relative position of the CDW and the lattice. It can be shown that the only Φ dependence in equation (17) is of the form $2|\Delta|^M(\cos M\Phi - 1)$. This implies a correction of the total energy per particle of order $(|\Delta|^2/\epsilon_F)(e|\Delta|/W)^{M-2}(M\Phi)^2/2$ for small Φ where W is the bandwidth. Using this value as a spring constant, the gap in the collective mode A_- is shown to be $\omega_T \approx \lambda^{1/2} M (e\Delta/W)^{M/2-1} \omega_Q$. We see that ω_T approaches zero rapidly for large M as the distinction between rational and irrational numbers becomes academic. On theoth er hand, the half filled case $M = 2$ is particularly unfavorable for the Fröhlich mechanism.[9] Here b_Q and b_{-Q} are in fact the same phonon. Only the A_+ mode exists and it is the usual optical phonon. The normal coordinate corresponding to the A_- mode is the acoustic phonon, which does not carry a current. In this case the conductivity is purely due to the single particle terms as discussed by Patton and Sham.[10]

(iv) Three dimensional ordering and related effects.

There exists experimental evidence that a class of Pt compounds,[1] $K_2Pt(CN)_4Br_{0.3} \cdot 3H_2O$, undergoes a Peierls distortion. In the organic compound TTF–TCNQ (tetrathiofulvalinium tetracyanoquinodimethane) both donor and acceptor chains are believed to be conducting at room temperature. At lower temperatures a Peierls distortion has been proposed.[2,11] If the charge transfer is incomplete, Q is likely to be incommensurate with the lattice. A transition occurs at ~60°K which we interpret as a three dimensional ordering temperature, below which both the electron and the hole chains distort with the same periodicity. The charge on each chain can be described by $\bar{\rho}_i + \delta\rho_i \cos(Qr + \Phi_i)$ where $\delta\rho_i/\bar{\rho}_i = \Delta_i/\epsilon_F^i\lambda_i$. The locking between the electron chain and the hole chain can easily be estimated from the Coulomb energy as a function of $(\Phi_1 - \Phi_2)$. For simplicity we assume that the chains have similar ϵ_F and λ and at low temperature $(2\Delta \gg \lambda\omega_Q)$ the gap ω_T is

$$\omega_T = \omega_Q Q r_\perp(\omega_p/4\epsilon_F)(K_0(Qr_\perp)/2\pi\lambda\epsilon_\perp)^{1/2} \quad (18)$$

where $\omega_p^2 = 4\pi e^2 n/m$, $r_\perp$ is the distance between the chains,[12] $\epsilon_\perp$ is the background dielectric constant and K_0 is the modified Bessel function. In TTF–TCNQ, ω_T is estimated to be ~0.1 ω_Q.

For the Pt compound locking between chains will not in itself give rise to a finite mode frequency. However the negatively charged Br ions provide a pinning energy in addition to the lattice commensurability energy. This pinning energy will be greatest if the Br ions are ordered and the Br concentration is one third that of Pt. In this case the Br vacancies have the same periodicity as the CDW and we estimate by analogy with equation (18) a value $\omega_T \approx 0.2\,\omega_Q$. If the Br ions are disordered or the concentration is 0.30, the pinning energy will be smaller. Using this estimaet of ω_T in equation (13) we obtain a large value of $\epsilon(\omega = 0)$ in fair agreement with the experimental[13] value of ~1000.[14] Furthermore as the temperature is raised the water molecules between chains can rotate more freely and $\epsilon_\perp$ can be expected to increase. This together with the decreasing $\Delta^{1/2}$ implies a smaller ω_T at higher temperature, leading to a larger $\epsilon(\omega = 0)$. A dielectric constant that increases with temperature has been observed experimentally.[13] On the other hand ω_{LO} is expected to remain temperature independent at $(1.5\lambda)^{1/2}\omega_Q$ until $2\Delta \lesssim \lambda\omega_Q$ at which point it decreases towards zero at T_c, according to mean field theory. Structure in the reflectivity has recently been observed around 50 cm^{-1} below 200K.[15] At higher temperature the mode will broaden and become unobservable when $\omega\tau \lesssim 1$.

Finally we would like to point out that very similar conclusions are expected for spin density waves (SDW). In fact, any mechanism which causes the electrons in a metal to condense into a charged lattice unrelated in structure or period to the underlying lattice will lead to similar effects. In particular, the Wigner lattice caused by long-range Coulomb effects in low-density metals, which is presumably antiferromagnetic and thus not in principle different from an SDW, would presumably be an example. The major difference is that the SDW arises out of direct electron–electron interaction. The effective mass $m^* = m$ and the full weight of the conduvtivity will be concentrated in a delta function at zero frequency. This has been pointed out for the three dimensional electron gas by Hopfield[16] and by Fedders and Martin.[17]

However, in three dimensions either an SDW or a CDW with phonon displacement must satisfy one extra condition not noted in reference 16 and 17. The Zittartz scattering of electrons within a bubble, which in our theory gives an exponentially small term (and which may actually give exactly zero if one follows the analog of two-fluid theory and keeps the gas of excitations stationary rather than assuming that they follow the velocity of the CDW) will not vanish unless there is a total energy gap E_g: i.e. unless the SDW or CDW lattice is a true insulator.

In conclusion, it may be useful to supplement our calculations with a physical description, somewhat speculative to be sure, of what we think is occurring in most of the " one-dimensional metals" so far observed. We feel that the reason for the rarity of these substances is that the Peierls distortion or other CDW is normally so strong in chain materials that even those with ostensibly partly-filled bands will undergo very large, frozen-in distortions which leave them in an insulating state with large energy gap. The commonest example is the half-filled band case, which has a commensurability energy as large as its condensation energy, and will probably never exhibit metallic behavior in one dimension.

If, however, because of incommensurability, and possibly also because of a rather delicate cancellation of attractive and repulsive interactions, the Peierls or other distortion has an intrinsic 'mean field' condensation energy of only a few hundred degrees, in a fairly large region, normally *below* T_{MF}, quasimetallic behavior may be observed. This conductivity, we feel, is dominated by the 'sliding conductivity'. of the ultra-low-frequency mode which we have discussed here. Its behavior is, however, enormously complicated by the fact that true ordering does not take place in one dimension and so all of the physical phenomenon are smeared out by the finite coherence length of the order. Eventually, when this coherence length becomes too long one of these phenomena pins down the sliding electron lattice:

(a) commensurability energy
(b) three-dimensional ordering, or
(c) 'sticking' at impurities.

The first two will give sharp, the last gradual transitions to a low-temperature insulating state with a large dielectric constant caused by the relatively weakly pinned sliding mode.

Acknowledgements – We wish to acknowledge stimulating discussions with many of our colleagues, especially B.I. Halperin, H.R. Zeller, A. Bloch, W.F. Brinkman, A.S. Barker, Jr. and L.J. Sham.

REFERENCES

1. ZELLER H.R., *Adv. Solid Phys.* 13, 31 (1973).
2. COLEMAN L.B., COHEN M.J., SANDMAN D.J., YAMAGISHI F.G., GARITO A.F. and HEEGER A.J., *Solid State Commun.* 12, 1125 (1973).
3. LEE P.A., RICE T.M. and ANDERSON P.W., *Phys Rev. Lett.* 31, 462 (1973).
4. FRÖHLICH H., *Proc. R. Soc., Lond.* A223, 296 (1954).
5. JEROME D., RICE T.M. and KOHN W., *Phys. Rev.* 158, 462 (1967).
6. ZITTARTZ J., *Phys. Rev.* 164, 575 (1967); ZITTARTZ J., *Phys. Rev.* 165, 605 (1968).
7. In these calculations we have neglected the dispersion of the phonon with wave vectors near Q. Since the resulting collective modes have substantial dispersion only phonons very close to Q are important and this is a reasonable approximation.
8. Interaction between chains will introduce dispersion in the collective mode in the transverse directions and remove the singularity in the density of states. There will be competition between impurity pinning and three dimensional ordering.
9. In an incommensurate situation close to half-filled there will be Umklapp scattering between the collective mode and low frequency phonons. Since Fröhlich conductivity is a form of phonon drag, such collisions will damp the effect.
10. PATTON B.R. and SHAM L.J., *Phys. Rev. Lett.* 31, 631 (1973).
11. BARDEEN J., *Solid State Commun.* 13, 357 (1973).
12. In a more accurate treatment one would integrate over the transverse charge distribution weighted by $K_0(Qr_\perp)$. For the purpose of obtaining a rough estimate we have measured $r_\perp$ between the centers of the molecules.
13. SHCHEGOLEV I.F., *Phys. Status Solidi* (a) 12, 9 (1972).
14. The interband contribution to $\epsilon(0)$ from the first nontrivial terms on the r.h.s. of equation (13) can be quite large. However, to account for the observed value of 1000 requires that $\Delta \approx 0.04$ eV, given $\omega_p = 2.88$ eV (see reference 1, p. 46). This seems to us much too small.
15. BRUESCH P., RICE M.J., STRASSLER S. and ZELLER H.R., preprint.
16. HOPFIELD J.J., *Phys. Rev.* 139, A419 (1965).
17. FEDDERS P.A. and MARTIN P.C., *Phys. Rev.* 143, 245 (1966).

Nach den neulich z.B. für verschiedenerlei organische Verbindungen aufgestellten eindimensionalen Metallmodellen spielt unterhalb eines Peierls-Übergangs die gekoppelte kollektive Elektron-Phonon-Ausbrietungsform bei der Leitfähigkeit eine wichtige Rolle. Die Formfrequenz geht bei $q = 0$ im Rahmen des Kolloidmodelles bzw. bei Fehlanpassung gegen 0 und ergibt die erstmals von Fröhlich angeregte unendliche Leitfähigkeit. Fremdatome, Gitteranpassung und räumliche Ordnung öffnen eine Lücke im Formspektrum. Die niederfrequente Leitfähigkeit sowie eine hohe Dielektrizitätskonstante werden vorausgesagt. Eine Spindichtewelle lässt ähnliche Effekte erwarten.

Electric field depinning of charge density waves

P. A. Lee and T. M. Rice
Bell Laboratories, Murray Hill, New Jersey 07974
(Received 9 October 1978)

The pinning of charge-density waves by impurities is considered in systems that exhibit at least short-range order in three dimensions. Impurities are classified into strong and weak with quite different pinning properties. The pinning of spin-density waves is weak and the phase values at impurity sites are almost random, in agreement with a recent experiment. The electric field required to depin the charge-density wave is estimated. The coupling between a drifting charge-density wave and carriers either from a remnant Fermi surface or thermal excitation is considered. Attention is focused on umklapp scattering of carriers by phasons as a coupling mechanism at finite temperature. The conductivity in the high-electric-field depinned limit can be large. Dislocations in the charge-density-wave lattice are examined with particular emphasis on the piecewise motion of the charge-density wave through the motion of dislocations. We also discuss the generation of dislocations by the analog of Frank-Read sources. The unusual nonlinear conductivity observed in $NbSe_3$ is interpreted in terms of depinning of charge-density waves. The possibility of observing similar effects in other systems is briefly examined.

I. INTRODUCTION

One of the most fascinating aspects of charge-density waves (CDW) is the possibility of carrying a current by drifting the electron fluid and the CDW, a possibility first recognized at the outset by Frohlich.[1] It is generally accepted[2-4] that because of impurity pinning a finite-strength electric field is required to dislodge the CDW. The oscillator strength in the linear electric conductivity is shifted to a finite frequency and the resulting optically active phase mode has been observed experimentally.[5,6] However, the competition between impurity pinning and the electric-field energy has not been examined in detail. Further impetus for research in this direction has been provided by the observation of nonlinear conductivity associated[7] with CDW formation[8,9] in $NbSe_3$. It is tempting to interpret the nonlinear conductivity as evidence of depinning of CDW. In this paper we study nonlinear conductivity associated with the depinning of the CDW.

We begin by elucidating the nature of impurity pinning in CDW systems. An extension of these ideas to the spin-density-wave state enables us to explain a recent observation[10] of a distribution of phases at impurity sites in Cr. In Sec. III we estimate the characteristic electric field to dislodge the CDW as a whole. In Sec. IV we consider the coupling of the drifting CDW and the free Fermi surface or thermally excited carriers. Some of these questions have been studied recently by Boriack and Overhauser.[4,11] We emphasize umklapp scattering by phasons as an effective coupling between the two systems, and write the phenomenological equation for the drifting of the coupled system. In Sec. V we consider the situation below the characteristic field and study the possibility of moving one part of the CDW relative to the rest. This naturally leads to the study of dislocations in the CDW lattice. We consider various mechanisms for the generation of such dislocations, especially the analogs of Frank-Read sources.[12] In Sec. VI we discuss the experimental observations on $NbSe_3$ in the light of our results.

There has been a considerable amount of work on nonlinear excitations in one-dimensional systems pinned by a periodic potential. These excitations are solitons and can be excited thermally[13,14] or created by quantum-mechanical tunneling in a large electric field.[15] However the soliton conduction mechanism does not permit one to get around the impurity-pinning problem.[16] As we shall see, dislocations in the CDW lattice may be thought of as generalization of the nonlinear excitation to three dimensions. Such excitations cannot be thermally generated in the three-dimensional system that we discuss in this paper and we consider the most likely source to be extrinsic, e.g., Frank-Read sources.

For the purpose of this paper it suffices to treat the CDW phenomenologically. The charge density is given by

$$\rho(\vec{r}) = \bar{\rho} + \rho_1 |\psi| \cos[Qz + \phi(\vec{r})], \tag{1.1}$$

where we have considered a single $\vec{Q}$ state in three-dimensional space and $\psi(= |\psi| e^{i\phi})$ is the CDW order parameter normalized to unity at $T = 0$. The phase variable $\phi(\vec{r})$ denotes the location of the CDW relative to the lab frame. Phenomenologically we think of the CDW as a charged lattice (e.g., a Wigner lattice). Therefore when it moves it carries a current

$$J=\rho_c\bar{\rho}e\dot{\phi}/Q. \quad (1.2)$$

The collective density ρ_c equals unity at $T=0$ for a Peierls insulator and not the amplitude of the charge modulation. It is progressively reduced for higher temperature in the manner discussed in Sec. IV. As the CDW moves the lattice distortion must move with it. As a result we can associate an effective mass m^* with the CDW.[1,2] The temperature dependence of the effective mass as well as ρ_c is calculated microscopically in a companion paper.[17] An electric field $\mathcal{E}$ in the z direction couples to the phase via the following additional term in the Hamiltonian

$$H'=\int d\vec{r}\,\frac{e\rho_{\text{eff}}\bar{\rho}\mathcal{E}_z\phi}{Q}, \quad (1.3)$$

where ρ_{eff} is an effective density to be discussed in Sec. IV. It is to be understood that the $\mathcal{E}_z$ field couples to the free-carrier density in the normal way.

II. COUPLING TO IMPURITIES

An impurity atom located at $\vec{r}_i$ can be described by a potential $v(\vec{r}-\vec{r}_i)$ which is the difference between the potential at the impurity site and the potential of the host atom. The interaction energy is

$$H_{\text{imp}}=\int d\vec{r}\,v(\vec{r}-\vec{r}_i)\rho(\vec{r})$$

$$=\rho_1|\psi|\,\text{Re}\int d\vec{r}\,v(\vec{r})e^{i\vec{Q}\cdot\vec{r}}e^{i(\vec{Q}\cdot\vec{r}_i+\phi)}$$

$$=\rho_1|\psi|v(Q)\cos[\vec{Q}\cdot\vec{r}_i+\phi(\vec{r}_i)]. \quad (2.1)$$

In a Peierls system $\rho_1/\bar{\rho}\approx\Delta_0/\lambda\epsilon_F$, where Δ_0 is the energy gap at $T=0$ and λ is the electron-phonon coupling constant. In a typical case $\rho_1/\bar{\rho}$ may be 0.1. For charged impurities, i.e., impurity from a different column of the Periodic Table, $v(Q)=[4\pi e^2/\epsilon_\infty(Q)]Q^{-2}$, where $\epsilon_\infty(Q)$ is the dielectric function which includes excitation across the Peierls gap. At a large wave vector Q such screening as well as carrier screening is small. Thus we may estimate that $\rho_1 v(Q)$ for charged impurities may be of the order of several tenths of eV. For isoelectronic impurities or impurities located away from the conducting chains [such as Br disorder in $K_2Pt(CN)_4Br_{0.3}\cdot nH_2O$ (KCP)] the impurity potential is considerably smaller.

The impurity coupling (2.1) has two consequences. First, the local phase $\phi(r_i)$ has certain preferred value; second, the linear coupling to the order parameter implied by Eq. (2.1) leads to a local enhancement of the ordered phase. This effect has been discussed by McMillan.[18] Let us consider a Ginzburg-Landau expansion for the order parameter

$$F=f_0\int d\vec{r}\,\mathcal{E}\Big(-t|\psi|^2+\tfrac{1}{2}|\psi|^4 + \xi_x^2\left|\frac{\partial\psi}{\partial x}\right|^2+\xi_y^2\left|\frac{\partial\psi}{\partial y}\right|^2+\xi_z^2\left|\frac{\partial\psi}{\partial z}\right|^2\Big), \quad (2.2)$$

where $t=-(T-T_c)/T_c$ and ξ_x, ξ_y, ξ_z are the coherence lengths. In all CDW's observed to date, with the notable exception of the $4Hb$ layered compounds,[19] the low-temperature state consists of CDW's with transverse ordering greater than the interchain or interlayer spacing. Such system must be considered as three dimensional as far as the Ginzburg-Landau expansion is concerned. Indeed by rescaling the length scale in the transverse directions, $x'=(\xi_z/\xi_x)x$ and $y'=(\xi_z/\xi_y)y$, Eq. (2.2) can be treated as an isotropic system:

$$F=f_0(\xi_x\xi_y/\xi_z^2)\times\int dx'dy'dz(-t|\psi|^2+\tfrac{1}{2}|\psi|^4+\xi_z^2|\nabla\psi|^2). \quad (2.3)$$

To this we add the pinning term due to a single impurity at the origin. From Eq. (2.1) we obtain

$$F_{\text{pin}}=-\text{Re}\int d\vec{r}\,\rho_1 v|\psi|e^{i[\phi(\vec{r})-\tilde{\phi}]}\delta(\vec{r}). \quad (2.4)$$

Let us choose the overall phase such that the solution far away from the impurity is real. Then $\tilde{\phi}$ is the preferred phase at the pinning site. For small v we can linearize

$$\psi(\vec{r})=t^{1/2}+\psi'$$

so that ψ' obeys

$$f_0(\xi_x\xi_y/\xi_z^2)[-\xi_z^2\nabla^2\psi'+2t(\psi'+\psi'^*)]=\rho_1 ve^{+i\tilde{\phi}}\delta(\vec{r}). \quad (2.5)$$

The solution of this equation is

$$\text{Re}\psi'=\epsilon\cos\tilde{\phi}(\xi_z/r)e^{-r/\xi(T)} \quad (2.6)$$

and

$$\text{Im}\psi'=\epsilon\sin\tilde{\phi}(\xi_z/r), \quad (2.7)$$

where

$$\epsilon=\rho_1 v/f_0\xi_x\xi_y\xi_z \quad (2.8)$$

and $\xi(T)=\xi_z t^{-1/2}$. The solution must be cut off at small r because the pinning potential is in reality not a δ function and more importantly because the length scale of the variation of ψ cannot be smaller than ξ_z, i.e., $\xi_z^2|\nabla\psi|^2$ must be less than or equal to $|\psi|^2$. This latter condition implies that Eqs. (2.6) and (2.7) are valid only for $r > \xi_z$. The condition of validity of the linearized solution is that $|\psi'(0)| < t^{1/2}$ or

$$\epsilon < t^{1/2}. \quad (2.9)$$

For $\epsilon \ll t^{1/2}$ we find that the phase at the impurity site is largely determined by the phase at infinity. The pinning potential is approximately $v\rho_1\cos\bar{\phi}$ with corrections of order $\epsilon v\rho_1$. In three dimensions the elastic energy cost increases with the size of the spatial variation about the impurity. As a result there is a minimum elastic energy that one must pay to interpolate the phase between $\bar{\phi}$ at the origin and zero at infinity, and that minimum energy is of order $f_0 t\xi_x\xi_y\xi_z\bar{\phi}^2$. On the other hand the energy to be gained from the impurity is $\rho_1 v t^{1/2}$. When Eq. (2.9) is satisfied the gain in impurity energy is simply not sufficient to overcome the elastic energy and the phase assumes its value at infinity everywhere.

While an individual weak impurity is unable to pin the phase as the preferred value, a collection of these weak impurities can still pin the overall phase of the CDW.[20,21] Such pinning is described as weak pinning by Fukuyama and Lee.[20] Basically the phase varies on a scale L much greater than the impurity spacing $n_i^{-1/3}$, where n_i is the impurity concentration. It gains energy from the fluctuation in the impurity potential of the order of

$$-v\rho_1|\psi|[(\xi_x\xi_y/\xi_z^2)L^3 n_i]^{1/2}.$$

It pays an elastic energy equal to $f_0|\psi|^2\xi_x\xi_y L$. The length L can be obtained by minimizing the free energy per unit volume:

$$L^{1/2} = \frac{4}{3}\left(\frac{f_0\xi_x\xi_y\xi_z|\psi|}{v\rho_1}\right)n_i^{-1/2}(\xi_x\xi_y)^{-1/2}. \quad (2.10)$$

On the other hand, if $\epsilon > t^{1/2}$ the linear solution described earlier breaks down. The order parameter then assumes an enhanced value at the origin which will saturate at some value which is relatively independent of t. At the same time the phase at the origin will be pinned at $\bar{\phi}$ and interpolates smoothly to the value at infinity. The pinning behavior of these strong impurities is quite different from the weak impurities.

Let us now make some estimates on the criterion given by Eq. (2.9). We will consider a quasi-one-dimensional problem. In this case $f_0 = \Delta_0^2/\epsilon_F\Omega$, where $\Omega = a_x a_y a_z$ is the volume of the unit cell. For a quasi-one-dimensional problem, if $\xi_x < a_x$, $\xi_y < a_y$, $f_0\xi_x\xi_y\xi_z$ should really be replaced by $f_0 a_x a_y \xi_z$. Using the relation $\xi_z/a_z \approx \epsilon_F/\Delta_0$, we obtain $f_0 a_x a_y \xi_0 \approx \Delta_0$ and the criterion (2.9) becomes simply $v\rho_1 > \Delta(T)$. Since $v\rho_1$ is estimated to be tenths of eV, for most systems we see that charged impurities will qualify as strong impurities whereas isoelectronic impurities will generally be weak except very near T_c.

The situation is quite different for spin-density waves (SDW) such as in chromium. In this case a charged impurity will not couple directly to the SDW but only to the second-order harmonic CDW that coexists with the SDW. Clearly this will lead to a much smaller coupling and to values which will be in the weak pinning regime. Furthermore the coupling will be of the form $\cos 2(\phi - \bar{\phi})$ for the CDW harmonic and this is analogous to the random-anisotropy problem,[22] rather than the random-field problem[20] discussed above. For both these reasons, it is to be expected that the SDW will not be strongly pinned to individual impurity sites even for dilute impurities. If so, the phase will not attain its preferred value $\bar{\phi}$ at individual sites. Instead, the values of the phase of the SDW will be random at individual impurity sites. This explains the rather unexpected observation of Teisseron *et al.*[10] that in the SDW phase of Cr doped with Ta to one part in 10^8, the spin density at individual Ta sites was random and did not take a unique value. This is exactly what happens in weak pinning where the individual impurities do not maximize (or minimize) the local density as in strong pinning, but rather the phase only pins to large-scale fluctuations in the impurity density.

III. ESTIMATES OF DEPINNING FIELD

Let us consider a large volume of CDW and ask what is the electric field required to dislodge the entire volume from the impurity pinning and move it bodily. An upper bound can be obtained by keeping the CDW rigid and simply comparing the total electric field energy with the pinning energy per unit volume. We have to treat the strong and weak impurities separately. From the arguments leading to Eq. (2.10) we see that for weak impurities the pinning energy per unit volume is given by

$$f_{\text{pin}} \approx -f_0|\psi|^2\xi_x\xi_y L/(L^3\xi_x\xi_y/\xi_z^2)$$
$$\approx |\psi|^2 f_0(\xi_z\xi_x\xi_y n_i)^2(\epsilon/|\psi|)^4. \quad (3.1)$$

The last factor $\epsilon/|\psi|$ is by definition less than unity. The quantity $\xi_z\xi_x\xi_y n_i$ is extremely small. For a quasi-one-dimensional system, taking $\xi_x = a_x$, $\xi_y = a_y$, and $\xi_z \approx 100\,a_z$, and an impurity concentration of one part in 10^5, $\xi_z\xi_x\xi_y n_i \approx 10^{-3}$. On the other hand, from Eq. (1.3) the electric-field energy per unit volume when the phase is advanced by 2π is given by

$$f(\mathcal{E}_z) = e\bar{\rho}\rho_{\text{eff}}\mathcal{E}_z(2\pi/Q). \quad (3.2)$$

Equating Eqs. (3.1) and (3.2), we obtain at low temperature ($\rho_{eff} \approx 1$)

$$e\mathcal{E}_z(2\pi/Q) \approx (\Delta_0^2/\epsilon_F)10^{-5}\epsilon^4 . \qquad (3.3)$$

Taking $\Delta_0^2/\epsilon_F \approx 0.1\ \Delta_0 \approx 10^{-3}$ eV and $\epsilon \approx 1$, this translates into a field $\approx 10^{-9}$ eV on the atomic scale, or 10^{-2} eV/cm. Based on these estimates we conclude that weak impurities can be depinned by extremely weak fields. The reason is that in three dimensions the domain size L is extremely large and a weak electric field provides an energy proportional to the large volume. The same reasoning leads us to believe that thermal depinning is unlikely in three-dimensional systems. As the temperature is raised the CDW remains pinned up to the CDW onset. There is no separate transition analogous to the spin-glass transition above which the local value of the phase becomes random.

The strong impurity has quite different pinning properties. In the presence of an electric field the phase ϕ_∞ far away from the impurity site will increase while the phase at the impurity site remains pinned at $\bar{\phi}$. This will cost elastic energy of the order of $|\psi(0)|^2 f_0 \xi_x \xi_y \xi_z (\phi_\infty - \bar{\phi})^2$. However the energy of the lowest-energy state must be periodic in $\phi_\infty - \bar{\phi}$. As ϕ_∞ continues to increase, the solution near the impurity site is metastable and eventually jump to the stable solution, providing phase slippage of 2π between the pinned phase and ϕ_∞. Such phase slippage may proceed by tunneling, thermal activation over barrier, or directly for a sufficiently large electric field. As an estimate for the depinning field we may balance the elastic energy per impurity with the electric field energy, setting

$$f_0 \xi_x \xi_y \xi_z n_i = e\mathcal{E}\rho_{eff}\bar{\rho}(2\pi/Q) . \qquad (3.4)$$

In a quasi-one-dimensional problem for n_i of the order of one part in 10^5, we obtain at low temperature $e\mathcal{E}_z(2\pi/Q) \approx \Delta_0 n_i \approx 10^{-7}$ eV or 1 eV/cm which is much larger than the depinning field for weak impurities.

Near T_c, as we remarked earlier, a local $|\psi|$ is induced at the impurity site. The elastic energy required for phase slippage is presumably a complicated function of Δ which is somewhere between a function linear in Δ (if we ignore the local enhancement of $|\psi|$) and a function independent of Δ. As we shall see in Sec. IV ρ_{eff} is linear in Δ near T_c. Thus the depinning field $\mathcal{E}_z$ should diverge as $\Delta^{-\eta}$ where $0<\eta<1$ as $T \to T_c$.

IV. INTERACTION BETWEEN THE DRIFTING CDW AND NORMAL CARRIERS

In the presence of a remnant Fermi surface, or of thermally excited carriers across the gap, it is necessary to clarify the relation between the drifting CDW and the quasiparticle contribution to the current. Let us go to a frame moving with the drift velocity $\vec{D}$ of the CDW so that the CDW is stationary (referred to as the CDW frame below). In this frame the single-particle energy $E(\vec{k})$ is the standard one, i.e.,

$$E(\vec{k}) = \zeta_{\vec{k}} \pm E_0(k) ,$$

where $\xi_k = \frac{1}{2}(\epsilon_{\vec{k}+\vec{Q}} - \epsilon_{\vec{k}})$ and $\zeta_k = \frac{1}{2}(\epsilon_{\vec{k}+\vec{Q}} + \epsilon_{\vec{k}})$ and $E_0^2(k) = \xi_k^2 + \Delta^2$. The occupation of the states will be determined by the balance between the external field on the one hand, and the relaxation to the lab frame on the other, as well as possible relaxation to the moving CDW frame. We follow Boriack and Overhauser[11] and make the simplifying approximation that the occupation (in the extended-zone scheme) is given by

$$f_{\vec{k}} = f(E_{\vec{k}}) - m(\vec{K} - \vec{D}) \frac{\partial E}{\partial \vec{k}} \frac{\partial f}{\partial E} , \qquad (4.1)$$

where $\vec{K}$ and $\vec{D}$ are in the z direction and m is the band mass. This can be viewed as an expansion of $f(E(\vec{k} - m(\vec{K} - \vec{D}))$, i.e., the single-particle distribution is centered at $\vec{K} - \vec{D}$ in the CDW frame. Since k transforms like a momentum under a Galilean transformation we see that $\sum_k k_z f_k = m(\vec{K} - \vec{D})$ in the CDW frame and is equal to $\vec{K}$ in the lab frame. Equation (4.1) is the usual approximation for the Boltzmann equation and leads to Matthiessen's rule, i.e., the additivity of various scattering processes.

The average velocity carried by such a state is given by (in the following we suppress the N^{-1} factor in front of $\sum_k$)

$$\vec{V} = \sum_k \frac{\partial E}{\partial k_z} f_k = -m(\vec{K} - \vec{D}) \sum_k \left(\frac{\partial E}{\partial k_z}\right)^2 \frac{\partial f}{\partial E} \qquad (4.2)$$

in the CDW frame. In the lab frame we have

$$\vec{V} = \vec{D}\rho_c + \vec{K}\rho_n , \qquad (4.3)$$

where

$$\rho_n = -m \sum_k \left(\frac{\partial E}{\partial k_z}\right)^2 \frac{\partial f}{\partial E} \qquad (4.4)$$

and $\rho_c = 1 - \rho_n$. This expression for ρ_n/m is the usual one that determines the plasma frequency of carriers in semiconductors. This must be the case because if $D=0$, i.e., if the CDW is pinned, this problem becomes identical to the usual semiconductor or semimetal.

It is instructive to look at Eq. (4.4) in a different way. Suppose there is no scattering. In the presence of an external electric field $\mathcal{E}$ in the z direction the crystal momentum is accelerated by $m\dot{\vec{K}} = e\vec{\mathcal{E}}$. Therefore the gain in momentum by the single particle after a time t is given by

$$m\sum_k \frac{\partial E}{\partial \vec{k}} f(\vec{k}-e\vec{\mathcal{E}}t) = -me\vec{\mathcal{E}}t\sum_k \left(\frac{\partial E}{\partial \vec{k}}\right)^2 \frac{\partial f}{\partial E} = e\vec{\mathcal{E}}t\rho_n .$$

The remainder of the momentum density fed into the system is then $e\vec{\mathcal{E}}t\bar{\rho}(1-\rho_n)$. This remainder must have gone into accelerating the CDW as a whole. Therefore the force field on the CDW is given by $(1-\rho_n)e\mathcal{E}$. It is natural to interpret $e\rho_c = e(1-\rho_n)$ as the fractional charge density associated with the condensate. To study the behavior of ρ_c near T_c it is convenient to use the relation

$$\frac{1}{mv_F^2} + \sum_{\vec{k},\sigma} \frac{\xi^2}{E_0^2(k)} \frac{\partial f}{\partial E} = \sum_{\vec{k},\sigma} \frac{\Delta^2}{E_0^3(k)} [1 - f(-\zeta + E_0) - f(\zeta + E_0)] .$$

For $\Delta \ll T$ we can expand

$$1 - f(-\zeta + E) - f(\zeta + E) \approx \tfrac{1}{2}\beta E_0$$

for regions in k space where $\zeta \ll T$. The remaining region gives negligible contribution and we find that ρ_c is linear in Δ near T_c.

Our expression for ρ_c is in agreement with Boriack and Overhauser's γ in the proper limit but it is in disagreement with Allender, Bray, and Bardeen[23] who obtained an answer analogous to the superfluid density in He and in which $\rho_c \sim \Delta^2$. The difficulty with their argument is that they work in the lab frame with a time-dependent CDW potential. They obtain an eigenvalue spectrum $\lambda'(\vec{E}) = E(\vec{k} - m\vec{D}) + \vec{D}\cdot\vec{k}$, which does not have the proper Galilean transformation properties of an energy. As Boriack and Overhauser[24] pointed out, the energy as defined by the mean value of $i\,\partial/\partial t$ is not λ' but $E(\vec{k} - m\vec{D}) + m\vec{D}\cdot\partial E/\partial\vec{k}$. Once this is properly taken into account, their result can be brought into agreement with Eq. (4.4).

In the presence of scattering there is an additional force on the condensate arising from the relative motion of the single particle and the CDW. Boriack and Overhauser[11] have considered the problem at low temperatures when impurity scattering is the dominant mechanism. At higher temperatures when the conductivity is temperature dependent we have to consider additional scattering mechanisms. This leads us to consider umklapp scattering of the single particles by phasons of the CDW. We assume that the phasons are in thermal equilibrium in the CDW frame. This is a good approximation for $kT \ll \omega_0$ (where ω_0 is the bare phonon frequency at Q) so that the mixing between the phason and the ordinary phonons is not strong. This problem is quite similar to ordinary umklapp scattering in polyvalent metals where Lawrence and Wilkins[25] have shown that umklapp scattering is dominant. The time rate of change in momentum in the single-particle system (which must go into accelerating the CDW) is given by

$$\frac{\Delta p}{\Delta t} = \sum_{k,k'} m(v_{k'} - v_k)W_{k,k'}\{\delta(E_k - E_{k'} - \omega_q)[f_k(1-f_{k'})(1+n_q) - (1-f_k)f_{k'}n_q] + (E_k - E_{k'} + \omega_q)[f_k(1-f_{k'})n_q - (1-f_k)f_{k'}(1+n_q)]\}, \tag{4.5}$$

where it is understood that $\vec{k}' = \vec{k} + \vec{q} + \vec{G}$ and $\vec{G}$ is a reciprocal-lattice vector associated with the CDW. The transition probability is given by

$$W_{k,k'} = (1 - m/m^*)(\omega_0/\omega_q)M^2 \tag{4.6}$$

and

$$M = gv_F q\Delta/E_0^2(k) . \tag{4.7}$$

In Eq. (4.6) the scattering rate $W_{k,k+q}$ contains the spectral weight[2] of the phase mode $(1 - m/m^*)$; ω_0 and ω_q are the bare phonon frequency and the phase mode frequency, respectively, and M is the matrix element between the phonon and the quasiparticle. It can easily be worked out in terms of the electron-phonon coupling constant g via the Bogoliubov operator and is given by Eq. (4.7).

Using the ansatz Eq. (4.1) in Eq. (4.5) we obtain the time rate of change in momentum per unit volume

$$\Delta p/\Delta t = (\bar{\rho}\rho_n/\tau)m(K - D) , \tag{4.8}$$

where

$$\frac{\rho_n}{\tau} = \sum_{kk'} (v_k - v_{k'})^2 W_{kk'} \times\left(\delta(E_k - E_{k'} - \omega_q)\frac{\partial f}{\partial E}(1 + n_q - f_{k'}) + \delta(E_k - E_{k'} + \omega_q)\frac{\partial f}{\partial E'}(1 + n_q - f_k)\right). \tag{4.9}$$

The order of magnitude of $1/\tau$ is estimated in the

Appendix. Since the phason is defined only for $q < \xi_0^{-1}$ it is clear that k must be restricted to be within ξ_0^{-1} of k_F as well. In a quasi-one-dimensional situation where part of the Fermi surface remains at low temperature, only the part of the Fermi surface that is within Δ of the gap will contribute to τ^{-1}. In this case τ^{-1} is estimated to be $\rho_n \tau^{-1} \sim \lambda (T^2/\omega_0) F$, where $\lambda = g^2/\omega_0 \epsilon_F$ is the dimensionless electron-phonon coupling constant and F is a geometric factor related to the fraction of the Fermi surface within Δ of the energy gap. Near T_c, $\rho_n \tau^{-1}$ goes to zero linearly with Δ.

The total force per unit volume accelerating the CDW is then given by

$$\vec{F}_D = e\bar{\rho}\rho_c \vec{\mathcal{E}} + \bar{\rho}\rho_n \tau^{-1} m(\vec{K} - \vec{D}) . \tag{4.10}$$

Since total momentum is conserved in the umklapp scattering, a similar term must appear in the acceleration of the quasiparticle,

$$\dot{\vec{K}} = -\tau_k^{-1}\vec{K} - \tau^{-1}(\vec{K} - \vec{D}) + e\vec{\mathcal{E}}/m , \tag{4.11}$$

where we have included a term which relaxes the momentum to the lab. The source of this relaxation may be impurity scattering or scattering by the ordinary phonons which are in equilibrium in the lab frame.

We can now study two limiting cases: (i) the CDW is pinned, so that $\vec{D} = 0$, and (ii) the high-field limit in which the CDW is depinned and the conductivity is again linear. In the first case we set $D = 0$ and solve Eq. (4.11) for $\vec{K}$. Inserting the result in Eq. (4.10), we obtain

$$\vec{F}_D = e\bar{\rho}\rho_{eff}\vec{\mathcal{E}} , \tag{4.12}$$

where $\rho_{eff} = \rho_c + \rho_n/(1 + \tau/\tau_k)$. Thus we see the coupling of the CDW to the electric field depends on the ratio τ/τ_k. Near T_c τ^{-1} approaches zero and $\vec{F}_D \approx e\rho_c \vec{\mathcal{E}}$, where ρ_c itself is linear in Δ. At low temperature, τ_k and τ may be comparable. If τ^{-1} dominates, F_D approaches $e\vec{\mathcal{E}}$. At very low temperature, τ^{-1} vanishes like T^2 and $\rho_{eff} \approx \rho_c$.

In the opposite limit we assume that the CDW is depinned. Then we can write the following equation of motion for the change in the collective contribution to the momentum $m^* \rho_c \vec{D}$:

$$\dot{\vec{D}} = -\frac{1}{\tau_D}\vec{D} - \frac{1}{\tau}\frac{\rho_n}{\rho_c}\frac{m}{m^*}(\vec{D} - \vec{K}) + \frac{e\vec{\mathcal{E}}}{m^*} , \tag{4.13}$$

where we have added a phenomenological decay time τ_D for the damping of the CDW to the lab frame. The source of damping may be mixing of the phason with the ordinary phonons which have a finite lifetime,[2] or it may be radiation of phasons at impurity sites, or it may be the kind of impurity damping discussed by Boriack and Overhauser.[11] Equations (4.11) and (4.13) can be solved to obtain

$$\frac{D - K}{K} = \left(\frac{1}{\tau_k}\frac{1}{m^*} - \frac{1}{\tau_D}\frac{1}{m}\right) \Big/ \left[\frac{1}{m}\left(\frac{1}{\tau^*} + \frac{1}{\tau_D}\right) + \frac{1}{m^*}\frac{1}{\tau}\right] , \tag{4.14}$$

$$\vec{K} = \frac{e\vec{\mathcal{E}}}{m}\left(\frac{1}{\tau^*} + \frac{1}{\tau_D} + \frac{m}{m^*}\frac{1}{\tau}\right) \Big/ \left[\frac{1}{\tau_k}\left(\frac{1}{\tau^*} + \frac{1}{\tau_D}\right) + \frac{1}{\tau_D}\frac{1}{\tau}\right] , \tag{4.15}$$

where $\tau^{*-1} = \tau^{-1}(\rho_n m/\rho_c m^*)$. The conductivity can be readily obtained by combining these equations with Eq. (4.3). It is interesting to point out here that the qualitative nature of the solution depends on whether the phason scattering is the dominant process. If $\tau_k^{-1} \gg \tau^{-1}$, i.e., the relaxation to the lab is the dominant process, e.g., when the conductivity is impurity dominated,[11] the conductivity is dominated by normal carriers if $\tau_k^{-1} \ll (m^*/m)\tau_D^{-1}$ and by the drifting CDW otherwise. A more interesting situation obtains if $\tau_k^{-1} \ll \tau^{-1}$ and $\tau_D^{-1} \ll \tau^{*-1}$. In this case we have $|\vec{D} - \vec{K}|/K \ll 1$ and

$$\vec{K} \approx \frac{e\vec{\mathcal{E}}\tau_k}{m}\left(\frac{1 + \rho_c/\rho_n}{1 + \tau_k \rho_c m^*/\tau_D \rho_n m}\right) . \tag{4.16}$$

The CDW and the normal carriers are drifting at similar rates. In particular the expression in parentheses in Eq. (4.16) may be of order unity leading to a total conductivity of the order of the normal conductivity in the absence of the CDW.

V. DISLOCATION IN THE CHARGE-DENSITY-WAVE LATTICE

Suppose the electric field is smaller than that required to move the CDW as a whole as discussed in Sec. IV. It may still be possible to move part of the CDW relative to the rest. This will require the presence of dislocations in the CDW lattice.

We shall restrict ourselves to the single-$\vec{Q}$ CDW state. The Burgers vector can only be parallel to the $\vec{Q}$ vector, i.e., in the z direction. Suppose we ignore amplitude fluctuations and consider a region in space far away from strong impurities. The free energy reduces to

$$F = f_0 \xi_x \xi_y |\psi|^2 \int dx'\, dy'\, dz\, (\nabla\phi)^2 . \tag{5.1}$$

For a straight line dislocation the solution is simply $\phi = \theta$, where θ is the angle in the plane normal to the dislocation. Then energy per unit length is

$$T = f_0 \xi_x \xi_y |\psi|^2 \ln(R/\xi_0) , \tag{5.2}$$

where R is a large-distance cutoff typically equal to the distance to the nearest dislocation with the opposite Burgers vector. It is instructive to examine two examples of dislocations. The first is a dislocation loop lying in the x-y plane. Since

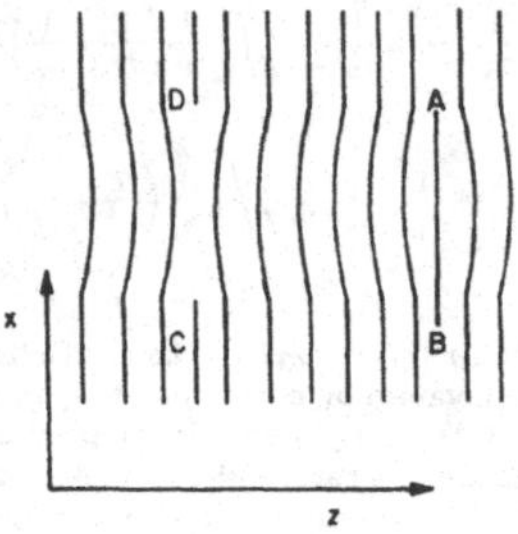

FIG. 1. Projection of two dislocation loops in the x-z plane. Solid lines are the contour of constant phase in units of 2π.

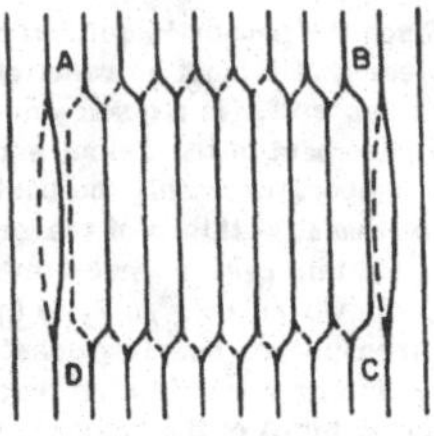

FIG. 2. Solid and dashed lines are the contours of constant phase for the layer above and below the plane in which the dislocation loop *ABCD* lies. The segments *AB* and *CD* are screw dislocations whereas *BC* and *AD* are edge dislocations.

the Burgers vector is in the z direction this is a purely edge-type dislocation. Its projection in the x-z plane is shown in Fig. 1. The line *AB* is the projection of a disk of an extra wavelength of the CDW. If we consider two loops with opposite Burgers vectors as shown in Fig. 1, clearly the phase within the volume bound by the two dislocation loops (the area *ABCD* in the projection) has slipped by 2π relative to the outside. An electric field will tend to pull the two loops apart in the z direction. Suppose the radius of the loop is R (in the scaled isotropic space) and the distance between the loops is z. The gain in energy from the electric field equals

$$\rho_{eff}\bar{\rho}\mathcal{E}_z\pi R^2 z(2\pi/Q)(\xi_x\xi_y/\xi_z^2) .$$

This is opposed by the elastic energy which causes an attraction between the two loops. This is analogous to the force between two parallel dislocations in a crystal. It can be evaluated in a straightforward way similar to the calculation in the limit when the crystal is replaced by an elastic continuum.[26] The attractive energy has a logarithmic dependence on the distance z, as in the solid, and takes the form

$$f_0\xi_x\xi_y 2\pi R|\psi|^2\ln(z/\xi_0) .$$

Minimizing the total energy we find that it has a maximum at a value z_0,

$$z_0=f_0\xi_z^2|\psi|^2/\rho_{eff}\bar{\rho}\mathcal{E}_z(2\pi/Q)R . \qquad (5.3)$$

For values of $z>z_0$ the electric field energy dominates and the disks will run away from each other whereas for $z<z_0$ the disk will collapse and annihilate.

A second example is a rectangular dislocation loop in the x-z plane such as *ABCD* shown in Fig. 2. The segments *AB* and *CD* parallel to z are pure screw dislocations. If we consider two loops spaced by y_0 in the y directions with opposite Burgers vectors, it is easy to see that the volume bound by the two loops has slipped by 2π relative to the rest of the CDW. It is particularly interesting to consider a layered system such that the CDW is weakly coupled in the y direction, i.e., $\xi_y < a_y$. In that case we can imagine the dislocation loop to be between layers. Then the dislocation loop describes the slippage of some layers relative to the bulk. The dislocation loop plays the role of a domain wall. Insofar as the stationary layer provides a periodic pinning potential to the slipped layer, the dislocation picture can be considered as a three-dimensional generalization of the soliton (or domain-wall) idea discussed by Rice *et al.*[13] in one dimension. Again, it is easy to show that for a sufficiently strong field the dislocation loop will expand in both the x and the y directions, thereby causing phase slippage and carrying a current.

However, examination of Eq. (5.3) shows that for electric fields too weak to depin the CDW as a whole, z_0 and R have to be enormous before the dislocations will grow and run away. Such large-scale objects preclude the possibility of thermally nucleating dislocation loops. In a real system there may be dislocation loops that were created in the process of condensation of the CDW. Generally, however, these dislocations can be used only once and are eliminated as they run into the surface of the sample or grain boundaries. This same problem arises in considering the shear in ordinary crystals. An ingenious proposal for dislocation sources was proposed by Frank and Read[12] and should be applicable to the CDW lattice as well. Let us consider a rectangular dislocation loop with sides x_0' and y_0' located in the x'-y' plane as shown in Fig. 3(a). We assume that the corners of the loop are pinned. A pinning mechanism may be regions in space where the CDW has reduced

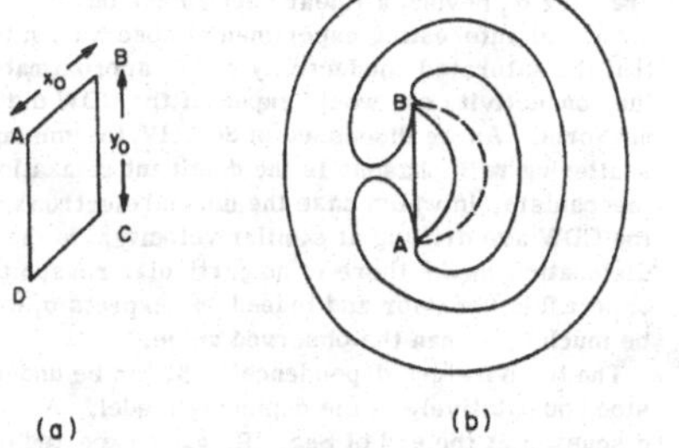

FIG. 3. (a) Dislocation loop in the x-y plane, segments of which can act as a Frank-Read source. (b) Shows the operation of a Frank-Read source. The dashed line is the semicircle that separates stable and unstable solutions.

amplitude which will attract the core of the dislocation line. Let us focus our attention on the segment AB lying along the x direction. Upon application of an electric field the segment will bow out as shown in Fig. 3(b). The balance is between the energy gained from the electric field and the elastic energy which is proportional to the length of the arc. The problem is analogous to that in crystals where the driving force is an applied stress and as in that case,[12,26] the boundary of the region of stable bowing of the segment AB occurs when the segment is a semicircle and the radius of the circle is just equal to one-half the length of the segment x_0', separating the stable and unstable solutions. The electric field required to produce an unstable solution is found[12,26] by balancing the energy gained from the field

$$\rho_{\rm eff}\bar{\rho}\,\mathcal{E}_z\left(\frac{2\pi}{Q}\right)\pi\left(\frac{x_0'}{2}\right)^2 y_0'\left(\frac{\xi_x\xi_y}{\xi_z^2}\right)$$

and the dislocation energy

$$f_0\xi_x\xi_y|\psi|^2\pi\left(\tfrac{1}{2}x_0'\right)\ln(x_0'/\xi_z)\,.$$

In terms of the length in real space $x_0=(\xi_x/\xi_z)x_0'$ and $y_0=(\xi_y/\xi_z)\,y_0'$, and using the quasi-one-dimensional expression for f_0 and ξ_z, we obtain the characteristic field

$$e\mathcal{E}_z\left(\frac{2\pi}{Q}\right)\approx\frac{f_0|\psi|^2}{\rho_{\rm eff}\bar{\rho}}\,\frac{\xi_x\xi_y}{x_0y_0}\ln\left(\frac{x_0\xi_x}{\xi_z}\right). \qquad (5.4)$$

For a quasi-one-dimensional system $f_0\approx\Delta_0^2/\epsilon_F\Omega$, ξ_x, ξ_y may be considerably less than a_x, a_y and for $x_0/a_x, y_0/a_y\approx 100$, the characteristic electric field may be quite small, ~1 eV/cm. For fields exceeding Eq. (5.4) the bowing out continues in the manner shown in Fig. 3(b), and the segment can operate as a source of dislocation loops expanding outward. At the same time the segment DC on the opposite side of the rectangle operates as a Frank-Read source in an identical manner. The two expanding loops then define an expanding domain of slipped phase.

As the dislocation loop expands it may encounter a strong impurity. The dislocation can get around the impurity in several ways. If the dislocation is screwlike it can cross-slip to a different glide plane. For dislocations that are edgelike it will have to climb (motion normal to the Burgers vector). In ordinary crystal, climbing requires migration of vacancies or interstitials. In a CDW, a climb can be accomplished by converting collective density to normal carriers. Such a conversion will take place when an edge dislocation hits a wall or grain boundary or when it is stopped by a strong impurity that it is unable to get around. The conversion of a normal electron to a collective density at one end of the sample and its transport by dislocation motion some distance down the sample constitutes a parallel channel of transport and the conductivity should be additive. The magnitude of the conductivity that this mechanism can provide is very complicated even to estimate, and Eq. (5.4) is to be understood as the minimum electric field required for the operation of a source of a particular dimension and hence the opening up of a new channel for conduction. As the electric field is increased, more and more Frank-Read sources operate, until a field is reached when the CDW is depinned and drifts as a whole, we then arrive at the situation described in Sec. III.

We should remark on an additional restriction for the Frank-Read source to operate, namely, that there must be no strong impurities within a radius of $\frac{1}{2}x_0'$ of the source. In the z direction this means a distance in real space of $z_0=(\xi_z/\xi_x)x_0$. This is the reason we restricted our attention to the segments AB and DC of the initial loop that lies along the x direction. In a layered system $\xi_y<\xi_x$ and hence there is a more severe restriction on the segments BC and AD to operate as Frank-Read sources.

VI. DISCUSSION

We would like to discuss the experimental results on $NbSe_3$ in light of the above theoretical considerations. It is found[7] that the nonlinear conductivity can be very well fitted by the form

$$\sigma(\mathcal{E})=\sigma_0+\sigma_1e^{-\mathcal{E}_0/\mathcal{E}}. \qquad (6.1)$$

Furthermore $\mathcal{E}_0$ is temperature dependent, behaving like $t^{-1/2}$ just below each of the two transition temperatures.[2] The minimum $\mathcal{E}_0$ is of the order of

1 eV/cm for the 144-K transition and 0.1 eV/cm for the 59-K transition. Below about 25 K, $\mathcal{E}_0$ increases abruptly and due to heating effects a voltage greater than 0.5 eV/cm cannot be applied. X-ray diffraction has been performed[9] with a current running through the sample, and no change in the superlattice period or intensity was observed. The extremely small electric field (10^{-8} eV on an atomic scale) indicates that we must be dealing with a phenomenon on a large scale. The natural explanation is that the CDW is being depinned by the electric field. Since the x ray is an instantaneous snapshot, it will not detect the small drift velocity of the CDW. The observation[7] that conductivity at microwave frequency approximates $\sigma_0+\sigma_1$ further confirms this view, as ac conductivity is not expected to be so strongly affected by the pinning. In a strictly one-dimensional problem the frequency-dependent conductivity is shown to be of the form $\sigma(\omega)\sim\exp(-\omega_0/\omega)$. In three dimensions the problem is more complicated and a detailed experimental determination of $\sigma(\omega)$ would be very interesting. Our study shows that the impurities can be divided into weak and strong ones and estimates made in Sec. III indicate that the weak impurities are easily depinned. The depinning field for the strong impurities was estimated to be of the order of 1 eV/cm, not too different from the measured value. (The higher temperature transition is expected to require a larger depinning field because Δ_0 is larger and ρ_{eff} is smaller, since a large portion of the Fermi surface survives the first transition.) However, we would expect a more sudden onset of extra conductivity, and there is no obvious way to understand the $\exp(-\mathcal{E}_0/\mathcal{E})$ dependence. One possibility may be that the CDW is broken up into grains (which may or may not be associated with grain boundaries of the real lattice). Each grain has its own depinning field, and an averaging over a distribution of depinning fields results in a more gradual nonlinear behavior, as discussed at the end of Sec. III.

In the dislocation model it is more natural to expect a distribution in the size x_0y_0 of the Frank-Read sources. A Poisson distribution of the size (resulting from a random distribution of dislocation pinning sites, for instance) can nicely account for the $\exp(-\mathcal{E}_0/\mathcal{E})$ behavior since x_0y_0 goes like $\mathcal{E}^{-1}$ according to Eq. (5.4). For a fixed density of strong impurities, Eq. (5.4) predicts an $\mathcal{E}_0$ that goes like $|\psi|$. However, from Eq. (2.9) we see that more and more impurities become strong impurities near T_c. The increase in strong impurity concentration will render some Frank-Read sources inoperative because the expanding dislocation line may run into the impurity before it becomes unstable. This will have the effect of increasing $\mathcal{E}_0$ beyond a linear dependence on $|\psi|$.

Another interesting experimental observation is that the saturated conductivity $\sigma_0+\sigma_1$ approximates the conductivity one would expect if the CDW did not form. As we discussed in Sec. IV, the umklapp scattering with phasons is the dominant relaxation mechanism, in which case the normal electron and the CDW are drifting at similar velocity. In the dislocation model there is no particular reason to expect this behavior and indeed one expects σ_1 to be much less than the observed value.

The temperature dependence of $\mathcal{E}_0$ can be understood qualitatively in the depinning model. As discussed at the end of Sec. III, $\mathcal{E}_0$ is expected to diverge near T_c even though the precise exponent is not known. Experimentally a $t^{-1/2}$ divergence is reported.[27] Below the 144-K transition $\mathcal{E}_0$ is observed to rise slowly even after the order parameter has apparently saturated. This can be explained by a temperature-dependent ρ_{eff} as given by Eq. (4.12) arising from the temperature dependence of the ratio τ/τ_k.

One of the most puzzling features of the experiment is the abrupt rise in $\mathcal{E}_0$ around 25 K. An explanation we would like to speculate upon here is that around 25 K there is a lock-in between the two apparently independent CDWs. The two CDW's have wave vectors $\vec{q}_1=(0, 0.243, 0)$ and $\vec{q}_2=(0.5, 0.263, 0.5)$. The harmonics $2q_1$ and $2q_2$ are almost identical up to a reciprocal-lattice vector and indeed a weak harmonic has been observed experimentally.[28] If $2q_1=2q_2$ there will be a term in the free energy $\Delta_1^2\Delta_2^{*2}$ which will tend to lock the relative phase of the two CDW's. Our speculation is that this lock-in occurs at around 25 K below which Δ_2 will have to move against Δ_1. Since Δ_1 is pinned by a much larger electric field we expect the depinning field to be dramatically increased. This model will predict a weak anomaly in the harmonic superlattice reflection around 25 K and that a field of several eV/cm which is big enough to depin the 144-K CDW will depin the CDW below 25 K. One unsatisfactory feature of this picture is that it is difficult to explain why the lock-in occurs only at 25 K when the order parameter has more or less saturated and does not occur closer to the onset of the second CDW.

If impurity pinning plays a strong role in the conductivity of $NbSe_3$ as we have suggested, there should be a correlation between the characteristic depinning field $\mathcal{E}_0$ and the concentration of strong impurities. In the depinning model we would predict that $\mathcal{E}_0$ should increase linearly with the strong impurity concentration. Recently it has been noted that $\mathcal{E}_0$ are different for samples with different room-temperature to low-temperature resistance ratios.[29] However, the resistance ratio is sensitive to both strong and weak impurity concentration. As a result, a quantitative correlation is not possible. One possible test is to dope the system with a small amount of charged impurities, such as Ti. Our picture predicts that even a small amount, of the order of one part in 10^4, will greatly reduce the nonlinear conductivity.

To summarize we have studied the possible nonlinear conductivity mechanisms for moving CDW in the presence of impurities. Those considerations should be applicable to all CDW systems, including layered compounds.[30] However, to exhibit an effect for modest electric fields requires very pure samples. So far only one system, $NbSe_3$, exhibits nonlinear conductivity. An electric field depinning of CDW is, to our knowledge, the only viable explanation for the unusual behavior in $NbSe_3$. As we see in this paper the qualitative features and the order of magnitude of the depinning fields are reasonably accounted for. It will be most interesting if the condensate can be made to move in other CDW systems. We have already seen that there is evidence that the SDW in chromium should be quite mobile; the problem is, of course, that one has to find means to couple to it. In other CDW systems, tetrathiafulvalene-tetracyanoquinodimethane (TTF-TCNQ) is not a suitable candidate, except perhaps for the narrow temperature region between 48 and 54 K,[29,31] because the CDW on oppositely charged chains provide a periodic pinning potential to each other. In KCP we expect a substantial amount of pinning due to disorder in the bromine sites, which leads to the surprisingly high pinned phase-mode frequency of ≈2.5 meV.[32] The depinning field is expected to be extremely large. A promising class of systems is the TTF halides and tetrathiafulvalene-thiocyanate [TTF-$(SCN)_{0.588}$].[33] However, disorder in the halides and in the SCN may also lead to a large depinning field. For the CDW in layered compounds our dislocation picture needs to be generalized to the state with three coexisting Q vectors. The basic physics is expected to remain the same and the layered compounds should be promising candidates for drifting CDW. We should mention the very interesting *4Hb* layered compounds,[19] where the CDW are apparently uncorrelated from layer to layer. In this case dislocation pairs may be created thermally within each layer.[34]

Note added in proof. Recent developments of note are (i) the report by Ong and collaborators [N. P. Ong, Bull. APS 24, 294 (1979) and N. P. Ong, J. W. Brill, J. C. Eckert, J. W. Savage, S. K. Khanna, and R. B. Somoano (unpublished)] that ϵ_0 scales as n_i^2 in samples doped with Ta in agreement with Eqs. (3.1) and (3.2); (ii) the observation of a threshold field in relatively pure $NbSe_3$ by Fleming and Grimes [R. M. Fleming and C. C. Grimes, Bull. APS 24, 386 (1979) and unpublished] which is in accord with the simple breakaway of the CDW as a whole from the pinning centers discussed in Sec. VI; and (iii) the observation of similar nonlinear conductivity below the structural phase transition in ZrV_2 and HfV_2 by V. M. Pan, I. E. Bulakh, A. L. Kasatkin, and A. D. Shevchenko, Pis'ma Zh. Eksp. Teor. Fiz. 27, 629 (1978) [JETP Lett. 27, 594 (1978)]. If our model is to be applicable to ZrV_2, an incommensurate CDW must be present also in this material, but to date no such CDW has been reported to our knowledge.

APPENDIX

In Eq. (4.9) the first term corresponds to emission of phonons and the second to absorption of phonons upon a transition from k to k'. By a transformation $k \to k'$ and we can show that the two terms are equal. Hence

$$\frac{\rho_n}{\tau} = 2 \sum_{kk'} (v_k - v_{k'})^2 W_{kk'} \delta(E_k - E_{k'} - \omega_q) \times \frac{\partial f}{\partial E}(1 + n_q - f_{k'}) . \qquad \text{(A1)}$$

To estimate the order of magnitude of τ^{-1} we observe that $\omega_q = sq$, where $s = (m/m^*)^{1/2} v_F$; the phason velocity is much lower than v_F and we may assume that for most of the scattering that contributes to the sum, $v_{k'} \approx -v_k \approx \xi/E_0$ and that $q \approx 2k$. The k sums can be converted to integrals over E and E', and we obtain upon using Eq. (4.7)

$$\frac{\rho_n}{\tau} = \frac{2}{\epsilon_F} \sum_{k_\perp} \int dE \frac{g^2 \xi^2 \Delta^2}{E_0^4} \frac{\omega_0}{\omega_{2k}} \frac{\partial f}{\partial E} \times [1 + n_{2k} - f(E - \omega_{2k})] . \qquad \text{(A2)}$$

Let us first examine the low-temperature limit $T \ll \Delta$. Noting that $\partial f/\partial E$ restricts E to be of order zero, we find the thermal factor $[1 + n_{2k} - f(E - \omega_{2k})]$ of order kT/ω_{2k} for $\omega_{2k} < kT$ and $\exp(-\omega_{2k}/T)$ for $\omega_{2k} \gg kT$. Further we note that phase mode q is restricted to less than ξ_0^{-1} and $\omega_q < \omega_0$. Thus if $kT < \omega_0$, only a fraction kT/ω_0 of the region in k space will contribute to the sum. Even for $\omega_0 > kT$, only regions of the Fermi surface within Δ of the gap will contribute, i.e., $E_0 \lesssim 2\Delta$. Equation (A2) is then estimated to be

$$\rho_n/\tau = (2g^2/\epsilon_F \omega_0) F T \min(1, kT/\omega_0) , \qquad \text{(A3)}$$

where F is a geometric factor which is roughly the area of the Fermi surface that is within Δ of the energy gap

$$F=\sum_{k_\perp}\int dE\,\frac{\xi^2\Delta^2}{E_0^4}\frac{\omega_0^2}{\omega_q^2}\Theta(2\Delta-E_0)\frac{\partial f}{\partial E}$$

$$\approx\sum_{k_\perp}\int dE\,\frac{\Delta^4}{E_0^4}\Theta(2\Delta-E_0)\frac{\partial f}{\partial E} \qquad \text{(A4)}$$

where Θ is the step function. At low temperature $\rho_n/\tau\sim\lambda FT^2/\omega_0$.

Next we consider the case near T_c, when $\Delta\ll T$. For the regions of the Fermi surface where $\xi\ll T$, $\partial f/\partial E\approx\beta E_0$ after we account for contribution for the electron and the hole pockets. Just as the estimate for ρ_c near T_c, it is clear that ρ_n/τ is linear in Δ just by scaling the integrals (A2).

---

[1] H. Frohlich, Proc. R. Soc. A 223, 296 (1954).

[2] P. A. Lee, T. M. Rice, and P. W. Anderson, Solid State Commun. 14, 703 (1974).

[3] J. B. Sokoloff, Phys. Rev. B 16, 3367 (1977).

[4] M. L. Bortack and A. W. Overhauser, Phys. Rev. B 16, 5206 (1977).

[5] P. Bruesch, S. Strassler, and H. R. Zeller, Phys. Rev. B 12, 219 (1975).

[6] J. E. Eldridge, Solid State Commun. 19, 607 (1976).

[7] N. P. Ong and P. Monceau, Phys. Rev. B 16, 3443 (1977).

[8] K. Tsutsumi, T. Takagaki, M. Yamamoto, Y. Shiozaki, M. Ido, T. Sambongi, K. Yamaya, and Y. Abe, Phys. Rev. Lett. 39, 1675 (1977).

[9] R. M. Fleming, D. E. Moncton, and D. B. McWhan, Phys. Rev. B 18, 5560 (1978).

[10] G. Teisseron, O. Berthier, P. Peretto, C. Benski, M. Robin, and S. Choulet, J. Magn. Magn. Mater. 8, 157 (1978).

[11] M. L. Bortack and A. W. Overhauser, Phys. Rev. B 17, 2395 (1978).

[12] F. C. Frank and W. T. Read, Phys. Rev. 79, 722 (1950); see, for example, F. Nabarro, *Theory of Crystal Dislocations* (Oxford University, New York, 1967), p. 709.

[13] M. J. Rice, A. R. Bishop, J. A. Krumhansl, and S. E. Trullinger, Phys. Rev. Lett. 36, 432 (1976).

[14] S. E. Trullinger, M. D. Miller, R. A. Guyer, A. R. Bishop, F. Palmer, and J. A. Krumhansl, Phys. Rev. Lett. 40, 206 (1978); R. A. Guyer and M. D. Miller, Phys. Rev. A 17, 1774 (1978).

[15] K. Maki, Phys. Rev. Lett. 39, 46 (1977).

[16] A. I. Larkin and P. A. Lee, Phys. Rev. B 17, 1596 (1978).

[17] T. M. Rice, P. A. Lee, and M. C. Cross (unpublished).

[18] W. L. McMillan, Phys. Rev. B 12, 1187 (1975); see also L. P. Gor'kov Pis'ma Zh. Eksp. Teor. Fiz. 25, 384 (1977) [JETP Lett. 25, 358 (1977)].

[19] F. J. DiSalvo, D. E. Moncton, J. A. Wilson, and S. Mahajan, Phys. Rev. B 14, 1543 (1976).

[20] L. J. Sham and B. R. Patton, Phys. Rev. B 13, 3151 (1976); Y. Imry and S. K. Ma, Phys. Rev. Lett. 35, 1399 (1975); H. Fukuyama and P. A. Lee, Phys. Rev. B 17, 535 (1978).

[21] D. E. Moncton, F. J. DiSalvo, J. D. Axe, L. J. Sham, and B. R. Patton, Phys. Rev. B 14, 3432 (1976).

[22] R. A. Pelcovits, E. Pytte, and J. Rudnick, Phys. Rev. Lett. 40, 476 (1978).

[23] D. Allender, J. W. Bray, and J. Bardeen, Phys. Rev. B 9, 119 (1974).

[24] M. L. Bortack and A. W. Overhauser, Phys. Rev. B 15, 2847 (1977).

[25] W. E. Lawrence and J. W. Wilkins, Phys. Rev. B 6, 4466 (1972).

[26] See, for example, J. Friedel, *Dislocations* (Addison-Wesley, Reading, Mass., 1964), p. 40.

[27] N. P. Ong, Phys. Rev. B 17, 3243 (1978).

[28] R. M. Fleming, D. E. Moncton, and D. B. McWhan (private communication).

[29] P. Bak and V. J. Emery, Phys. Rev. Lett. 36, 978 (1976).

[30] J. A. Wilson, F. J. DiSalvo, and S. Mahajan, Adv. Phys. 24, 117 (1975).

[31] W. D. Ellenson, S. M. Shapiro, G. Shirane, and A. F. Garito, Phys. Rev. B 16, 3244 (1977).

[32] K. Carneiro, G. Shirane, S. A. Werner, and S. Kaiser, Phys. Rev. B 13, 4258 (1976).

[33] F. Wudl, D. E. Schafer, W. M. Walsh, Jr., L. W. Rupp, Jr., F. J. DiSalvo, J. V. Waszczak, M. L. Kaplan, and G. A. Thomas, J. Chem. Phys. 66, 377 (1977); R. B. Somoano, A. Gupta, V. Kadek, M. Novotny, M. Jones, T. Datta, R. Deck, and A. M. Hermann, Phys. Rev. B 15, 595 (1977).

[34] J. M. Kosterlitz and D. J. Thouless, J. Phys. C 6, 1181 (1973).

Sliding charge density waves

from P. A. Lee

THE idea of a charge density wave transition goes back to Peierls[1] and, independently, to Fröhlich[2] in 1954. They reasoned that a one-dimensional metal can lower its energy by distorting the lattice and forming a gap at the Fermi surface, thereby making a transition into an insulating state. The term 'charge density wave' refers to the fact that the lattice and the electron charge density form a new periodic structure, with a wavelength λ that is longer than the original lattice period a. Fröhlich further reasoned that if the charge density wave is incommensurate, that is, a/λ is not a simple rational fraction like 1/2 or 2/3, then the entire charge density wave structure can slide through the lattice, thereby contributing to the electrical conductivity. Since these ideas were originally formulated in terms of one-dimensional systems, for many years they were considered with nothing more than theoretical curiosity.

In the early 1970s the interests of many physicists turned to quasi one- and two-dimensional systems. The existence of a charge density wave ground state in these systems turns out to be the rule rather than the exception, and a large number of examples have been discovered. Experimentally the onset of the charge density wave is signaled by a rise in the resistivity at some temperature T_c. In quasi one-dimensional systems a transition to an insulating state often occurs, whereas in the layered compounds the system often stays metallic. The most direct microscopic evidence for a charge density wave transition is the appearance of new superlattice spots associated with the new lattice periodicity λ which is detected by X-ray, neutron or electron diffraction. Prominent examples of charge density wave systems are the organic charge transfer salts[3], such as tetrathiafulvalene-tetracyanoquinodimethane (TTF-TCNQ) and layered compounds[4] formed from transition metals and chalcogens, such as TaS_2 and $NbSe_2$.

However, until 1976, none of the charge density wave systems had exhibited the sliding conductivity envisioned by Fröhlich. This is because the charge density wave was either commensurate or pinned by impurities. Impurities are sensitive to the charge density oscillations and can lock the charge density wave in place, a phenomenon very analogous to static friction[5]. As in friction, a sufficiently large force should dislodge the charge density wave. Thus the observation[6,7] in 1976 of nonlinear electrical conductivity in $NbSe_3$ was greeted with a good deal of interest. In $NbSe_3$ there are two transitions at 59K and 144K; below each of these temperatures the resistivity rises. It was found that beyond a certain threshold electric field[8], which can be as low as a few meV per cm for the low-temperature transition and 0.1 eV per cm for the higher-temperature transition, the resistivity becomes nonlinear and begins to drop, until it saturates to a value close to that expected if the charge density wave transition had not occurred.

The conductivity is also found to be highly frequency dependent, rising with increasing frequency until it coincides with the high field d.c. value at ~ 100 MHz (refs 7 and 9). It soon became clear that the electric field is not destroying the charge density wave, because an X-ray diffraction experiment shows that the superlattice spots are not affected by a current through the sample even when the current exceeds the linear threshold[10]. Diffraction experiments reveal incommensurate lattice distortions with periods λ such that $a/\lambda = 0.24$ and 0.26 for the upper and lower transitions. These data are consistent with an interpretation in terms of impurity pinning of the charge density wave[11]. The charge density wave can undergo small oscillation, thus accounting for the frequency-dependent conductivity. The pinning frequency is very low, possibly because the impurity concentration in $NbSe_3$ is very small. What makes $NbSe_3$ special among charge density wave systems in this respect is not entirely understood at present. The weakness of the pinning also manifests itself in the low-frequency dielectric constant, which has been measured to be greater than 10^8 (see ref. 9). Presumably a small electric field is sufficient to overcome the weak pinning so that the charge density wave can slide in the way envisioned by Fröhlich — an interpretation supported by an experiment in which impurities were introduced into $NbSe_3$ and the threshold electric field was found to increase[12]. Instead of classical depinning, an alternative mechanism involving the quantum mechanical tunneling of macroscopic segments (up to several microns in length) of charge density waves has been proposed[13], but all existing theories are based on the collective motion of the charge density wave ground state described by Fröhlich.

Further study of $NbSe_3$ yielded more surprises. When the current exceeds the threshold, voltage noise across the sample greatly increases. Superimposed on this background noise are periodic components with well defined frequencies[8], which scale linearly with the excess nonlinear current[14] and are typically in the range of kHz to MHz. A model for the noise is that it arises when the charge density wave moves through the impurities or the lattice, much like a particle rolling down a washboard. If certain assumptions are made about the charge being carried by the sliding charge density wave (a quantity under debate at present), the frequency of the noise can be accounted for[14].

Recently there was a report of dark-field electron microscopy on $NbSe_3$ (ref. 15). An image of an area of the sample is formed using the superlattice reflection alone; thus any feature in the imaging is due to the charge density wave formation. It is found that stripes approximately 200Å wide run along the conducting axis, and across the stripes are bands of lighter and darker intensities, suggestive of a Moiré pattern. It appears that the charge density wave structure may itself be broken into domains and that more than one superlattice wave vector is possible, so that interference between them forms the Moiré pattern. What is even more intriguing is that at low temperatures the bands are found to shift in time, so that the entire pattern shimmers on a time scale of a fraction of a second. This suggests some time-dependent motion of the charge density wave, but it is occurring on such a slow time scale that its connection with the nonlinear and frequency-dependent conductivity is unclear.

It is rare to find a material like $NbSe_3$ that exhibits such a rich variety of unusual properties. At present the idea of a sliding charge density wave provides a framework for the qualitative understanding of many of these properties, but clearly much work remains to be done before a quantitative understanding is achieved.

P.A. Lee is at the Bell Laboratories, Murray Hill, New Jersey.

1. Peierls, R. E. *Quantum Theory of Solids* (Oxford University Press, London, 1955).
2. Fröhlich, H. *Proc. R. Soc.* **A223**, 296 (1954).
3. *The Physics and Chemistry of Low Dimensional Solid* (ed. Alcacer, L., 1979).
4. Wilson, J. A. *et al. Adv. Phys.* **24**, 117 (1975).
5. Lee, P. A. *et al. Solid State Commun.* **14**, 703 (1974).
6. Monceau, P. *et al. Phys. Rev. Lett.* **37**, 602 (1976).
7. Ong, N. P. & Monceau, P. *Phys. Rev.* **B16**, 3443 (1977).
8. Flemming, R. M. & Grimes, C. C. *Phys. Rev. Lett.* **42**, 1423 (1979).
9. Grüner, G. *et al. Phys. Rev. Lett.* **45**, 935 (1980).
10. Fleming, R. M. *et al. Phys. Rev.* **B18**, 5560 (1978).
11. Lee, P. A. & Rice, T. M. *Phys. Rev.* **B19**, 3970 (1979).
12. Ong, N. P. *et al. Phys. Rev. Lett.* **42**, 811 (1979).
13. Bardeen, J. *Phys. Rev. Lett.* **45**, 1978 (1980).
14. Monceau, P. *et al. Phys. Rev. Lett.* **45**, 43 (1980).
15. Fung, K. K. & Steeds, J. W. *Phys. Rev. Lett.* **45**, 1696 (1980).

0028-0836/81/190011-02$01.00

Exact Results in the Kondo Problem. II. Scaling Theory, Qualitatively Correct Solution and Some New Results on One-Dimensional Classical Statistical Models

P. W. Anderson and G. Yuval

Bell Telephone Laboratories, Murray Hill, New Jersey 07974

and

Cavendish Laboratory, Cambridge University, Cambridge, England*

and

D. R. Hamann

Bell Telephone Laboratories, Murray Hill, New Jersey 07974

(Received 10 September 1969)

The simplest Kondo problem is treated exactly in the ferromagnetic case, and given exact bounds for the relevant physical properties in the antiferromagnetic case, by use of a scaling technique on an asymptotically exact expression for the ground-state properties given earlier. The theory also solves the $n=2$ case of the one-dimensional Ising problem. The ferromagnetic case has a finite spin, while the antiferromagnetic case has no truly singular $T\to 0$ properties (e.g., it has finite χ).

I. INTRODUCTION

A previous paper[1] showed that the simplest Kondo problem is equivalent to a certain class of problems in the classical statistical mechanics of one-dimensional systems. One limit of the problem of the Anderson model of a magnetic impurity also leads to the same classical problem.[2] This problem was stated in Ref. 1 as the statistical mechanics of a set of alternating hard rods on a line interacting via logarithmic ("two-dimensional Coulomb") potentials:

$$\langle 0|e^{-\beta\mathcal{H}}|0\rangle = Z' = \sum_n \left(\frac{J_\perp}{2}\right)^{2n} \int_0^\beta d\beta_{2n} \int_0^{\beta_{2n}-\tau} d\beta_{2n-1} \times \cdots \int_0^{\beta_2-\tau} d\beta_1 \exp\sum_{i>j} (-1)^{i-j}(2-\epsilon) \times \ln\left(\frac{\beta_i-\beta_j}{\tau}\right). \quad (1)$$

Here $\mathcal{H}$ is the Hamiltonian of the Kondo system:

$$\mathcal{H} = \text{K. E.} + 2J_\perp(S_+s_- + S_-s_+) + J_z s_z S_z\,, \quad (2)$$

S being the local spin ($S=\frac{1}{2}$) and s the spin of the free electrons at the local site. τ is a cutoff of order $1/E_F$, and

$$\epsilon = 8\delta/\pi - 8\delta^2/\pi^2 \simeq 2J_z\tau\,, \quad (3)$$

where δ is the scattering phase shift of antiferromagnetic sign caused by the $J_z s_z S_z$ term. $|0\rangle$ is the unperturbed ground state of $\mathcal{H}_0 = \text{K. E.} + JS_z s_z$.

Since we depend so completely here on the result Eq. (1) of Ref. 1, let us outline the argument of that paper and try to clarify the meaning of Eq. (1). In that paper we go to the Feynman space-time formalism (actually space-imaginary time = temperature) and, since the Kondo problem of the magnetic impurity treats only a single-point impurity, the question reduces to a sum over paths in only the one ("time") dimension. In addition, the perturbation [which we take as the $J_\perp$ term of Eq. (2)] has the effect of flipping the local spin at each application, so that the problem reduces to calculating the amplitude for a succession of spin flips at times $\beta_1, \beta_2, \ldots$ and the sum over histories is just the sum over all possible numbers and positons of flips. Thus, formally we can write a ground-state average such as Eq. (1) as a grand partition function of an effective one-dimensional gas of spin flips. The one difficult step of Ref. (1) is that of showing that the effective interaction in this one-dimensional gas is a simple logarithmic pair interaction, and it is only at long enough distances $\beta_i-\beta_j$ that the proof we used is precise. Fortunately, the classic Kondo problem has been defined as the calculation of the limiting behavior for small J, in which case the gas of spin flips becomes increasingly rarified and the behavior for large distances must be controlling; as we shall see, this statement has a precise meaning in the context of the present paper. This corresponds to the fact that the Kondo effect has always been assumed to involve only electrons near the Fermi surface. But because of the singular nature at small $\beta_i-\beta_j$ of the asymptotic expression valid for large $\beta_i-\beta_j$, at several stages of the problem the behavior for small $\beta_i-\beta_j$ must be handled in some way such as to avoid ultraviolet divergences; and cutoffs of various shapes [of which Eq. (1) is an example] must be introduced. It is easy to see that such a cutoff must be present physically; energy bands are not infinitely wide, and J's involve form factors of physical wave functions. In the derivation of Eq. (1), the "origin-to-origin unperturbed Green's function"

$$G_0(t) = \langle 0 | T\psi(0)\psi^\dagger(t) | 0 \rangle ,$$

where ψ is the normalized wave function coupled to the local spin, enters. G_0 behaves like τ/t at large times, where τ is the density of states $\sim 1/E_F$, and has an easily computed cutoff at short times for any given ψ and band structure. Of Eq. (1) we know only that the relevant τ is closely related to the cutoff in G_0, a relationship which may be computed precisely only in certain limiting cases.

Let us also remark that only a slight modification of Eq. (1) gives the finite-temperature partition function of the Kondo problem, which will be the subject of a later paper. Essentially, one replaces $\ln[(\beta_i - \beta_j)/\tau]$ by

$$\ln\{(\beta/2\pi\tau)\sin[(\beta_i - \beta_j)/2\pi\beta]\} ,$$

where $\beta = 1/T$.[3]

One important thing to note is that a transformation $S_x \to -S_x$, $S_y \to -S_y$, $S_z \to +S_z$ leaves the dynamics of the spin unchanged (it is simply a proper rotation of the coordinate system) so that the sign of $J_\pm$ is irrelevant. The sign of J_z, and thus of ϵ, determines whether the coupling is ferromagnetic or antiferromagnetic. As we have it, $\epsilon > 0$ is antiferromagnetic. Thus by varying ϵ with $J_\pm$ fixed (i.e., varying the effective "temperature" of the statistical problem) we can go continuously from ferromagnetic to antiferromagnetic coupling. Manifestly, the quantity (1) is a function only of three parameters, β/τ, $J_\pm\tau$, and $\epsilon \simeq 2J_z\tau$. As $\beta \to \infty$ we expect $Z = e^{\beta F}$, and $F\tau$ is now a function only of $J_\pm\tau$ and ϵ. We may think of these two parameters as the exponential of the chemical potential for spin flips, and as the effective temperature of the classical problem (which is to be carefully distinguished from the real temperature β^{-1}, which is the inverse of the "volume" and goes to zero as the length of the line increases to ∞). The quantity F is the negative of the ground-state energy relative to that of $\mathcal{H}_0$; in the one-dimensional problem, however, it plays the role of a "pressure," conjugate to the "volume" β. All $\beta \to \infty$ problems may be plotted on a diagram [Fig. 1(a)] on which two lines, radiating from $\epsilon = 0$ to left and right, represent the manifold of physical (isotropic) Kondo problems of ferromagnetic and antiferromagnetic sign.

It is commonly believed that the ferromagnetic Kondo system has a mean spin moment at absolute zero, while the antiferromagnetic one does not. It is easy to see that the possession of a spin moment at 0 corresponds to a long-range order of the classical system in which + and − charges are all associated in pairs pointing in one direction, either left or right. This type of order is clearer if we partially integrate the interaction twice and write Eq. (1) as an integral over all possible paths of a function $S_z(\beta')$:

$$\langle 0 | e^{-\beta\mathcal{H}} | 0 \rangle = \int d(\text{paths}) \exp\left[\frac{2-\epsilon}{2}\int_0^\beta\int_0^\beta d\beta'\, d\beta''\, \frac{S_z(\beta')S_z(\beta'')}{(\beta'-\beta'')^2+\gamma^2} - (\ln J_\pm)\times(\text{number of jumps})\right]. \quad (4)$$

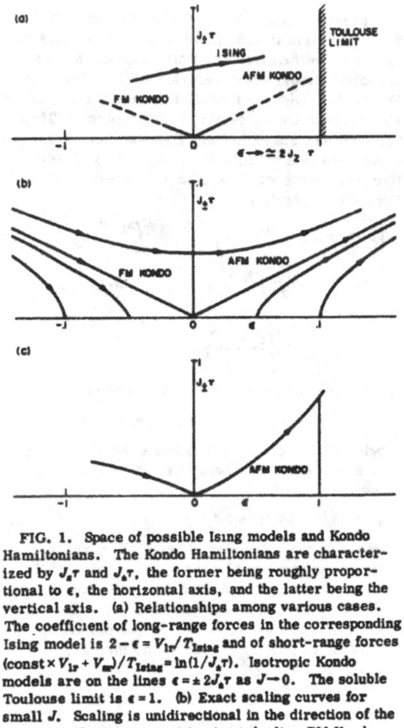

FIG. 1. Space of possible Ising models and Kondo Hamiltonians. The Kondo Hamiltonians are characterized by $J_z\tau$ and $J_\pm\tau$, the former being roughly proportional to ϵ, the horizontal axis, and the latter being the vertical axis. (a) Relationships among various cases. The coefficient of long-range forces in the corresponding Ising model is $2-\epsilon = V_{lr}/T_{Ising}$ and of short-range forces $(\text{const}\times V_{lr} + V_{sr})/T_{Ising} = \ln(1/J_\pm\tau)$. Isotropic Kondo models are on the lines $\epsilon = \pm 2J_\pm\tau$ as $J \to 0$. The soluble Toulouse limit is $\epsilon = 1$. (b) Exact scaling curves for small J. Scaling is unidirectional in the direction of the arrows. The Ising transition is at the line FM Kondo. (c) Approximate scaling curves for strong interactions according to "upper limit" of Fig. 5. "Best guess" would be almost indistinguishably lower.

Here $S_z(\beta)$ is a function of the form of Fig. 2; it takes on only the values $\pm\frac{1}{2}$ and jumps (with either a minimum jump time or a form factor of order τ) between these two values at will. Then the long-range order which implies magnetization is the long-range order of $S_z(\beta')$. This last expression is, in turn, essentially equivalent to an Ising model with a long-range ferromagnetic interaction with form $1/(i-j)^2$ and strength $(1-\epsilon/2)$, and a short-range ferromagnetic interaction $\ln(1/J_\pm\tau)$ + const.

It may be best to make a pedagogic point about the meaning of the asymptotic validity of (1) on this Ising-model version of the problem. It is recognized universally, for sound but not generally explicitly stated reasons, that the qualitative behavior, at least, of such Ising models is entirely controlled by the long-range forces. The basic reason is that short-range forces cannot lead to long-range order, or even to short-range order decaying more slowly than exponentially. Thus, the nature of all long-range singularities (corresponding to low-temperature or low-frequency singularities in the Kondo effect) is entirely determined by the long-range interaction, for which the asymptotic theory that we use is accurate.

Incidentally, a theorem of Griffith[4] shows that the correlations cannot be weaker than the long-range forces themselves in the Ising model, thus, than $1/\beta^2$. We shall show that this is indeed the correlation behavior in the antiferromagnetic case,

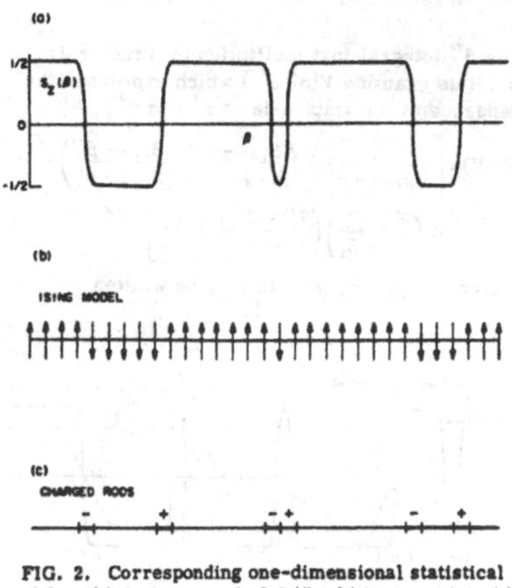

FIG. 2. Corresponding one-dimensional statistical models. (a) Path integral of $S_z(\beta)$, (b) Ising model, (c) charged rod model. Corresponding configurations as a function of β are shown.

since we derive also an upper limit to the correlations in that case.

Dyson has been unable to determine whether or not this Ising model has a phase transition.[5] One purpose of this paper is to show that it does [actually on the original model (1)] and that at least for small $J_\pm$ this does indeed occur at the ferromagnetic-antiferromagnetic boundary point $\epsilon = 0$. We also throw considerable further light on the Kondo problem (though, unfortunately, without achieving a completely exact solution of the interesting antiferromagnetic case) by showing that there is a rigorous scaling technique which reduces the two-dimensional manifold of Fig. 1 to a one-dimensional one. In particular, we can map the entire ferromagnetic line on the point at the origin ($\epsilon = 0$), thus solving the ferromagnetic case really exactly and showing that it has finite magnetization. All cases to the left of the ferromagnetic ones can be mapped on the $J_\pm = 0$ line, and thus are soluble and ordered.

The same scaling laws map all antiferromagnetic cases onto each other, the scale factor being the Kondo temperature. Unfortunately, the direction is in the sense of increasing $J_\pm\tau$, which carries one eventually into the region where the scaling equations are form-factor dependent numerically. Nonetheless, the meaning of the Kondo temperature, the fact that the state is unpolarized, and the connection with perturbation theory are all clear. Inequalities can be found which show that the renormalization procedure is qualitatively valid up to $\epsilon = 1$. Thus, since all cases with $\epsilon = 1$ are exactly soluble (a remark due independently to Toulouse[6]), we can give a solution with parameters correct to logarithmic accuracy. We believe this solution to be at least qualitatively right; in particular, it gives correlation functions which indicate nonsingular properties at absolute zero, in contradiction to most previous theories.[7]

II. SCALING THEORY OF THE COULOMB GAS MODEL

The technique we use is a "renormalization" of the cutoff parameter τ, which leads to a set of scaling laws connecting a given problem to ones with different parameter values. We show that all pairs of flips closer than $\tau_1 > \tau$ can be eliminated, leaving a problem of precisely the same form but with modified parameters $\tilde{J}_\pm\tau_1$ and $\tilde{\epsilon}(\tau_1)$, and a modified F:

$$Z'(J_\pm\tau, \epsilon) = Z'(\tilde{J}_\pm\tau_1, \tilde{\epsilon})\exp\Delta F_{\tau_1}\beta \,. \tag{5}$$

These scaling laws are exact for small $J_\pm\tau$ and ϵ, and are subject to precise inequalities in any case.

First we observe that if $J_\pm\tau$ is small enough, there will be few spin flips + or −: Our "Coulomb gas" of spin flips is a rarified one. On the other hand, in the presence of the strong $\ln(\beta_i - \beta_j)$ interaction, those which are present will tend to appear as close pairs. Clusters of more than two will not form with high probability because the + and − members of a pair attract and repel a third flip with equal intensity:

$$-\ln(\beta_i + \Delta\beta - \beta_j) + \ln(\beta_i - \beta_j) \simeq -[\Delta\beta/(\beta_i - \beta_j)] .$$

Thus, pairs attract singles with a weak $1/x$ potential, which cannot overwhelm the large $(-\ln J_{\perp})$ extra energy required to make a group of three. Thus, close pairs of flips form reasonably self-contained systems whose behavior is mostly determined by the internal force between the pair. It is easy to calculate that the mean distance between pairs is

$$\frac{1}{l_0} = N = \frac{1}{\beta}\frac{\partial \ln Z'}{\partial \mu} = \frac{\partial E_g}{\partial[\ln(\frac{1}{2}J_{\perp})^2]} , \tag{6}$$

where we note (see Ref. 1) that Z' as defined in Eq. (1) is a constant times $e^{-\beta E_g}$, and E_g is the ground-state energy relative to $\langle 0|\mathcal{H}|0\rangle$. Second-order perturbation theory gives an estimate for E_g adequate for our purposes:

$$E_g \simeq \tfrac{1}{4}J_{\perp}^2\tau , \quad l_0 \sim 2\tau/(J_{\perp}\tau)^2 . \tag{7}$$

We can also estimate that the separation of spin flips tends to be of the order

$$\bar{X} \sim \tau/\epsilon , \tag{8}$$

although the actual mean value is dependent on the precise treatment of long-range forces and is not well defined. In any case for ϵ and $J_{\perp}\tau$ small

$$l_0 \gg \bar{X} \gg \tau . \tag{9}$$

The physical picture we have is of many close pairs of flips which change the mean magnetization slightly, interspersed between pairs of isolated flips which are real reversals of M over a larger timescale. This suggests that we might consider the isolated flips as operating in a medium where the close pairs merely modify the mean magnetization (see Fig. 3). It is this idea which we now give a rigorous form.

Consider only "close pairs," so close that their separation is between the limits $\tau \le \beta_{i+1} - \beta_i < \tau + d\tau$. If $d\tau$ is infinitesimal, such a pair occurs very rarely, so we may completely neglect the possibility of two successive close pairs occurring. Then we can rewrite the integrals in Eq. (1) as follows:

$$\cdots \int_0^{\beta_5-\tau} d\beta_4 \int_0^{\beta_4-\tau} d\beta_3 \int_0^{\beta_3-\tau} d\beta_2 \int_0^{\beta_2-\tau} d\beta_1$$

$$= \int_0^{\beta_5-\tau-d\tau} d\beta_4 \int_0^{\beta_4-\tau-d\tau} d\beta_3$$

$$\times \int_0^{\beta_3-\tau-d\tau} d\beta_2 \int_0^{\beta_2-\tau-d\tau} d\beta_1 \qquad \text{(all "free")}$$

$$+ \int_0^{\beta_5-\tau-d\tau} d\beta_4 \int_0^{\beta_4-\tau-d\tau} d\beta_3$$

$$\times \int_0^{\beta_3-\tau} d\beta_2 \int_{\beta_2-\tau-d\tau}^{\beta_2-\tau} d\beta_1 \qquad \text{(1, 2 paired)}$$

$$+ \int_0^{\beta_5-\tau-d\tau} d\beta_4 \int_0^{\beta_4-\tau} d\beta_3$$

$$\times \int_{\beta_3-\tau-d\tau}^{\beta_3-\tau} d\beta_2 \int_0^{\beta_2-\tau-d\tau} d\beta_1 \qquad \text{(2, 3 paired)}$$

$$+ \cdots , \tag{10}$$

and between any pair of free flips (such as 4 and 1 in the second version above) there may be only one close pair; two only occur $\sim (d\tau)^2 = 0$.

We rearrange the sum (10) in a familiar way: We group together all terms with $2n$ free β's, between each pair of which there may be either zero or one close pair with separation $\sim \tau$. Consider one particular pair of free β's, β_i and β_{i+1}. We now have

$$Z = \sum_n \int^{\beta} \cdots \int_0^{\beta_{i+2}-\tau-d\tau} d\beta_{i+1} \int_0^{\beta_{i+1}-\tau-d\tau} d\beta_i$$

$$\cdots (\tfrac{1}{2}J_{\perp})^{2n} \exp[-V_0(\beta_{2n} \cdots \beta_{i+1}, \beta_i \cdots)]$$

$$\times \prod_i \{1 + \tfrac{1}{4}J_{\perp}^2 \int_{\beta_i+2\tau}^{\beta_{i+1}} d\beta' \int_{\beta'-\tau-d\tau}^{\beta'-\tau} d\beta'' \exp[-V$$

$$\times (\beta' - \beta_{2n}, \beta' - \beta_{2n-1} \cdots, \beta'' - \beta_2 \cdots, \beta' - \beta'')]\} . \tag{11}$$

The β'' integral just multiplies by a factor $d\tau$. Now let us examine $V(\beta', \beta'')$ which expresses the dependence of the amplitude on β' and β'':

$$\exp(-V) = \left[\left(\frac{\beta_{i+1} - \beta''}{\beta_{i+1} - \beta'}\right)\left(\frac{\beta_{i+2} - \beta'}{\beta_{i+2} - \beta''}\right)\left(\frac{\beta_{i+3} - \beta''}{\beta_{i+3} - \beta'}\right)\cdots \right.$$

$$\left. \times \left(\frac{\beta' - \beta_i}{\beta'' - \beta_i}\right)\left(\frac{\beta'' - \beta_{i-1}}{\beta' - \beta_{i-1}}\right)\cdots\right]^{2-\epsilon} .$$

We have $\beta' - \beta'' = \tau$, so this may be written

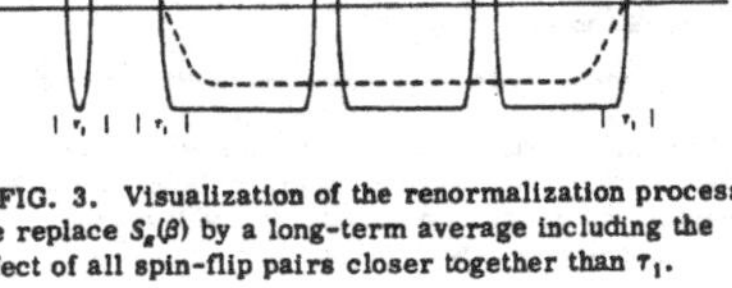

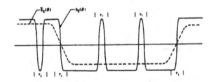

FIG. 3. Visualization of the renormalization process: We replace $S_z(\beta)$ by a long-term average including the effect of all spin-flip pairs closer together than τ_1.

$$\exp(-V)=\left[\left(1+\frac{\tau}{\beta_{i+1}-\beta}\right)\left(1+\frac{\tau}{\beta''-\beta_i}\right)\right.$$
$$\left.\times\left(1+\frac{\tau}{\beta_{i+2}-\beta'}\right)^{-1}\left(1+\frac{\tau}{\beta''-\beta_{i-1}}\right)^{-1}\cdots\right]^{2-\epsilon} \tag{12}$$

and our basic approximation is simply to write the integral of this over the permitted range of β' as (we denote the integral by an average $\langle\ \rangle$ to save writing)

$$\langle\exp(-V)-1\rangle\cong(2-\epsilon)\tau\left\langle\frac{1}{\beta_{i+1}-\beta'}+\frac{1}{\beta''-\beta_i}\right.$$
$$\left.-\frac{1}{\beta_{i+2}-\beta'}-\frac{1}{\beta''-\beta_{i-1}}+\cdots\right\rangle. \tag{13}$$

Let us make some comments about this approximation, which is much better than it looks. (1) It obviously correctly reproduces the dependence on all β's far from i. (2) The alternating series are rapidly converging as is the product in Eq. (12). (3) For small ϵ even $\tau/(\beta_i-\beta')$ is of order ϵ and thus small. (4) All intervals between β's are $>\tau$ so that $\exp(-V)<4$ and that occurs only with very low probability. In the Appendix, we consider the validity of the approximations in great detail and show that Eq. (12) is exact in a certain asymptotic sense, and that rigorous limits may be set on the behavior.

Inserting Eq. (13) in Eq. (10), we get

$$Z=\sum_n\int_0^{\beta}\cdots\int_0^{\beta_{i+2}-\tau-d\tau}d\beta_{i+1}\cdots(\tfrac{1}{2}J_{\pm})^{2n}\exp(-V_0)$$
$$\times\prod_i\left[1-(\tfrac{1}{2}J_{\pm})^2d\tau\left\{(\beta_{i+1}-\beta_i-3\tau)\right.\right.$$
$$\left.\left.\times\left(1+(2-\epsilon)\tau\left\langle\sum\pm\frac{1}{\beta_n-\beta'}\right\rangle\right)\right\}\right] \tag{14}$$

Since $d\tau$ is infinitesimal, we may exponentiate the $J_{\pm}^2 d\tau$ terms. The factor involving $\beta_{i+1}-\beta_i$ gives us

$$Z=\exp[(\tfrac{1}{2}J_{\pm})^2\beta d\tau]\tilde{Z}, \tag{15}$$

where $\tilde{Z}$ is formally the same as in Eq. (1) except for modifications of the amplitudes. Calculating these modifications, we distinguish two regions. First, there is the region of small $J_{\pm}\tau$ and ϵ, where $\beta_{i+1}-\beta_i$ is in general $\sim l_0\gg\tau$, and we neglect corrections of order $\tau/\Delta\beta$. Then we have very simply

$$\int\frac{d\beta'}{\beta_n-\beta'}\cong\ln\frac{\beta_n-\beta_{i+1}}{\beta_n-\beta_i}, \tag{16}$$

with each interval $i-i+1$ contributing two terms, one for each end; counting in the ones for $n-n\pm1$ we get four in all. Then we neglect τ relative to the interval and get

$$\tilde{Z}=\sum_n(\tfrac{1}{2}J_{\pm})^{2n}\prod_m\int_0^{\beta_{m+1}-\tau-d\tau}d\beta_m$$
$$\times\exp\{(2-\epsilon)n\ln\tau+[2-\epsilon-J_{\pm}^2\tau d\tau(2-\epsilon)]$$
$$\times\sum_{i>j}(-1)^{i-j}\ln(\beta_i-\beta_j)\}. \tag{17}$$

Clearly, $\tilde{Z}$ is of the appropriate form of a modified Z with a new interaction $\epsilon-\tilde{\epsilon}=\epsilon+d\epsilon$

$$d\epsilon=(2-\epsilon)J_{\pm}^2\tau d\tau=(2-\epsilon)(J_{\pm}\tau)^2d(\ln\tau); \tag{18}$$

Although apparently $J_{\pm}$ is unmodified, we know that, in fact, there will be fewer flips per unit β in $\tilde{Z}$ than in Z. The cutoff has been altered, while the dimensional coefficient $\tau^{(2-\epsilon)n}$ in the amplitude is unchanged. In order to reduce $\tilde{Z}$ to the form of Z in Eq. (1), we must change

$$J_{\pm}\tau-\tilde{J}_{\pm}\tilde{\tau}=J_{\pm}\tau[(\tau+d\tau)/\tau]^{\epsilon/2} \tag{19}$$

and then it may be verified that

$$\tilde{Z}=Z[\beta/\tilde{\tau},\ (\tilde{J}_{\pm}\tilde{\tau}),\ \tilde{\epsilon}]\ . \tag{20}$$

Equation (19) may be written in differential form:

$$d(J_{\pm}\tau)=\tfrac{1}{2}\epsilon(J_{\pm}\tau)d(\ln\tau)\ . \tag{21}$$

Equations (15), (18), and (21) are the basic scaling laws which solve the problem in the small ϵ, $J_{\pm}$ case.

Let us add a few more words about the validity of Eqs. (15), (18), and (21). It may appear at first that the use of $d\tau$ infinitesimal may make the computation sensitive to the region near the cutoff. Note, however that in the basic laws, Eqs. (18) and (21), the infinitesimal is $d(\ln\tau)$ multiplied by $(J_{\pm}\tau)^2$. Thus, so long as the latter is even reasonably small, rather large jumps in $\ln\tau$, and thus large *factors* in τ, can be considered to be infinitesimal. (This was, in fact, the route by which we arrived originally at these equations.) Thus the first few steps for any true Kondo problem (defined as small $J\tau$) are cutoff independent. After the first step, the cutoff is an artifact of the method and may be chosen at will as in the Appendix. Note also that the small factors need only be $J_{\pm}\tau$ in all equations; finite ϵ is still treated as accurately as desired. Thus the scaling laws are exact throughout the lower region of Fig. 1.

Note that Eqs. (21) and (18) are compatible in the isotropic case $J_z=J_{\pm}$, $\epsilon\cong 2J_z\tau\cong 2J_{\pm}\tau$, and where ϵ and $J_{\pm}\tau$ are small. Thus ϵ and $J_{\pm}\tau$ scale together, and, as should not be entirely unexpected, the isotropic case remains isotropic at every time scale. For anisotropic cases,

$$d\epsilon/d(J_{\pm}\tau)\cong 4J_{\pm}\tau/\epsilon,\quad \epsilon^2-4J_{\pm}^2\tau^2=\text{const.} \tag{22}$$

The scaling lines are a set of hyperbolas with the isotropic cases as asymptotes [see Fig. 1(*b*)]. All ferromagnetic cases below the isotropic one scale onto the case $J_\pm\tau = 0$, which is manifestly ordered, thus locating the transition line at the ferromagnetic case [or above, but we shall argue that all antiferromagnetic cases are disordered and that the two scale into each other according to Fig. 1(b)].

Before going on to further results let us discuss the regime where ϵ and $J_\pm\tau$ are no longer negligible compared to 1. It turns out that for many reasons it is not possible to follow the renormalization in complete detail when it reaches this region (as it obviously shall for the antiferromagnetic cases from Fig. 1). It is not, in fact, even very necessary to do so. What is necessary is to show that there is no tendency for the renormalization process to stop short of $\epsilon = 1$, the case where an exact solution exists as we shall see. To alter the actual parameters by numerical constants which are not exponentially large is physically almost irrelevant. Thus the essential thing is to bound the corrections $d\epsilon/d\tau$ and $dJ_\pm\tau/d\tau$ above (below) some finite numbers. This program is carried out in the Appendix.

III. RESULTS: FERROMAGNETIC CASE

The simple limiting equations (15), (18), (22) suffice to solve the ferromagnetic weak-coupling case. We start at an isotropic case $\tau = \tau_0$, $\epsilon = \epsilon_0$, $J_\pm\tau = \frac{1}{2}\epsilon_0 \ll 1$. Equation (18) gives us

$$4\,d\epsilon/\epsilon^2(2-\epsilon) = d(\ln\tau)\,,$$

and neglecting ϵ relative to 2,

$$\ln(\tau/\tau_0) = 2/|\epsilon| - 2/\epsilon_0\,. \qquad (23)$$

At the same time Eq. (21) gives

$$\ln(\tau/\tau_0) = 1/J_\pm\tau - 2/\epsilon_0 = 1/J_\pm\tau - 1/(J_\pm\tau)_0\,. \qquad (24)$$

Thus the renormalization takes place toward the origin; events on an ever larger time scale occur according to ever weaker interactions. Equation (24) says that the number of flips farther apart than τ decreases logarithmically with τ; this means an extremely slow decay of fluctuations, but in the end the state is polarized.

We expand on the nature of our solution: At a stage at which we have eliminated all pairs closer together than $\tau \gg \tau_0$, the remaining pairs occur at a mean separation of order $l \sim \tau/\epsilon^2(\tau) \sim \tau\ln^2(\tau/\tau_0)$ and have a length $\sim \tau/\epsilon \sim \tau\ln(\tau/\tau_0)$. Thus averaging over lengths of order $\tau\ln^2(\tau/\tau_0)$ we would see a mean polarization of order

$$\epsilon(\tau) \simeq \frac{2}{\ln(\tau/\tau_0)} \simeq \frac{2}{\ln(l/\tau_0) - 2\ln\ln(l/\tau_0)}\,.$$

We have supposed it permissible to rotate in the complex time plane and to interpret this as the qualitative behavior of the time spin-spin correlation function.

$F = E_g$ can be obtained using Eq. (15) with (24).

$$-E_g = F = \tfrac{1}{4}\int_{\tau_0}^{0} J_\pm^2\tau^2\frac{d\tau}{\tau^2} = \tfrac{1}{4}\int_{(J\tau)_0}^{0}\frac{d(J_\pm\tau)}{d}$$

$$= \frac{1}{2\tau_0}\exp\left(\frac{2}{(J_\pm\tau)_0}\right)\int_{(J\tau)_0} dx\,\exp\left(-\frac{2}{x}\right). \qquad (25)$$

This expression has the interesting property that while it is very nonanalytic at $J\tau = 0$, it has a perfectly innocent-seeming asymptotic series at that point which agrees term by term with perturbation theory; as Kondo[8] has noted, perturbation theory gives no logarithmically singular terms in the series for the energy (relative to E_0 of course):

$$E_g = -\left[(\tfrac{1}{2}J_\pm)^2\tau_0\right]\left[1 - 2J_\pm\tau_0 + 6(J_\pm\tau_0)^2 + \cdots\right]\,. \qquad (26)$$

Equation (26) is, as we shall see, also the correct asymptotic series in the antiferromagnetic case with appropriate sign changes, but does not in that case represent the answer adequately: A Stokes phenomenon has intervened. The situation here is almost a classic case of the dangers of relying on perturbation-theory methods and of the complicated analytic behavior which may underly the simple and well-known fact that perturbation series are usually asymptotic.

IV. RESULTS: ANTIFERROMAGNETIC CASE AND TOULOUSE LIMIT

As already noted, the antiferromagnetic (AF) case cannot quite be settled in the same conclusive way. In the ferromagnetic case, as we scale $\tau \to \infty$, $\epsilon \to 0$, and $J_\pm \to 0$. All our expressions become more and more accurate and we have a full and exact solution so long as we start from small values of these parameters. In the AF case, ϵ and $J_\pm$ increase starting from any values, no matter how small, and there is no case for which we cannot eventually find a timescale for which $\epsilon \sim 1$. In fact, this is the fundamental expression of the Kondo phenomenon: Solving Eqs. (18) and (21) approximately by neglecting ϵ relative to 2, we have $\epsilon^2 - (2J_\pm\tau)^2 = \text{const} = 0$ for the isotropic case, and

$$2/\epsilon_0 - 2/\epsilon(\tau) = \ln(\tau/\tau_0)\,. \qquad (27)$$

Defining τ_K as the point where $\epsilon = 1$, we have

$$2/\epsilon_0 - 2 = \ln(\tau_K/\tau_0)$$

$$\text{or}\quad e^2(\tau_K/\tau_0) = e^2/\epsilon_0\,, \qquad (28)$$

the familiar expression for $1/T_K$, T_K being the Kondo temperature. Thus the meaning of the Kondo temperature is merely that it is the scale of "time" = temperature at which the system behaves as though it were strongly coupled. Incidentally, the fact that our equations are exact as $J_s\tau \to 0$ and give a vertical slope at all points except $\epsilon = 0$ on that line shows that the scaling curve cannot reenter the horizontal axis.

Why is $\epsilon = 1$ important? Because it is equivalent to a trivially soluble problem.[8] Consider the Hamiltonian

$$\mathcal{H}_1 = \sum_k \epsilon_k n_k + V_s \sum_k (c_d^\dagger c_k + c_k^\dagger c_d) = \mathcal{H}_0 + V. \quad (29)$$

Here n_k, and c_k, $c_k^\dagger$ are Fermi operators for free spinless electrons, while $c_d^\dagger$ is the same for a local resonant state. This can be solved completely in well-known fashion (it is the "Lee model") as we shall do shortly. On the other hand, it is also possible to calculate the quantity

$$\langle 0|e^{-\beta\mathcal{H}_1}|0\rangle = \langle 0|e^{-\beta\mathcal{H}_0} T \exp\left(-\int_0^\beta V d\beta\right)|0\rangle$$

$$= \sum_n V_s^{2n} \int_0^\beta d\beta_{2n} \int_0^{\beta_{2n}} d\beta_{2n-1} \cdots \int_0^{\beta_2} d\beta_1 G(\beta_1 \cdots \beta_{2n}). \quad (30)$$

Here G is easily shown to be

$$G = \det_{ij} G_0(\beta_{2i} - \beta_{2j-1}), \quad (31)$$

where G_0 is the free-particle Green's function, and since we may assume a cutoff form for G_0

$$G_0(\beta) = \tau/\beta, \qquad \beta > \tau \quad (32)$$

this determinant can be evaluated as a Cauchy determinant, giving

$$\langle 0|e^{-\beta\mathcal{H}_1}|0\rangle$$
$$= \sum_n V_s^{2n} \int\int\int\int \exp\left((-1)^{i-j} \sum_{i>j} \ln \frac{\beta_i - \beta_j}{\tau}\right)$$
$$= Z'(\beta/\tau, 2V_s\tau, \epsilon = 1). \quad (33)$$

(We need not specialize to Eq. (32); Mushkelishvili methods[9] give us the form (33) for any reasonable G_0.)

The ground-state energy of Eq. (29), which is the Hamiltonian of a resonance of width $\Delta = V_s^2\tau$ at the Fermi surface, is easily calculated to be

$$E_{gT} = V_s^2\tau \ln(V_s^2\tau^2) + \text{cutoff-dependent terms}$$
$$\cong V_s^2\tau \ln(V_s^2\tau^2) - (1/\tau)\cot^{-1}(1/V_s^2\tau^2)$$
$$- V_s^2\tau \ln(1 + V_s^4\tau^4)^{1/2}, \quad (34)$$

the latter for the simple assumption of a constant density of states, $G_0 = (\tau/\beta)(1 - e^{-\beta/\tau})$, which is quite close enough to our assumption of the Appendix.

The $S_z - S_z$ correlation function of the real Kondo system with $\epsilon = 1$ obviously corresponds to the $n_d - n_d$ correlation function of Eq. (29). This correlation function is just

$$G_d^2(t) \simeq (\Delta/t)^2.$$

A $1/t^2$ correlation function corresponds to

$$\chi(T=0) = (1/T)\int_0^{1/T} S_z(0) S_z(t)\, dt = \text{finite} \quad .$$

Our scheme, then, is to scale by means of the basic equations (15), (18), (21) (or their more accurate counterparts from the Appendix)

$$dF = [(2-\epsilon)/(2-\epsilon_0)](\tfrac{1}{2}J_s)\, d\tau, \quad (35)$$

$$d\epsilon = [(2-\epsilon)^2/(2-\epsilon_0)] J_s^2 \tau^2 (d\tau/\tau), \quad (36)$$

$$d\ln(J_s^2\tau^2) = \epsilon\,(d\tau/\tau). \quad (37)$$

These equations must be solved starting from their weak-coupling asymptotes [Eq. (27)]. Once integrated up to $\epsilon = 1$, at which we obtain $J_s\tau = 0.783$, we obtain the total ground-state energy by adding the Toulouse result Eq. (34)

$$E_g \doteq -\int_{\tau_0}^{\epsilon(\tau)=1} \left(\frac{2-\epsilon}{2-\epsilon_0}\right)\left(\frac{J_s\tau}{2}\right)^2 \frac{d\tau}{\tau^2} - E_{gT}\left(\tau_\kappa, \frac{J_s\tau_\kappa}{2}\right). \quad (38)$$

Both the upper limit and the Toulouse term give results of the order of T_κ, while the lower limit term is the series [Eq. (26)]. Further numerical results will be given in succeeding communications.

In what sense is this a complete solution of the antiferromagnetic problem? We believe it is so in a very real sense, just as Fermi liquid theory solves most problems of pure metals even without giving precise numerical parameters. We establish the scaling factor to logarithmic accuracy only, but what is important is to prove that it exists: that all magnetic impurities at *some* scale behave just like ordinary ones.

V. CONCLUSION

In conclusion let us, for one thing, make some remarks about experimental comparisons. The most interesting question on the Kondo effect[10] has been from the start whether it did or did not fit into the structure of usual Fermi gas theory: In particular, does a true infrared singularity occur as in the x-ray problem,[8] or does the Kondo impurity obey phase-space arguments as $T \to 0$ and give no energy dependences more singular than E^2 (or T^2), and is $\chi(T=0)$ finite? The result we find is that the usual antiferromagnetic case in fact *does* fit after the time scale has been revised to τ_κ, i.e., that it behaves like a true bound singlet as was conjectured originally by Nagaoka.[7] Thus, experimental results giving singular behavior are, after all, as suggested by Star,[11] probably interaction effects. This is a satisfactory situation from the point of view of many-body theory but a highly unsatisfactory one from the experimentalists' side.

The ferromagnetic case, on the other hand, exhibits strange enough behavior to satisfy the most particular, with finite S, zero effective J, and logarithmic correlation functions.

The one-dimensional Ising model with $n = 2$ is solved en route and Thouless's conjecture[12] that a finite magnetization jump occurs at T_c is verified. Nagle and Bonner[13] have, by means of numerical extrapolations, calculated an approximate transition temperature for this case with no additional nearest-neighbor force (i.e., $J_\pm \tau \simeq 1$) which is consistent with Fig. 1(c). Nagle and Bonner's estimate of the critical exponent β of 0 is consistent with our finite jump of M, also.

But perhaps the most interesting implications of this work are purely theoretical: First, it represents a soluble case of a true many-body problem where one can gain insight into the reality behind approximations of many different kinds; and second, it may throw a very great deal of light onto the formal theory of scaling laws in statistical mechanics.

ACKNOWLEDGMENTS

We should like particularly to acknowledge extensive and helpful discussions on these questions with P. Nozières and J. J. Hopfield, and the use of their work prior to publication.

APPENDIX: SOFT CUTOFFS – AN APPROXIMATE THEORY AND EXACT LIMITS IN THE STRONG-COUPLING CASE

The "sharp" cutoffs we have used in the bulk of the paper are a strictly artificial representation of the actual physics. The physical effect in any reasonable band structure of bringing two flips together is not to reduce the amplitude to zero but to cancel out the effects of the two flips leaving unit amplitude. (This is just a statement about the algebra of the spin operators at equal times.) Thus, in the initial problem, the cutoff at τ should essentially be to unit amplitude: For $[\tau/(\beta_i - \beta_{i+1})]^{2-\epsilon}$, we should substitute some function such as is shown in Fig. 4(c). For definiteness we pick the simplified "flat-top" cutoff also shown in Fig. 4(b).

$$\varphi_{ft}(\beta) = 1, \qquad \beta < \tau$$
$$= (\tau/\beta)^{2-\epsilon}, \quad \beta > \tau . \tag{A1}$$

The quantity Z' we wish to evaluate, then, now has integrals covering the full region, but whenever any two arguments β_i, β_j come closer than τ we replace the $[(\beta_i - \beta_j)/\tau]$ factor by unity:

$$Z' = \sum_n (\tfrac{1}{2}J_\pm)^{2n} \int_0^\beta d\beta_{2n} \int_0^{\beta_{2n}} d\beta_{2n-1} \cdots \int_0^{\beta_{2i+1}} d\beta_{2i}$$
$$\times \cdots \int_0^{\beta_2} d\beta_1 \times \prod_{i>j} [\varphi(\beta_i - \beta_j)]^{(-1)^{i-j+1}} . \tag{A2}$$

Now the process of scaling the cutoff τ may be carried out by making a small change in φ:

$$\varphi = \varphi' + d\varphi , \tag{A3}$$

where for the flat-top function we choose to use

$$\begin{aligned} \varphi'_{ft} &= 1 - (2-\epsilon)\, d\tau/\tau, \quad && \beta < \tau + d\tau \\ &= (\tau/\beta)^{2-\epsilon}, && \beta > \tau + d\tau \\ d\varphi &= (2-\epsilon)\, d\tau/\tau, && \beta < \tau + d\tau \\ &= 0, && \beta > \tau + d\tau . \end{aligned} \tag{A4}$$

[In general, of course, we are simply scaling the parameter in φ. The sharp cutoff function used in the text can be considered from this point of view, see Fig. 4(a).]

Now we insert (A3) into (A2) and rearrange in just the same way, keeping all the terms in which

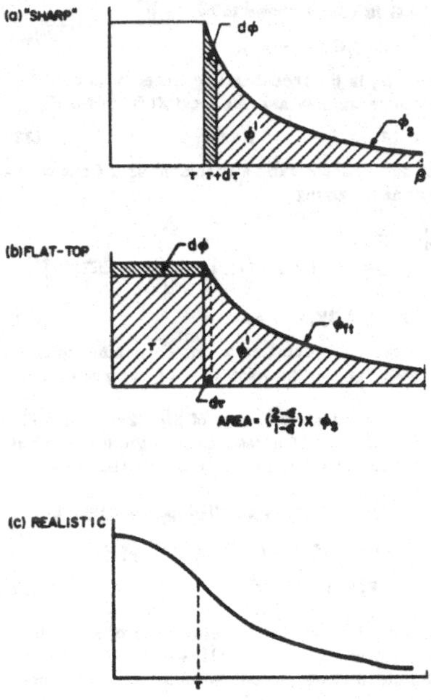

FIG. 4. Cutoff functions. (a) The sharp cutoff φ_s used in the main text and the change in φ_s on a modification of the cutoff by $d\tau$, (b) the flat-top cutoff used in the Appendix and the corresponding rescaling, (c) a possible realistic cutoff function.

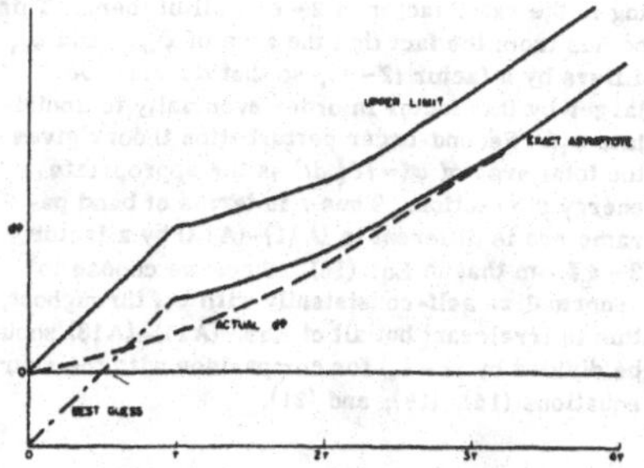

FIG. 5. Changes $d\psi$ in the cutoff function upon modification of τ to $\tau + d\tau$ (arbitrary scale; we have chosen $\epsilon = 1$ for illustration). Dotted curve matches exact low-β behavior and exact asymptote; upper limit is a self-reproducing approximation for $d\psi$ which is definitely above the exact curve; and best guess is an approximate self-reproducing estimate.

φ' appears $2n$ times as the new $2n$th term, and in this new term $d\varphi$ may either enter between any pair of β's, β_{2i} and β_{2i+1}, or not. The cutoff properties are now such that it is not impossible, but simply negligibly rare, that β_{2i} and β_{2i+1} are closer together than τ, and certainly even rarer that β_{2i-1} and β_{2i+2} are also within τ of β_{2i+1} or β_{2i}, respectively. Thus without fear of appreciable error and noting that the error is always in the direction of *over*estimating correlations, we expand the dependence on distant β's as in the text, and focus on a particular pair β_{2i} and β_{2i+1} of interest:

$$Z' = \sum (\tfrac{1}{2} J_\bot)^2 \int \cdots \int_{\beta_{2i-1}}^{\beta_{2i+2}} d\beta_{2i+1} \int_{\beta_{2i-1}}^{\beta_{2i+1}} d\beta_{2i} \times [\varphi'(\beta_{2i+1} - \beta_{2i})(1 + d\psi(\beta_{2i+1} - \beta_{2i}))] \quad , \quad \text{(A5)}$$

where $d\psi$ is the function which results from having a pair of flips β', β'' with a factor $d\varphi(\beta'' - \beta')$ between β_{2i+1} and β_{2i}. We can establish three facts about the function $d\psi$:

$$d\psi(\beta) > 0 \ , \qquad \text{(A6a)}$$

$$d\psi \le \left(\frac{J_\bot}{2}\right)^2 \frac{d\tau}{\tau} \beta^2 \left(\frac{2-\epsilon}{2}\right), \quad \beta \to 0 \qquad \text{(A6b)}$$

$$d\psi \sim dF \times \beta - d\epsilon \ln\beta - d\gamma + O(1/\beta), \quad \beta \to \infty. \qquad \text{(A6c)}$$

We sketch the behavior of $d\psi$ in Fig. 5, according to the calculation we shall do shortly, and also the behavior of $d\psi$ which corresponds to the approximations we shall use: (1) We keep only the first three terms of the asymptotic series, corresponding to the three scaling laws we have already introduced, of F, ϵ, and $J_z\tau$; and (2) we go on to the next stage with the same $\varphi'(\beta_{2i+1} - \beta_{2i})/\tau$, i.e., we get $d\varphi = \text{const} + dF\beta$ inside τ and $=$ (A6c) outside. It is obvious that this can be made a *lower* limit for $d\psi$, and thus a state which has *stronger* correlation, by setting $d\gamma = 0$. This should be a fair estimate as well. Thus we do not falsify any long-range low-frequency singular behavior by Griffith's inequalities. A limit in the other direction is shown in Fig. 5; numerical calculations show that the two are indistinguishable, in fact.

Next we establish (A6). Equation (A6a) is obvious. Let us, for the other two, write out $d\psi$ in Eq. (A5):

$$d\psi(\beta) = \left(\frac{J_\bot}{2}\right)^2 \int_0^\beta d\beta'' \int_0^{\beta''} d\beta' \times \frac{\varphi(\beta - \beta'')\, d\varphi(\beta'' - \beta')\, \varphi(\beta')}{\varphi(\beta - \beta')\, \varphi(\beta'')} \ . \qquad \text{(A7)}$$

For (A6b) we note that when $\beta < \tau$ all functions except $d\varphi$ are unity and the integrals give simply the total area $\frac{1}{2}\beta^2$.

To establish the asymptotic behavior it suffices to go to $\beta \gg \tau$ in Eq. (A7). We then write

$$d\psi = \left(\frac{J_\bot}{2}\right)^2 \frac{d\tau}{\tau} (2-\epsilon) \int_0^\beta d\beta'' \times \int_{\beta''-\tau \text{ or } 0}^{\beta''} d\beta' \left[1 + \left(\frac{\varphi(\beta - \beta'')}{\varphi(\beta - \beta')} \frac{\varphi(\beta')}{\varphi(\beta'')} - 1\right)\right] \ .$$

The "1" gives

$$(\tfrac{1}{2} J_\bot)^2 (d\tau/\tau)(2-\epsilon)[\tau(\beta - \tau) + \tfrac{1}{2}\tau^2] \ . \qquad \text{(A8)}$$

The second term contains, first, the expansion which is valid throughout the interior of the interval,

$$\left(\frac{J_\bot}{2}\right)^2 \frac{d\tau}{\tau} (2-\epsilon) \int_\tau^{\beta-\tau} d\beta'' \int_{\tau \text{ or } \beta''-\tau}^{\beta''} d\beta' \times (2-\epsilon)(\beta'' - \beta') \left(\frac{1}{\beta - \beta'} + \frac{1}{\beta'}\right) + O\left(\frac{1}{\beta^2}\right) \ ,$$

which may be evaluated by assuming τ always small relative to β's:

$$= -\left(\frac{J_\bot}{2}\right)^2 \tau\, d\tau \frac{(2-\epsilon)^2}{2} \times 2 \ln\left(\frac{\beta - \tau/2}{\tau}\right) \ . \qquad \text{(A9)}$$

Finally, we have contributions from the ends of the interval which are again of order $1/\beta$. Neglecting all $1/\beta$ contributions we get, adding (A8) and (A9),

$$d\psi \sim (2-\epsilon)(J_\bot/2)^2\, d\tau [\beta - (2-\epsilon)\tau \ln(\beta/\tau) - \tau/2] \ . \qquad \text{(A10)}$$

We have therefore the new set of scaling laws, setting $(1+d\psi)\simeq e^{d\phi}$, and carrying out the argument as in the main text:

$$dF=(\tfrac{1}{2}J_\pm)^2\,d\tau(2-\epsilon)\,, \tag{A11}$$

$$d\epsilon=\tfrac{1}{2}(2-\epsilon)^2J_\pm^2\tau\,d\tau\quad, \tag{A12}$$

$$d\ln(J_\pm^2)=-\tfrac{1}{2}(J_\pm)^2\tau\,d\tau$$

$$+\text{rescaling correction as in text ("best")}$$

$$=0+\text{rescaling ("upper limit").} \tag{A13}$$

We use this latter estimate in the text because it is simpler.

One feature of these laws which is a bit surprising is the extra factor of $2-\epsilon$ in all of them. This comes from the fact that the area of φ_{sharp} and φ_{ft} differs by a factor $(2-\epsilon)$, so that $d\varphi$ must be larger by this factor in order eventually to annihilate φ_{ft}. Second-order perturbation theory gives the total area of $\varphi(=\int G_0^2\,dt)$ as the appropriate energy correction. Thus τ in terms of band parameters is different in (A11)–(A13) by a factor $2-\epsilon$ from that in Eq. (15). Since we choose to renormalize self-consistently with φ_{ft} throughout, this is irrelevant but all of Eqs. (A11)–(A13) should be divided by $(2-\epsilon_0)$ for comparison with the sharp equations (15), (18), and (21).

---

*Work at the Cavendish Laboratory supported in part by the Air Force Office of Scientific Research Office of Aerospace Research, U.S. Air Force, under Grant No. 1052-69.

[1] P. W. Anderson and G. Yuval, Phys. Rev. Letters 23, 89 (1969); G. Yuval and P. W. Anderson, Phys. Rev. B 1, 1522 (1970).

[2] D. R. Hamann, Phys. Rev. Letters 23, 95 (1969).

[3] Two of us (G. Yuval and P. W. Anderson) had arrived at this expression, and also K.-D. Schotte has derived it in a preprint independently, using Tomonaga oscillator methods.

[4] R. B. Griffith, J. Math. Phys. 8, 478 (1967).

[5] F. J. Dyson, Commun. Math. Phys. 12, 91 (1969); and to be published.

[6] G. Toulouse (private communication). This result was discovered independently by one of the authors.

[7] H. Suhl, Phys. Rev. 138, A515 (1965); Physics 2, 69 (1968); Phys. Rev. 141, 483 (1966); Physics 3, 17 (1967); A. A. Abrikosov, *ibid.* 25, 61 (1965); P. W. Anderson, Phys. Rev. 164, 352 (1967); J. Appelbaum and J. Kondo, *ibid.* 170, 542 (1968); Y. Nagaoka, Progr. Theoret. Phys. (Kyoto) 37, 13 (1967).

[8] J. Kondo, Progr. Theoret. Phys. (Kyoto) 40, 683 (1968); K. Yosida, Phys. Rev. 147, 223 (1966).

[9] P. Nozières and C. de Dominicis, Phys. Rev. 178, 1097 (1969).

[10] P. W. Anderson, in *The Many-Body Problem*, edited by R. Balian and C. De Witt (Gordon and Breach, New York, 1968).

[11] W. M. Starr, in *Proceedings of the Eleventh International Conference on Low Temperature Physics, St. Andrews, Scotland*, 1968, edited by V. F. Allen, D. M. Finlayson, and D. M. McCall (University of St. Andrews Printing Dept., St. Andrews, Scotland, 1969), pp. 1250, 1280.

[12] D. J. Thouless (unpublished).

[13] J. F. Nagle and J. C. Bonner (unpublished).

ERRATA - P. W. Anderson, G. Yuval and D. R. Hamann Phys. Rev. B 1, 4464-4473 (1970).

1. Third line of Eq. (11) should read:

$$\times \prod_i \{1+\tfrac{1}{4}J_\pm^2\int_{\beta_i+2\tau}^{\beta_{i+1}-\tau} d\beta'\int_{\beta'-\tau-d\tau}^{\beta'-\tau} d\beta''\exp[-V$$

2. First line of Eq. (12) should read:

$$\exp(-V)=\left[\left(1+\frac{\tau}{\beta_{i+1}-\beta'}\right)\left(1+\frac{\tau}{\beta''-\beta_i}\right)\right.$$

3. Line preceding Eq. (14) should read:

Inserting Eq. (13) in Eq. (11), we get...

4. Second line of Eq. (14) should read:

$$\times\prod_i\left[1+(\tfrac{1}{2}J_\pm)^2\,d\tau\left\{(\beta_{i+1}-\beta_i-3\tau)\right.\right.$$

5. Eq. (16) should read:

$$\int\frac{d\beta'}{\beta_n-\beta'}=-\ln\frac{\beta_n-\beta_{i+1}}{\beta_n-\beta_i}$$

6. Second line of Eq. (17) should read:

$$\times\exp\{(2-\tilde{\varepsilon})\,n\ln\tau+[2-\varepsilon-J_\pm^2\tau d\tau(2-\varepsilon)]$$

7. Second line of Eq. (28) should read:

$$e^2(\tau_K/\tau_0)=e^{2/\varepsilon_0}$$

A poor man's derivation of scaling laws for the Kondo problem

P. W. ANDERSON
Cavendish Laboratory,† Cambridge, England
and
Bell Telephone Laboratories, Murray Hill, New Jersey, USA
MS. received 28th April 1970

Abstract. The scaling laws derived by a complicated space-time approach for the Kondo problem in previous work are rederived by a 'cutoff renormalization' technique used previously in the theory of superconductivity.

In a series of recent papers (Anderson and Yuval 1969, 1970, Anderson *et al.* 1970a, b) we showed the equivalence of an anisotropic Kondo problem to a certain kind of one dimensional statistical problem, and solved that problem by deriving scaling laws connecting solutions for different sets of parameters with each other, and specifically with one or another soluble case.

Here we show that these scaling laws are derivable directly and easily in the Kondo problem itself. As in the previous work, the solution of the ferromagnetic case becomes trivial, but the antiferromagnetic one may require, for some purposes, solution of an auxiliary problem. As in previous methods, the parameters of the auxiliary problem are not perfectly defined, but it is clear that that solution behaves like a simple bound state with no singular properties.

The method is based on a technique used in the theory of the 'Coulomb pseudopotential' in superconductivity (Morel and Anderson 1962, Schrieffer 1964). In that problem it was convenient to confine one's interest to the region within a Debye energy of the Fermi surface so what was done was to calculate an effective interaction or 'pseudopotential' which was equivalent, acting in this limited region of momentum space, to the real potential in the full region. Here our technique will be the same: to eliminate successively the higher energy regions in favour of an effective interaction.

Consider the Dyson equation for the scattering matrix T

$$T(\omega) = V_{\text{int}} + V_{\text{int}}\, G_0(\omega)\, T(\omega) \tag{1}$$

where

$$V_{\text{int}} = \frac{J_{\pm}}{2}(S_+s_- + S_-s_+) + J_z S_z s_z \tag{2}$$

is the anisotropic Kondo interaction. ($S_{\pm} = S_x \pm iS_y$). G_0 is the unperturbed Green function for propagation of free electrons in the band

$$G_0 = (\omega - \mathscr{H}_0)^{-1} \tag{3}$$

$$\mathscr{H}_0 = \sum_{\epsilon_k = E_F - E_c}^{E_F + E_c} \epsilon_k n_{k\sigma} \tag{4}$$

† Work at the Cavendish Laboratory supported in part by the Air Force Office of Scientific Research Office of Aerospace Research, US Air Force, under Grant Number 1052-69.

and for simplicity this band is assumed symmetric about the Fermi surface with a sharp cutoff at E_c.

(1) is the exact definition of the matrix T, which is not assumed to have single particle scattering character or any other limitation. We are interested in ω near the ground state in energy, so not a great many excited particles will in fact be present. It is an exact identity that

$$G = G_0 + G_0 T G_0 \tag{5}$$

and the lowest pole of G, and thus of T, determines the ground state. G is here *not* a many body Green function in the usual Matsubara sense but the actual resolvent operator $(\omega - \mathscr{H})^{-1}$.

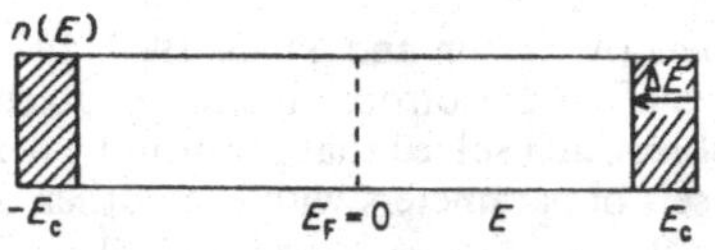

Figure 1.

We now attempt to reduce the cutoff energy E_c to $E_c - \Delta E$, $\Delta E \ll E_c$ (see figure 1). States near E_c are not highly occupied in the ground state and ΔE is arbitrarily small, so it is certainly rigorous to study only states with few or no excited electrons k'' with $E_c > |\epsilon_{k''}| > E_c - \Delta E$ (we shift energy zero to $E_F = 0$). We now try to take into account explicitly all scatterings from our low energy state of interest into $E_c > \epsilon_{k''} > E_c - \Delta E$ by introducing a projection operator $P_{\Delta E}$ which projects *on to states containing one or more such particles*. We then simply resolve (1)

$$T = V + VP_{\Delta E}G_0 T + V(1 - P_{\Delta E})\, G_0 T$$

and resubstituting (1) in the second term we get

$$\begin{aligned} T &= V + VP_{\Delta E}G_0 V + (V + VP_{\Delta E}G_0 V)(1 - P_{\Delta E})\, G_0 T + VP_{\Delta E}G_0 VP_{\Delta E}G_0 T \\ &= V' + V'(1 - P_{\Delta E})\, G_0 T \end{aligned} \tag{6}$$

where

$$V' = V + VP_{\Delta E}G_0 V \tag{7}$$

In getting (6) we have neglected (relative to 1)

$$VP_{\Delta E}G_0 VP_{\Delta E}G_0 \simeq (\Delta E)^2\, \rho^2(E)\, V^2 / E_c^2 \tag{8}$$

which is proportional to $(\Delta E)^2$ and may thus be made small.

Equations (5), (6) and (7) are equivalent to the statement that the original problem with cutoff at E_c is exactly (to order $(\Delta E)^2$) equivalent to a new problem in which, according to (6), we may cut out all matrix elements of T (and therefore of V) leading into the ΔE subspace. By multiplying (6) by $(1 - P)$ on the left and right, we eliminate all matrix elements outside the new cutoff, and have a complete new physical problem as far as matrix elements of G within the remaining subspace are concerned. The new potential V', according to (7), is changed from V by an amount

$$\mathrm{d}V = VP_{\Delta E}G_0 V \tag{9}$$

proportional to ΔE (The projection operators $(1 - P)$ on the left and right may be understood of course.).

Clearly if we can evaluate (9) we are left with a new problem to which the same trick may be applied and in the end we get a differential equation relating sets of equivalent problems. Let us then insert (2) in (9). Formally,

$$\mathrm{d}V = \sum_{k_1,\sigma_1}^{|\epsilon_{k_1}|<E_c-\Delta E} \sum_{k_2,\sigma_2}^{|\epsilon_{k_2}|<E_c-\Delta E} \sum_{k,\sigma}^{E_c>|\epsilon_k|>E_c-\Delta E} \frac{1}{\omega - E_c - |\epsilon_{k_1}|}$$

$$\times \left[(C^+_{k_2\sigma_2}C_{k\sigma})(C^+_{k\sigma}C_{k_1\sigma_1})\left\{\frac{J_\pm}{2}(S_+(s_-)_{\sigma_2\sigma} + S_-(s_+)_{\sigma_2\sigma}) + J_zS_z(s_z)_{\sigma_2\sigma}\right\}\right.$$

$$\times \left\{\frac{J_\pm}{2}(S_+(s_-)_{\sigma\sigma_1} + S_-(s_+)_{\sigma\sigma_1}) + J_zS_z(s_z)_{\sigma\sigma_1}\right\} + C^+_{k\sigma}C_{k_2\sigma_2}C^+_{k_1\sigma_1}C_{k\sigma}\left\{\frac{J_\pm}{2}(S_+(s_-)_{\sigma\sigma_2}\right.$$

$$\left.\left. + S_-(s_+)_{\sigma\sigma_2}) + J_zS_z(s_z)_{\sigma\sigma_2}\right\}\left\{\frac{J_\pm}{2}(S_+(s_-)_{\sigma_1\sigma} + S_-(s_+)_{\sigma_1\sigma}) + J_zS_z(s_z)_{\sigma_1\sigma}\right\}\right].$$

This formidable expression has already had some simplifications carried out. It is assumed that k_1 is not excited in the initial state in order to insert $|\epsilon_{k_1}|$; ΔE is small so $|\epsilon_k| \simeq E_c$, and we know k is unexcited. Actually the number of excited particles is relatively $\simeq 1/\text{total volume} \lll 1$.

To further simplify we set $C^+_{k\sigma}C_{k\sigma} = 1$ or 0 for k below or above the Fermi sea, and sort out the algebra of the s by summing over σ and using

$$s_+s_- = \tfrac{1}{2} + s_z \qquad s_-s_+ = \tfrac{1}{2} - s_z$$

$$s_-s_z = -s_zs_- = \tfrac{1}{2}s_- \qquad s_+s_z = -\tfrac{1}{2}s_+ = -s_zs_+$$

$$s_zs_z = \tfrac{1}{4}$$

(specializing also to $|S| = \frac{1}{2}$). This gives us

$$\mathrm{d}V = \sum_{k_1,\sigma_1}\sum_{k_2,\sigma_2}\frac{\rho\Delta E}{\omega - E_c - |\epsilon_{k_1}|}\left[C^+_{k_2\sigma_2}C_{k_1\sigma_1}\left\{\delta_{\sigma_1\sigma_2}\left(\frac{J^2_\pm}{8} + \frac{J^2_z}{16}\right) - \frac{J^2_\pm S_z(s_z)_{\sigma_2\sigma_1}}{2}\right.\right.$$

$$\left. - \frac{J_\pm J_z}{4}(S_+(s_-)_{\sigma_2\sigma_1} + S_-(s_+)_{\sigma_2\sigma_1})\right\} + C_{k_2\sigma_2}C^+_{k_1\sigma_1}\left\{\delta_{\sigma_1\sigma_2}\left(\frac{J^2_\pm}{8} + \frac{J^2_z}{16}\right) + \frac{J^2_\pm S_z(s_z)_{\sigma_1\sigma_2}}{2}\right.$$

$$\left.\left. + \frac{J_\pm J_z}{4}(S_+(s_-)_{\sigma_1\sigma_2})\right\}\right]. \tag{10}$$

The reason for the solubility of the Kondo problem is that (10) has virtually the same form as (2). To reduce it to exactly the same form we have to make some approximations. First a very good one: the $\delta_{\sigma_1\sigma_2}$ terms for $k_1 \neq k_2$ are simply

$$\frac{1}{16}(J^2_z + 2J^2_\pm)\,C^+_{k_1\sigma_1}C_{k_2\sigma_1}\,\rho\Delta E\left\{\frac{|\epsilon_{k_1}| - |\epsilon_{k_2}|}{(\omega - E_c - |\epsilon_{k_1}|)(\omega - E_c - |\epsilon_{k_2}|)}\right\} \tag{11}$$

which is an ordinary scattering potential which is small, and actually zero at the Fermi surface and on the average. Clearly (11) will not contribute significantly to the infrared divergences which are the heart of the Kondo problem, so hereafter we neglect it. The $k_1 = k_2$ terms, however, lead to a significant result

$$\mathrm{d}V_0 = \frac{\rho\Delta E}{8}(J^2_z + 2J^2_\pm)\sum_{\epsilon_k=0}^{E_c}(\omega - E_c - |\epsilon_k|)^{-1}. \tag{12}$$

This term is not of a form similar to the unperturbed Hamiltonian; it is simply a constant number, a shift of the zero of the energy scale. It represents the net shift of total ground state energy due to scatterings into the ΔE subspace. It would be perfectly consistent to retain it by adding a constant to the interaction V but it is more convenient to incorporate it in $\mathscr{H}_0$ and thus in G_0 at each stage

$$\mathscr{H}'_0 = \mathscr{H}_0 + \mathrm{d}V_0$$

$$G'_0 = (\omega - \mathscr{H}_0 - \mathrm{d}V_0)^{-1}.$$

Since dV_0 is simply a numerical constant (not a particle selfenergy!) it has no effect to include this ground state energy change except to redefine ω. Thus we may sum up the total ground state energy shift at any stage as

$$\Delta(E_g) = \int_{E_c}^{E_c^0} dE \left(\frac{(J_z\rho)^2}{8} + \frac{(J_\pm\rho)^2}{4}\right) \ln\left(\frac{2E - \omega - \Delta}{E - \omega - \Delta}\right) \simeq \ln 2 \int_{E_c}^{E_c^0} \left(\frac{(J_z\rho)^2}{8} + \frac{(J_\pm\rho)^2}{4}\right) dE. \tag{13}$$

This is the first of the scaling laws and corresponds to (15) of our previous paper (Anderson *et al.* 1970).

We are left with the spin dependent terms. The simplest and most important case occurs when $J\rho \ll 1$. If we look back at (8) we will see that the limit on $\Delta E/E_c$ is set by $1/J\rho$. If ΔE may be set fairly large (we call this 'renormalization by leaps and bounds') the relevant values of $|\epsilon_{k_1}|$ are small compared with E_c, and it may be very accurately stated that

$$\frac{dV}{\Delta E} = \frac{\rho}{\omega - E_c + \Delta(E_c)} \sum_{k_1\sigma_1 k_2\sigma_2} C^+_{k_2\sigma_2} C_{k_1\sigma_1} \Big\{ -J_\pm^2 S_z(s_z)_{\sigma_2\sigma_1} - \frac{J_\pm J_z}{2} (S_+(s_-)_{\sigma_2\sigma_1} + S_-(s_+)_{\sigma_2\sigma_1}) \Big\}.$$

This equation shows that the change in the effective perturbation V which is caused by changing the cutoff energy E_c by a small amount is formally equivalent to an anisotropic exchange interaction, and so the effect is merely to change the parameters of the original interaction. This is why we call the procedure 'scaling laws', because what we have done is to show that a problem with one set of parameters is entirely equivalent to that with another set. This leaves us with the hope, usually fulfilled, that by following the scaling procedure out repeatedly we can eventually reach a region where a solution by some other method may be found. The differential equations connecting the three parameters E_c, J_z and $J_\pm$ are, then,

$$\frac{dJ_z}{dE_c} = -\frac{\rho}{\omega - E_c + \Delta} J_\pm^2 \tag{14}$$

$$\frac{dJ_\pm}{dE_c} = -\frac{\rho}{\omega - E_c + \Delta} J_z J_\pm. \tag{15}$$

These are the two primary scaling laws corresponding to (18) and (21) of the previous paper (Anderson *et al.* 1970). The most important conclusion is obtained by dividing (14) by (15) and integrating the resulting equation to obtain

$$J_z^2 - J_\pm^2 = \text{const} \tag{16}$$

the set of hyperbolic curves connecting different cases in the neighbourhood of the origin (see figure 2). Equation (16) shows that ferromagnetic cases on or to the right of the isotropic line are solved trivially by scaling to $J_\pm = 0$, as discussed previously and in agreement with Mattis' general theorem (Mattis 1967)† and with the results of Shiba (preprint). All antiferromagnetic cases with $J\rho \ll 1$, and ferromagnetic cases on the antiferromagnetic side of the isotropic line, scale *away* from small J in the end, and, no matter how weak the original coupling, eventually become equivalent to large $J\rho$. (Note that the sign of (14) shows that J_z always increases; ferromagnetic is J_z negative; while (15) changes sign at $J_z = 0$.) Fuller implications of the scaling laws for small J, especially in regard to the Ising model, are being presented elsewhere.

† Dr Mattis has kindly shown us a generalization to anisotropic exchange which shows that the ferromagnetic isotropic case is indeed the boundary.

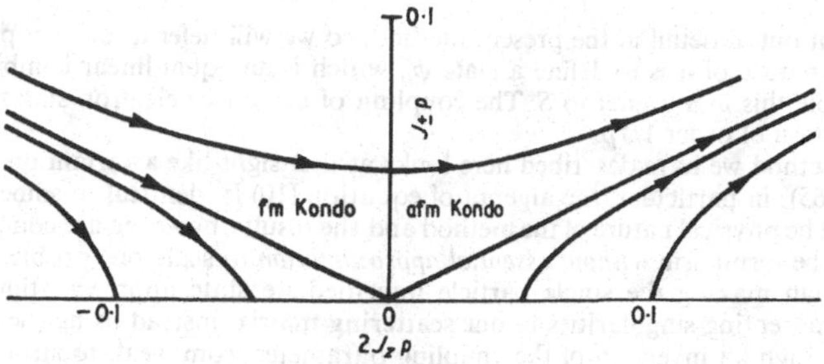

Figure 2.

The interesting antiferromagnetic case, as shown by our previous work, is unfortunately the case in which, as E_c decreases, (14) and (15) become less accurate. In this case the ω and k dependence in (10) becomes increasingly less negligible. Leaving that dependence in, we obtain

$$dV = -\rho\Delta E \sum_{k_1\sigma_1k_2\sigma_2}\left\{\frac{J_\pm^2}{2}S_z(s_z)_{\sigma_2\sigma_1} - \frac{J_\pm J_z}{4}(S_+(s_-)_{\sigma_2\sigma_1} + S_-(s_+)_{\sigma_2\sigma_1})\right\}$$
$$\times\left\{\frac{1}{\omega - E_c + \Delta - |\epsilon_{k_1}|} + \frac{1}{\omega - E_c + \Delta - |\epsilon_{k_2}|}\right\}C^+_{k_2\sigma_2}C_{k_1\sigma_1}. \tag{16}$$

Since in general the denominators are negative, this means that the increase in J is slightly greater at the centre of the band where $|\epsilon_k| = 0$. Basically, (14) and (15) get multiplied by somewhat k dependent functions, which may in the long run lead to rather large k variation of J, especially since the approximate solution of (14) and (15) neglecting the k variation diverges: for the isotropic case, for instance

$$\frac{dJ}{\rho J^2} = \frac{dE_c}{E_c - \omega - \Delta}$$
$$\frac{1}{\rho J_0} - \frac{1}{\rho J} \simeq \ln\left(\frac{E_c^0}{E_c - \omega - \Delta}\right)$$

and at some $\omega < -\Delta$ (which, presumably, is the binding energy of the ground state) given by

$$\omega_0 \simeq -\Delta(0) - E_c^0 \exp(-1/\rho J_0) = -\Delta(0) - E_K \tag{17}$$

$\rho J \to \infty$. Near this point

$$\frac{1}{\rho J} = \ln\frac{E_c - \omega - \Delta}{E_K}. \tag{18}$$

While this state of affairs is highly satisfactory from the point of view of finding the binding energy, it does not leave us happy about being able to define the properties of the low states satisfactorily. Continuing ω beyond the pole leads to great complications, and it is more satisfactory to abandon the scaling procedure at some intermediate stage at which $J\rho$ is fairly large at the Fermi surface but not infinite, and to treat the problem the other way round: as a perturbation theory in $1/J\rho$. It will be seen, however, that the complication of (16) is such that it is not easy to define precisely the exact relationship between $J\rho$ and E_c when $J\rho$ is fairly large, just as in the original method cutoff shape dependence left us uncertain as to precise numerical results.

The solution of this strong coupling case is, as in the previous method, complicated without being particularly edifying. It is our impression that the solution in the other method can be given more rigorous limits, and that therefore it is not particularly urgent to carry it out in detail in the present method, so we will defer it for later publication. The essential nature of it is to define a state ϕ_0 which is an equal linear combination of all k, and couple this in a singlet to S. The coupling of ϕ_0 to free electron states is then a small perturbation of order $1/J\rho$.

The method we have described here looks at first sight like a variant on Suhl's method; (Suhl 1965); in particular, the algebra of equation (10) is identical to some of his manipulations. The physical nature of the method and the results, however, are completely different. We use the formalism *without essential approximation* to scale one problem onto another, rather than making the single particle intermediate state approximation; and we stop short of accepting singularities in our scattering matrix, instead using the method only to scale through an inversion of the coupling parameter from weak to strong. The essential nature of the difficulties of the Kondo problem is thus made clear: one is trapped between a region with all the complications of a true phase transition, and another where the perturbation theory must be done in the inverse of the obvious coupling parameter.

References

ANDERSON, P. W., and YUVAL, G., 1969, *Phys. Rev. Lett.*, **23**, 89.
—— 1970, *Phys. Rev.*, **131**, 1522.
ANDERSON, P. W., YUVAL, G., and HAMAN, D., 1970a, *Phys. Rev.*, **131**, 4464.
—— 1970b, *Solid St. Commun.*, **8**, 1033.
MATTIS, D. C., 1967, *Phys. Rev. Lett.*, **19**, 1478
MOREL, P., and ANDERSON, P. W., 1962, *Phys. Rev.*, **125**, 1263.
SCHRIEFFER, J. R., 1964, *The Theory of Superconductivity* (New York: Benjamin), p. 186.
SUHL, H., 1965, *Phys. Rev.*, **138**, A515.

ERRATA - P. W. Anderson, J. Phys. C 3, 2436-2441 (1970).

1. Last line of Eq. (10) should read:

$$+ \frac{J_\pm J_z}{4} \left(S_+(s_-)_{\sigma_1\sigma_2} + S_-(s_+)_{\sigma_1\sigma_2}\right)\Big\}\Big] .$$

2. Eq. (11) should read:

$$\frac{1}{16}\left(J_z^2+2J_\pm^2\right) c^+_{k_1\sigma_1} c_{k_2\sigma_2}$$

$$\rho\Delta E \left\{ \frac{|\varepsilon_{k_1}|-|\varepsilon_{k_2}|}{(\omega-E_c-|\varepsilon_{k_1}|)(\omega-E_c-|\varepsilon_{k_2}|)} \right\}$$

3. Eq. (13) should read:

$$\Delta E_g = -\int_{E_c}^{E_c^0} dE \left(\frac{(J_z\rho)^2}{8} + \frac{(J_\pm\rho)^2}{4} \right) \ell n \left(\frac{2E-\omega-\Delta}{E-\omega-\Delta} \right)$$

$$\simeq -\ell n 2 \int_{E_c}^{E_c^0} \left(\frac{(J_z\rho)^2}{8} + \frac{(J_\pm\rho)^2}{4} \right) dE$$

Scaling Theory of the Asymmetric Anderson Model

F. D. M. Haldane[(a)]
Bell Laboratories, Murray Hill, New Jersey 07974, and Physics Department, Princeton University, Princeton, New Jersey 08540
(Received 9 September 1977; revised manuscript received 30 November 1977)

A scaling theory is used to show that for temperatures $T \ll U$, the properties of the asymmetric Anderson model ($U \gg |E_d|, \Delta$) are universal functions of the scaling invariants Δ and $E_d^* = E_d + (\Delta/\pi)\ln(W_0/\Delta)$, where W_0 is the conduction electron bandwidth or U, whichever is smaller. Crossovers between various regimes of simple behavior as the temperature changes are described. $|E_d^*| \lesssim \Delta$ is identified as the criterion for a "mixed-valence" ground state, where the susceptibility $\approx \Delta^{-1}$. For $-E_d^* \gg \Delta$, there is a local-moment regime with a Kondo temperature $T_K \approx \Delta \exp(\pi E_d^*/2\Delta)$.

There has been recent interest in the asymmetric Anderson model[1] ($U \gg |E_d|, \Delta$) in connection with the theory of "mixed-valence" rare-earth materials.[2-4] The numerical renormalization-group technique pioneered by Wilson[5] allows the thermodynamic properties to be calculated,[3] but the parameter space is large. Analytic results can clarify the dependence of physical properties on the model parameters, and provide a framework for the numerical exploration of the "crossovers" between limits describable by a simple effective Hamiltonian. This Letter reports a *scaling property* of the asymmetric Anderson model; that is, universality of model properties as functions of the scaling invariants Δ and $E_d^* = E_d + (\Delta/\pi)\ln(W_0/\Delta)$, where $W_0 \approx U$ or the conduction electron bandwidth, whichever is smaller. The scaling equations also allow a simple description of the temperature dependence of physical properties.

The (nondegenerate) Anderson model is characterized by the parameters E_d, U, and $\Delta(\omega)$, and is

$$H = H^0 + E_d \sum_\sigma n_{d\sigma} + U n_{d\uparrow} n_{d\downarrow} + \sum_{k\sigma} V_{kd}\, c_{k\sigma}^\dagger c_{d\sigma} + \text{H.c.},$$
$$H^0 = \sum_{k\sigma} \epsilon_k n_{k\sigma}; \quad \Delta(\omega) = \sum_k |V_{kd}|^2 \delta(\omega - \epsilon_k). \qquad (1)$$

$\Delta(\omega)$ is essentially characterized by Δ $[=\Delta(0)]$, and a bandwidth W where $\Delta(\omega)\approx\Delta$ for $|\omega|\ll W$ and $\Delta(\omega)\approx 0$ for $|\omega|\gg W$. The limit $U\gg\Delta$ may be investigated by a perturbation expansion in Δ. If $W\gg U$ the expansion is independent of W, but for $U\gg W\gg |E_d|$, logarithmic dependence on W appears in each order; for $W\gg T$, the leading terms in the expansion for the impurity susceptibility are

$$\chi=\frac{1}{6T}\left[1+\frac{\Delta}{3\pi T}\ln\left(\frac{T}{W}\right)+\ldots\right] \quad (T\gg |E_d|), \quad (2)$$

$$\chi=\frac{\Delta}{2\pi E_d^2}\left[1+\frac{2\Delta}{\pi E_d}\ln\left(\frac{E_d}{W}\right)+\ldots\right] \quad (E_d\gg T), \quad (3)$$

$$\chi=\frac{1}{4T}\left[1+\frac{2\Delta}{\pi E_d}+\frac{1}{2}\left(\frac{2\Delta}{\pi E_d}\right)^2\ln\left|\frac{T^2}{WE_d}\right|+\ldots\right] \quad (-E_d\gg T). \quad (4)$$

These expansions are ultraviolet divergent as $W\to\infty$; however, for $W\gg U$, they are given by equivalent expressions where a quantity of order U replaces W. (The effects of processes involving high-energy conduction-band states cancel exactly in the noninteracting limit $U=0$; this cancellation remains for conduction band energies $\gg U$.) Perturbation theory indicates that states with energies in the range $|E_d|, T\ll|\omega|\ll U$ play an important role in low-energy processes as virtually excited intermediate states.

Logarithmic dependence on a high-energy cutoff is the hallmark of a *scaling property*, where the physics at low energies depends not on the "bare" parameters, but on renormalized parameters that take into account the effect of high-energy intermediate states. Such quantities may be identified as the invariants of a *scaling transformation:* If the cutoff W is reduced to $W-|dW|$ by integrating out states with energies $W-|dW|<|\omega|<W$, the bare parameters are renormalized, but the low-energy physics is unchanged, and thus depends on the *scaling invariants*. Of course, such a truncation of the conduction band not only renormalizes the bare parameters, but also generates both new couplings and retardation.[6] However, retardation should not affect processes with energies $|\omega|\ll W$, and the new couplings should be "irrelevant" in that they vanish in the limit $W\to\infty$. In this limit, a truncation procedure that generates a new effective Hamiltonian with renormalized parameters does so as a consequence of an intrinsic scaling property of the model.

The scaling equations may be derived in a manner reminiscent of Anderson's "poor man's" treatment[7] of the Kondo problem. Divide $\Delta(\omega)$ into $\Delta((1+\lambda)\omega)-\lambda\omega\Delta'(\omega)$, where λ is a positive infinitesimal, and $\Delta'(\omega)$ is the derivative. The positive quantity $-\lambda\omega\Delta'(\omega)$ represents the contribution to $\Delta(\omega)$ from high-energy states which are to be integrated out, preserving the *form* of $\Delta(\omega)$, but changing its *scale*. If $U\gg W\gg|E_d|$, hybridization with these states renormalizes the d-orbital states $|0\rangle$ and $|1\sigma\rangle$, but the state $|2\rangle$ is decoupled from the conduction band. Particle states in the cutoff region are labeled k^+, hole states k^-; to lowest order in λ $(=-d\ln W)$ the transformation is

$$|\tilde{0}\rangle=|0\rangle-\sum_{k=k^-,\sigma}\frac{V_{kd}^*}{|\epsilon_k|+E_1-E_0}c_{k\sigma}|1\sigma\rangle, \quad (5)$$

$$\tilde{E}_0=E_0-\sum_{k=k^-,\sigma}\frac{|V_{kd}|^2}{|\epsilon_k|+E_1-E_0}, \quad (6)$$

$$|1\tilde{\sigma}\rangle=|1\sigma\rangle-\sum_{k=k^+}\frac{V_{kd}}{|\epsilon_k|+E_0-E_1}c_{k\sigma}^\dagger|0\rangle, \quad (7)$$

$$\tilde{E}_1=E_0-\sum_{k=k^+}\frac{|V_{kd}|^2}{|\epsilon_k|+E_0-E_1}. \quad (8)$$

The scaling equation for E_d $(\equiv E_1-E_0)$ is thus

$$\frac{dE_d}{d\ln W}=\frac{1}{\pi}\int_0^\infty d\omega\left(\frac{2\omega\Delta'(-\omega)}{\omega+E_d}+\frac{\omega\Delta'(\omega)}{\omega-E_d}\right). \quad (9)$$

Since $\Delta'(\omega)\approx 0$ unless $|\omega|\approx W$, (9) simplifies to

$$dE_d/d\ln W=-\Delta(0)/\pi+O(E_d\Delta(W)/W). \quad (10)$$

The transformation of $\Delta(0)$ is found from the renormalization of V_{kd} when $\epsilon_k=0$:

$$\tilde{V}_{kd}=\langle\tilde{0}|c_{k\sigma}H|1\tilde{\sigma}\rangle(\langle\tilde{0}|\tilde{0}\rangle\langle 1\tilde{\sigma}|1\tilde{\sigma}\rangle)^{-1/2}, \quad (11)$$

$$d\Delta(0)/d\ln W=O(\Delta(0)\Delta(W)/W). \quad (12)$$

In the limit $W\to\infty$, the right-hand side of (12) vanishes, and $\Delta(0)$ is unrenormalized; however, (10) is nontrivial, and E_d is strongly renormalized by scaling. This is because the state $|0\rangle$ can hybridize with both $c_{k\uparrow}|1\uparrow\rangle$ and $c_{k\downarrow}|1\downarrow\rangle$, while (as $|2\rangle$ is decoupled) $|1\sigma\rangle$ only mixes with $c_{k\sigma}^\dagger|0\rangle$; E_0 is thus reduced by twice as much as E_1, and E_d rises as scaling proceeds. This feature is absent if $W\gg U$; a similar derivation in that case produces no nontrivial scaling equations. It would be pointless to be more precise about the terms $O(\Delta(W)/W)$; not only are they "irrelevant" (when they are significant, so are new couplings and retardation), but they depend on the detailed form of $\Delta(\omega)$ when $|\omega|\simeq W$—note that the nontrivial term in (10) involves $\Delta(0)$, independent of its form in the cutoff region. Essentially equivalent equations have recently been independently reported by Jefferson,[4] though they are phrased in somewhat different notation, and include "irrelevant" terms derived using a particular cutoff prescription.

The scaling invariant obtained by integrating (10) is $E_d^* = E_d + (\Delta/\pi)\ln(W_0/\Delta)$, where E_d and W_0 are the initial or "bare" values.[8] (If initially $W \gg U$, renormalization of E_d only begins when W has been scaled down to $W \approx W_0 \approx U$.) As W is reduced, E_d rises along the *scaling trajectory*

$$E_d(W) = E_d^* - (\Delta/\pi)\ln(W/\Delta). \tag{13}$$

The scaling trajectories (13) are plotted in Fig. 1.

The scaling laws were derived assuming that particle states in the cutoff were empty and hole states full. For $W < T$, the scaling laws change,[9] and further scaling produces no renormalization of E_d. For $T \gg \Delta$, the physics is essentially atomic, and is thus described by a free orbital with a temperature-dependent level $E_d(T)$, given by setting $W = T$ in (13).

For $E_d^* \gg \Delta$, scaling stops when $W \simeq E_d(W) \simeq T^*$ ($\gg \Delta$), where

$$T^* + (\Delta/\pi)\ln(\alpha T^*/\Delta) = E_d^*; \tag{14}$$

α [$\approx O(1)$] is a universal number characteristic of the crossover.[10] Reducing W below T^* produces no further renormalization[9] since the states $|1\sigma\rangle$ become decoupled from low-energy processes. At temperatures below T^*, charge fluctuations are frozen out, $\langle n_d \rangle \approx 0$, and the impurity susceptibility is given by perturbation theory:

$$\chi(T) = \frac{1}{4T}\left(\frac{2\exp(-T^*/T)}{1+2\exp(-T^*/T)}\right) + \frac{\Delta}{2\pi T^{*2}}. \tag{15}$$

Below some temperature T_{FL}, (15) is dominated by the second, temperature-independent term, characteristic of a Fermi liquid. The effective Curie constant $T\chi$ is shown schematically as a function of temperature in Fig. 2(a). It is a measure of the effective degeneracy of the impurity orbital, and rises from $\frac{1}{8}$ (fourfold degeneracy) for $T \gg U$, to $\frac{1}{6}$ (triplet) for $U \gg T \gg T^*$; below T^* it falls to zero (singlet), becoming linear in the Fermi-liquid regime below T_{FL}.

When $|E_d^*| \lesssim \Delta$, the orbital retains effective triplet degeneracy till the crossover temperature $T \approx \Delta$ when irrelevent terms grow and scaling breaks down. The system goes directly into a Fermi-liquid regime; $\langle n_d \rangle$ remains substantially nonintegral at $T = 0$, and this regime may be de-

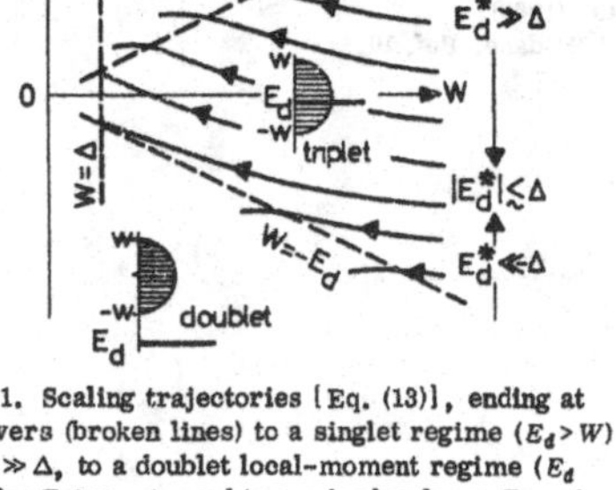

FIG. 1. Scaling trajectories [Eq. (13)], ending at crossovers (broken lines) to a singlet regime ($E_d > W$) for $E_d^* \gg \Delta$, to a doublet local-moment regime ($E_d < -W$) for $E_d^* \ll -\Delta$, and to a mixed-valence Fermi-liquid regime for $|E_d^*| \lesssim \Delta$.

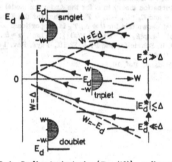

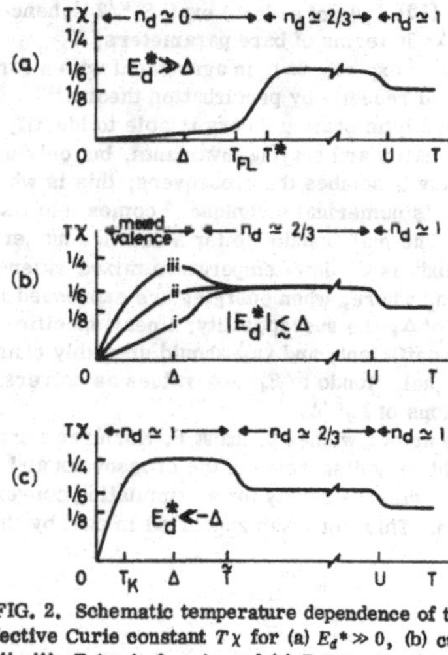

FIG. 2. Schematic temperature dependence of the effective Curie constant $T\chi$ for (a) $E_d^* \gg 0$, (b) curves i, ii, iii, $E_d^* \approx \Delta, 0, -\Delta$, and (c) $E_d^* \ll -\Delta$, showing crossovers between atomic regimes where the effective degeneracy is fourfold ($T\chi = \frac{1}{8}$), triplet ($T\chi = \frac{1}{6}$), doublet ($T\chi = \frac{1}{4}$), and singlet ($T\chi = 0$). At low temperatures there is a crossover to a Fermi-liquid regime where $T\chi$ is linear in T. Note that though the temperature scale is drawn linearly to emphasize this Fermi-liquid behavior, T_K, Δ, $\tilde{T}$, and T^* indicate *scales* of temperature which may differ by many orders of magnitude.

scribed as one of "mixed valence." (Note that $|E_d^*|\lesssim\Delta$ is the criterion for mixed valence, *not* $|E_d|\lesssim\Delta$ as commonly supposed[2-4]; depending on W_0, E_d^* may be arbitrarily larger than E_d.) For $T<\Delta$, χ will be of the order of the inverse of the crossover temperature (i.e., $\chi\approx\Delta^{-1}$ as predicted by Varma and Yafet[2]). Since this is a crossover region between two simple limits ("integral valence"), low-temperature properties will be sensitive—though universal—functions of E_d^*/Δ.

When $-E_d^*\gg\Delta$ [Fig. 2(c)], renormalization stops when the state $|0\rangle$ is decoupled at $W\approx -E_d(W)\approx\tilde{T}$ ($\gg\Delta$), where

$$\tilde{T}-(\Delta/\pi)\ln(\tilde{\alpha}\tilde{T}/\Delta)=-E_d^*; \tag{16}$$

$\tilde{\alpha}$ is analogous to α in (14).[10] For $T<\tilde{T}$, charge fluctuations are frozen out leaving $\langle n_d\rangle\simeq 1$, and a local moment. The Schrieffer-Wolff transformation[11] to a Kondo model is then valid, giving $(J\rho)^{\text{eff}}=-2\Delta/\pi\tilde{T}$, $D^{\text{eff}}\approx\tilde{T}$. Below a Kondo temperature T_K, given by[5] $D(|J\rho|)^{1/2}\exp(1/J\rho)$, the local moment is quenched, leaving a Fermi liquid. From (16), T_K is of order $\Delta\exp(\pi E_d^*/2\Delta)$, hence, $T_K\ll\Delta$; in terms of bare parameters, $T_K\approx(W_0\Delta)^{1/2}\exp(\pi E_d/2\Delta)$, in agreement with a form obtained recently by perturbation theory.[12]

An analytic scaling theory is able to identify universality and scaling invariants, but only qualitatively describes the crossovers; this is where Wilson's numerical technique[5,3] comes into its own. The main candidate for a detailed numerical study is the low-temperature mixed-valence region, where, when energies are expressed in units of Δ, the susceptibility, linear specific heat coefficient, and $\langle n_d\rangle$ should smoothly change from their Kondo to $E_d^*\gg\Delta$ values as universal functions of E_d^*/Δ.

The author wishes to thank P. Nozières for an enlightening discussion of the crossovers and H. R. Krishna-murthy for a stimulating conversation. This work was supported in part by the National Science Foundation under Contract No. DMR 7600866A01 at Princeton University. The hospitality of the Aspen Center for Physics and of Professor S. Doniach at Stanford University's Department of Applied Physics is gratefully acknowledged.

[(a)]Current address: Institut Laue-Langevin, B.P. 156X, 38042 Grenoble, France.

[1]P. W. Anderson, Phys. Rev. 124, 41 (1961).

[2]C. M. Varma and Y. Yafet, Phys. Rev. B 13, 2950 (1975).

[3]H. R. Krishna-murthy, K. G. Wilson, and J. W. Wilkins, in *Valence Instabilities and Related Narrow Band Phenomena*, edited by R. D. Parks (Plenum, New York, 1977), p. 177.

[4]J. H. Jefferson, J. Phys. C 10, 3589 (1977).

[5]K. G. Wilson, Rev. Mod. Phys. 47, 773 (1975).

[6]P. Nozières, J. Phys. (Paris), Colloq. 37, C1-271 (1976).

[7]P. W. Anderson, J. Phys. C 3, 2346 (1970).

[8]W_0 may be precisely defined by the value W in the perturbation expansion (2).

[9]Notionally, "scaling laws" may be derived in the limits $\omega\ll|E_d|$ or $\omega\ll T$, provided $\omega\gg\Delta$, by using a similar procedure to (5)–(12). The resulting equations are "trivial" in the manner of (12), as states in this energy range do not cause any renormalization of E_d or Δ. In the crossover regions $W\approx|E_d|$ or $W\approx T$, such "scaling equations" are not well defined, as they depend on the cutoff structure.

[10]α and $\tilde{\alpha}$ relate $E_d(T=0)$ ($=T^*$ or $-\tilde{T}$) to $E_d(T)$ at temperatures above the crossover. They may be obtained by detailed comparison of the perturbation expansions when T is above and below the crossover {F. D. M. Haldane, to be published; comparison of perturbation theory in Δ for the Anderson model with $W=\infty$ and E_d+U, $-E_d\gg T$, with that for the Kondo model shows that for $U\gg\Delta$, T_K is proportional to $(U\Delta)^{1/2}\exp[E_d(E_d+U)/(2\Delta U/\pi)]$}. In fact, $\alpha=\tilde{\alpha}$.

[11]J. R. Schrieffer and P. A. Wolff, Phys. Rev. 149, 491 (1966).

[12]Haldane, Ref. 10.

Ordering, metastability and phase transitions in two-dimensional systems

J M Kosterlitz and D J Thouless
Department of Mathematical Physics, University of Birmingham, Birmingham B15 2TT, UK

Received 13 November 1972

Abstract. A new definition of order called topological order is proposed for two-dimensional systems in which no long-range order of the conventional type exists. The possibility of a phase transition characterized by a change in the response of the system to an external perturbation is discussed in the context of a mean field type of approximation. The critical behaviour found in this model displays very weak singularities. The application of these ideas to the *xy* model of magnetism, the solid–liquid transition, and the neutral superfluid are discussed. This type of phase transition cannot occur in a superconductor nor in a Heisenberg ferromagnet, for reasons that are given.

1. Introduction

Peierls (1935) has argued that thermal motion of long-wavelength phonons will destroy the long-range order of a two-dimensional solid in the sense that the mean square deviation of an atom from its equilibrium position increases logarithmically with the size of the system, and the Bragg peaks of the diffraction pattern formed by the system are broad instead of sharp. The absence of long-range order of this simple form has been shown by Mermin (1968) using rigorous inequalities. Similar arguments can be used to show that there is no spontaneous magnetization in a two-dimensional magnet with spins with more than one degree of freedom (Mermin and Wagner 1966) and that the expectation value of the superfluid order parameter in a two-dimensional Bose fluid is zero (Hohenberg 1967).

On the other hand there is inconclusive evidence from the numerical work on a two-dimensional system of hard discs by Alder and Wainwright (1962) of a phase transition between a gaseous and solid state. Stanley and Kaplan (1966) found that high-temperature series expansions for two-dimensional spin models indicated a phase transition in which the susceptibility becomes infinite. The evidence for such a transition is much stronger for the *xy* model (spins confined to a plane) than for the Heisenberg model, as can be seen from the papers of Stanley (1968) and Moore (1969). Low-temperature expansions obtained by Wegner (1967) and Berezinskii (1970) give a magnetization proportional to some power of the field between zero and unity, and indicate the possibility of a sharp transition between such behaviour and the high-temperature régime where the magnetization is proportional to the applied field.

In this paper we present arguments in favour of a quite different definition of long-range order which is based on the overall properties of the system rather than on the behaviour of a two-point correlation function. In the cases we consider, the usual correlation function, such as the spin–spin correlation function, vanishes at any finite temperature in contrast to the corresponding cases in three dimensions. This type of long-range order, which we refer to as topological long-range order, may exist for the two-dimensional solid, neutral superfluid, and for the xy model, but not for a superconductor or isotropic Heisenberg model. In the case of a solid, the disappearance of topological long-range order is associated with a transition from an elastic to a fluid response to a small external shear stress, while for a neutral superfluid it is associated with the instability of persistent currents. Recently, Berezinskii (1971) has put forward similar arguments, but there are some important differences in our results. A brief account of this theory has already been given (Kosterlitz and Thouless 1972).

The definition of long-range order which we adopt arises naturally in the case of a solid from the dislocation theory of melting (Nabarro 1967). In this theory, it is supposed that a liquid close to its freezing point has a local structure similar to that of a solid, but that in its equilibrium configurations there is some concentration of dislocations which can move to the surface under the influence of an arbitrarily small shear stress, and so produce viscous flow. In the solid state there are no free dislocations in equilibrium, and so the system is rigid. This theory is much easier to apply in two dimensions than in three since a dislocation is associated with a point rather than a line.

Although isolated dislocations cannot occur at low temperatures in a large system (except near the boundary) since their energy increases logarithmically with the size of the system, pairs of dislocations with equal and opposite Burgers vector have finite energy and must occur because of thermal excitation. Such pairs can respond to an applied stress and so reduce the rigidity modulus. At sufficiently high temperatures, the largest pairs become unstable under an applied shear stress and produce a viscous response to the shear.

The presence or absence of free dislocations can be determined in the following manner. We suppose that the system has a fair degree of short-range order so that a local crystal structure can be identified. To be definite we take the crystal structure to be a simple square lattice with spacing $\boldsymbol{a}$. Using the local order, we attempt to trace a path from atom to atom which, in the perfect crystal, would be closed. Mermin (1968) has commented that the thermal motion does not necessarily destroy the correlation in orientation of the crystal axes at large distances, but even if this is destroyed, the direction taken in one region can be defined in terms of that taken in a previous region in the same neighbourhood. Local defects can be avoided by small deformations of the path. This procedure is possible provided there are no grain boundaries. If there are free dislocations present, the number of dislocations contained within the region surrounded by the contour will be proportional to the area of the region. Since the Burgers vectors of the individual dislocations can point in two possible directions, the average total Burgers vector will be proportional to the square root of the area. The path will fail to close by an amount proportional to L, the length of the path. If there are only pairs of dislocations present, only those pairs which are cut by the contour will contribute to the total Burgers vector. The number of pairs cut by the path is proportional to Lr where r is the mean separation of the pairs. Averaging over the possible orientations of the individual Burgers vectors, we see that the path will fail to close by an amount proportional to $(Lr)^{1/2}$. This allows us to determine the presence or absence of topological long-range order in the system.

We can obtain estimates of the transition temperature for the systems discussed by arguments similar to those used by Thouless (1969) for a one-dimensional Ising model with an interaction falling off at large distances as r^{-2}. At large distances from a dislocation (or vortex in the case of the xy model and neutral superfluid) the strain produced is inversely proportional to the distance and so the energy of a single dislocation depends logarithmically on the size of the system. In fact, the energy of an isolated dislocation with Burgers vector of magnitude b in a system of area A is (Friedel 1964)

$$E = \frac{\nu b^2(1+\tau)}{4\pi} \ln \frac{A}{A_0} \tag{1}$$

where ν and τ are the two-dimensional rigidity modulus and Poisson's ratio respectively, and A_0 is an area of the order of b^2. The entropy associated with a dislocation also depends logarithmically on the area, and, since there are approximately A/A_0 possible positions for the dislocation, the entropy is

$$S = k_B \ln \frac{A}{A_0} + O(1) \tag{2}$$

where k_B is the Boltzmann constant.

Since both energy and entropy depend on the size of the system in the same way, the energy term will dominate the free energy at low temperatures, and the probability of a single dislocation appearing in a large system will be vanishingly small. At high temperatures, dislocations will appear spontaneously when the entropy term takes over. The critical temperature at which a single dislocation is likely to occur is the temperature at which the free energy changes sign, namely (Kosterlitz and Thouless 1972)

$$k_B T_c = \frac{\nu b^2(1+\tau)}{4\pi} \tag{3}$$

Identical considerations for the xy model give

$$k_B T_c = \pi J \tag{4}$$

where J is the spin–spin coupling constant and

$$k_B T_c = \frac{\pi \hbar^2 \rho}{2m} \tag{5}$$

for the neutral superfluid where ρ is the density of particles per unit area and m is the effective atomic mass which is not necessarily the same as the atomic mass for a particle moving on a substrate.

That these estimates are upper bounds can be seen by the following argument. Although the formation of isolated dislocations will not occur at low temperatures, there can always be production of a pair of dislocations with equal and opposite Burgers vector, since the strain produced by such a pair falls off sufficiently rapidly at large distances so that the energy is finite. The critical temperatures as calculated above are the temperatures at which a pair of dislocations (or vortices) will dissociate, ignoring the effect of other pairs in the system. These other pairs will relax in the field of the first pair thereby renormalizing the rigidity modulus etc. downwards and consequently reducing the critical temperature. Most of the rest of the paper is devoted to a detailed study of this phenomenon in the systems discussed.

2. A model system

The statistical problem we are faced with is essentially that of a two-dimensional gas of particles with charges $\pm q$ interacting via the usual logarithmic potential, the number of particles being constrained only by the requirement that the system has overall electrical neutrality. The only essential modification necessary to treat the dislocation problem is to allow for the two different possible orientations of the Burgers vector. The hamiltonian for such a system is

$$H(\boldsymbol{r}_1 \ldots \boldsymbol{r}_N) = \tfrac{1}{2} \sum_{i \neq j} U(|\boldsymbol{r}_i - \boldsymbol{r}_j|) \tag{6}$$

where

$$U(|\boldsymbol{r}_i - \boldsymbol{r}_j|) = -2q_i q_j \ln\left|\frac{\boldsymbol{r}_i - \boldsymbol{r}_j}{r_0}\right| + 2\mu \qquad r > r_0$$
$$= 0 \qquad r < r_0 .$$

Here q_i and $\boldsymbol{r}_i$ are the charge and position of the ith particle, 2μ is the energy required to create a pair of particles of equal and opposite charge a distance r_0 apart, and r_0 is some suitable cutoff to avoid spurious divergences at small separations. We expect that the cutoff r_0 is of the order of the particle diameter or, for a lattice, the lattice spacing. In such a lattice near the critical point, we will be able to replace sums over the lattice sites by an integral over the whole system since the important contributions to the sum will come from the long-range part of the interaction, that is $|\boldsymbol{r}_i - \boldsymbol{r}_j| \gg r_0$.

To obtain a tractable theory, we further assume that the chemical potential μ is sufficiently large that there are very few particles present in the system. Since we have such a dilute gas, the configurations of least energy are those where equal and opposite charges are closely bound in dipole pairs, well separated from one another. At low temperatures, therefore, the fluctuations of charge within a given region are restricted, while at high temperatures charges may occur in isolation so much larger fluctuations may occur. For such configurations, we see that the mean square separation of the particles making up a dipole pair (ignoring for the present interactions between the pairs) is

$$\langle r^2 \rangle = \frac{\int_{r_0}^{\infty} \mathrm{d}r\, r^3 \exp\{-2\beta q^2 \ln(r/r_0)\}}{\int_{r_0}^{\infty} \mathrm{d}r\, r \exp\{-2\beta q^2 \ln(r/r_0)\}} = r_0^2 \frac{\beta q^2 - 1}{\beta q^2 - 2} \tag{7}$$

where $\beta = 1/k_B T$. The probability of finding a pair within a given area is found by summing $\exp\{-2\beta\mu - 2\beta q^2 \ln(|\boldsymbol{r}_i - \boldsymbol{r}_j|/r_0)\}$ over all values of $\boldsymbol{r}_i$ and $\boldsymbol{r}_j$ in the area. The double sum can be replaced by a double integral if we normalize by a factor of order r_0^{-4}, so that the mean separation d between such pairs is given in the same approximation by

$$\frac{1}{d^2} \approx \frac{\mathrm{e}^{-2\beta\mu}}{r_0^4} \int \mathrm{d}^2\boldsymbol{r} \exp\{-2\beta q^2 \ln(r/r_0)\} + O(\mathrm{e}^{-4\beta\mu})$$
$$= \frac{\pi}{r_0^2} \mathrm{e}^{-2\beta\mu} \frac{1}{\beta q^2 - 1} . \tag{8}$$

Here we can see the necessity for a large chemical potential μ, since already in calculating d^{-2}, any terms beyond first order in $\exp(-2\beta\mu)$ become intractable. Thus

$$\langle (r/d)^2 \rangle \approx \frac{\pi \mathrm{e}^{-2\beta\mu}}{\beta q^2 - 2} \ll 1 \qquad \text{for} \qquad \beta q^2 < 2 . \tag{9}$$

We can already see that a phase transition to a conducting state will take place at a temperature T_c given by

$$k_B T_c \approx \tfrac{1}{2}q^2 \tag{10}$$

since, in this simple approximation the polarizability, which is proportional to $\langle r^2 \rangle$, diverges. The closely bound dipole pairs will separate to give a uniform two-dimensional plasma of oppositely charged particles. The free energy of an isolated charge is

$$F = E - TS \approx \tfrac{1}{2}q^2 \ln(R^2/r_0^2) - k_B T \ln(R^2/r_0^2) \tag{11}$$

where R is the radius of the system. Thus we see that isolated charges can appear spontaneously when the temperature reaches T_c as given by equation (10). We are therefore particularly interested in values of βq^2 near 2.

Hauge and Hemmer (1971) have investigated the two-dimensional Coulomb gas under very different conditions. In the limit of vanishing particle size, they find a transition at $kT = q^2$ (in our units), the temperature at which pairs begin to form as $kT \to q^2$ from above. We look at a different temperature range and find a transition near $kT = \frac{1}{2}q^2$, when the largest pairs dissociate as the transition is approached from below. The main difference between the two models is the presence of a finite cutoff in this paper, which makes the potential well behaved at small distances, while Hauge and Hemmer allow the potential to be singular at the point of closest approach.

The divergence of the polarizability as the transition temperature is approached from below suggests that the most important effect of the interactions between different pairs may be described by the introduction of a dielectric constant. The field of a pair separated by a small distance of the order of r_0 does not extend much beyond the distance of separation of the two charges and from equation (8) it is very unlikely that there will be another pair in their immediate vicinity. Thus the dielectric constant appropriate for the expression of the energy of such a pair is unaffected by the polarizability of the other pairs. It is only for charges separated by an amount greater than the mean separation d of pairs that the energy is modified by the presence of other smaller pairs lying within the range of the field. The effective dielectric constant, therefore, becomes larger as we consider larger and larger pairs, so that the problem becomes a sort of iterated mean field approximation. We are therefore faced with problems similar to those which led to rescaling in the paper of Anderson *et al* (1970).

Within the range of a pair with large separation $r \gg r_0$ it is expected that there will be other pairs with separation up to $r \exp(-\beta\mu)$. The effect of these will be to reduce the interaction energy of the large pair by introducing an effective dielectric constant $\epsilon(r)$. To calculate this dielectric constant we consider pairs with separations lying in a small interval between r and $r + \mathrm{d}r$ with $\mathrm{d}r \ll r$, and consider terms to first order only in $\mathrm{d}r$. Using the standard methods of linear response theory, we apply a weak electric field E at an angle θ to the line joining the two charges. The polarizability per pair is given by

$$p(r) = q\frac{\partial}{\partial E}\langle r\cos\theta\rangle\bigg|_{E=0} \tag{12}$$

where the average is taken over the annulus $r < r' < r + \mathrm{d}r$ with Boltzmann factor

$$\exp\{-\beta U(r)\} = \left(\frac{r}{r_0}\right)^{-2\beta q^2/\epsilon(r)} \exp(\beta Eqr\cos\theta). \tag{13}$$

Although formally the factor $\exp(\beta Erq\cos\theta)$ will cause a divergence in the integrals of equation (12), we can argue that we can carry out the differentiation with respect to E

and take the limit $E \to 0$ before integrating as follows. Inside the medium the effective field is set up by the two charges, and falls off rapidly at distances greater than the separation between the charges and so has a finite range. Moreover, since the dipole pairs within this field have a separation very much less than the separation of the two charges of interest, the field experienced by the dipole pairs is effectively constant over them. Thus, our method of calculating the polarizability of a given pair from equation (12) is justified. We easily obtain

$$p(r) = \tfrac{1}{2}\beta q^2 r^2. \tag{14}$$

The density of such pairs is, by the same argument that leads to equation (8),

$$\mathrm{d}n(r) = \frac{1}{r_0^4}\int_0^{2\pi} \mathrm{d}\theta \int_r^{r+dr} \mathrm{d}r'\, r' \exp\{-\beta U_{\mathrm{eff}}(r')\} + \mathrm{O}(\mathrm{e}^{-4\beta\mu}) \tag{15}$$

where $U_{\mathrm{eff}}(r)$ is the energy of two charges separated by r in such a dielectric medium. The terms $\mathrm{O}(\mathrm{e}^{-4\beta\mu})$ correspond to the fact that in the normalization of the probability for seeing such pairs there are terms corresponding to seeing no pairs, one pair etc, all of which we ignore except for the no-pair term. Another complication arises at this point because the energy $U_{\mathrm{eff}}(r)$ is given by

$$U_{\mathrm{eff}}(r) = 2\mu + 2q^2 \int_{r_0}^{r} \frac{\mathrm{d}r'}{r'\epsilon(r')}\,. \tag{16}$$

As it stands, we cannot evaluate equation (16) because of the unknown function $\epsilon(r)$. However, assuming that $\mathrm{d}\epsilon^{-1}(r)/\mathrm{d}r$ is sufficiently small, we can write

$$U_{\mathrm{eff}}(r) = \frac{2q^2 \ln (r/r_0)}{\epsilon(r)} \tag{17}$$

and check the consistency of this assumption *a posteriori*. We finally obtain the susceptibility due to these pairs

$$\mathrm{d}\chi(r) = \pi\beta q^2 \mathrm{e}^{-2\beta\mu} \left(\frac{r}{r_0}\right)^{-2\beta q^2/\epsilon(r) + 4} \mathrm{d}\left(\ln \frac{r}{r_0}\right). \tag{18}$$

Lastly, we change variables to

$$x = \ln \frac{r}{r_0} \qquad \text{and} \qquad y(x) = \frac{2\beta q^2}{\epsilon(x)} - 4$$

to obtain the differential equation

$$\frac{\mathrm{d}y}{\mathrm{d}y} = -\pi^2 \mathrm{e}^{-2\beta\mu}(y + 4)^2\ \mathrm{e}^{-x}) \tag{19}$$

which is subject to the boundary conditions

$$y(0) = 2\beta q^2 - 4 \qquad \text{and} \qquad y(\infty) = \frac{2\beta q^2}{\epsilon} - 4 \tag{20}$$

where ϵ is the macroscopic dielectric constant. Since, as discussed previously, when $x = 0$, $r = r_0$ so that the interaction of a pair separated by r_0 is unaffected by any other pairs, so that $\epsilon(x = 0) = 1$. Similarly, as $x \to \infty$, the effective interaction is reduced by a factor $\epsilon(x = \infty)$, which must be the same as the dielectric constant as measured by application of an external field to a macroscopic region of the medium. We can simplify equation (19) further by noticing that, since the derivative dy/dx is proportional to $e^{-2\beta\mu}$, the difference $y(0) - y(\infty)$ for $y(0) > y_c(0)$ must also be very small (in fact $y(0) - y(\infty) < O(e^{-\beta\mu})$) so that we can replace the factor $(y + 4)^2$ in equation (19) by $(y(0) + 4)^2$.

The problem is now reduced to solving the equation

$$\frac{d\tilde{y}}{d\tilde{x}} = -\exp(-\tilde{x}\tilde{y}) \tag{21}$$

where we have rescaled $x = (\pi e^{-\beta\mu}(y(0) + 4))^{-1}\,\tilde{x}$ and $y = (\pi e^{-\beta\mu}(y(0) + 4))\,\tilde{y}$ to eliminate the constant multiplying the exponential. This equation does not possess an analytic solution but the solutions fall into two classes (see Appendix)

$$\begin{aligned} &\text{(i)} && \frac{q^2}{k_B T\epsilon(T)} - 2 > 0 && \text{for } T < T_c \\ &\text{(ii)} && \frac{q^2}{k_B T\epsilon(T)} - 2 \to -\infty && \text{for } T > T_c . \end{aligned} \tag{22}$$

We interpret this sudden change in the behaviour of $\epsilon(T)$ as T passes through T_c as a phase transition to the conducting state where the pairs become unbound. The critical temperature T_c is given by

$$\frac{q^2}{k_B T_c} = 2\left[1 + \left\{\tilde{y}_c(0)\,\pi\exp\left(\frac{-\mu}{k_B T_c}\right)\right\}\right] \tag{23}$$

and the critical value of the dielectric constant at the transition temperature is given by

$$\frac{q^2}{k_B T_c} - 2\epsilon(T_c) = 0. \tag{24}$$

To determine the values of T_c and ϵ_c and to investigate the nature of the phase transition, it is necessary to find an approximate solution of equation (21). This can be done in two limiting cases, $T \ll T_c$ and $T \lesssim T_c$. In the first case, $y(0) \approx y(\infty) > 0$ so we can immediately integrate equation (21) by replacing $y(x)$ by $y(0)$ in the exponential to obtain

$$\epsilon \approx 1 + \frac{\pi^2\beta q^2 e^{-2\beta\mu}}{\beta q^2 - 2} . \tag{25}$$

In the second case $T \lesssim T_c$, equation (21) may be solved approximately by the methods described in the Appendix to obtain $\tilde{y}_c(0) = 1{\cdot}3$ so that

$$\frac{q^2}{k_B T_c} \approx 2\{1 + 1{\cdot}3\pi\exp(-\mu/k_B T_c)\} \tag{26}$$

and

$$\left(\frac{q^2}{k_B T\epsilon(T)} - 2\right) \underset{T\to T_c}{\sim} \exp\left\{-\left(\ln\frac{T_c}{T_c - T}\right)^{1/2}\right\} . \tag{27}$$

This singularity in $\epsilon(T)$ is of a most unusual type but, since we have used a mean field theory, this form of the singularity is unlikely to be the correct one. Note in particular that $\epsilon(T_c)$ is finite so that the susceptibility does not diverge at the critical temperature as one would expect from the simple arguments presented earlier.

We are now in a position to check our assumption that $d\epsilon^{-1}(r)/dr$ is sufficiently small to do the integration in equation (16). Clearly, for $T \ll T_c$, the replacement of $\epsilon^{-1}(r)$ by a constant is valid. The most unfavourable case is for the critical trajectory when $y(\infty) = 0$ and $d\epsilon^{-1}(r)/dr$ is largest. The condition for the replacement to be valid is

$$\frac{d\epsilon^{-1}(r)}{dr} \underset{r\to\infty}{\ll} \frac{1}{r\ln r}\,. \tag{28}$$

The critical solution of equation (21) is, as can be verified by direct substitution

$$\tilde{y}(x) \underset{x\to\infty}{\sim} \frac{2\ln\tilde{x}}{\tilde{x}} - \frac{\ln\ln\tilde{x}}{\tilde{x}} + \ldots \tag{29}$$

Remembering that $\tilde{x} \propto \ln r/r_0$, this gives

$$\frac{d\epsilon^{-1}(r)}{dr} \underset{r\to\infty}{\sim} \frac{1}{r\ln r}\,\frac{\ln\ln r}{\ln r} \ll \frac{1}{r\ln r}\,. \tag{30}$$

The free energy has a singularity at the critical temperature but it is so weak that any observable except the dielectric constant is finite and all derivatives are finite and bounded for $T < T_c$. Not surprisingly, this is in agreement with the results obtained by Anderson and Yuval (1971) for the one-dimensional Ising model with inverse square interaction, which can be treated with our methods to obtain most of their results. The energy of a pair of charges is

$$U = \frac{q^2\int_0^\infty dx\, x\,\epsilon^{-1}(x)\exp\{-2(\beta q^2\epsilon^{-1}(x) - 1)x\}}{\int_0^\infty dx \exp\{-2(\beta q^2\epsilon^{-1}(x) - 1)x\}} \tag{31}$$

where we have substituted $x = \ln(r/r_0)$. Since the specific heat C is proportional to $dU/d\beta$, any singularities in C will show up in integrals of the form

$$\int_0^\infty dx\, x^n \frac{\partial\epsilon^{-1}(x)}{\partial\beta}\exp\{-2(\beta q^2\epsilon^{-1}(x) - 1)x\} \approx \int_0^\infty dx\, x^n\, e^{-\alpha x}\frac{\partial y(x)}{\partial y(0)} \tag{32}$$

where we have used the fact that $\beta q^2\epsilon^{-1}(x) - 1 \approx \text{const} > 0$ for $T < T_c$. This integral will certainly be finite except possibly at $T = T_c$, in which case we want the asymptotic form of $y(x)$ as given by equation (29), since it is clear that $\partial y(x)/\partial y(0)$ is singular only at infinity.

Differentiating equation (21) with respect to $\tilde{y}(0)$ and using equation (29) we obtain

$$\frac{\partial}{\partial\tilde{x}}\left(\frac{\partial\tilde{y}(\tilde{x})}{\partial\tilde{y}(0)}\right) \underset{x\to\infty}{\sim} \frac{\partial\tilde{y}(\tilde{x})}{\partial\tilde{y}(0)}\,\frac{\ln\tilde{x}}{\tilde{x}} \tag{33}$$

which has the solution

$$\frac{\partial\tilde{y}(\tilde{x})}{\partial\tilde{y}(0)} \underset{x\to\infty}{\sim} \exp\{\tfrac{1}{2}(\ln\tilde{x})^2\}. \tag{34}$$

The integrals of the form (32) in the expression for C are thus all finite since the $e^{-\alpha x}$ factor ensures convergence. We can similarly show that all derivatives of C of finite order are finite and bounded at the critical temperature, which strongly indicates that C is analytic there.

It must be borne in mind that mean field approximations are notoriously bad in predicting the form of the singularities in specific heats etc, since such a theory ignores fluctuations in the internal field. We therefore expect that our value of T_c is an over-estimate and that in an exact treatment there may be a weak singularity in C.

Just above the critical temperature, the largest pairs will dissociate and be able to carry charge so that the medium will conduct. Assuming that the mobility of a free charge behaves smoothly at T_c, the behaviour of the DC conductivity as $T \to T_c$ from above will be determined by that of the density of dissociated pairs $\delta n(T)$. This number may be estimated by calculating the number of pairs whose separation is less than R, where R is defined by $y(R) = 0$. When the separation r increases so that $y(r) < 0$. the trajectory $y(x)$ ($x = \ln(r/r_0)$) falls off rapidly to minus infinity (see figure A1). We interpret this failure of the theory as the complete dissociation of such a pair. We find that R is given by (see Appendix)

$$\ln\frac{R}{r_0} \underset{T\to T_c^+}{\sim} \exp\left\{\left(\ln\frac{1}{T-T_c}\right)^{1/2}\right\}. \tag{35}$$

The density of pairs with separation less than R is

$$\begin{aligned} n(R) &\approx \frac{2\pi}{r_0^2}\int_{r_0}^{R}\frac{\mathrm{d}r}{r_0}\left(\frac{r}{r_0}\right)^{-(2\beta q^2/\epsilon(r)-1)} \\ &\approx \frac{\pi}{r_0^2}\,\frac{1-(R/r_0)^{-2(\beta q^2-1)}}{\beta q^2-1} \end{aligned} \tag{36}$$

where we have taken $\epsilon(r) \approx 1$ throughout the region of integration. This is justified here because $\{\beta q^2/\epsilon(r)\} - 1 \approx 1$ near T_c for all r. Thus, the density of dissociated pairs $\delta n(T)$ is given by

$$\ln\frac{\delta n(T)}{n} \underset{T\to T_c^+}{\sim} -\exp\left\{\left(\ln\frac{1}{T-T_c}\right)^{1/2}\right\} \tag{37}$$

where $2n$ is the total density of charges. Also

$$\frac{\mathrm{d}}{\mathrm{d}T}\ln\frac{\delta n(T)}{n} \underset{T\to T_c^+}{\sim} \frac{\exp\left([\ln\{1/(T-T_c)\}]^{1/2}\right)}{T-T_c} \tag{38}$$

so that

$$\frac{\delta n(T)}{n} \underset{T\to T_c^+}{\to} 0 \tag{39}$$

and

$$\frac{\mathrm{d}^r}{\mathrm{d}T^r}\,\frac{\delta n(T)}{n} \underset{T\to T_c^+}{\to} 0\,. \tag{40}$$

Thus, the DC conductivity tends to zero with all derivatives zero as the critical temperature is approached from above.

To conclude this section on the model system, we would like to point out that the assumption of a very dilute system ($e^{-2\beta\mu} \ll 1$) is not necessarily valid in a real system. However, we expect that the qualitative arguments will go through even in such a case and the general form of the results will be unchanged. We can imagine increasing the cutoff r_0 to some value R_0 such that the energy of two charges a distance R_0 apart is $2\mu(R_0)$ where $\exp\{-2\mu(R_0)\beta\} \ll 1$. For charges further apart than R_0, we can use the theory as outlined previously. The boundary conditions given by equation (20) will be changed to

$$y(0) = \frac{2q^2}{k_B T\epsilon(R_0)} - 4 \tag{41}$$

with $\epsilon(R_0)$ an unknown function. The critical temperature and the dielectric constant will now be determined in terms of $\epsilon(R_0)$ and $\mu(R_0)$. To determine these two quantities, a more sophisticated treatment is required, but we expect that the behaviour of the dielectric constant and specific heat at the critical temperature will be unchanged.

3. The two-dimensional *xy* model

The two-dimensional *xy* model is a system of spins constrained to rotate in the plane of the lattice which, for simplicity, we take to be a simple square lattice with spacing $\boldsymbol{a}$. The hamiltonian of the system is

$$H = -J\sum_{\langle ij\rangle} \boldsymbol{S}_i . \boldsymbol{S}_j = -J\sum_{\langle ij\rangle} \cos(\phi_i - \phi_j) \tag{42}$$

where $J > 0$ and the sum $\langle ij\rangle$ over lattice sites is over nearest neighbours only. We have taken $|\boldsymbol{S}_i| = 1$ and ϕ_i is the angle the ith spin makes with some arbitrary axis. Only slowly varying configurations, that is, those with adjacent angles nearly equal, will give any significant contribution to the partition function so that may expand the hamiltonian up to terms quadratic in the angles.

It has been shown by many authors (Mermin and Wagner 1966, Wegner 1967, Berezinskii 1970) that this system does not have any long-range order as the ground state is unstable against low-energy spin-wave excitations. However, there is some evidence (Stanley 1968, Moore 1969) that this system has a phase transition, but it cannot be of the usual type with finite mean magnetization below T_c. As we shall show, there exist metastable states corresponding to vortices which are closely bound in pairs below some critical temperature, while above this they become free. The transition is characterized by a sudden change in the response to an applied magnetic field.

Expanding about a local minimum of H

$$H - E_0 \approx \tfrac{1}{2}J\sum_{\langle ij\rangle}(\phi_i - \phi_j)^2 = J\sum_{\boldsymbol{r}}(\Delta\phi(\boldsymbol{r}))^2 \tag{43}$$

where Δ denotes the first difference operator, $\phi(\boldsymbol{r})$ is a function defined over the lattice sites, and the sum is taken over all the sites. If we consider the system in the configuration of figure 1, its energy is, from equation (43)

$$H - E_0 \approx \pi J \ln\frac{R}{a} \tag{44}$$

where R is the radius of the system. Thus we have a slowly varying configuration, which we shall call a vortex, whose energy increases logarithmically with the size of the system.

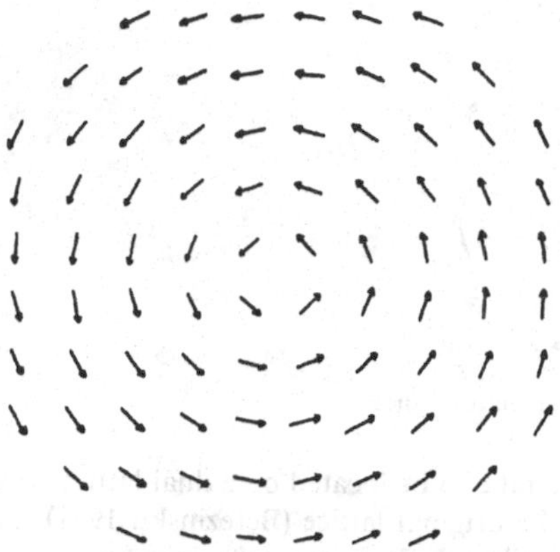

Figure 1. An isolated vortex in the xy model.

From the arguments of the Introduction, this suggests that a suitable description of the system is to approximate the hamiltonian by terms quadratic in $\Delta\phi(\boldsymbol{r})$ and split this up into a term corresponding to the vortices and another to the low-energy excitations (spin waves).

We extend the domain of $\phi(\boldsymbol{r})$ to $-\infty < \phi(\boldsymbol{r}) < \infty$ to allow for the fact that, in the absence of vortices, $\langle(\phi(\boldsymbol{r}) - \phi(\boldsymbol{r}'))^2\rangle$ increases like $\ln(|\boldsymbol{r} - \boldsymbol{r}'|)$ (Berenzinskii 1971). Thus, at large separations, the spins will have gone through several revolutions relative to one another. If we now consider a vortex configuration of the type of figure 1, as we go round some closed path containing the centre of the vortex, $\phi(\boldsymbol{r})$ will change by 2π for each revolution. Thus, for a configuration with no vortices, the function $\phi(\boldsymbol{r})$ will be single-valued, while for one with vortices it will be many-valued. This may be summarized by

$$\oint \Delta\phi(\boldsymbol{r}) = 2\pi q \qquad q = 0, \pm 1, \pm 2 \ldots \tag{45}$$

where the sum is over some closed contour on the lattice and the number q defines the total strength of the vortex distribution contained in the contour. If a single vortex of the type shown in figure 1 is contained in the contour, then $q = 1$.

Let now $\phi(\boldsymbol{r}) = \psi(\boldsymbol{r}) + \bar{\phi}(\boldsymbol{r})$, where $\bar{\phi}(\boldsymbol{r})$ defines the angular distribution of the spins in the configuration of the local minimum, and $\psi(\boldsymbol{r})$ the deviation from this. The energy of the system is now

$$H - E_0 \approx J\sum_{\boldsymbol{r}} (\Delta\psi(\boldsymbol{r}))^2 + J\sum_{\boldsymbol{r}} (\Delta\bar{\phi}(\boldsymbol{r}))^2 \tag{46}$$

where

$$\oint \Delta\psi(\boldsymbol{r}) = 0 \qquad \text{and} \qquad \oint \Delta\bar{\phi}(\boldsymbol{r}) = 2\pi q. \tag{47}$$

The cross term vanishes because of the condition (47) obeyed by $\psi(\boldsymbol{r})$. Clearly the configuration of absolute minimum energy corresponds to $q = 0$ for every possible contour when $\bar{\phi}(\boldsymbol{r})$ is the same for all lattice sites. We see from equation (45) that, if we shrink the contour so that it passes through only four sites as in figure 2, we will obtain the strength

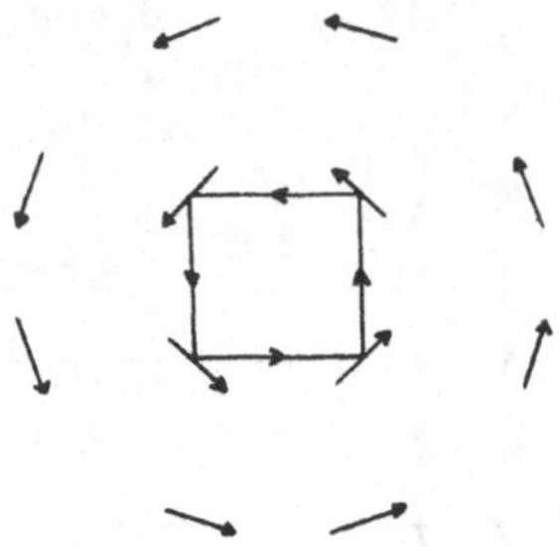

Figure 2. Contour around centre of vortex.

of the vortex whose centre we can take to be located on a dual lattice whose sites r^* lie at the centres of the squares of the original lattice (Berezinskii 1971). This procedure enables us to define the vortex distribution function $\rho(r^*)$ given by

$$\rho(r^*) = \sum_e q_e \, \delta_{r^*, r_e^*} . \tag{48}$$

Going now to a continuum notation for convenience, using equation (47), we can easily find the equation obeyed by $\bar{\phi}(r)$

$$\nabla^2 \bar{\phi}(r) = 2\pi\rho(r). \tag{49}$$

In terms of ρ, the energy of the system in a given configuration is

$$H - E_0 = J \int \mathrm{d}^2 r (\nabla\psi)^2 - 4\pi^2 J \iint \mathrm{d}^2 r \, \mathrm{d}^2 r' \, \rho(r) \, g(r - r') \, \rho(r')$$
$$+ 2\pi J \iint \mathrm{d}^2 r \, \mathrm{d}^2 r' \, \rho(r) \, \rho(r') \ln \frac{R}{r_0} \tag{50}$$

where R is the radius of the system, r_0 is a cutoff of the order of the lattice spacing a and $g(r)$ is the Green function of the square lattice defined so that $g(0) = 0$ (Spitzer 1964). The last term of equation (50) requires that $\Sigma q_e = 0$ which is the condition that the total vorticity of the system vanishes, and corresponds to the requirement of electrical neutrality in the model system. The first term produces the spin-wave excitations and is responsible for destroying long-range order, and the second term is the interaction energy of the vortices which cause the phase transition.

If we ignore all vortex configurations, we obtain for the spin–spin correlation function

$$\langle S_i . S_j \rangle = \langle \exp\{\mathrm{i}(\phi_i - \phi_j)\}\rangle = \exp\left(-\frac{k_B T}{2J} g(r_i - r_j)\right) \tag{51}$$

where (Spitzer 1964)

$$g(r) \underset{|r| \gg a}{\approx} \frac{1}{2\pi} \ln \left| \frac{r}{r_0} \right| \tag{52}$$

and

$$\frac{r_0}{a} = \frac{\mathrm{e}^{-\gamma}}{2\sqrt{2}} \qquad (\gamma = \text{Euler's constant}).$$

Thus

$$\langle S_i S_j \rangle \underset{|r_i - r_j| \gg a}{\approx} \left| \frac{r_i - r_j}{r_0} \right|^{-k_B T/4\pi J} \tag{53}$$

and we can see that the spin-wave excitations are responsible for destroying any long-range order in the system, but have nothing to do with any phase transition which may occur.

The energy E_v of the vortex configuration is now simply

$$E_v \underset{|r_i - r_j| \gg a}{\approx} -2\pi J \sum_{i \neq j} q_i q_j \ln \left| \frac{r_i - r_j}{r_0} \right| \tag{54}$$

where q_i is the strength of the ith vortex whose centre is located at r_i. We no longer distinguish between the original and dual lattices. This asymptotic expression for the energy is in fact a very good estimate even down to $|r| = a$ (Spitzer 1964) so we can use it for all $|r| > a$. Since in our approximation, the spin-waves and vortices do not interact with one another, the problem is now reduced to that of the model system with the vortices playing the rôle of the charged particles and

$$E_v = -2\pi J \sum_{i \neq j} q_i q_j \ln \left| \frac{r_i - r_j}{a} \right| - 2\pi J \sum_i q_i^2 \ln \frac{r_0}{a} \qquad \text{for } |r_i - r_j| > a$$

$$= 0 \qquad \text{otherwise.} \tag{55}$$

The chemical potential μ of a single vortex of unit strength is

$$\mu = -2\pi J \ln \frac{r_0}{a} = 2\pi J(\gamma + \tfrac{3}{2} \ln 2) \tag{56}$$

which is to be compared with the exact expression for the energy of two vortices separated by a single lattice spacing

$$\mu = \pi^2 J \tag{57}$$

which is obtained from equation (50) using the exact value of $g(a) = \frac{1}{4}$. The critical temperature T_c at which the transition takes place is given by the solution of the equation

$$\frac{\pi J}{k_B T_c} - 1 \approx \pi\, \tilde{y}_c(0) \exp\left(\frac{-\pi^2 J}{k_B T_c}\right) \tag{58}$$

$$\approx 0{\cdot}12.$$

Below T_c, the vortices will be bound in pairs of zero total vorticity, while above T_c they are free to move to the surface under the influence of an arbitrarily weak applied magnetic field, thereby causing a sudden change in the form of the response to the applied field.

One way of seeing the effect of the transition is to consider the spin–spin correlation function $\langle S_i . S_j \rangle$. Taking the vortices into account, this is modified to

$$\langle S_i . S_j \rangle = \langle \exp\{i(\psi_i - \psi_j)\}\rangle \langle \exp\{i(\bar{\Phi}_i - \bar{\Phi}_j)\}\rangle. \tag{59}$$

Since the vortices and spin waves do not interact in our approximation, the two averages are taken independently. The average $\langle \exp\{i(\psi_i - \psi_j)\}\rangle$ is taken over all possible values of ψ_i, and is given by equation (53) and $\langle \exp\{i(\bar{\Phi}_i - \bar{\Phi}_j)\}\rangle$ is taken over all possible vortex distributions. In our mean field approximation, we can estimate the latter term as follows

$$\langle \exp\{i(\bar{\Phi}_i - \bar{\Phi}_j)\}\rangle = \langle \bar{S}_i . \bar{S}_j \rangle \underset{|r_i - r_j| \to \infty}{\equiv} K(T) \tag{60}$$

where $\bar{S}_i$ is the spin on the ~~ith~~-site in the absence of spin-wave excitations. Thus,

$$K(T) = \bar{\sigma}^2 \tag{61}$$

where $\bar{\sigma}$ is the mean magnetization of the metastable configurations without the spin-wave excitations. Thus, $\bar{\sigma}$ corresponds exactly to ϵ^{-1} of the model system so that $K(T) = O(1)$ for $T < T_c$ and vanishes discontinuously at $T = T_c$ according to equation (27).

Since the spin waves and vortices do not interact, the system of vortices alone is equivalent to a system of freely moving, straight parallel current-carrying wires whose number is not conserved, located at the centres of the vortices. The sign of the forces between the wires is reversed, and $I_i = q_i\sqrt{2\pi J}$, I_i being the current in the ith wire in suitable units. The chemical potential $\mu = \pi^2 J$ is the energy required to 'create' a wire with no current. The sum of the currents in the wires is contained to be zero, corresponding to the condition $\Sigma q_i = 0$. This analogy tells us that, on applying a magnetic field in the plane of the system. the vortices will tend to move at right angles to the field, the direction determined by the sign of q_i.

4. Two-dimensional crystal

For simplicity, we consider a crystal with a square lattice of spacing a, and when possible as a continuum with a short distance cutoff r_0 of the order of the lattice spacing a. As far as any critical phenomena are concerned, this will make no essential difference. The short-distance behaviour will affect quantities like the effective chemical potential. but for our purposes we require only that this is sufficiently large. We can therefore use standard linear elasticity theory to describe our crystal (Landau and Lifshitz 1959).

The stress σ_{ij} is related to the strain u_{ij} for an isotropic medium by

$$\sigma_{ij} = 2\nu u_{ij} + \lambda\delta_{ij}u_{kk} \tag{62}$$

and the internal energy by

$$U = \tfrac{1}{2}\int d^2\boldsymbol{r}\, u_{ij}\sigma_{ij}. \tag{63}$$

As mentioned in the Introduction. there is no long-range order of the usual type in this system, exactly as for the xy model of the previous section. Writing the energy in terms of the displacement field $\boldsymbol{u}(\boldsymbol{r})$, which describes the displacement of the lattice sites from their equilibrium positions, and expanding to terms quadratic in $\boldsymbol{u}(\boldsymbol{r})$, we can show that the mean square deviation of a site from its equilibrium position increases logarithmically with the size of the system (Peierls 1935. Berezinskii 1970). This destruction of long-range order is caused by the low-energy phonon modes, and the phase transition from the solid to the liquid state by the dislocation configurations. Since in our approximation, the phonons and dislocations do not interact, we may treat them separately. As in the magnetic case, we may decompose the displacement field as

$$\boldsymbol{u}(\boldsymbol{r}) = \boldsymbol{v}(\boldsymbol{r}) + \bar{\boldsymbol{u}}(\boldsymbol{r}) \tag{64}$$

where

$$\oint \boldsymbol{v}(\boldsymbol{r})\, dl = 0 \qquad \text{and} \qquad \oint \bar{\boldsymbol{u}}(\boldsymbol{r})\, dl = \boldsymbol{b} \tag{65}$$

where the integral is taken round some closed contour and $\boldsymbol{b}$ is the total Burger's vector of the dislocation distribution contained within the contour. $\boldsymbol{v}(\boldsymbol{r})$ is the displacement field of the phonons and $\bar{\boldsymbol{u}}(\boldsymbol{r})$ that of the dislocations. If the contour in equation (65) is taken round only one dislocation it gives the Burger's vector $\boldsymbol{b}$ of that dislocation. For simplicity we shall assume that $|\boldsymbol{b}|$ is the same for all dislocations and the most likely value of $|\boldsymbol{b}|$ will be the smallest possible, that is $|\boldsymbol{b}| \approx a$.

The most convenient way of treating a medium with dislocations is to introduce a stress function $\chi(\boldsymbol{r})$ which is related to the stress $\sigma_{ij}(\boldsymbol{r})$ by (Friedel 1964, Landau and Lifshitz 1959)

$$\sigma_{ij}(\boldsymbol{r}) = \epsilon_{ik}\epsilon_{jl}\frac{\partial^2\chi(\boldsymbol{r})}{\partial x_k \partial x_l} \tag{66}$$

where

$$\epsilon_{12} = +1, \qquad \epsilon_{21} = -1, \qquad \epsilon_{ij} = 0 \text{ otherwise}.$$

We can then define a source function $\eta(\boldsymbol{r})$ describing the distribution of dislocations so that $\chi(\boldsymbol{r})$ obeys the equation

$$\nabla^4\chi(\boldsymbol{r}) = K\eta(\boldsymbol{r}) \tag{67}$$

where

$$K = \frac{4\nu(\nu+\lambda)}{2\nu+\lambda} = 2\nu(1+\tau). \tag{68}$$

$\eta(\boldsymbol{r})$ is given by (Friedel 1964)

$$\eta(\boldsymbol{r}) = \sum_{\alpha} \epsilon_{ij}\, b_j^{(\alpha)} \frac{\partial}{\partial x_i^{(\alpha)}} \delta^{(2)}(\boldsymbol{r} - \boldsymbol{r}^{(\alpha)}) \tag{69}$$

where $\boldsymbol{r}^{(\alpha)}$ and $\boldsymbol{b}^{(\alpha)}$ are the position and Burger's vector respectively of the αth dislocation.

Having set up such a formalism, we can investigate the response of a medium containing dislocations to an applied stress in exact analogy to the electrostatic case. In our case, σ_{ij} and u_{ij} correspond respectively to the electric field $\boldsymbol{E}$ and displacement field $\boldsymbol{D}$. The strain energy of the medium due to the dislocations is

$$\begin{aligned} U &= \tfrac{1}{2}\int d^2r\, \chi(\boldsymbol{r})\,\eta(\boldsymbol{r}) \\ &= \frac{1}{2}K\int d^2r\, d^2r'\, \eta(\boldsymbol{r})\, g(\boldsymbol{r}-\boldsymbol{r}')\,\eta(\boldsymbol{r}') + O\left(\sum_{\alpha\beta} \boldsymbol{b}^{(\alpha)}\cdot\boldsymbol{b}^{(\beta)} \ln\frac{R}{a}\right) \end{aligned} \tag{70}$$

where $g(\boldsymbol{r})$ is the Green function of equation (67)

$$g(\boldsymbol{r}) \underset{|\boldsymbol{r}|\gg a}{\approx} \frac{1}{8\pi} r^2 \ln\left|\frac{\boldsymbol{r}}{r_0}\right| \tag{71}$$

and

$$g(0) = 0.$$

Using those equations, we can immediately see that the energy of an isolated dislocation increases logarithmically with the area of the system. The strain energy of a pair of dislocations with equal but oppositely directed Burger's vectors is easily found to be (Friedel 1964)

$$U_{\text{pair}}(\boldsymbol{r}) \underset{|\boldsymbol{r}|\gg a}{\approx} \frac{Kb^2}{4\pi}\left(\ln\left|\frac{\boldsymbol{r}}{a}\right| - \tfrac{1}{2}\cos 2\theta\right) + 2\mu \tag{72}$$

where $|\boldsymbol{r}|$ is the separation of the two dislocations, θ the angle between $\boldsymbol{b}$ and $\boldsymbol{r}$, and 2μ the energy required to create two dislocations one lattice spacing apart. This result implies that, at low temperatures, the dislocations tend to form closely bound dipole pairs. Clearly, the condition $\Sigma \boldsymbol{b}^{(\alpha)} = 0$ must be satisfied so that at low temperatures the energy of the system is finite, corresponding to the condition of electrical neutrality in the model system.

The next step is to identify a quantity analogous to the dielectric constant. Consider the stress function $\chi^{(\alpha)}(\boldsymbol{r})$ due to the αth dipole with source function $\eta^{(\alpha)}(\boldsymbol{r})$, (cf Jackson 1962)

$$\chi^{(\alpha)}(\boldsymbol{r}) = K \int d^2\boldsymbol{r}'\, \eta^{(\alpha)}(\boldsymbol{r}')\, g(\boldsymbol{r} - \boldsymbol{r}^{(\alpha)} - \boldsymbol{r}') \tag{73}$$

where $\boldsymbol{r}^{(\alpha)}$ denotes the centre of the αth dipole. Assuming the dipoles are very small, we expand in powers of $\boldsymbol{r}'$ for $|\boldsymbol{r}'| \ll |\boldsymbol{r} - \boldsymbol{r}^{(\alpha)}|$

$$\chi^{(\alpha)}(\boldsymbol{r}) = K \int d^2\boldsymbol{r}'\, \eta^{(\alpha)}(\boldsymbol{r}')\, g(\boldsymbol{r} - \boldsymbol{r}^{(\alpha)}) + \tfrac{1}{2}K \int d^2\boldsymbol{r}'\, \eta^{(\alpha)}(\boldsymbol{r}')\, x_i'\, x_j' \frac{\partial^2 g(\boldsymbol{r} - \boldsymbol{r}^{(\alpha)})}{\partial x_i^{(\alpha)} \partial x_j^{(\alpha)}} + \ldots \tag{74}$$

since the linear term vanishes for a dipole. We can now define the macroscopic stress function by averaging over a region ΔA such that $\langle r^2 \rangle \ll \Delta A \ll A$, where $\langle r^2 \rangle$ is the mean square separation of the dipole and A is the area of the system.

$$\chi(\boldsymbol{r}) \approx K \int d^2\boldsymbol{r}'\eta(\boldsymbol{r}')g(\boldsymbol{r} - \boldsymbol{r}') + K \int d^2\boldsymbol{r}' C_{ij}(\boldsymbol{r}') \frac{\partial^2 g(\boldsymbol{r} - \boldsymbol{r}')}{\partial x_i' \partial x_j'} \tag{75}$$

where

$$\eta(\boldsymbol{r}) = \langle \sum_\alpha \int d^2\boldsymbol{r}'\eta^{(\alpha)}(\boldsymbol{r}')\delta(\boldsymbol{r} - \boldsymbol{r}^{(\alpha)})\rangle_{\Delta A}\, n(\boldsymbol{r}) \tag{76}$$

and

$$C_{ij}(\boldsymbol{r}) = \tfrac{1}{2}\langle \sum_\alpha \int d^2\boldsymbol{r}'\eta^{(\alpha)}(\boldsymbol{r}')x_i'x_j'\delta(\boldsymbol{r} - \boldsymbol{r}^{(\alpha)})\rangle_{\Delta A}\, n(\boldsymbol{r})$$

where $n(\boldsymbol{r})$ is the density of dipoles.

Simple tensor analysis shows that, for an isotropic medium, C_{ij} is given in terms of the applied stress σ_{ij} by

$$C_{ij} = \epsilon_{ik}\epsilon_{kl}(C_1\sigma_{kl} + C_2\delta_{kl}\sigma_{mm}) + O(\sigma^2) \tag{77}$$

where C_1 and C_2 are to be determined by a model calculation. From equation (75), we find after an integration by parts

$$\chi(\boldsymbol{r}) = \frac{K}{1 - K(C_1 + C_2)} \int d^2\boldsymbol{r}'\eta(\boldsymbol{r}')g(\boldsymbol{r} - \boldsymbol{r}'). \tag{78}$$

Thus, the effect of the dislocation pairs is to renormalize $K^{-1} \to K^{-1} - (C_1 + C_2)$, and we immediately see that the quantity corresponding to the dielectric constant $\epsilon(r)$ is $\epsilon(r) = 1 - K\{C_1(r) + C_2(r)\}$. Using simple linear response theory as in the model system, we obtain

$$\begin{aligned} C_1(r) &= -\tfrac{1}{4}\beta b^2 n\langle r^2\rangle \\ C_2(r) &= \tfrac{1}{8}\beta b^2 n\langle r^2 \cos 2\theta\rangle \end{aligned} \tag{79}$$

where the averages are to be taken with Boltzmann factor $\exp\{-\beta U_{\text{eff}}(r)\}$ and

$$U_{\text{eff}}(r) = \frac{b^2 K}{4\pi\epsilon(r)}\left(\ln\left|\frac{r}{r_0}\right| - \tfrac{1}{2}\cos 2\theta\right) + 2\mu. \tag{80}$$

Provided we choose the cutoff r_0 sufficiently large so that our continuum approximation holds, and that μ is large enough, we only have to show that the average over θ does not change the analogy with the model system of § 2. We can easily show that equation (19) is modified to

$$\frac{\mathrm{d}y}{\mathrm{d}x} = -\pi^2 \mathrm{e}^{-2\beta\mu}(y+4)^2\left(1 - \frac{I_1(2+\frac{1}{2}y)}{2I_0(2+\frac{1}{2}y)}\right)\mathrm{e}^{-xy} \tag{81}$$

with

$$y(x) = \frac{\beta b^2}{4\pi K\epsilon(x)} - 4$$

and $I_\nu(z)$ is the νth-order modified Bessel function (Ambramowitz and Stegun 1965). Provided μ is sufficiently large, $y(x)$ is very small for all x, so that the Bessel functions occurring in equation (81) are essentially constant. Thus we can immediately apply all the results of § 2 to this case.

5. Neutral superfluid

The same type of argument can be applied to a neutral superfluid in two dimensions. All that is required is that Bose condensation should occur in small regions of the system, so that locally a condensate wavefunction can be defined. The argument of Hohenberg (1967) shows that the phase of the condensate wavefunction fluctuates over large distances so that the type of order defined by Penrose and Onsager (1956) cannot occur. If a condensate wavefunction can be defined in a local region, it should be possible to explore the variation of its phase from one region to a neighbouring region. The total vorticity within a region is found by the change of phase along the boundary, divided by 2π. Just as the energy of a vortex in the magnetic system or dislocation in the crystal, so the energy of a superfluid vortex increases logarithmically with the size of the system. At low temperatures, there will be no free vortices, only clusters of zero total vorticity. Neglecting the interaction between clusters, the critical temperature is given by equation (5). The analogy with the model system is close as there is only one type of vortex which may have either sign. As in this model, we expect the interaction between vortex clusters to lower the critical temperature from the estimate of equation (5). The lattice Bose gas has been discussed in detail by Berezinskii (1971) who finds the parameters of the Bose gas to be related to those of the magnetic system by

$$J = \frac{\rho\hbar^2}{m} \quad \text{and} \quad r_0 \approx 1{\cdot}28\frac{\hbar c}{T}$$

where c is the speed of sound. Using these parameters, we can estimate the value of T_c from equation (58) and the superfluid density $\rho_s(T_c)$ which is nonzero in contrast to the result of Berezinskii.

Below the critical temperature, for a lattice Bose system with periodic boundary conditions, superfluid flow is stable in the thermodynamic limit, since a state with flow is one in which the phase of the order parameter changes by a multiple of 2π round the system. The system can change from one such state to another by the creation and separation of a pair of vortices, one of which passes right round the system before recombining with the other, but there is a high energy barrier which prevents this process. The different states of superfluid flow are topologically distinct states which do not make transitions between one another at low temperatures. The phase transition is characterized by these states becoming mutually accessible above the transition temperature, so that the flow states are no longer metastable.

Experiments on the onset of superfluid flow for thin films—see, for example, Symonds *et al* (1966), Chester *et al* (1972) and Herb and Dash (1972)—show a strong depression of the temperature as the film becomes thinner. According to the considerations of this paper the critical temperature should be given by

$$k_B T_c = \frac{\pi \hbar^2 \rho_s}{2m^*} \tag{82}$$

where ρ_s is the density of superfluid particles per unit area and m^* is the effective mass of the helium atoms in the film. If ρ_s is taken to be the bulk value multiplied by the film thickness and m^* is taken to be unaffected by the substrate, and so equal to the atomic mass, this formula gives too high a value for T_c, so the effect of the boundaries in reducing ρ_s must be important. The value of ρ_s at the onset temperature must be nonzero; for a film thickness 1·5 nm at 1·5 K the superfluid density given by equation (81) is about 0·22 times the particle density, if m^* is taken equal to the atomic mass. It is not surprising that the specific heat maximum occurs at a higher temperature than the onset of superfluidity, since a considerable degree of short-range order is necessary before the considerations of this paper have any relevance. De Gennes has made a relevant comment on this matter (see Symonds *et al* 1966).

For a charged superfluid (superconductor) the argument cannot be carried through because, as a result of the finite penetration depth λ, the energy of a single flux line is finite. The circulating current density inside a thin superconducting film of thickness $d \ll \lambda$ is (Pearl 1964)

$$\begin{aligned} J(r) &\approx \frac{\hbar c^2}{4e} \frac{d}{\lambda^2} \frac{1}{r} & r &\ll \frac{\lambda^2}{d} \\ &\approx \frac{\hbar c^2}{4\pi e} \frac{1}{r^2} & r &\gg \frac{\lambda^2}{d} \end{aligned} \tag{83}$$

where r is the distance from the flux line. The repulsion energy between two vortices with opposite circulation falls off like r^{-1} instead of increasing as $\ln r$ for large separations. The self energy of a single flux line is thus determined by the current near the flux line and is approximately (De Gennes 1966)

$$U \approx \left(\frac{\hbar c}{4e}\right)^2 \frac{d}{\lambda^2} \ln \frac{\lambda^2}{d\xi} \tag{84}$$

where ξ is the 'hard core' radius of the flux line.

6. Isotropic Heisenberg model in two dimensions

The isotropic Heisenberg model for a two-dimensional system of spins is quite different from the *xy* model. To show the nature of this difference, we consider a large system with periodic boundary conditions, but similar considerations apply to other boundary conditions. In the xy model at low temperatures, the direction of magnetization in a region is defined by a single angle ϕ which varies slowly in space. Although the angle ϕ fluctuates by a large amount in a large system, the number of multiples of 2π it changes by on a path that goes completely round the system is a topological invariant, so that

$$n_x = \frac{1}{2\pi}\int_0^{L_x} \frac{\partial \phi}{\partial x}\,\mathrm{d}x \qquad n_y = \frac{1}{2\pi}\int_0^{L_y} \frac{\partial \phi}{\partial y}\,\mathrm{d}y \tag{85}$$

are numbers defining a particular metastable state. Transitions can only take place from one metastable state to another if a vortex pair is formed, the two vortices separate and recombine after one has gone right round the system. Such a process will cause a change of one in either n_x or n_y, but there is a logarithmically large energy barrier to prevent such a transition.

In the case of the isotropic Heisenberg model, the direction of magnetization is defined by two polar angles θ and ϕ. A quantity such as

$$\frac{1}{2\pi}\int_0^{L_x} \frac{\partial \phi}{\partial x}\,\mathrm{d}x$$

is not a topological invariant. A twist of the angle ϕ by 2π across the system can be changed continuously into no twist by changing the other polar angle θ (which we take to be the same everywhere) from $\frac{1}{2}\pi$ to zero. There is in fact a single topological invariant for the Heisenberg model in two dimensions, which is

$$N = \frac{1}{4\pi}\iint \sin\theta\left(\frac{\partial\theta}{\partial x}\frac{\partial\phi}{\partial y} - \frac{\partial\theta}{\partial y}\frac{\partial\phi}{\partial x}\right)\mathrm{d}x\,\mathrm{d}y. \tag{86}$$

If we regard the direction of magnetization in space as giving a mapping of the space on to the surface of a unit sphere, this invariant measures the number of times the map of the space encloses the sphere. This invariant is of no significance in statistical mechanics, because the energy barrier separating configurations with different values of N is of order unity. To show this, we consider how a configuration with N equal to unity can be transformed continuously into one with N equal to zero.

A simple example of a configuration with $N = 1$ is one in which θ is a continuous function of $r = (x^2 + y^2)^{1/2}$, equal to π for r greater than some value a, and equal to zero at the origin. The angle ϕ is assumed to be equal to $\tan^{-1}(y/x)$. The energy of a slowly varying configuration is proportional to

$$\tfrac{1}{2}\iint ((\nabla\theta)^2 + \sin^2\theta(\nabla\phi)^2)\,\mathrm{d}x\,\mathrm{d}y$$

which, for the configuration described above

$$= \pi\int_0^a \left\{\left(\frac{\mathrm{d}\theta}{\mathrm{d}r}\right)^2 + \frac{1}{r^2}\sin^2\theta\right\} r\,\mathrm{d}r. \tag{87}$$

Even if θ varies linearly between the origin and $r = a$, this integral is finite and independent of a. The configuration can be changed continuously into a configuration with $N = 0$ by letting a tend to zero. Of course, for small values of a, this expression for the energy is invalid, but the number of spins in a disc of radius a is then small so that any energy barrier is small.

We must conclude that there is no topological long-range order in the Heisenberg model in two dimensions. Supposing that the occurrence of a phase transition is really connected with the existence of long-range order we would conjecture that the xy model has a phase transition but the Heisenberg model does not. Such a conjecture does not seem to be in conflict with the evidence provided by the analysis of power series for the models by Stanley (1968) and Moore (1969).

Acknowledgments

The authors would particularly like to thank Professor T H R Skyrme for the solution of the differential equation and other members of the Department of Mathematical Physics for many useful and illuminating discussions, especially Dr R K Zia and Dr J G Williams. They would also like to thank Professor P C Martin of Harvard University for drawing their attention to the work of Berezinskii, and for a helpful discussion.

Appendix. Solution of $dy/dx = -e^{-xy}$

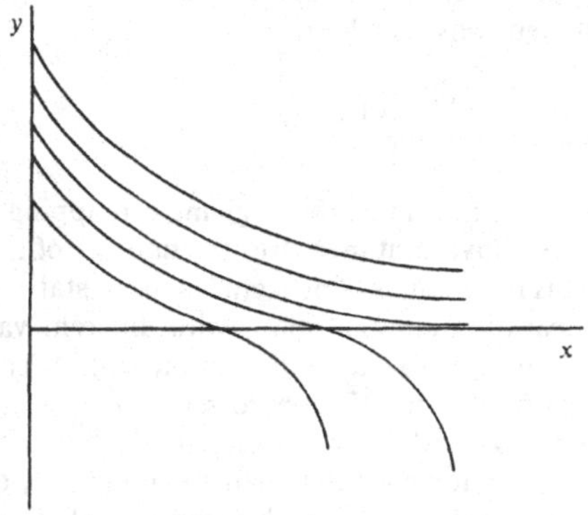

Figure A1. Trajectories of $dy/dx = e^{-xy}$.

From the method of isoclines, we see that the trajectories $y(x)$ behave as plotted in figure A1 with the singular point at infinity. It is easy to see that the solutions fall into two classes

$$\begin{array}{lll} \text{(i)} & y(\infty) \geqslant 0 & y(0) \geqslant y_c(0) \\ \text{(ii)} & y(\infty) \to -\infty & y(0) < y_c(0) \end{array} \qquad (\text{A.1})$$

corresponding to equation (22). For convenience, we make a transformation of variables to brong the singular point to the origin by defining

$$z = \frac{1}{xy} \qquad w = xy - \ln\frac{x}{y} \qquad (\text{A.2})$$

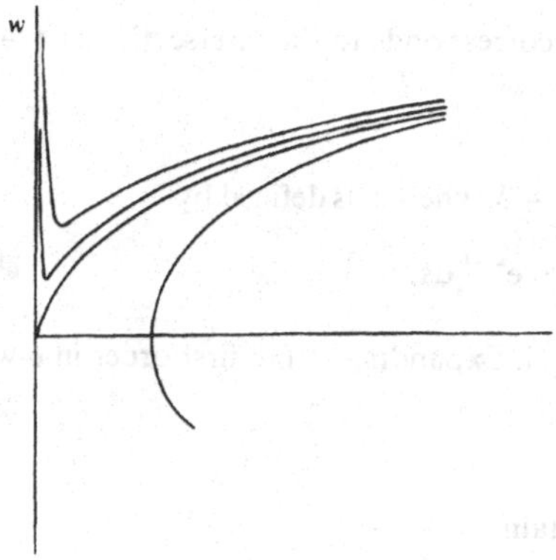

Figure A2. Trajectories in transformed variables.

so that

$$\frac{dw}{dz} = -\frac{1}{z^2} + \frac{1}{z}\coth \tfrac{1}{2}w. \tag{A.3}$$

The trajectories $w(z)$ have the form displayed in figure A2, where the trajectories corresponding to $y(0) \geqslant y_c(0)$ are those with $w(z) \geqslant 0$. We can find approximate solutions in the three regions

(i) $w \sim 0, \quad z \sim 0$

(ii) $w \to \infty, \quad z \sim 0$

(iii) $w \to \infty, \quad z \to \infty$

and match the solutions at $w = 2$. In region (i)

$$\frac{dw}{dz} \approx -\frac{1}{z^2} + \frac{2}{wz}. \tag{A.4}$$

Changing variables to $2t = (1/z) - w$ we obtain

$$z(t) \approx e^{t^2} \int_t^{t_1} e^{-s^2}\, ds \tag{A.5}$$

where $t_1 \to \infty$ as $y(0) \to y_c(0)$ from above. In regions (ii) and (iii)

$$w(z) \approx \frac{1}{z} + \ln z + 2 \ln y \tag{A.6}$$

where $y = y(\infty)$ in (ii) and $y = y(0)$ in (iii).

We can make the best match at $w(z) = 2$, where the gradients in the three regions are equal. However, the numbers are not too good because $\coth 1 \approx \frac{4}{3}$, while we have taken $\coth \frac{1}{2}w = 1$ in regions (ii) and (iii). The points at which we match are defined by the intersections of the curves

$$z(t) = \frac{1}{2(t+1)} \quad \text{and} \quad z(t) = e^{t^2} \int_t^{t_1} e^{-s^2}\, ds. \tag{A.7}$$

The match between regions (i) and (ii) corresponds to the intersection at $t \to \infty$, that is

$$t \approx t_1 - \frac{\ln t_1}{2t_1}. \tag{A.8}$$

At the other intersection, we put $t = t_c + \delta$, where t_c is defined by

$$z_c(t_c) = \frac{1}{2(t_c + 1)} = e^{t_c^2} \int_{t_c}^{\infty} e^{-s^2}\, ds. \tag{A.9}$$

Numerical calculations give $t_c \approx -0{\cdot}84$. Expanding $z(t)$ to first order in δ we obtain

$$t \approx t_c + \frac{(t_c + 1)\, e^{t_c^2}}{(-2t_c - 1)} \frac{e^{-t_1^2}}{t_1}. \tag{A.10}$$

Using equations (A.6) to (A.10), we obtain

$$y(\infty) \approx \sqrt{2t_1}\, e^{-t_1}$$

$$y(0) - y_c(0) \approx \frac{(1 + t_c)^2 \exp(t_c^2 - 2t_c)}{2y_c(0)} \frac{e^{-t_1^2}}{t_1} \tag{A.11}$$

where

$$y_c^2(0) = 2(t_c + 1)\, e^{-2t_c} \approx 1{\cdot}7. \tag{A.12}$$

Taking logarithms of equation (A.11) eliminating t_1, we obtain the leading singularity in $y(\infty)$ for $y(0) \gtrsim y_c(0)$

$$\ln y(\infty) \sim -\left(\ln \frac{1}{y(0) - y_c(0)}\right)^{1/2}. \tag{A.13}$$

Substituting the expressions for $y(\infty)$ and $y(0)$ given by equation (20), and ignoring all but the most singular terms, we immediately obtain equation (27).

Just above the critical temperature where $y(0) \leqslant y_c(0)$, the appropriate solution of equation (A.4) is

$$z(t) = e^{t^2} \int_t^{\infty} e^{-s^2}\, ds + K e^{t^2} \tag{A.14}$$

where $K \geqslant 0$. In the region $w \to -\infty$, $z \to +\infty$

$$w(z) \approx \frac{1}{z} - \ln z - 2 \ln X \tag{A.15}$$

where X is defined by $y(X) = 0$. We can now carry out exactly the same procedure as above to estimate the behaviour of X for $y(0) \to y_c(0)$ when $K \to 0^+$ by matching the solutions (A.6), (A.14) and (A.15) at $w = \pm 2$. We find

$$\ln X \underset{y(0) \to y_c(0)}{\sim} \{-\ln(y_c(0) - y(0))\}^{1/2}. \tag{A.16}$$

Using the expressions for $y(0)$ and $y_c(0)$ of equation (20), we immediately obtain equation (35).

References

Abramowitz M and Stegun I A 1965 *Handbook of Mathematical Functions* (New York: Dover) p 376
Alder B J and Wainwright T A 1962 *Phys. Rev.* **127** 359–61
Anderson P W and Yuval G 1971 *J. Phys. C: Solid St. Phys.* **4** 607–20

Anderson P W, Yuval G and Hamann D R 1970 *Phys. Rev.* B **1** 4464–73
Berezinskii V L 1970 *Sov. Phys.–JETP* **32** 493–500
—— 1971 *Sov. Phys.–JETP* **34** 610–6
Chester M, Yang L C and Stephens J B 1972 *Phys. Rev. Lett.* **29** 211–4
Friedel J 1964 *Dislocations* (Oxford: Pergamon) p 40
De Gennes P G 1966 *Superconductivity of Metals and Alloys* (New York: Benjamin) p 63
Hauge E H and Hemmer P C 1971 *Phys. Norvegica* **5** 209–17
Herb J A and Dash J G 1972 *Phys. Rev. Lett.* **29** 846–8
Hohenberg P C 1967 *Phys. Rev.* **158** 383–6
Jackson J D 1962 *Classical Electrodynamics* (New York: Wiley) 103–22
Kosterlitz J M and Thouless D J 1972 *J. Phys. C: Solid St. Phys.* **5** 124–6
Landau L D and Lifshitz E M 1959 *Theory of Elasticity* (London: Pergamon) pp 1–20
Mermin N D 1968 *Phys. Rev.* **176** 250–4
Mermin N D and Wagner H 1966 *Phys. Rev. Lett.* **22** 1133–6
Moore M A 1969 *Phys. Rev. Lett.* **23** 861–3
Nabarro F R N 1967 *Theory of Crystal Dislocations* (Oxford: Pergamon) pp 688–90
Pearl J 1964 *Appl. Phys. Lett.* **5** 65–6
Peierls R E 1934 *Helv. Phys. Acta.* **7** Suppl II 81–3
—— 1935 *Ann. Inst. Henri Poincaré* **5** 177–222
Penrose O and Onsager L 1956 *Phys. Rev.* **104** 576–84
Spitzer F 1964 *Principles of Random Walk* (Princeton: Van Nostrand) pp 148–51
Stanley H E 1968 *Phys. Rev. Lett.* **20** 589–92
Stanley H E and Kaplan T A 1966 *Phys. Rev. Lett.* **17** 913–5
Symonds A J, Brewer D F and Thomason A L 1966 in *Quantum Fluids* ed D F Brewer (Amsterdam: North-Holland) pp 267–70
Thouless D J 1969 *Phys. Rev.* **187** 732–3
Wegner F 1967 *Z. Phys.* **206** 465–70

Scaling Theory of Localization: Absence of Quantum Diffusion in Two Dimensions

E. Abrahams
Serin Physics Laboratory, Rutgers University, Piscataway, New Jersey 08854

and

P. W. Anderson,[a] D. C. Licciardello, and T. V. Ramakrishnan[b]
Joseph Henry Laboratories of Physics, Princeton University, Princeton, New Jersey 08540
(Received 7 December 1978)

Arguments are presented that the $T=0$ conductance G of a disordered electronic system depends on its length scale L in a universal manner. Asymptotic forms are obtained for the scaling function $\beta(G)=d\ln G/d\ln L$, valid for both $G \ll G_c \simeq e^2/\hbar$ and $G \gg G_c$. In three dimensions, G_c is an unstable fixed point. In two dimensions, there is no true metallic behavior; the conductance crosses over smoothly from logarithmic or slower to exponential decrease with L.

Scaling theories of localization have been discussed by Thouless and co-authors[1-3] and by Wegner.[4] Recently Schuster,[5] using methods related to those of Aharony and Imry,[6] has proposed a close relationship of the localization problem to a dirty XY model of the same dimensionality. Wegner has proposed a scaling for dimensionality $d \neq 2$ of the conductivity

$$\sigma \sim (E-E_c)^{(d-2)\nu}, \tag{1}$$

while Schuster identifies ν as the correlation-length exponent of the XY model $-\frac{1}{2}$ at $d>4$. The latter proposes a universal jump of conductivity for $d=2$ given by $e^2/\hbar\pi^2$. This is not inconsistent with the results of Wegner[4] and is in rough agreement with the early calculations of Ref. 2. It has not been clear how (1) could be reconciled with the physical ideas of Mott[7] as related to the beginning of a scaling theory by Thouless.[3] We here develop a renormalization-group scheme based on the Mott-Thouless arguments, which in many essential ways agrees with Wegner's results, and in other ways severely disagrees. In particular, we recover (1) for $d>2$, where ν is the localization-length exponent below E_c. This is in clear contradiction to Mott,[7] who argues that in all cases the conductivity jumps to zero at E_c. At $d=2$, we find no jump in σ but a steep crossover from exponential to very slow dependence on size. There is *no* true metallic conductivity. These results were presaged by Thouless and coworkers[8,9] to some extent, with Ref. 8 indicating a transition region for three dimensions, and Ref. 9 a size-dependent minimum metallic conductivity.

Our ideas are based on the relationship[1] between conductance as determined by the Kubo-Greenwood formula and the response to perturbation of boundary conditions in a finite sample described by Thouless and co-workers[3]

$$\frac{``V"}{W}=\frac{\Delta E}{dE/dN}=\frac{2\hbar}{e^2}G=\frac{2\hbar}{e^2}\sigma L^{d-2}. \tag{2}$$

Here G is the conductance (*not* conductivity σ) of a hypercube of size L^d [here $L \ggg L_0$ (L_0 = microscopic size)], dE/dN is the mean spacing of its energy levels, and ΔE is the geometric mean of the fluctuation in energy levels caused by replacing periodic by antiperiodic boundary conditions. Actually, when "V/W" is relatively large, it is hard to match the energy levels and, in fact, ΔE is defined using the curvature for small χ when we replace periodic $\psi(n+1)=\psi(1)$ by $\psi(n+1)=e^{i\chi} \times \psi(1)$ boundary conditions. This procedure is valid throughout the range of interest.[3] We will comment on the validity of (2) in a fuller paper, but here we add the following remarks. The equivalence of the Kubo-Greenwood formula and the breadth ν of the distribution of ΔE as described in Ref. 3a is not quite precisely provable but does not depend as stated in that reference on independence of momentum matrix elements $p_{\alpha\beta}$ and energy difference $E_\alpha - E_\beta$ between two states, only on a uniform distribution of those $E_\alpha - E_\beta$ which have large $p_{\alpha\beta}$.

Our scaling theory depends on the following ideas.

(I) We define a generalized dimensionless conductance which we call the "Thouless number" as a function of scale L:

$$g(L)=\frac{\Delta E(L)}{dE(L)/dN}\left(=\frac{G(L)}{e^2/2\hbar}\right), \tag{3}$$

where we now contemplate a small *finite* hypercube of size L. In the case $L \gg l$, the mean free path, we may use (2) to define a conductance $G(L)$ which is not related directly to the macroscopic conductivity but is a function of L, and is defined by (3) from the average of the Thouless energy-level differences at scale L. When $L < l$, there is phase coherence on a scale L and g is no longer given by (3) but it can be shown that $(e^2/\hbar)g = G$ can be defined as the conductance of a hypercube imbedded in a perfect crystal.

(II) We remark that $g(L)$ is the relevant dimensionless ratio which determines the change of energy levels when two hypercubes are fitted together. This is the hypothesis of Thouless and can be justified in several ways on physical grounds. For instance, once $L >$ mean free path, the phase relationships for an arbitrary integration of the wave equation across the cube are as random from one side to another as those between wave functions on different cubes. This could be shown to be related to Wegner's "neglect of eigenvalues far from E_F" by a scaling argument. In this limit $g(L)$ represents [as indicated in (2)] the "V/W" of an equivalent many-level Anderson model where each block has $(L/L_0)^d$ energy levels and a width of spectrum

$$W = (dE/dN)(L/L_0)^d .$$

We cannot see how any statistical feature of the energy levels other than this coupling/granularity ratio can be relevant.

(III) We then contemplate combining b^d cubes into blocks of side bL and computing the new $\Delta E'/(dE/dN)'$ at the resulting scale bL. The result will be

$$g(bL) = f(b, g(L)), \tag{4}$$

or in continuous terms

$$d\ln g(L)/d\ln L = \beta(g(L)). \tag{5}$$

The scaling trajectory has only one parameter, g.

(IV) At large and small g we can get the asymptotics of β from general physical arguments. For large g, macroscopic transport theory is correct and, as in (2),

$$G(L) = \sigma L^{d-2},$$

so that

$$\lim_{g\to\infty} \beta_d(g) = d - 2. \tag{6}$$

For small g ($V/W \ll 1$), exponential localization is surely valid and therefore g falls off exponentially:

$$g = g_a e^{-\alpha L}.$$

Thence

$$\lim_{g\to\infty} \beta_d(g) = \ln[g/g_a(d)]. \tag{7}$$

Here g_a is a dimensionless ratio of order unity.

From the asymptotics (6) and (7), we may sketch the universal curve $\beta_d(g)$ in $d = 1, 2, 3$ dimensions (Fig. 1). The central assumption of Fig. 1 is continuity: Since β represents the blocking of finite groups of sites, it can have no built-in singularity, and hence it would be unreasonable for it to have the cusp indicated by the dashed line: This is the curve which would be required to give the Mott-Schuster jump in conductivity for $d = 2$. The only singularities then, must be fixed points $\beta = 0$. Physically, it is also certain that β is monotonic in g, since smaller V/W surely always means more localization.

In constructing Fig. 1, we have used perturbation theory in V/W which shows that the first deviation of β from $\ln(g/g_a)$ is *positive*, with

$$\beta = \ln(g/g_a)[1 + \alpha g + \sim g^2 + \dots], \tag{8}$$

since this is essentially just the "locator" perturbation series first discussed by Anderson.[10] The steepening of the slope of β given by (8) makes $\nu \lesssim 1$, as we shall see, for $d = 3$ or greater.

For large g, we suppose that β may be calculated as a perturbation series in $W/V = g^{-1}$:

$$dg/d\ln L = g(d - 2 - a/g + \dots). \tag{9}$$

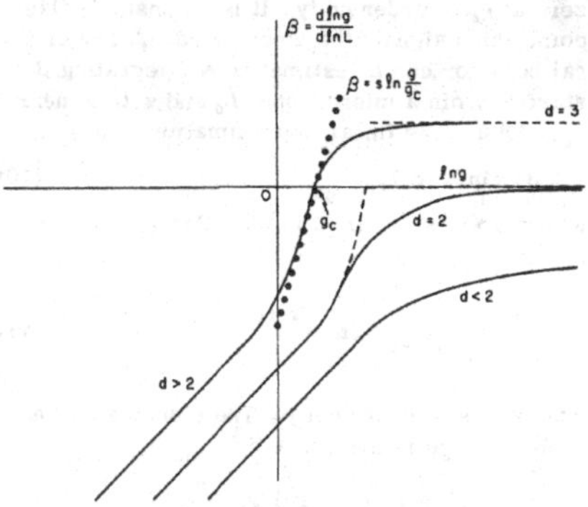

FIG. 1. Plot of $\beta(g)$ vs $\ln g$ for $d > 2$, $d = 2$, $d < 2$. $g(L)$ is the normalized "local conductance." The approximation $\beta = s\ln(g/g_c)$ is shown for $g > 2$ as the solid-circled line; this unphysical behavior necessary for a conductance jump in $d = 2$ is shown dashed.

The first correction term in this series may be estimated in perturbation theory by considering backscattering processes of the sort first discussed by Langer and Neal[11] in their analysis of the dependence of resistivity on impurity concentration.

The use of Langer and Neal involves a rather subtle question. Converting the calculation of Langer and Neal to dimensionality 2 in particular, one obtains

$$g(L) = g_0 - g_1 \ln\Lambda,$$

where Λ is a length cutoff for a certain divergent integral of second order in the density of scatterers. Langer and Neal assume this cutoff is l, the mean free path; for scales $L < l$, it should, of course, be L and we obtain just the result expected from (9), $d\ln g/d\ln L \sim -g^{-1}$. On the other hand, our universality argument seems to require $L > l$. We have restudied the cutoff question and will show in a fuller paper that their cutoff is not correct.

On the other hand, we have been unable to show definitively that the mean free path does *not* represent a relevant scale for the problem, since once $L > l$, we find, for example, that the coefficient of $\ln L$ depends on l. We must rely rather on our general arguments from continuity and regularity, and an intuition that only g is relevant, to suppose that a series development of β in g^{-1} should exist, once $L \gg l$.

The consequences of Fig. 1 and Eqs. (8) and (9) are as follows: For $d > 2$, the β function has a zero at g_c of order unity. It is an unstable fixed point which signals the mobility edge. The critical behavior can be estimated by integrating β starting from a microscopic L_0 and with g_0 near g_c. We use the linear approximation

$$\beta = s \ln(g/g_c), \tag{10}$$

where $s > 1$, since $\alpha > 0$ in (8). For $g_0 > g_c$, we obtain

$$\sigma = A\frac{e^2}{\hbar}\,\frac{g_c}{L_0^{d-2}}\left(\ln\frac{g_0}{g_c}\right)^{(d-2)/s}, \tag{11}$$

where A is of order unity. The distance to the mobility edge is measured by

$$\epsilon = \ln(g_0/g_c) \approx (g_0 - g_c)/g_c, \tag{12}$$

and the factor $Ae^2/\hbar L_0^{d-2}$ in (11) is the Mott conductivity which here appears in the scaling form proposed by Wegner.

In the localized regime ($g_0 < g_c$), we get

$$g \approx g_c \exp(-A|\epsilon|^{1/s} L/L_0), \tag{13}$$

so that the exponent of the localization length is the inverse slope of β at g_c,

$$\nu = 1/s. \tag{14}$$

These results again agree with those of Wegner.

In two dimensions, we have a strikingly different picture (see Fig. 1). Instead of a sharp mobility edge there is *no* critical g_c where $\beta(g_c) = 0$, but β is *always negative* so that in all cases $g(L \to \infty) = 0$. Instead of a sharp universal minimum metallic conductivity, there is a universal crossover from logarithmic to exponential behavior which for many experimental purposes may resemble a sharp mobility edge fairly closely. If we extrapolate the form we would deduce from Langer's perturbation-theory calculation, on the "extended" side of the crossover

$$g \simeq g_0 - Ag_c \ln(L/L_0) \tag{15}$$

the conductivity decreases logarithmically with scale until $g = g_c$ at the scale L_l, where

$$\frac{L_l}{L_0} = \exp\left[\frac{1}{A}\left(\frac{g_0}{g_c} - 1\right)\right]. \tag{16}$$

From this point on g decreases exponentially with L, the localization length being of order L_l as given by (16). This type of behavior was already anticipated from computer studies,[9] but the nature of the actual solution is surprising to say the least, as well as the fact that it appears to have been anticipated in terms of a divergence of perturbation theory in the weak-coupling limit by Langer and Neal.[11]

This work is supported in part by the National Science Foundation Grants No. DMR 78-03015 and No. DMR 76-23330-A-1, and by the U. S. of Naval Research.

[(a)] Also at Bell Laboratories, Murray Hill, N. J. 07974.

[(b)] On leave from Indian Institute of Technology, Kanpur, India.

[1] J. T. Edwards and D. J. Thouless, J. Phys. C 5, 807 (1972).

[2] D. C. Licciardello and D. J. Thouless, J. Phys. C 8, 4157 (1975).

[3a] D. J. Thouless, Phys. Rep. 13C, 93 (1974).

[3b] D. J. Thouless, to be published.

[4]F. J. Wegner, Z. Phys. 25, 327 (1976).
[5]H. G. Schuster, Z. Phys. 31, 99 (1978).
[6]A. Aharony and Y. Imry, J. Phys. C 10, L487 (1977).
[7]N. F. Mott, *Metal Insulator Transitions* (Taylor and Francis, London, 1974).
[8]B. J. Last and D. J. Thouless, J. Phys. C 7, 699 (1974).
[9]D. C. Licciardello and D. J. Thouless, J. Phys. C 11, 925 (1978).
[10]P. W. Anderson, Phys. Rev. 109, 1492 (1958).
[11]J. S. Langer and T. Neal, Phys. Rev. Lett. 16, 984 (1966).

Scaling theory of the metal-insulator transition in amorphous materials

W. L. McMillan
Department of Physics and Materials Research Laboratory, University of Illinois at Urbana-Champaign, Urbana, Illinois 61801
(Received 17 July 1980; revised manuscript received 22 December 1980)

A scaling model is presented for the metal-insulator transition in amorphous materials which includes a localization, correlation, and screening. The model predicts a continuous phase transition at zero temperature with localized states and a correlation gap in the insulating phase.

Recent experiments have indicated that the metal-insulator transition in several amorphous or disordered materials [a-$Ge_{1-x}Au_x$,[1,2] granular alumina,[3] and crystalline P-doped Si (Ref. 4)] is continuous. In a-$Ge_{1-x}Au_x$,[1] in particular, measurements of conductivity versus temperature show that the electronic states are extended in the metallic phase and localized in the insulating phase; the transition involves localization and therefore has something of the character of the Anderson transition.[5] However, tunneling experiments on a-$Ge_{1-x}Au_x$ (Ref. 2) and granular alumina[3] find a giant zero-bias anomaly in the one-electron density of states of the metal due to electron-electron interaction and also, apparently, a correlation gap in the insulating state. Thus the transition also has something of the character of a Mott transition.[6]

In this paper I develop a scaling theory of the metal-insulator transition in amorphous materials including the effects of localization and correlation. This theory is an extension of the scaling theory of the Anderson transition by Abrahams *et al.*[5] which treats localization within a one-electron model. Correlation effects are included by adapting a weak-coupling approximation due to Altshuler and Aronov[7] and Altshuler *et al.*[7] and by screening the electron-electron interaction within linear response theory. The effects of correlation are cast into renormalization-group language using an "exact eigenstates" method.[8] Included in the theory are one-electron localization, many-body localization, screening, and interaction effects on the one-electron density of states and the correlation gap.

We consider a model for the motion of electron wave packets of length scale L. We assume that the single-particle motion is diffusive with diffusion constant D_L for wave packets of radius L. The wave packets are to be made up of eigenstates $\phi_l(x)$ with eigenvalue $\hbar\omega_l$ of the one-electron problem and chosen to have minimum energy spread F_L. There is a characteristic energy $\hbar D_L/L^2$ associated with D_L which turns out to be the minimum energy spread F_L (within a constant of order unity). We therefore define $F_L = \hbar D_L/L^2$. One can verify this relationship using the exact eigenstates method[8] which is discussed below. We write down an unnormalized wave function

$$\psi(x) = \sum_l \phi_l^*(x_0)\phi_l(x)\exp[-\hbar^2(\omega_l-\omega)^2/4F^2]$$

which is centered at x_0 with mean energy $\hbar\omega$ and energy spread F. The mean-square radius is easily found to be

$$\langle(\overline{x}-\overline{x}_0)^2\rangle = 3\hbar D/\sqrt{2\pi}F$$

in three dimensions. I have also verified this relationship by simulation. This relationship provides a fundamental connection between length scale L and energy scale F_L. We use a renormalization-group approach and adopt a model at length scale L which retains one-electron states within energy F_L of the Fermi energy E_F. Roughly speaking, quantum states ϕ_l with energy $|\hbar\omega_l - E_F| > F_L$ will have been "integrated out" by including their contributions to various physical quantities. More concisely, we will integrate out transitions between quantum state ϕ_l and ϕ_m so that at length scale L transition between states such that $\hbar|\omega_l-\omega_m| > F_L$ have been removed; we actually remove matrix elements from the Hamiltonian, not quantum states. A second physical parameter of the system is the one-electron density of states (of one spin and for a unit volume) N_L at the Fermi energy; we find a second energy scale $E_L = 1/N_L L^3$. We write for the interaction between electrons $U(r) = e^2/\epsilon_L r$, where ϵ_L is the long-wavelength dielectric constant; ϵ_L contains the screening contributions from the quantum states which have been integrated out. The dielectric constant is, of course, a function of wave number q but it is approximately constant for $qL < 1$. The third energy scale is $U_L = e^2/L\epsilon_L$. We believe that the three physical quantities D_L, N_L, and ϵ_L or, equivalently, the three energy scales F_L, E_L, and U_L are sufficient to describe the Hamiltonian of the system at length scale L. We now define two dimensionless parameters from ratios of the energy scales. We first define a conductivity parameter $\sigma_L = 2e^2 N_L d_L$ such that the physical conductivity at zero temperature is $\sigma = \lim_{L\to\infty}\sigma_L$. Following Abrahams *et al.*[5] we define a dimensionless conductance

$$g_L = 8\sqrt{2}\pi\hbar L\sigma_L/e^2 = 16\sqrt{2}\pi F_L/E_L \ .$$

The second dimensionless parameter is the dimensionless interaction strength $\lambda_L = 16U_L/E_L$. We derive below the renormalization-group (RG) equations for g_L and λ_L.

We will use the exact eigenstates method[8] to calculate various physical quantities. We assume a one-electron Hamiltonian with eigenstates $\phi_l(x)$ normalized in unit volume and eigenvalues $\hbar\omega_l$. We require certain averages of the matrix element

$$M_{lm}(q) = \int d^3x\, \phi_l^*(x) e^{-i\vec{q}\cdot\vec{x}} \phi_m(x) \ . \quad (1)$$

which we derive as follows. We make up a wave packet

$$\psi(x,t) = \sum_l \phi_l^*(x_0)\phi_l(x)\frac{e^{-i\omega_l t}}{\sqrt{n}} \quad (2)$$

from n quantum states and assume diffusive motion of the wave packet

$$|\psi(x,t)|^2 = \frac{\exp[-(x-x_0)^2/4Dt]}{(4\pi Dt)^{3/2}} \ . \quad (3)$$

Fourier-transforming both sides of Eq. (3) in space $\int d^3x\, e^{-i\vec{q}\cdot(\vec{x}-\vec{x}_0)}$ and time $\int_0^\infty dt\, e^{i\omega t - \eta t}$, space averaging $\int d^3x_0$, and taking the imaginary part we find the average of the matrix element squared over all states with fixed energy difference $\hbar(\omega_l - \omega_m)$ (Ref. 7)

$$|M_{lm}(q)|^2 = (Dq^2/\pi N_L\hbar)/[(\omega_l-\omega_m)^2 + D^2q^4] \ . \quad (4)$$

We now change length scale from L to L' by changing the energy cutoff from $F_L = \hbar D_L/L^2$ to $F_{L'} = \hbar D_{L'}/L'^2$; to lowest order in λ and g^{-1}, D_L is independent of L and $d\ln F/d\ln L = -2$. We change energy scale from F_L to $F_{L'}$ by including the contribution to all physical quantities from transitions between states l and m such that $F_{L'} < \hbar|\omega_l - \omega_m| < F_L$. The contribution to the long-wavelength dielectric constant is

$$d\epsilon = \lim_{q\to 0} \frac{16\pi e^2}{q^2} \sum_{lm}{}' |M_{lm}(q)|^2 \frac{f_l(1-f_m)}{\hbar(\omega_m-\omega_l)} \ , \quad (5)$$

where f_l is the occupation number of the lth state and the summation is over all states such that $F_{L'} < \hbar|\omega_l - \omega_m| < F_L$. Performing the integrals we find

$$\frac{d\ln\epsilon_L}{d\ln L} = +2\lambda_L \ . \quad (6)$$

We next calculate the renormalization of the density of states which is given by $(1 + d\Sigma_m/dh\omega_m)^{-1}$ where Σ_m is the self-energy of the mth state. We use the Hartree-Fock approximation. The Hartree term screens the one-electron potential and the contribution from the exchange term is

$$d\left(1 + \frac{d\Sigma_m}{d\hbar\omega_m}\right) = -\sum_l{}' \sum_q U_q \frac{d}{d\hbar\omega_m}|M_{lm}(q)|^2 f_l \ , \quad (7)$$

where the sum is restricted by $F_{L'} < \hbar|\omega_l - \omega_m| < F_L$ and $U_q = 4\pi e^2/q^2\epsilon_L$ is the Fourier transform of the interaction. Performing the integrals we find

$$\frac{d\ln N_L}{d\ln L} = -\frac{\lambda_L}{g_L} \ . \quad (8)$$

Finally we need to calculate the renormalizaton of the diffusion constant. Instead of developing the exact eigenstates formalism to calculate the conductivity we will adapt the perturbative calculation of Altshuler and Aronov[7] to our purposes. These authors find for the renormalizaton of the zero frequency, zero-temperature conductivity

$$\frac{d\sigma}{\sigma} = \frac{4\pi}{3}\int_{F_{L'}}^{F_L} E\,dE \sum_q U_q \mathrm{Re}(iE + D_q^2)^{-3} \ . \quad (9)$$

The perturbative calculation yielded limits on the energy integral from zero to infinity and we have modified that result by summing only over transitions between $F_{L'}$ and F_L. Carrying out the integrals we find

$$\frac{d\ln\sigma_L}{d\ln L} = -\frac{2\lambda_L}{g_L} \ . \quad (10)$$

From the definition of σ_L we have

$$\frac{d\ln D_L}{d\ln L} = \frac{d\ln\sigma_L}{d\ln L} - \frac{d\ln N_L}{d\ln L} \quad (11)$$

so that Eqs. (8), (10), and (11) determine the renormalization of the diffusion constant due to interactions. There is also a one-electron contribution to the renormalization of the diffusion constant which, according to Abrahams *et al.*,[5] is of the form $-c/g_L$ with c unknown. The above results are valid within perturbation theory, $\lambda_L \ll 1$, $1/g_L \ll 1$.

Assembling these results we find

$$\begin{aligned}
&\frac{d\ln N}{d\ln L} = -\frac{\lambda}{g}, \quad \frac{d\ln\epsilon}{d\ln L} = 2\lambda \ ,\\
&\frac{d\ln D}{d\ln L} = -\frac{(c+\lambda)}{g}, \quad \frac{d\ln E}{d\ln L} = -3 + \frac{\lambda}{g} \ ,\\
&\frac{d\ln F}{d\ln L} = -2 - \frac{(c+\lambda)}{g}, \quad \frac{d\ln U}{d\ln L} = -1 - 2\lambda \ ,\\
&\frac{d\ln g}{d\ln L} = 1 - \frac{(c+2\lambda)}{g} \ ,\\
&\frac{d\ln\lambda}{d\ln L} = 2 - 2\lambda - \frac{\lambda}{g} \ .
\end{aligned} \quad (12)$$

Equations (12) are the differential equations of the

two-parameter renormalization group, derived within weak coupling ($\lambda << 1$, $g^{-1} << 1$). The first correction term in the conductance equation is the one-electron localization correction of Abrahams *et al.*[5]; we interpret the second term as a many-body localization correction. We feel comfortable extrapolating the RG equations to intermediate coupling to study the phase transition.

We now study the properties of the RG equations. We first study the RG trajectories in parameter space to locate the fixed points representing phase transitions and stable phases. The constant c is unknown; however, we expect the qualitative behavior of the theory to be insensitive to the value of the constant. We will work through the theory with the value $c=\frac{7}{4}$, chosen so that one-particle localization and many-body localization are of equal importance at the phase transition. With this value of c the flow diagram is given in Fig. 1. There is a fixed point at $(g,\lambda)=(\frac{7}{4},0)$ which represents the Anderson transition studied by Abrahams *et al.*[5] The trajectory to the right is toward $(\infty,0)$ which represents the noninteracting conducting state with extended states; the trajectory to the left is toward $(0,0)$ which represents the noninteracting insulating state with localized quantum states. The Anderson-transition fixed point is unstable with respect to interactions and does not represent an observable metal-insulator transition (according to the present theory) although it can represent a mobility edge far from the Fermi energy. The noninteracting insulator and conducting states are not physically observable states. The fixed point at $(\frac{7}{2},\frac{7}{8})$ represents the metal-insulator transition in the interacting system and the two trajectories flowing into it are the critical surface separating parameter space into two regions, conducting on the right and insulating on the left. On the right the flow is toward $(\infty,1)$ which represents an observable phase which we call the amorphous conductor phase. On the left the flow approaches the λ axis as g goes to zero which represents an observable phase which we call the amorphous insulator phase. We show below that these two phases have unique, universal properties which depend only on a length scale ξ and an energy scale Δ.

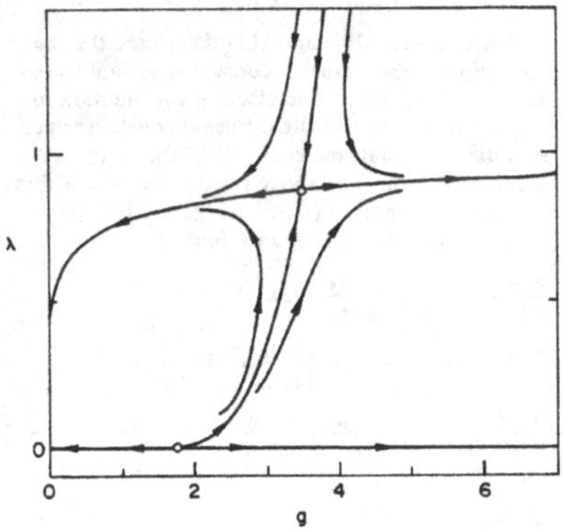

FIG. 1. Flow diagram of the two-parameter renormalization group showing trajectories in parameter space with the two fixed points marked by open circles.

We now study the fixed point at $(\frac{7}{2},\frac{7}{8})$ and develop the scaling theory. We start the calculation at microscopic length scale "a" with properties N_a, etc. If we start on the critical surface we find $E_L=E_a(a/L)^\eta$, $F_L=F_a(a/L)^\eta$, $U_L=U_a(a/L)^\eta$, $D_L=D_a(a/L)^{\eta-2}$, $\epsilon_l=\epsilon_a(a/L)^{1-\eta}$, $N_L=N_a(a/L)^{3-\eta}$, $\sigma_L=\sigma_a(a/L)$, with $\eta=\frac{11}{4}$. We now linearize the differential equations near the fixed point and find a single positive eigenvalue of $\sqrt{17/8}-\frac{1}{2}=0.96$. Making the usual renormalization-group arguments[9] the correlation length ξ (at which the system crosses over from the critical regime to the conducting or insulating regime) is proportional to $|x-x_c|^{-\nu}$ where $\nu^{-1}=0.96$ and x is the composition with x_c the critical composition. We assume here that composition is the external control parameter that drives the system through the transition (at zero temperature). Thus we find two critical exponents η, relating energy scale and length scale in the critical regime, and ν, relating correlation length and composition. Both exponents depend upon the value of the constant c and upon the way in which the theory is extrapolated from weak coupling and therefore the numerical values should not be taken seriously. We have $1<\eta<3$ and we expect $\nu\approx 1$. The energy scale of the system as it crosses over from the critical regime to the conducting or insulating regime is

$$\Delta=F_\xi=\hbar D_\xi/\xi^2=(\hbar D_a/a^2)(a/\xi)^\eta \ .$$

We now discuss the properties of the system in the three regimes: critical, conducting, and insulating. In what follows we will estimate the order of magnitude of various quantities and will omit constants of order unity. In order to estimate properties of the system at finite energy E, frequency ω, or temperature T, we stop the renormalization group at a length scale such that the energy scale is E or $\hbar\omega$ or kT. In the critical regime at finite temperature we stop at a length scale $L_T=a(F_a/kT)^{1/\eta}$ and find for the conductivity

$$\sigma(T)=\sigma_a(a/L_T)=\sigma_a(kT/F_a)^{1/\eta} \ .$$

Similarly at finite frequency $\sigma(\omega)=\sigma_a(\hbar\omega/F_a)^{1/\eta}$. The one-electron density of states versus energy is

$N(E) = N_a(E/F_a)^{-1+3/\eta}$. The Coulomb interaction is partially screened $U(r) = (e^2/\epsilon_a)(a/r)^\eta$. In the conducting regime the solution to the RG equations are $\lambda = 1$, $g_L = 3.5(1+L/\xi)$. We find

$$\sigma_L = 0.1(e^2/\hbar\xi)(1+\xi/L)$$

and energy scale $F_L = \Delta(\xi/L)^2$. The length scale at finite temperature is $L_T = \xi\sqrt{\Delta/kT}$ and the conductivity at finite temperature is therefore

$$\sigma(T) = 0.1(e^2/\hbar\xi)(1+\sqrt{kT/\Delta}) \ .$$

Solving the differential equation for the density of states we find $N_L = N_\xi(1+\xi/L)/2$ so that the density of states versus energy is

$$N(E) = N_a(a/\xi)^{3-\eta}(1+\sqrt{E/\Delta}) \ .$$

The dielectric constant is

$$\epsilon_L = \epsilon_a(\xi/a)^{\eta-1}(L/\xi)^2$$

and

$$\epsilon_q = \epsilon_a(\xi/a)^{\eta-1}/(q\xi)^2 \ ;$$

this implies exponential screening with a screening length ξ. The theory behaves sensibly in the insulating regime. The conductance goes to zero at a finite length scale ξ which we interpret as the localization length. The dielectric constant goes to a constant $\epsilon = (\epsilon_a)(\xi/a)^{\eta-1}$ and the Coulomb energy $U_\xi = \Delta = U_a(a/\xi)^\eta$, which is the relevant energy scale in the insulating phase, is finite. The density of states goes sharply to zero and there is a correlation energy gap equal to Δ. The conductivity goes quickly (exponentially) to zero for $kT < \Delta$. All these properties are undoubtedly characteristic of the amorphous insulating phase. We do not, of course, expect a sharp energy gap in the density of states; we expect band tailing below Δ. At present we have no viable microscopic model of the amorphous insulating phase; the theory presented above in deriving the RG equations is a primitive microscopic theory of the amorphous conducting phase.

There is a characteristic conductivity which separates normal metallic behavior from the amorphous conductor behavior. In the normal metallic regime the conductivity is $e^2k_F^2l/3\pi^2\hbar$, where k_F is the Fermi wave number and l the mean free path; this description breaks down for mean free path less than atomic spacing "a" (we assume a macroscopically homogeneous material). Since k_Fa is a typically π, the normal-metal regime is for $\sigma > e^2/3\hbar a$. The conductivity in the amorphous conducting regime is $0.1e^2/\hbar\xi$ which should apply when $\xi > a$ or $\sigma < 0.1e^2/\hbar a$. Therefore we expect a crossover from normal metallic behavior, with σ decreasing with increasing temperature, to amorphous metallic behavior, with σ increasing with increasing temperature for $\sigma \approx 0.2e^2/\hbar a$ which is a resistivity of about 200 $\mu\Omega$ cm for metals.

The theory predicts a strong universality: all metal-insulator transitions in (macroscopically homogeneous) disordered materials should have the same exponents. Further, since the only relevant parameters are the length scale ξ and energy scale Δ, both the amorphous insulator and amorphous conductor phases should obey a law of corresponding states. The materials must be homogeneous on a length scale larger than ξ to avoid the complication of a classical percolation problem. I see no reason why the theory should not be applicable to granular materials provided the correlation length is larger than the grain size and the material is homogeneous; presumably, the critical region will be narrower for granular materials.

It is clearly desirable to have a more detailed picture of the conducting and insulating phases than that presented here. One wants to use the renormalization-group approach for $L < \xi$ to find the Hamiltonian parameters for $L = \xi$ and then develop a microscopic theory using the Hamiltonian. The physical picture of the conducting phase is pretty clear. There is electron-hole symmetry (in both phases). There is a square-root anomaly in the one-electron density of states which is a precursor to the opening of the correlation gap at the transition. The screening length is ξ and the electron-electron interaction is strong and is the dominant interaction. A somewhat more sophisticated calculation of the self-energy using the screened exchange approximation shows that the quasiparticle approximation is valid for energies much less than Δ but breaks down for $E \approx \Delta$ due to electron-electron scattering. The conductivity mechanism at low temperatures is quantum diffusion of quasiparticles; the quantum states are, or course, extended in the Anderson sense. Since the density-of-states renormalization for each spin direction follows the Fermi energy for that spin, the spin susceptibility is not renormalized. In the insulating phase it is clear that there is a correlation gap but there is, at present, no satisfactory microscopic model of that phase.

The present experimental situation is as follows. The metal-insulator transition appears to be continuous in a-$Ge_{1-x}Au_x$,[1] granular alumina,[3] and, possibly, phosphorus-doped silicon.[4] Certainly one can make samples with very small conductivities; it is not clear at what point inhomogeneities and classical percolation become imporatnt. The observation of correlation effects by tunneling experiments[2] in a-$Ge_{1-x}Au_x$ motivated the present theoretical work. More recent tunneling experiments on granular alumina[3] have shown that the square-root anomaly in the density of states scales with the conductivity over several decades and that $\eta \approx 2$; this is the first real test of the scaling theory. Tunneling experiments[2] in the insulating phase indicate an absence of available states at low energy and thus an energy gap; however, a quantitative interpretation of the experiment requires an understanding of field penetration into the insulator which requires a detailed model of the insulating phase. The conductivity crosses over from an exponential dependence (with a fractional inverse power of T) at low temperature to an algebraic dependence at high temperature. The crossover temperature is a characteristic energy which might be interpreted as an energy gap; however, one could alternatively interpret this behavior as a crossover from variable-ranged hopping[6] at low temperature to nearest-neighbor hopping at high temperature. Thus, the experimental evidence for an energy gap is ambiguous. In summary, the experimental evidence is strong that correlation and localization effects are important near the metal-insulator transition and one scaling prediction has been tested experimentally. However, many of the predictions of the theory await experimental confirmation.

ACKNOWLEDGMENTS

The author would like to thank B. Dodson, R. C. Dynes, and J. Mochel for extensive discussions of their experiments and P W. Anderson and P. A. Lee for theoretical discussions. I would also like to acknowledge support by the National Science Foundation under Grant No. NSF-DMR-77-27091.

---

[1]B. M. Dodson, W. L. McMillan, J. M. Mochel, and R. C. Dynes, Phys. Rev. Lett. 46, 46 (1981).
[2]W. L. McMillan and J. Mochel, Phys. Rev. Lett. 46, 556 (1981).
[3]R. C. Dynes and J. Garno, Phys. Rev. Lett. 46, 137 (1981).
[4]T. F. Rosenbaum, K. Andres, G. A. Thomas, and R. N. Bhatt, Phys. Rev. Lett. 45, 1723 (1980).
[5]E. Abrahams, P. W. Anderson, D. C. Licciardello, and T. V. Ramakrishnan, Phys. Rev. Lett. 42, 673 (1979).
[6]N. F Mott, *Metal-Insulator Transitions* (Taylor and Francis, London, 1974).
[7]B. L. Altshuler and A. G. Aronov, Zh. Eksp. Teor. Fiz. 77, 2028 (1979) [Sov. Phys. JETP 50, 968 (1979)]; B. L. Altshuler, A. G. Aronov, and P. A. Lee, Phys. Rev. Lett. 44, 1288 (1980).
[8]E. Abrahams, P. W. Anderson, P. A. Lee, and T. V. Ramakrishnan (unpublished).
[9]K. Wilson and J. Kogut, Phys. Rep. C 12, 75 (1974).

PHYSICAL REVIEW B VOLUME 14, NUMBER 3 1 AUGUST 1976

Quantum critical phenomena*

John A. Hertz
The James Franck Institute and The Department of Physics, The University of Chicago, Chicago, Illinois 60637
(Received 8 September 1975)

This paper proposes an approach to the study of critical phenomena in quantum-mechanical systems at zero or low temperatures, where classical free-energy functionals of the Landau-Ginzburg-Wilson sort are not valid. The functional integral transformations first proposed by Stratonovich and Hubbard allow one to construct a quantum-mechanical generalization of the Landau-Ginzburg-Wilson functional in which the order-parameter field depends on (imaginary) time as well as space. Since the time variable lies in the finite interval $[0, -i\beta]$, where β is the inverse temperature, the resulting description of a d-dimensional system shares some features with that of a $(d+1)$-dimensional classical system which has finite extent in one dimension. However, the analogy is not complete, in general, since time and space do not necessarily enter the generalized free-energy functional in the same way. The Wilson renormalization group is used here to investigate the critical behavior of several systems for which these generalized functionals can be constructed simply. Of these, the itinerant ferromagnet is studied in greater detail. The principal results of this investigation are (i) at zero temperature, in situations where the ordering is brought about by changing a coupling constant, the dimensionality which separates classical from nonclassical critical-exponent behavior is not 4, as is usually the case in classical statistics, but $4 - z$ dimensions, where z depends on the way the frequency enters the generalized free-energy functional. When it does so in the same way that the wave vector does, as happens in the case of interacting magnetic excitons, the effective dimensionality is simply increased by 1; $z = 1$. It need not appear in this fashion, however, and in the examples of itinerant antiferromagnetism and clean and dirty itinerant ferromagnetism, one finds $z = 2$, 3, and 4, respectively. (ii) At finite temperatures, one finds that a classical statistical-mechanical description holds (and nonclassical exponents, for $d < 4$) very close to the critical value of the coupling U_c, when $(U - U_c)/U_c \ll (T/U_c)^{2/z}$ $z/2$ is therefore the quantum-to-classical crossover exponent.

I. INTRODUCTION

The spectacularly successful analysis of critical phenomena in a wide variety of systems using Wilson's renormalization-group ideas[1] has hitherto been limited to *classical* statistical-mechanical models. Such a description is appropriate whenever the critical temperature is finite, provided one is close enough to the instability. Then, when all fluctuation modes have characteristic energies $\ll kT_c$, classical statistics are appropriate. However, one can also think about a phase transition in a zero-temperature system which occurs when, say, a coupling constant reaches a certain threshold. In this case, none of the fluctuation modes have thermal energies, and their statistics will be highly nonclassical. By the same token, in the same system at a finite but low temperature, one should expect quantum effects to be dominant except in a narrow range of coupling strengths near the critical value. (By low temperature, I mean kT much less than characteristic microscopic energies, such as the Fermi energy, bandwidth, Coulomb or exchange energies, etc.)

In addition to quantum effects at low or zero temperature in the equilibrium correlation functions and static-response coefficients, we should expect quite different dynamical properties. In the classical case, one can study dynamical critical phenomena using time-dependent Landau-Ginzburg equations or generalizations thereof.[2] These equations contain as parameters transport coefficients whose existence depends on the presence of collisions to maintain local thermal equilibrium. In a zero-temperature problem, by contrast, there are no collisions, and consequently no transport coefficients and no time-dependent Landau-Ginzburg equations. Similarly, at low T, the dynamics will be effectively collisionless except very close to the critical coupling.

One feature of the classical problem is the separability of the statics and the dynamics—the former may be solved independently of the latter. We shall see here that this, too, breaks down in systems where quantum mechanics is important. Statistics and dynamics are then inextricably connected, and one has to solve for both equilibrium and nonequilibrium properties together in the same formalism, rather than doing the dynamics afterwards. This complication is offset, however, by the fact that the formalism we shall use makes this unified approach the straightforward and natural one.

Our principal formal tool for setting up this

class of problems is the functional-integral transformation of Stratonovich and Hubbard.[3] It allows one to construct an exact quantum generalization of the Landau-Ginzburg-Wilson (LGW) free-energy functional used in classical problems. The precise form of this functional will depend on the character of the dynamics of the system in question, but all quantum functionals share the feature that the order-parameter field depends on time as well as space. The time variable is, as one might expect in a quantum-statistical problem, imaginary and in the interval $[0, -i\beta]$ $(\beta = 1/kT)$. The Fourier transform of the order parameter (in terms of which it is usually simpler to write the functional) therefore is a function of frequency as well as wave vector, and the frequencies which occur are the (Bose) Matsubara frequencies $i\omega_n = 2\pi i n/\beta$. Section III is devoted to a discussion of the derivation of this functional for the problem of interacting paramagnons in itinerant ferromagnetism.

With this as a starting point, in Sec. III we apply the renormalization group and study the evolution of the parameters in the functional as high wave numbers *and high frequencies* are scaled out of the problem. We show that the quantum LGW functional has a stable Gaussian fixed point under the renormalization group at zero temperature for $d \geq 1$. The $T=0$ critical exponents are thus mean-field-like. We then examine the instability of this fixed point at $T \neq 0$ and calculate the crossover exponent which characterizes the eventual switch to a non-Gaussian fixed point and non-mean-field exponents. In Sec. IV we discuss the utility of approximate solutions of the renormalization-group equations as a substitute for more conventional perturbation-theoretical techniques in problems like this, and examine the effect of the hitherto ignored higher-order terms in the generalized free-energy functional on such solutions. In Sec. V we introduce and apply the renormalization group to several other models in which quantum effects can be important—itinerant antiferromagnetism, interacting magnetic excitons, and the paramagnon problem in the presence of impurities. Finally, Sec. VI is devoted to a somewhat different version of the quantum renormalization group in which all frequency components of the order parameter $\Psi(q, \omega)$ with same q are scaled out of the problem together at each stage of the renormalization-group operation. This procedure is different from that mentioned above, where one scales out high ω and high q together, in that time and space are no longer treated on the same footing. The results are the same, however, and this formulation does have the advantage that it, in principle, allows one to follow the crossover from quantum to classical scaling continuously.

An abbreviated account of part of this work was presented earlier.[4]

II. GENERALIZED LGW FUNCTIONAL FOR INTERACTING PARAMAGNONS

The application of the Stratonovich-Hubbard transformation to itinerant ferromagnetism has been discussed extensively in the literature.[5-7] Here we only outline the steps involved in generating the free-energy functional. We start with a Hubbard interaction Hamiltonian,[8] written in terms of charge- and spin-density variables:

$$H' = U \sum_i n_{i\uparrow} n_{i\downarrow}$$

$$= \frac{U}{4} \sum_i (n_{i\uparrow} + n_{i\downarrow})^2 - \frac{U}{4} \sum_i (n_{i\uparrow} - n_{i\downarrow})^2 . \quad (2.1)$$

Statistical mechanics requires knowledge of matrix elements of the operator

$$e^{-\beta H} = e^{-\beta H_0} T \exp\left(-\int_0^\beta d\tau H'(\tau)\right) . \quad (2.2)$$

The Stratonovich-Hubbard transformation applies the identity

$$e^{a^2/2} = \int_{-\infty}^{\infty} \frac{dx}{\sqrt{2\pi}} e^{-x^2/2 - ax} \quad (2.3)$$

to (2.2) for each imaginary time τ between 0 and β and for every site in the lattice, with the result that

$$Z = \mathrm{Tr}\, e^{-\beta H}$$

$$= Z_0 \int \delta\Psi \exp\left(-\frac{1}{2}\int_0^\beta d\tau \sum_i \Psi_i^2(\tau)\right)$$

$$\times \left\langle \mathrm{Tr}\, T \exp\left(-\int_0^\beta d\tau \sum_{i\sigma} \sigma V_i(\tau) n_{i\sigma}(\tau)\right)\right\rangle_0 . \quad (2.4)$$

Here $V_i(\tau) = (\frac{1}{2}U)^{1/2}\Psi_i(\tau)$ is a time-dependent magnetic field acting on site i at "time" τ, and Z_0 is the partition function of the noninteracting system. [In addition to V, there should be another field in the exponential inside the expectation value in (2.4), coupled to the charge density. We ignore it here, since we want to concentrate on the spin fluctuations and expect that charge-density fluctuations will be relatively unimportant.] The expectation value in (2.4) can be expressed in terms of the electron Green's functions of the noninteracting system,

$$G^0_{ij}(\tau,\tau')=\frac{1}{\beta}\sum_{kn}\frac{\exp[i\vec{k}\cdot(\vec{R}_i-\vec{R}_j)-iE_n\tau]}{iE_n-\epsilon_k}, \quad (2.5)$$

so that

$$Z=Z_0\int\delta\Psi\exp\left(-\frac{1}{2}\int_0^\beta d\tau\sum_i\Psi_i^2(\tau)+\sum_\sigma \mathrm{Tr}\ln(1-\sigma VG^0)\right), \quad (2.6)$$

where the matrix V has elements $V_{ij}(\tau,\tau')=V_i(\tau)\delta_{ij}\delta(\tau-\tau')$.

The exponential in Eq. (2.6) is then a formally exact free-energy functional $\Phi[\Psi]$ in which $\Psi_i(\tau)$ [or, in a continuum limit, $\Psi(x,\tau)$] is the order-parameter field. To do much with it, it is generally advantageous to expand it in a power series in Ψ, leading to an expression of the general form

$$\Phi[\Psi]=\frac{1}{2}\sum_{q\omega}v_2(q,\omega)|\Psi(q,\omega)|^2+\frac{1}{4\beta N}\sum_{q_i\omega_i}v_4(q_1\omega_1,q_2\omega_2,q_3\omega_3,q_4\omega_4)\Psi(q_1,\omega_1)\Psi(q_2,\omega_2)\Psi(q_3,\omega_3)\Psi(q_4,\omega_4)\times\delta\left(\sum_{i=1}^4 q_i\right)\delta\left(\sum_{i=1}^4\omega_i\right)+\cdots$$
$$+\frac{1}{m(\beta N)^{m/2-1}}\sum_{q_i,\omega_i}v_m(q_1\omega_1,\ldots,q_m\omega_m)\prod_{i=1}^m\Psi(q_i,\omega_i)\,\delta\left(\sum_{i=1}^m q_i\right)\delta\left(\sum_{i=1}^m\omega_i\right)+\cdots. \quad (2.7)$$

Clearly the form is analogous to that discussed by Wilson for classical statistics. The effect of quantum mechanics can be traced to the noncommutativity of H_0 and H', which forced us to write $e^{-\beta H}$ in the interaction representation (2.2), requiring the functional averaging identity (2.3) to be applied for each time τ. This makes the order parameter time dependent, with the consequence that (Matsubara) frequencies appear in (2.7) on the same footing as wave vectors. It is as if another dimension were added to the system, but, except at zero temperature, the extent of the system in the extra dimension is finite. We shall examine the effects of the consequent finite spacing between Matsubara frequencies at the end of Sec. III.

The coefficients v_m in (2.7) (irreducible bare-m-point vertices in a diagrammatic perturbation-theoretic development) can in principle be evaluated in terms of the band propagators (2.5); v_m is just proportional to a loop of m electron propagators, with four-momentum transfers $q_1,\omega_1,\ldots,q_m,\omega_m$ between propagator lines.[7] The quadratic coefficient v_2, which we will want to examine here, has an extra term of unity because of the Gaussian weight factor in the functional integral,

$$v_2(q,\omega)=1-U\chi_0(q,\omega), \quad (2.8)$$

where χ_0 is the function evaluated by Lindhard for a free-electron model,[9]

$$\chi_0(q,\omega)=-\frac{1}{\beta}\sum_{kn}G(k,iE_n)G(k+q,iE_n+\omega)$$
$$=-\sum_k\frac{f(\epsilon_k)-f(\epsilon_{k+q})}{\epsilon_k-\epsilon_{k+q}+\omega}. \quad (2.9)$$

For small q and small ω/qv_F, it has the expansion[10]

$$\chi_0(q,i\omega_m)=N(E_F)[1-\tfrac{1}{3}(q/2k_F)^2-\tfrac{1}{2}\pi(|\omega_m|/qv_F)+\cdots]. \quad (2.10)$$

As long as we are near the ferromagnetic instability $UN(E_F)=1$, this long-wavelength, low-frequency form of χ_0 gives a good representation of the paramagnon propagator or its spectral weight function

$$\mathrm{Im}\chi(\omega+i\delta)=\mathrm{Im}\left(\frac{\chi_0(q,\omega)}{1-U\chi_0(q,\omega)}\right)\approx\frac{(2/\pi)v_FqN(E_F)\omega}{\omega^2+\{(2/\pi)v_Fq[1-UN(E_F)+\frac{1}{3}(q/2k_F)]\}^2} \quad (2.11)$$

for frequencies where most of the spectral weight lies, since $\mathrm{Im}\chi$ is sizable mostly near a frequency

$$\omega_q=(2/\pi)v_Fq[1-UN(E_F)+\tfrac{1}{3}(q/2k_F)^2]\ll v_Fq. \quad (2.12)$$

The higher-order loops v_m, $m\geq 4$, are complicated functions of the q_i and ω_i, but they share with v_2 the fact that they vary with any q on a scale of $2k_F$ and with ω on a scale of E_F. Put physically, the force between paramagnons has a range $\approx (2k_F)^{-1}$ in space and a retardation $\approx 1/E_F$ in time. When all of the q_i and ω_i vanish, v_m is simply proportional to the $(m-2)$nd derivative of the band density of states at E_F.[7]

In this section we will see an approximation to the full functional in which we use the expansion (2.10) in v_2, ignore all q and ω dependence on v_4, and discard higher-order vertices completely. The finite range and retardation effects will be simulated by cutting off all q sums at $\approx 2k_F$ and all ω sums at $\approx E_F$. Choosing units appropriately, we can write our approximate functional as

$$\Phi[\Psi]=\frac{1}{2}\sum_{q\omega}\left(r_0+q^2+\frac{|\omega|}{q}\right)|\Psi(q,\omega)|^2$$
$$+\frac{u_0}{4N\beta}\sum_{q_i\omega_i}\Psi(q_1,\omega_1)\Psi(q_2,\omega_2)\Psi(q_3,\omega_3)$$
$$\times\Psi(-q_1-q_2-q_3,-\omega_1-\omega_2-\omega_3), \tag{2.13}$$

where, in terms of microscopic parameters,

$$r_0=1-UN(E_F), \tag{2.14a}$$
$$u_0=\tfrac{1}{12}U^2N''(E_F). \tag{2.14b}$$

This Φ is of almost the same form as the classical LGW functional, except for the presence of the frequency-dependent term in v_2, which contains the essential information about the dynamics. It tells us that the decay mechanism for the paramagnon excitations is Landau damping—the lifetime of a free particle-hole pair of total momentum q is $(v_Fq)^{-1}$, and the correlations enhance this lifetime (for small q) by a factor $[1-UN(E_F)]^{-1}$, as reflected in (2.11). This is why ω enters (2.13) in the form $|\omega|/q$. If the dynamics were different, this term would have a different form. We shall examine examples with different dynamics in Sec. V.

Our functional (2.13) therefore describes a set of interacting, weakly-Landau-damped excitations. Terms of higher order in Ψ, as well as higher-order expansions of the coefficients included here in powers of q and ω, contain no essential new physics and in fact are "irrelevant" to the zero-temperature critical behavior in the sense described by Wilson.[1]

It is also possible to write the generalized LGW functional in a form which preserves the rotational invariance of the original Hamiltonian by using a vector paramagnon field $\vec{S}$ in place of Ψ. We shall not dwell at length on the formal derivation of this functional, since this aspect of the problem has been discussed elsewhere.[6,11] Our emphasis is on the form of the coefficients to order S^4 and their physical implications.

The starting point lies in expressing the interaction Hamiltonian as

$$H'=\tfrac{1}{2}U\sum_i(n_{i\uparrow}+n_{i\downarrow})-\tfrac{2}{3}U\sum_i\vec{S}_i\cdot\vec{S}_i \tag{2.15}$$

instead of (2.1), which only has $S_i^zS_i^z$ terms. Then the application of the identity (2.3) leads to an expression for the partition function [cf. (2.6)]

$$z=z_0\int\delta S\exp\left(-\frac{1}{2}\sum_i\int_0^\beta d\tau\vec{S}_i^2(\tau)+\mathrm{Tr}\ln(1-VG^0)\right) \tag{2.16}$$

in which V and G are matrices in spin space as well as in space-time indices, and the Tr indicates a trace over both spin and space-time indices. Explicitly, G^0 is the spin diagonal with elements (2.5), and

$$\langle i,\tau,m|V|j,\tau',m'\rangle$$
$$=(\tfrac{1}{3}U)^{1/2}\delta_{ij}\delta(\tau-\tau')\vec{S}_i(\tau)\cdot\langle m|\vec{\sigma}|m'\rangle. \tag{2.17}$$

When the Tr ln in (2.16) is expanded in powers of V, the quadratic term of the exponent becomes

$$\frac{1}{2}\sum_{q\omega\alpha}[1-\tfrac{2}{3}U\chi_0(q,\omega)]|S_\alpha(q,\omega)|^2 \tag{2.18}$$

and the fourth-order term looks like

$$\frac{1}{4\beta N}\sum_{\substack{q_i,\omega_i\\ \alpha\beta\gamma\delta}}v_4^{\alpha\beta\gamma\delta}(\{q_i,\omega_i\})$$
$$\times S_\alpha(q_1\omega_1)S_\beta(q_2\omega_2)S_\gamma(q_3\omega_3)S_\delta(q_4\omega_4)$$
$$\times\delta\left(\sum_{i=1}^4 q_i\right)\delta\left(\sum_{i=1}^4\omega_i\right), \tag{2.19}$$

where $v_4^{\alpha\beta\gamma\delta}$ is proportional to the v_4 which appeared in the scalar description [Eq. (2.7)],

$$v_4^{\alpha\beta\gamma\delta}=\tfrac{1}{2}(\tfrac{2}{3})^2v_4\,\mathrm{Tr}(\sigma^\alpha\sigma^\beta\sigma^\gamma\sigma^\delta). \tag{2.20}$$

The important point is that the dependence of the quadratic and quartic coefficients on wavevector and frequency arguments is the same as in the scalar case, and the dependence on the polarization labels follows simply from the Pauli spin algebra. So in order that four-paramagnon modes have nonvanishing interactions, the Pauli matrices corresponding to their polarizations must multiply to give the unit matrix. One way to do this is to have $\alpha=\beta$ and $\gamma=\delta$; this part of the interaction is then of the form $(\vec{S}\cdot\vec{S})(\vec{S}\cdot\vec{S})$, as occurs in the usual LGW functional for a vector field. But one can also have $\alpha=\gamma$, $\beta=\delta$ or $\alpha=\delta$,

$\beta=\gamma$, leading to a part of $v_4^{\alpha\beta\gamma\delta}$ proportional to $\delta_{\alpha\delta}\delta_{\beta\gamma}-\delta_{\alpha\gamma}\delta_{\beta\delta}=\epsilon_{\alpha\beta\mu}\epsilon_{\gamma\delta\mu}$. This means there is also a part of the S^4 interaction of the form $(\vec{S}\times\vec{S})\cdot(\vec{S}\times\vec{S})$. This term vanishes if all frequency and wave-vector dependence of v_4 is ignored, since we could then write its contribution to Φ as something proportional to

$$v_4\sum_i\int_0^\beta d\tau[\vec{S}_i(\tau)\times\vec{S}_i(\tau)]^2=0\,. \qquad (2.21)$$

Nonlocal or retarded terms must therefore be retained in order to see any effects of these parts of the interaction. Such terms, however, are beyond the scope of the present discussion, in which nonlocal effects in v_4 are irrelevant to the phenomena of interest. Accordingly, our vector paramagnon LGW functional is

$$\Phi[\vec{S}]=\frac{1}{2}\sum_{q\omega\alpha}\left(r_0+q^2+\frac{|\omega|}{q}\right)|S_\alpha(q,\omega)|^2 + \frac{u_0}{4\beta N}\sum_{\substack{q_i\omega_i\\ \alpha\beta}}S_\alpha(q_1\omega_1)S_\alpha(q_2\omega_2)S_\beta(q_3\omega_3) \times S_\beta(-q_1-q_2-q_3,-\omega_1-\omega_2-\omega_3)\,. \qquad (2.22)$$

The parameters r_0 and u_0 differ from their scalar problem counterparts [as in (2.18) and (2.20)], but this point will not be important here.

There is nothing really new in this section. I have simply collected from various sources the points relevant to establishing the basis of the model functional. The purpose of doing so was purely pedagogical.

III. RENORMALIZATION-GROUP TRANSFORMATION

Beal-Monod was the first to apply the renormalization group to the quantum functional (2.13).[12] She noted that the time acted like an extra dimension, and asserted that the critical behavior would be just the same as that of a $(d+1)$-dimensional system. This is not true, however. The frequency enters (2.13) in the form of a term in Φ proportional to $|\omega|/q$, which is quite different from the way the wave vector occurs. This anisotropy partially destroys the analogy between the present problem and a $(d+1)$-dimensional classical LGW problem, and renders her conclusion invalid. We will see here that it is necessary to generalize the Wilson scaling procedure somewhat to deal with the anisotropic coupling.

The general idea of the renormalization operation is the same as Wilson's. There are three steps: (a) Terms in Φ which have the wave vectors or frequencies of some of the Ψ fields in an "outer shell" are eliminated from the functional integral by carrying out the integration over these $\Psi(q,\omega)$, while holding fixed the $\Psi(q,\omega)$ with smaller q or ω. (b) The variables q and ω, which in the remaining functional integral run up only to a cutoff less than the original one, are rescaled, so that they once again take on the range of values they had in the original problem, before step (a). (c) The fields Ψ are rescaled, so that in terms of the new fields and the rescaled q and ω, the terms with q^2 and $|\omega|/q$ in the quadratic part of Φ look just like those in the original functional. That is, the coefficients of q^2 and $|\omega|/q$ in (2.13) must remain at unity under the group transformation. The only difference will turn out to be that in step (b), q and ω must be rescaled differently, as a consequence of the anisotropy of the functional in the "extra dimension."

To see why this happens, let us try doing the scaling isotropically. We use, as we will throughout this paper, a scaling procedure in which only an infinitesimally thin shell of Ψ's is removed at each stage of the renormalization procedure.[13]

Suppose that in step (a) we have removed Ψ's with $e^{-l}<q<1$ and $e^{-l}<|\omega|<1$, with l infinitesimal. This will affect the quadratic term of (2.12) in two ways: (i) r_0 will be changed to a new value r_0' (the change is of order l), and (ii) the sums on q and ω now have q and $|\omega|$ less than e^{-l}. It is easier to keep track of the manipulations we make if we write the sums as integrals, so the quadratic term (call it Φ_2 originally) now looks like

$$\Phi_2'=\tfrac{1}{2}\beta N\int_0^{e^{-l}}\frac{d^dq\,d\omega}{(2\pi)^{d+1}}\left(r_0'+q^2+\frac{|\omega|}{q}\right)|\Psi(q,\omega)|^2\,. \qquad (3.1)$$

Rescaling q and ω (step b) by letting

$$q'=qe^l,\quad \omega'=\omega e^l, \qquad (3.2)$$

we can write Φ_2' as

$$\Phi_2'=\tfrac{1}{2}\beta Ne^{-(d+1)l}\int_0^1\frac{d^dq'\,d\omega'}{(2\pi)^{d+1}} \times\left(r_0'+q'^2e^{-2l}+\frac{|\omega'|}{q'}\right)|\Psi(q'e^{-l},\omega'e^{-l})|^2\,. \qquad (3.3)$$

Step (c) would then have us define a $\Psi'(q',\omega')$, proportional to $\Psi(q'e^{-l},\omega'e^{-l})$ with the coefficient of proportionality chosen so that the coefficients of q'^2 and $|\omega|/q$ are both unity. But this cannot be dome, since the two terms in (3.3) have different coefficients ($e^{-(d+3)l}$ and $e^{-(d+1)l}$, respectively), so any redefinition of Ψ, which multiplies both terms, cannot make *both* coefficients unity. One way to proceed would be to settle for keeping one

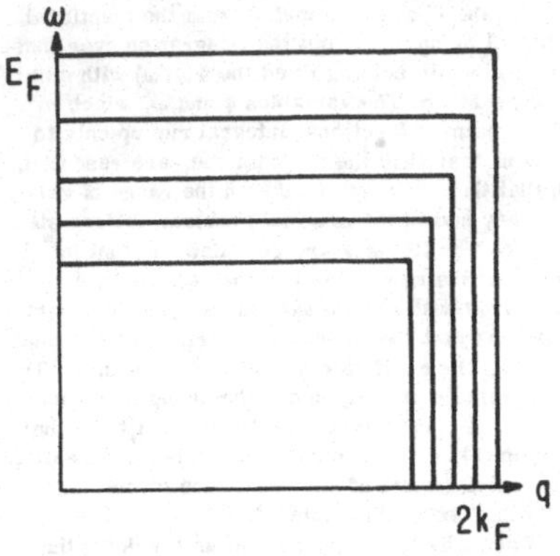

FIG. 1. Scaling procedure (3.4) in q and ω space.

of the coefficients, say that of q'^2, fixed, and letting the other one vary as dictated by that transformation. Here we shall get around the difficulty in a different way, however.

Instead of (3.2), we choose a more general scaling,

$$q'=qe^l, \quad \omega'=\omega e^{zl}, \tag{3.4}$$

that is, we scale down at different rates in wave number and frequency. This is pictured schematically in Fig. 1. Then instead of (3.3) we have

$$\Phi_2'=\tfrac{1}{2}\beta Ne^{-(d+z)l}\int_0^1\frac{d^dq'\,d\omega'}{(2\pi)^{d+1}} \times\left(r_0'+q'^2e^{-2l}+\frac{|\omega'|e^{-zl}}{q'e^{-l}}\right)|\Psi(q'e^{-l},\omega'e^{-zl})|^2. \tag{3.5}$$

It is apparent that if we choose $z=3$, the coefficient of $|\omega'|/q'$ inside the large parentheses will be e^{-2l}, the same as the coefficient of q'^2. Then a rescaling of Ψ can make the total coefficient of both of them unity. One demands

$$|\Psi'(q',\omega')|^2=e^{-(d+z)l}\,e^{-2l}|\Psi(q'e^{-l},\omega'e^{-zl})|^2$$

or

$$\Psi'(q',\omega')=e^{-(d+z+2)l/2}\Psi(q,\omega). \tag{3.6}$$

Then

$$\Phi_2'=\frac{\beta N}{2}\int_0^1\frac{d^dq'\,d\omega'}{(2\pi)^{d+1}}\left(r_0'+q'^2+\frac{|\omega'|}{q'}\right)|\Psi'(q',\omega)|^2, \tag{3.7}$$

and we see that under the infinitesimal generator of the renormalization group, $r_0\to r(l)=r_0'e^{2l}$. This is the same behavior found for a classical functional. Under this transformation, however, the quartic term will become

$$\Phi_4'=\tfrac{1}{2}(\beta N)^2u_0'\int_0^{e^{-l}}\prod_{i=1}^{4}\frac{d^dq_i}{(2\pi)^d}\int_0^{e^{-zl}}\prod_{i=1}^{4}\frac{d\omega_i}{2\pi}\Psi(q_1,\omega_1)\Psi(q_2,\omega_2)\Psi(q_3,\omega_3)\Psi(-q_1-q_2-q_3,-\omega_1-\omega_2-\omega_3)$$

$$=\exp[-\tfrac{3}{2}(d+z)l+4(d+z+2)l]\,\tfrac{1}{4}\,u_0'(\beta N)^2\int_0^1\prod_{i=1}^{4}\frac{d^dq_i'\,d\omega_i'}{(2\pi)^{d+1}}\Psi'(q_1',\omega_1')$$

$$\times\Psi'(q_2',\omega_2')\Psi'(q_3',\omega_3')\Psi'(-q_1'-q_2'-q_3',-\omega_1'-\omega_2'-\omega_3'), \tag{3.8}$$

where u_0' is different from u_0 because of the elimination of the shell variables. Hence u must transform according to

$$u_0\to u(l)=u_0'e^{[4-(d+z)]l}$$
$$=u_0'e^{\epsilon l}, \tag{3.9}$$

where

$$\epsilon=4-(d+z). \tag{3.10}$$

As in the classical case, integrating out the Ψ's with q's in the shell will lead to a u_0' different from u_0 by something of order u_0^2. Hence for small u the change in u from rescaling, which is linear in u, will be dominant. We will have

$$\frac{du}{dl}=\epsilon u+O(u^2), \tag{3.11}$$

so that the Gaussian fixed point, with $u=0$, will be stable if ϵ is negative, that is, if $d>4-z$. One way of putting this is to say that the effective dimensionality is increased by z. In the present example, then, where $z=3$, we should expect a stable Gaussian fixed point and Landau exponents for $d>1$. This result is the central point of this paper.

One should be careful to note, however, that

our model functional was derived from the analytic features of a three-dimensional electron gas. Actually, one- and two-dimensional electron gases do not have Lindhard functions which behave like (2.10). In two dimensions the coefficient of q^2 vanishes,[14] while in one dimension χ_0 has its maximum at $q = 2k_F$, not $q = 0$, indicating that the dominant fluctuations are nearly *antiferromagnetic* paramagnons. It is nevertheless interesting to think about the model (2.13) in arbitrary dimensionality anyway.

In order to complete the derivation of the renormalization-group equations for this problem, it is now necessary only to do the integration out of the outer-shell fields $\Psi(q, \omega)$, with $e^{-l} < q < 1$ and $e^{-zl} < |\omega| < 1$. (We continue to write the dynamical exponent as z even though we know it to be 3 in this problem because other problems will have other values of z, but will be otherwise very similar.) Consider the quartic term in the functional. The important terms to consider (to lowest order in u) are those with two q_i or ω_i in the outer shell, and where these four-momenta are equal and opposite. (This is because they are positive definite.) Since there are $\frac{1}{2}(4\times 3) = 6$ ways to pick this pair of Ψ's, we can write

$$\Phi_4 = \frac{u_0}{4\beta N} \sum{}'' \prod_{i=1}^{4} \Psi(q_i, \omega_i)\, \delta\Big(\sum_{i=1}^{4} q_i\Big)\, \delta\Big(\sum_{i=1}^{4} \omega_i\Big) + \frac{3u_0}{2\beta N} \sum{}'' |\Psi(q, \omega)|^2 \sum{}' |\Psi(q, \omega)|^2 , \tag{3.12}$$

where a prime on a sum indicates summing on q's and ω's in the shell and a double prime indicates summing on q's and ω's not in the shell. Errors we make in this approximation will first appear in terms of order Ψ^6 in the new functional[13] and are therefore irrelevant here.

The part of the functional integration over the shell variables is now a product of independent Gaussian integrals

$$\int \frac{d\Psi(q, \omega)\, d\Psi(-q-\omega)}{2\pi} \exp\left[-\frac{1}{2}\left(r_0 + q^2 + \frac{|\omega|}{q} + \frac{3u_0}{\beta N} \sum_{q'\omega'}{}'' |\Psi(q', \omega')|^2\right) |\Psi(q, \omega)|^2\right] = \left(r_0 + q^2 + \frac{|\omega|}{q} + \frac{3u_0}{\beta N} \sum_{q\omega}{}'' |\Psi(q, \omega)|^2\right)^{-1} . \tag{3.13}$$

Hence

$$Z = \int \delta\Psi \exp\Bigg[-\frac{1}{2} \sum_{q,\omega}{}'' \left(r_0 + q^2 + \frac{|\omega|}{q}\right) |\Psi(q, \omega)|^2 - \frac{1}{4} \frac{u_0}{\beta N} \sum{}'' \prod_{i=1}^{4} \Psi(q_i, \omega_i)\, \delta\Big(\sum_{i=1}^{4} q_i\Big)\, \delta\Big(\sum_{i=1}^{4} \omega_i\Big) - \frac{1}{2} \sum_{q'\omega'}{}' \ln\left(r_0 + q'^2 + \frac{|\omega'|}{q'} + \frac{3u_0}{\beta N} \sum_{q\omega}{}'' |\Psi(q, \omega)|^2\right)\Bigg] . \tag{3.14}$$

The expansion of the ln to fourth order in Ψ then gives a change in r_0,

$$r_0 \to r_0' = r_0 + \frac{3u_0}{\beta N} \sum_{q\omega}{}' \left(r_0 + q^2 + \frac{|\omega|}{q}\right)^{-1} , \tag{3.15}$$

and a change in u_0,

$$u_0 \to u_0' = u_0 - \frac{9u_0^2}{\beta N} \sum_{q\omega}{}' \left(r_0 + q^2 + \frac{|\omega|}{q}\right)^{-2} . \tag{3.16}$$

In a model with a wave-vector cutoff at $2k_F$ $(=1)$ and a frequency cutoff at E_F $(=1)$ the (primed) sums of q and ω in (3.15) and (3.16) are over an L-shaped region of space, as shown in Fig. 1. We have

$$\frac{1}{\beta N} \sum_{q\omega}{}' = 2 \int_{e^{-zl}}^{1} \frac{d\omega}{2\pi} \int_0^1 q^{d-1}\, dq \int \frac{d\Omega_d}{(2\pi)^d} + \int_{-1}^{1} \frac{d\omega}{2\pi} \int_{e^{-l}}^{1} q^{d-1}\, dq \int \frac{d\Omega_d}{(2\pi)^d} . \tag{3.17}$$

This first term is the integration along the horizontal strip ($|\omega| \approx 1$, $0 < q < 1$) and the second is the integration along the vertical strip ($q \approx 1$, $0 < |\omega| < 1$). (Ω_d is the solid angle in d dimensions.) Then ($m = 1, 2$)

$$\frac{1}{\beta N} \sum_{q\omega}{}' \left(r_0 + q^2 + \frac{|\omega|}{q}\right)^{-m} = \frac{2z\Omega_d l}{(2\pi)^{d+1}} \int_0^1 q^{d-1} \left(r_0 + q^2 + \frac{1}{q}\right)^{-m} dq + \frac{2\Omega_d l}{(2\pi)^{d+1}} \int_0^1 d\omega\, (r_0 + 1 + \omega)^{-m} . \tag{3.18}$$

Thus using (3.18) in (3.15) and (3.16), and combining this information with what we discussed earlier about the changes of r and u from rescaling, leads to the renormalization-group equations

$$\frac{dr}{dl}=2r+3C_d u\left[\ln\left(\frac{r+2}{r+1}\right)+z\int_0^1\frac{x^d\,dx}{x^3+rx+1}\right], \quad (3.19)$$

$$\frac{du}{dl}=\epsilon u-9C_d u^2\left[\frac{1}{(r+1)(r+2)}+z\int_0^1\frac{x^{d+1}\,dx}{(x^3+rx+1)^2}\right], \quad (3.20)$$

where $C_d=2\Omega_d/(2\pi)^{d+1}$. In each equation, the first term comes from the rescaling and the second term from the elimination of the outer shell. Within each second term, the first contribution is from the integration along the horizontal strip.

The fact that (3.19) and (3.20) look somewhat messier than their counterparts in the classical problem is just a consequence of the way we chose the cutoffs in our problem—the q cutoff and ω cutoff were independent of each other, so the integration over the shell variables was along two distinct strips. We could alternatively make a model with different cutoffs that simplify the algebra. One simple choice is to choose a q- and ω-dependent cutoff which excludes all $\Psi(q,\omega)$ for which

$$q^2+|\omega|/q>1\,. \quad (3.21)$$

The region of modes in q and ω space included in this model is shown graphically in Fig. 2. The ω cutoff is q dependent,

$$\omega_c(q)=q-q^3\,. \quad (3.22)$$

The outer shell now becomes a strip along the curve $\omega_c(q)$ defined by

$$e^{-2l}<q^2+|\omega|/q\le 1 \quad (3.23)$$

and

$$\frac{1}{\beta N}\sum_{q\omega}{}'=C_d\int_0^{e^{-l}}q^{d-1}\,dq\int d\omega\,\theta\left(1-q^2-\frac{|\omega|}{q}\right)\times\theta\left(q^2+\frac{|\omega|}{q}-e^{-2l}\right)$$
$$=\frac{2lC_d}{d}\equiv K_d l\,. \quad (3.24)$$

Everywhere along the strip, by the terms of this model, $r_0+q^2+|\omega|/q=r_0+1$, so we get the simpler renormalization-group equations

$$\frac{dr}{dl}=2r+\frac{3K_d u}{1+r}, \quad (3.25)$$

$$\frac{du}{dl}=\epsilon u-\frac{9K_d u^2}{(1+r)^2}, \quad (3.26)$$

which have a form identical to those obtained in

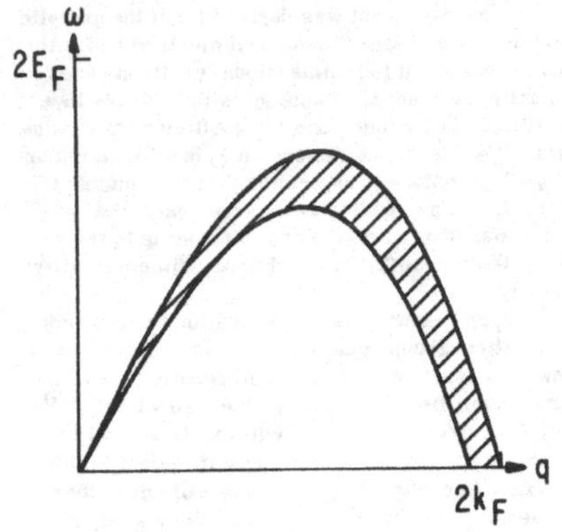

FIG. 2. Region of (q,ω) space contained in the model defined by (3.21) and (3.22). The hatched region is scaled out at each step.

the classical LGW problem, except that ϵ (and K_d) are different. In fact, the seemingly arbitrary cutoff (3.20) is not as contrived as it sounds, since we know that for small q, the microscopic random-phase-approximation (RPA) spectral weight function cuts off abruptly at[9] $\omega>qv_F$, although the Lorentzian spin-fluctuation model (2.11) does not. This cutoff enables the full RPA χ to satisfy the f-sum rule. We can then think of the model with the odd-cutoff rule (3.21) or (3.22) as enforcing an f-sum-rule constraint for small q. This may be irrelevant to critical behavior, but it is certainly no more artificial than the original cutoff procedure.

We next summarize the consequences of the renormalization-group equations (3.19) and (3.20) or (3.25) and (3.26). As we mentioned earlier, for $d+z>4$, i.e., $d>1$, ϵ is negative and u is irrelevant, since $u(l)$ dies exponentially with l. The critical properties of the model are therefore Landau-like, and RPA theory is qualitatively correct.[15] For $d<1$ (admittedly a case of only formal relevance) the group equations are correct to first order in ϵ, and corrections to Landau exponents may be calculated from them in the usual way, to order ϵ. At $d=1$, u is marginal, and one obtains logarithmic corrections to power-law critical behavior. That this is true follows trivially in the case of Eqs. (3.25) and (3.26) (the case of the odd cutoff) because their form is just that studied by Fisher *et al.*[16] It is not quite so obvious in the other cutoff model, but I prove it in the appendix.

The results of this section are not dependent on the scalar-field description of the interacting paramagnons. Because the vector-field functional of Eq. (2.22) differs from the scalar version (2.13) only in the number of spin components, it is straightforward to generalize all of the preceding arguments, in direct analogy to the Wilson theory for a vector field. The only difference is that the factor 3 in Eq. (3.19) or (3.25) becomes $n+2=5$, and the factor 9 in (3.20) or (3.26) becomes $n+8=11$. These changes, of course, do not alter the value of ϵ or influence the relevance or irrelevance of the parameters r or u.

Although in this paper we shall not try to do any better than first order in ϵ, it is worth remarking that in generating renormalization-group equations which are correct to order ϵ^2, one finds corrections to the coefficient of q^2 in Φ_2 (which leads to a nonzero η), but no corrections to the $|\omega|/q$ term. Consequently, $z=3-\eta$.

We conclude this section by looking at this problem at finite temperature. The Matsubara frequencies then no longer form a continuum, but are spaced by $2\pi T$. This is of little consequence initially if T is much less than the original frequency cutoff ($\approx E_F$), but as we remove high frequencies from the problem, we eventually reach a point where only a few Matsubara frequencies remain, and we can no longer approximate them very well by a continuum. Finally, only the $\omega=0$ terms will remain in the functional, and we will arrive at the classical LGW problem. Beyond this point, the renormalization-group equations will be of Wilson form, with $\epsilon=4-d$, and close enough to the critical coupling r_c the exponents will be nonclassical. On the other hand, if T is very low, this true critical region will be very narrow. The problem can be phrased in terms of the standard crossover language,[17] where the temperature is the symmetry-breaking parameter.

Actually, one encounters precisely the same sort of situation in a finite classical system, where the wave vectors q also have a finite spacing $2\pi/L$ between them (L is the linear size of the system). The scaling naturally breaks down when the size of the Kadanoff cell exceeds that of the system. Our problem here is analogous to that of a $(d+1)$-dimensional system which is finite (length β) in the extra dimension. Our crossover is analogous to that which occurs between $(d+1)$-dimensional and d-dimensional critical behavior when the correlation length exceeds the length of the system in the finite dimension.

To make this idea more quantitative, note that when the scaling parameter in the renormalization group equations has value l, frequencies between $E_F e^{-zl}$ and E_F have been removed from the problem. (We write the original frequency cutoff E_F explicitly in this section.) Thus the quantum scaling stops, roughly, when just one finite Matsubara frequency remains. That is, $l=\hat{l}$, where

$$2\pi T = E_F e^{-z\hat{l}} .$$

The maximum wave vector left in the problem is then

$$q_c \approx k_F e^{-\hat{l}} = k_F (2\pi T/E_F)^{1/z} . \qquad (3.27)$$

For $q>q_c$ and $\omega>T$, then, the fluctuations of $\Psi(q,\omega)$ are governed by the quantum renormalization-group equations derived above. In three dimensions, our analysis shows that RPA theories will be qualitatively valid. One way to think of q_c is as the inverse of the length over which one has to average microscopic quantities in order to be able to treat them as classical thermodynamic variables.

In the remaining corner of q, ω space, fluctuations will be classical in nature. Figure 3 shows the classical and quantum regions of q, ω space at a particular $T \ll E_F$. It is not obvious from this discussion how to deal with the dynamics when all of the finite Matsubara frequencies have disappeared. This is because I have been too cavalier in treating the ω's as a continuum until all but the very last one was gone. We will see how to do better than this in Sec. VII.

The crossover phenomenon is most simply discussed as follows (we talk only about the case where $\epsilon<0$ before the crossover): For $1 \ll l < \hat{l}$, r grows as

$$r(l) = \bar{r}_0 e^{2l} \qquad (3.28)$$

($\bar{r}_0$ differs from r_0 because of the effects of some transient terms in the solutions of the renormalization-group equations). When $r(l)$ gets to unity, we stop the scaling, since we have scaled the problem into one with a small $[O(1)]$ correlation length, which can be treated by perturbation theory. Thus if $r(l)$ gets to unity before l gets to $\hat{l}$, the transition to the classical LGW functional and the Wilson renormalization-group equations never gets a chance to happen. The critical exponents will deffer from their Landau values, then, only if $r(\hat{l})<1$, i.e.,

$$\bar{r}_0 < (2\pi T/E_F)^{2/z} . \qquad (3.29)$$

In the usual terminology,[16] $\frac{1}{2}z$ is the crossover exponent.

Figure 4 illustrates the consequences of this effect. At each T, we assume we have a different critical Hubbard coupling strength $U_c(T)$; the system is ferromagnetic for $U>U_c(T)$. Stoner theory predicts $U_c(T)=U_c(0)+\alpha T^2$, or $T_c(U) \propto [U-U_c(0)]^{1/2}$. As we approach the transition line at fixed T,

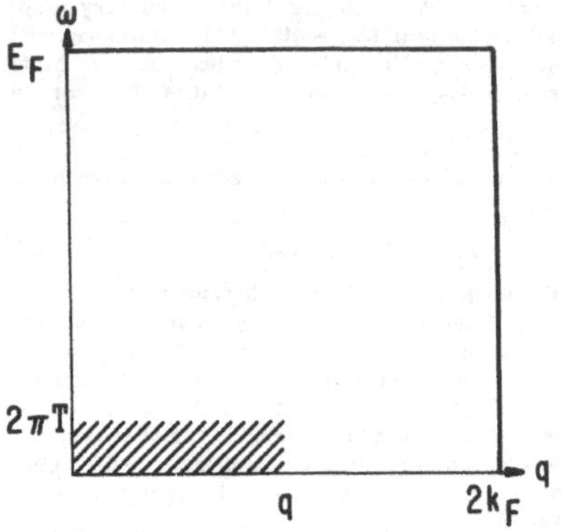

FIG. 3. Regions of the (q,ω) plane where correlation functions are dominated by classical (hatched) and quantum (unhatched) effects.

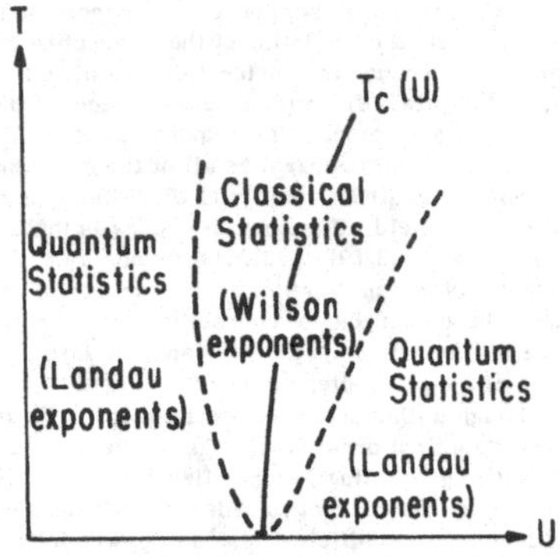

FIG. 4. Crossover diagram showing whether long-wavelength low-frequency critical properties are determined by classical or quantum renormalization groups, as a function of temperature T and coupling constant U, for itinerant ferromagnetic models (2.13) or (2.22).

varying U, $\bar{r}_0 \propto [U - U_c(T)]/U_c(T)$. From (3.29) the crossover occurs when $\bar{r}_0 \approx (T/E_F)^{2/z}$, so exponents are effectively Landau-like outside and Wilson-like inside the region between the dotted lines in Fig. 4.

Since $z>1$, the temperature region in which, for a given U, the classical LGW functional becomes relevant has a width at least of the order of T_c itself. The quantum-to-classical crossover is therefore more easily observed by sitting at a fixed low T and varying U by alloying or pressure. Note also that just below $U_c(0)$, one may be able to pass from a quantum region to a classical one and back to a quantum one as T is varied.

Another observation worth making is that in the case of fixed $U = U_c(0)$, where $T_c = 0$, as one varies T down to zero, one approaches the instability within a classical region. Therefore non-Landau (Wilson) critical exponents should characterize this transition. This result depends on the fact that $z>1$, so that the paramagnetic crossover boundary in Fig. 4 moves initially toward the left as T increases from zero.

One may also note that the quantum renormalization-group procedure for $l<\hat{l}$ provides a way of explicitly deriving the LGW functional microscopically. One will usually want to transform back to the original scale, letting the fact that the momentum cutoff in this functional is generally much smaller than the microscopic characteristic inverse length, $q_c \ll k_F$, appear explicitly.

Finally, the description of the quantum-classical crossover is independent of the number of components n of the order parameter (for physically relevant dimensionalities). This is because the quantum problem has $\epsilon<0$ with ϵ independent of n, and the determination of the crossover value of l requres only asking when the renormalization-group equations in the quantum region break down. (Inside the classical region, critical indices will depend on n, but that is not our concern here.)

IV. APPROXIMATE SOLUTIONS OF THE RENORMALIZATION-GROUP EQUATIONS

In Sec. III, we looked at the properties of the renormalization group we had derived insofar as they bore on the critical properties of the model. Here we show how we can make approximate solutions of these equations to give more qualitative information about the nature of the correlation functions. The form of these solutions is similar, but not identical to that obtained by simple perturbation theory in U. We start with the model (2.13) and its associated renormalization-group equations (3.25) and (3.26). As explained in Ref. 7, this model affords a credible description of weak itinerant ferromagnetism and strong paramagnetism when the band density of states is not too rapidly varying near E_F. Equivalently, one must have u_0 (2.14b) fairly small. (If the density of states is rapidly varying, one must keep higher-order terms than $u_0\Psi^4$.)

We deal with the physically relevant case of $\epsilon(=-2)<0$, so the Gaussian fixed point is stable. Our approximation procedure is to linearize the general renormalization-group equations around this fixed point and solve the subject to the initial conditions $r(0)=r_0$, $u(0)=u_0$. We are therefore ignoring the effects of Ψ^6 and higher-order terms which are generated in the exact renormalized LGW functional as transient terms but die out (faster than r and u) for large l. The only kinds of terms we include are those which are present in the original functional. For u_0 small, we can ignore the second term in the equation for du/dl, giving

$$u(l)=u_0e^{-|\epsilon|l}. \tag{4.1}$$

In (3.25) for dr/dl, we ignore the r in the denominator of the second term since corrections to this would be $O(ru)$. We therefore have to solve

$$\frac{dr}{dl}-2r=3K_d u_0 e^{-|\epsilon|l}. \tag{4.2}$$

The solution is standard:

$$r(l)=[r_0+3K_du_0/(2+|\epsilon|)]e^{2l}$$
$$-[3K_du_0/(2+|\epsilon|)]e^{-|\epsilon|l}. \tag{4.3}$$

To apply this to the zero-temperature problem, we scale (that is, carry out the renormalization-group transformations) (4.1) and (4.3) until $r(l)=1$. Beyond this point, the problem is one whose Kadanoff cell size exceeds the correlation length, so it can be dealt with in perturbation theory. We have $r(l)=1$ when

$$l=l_1=\tfrac{1}{2}\ln\left[\left(\frac{1}{r_0+3K_du_0/2+|\epsilon|}\right)^{-1}\right], \tag{4.4}$$

provided this $l_1 \gg 1$ [that is, $r_0+3K_du_0/(2+|\epsilon|) \ll 1$]. At this l the correlation length is the Kadanoff cell size, that is, the factor by which distances have been scaled, which is e^l. Thus

$$\xi=(r_0+3K_du_0/2+|\epsilon|)^{-1/2}\equiv\bar{r}_0^{-1/2}. \tag{4.5}$$

This just tells us that the effect of the anharmonicity is to shift the instability point from $r_0=0$ to $\bar{r}_0=0$. Such a result could be obtained straightforwardly by perturbation theory in u_0, but it is enlightening to see how it emerges from the linearized renormalization-group analysis.

We can also ask about the finite-temperature problem and the form of the LGW functional after the quantum-to-classical crossover. The crossover occurs when $e^{-zl}\approx T/E_F$, i.e., at l equal to

$$\hat{l}=z^{-1}\ln(E_F/T)\gg 1. \tag{4.6}$$

Thus the Landau-Ginzburg parameters at this point are

$$r(\hat{l})=\bar{r}_0(E_F/T)^{2/z}, \tag{4.7}$$

$$u(\hat{l})=u_0(T/E_F)|\epsilon|/z. \tag{4.8}$$

When this problem is expressed back in the scale of the original one, we simply have a problem with $r_{\rm eff}=\bar{r}_0$, $u_{\rm eff}=u_0$, and a cutoff at $q_c\approx k_F(T/E_F)^{1/z}$. Another way to express this result is to say that perturbation theory is sufficient to calculate the change in effective Landau-Ginzburg parameters which comes from scaling the finite Matsubara frequencies out of the problem (provided the quantum region ϵ is negative). The situation would be more complicated (and more interesting) if the Gaussian fixed point were not stable for $l<\hat{l}$. A hint of what can happen then can be seen in the $\epsilon=0$ case discussed in the Appendix.

It is not difficult to generalize this discussion to a generalized LGW functional with local couplings of all (even) orders in Ψ, such as appear in (2.7), rather than just a Ψ^4 term. We then need an infinite set of renormalization-group equations, rather than just two of them. To derive them, we follow the same argument we used in Sec. III. The nth-order anharmonic term in Φ has the form

$$\Phi_n=\frac{u_0^{(n)}}{n!}(\beta N)^{1-n/2}$$
$$\times\sum_{q_i\omega_i}\prod_{i=1}^{n}\Psi(q_i,\omega_i)\,\delta\Big(\sum_{i=1}^{n}q_i\Big)\,\delta\Big(\sum_{i=1}^{n}\omega_i\Big). \tag{4.9}$$

Under the rescaling of wave vectors, frequencies, and fields (3.4) and (3.6), this becomes

$$\Phi_n=\frac{u_0^{(n)}}{n!}(\beta N)^{1-n/2}(e^{(d+z+2)l/2})^n(e^{-(d+z)l})^{(n-1)}$$
$$\times\sum_{q_i\omega_i}\sum_{i=1}^{n}\Psi'(q_i',\omega_i')\,\delta\Big(\sum_{i=1}^{n}q_i'\Big)\,\delta\Big(\sum_{i=1}^{n}\omega_i'\Big). \tag{4.10}$$

(The first exponential factor comes from the n fields rescaled, the second from the rescaling of the variables of integration.) Thus the rescaling gives a contribution to

$$\left(\frac{du^{(n)}}{dl}\right)_{\rm rescaling}=\epsilon_n u^{(n)}, \tag{4.11}$$

where

$$\epsilon_n=n-(d+z)(\tfrac{1}{2}n-1). \tag{4.12}$$

The fact that all ϵ_n are negative for $n\geq 4$ means that the Gaussian fixed point $u^{(n)}=0$ is at least metastable. We shall not consider the possibility of a different true stable fixed point.

We turn then to the elimination of the shell fields $\Psi(q,\omega)$ with $e^{-l}<q\leq 1$ and $e^{-zl}<\omega\leq 1$. In each term (4.9) we separate out the terms with factors $\Psi(q,\omega)\Psi(-q,-\omega)$, with q and ω in the outer shell. As in previous discussion, the "paired" terms like this are important because they are positive definite, so they dominate those in which only one Ψ has its (q, ω) in the shell as well as those with two Ψ's with arguments in the shell but not equal and opposite. The fact that the shell is thin, on the other hand, permits us to ignore any terms with more than two Ψ's with arguments in the shell; if l is the thickness of the shell, they contribute only to order l^2. Furthermore, there are $\frac{1}{2}n(n-1)$ ways to pick the pair. Thus the part of Φ_n involving shell variables is

$$\Phi_n' = \frac{u_0^{(n)}}{2(n-2)!}(\beta N)^{1-n/2} \sum_{q_i\omega_i}'' \prod_{i=1}^{n-2} \Psi(q_i, \omega_i) \times \delta\left(\sum_{i=1}^{n-2} q_i\right)\delta\left(\sum_{i=1}^{n-2} \omega_i\right) \sum_{q\omega}' |\Psi(q,\omega)|^2 . \tag{4.13}$$

Now the coefficient of the shell Ψ's is no longer just a quadratic function of the nonshell Ψ's, but includes the sum of all of the expressions like (4.14) over all n. In place of (3.14) we find

$$Z = \int \delta\Psi \exp\left\{-\frac{1}{2}\sum_{q\omega}''\left(r_0 + q^2 + \frac{|\omega|}{q}\right)|\Psi(q,\omega)|^2 - \sum_{n=4}^{\infty} \frac{u_0^{(n)}}{n!}(\beta N)^{1-n/2} \sum_{q_i\omega_i}'' \prod_{i=1}^{n} \Psi(q_i,\omega_i)\delta\left(\sum_{i=1}^{n} q_i\right)\delta\left(\sum_{i=1}^{n}\omega_i\right) \right.$$
$$\left. - \frac{1}{2}\sum_{q'\omega'} \ln\left[r_0 + q'^2 + \frac{|\omega'|}{q'} + \sum_{n=2}^{\infty}\frac{u_0^{(n+2)}}{n!}(\beta N)^{-n/2} \sum_{q_i,\omega_i}'' \prod_{i=1}^{n}\Psi(q_i,\omega_i)\delta\left(\sum_{i=1}^{n} q_i\right)\delta\left(\sum_{i=1}^{n}\omega_i\right)\right]\right\}. \tag{4.14}$$

(In this section, all sums on n are taken over *even* n only.) In words, the new functional has two parts, the first of which is just the old functional restricted to the nonshell Ψ's and the second of which is an average over the shell of the logarithm of the inverse RPA propagator plus the second functional derivative of the original functional with respect to the shell Ψ's.

As in the preceding discussion of the Ψ^4-only problem, we linearize the group equations around the Gaussian fixed point $u^{(n)} = 0$, and solve them subject to the initial conditions $u^{(n)}(0) = u_0^{(n)}$. Things simplify enormously on linearization. On expanding the logarithm, we get a change in $u^{(n)}$ proportional to $u^{(n+2)}$,

$$\delta u^{(n)} = \tfrac{1}{2}u^{(n+2)}(\beta N)^{-1}\sum_{q'\omega'}'\left(q^2 + \frac{|\omega|}{q}\right)^{-1}$$
$$= \tfrac{1}{2}lC_d\left(\ln 2 + z\int_0^1 \frac{x^d\,dx}{x^3+1}\right)u^{(n+2)}$$
$$\equiv \tfrac{1}{2}lC_d\lambda u^{(n+2)} . \tag{4.15}$$

The other cutoff model (3.21) gives

$$\delta u^{(n)} = \tfrac{1}{2}lK_d u^{(n+2)} . \tag{4.16}$$

We shall use (4.16) rather than (4.15); the difference is only quantitative. Combining the changes in $u^{(n)}$ from rescaling (4.11) with (4.16) gives the renormalization-group equations

$$\frac{du^{(n)}}{dl} = \epsilon_n u^{(n)} + \tfrac{1}{2}K_d u^{(n+2)} . \tag{4.17}$$

We solve them by Laplace transform. Define

$$\bar{u}^{(n)}(S) = \int_0^\infty dl\, e^{-lS}u^{(n)}(l) \tag{4.18}$$

and take the Laplace transform of (4.17),

$$S\bar{u}^{(n)}(S) - u_0^{(n)} = \epsilon_n \bar{u}^{(n)}(S) + \tfrac{1}{2}K_d \bar{u}^{(n+2)}(S) . \tag{4.19}$$

Then

$$\bar{u}^{(2)}(S) = \frac{u_0^{(2)}}{S-\epsilon_2} + \frac{\frac{1}{2}K_d\bar{u}^{(4)}(S)}{S-\epsilon_2}$$
$$= \frac{u_0^{(2)}}{S-\epsilon_2} + \frac{\frac{1}{2}K_d u_0^{(4)}}{(S-\epsilon_2)(S-\epsilon_4)} + \frac{(\frac{1}{2}K_d)^2 u_0^{(6)}}{(S-\epsilon_2)(S-\epsilon_4)(S-\epsilon_6)} + \cdots . \tag{4.20}$$

In taking the inverse Laplace transform

$$r(l) = u^{(2)}(l)$$
$$= \frac{1}{2\pi i}\int_C dS\, e^{lS}\left(\frac{u_0^{(2)}}{S-\epsilon_2} + \frac{\frac{1}{2}K_d u_0^{(4)}}{(S-\epsilon_2)(S-\epsilon_4)} + \cdots\right), \tag{4.21}$$

we are interested only in the large-l behavior, so we need worry only about the poles at the largest ϵ_n, that is, $\epsilon_2 = 2$. Then

$$r(l) \sim e^{2l}\left[u_0^{(2)} + \frac{\frac{1}{2}K_d u_0^{(4)}}{\epsilon_2-\epsilon_4} + \frac{(\frac{1}{2}K_d)^2 u_0^{(6)}}{(\epsilon_2-\epsilon_4)(\epsilon_2-\epsilon_6)} + \cdots\right]. \tag{4.22}$$

But since $\epsilon_n - \epsilon_{n+2} = d + z - 2 \equiv \Delta$, this is

$$r(l) \sim e^{2l}\left(u_0^{(2)} + \frac{\frac{1}{2}K_d u_0^{(4)}}{\Delta} + \frac{(\frac{1}{2}K_d)^2 u_0^{(6)}}{2\Delta^2} + \frac{(\frac{1}{2}K_d)^3 u_0^{(8)}}{3\times 2\Delta^3} + \cdots\right)$$
$$= e^{2l}\sum_{n=0}^{\infty}\left(\frac{K_d}{2\Delta}\right)^{n/2}\frac{u_0^{(n)}}{(\frac{1}{2}n)!} . \tag{4.23}$$

There is a simple way of looking at this result.[7] The coefficient of e^{2l} is just the average of the second derivative of the function

$$u(x)=\sum_{n=0}^{\infty}\frac{1}{n!}u_0^{(n)}x^n \tag{4.24}$$

over a Gaussian distribution of x with variance K_d/Δ. To see this, just evaluate

$$\langle u''(x)\rangle=\sum_{n=2}^{\infty}\frac{1}{(n-2)!}u_0^{(n)}\langle x^{n-2}\rangle$$

$$=\sum_{n=0}^{\infty}\frac{1}{n!}u_0^{(n+2)}\langle x^n\rangle$$

$$=\sum_{n=0}^{\infty}\frac{1}{n!}u_0^{(n+2)}(n-1)!!\left(\frac{K_d}{\Delta}\right)^{n/2}$$

$$=\sum_{n=0}^{\infty}\frac{u_0^{(n+2)}}{2^{n/2}(\frac{1}{2}n)!}\left(\frac{K_d}{\Delta}\right)^{n/2}, \tag{4.25}$$

which is just what appears in (4.23). Hence the arguments which led to (4.5) give a correlation length

$$\xi=(\langle u''\rangle)^{-1/2}=[r_0+\langle(u''-r_0)\rangle]^{-1/2}. \tag{4.26}$$

A dimensionless parameter measuring the deviation from the pure RPA result is

$$\sigma=\left(\frac{K_d}{\Delta}\right)^{1/2}=\left(\frac{4\Omega_d}{(2\pi)^{d+1}d(d+z-2)}\right)^{1/2}. \tag{4.27}$$

In three dimensions, $\sigma=0.0518$.

V. OTHER MODELS

It is apparent from the discussion in Sec. III that the value of the dynamical exponent z is crucial in determining the qualitative structure of the renormalization-group equations, and that its value (i.e., 3) in the paramagnon problem is a consequence of the fact that frequency occurs in the generalized LGW functional in the form of a term in the quadratic part of Φ proportional to $|\omega|/q$. In this section we examine some other systems, finding their z's and discussing the consequences for the $T=0$ critical behavior and the low-T crossover to normal critical exponents. We shall not derive the quantum LGW functionals for these systems; rather, we shall appeal to the physical interpretation of the LGW coefficients to argue what qualitative form they should have.

Example 1: dirty itinerant ferromagnet. Fulde and Luther have shown how impurities lead to spin diffusion in the RPA,[18] that is, the spin-fluctuation spectral weight function has a form

$$\chi''(q,\omega)=\chi(q)Dq^2\omega/[\omega^2+(Dq^2)^2] \tag{5.1}$$

when $ql\ll 1$, where l is the electronic mean free path. This should be contrasted with the Landau-damping form (2.11). We can incorporate these effects into our formalism by using a quadratic part of Φ of the form

$$\Phi_2=\frac{1}{2}\sum_{q\omega}\left(r_0+q^2+\frac{|\omega|}{D_0q^2}\right)|\Psi(q,\omega)|^2 \tag{5.2}$$

instead of the expression in (2.13). The imaginary part of the reciprocal of the coefficient in (5.2) is then of the form (5.1). The only significant difference between this problem and the previous one is in the rescaling. When $q\to q'=qe^l$ and we let $\omega\to\omega'=\omega e^{zl}$, $q^2=q'^2e^{-2l}$ and $|\omega|/D_0q^2=(|\omega'|/D_0q'^2)e^{(z-2)l}$; thus we must choose $z=4$ to make the two coefficients identical, so that they can be made equal to unity by a scale change in Ψ. Therefore one expects Landau critical exponents in any positive dimensionality for this problem.

There will also be minor quantitative differences in the form of the group equations for this case, since in eliminating the shell variables, one now encounters integrations of the form ($D_0=1$)

$$I_m=\frac{1}{\beta N}\sum_{q\omega}{}'\left(r_0+q^2+\frac{|\omega|}{q^2}\right)^{-m} \tag{5.3}$$

instead of (3.18). The second terms in the brackets in (3.19) and (3.20) are therefore replaced by

$$z\int_0^1\frac{x^{d+1}dx}{x^4+rx^2+1}\quad\text{and}\quad z\int_0^1\frac{x^{d+3}dx}{(x^4+rx^2+1)^2}, \tag{5.4}$$

respectively. [If we construct a model with a cutoff in analogy to (3.22), we get equations like (3.25) and (3.26), except that K_d is multiplied by a factor $d/(d+1)$.]

Actually, the introduction of randomness should actually be regarded as having a more fundamental effect on the LGW functional than just making the dynamics diffusive. The parameters r_0 and u_0 should become random functions of position with specified probability distributions, as in the work of Lubensky and Harris[19] and of Grinstein and Luther.[19] However, their work indicates that this has little effect on critical properties when the order parameter is a three-vector, as it really should be in this case.

Example 2: itinerant antiferromagnet. In this case the instability is at a finite wave vector Q, so fluctuations near this wave vector have no special q dependence like those imposed by rotational invariance on the ferromagnetic fluctuations of long wavelength in the previous examples. We characterize their decay by a single relaxation time τ, and express this in writing the quadratic part of Φ as

$$\Phi_2=\frac{1}{2}\sum_{q\omega}(r_0+q^2+|\omega|\tau)|\Psi(q,\omega)|^2. \tag{5.6}$$

In order that the second and third terms here scale in the same way, we must take $z=2$. The Gaussian fixed point will therefore be stable in greater than two dimensions in this case. There will also be minor changes in the form of the group equations, as in the previous example, but we will not write these here. This model also describes incipient charge-density wave or superconducting fluctuations.

The order parameter should, strictly speaking, be a vector in this problem as well. In fact, if there are m inequivalent values of Q at which the instability can happen, Ψ should be a ($3m$)-component vector. Similarly, for a charge-density wave instability, where the charge density is a scalar, the order parameter has m components if there are m inequivalent values of the instability wave vector, and for a superconductor, Ψ has two components. But, again, the points we discussed in Sec. III do not depend on the number of components of Ψ.

One must use extreme caution in applying the model (5.6) to one-dimensional metallic models, or to higher-dimensional ones with one-dimensional features such as flat pieces of Fermi surface. In these cases, the coefficients r_0, u_0, and all higher-order $u^{(n)}$ have a singular temperature dependence as $T \to 0$, and each $u^{(n)}$ is more singular than $u^{(n-1)}$. One therefore may not truncate Φ at any finite order at low temperatures.

Example 3. singlet-ground-state magnet (singlet-singlet model). In these systems, a nonvanishing matrix element of the z component of the total atomic angular momentum between two crystal-field–split levels leads to a magnetic state for sufficient exchange strength.[20] (The problem is isomorphic to that of an Ising model in a transverse field equal to the crystal-field splitting.) Even in the absence of magnetic order the exchange allows the crystal-field excitons to propagate in the lattice; the excitation structure is reflected in the RPA susceptibility

$$\chi(q,\omega) \propto \frac{1}{\omega^2 - \omega^2(q)} \quad (\omega \text{ real}), \tag{5.7}$$

where $\omega^2(q)$ is an even function of q whose q-independent part goes to zero as the instability is approached. This is often described as a soft magnetic exciton. In properly chosen units, and for sufficiently small q,

$$\omega^2(q) = r_0 + q^2 . \tag{5.8}$$

We can put this information into a generalized LGW functional by writing a quadratic part of Φ,

$$\Phi_2 = \frac{1}{2}\sum_{q\omega} (r_0 + q^2 + |\omega|^2)\,|\Psi(q,\omega)|^2 . \tag{5.9}$$

[The change in relative sign of q^2 and ω^2 between (5.7) and (5.9) is because the ω's in (5.9) are (imaginary) Matsubara frequencies.] This heuristic procedure has been justified microscopically by Klenin and the author.[21] The quadratic term looks like (5.9), and at zero temperature the quartic term is of the form we have been using here.[21] At finite T, there are anomalous singular quartic terms proportional to $e^{-\text{const}/T}$, which we can ignore here for very low temperatures. In this problem, then, time acts just like another dimension [(5.9) is Lorentz invariant], the dynamical exponent $z=1$, and three dimensions is the dividing line between Landau and Wilson critical exponents.[22,23]

The singlet-triplet model, which has very different critical dynamics in the classical statistical region at finite temperature because of its rotational symmetry,[24] is not expected to behave very differently from the singlet-singlet model in the quantum scaling region, since as we emphasized above, the vector or scalar character of the order parameter is irrelevant to the critical dimensionality and crossover index.

One way to look at this problem or the preceding example is to think of the frequency label of Ψ as labeling different components of an infinite-dimensional vector field,

$$\Phi_2 = \frac{1}{2}\sum_{qm} (r_0^m + q^2)\,|\Psi_m(q)|^2 , \tag{5.10}$$

with $r_0^m = r_0 + |\omega_m|^2$ in this example, or $r_0^m = r_0 + |\omega_m|$ in the antiferromagnet. (These frequencies are measured in units of the high-frequency cutoff.) This point has been made independently by Young.[23] In carrying out the renormalization group on this anisotropic classical problem, initially the behavior of the group transformations is as if the spin dimensionality is infinite, provided that the anisotropy is small. As the scaling proceeds, however, all m values except $m=0$ become irrelevant, and there is an eventual crossover to a scalar-field problem. The crossover exponent can be extracted simply in this picture by the same sort of arguments we used in obtaining (3.29). The scaling stops when $r^0(l) = r_0^0 e^{2l} \approx 1$, and if the other $r^m(l) = r_0^m e^{2l}$ have not reached unity well before this, the critical behavior will not be characteristic of a scalar field. Whether this has happened depends then on whether $r_0^m \gg r_0^0$, that is, $|\omega_m|^2 \gtrless r_0$ (example 3) or $|\omega_m| \gtrless r_0$ (example 2). This gives a crossover exponent of 1 for the antiferromagnet and $\frac{1}{2}$ for the singlet-ground-state problem, in agreement with the prediction $\frac{1}{2}z$ obtained from the $(d+1)$-dimensional scaling procedure in Sec. III.

The crossover diagram looks slightly different

from Fig. 4 in this case, and is shown in Fig. 5. From mean-field theory, we find that the curve $T_c(J)$ rises from zero at the critical exchange J_c with all of its derivatives infinite. Since the crossover exponent is $\frac{1}{2}z = \frac{1}{2}$, the crossover boundaries approach the $T = 0$ instability point with infinite slope, in contrast with the itinerant ferromagnetic case of Fig. 4, where they came in with zero slope. In both cases, however, the approach to zero temperature with coupling fixed so that $T_c = 0$ is in a classical statistical region.

This is not necessarily universally true, however, as the following example shows: Suppose that we designate the variable coupling parameter (U or J in previous examples) by x, i.e., $\bar{r} = x - x_c$. Suppose further that the phase boundary $T_c(x)$ rises from zero at $x_c(0)$ with a power law $[x - x_c(0)]^{\bar{\beta}}$. Then if the crossover exponent ϕ is greater than $\bar{\beta}$ (or equivalently, if $z < 2/\bar{\beta}$) the paramagnetic crossover boundary will move leftward from x_c as T rises, as in the previous examples, but if $\phi < \bar{\beta}$, it will begin by moving rightward instead. The simplest way to see this is to look at the crossover boundaries as functions of T,

$$x_\pm(T) = x_c(T) \pm \alpha T^{2/z} = x_c(0) + \alpha' T^{1/\bar{\beta}} \pm \alpha T^{2/z} \tag{5.11}$$

(α and α' are constants). If $2/z < 1/\bar{\beta}$, the $T^{1/\bar{\beta}}$ term is initially negligible relative to the $T^{2/z}$ term, so the net sign is necessarily negative for the paramagnetic crossover boundary. But if $2/z > 1/\bar{\beta}$, the $T^{2/z}$ term is a small correction on the $T^{1/\bar{\beta}}$ piece so the total term (5.11) is necessarily positive for small enough T. The latter

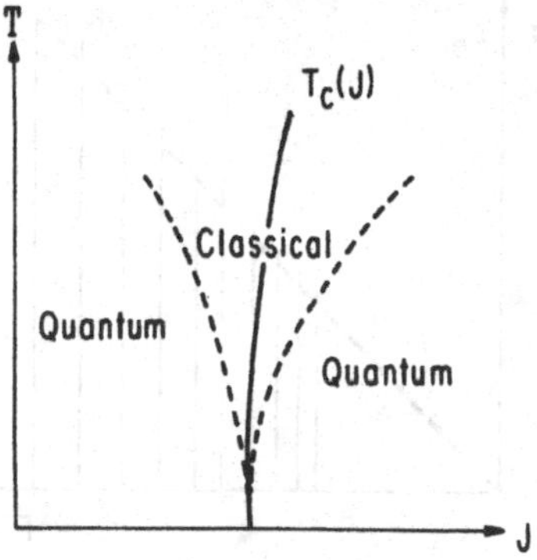

FIG. 5. Crossover diagram for the singlet-ground-state ferromagnet (Ising model in a transverse field). (J is the exchange coupling.)

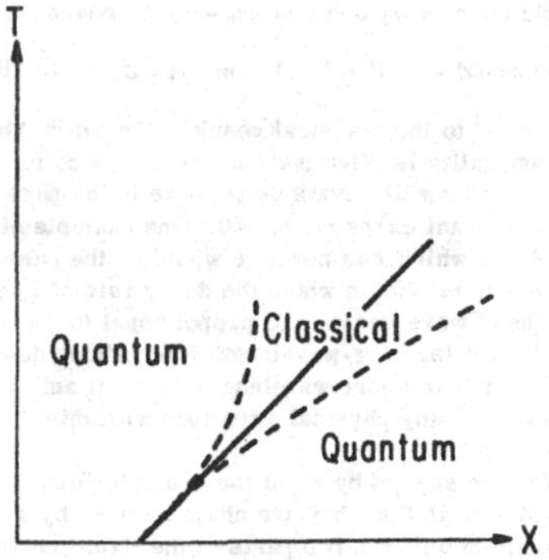

FIG. 6. Crossover diagram for a hypothetical system with $\beta = 1$ and $\varphi = \frac{1}{2}$. X is a coupling-strength parameter.

situation is shown in Fig. 6, where we take $\bar{\beta}$ and $z = 1$ for illustrative purposes.

One sees that here at $T = 0$ critical point [with $x = x_c(0)$] is approached entirely in the *quantum* region as T is lowered to zero. Furthermore, in this case the width of the temperature region in which classical statistics hold (for a fixed x) also becomes negligibly small relative to T_c itself, unlike the previously encountered situations, where no crossover could be observed as a function of T for temperatures near T_c. The nature of the phase transition when $T_c = 0$ and the existence or nonexistence of a crossover near T_c as T is varied both depend on the relative magnitudes of φ and β.

One can generalize the discussion of this section to any quantum LGW functional with a Ψ^4 interaction which is local in space and time and whose quadratic part is of the form

$$\Phi_2 = \sum_{q\omega}\left(r_0 + q^\sigma + \frac{|\omega|^m}{q^{m'}}\right)|\Psi(q,\omega)|^2 . \tag{5.12}$$

Under the rescaling transformation, this becomes

$$\Phi_2 = e^{-(d+z)l}\sum_{q'\omega'}\left(r_0 + q'^\sigma e^{-\sigma l} + \frac{|\omega'|^m}{(q')^{m'}}e^{(m'\sigma - mz)l}\right) \times |\Psi(q'e^{-l}, \omega' e^{-zl})|^2 , \tag{5.13}$$

so the fields must rescale like

$$\Psi'(q',\omega') = e^{-(d+z+\sigma)l/2}\Psi(q'e^{-l}, \omega' e^{-zl}) \tag{5.14}$$

and z must be chosen to satisfy $\sigma = mz - m'\sigma$, i.e.,

$$z = (\sigma + m')/m . \tag{5.15}$$

When the rescaled fields are substituted into the

quartic term in Φ, one finds $u_0 \to u_0 e^{\epsilon l}$, where

$$\epsilon = 2\sigma - d - z = (2 - 1/m)\sigma - m'/m - d, \quad (5.16)$$

in contrast to the classical result $\epsilon = 2\sigma - d$.[18] The dimensionality is effectively increased by z, as before, and z will always be positive in the physically relevant cases m, $m' \geq 0$. One example of a model in which z is negative would be the case with $m = 1$, $m' < 0$, in which the decay rate of fluctuations of wave vector q is proportional to $q^{-|m'|}$—that is, long-wavelength fluctuations decay faster than short-wavelength ones. (I am not aware of any physical situations with this property.)

A feature shared by all of the examples discussed here is that they are characterized by a competition between two parts of the Hamiltonian, one of which wants the system to order and the other of which minimized when there is no long-range order. If the former is just barely strong enough to counteract the latter, T_c will be much lower than the characteristic energies of either part, and many finite-ω Ψ's must be integrated out of the problem before it can be cast into classical LGW form. Our discussions here are relevant to any such situation. An example of a system where these effects are negligible is an ordinary Heisenberg magnet, since there T_c is of the order of the only characteristic microscopic energy J, hence near T_c no modes have characteristic energies $\gg T$.

VI. ALTERNATIVE FORMULATION OF THE RENORMALIZATION GROUP

In this section, we set up the renormalization group in a diagrammatic language[1,25] and perform the scaling in a different way. We will scale out shell variables explicitly only in q; the scaling in frequency will be taken care of implicitly by Bose functions which appear in the integrals. We start again with the scalar paramagnon problem. The elimination of the shell variables is expressed as a Hartree (single-loop) self-energy correction to the free propagator $(r_0 + q^2 + |\omega|/q)^{-1}$ and internal-loop corrections to the four-point interaction vertex u_0 (Fig. 7). In both of these corrections, the internal loops have their momenta between e^{-l} and 1. If we also restricted their Matsubara frequencies to lie between e^{-zl} and 1, these self-energy and vertex corrections would lead to the terms in brackets in (3.19) and (3.20). Here, however, we choose to sum over all frequencies. The scaling is thus in vertical strips in (q, ω) space (Fig. 8). We therefore have a self-energy correction

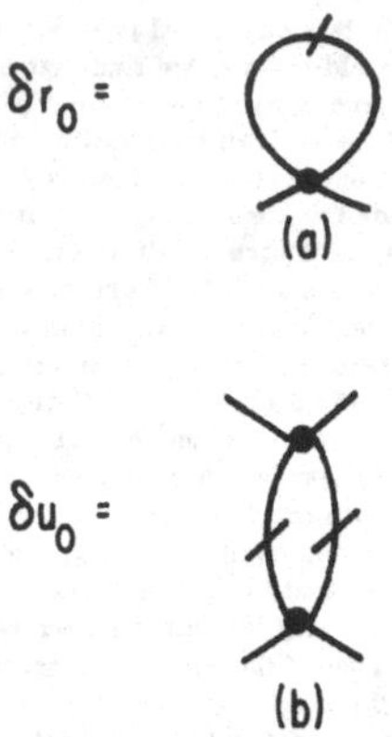

FIG. 7. Lowest-order diagrams for the change in parameters r (a) and u (b) under the renormalization group. A bar through a propagator indicates that its q is in the outer shell $e^{-l} < q \leq 1$.

$$\delta r = \frac{3u_0}{\beta N} \sum_{q\omega}{}' \chi(q, \omega) \quad (6.1)$$

and a vertex correction

$$\delta u = -\frac{9u_0^2}{\beta N} \sum_{q\omega}{}' \chi^2(q, \omega), \quad (6.2)$$

where $\chi(q, \omega) = (r_0 + q^2 + |\omega|/q)^{-1}$ is the free-paramagnon propagator. (Here the prime on the

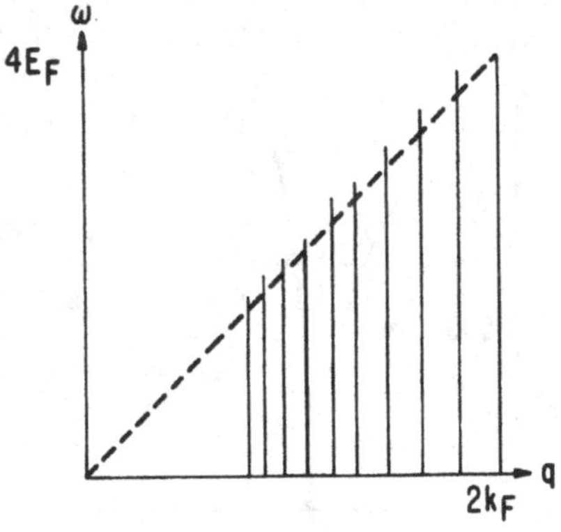

FIG. 8. Scaling procedure used in Sec. VII. The dashed line marks the f-sum-rule cutoff imposed to make (7.4) finite.

sum indicates that q's, but not ω's, lie in the shell $e^{-l} < q \leq 1$.) The Matsubara sums can be converted to integrals over real frequency, using the spectral representation

$$\chi(q, \omega_m) = \int_{-\infty}^{\infty} \frac{d\omega}{\pi} \frac{\chi''(q, \omega)}{\omega - \omega_m}, \tag{6.3}$$

so that

$$\delta r = \frac{3u_0}{N} \sum_q \int_{-\infty}^{\infty} \frac{d\omega}{\pi} \chi''(q, \omega)[n(\omega) + \tfrac{1}{2}] \tag{6.4}$$

and

$$\delta u = -\frac{9u_0^2}{N} \sum_q{}' \int_{-\infty}^{\infty} \frac{d\omega\, d\omega'}{\pi^2} \frac{\chi''(q, \omega)\chi''(q, \omega')}{\omega' - \omega} \times [n(\omega) - n(\omega')]. \tag{6.5}$$

The first of these is formally logarithmically divergent as it stands, since the Bose function approaches zero and χ'' goes like $1/\omega$ for large ω. We remedy this deficiency by noting that the true spectral weight function from which our model (2.11) was derived has an abrupt cutoff when $\omega = qv_F$. We therefore cut the integrals off at $\omega = q = 1$.

Two limiting cases are apparent: In the high-T limit, the Bose function may be approximated by T/ω for $\omega < 1$. Then the frequency integral (6.4) just gives $T\chi(q, 0)$, the equal-time correlation function in this classical limit. Similarly, the frequency integral in the vertex correction (6.5) is just proportional to $T\chi^2(q, 0)$. Only static susceptibilities enter the renormalization-group equations. In this way, the dynamics become irrelevant to the static critical behavior in the high-temperature problem.

In the low-T limit, the Bose function is $-\theta(-\omega)$, or $n(\omega) + \frac{1}{2} = \frac{1}{2}\,\mathrm{sgn}\omega$. Here we recover the quantum limit we have discussed above, except for numerical factors which are a consequence of the cutoff model we use here. (It is different from both of the models described in Sec. III.) We obtain

$$\delta r = \frac{3u_0}{N} \sum_q{}' \int_{-\infty}^{\infty} \frac{d\omega}{2\pi} \chi''(q, \omega)\,\mathrm{sgn}\omega \tag{6.6}$$

and

$$\delta u = -\frac{9u_0^2}{N} \sum_q{}' \int_{-\infty}^{\infty} \frac{d\omega\, d\omega'}{2\pi^2} \frac{\chi''(q, \omega)\chi''(q, \omega')}{\omega' - \omega} \times (\mathrm{sgn}\omega - \mathrm{sgn}\omega'). \tag{6.7}$$

These answers are just what one gets by treating the Matsubara frequencies in (6.1) and (6.2) as a continuum and integrating from zero to unity.

The point to note here is that we can in principle deal with the more general expressions (6.4) and (6.5). But as we scale down in frequency, the argument of the Bose functions must also be scaled by a factor e^{-zl}. Explicitly, the renormalization-group equations are

$$\frac{dr}{dl} = 2r + 3uC_d \int_{-1}^{1} d\omega\, \chi''(1, \omega)[n(\omega e^{-zl}) + \tfrac{1}{2})], \tag{6.8}$$

$$\frac{du}{dl} = \epsilon(l)u - 9u^2 C_d \int_{-1}^{1} \frac{d\omega_1\, d\omega_2}{\pi} \chi''(1, \omega_1)\chi''(1, \omega_2) \times \left(\frac{n(\omega_1 e^{-zl}) - n(\omega_2 e^{-zl})}{\omega_2 - \omega_1}\right). \tag{6.9}$$

Unlike previous renormalization-group equations, they depend unavoidably on l on their right-hand sides, through $\epsilon(l)$ and the Bose factors. The l dependence of ϵ has the following origin: In the diagrammatic perturbation theory, each successive order in u_0 involves two propagators and one integration over k and summation over ω. The rules are given in the original scale with the unrenormalized propagators. Now in terms of the rescaled wave vectors and frequencies (q', ω') the propagator is

$$\chi(q, \omega) = [r_0 + (q'^2 + |\omega|/q')e^{-2l}]^{-1}. \tag{6.10}$$

But we want to write things in terms of the renormalized propagator

$$\tilde{\chi}(q', \omega') = (r_0' e^{2l} + q'^2 + |\omega'|/q')^{-1}, \tag{6.11}$$

which differs from (6.10) by a factor e^{2l}. This, together with the phase-space renormalization, forces us to renormalize the interaction to $\tilde{u}_0$, so that the physically relevant quantity

$$u_0 \sum_{q\omega} \chi^2 = \tilde{u}_0 \sum_{q'\omega'} \tilde{\chi}^2 \tag{6.12}$$

remains invariant. For example, in the diagram of Fig. 7(b), we want to renormalize the interaction so that

$$u_0 \sum_{q\omega}{}' \chi^2(q, \omega) = u_0 \sum_q{}' \int \frac{d\omega_1\, d\omega_2}{\pi^2} \chi''(q, \omega_1)\chi''(q, \omega_2) \times \left(\frac{n(\omega_1) - n(\omega_2)}{\omega_2 - \omega_1}\right) \tag{6.13}$$

remains constant. One can see that this renormalization depends on the temperature. For the high-T case, this just becomes

$$Tu_0 \sum_q \chi^2(q, 0) = Tu_0 e^{-dl} \sum_{q'}{}' [e^{2l}\tilde{\chi}(q', 0)]^2 = T^2 \tilde{u}_0 \sum_{q'}{}' \tilde{\chi}^2(q', 0), \tag{6.14}$$

so we must choose $\tilde{u}_0 = u_0 e^{(4-d)l}$. Then $\epsilon = 4 - d$, as in Wilson theory. But at zero T, (6.13) is

$$u_0 e^{-dl} \sum_q{}' \int \frac{e^{-2zl}\, d\omega_1'\, d\omega_2'}{2\pi^2} [e^{2l}\chi''(q,\omega_1')][e^{2l}\chi''(q,\omega_2')] \left(\frac{\mathrm{sgn}\omega_1' - \mathrm{sgn}\omega_2'}{e^{-zl}(\omega_1' - \omega_1')}\right)$$

$$= \bar{u}_0 \sum_q{}' \int \frac{d\omega_1'\, d\omega_2'}{2\pi^2} \bar{\chi}''(q,\omega_1')\bar{\chi}''(q,\omega_2') \left(\frac{\mathrm{sgn}\omega_1' - \mathrm{sgn}\omega_2'}{\omega_2' - \omega_1'}\right), \tag{6.15}$$

so that $\bar{u}_0 = u_0 e^{(4-d-z)l}$, or $\epsilon = 4 - d - z$. That is, ϵ changes from $4-d-z$ to $4-d$ as l passes through $\hat{l}$ (3.28). In the intermediate region, one can formally determine $\epsilon(l)$ by defining

$$\bar{u}_0 = u_0 \exp\left(\int_0^l \epsilon(l')\, dl'\right).$$

That is,

$$\epsilon(l) = 4 - d - z + \frac{\int d\omega_1\, d\omega_2\, \chi''(\omega_1)\chi''(\omega_2)\{[n(\omega_1 e^{-zl}) - n(\omega_2 e^{-2l})]/(\omega_2 - \omega_1)\}}{\int d\omega_1\, d\omega_2\, \chi''(\omega_1)\chi''(\omega_2)\{[n(\omega_1) - n(\omega_2)]/(\omega_2 - \omega_1)\}}, \tag{6.16}$$

where $\chi''(\omega)$ means $\chi''(1,\omega)$. In either limit, $l \ll \hat{l}$, we could of course have found $\epsilon(l)$ without recourse to analysis of any particular diagram, since for $l \ll \hat{l}$ the Matsubara frequencies effectively form a continuum and the sum of ω may be approximated as an integral, while for $l \gg \hat{l}$ only the $\omega = 0$ terms matter. But the full formalism [(6.8) and (6.9), with (6.15)] gives one a well-defined, if messy, calculational procedure for treating the crossover region.

It is also worth pointing out in passing, although this limit is not the subject of this paper, that in the high-temperature problem the procedure outlined here gives one a handle on the dynamical problem without recourse to the Langevin-equation approach generally used to discuss critical dynamics. The results are the same, of course. The perturbation series which comes out of the generalized LGW functional here is equivalent, after analytic continuation of frequencies to the real axis, to the perturbative solutions of the Langevin equations of time-dependent Landau-Ginzberg theory,[2] provided that the Bose functions are always replaced by their classical limits T/ω.

ACKNOWLEDGMENTS

I am grateful to S. Doniach for his hospitality at Stanford, where this work was started, and to both him and G. Mazenko for enlightening discussions of this and related problems.

APPENDIX: THE ONE-DIMENSIONAL PARAMAGNON MODEL

The solution of the problem expressed by (2.13) on its borderline dimensionality (1) makes a nice example of how the quantum renormalization works, even though, as I mentioned in Sec. III, it has nothing to do with one-dimensional interacting electrons. Part of the model, of course, is the specification of the cutoff, and here we use the first of the two schemes discussed in Sec. III, in which the wave-vector and frequency cutoffs are each taken (independently) to be unity. The renormalization-group equations come out slightly different from (3.18) and (3.19) because it does not make much sense to talk about a solid angle in one dimension. Instead of (3.16) we have

$$\frac{1}{\beta N}\sum_{q\omega}{}' = 2\int_{e^{-zl}}^{1} \frac{d\omega}{2\pi} \int_{-1}^{1} \frac{dq}{2\pi} + 2\int_{-1}^{1} \frac{d\omega}{2\pi} \int_{e^{-l}}^{1} \frac{dq}{2\pi}, \tag{A1}$$

so

$$\frac{1}{\beta N}\sum_{q\omega}{}' \left(r_0 + q^2 + \frac{|\omega|}{q}\right)^{-m} = \frac{zl}{\pi^2}\int_0^1 dq \left(r_0 + q^2 + \frac{1}{q}\right)^{-m} + \frac{l}{\pi^2}\int_0^1 \frac{d\omega}{(r_0 + 2 + \omega)^m}. \tag{A2}$$

Thus (3.18) and (3.19) become ($\epsilon = 0$, $z = 3$)

$$\frac{dr}{dl} = 2r + \frac{3u}{\pi^2}\left[\ln\left(\frac{r+2}{r+1}\right) + 3\int_0^1 \frac{x^2\, dx}{(x^3 + rx + 1)^2}\right], \tag{A3}$$

$$\frac{du}{dl} = \frac{9u^2}{\pi^2}\left(\frac{1}{(r+1)(r+2)} + 3\int_0^1 \frac{x^2\, dx}{(x^3 + rx + 1)^2}\right). \tag{A4}$$

For small r and u we can ignore the r in the denominator of the equation for du/dl, so that

$$\frac{du}{dl} = -\frac{9u^2}{\pi^2}, \tag{A5}$$

whose solution is

$$u(l) = (u_0^{-1} + 9l/\pi^2)^{-1}. \tag{A6}$$

We then substitute this expression into (A3), and keep terms to first order in r on the right-hand side.[16] The expansion of the integral gives

$$\int_0^1 \frac{x\,dx}{1+x^3+rx} = \int_0^1 \frac{x\,dx}{1+x^3} - \tfrac{1}{6}r + O(r^2)\,, \qquad \text{(A7)}$$

so

$$\frac{dr}{dl} = 2r + \frac{3}{\pi^2(u_0^{-1}+9l/\pi^2)}(\lambda - r)\,, \qquad \text{(A8)}$$

where

$$\lambda = \ln 2 + \int_0^1 \frac{x\,dx}{x^3+1}\,. \qquad \text{(A9)}$$

The solution of (A8), ignoring transients which decay as $1/l$, is

$$r(l) = \bar{r}_0 e^{2l}[l_0/(l_0+l)]^{1/3}\,, \qquad \text{(A10)}$$

where $l_0 = \pi^2/9u_0$. As in Sec. IV, $\bar{r}_0$ differs from r_0 because of the transients. The linearized equation is valid up to the $l = l_1$ where $r = 1$, but we need go no further, since at this point the effective Kadanoff cell size is equal to the correlation length, and perturbation theory will suffice to finish the problem. To find the correlation length, then, we set $r(l_1) = 1$ in (A10) and use $e^{l_1} = \xi$. Then

$$\xi = \bar{r}_0^{-1/2}(\ln 1/r_0)^{1/6}\,, \qquad \text{(A11)}$$

which is exactly the $\epsilon = 0$ result found from the more usual sort of renormalization-group equations (like 3.25 and 3.26).[16] This is not surprising, since the linearized equations in the two cases are the same.

This analysis tells us that there is an ordered state at $T = 0$ if the coupling strength is large enough. It is interesting to look at this case ($\bar{r}_0 < 0$) at finite T, since we know there can then be no order. When the last finite Matsubara frequency is scaled out, $l = \hat{l} = \frac{1}{3}\ln(E_F/T)$, so ($\hat{l} \gg l_0$)

$$r(\hat{l}) = -|\bar{r}_0|\left(\frac{E_F}{T}\right)^{2/3}\left(\frac{3l_0}{\ln(E_F/T)}\right)^{1/3}\,, \qquad \text{(A12)}$$

$$u(\hat{l}) = \frac{\pi^2}{3\ln(E_F/T)}\,. \qquad \text{(A13)}$$

I have not found any reasonable way to use the renormalization group (or at least its momentum-space version) on the classical one-dimensional LGW problem that remains at this point. However, this problem has been studied using other methods.[26] One finds, for $|r| \gg u$, a correlation length

$$\begin{aligned}\xi &= |r(l)|^{-1/2}\exp[|r(l)|^{3/2}/u(l)]\\ &= |\bar{r}_0|^{-1/2}\left(\frac{T}{E_F}\right)^{1/3}\left(\frac{\ln(E_F/T)}{3l_0}\right)^{1/6}\\ &\quad\times\exp\left[\frac{(3|\bar{r}_0|)^{3/2}}{\pi^2}\left(\frac{E_F}{T}\right)\left(l_0\ln\frac{E_F}{T}\right)^{1/2}\right]. \qquad \text{(A14)}\end{aligned}$$

However, this is the correlation length in a problem whose scale differs from that of the original problem by a factor $e^l = (E_F/T)^{1/3}$. Thus the true ξ is larger by a factor e^{2l},

$$\begin{aligned}\xi &= |\bar{r}_0|^{-1/2}\left(\frac{3u_0\ln(E_F/T)}{\pi^2}\right)^{1/6}\\ &\quad\times\exp\left[\frac{\sqrt{3}|\bar{r}_0|^{3/2}}{\pi u_0^{1/2}}\left(\frac{E_F}{T}\right)\left(\ln\frac{E_F}{T}\right)^{1/2}\right]. \qquad \text{(A15)}\end{aligned}$$

The quantum effects manifest themselves in the fractional powers of $\ln(E_F/T)$ which produce corrections to classical one-dimensional LGW behavior.

---

*Work supported by the NSF through Grant GH 40883 and NSF-MRL at the University of Chicago.
†Alfred P. Sloan Foundation Fellow.

[1]K. Wilson and J. Kogut, Phys. Rep. C 12, 75 (1974); S. Ma, Rev. Mod. Phys. 45, 589 (1973); M. E. Fisher, *ibid.* 46, 597 (1974).
[2]B. I. Halperin, P. C. Hohenberg, and S. Ma, Phys. Rev. Lett. 29, 1548 (1972); Phys. Rev. B 10, 137 (1974); B. I. Halperin, P. C. Hohenberg, and E. Siggia, Phys. Rev. Lett. 28, 548 (1974); S. Ma and G. Mazenko, *ibid.* 33, 1384 (1974); Phys. Rev. B 11, 4077 (1975); R. Freedman and G. Mazenko, Phys. Rev. Lett. 34, 1575 (1975).
[3]R. L. Stratonovich, Dokl. Akad. Nauk SSSR 2, 1097 (1957) [Sov. Phys.-Doklady 2, 416 (1957)]; J. Hubbard, Phys. Rev. Lett. 3, 77 (1959).
[4]J. Hertz, AIP Conf. Proc. 24, 298 (1975).
[5]S. Q. Wang, W. E. Evenson and J. R. Schrieffer, Phys. Rev. Lett. 23, 92 (1969); J. R. Schrieffer, *ibid.* 23, 92 (1969); J. R. Schrieffer, W. E. Evenson and S. Q. Wang, J. Phys. (Paris) 32, C1, (1971).
[6]J. R. Schrieffer (unpublished); W. E. Evenson, J. R. Schrieffer, and S. Q. Wang, J. Appl. Phys. 41, 1199 (1970).
[7]J. A. Hertz and M.A. Klenin, Phys. Rev. B 10, 1084 (1974).
[8]J. Hubbard, Proc. R. Soc. A 276, 238 (1963).
[9]J. Lindhard, Dan. Vidensk. Selsk. Mat.-Fys. Medd. 28, 8 (1954).
[10]S. Doniach and S. Engelsberg, Phys. Rev. Lett. 17, 750 (1966).
[11]M. T. Beal-Monod, K. Maki, and J. P. Hurault, J. Low Temp. Phys. 17, 439 (1974).
[12]M. T. Beal-Monod, Solid State Commun. 14, 677 (1974).
[13]F. J. Wegner and A. Houghton, Phys. Rev. A 8, 401 (1973); J. Hertz, J. Low Temp. Phys. 5, 123 (1971).
[14]C. Kittel, in *Solid State Physics*, edited by H. Ehrenreich, F. Seitz, and D. Turnbull (Academic, New York, 1969), Vol. XXII, p. 14.
[15]This result was also obtained in a different way by M. T. Beal-Monod, J. Low Temp. Phys. 17, 467 (1974), and M. T. Beal-Monod and K. Maki, Phys. Rev. Lett. 34, 1461 (1975).
[16]M. E. Fisher, S. Ma, and B. G. Nickel, Phys. Rev.

Lett. 29, 917 (1972); see also F. J. Wegner and E. K. Riedel, Phys. Rev. B 7, 248 (1973).

[17]E. K. Riedel and F. J. Wegner, Z. Phys. 225, 195 (1969); Phys. Rev. Lett. 24, 730 (1970).

[18]P. Fulde and A. Luther, Phys. Rev. 170, 570 (1968).

[19]T. Lubensky, Phys. Rev. B 11, 3573 (1975); T. Lubensky and A. B. Harris, AIP Conf. Proc. 24, 311 (1975); G. Grinstein, *ibid.* 24, 313 (1975); G. Grinstein and A. Luther (unpublished).

[20]See, e.g., P. Fulde and I. Peschel, Adv. Phys. 21, 1 (1972).

[21]M. A. Klenin and J. A. Hertz, AIP Conf. Proc. 24, 242 (1975), and unpublished.

[22]This was also noticed for this system by J. Lajzerowicz and P Pfeuty, Phys. Rev. B 11, 4560 (1975).

[23]A. P. Young, Oxford report (unpublished).

[24]P C. Hohenberg and J. Swift (unpublished).

[25]J. Rudnick, Phys. Rev. B 11, 363 (1975) has used a similar diagrammatic procedure in the classical LGW problem.

[26]D. J. Scalapino, M. Sears, and R. Ferrell, Phys. Rev. B 6, 3409 (1972); J. Krumhansl and J. R. Schrieffer, *ibid.* 11, 3535 (1975).

INDEX

Printed in the United States
by Baker & Taylor Publisher Services